# A Textbook of
# Biotechnology



**Dr. R.C. DUBEY**
M.Sc., Ph.D., FNRS, FBS, FPSI
Professor
*Department of Botany and Microbiology*
*Gurukul Kangri University*
*Haridwar (Uttarakhand)*

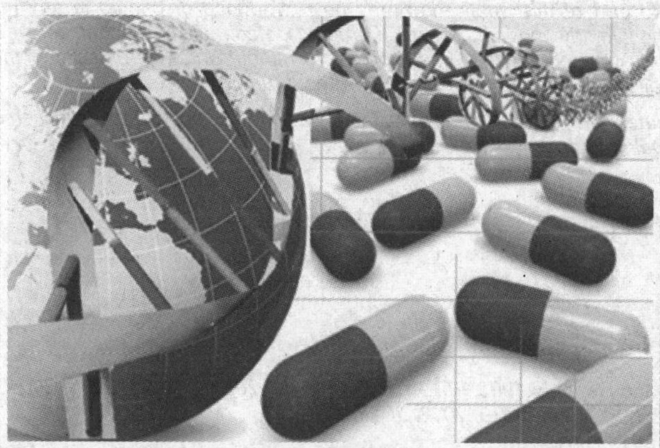

**S. CHAND**
**PUBLISHING**

**S Chand And Company Limited**
**(ISO 9001 Certified Company)**

**S Chand And Company Limited**
**(ISO 9001 Certified Company)**
***Head Office:*** D-92, Sector–2, Noida – 201301, U.P. (India), Ph. 91-120-4682700
***Registered Office:*** A-27, 2nd Floor, Mohan Co-operative Industrial Estate,
New Delhi – 110 044, Phone: 011-49731800
www.**schandpublishing.com;** e-mail: **info@schandpublishing.com**

**Marketing Offices:**

Chennai : Ph: 23632120; chennai@schandpublishing.com
Guwahati : Ph: 2738811, 2735640; guwahati@schandpublishing.com
Hyderabad : Ph: 40186018; hyderabad@schandpublishing.com
Jalandhar : Ph: 4645630; jalandhar@schandpublishing.com
Kolkata : Ph: 23357458, 23353914; kolkata@schandpublishing.com
Lucknow : Ph: 4003633; lucknow@schandpublishing.com
Mumbai : Ph: 25000297; mumbai@schandpublishing.com
Patna : Ph: 2260011; patna@schandpublishing.com



*First Edition 1993*
*Subsequent Editions and Reprints 1993, 95, 96, 98, 99, 2001, 2002, (Twice), 2003, 2004, 2005, 2006, 2007, 2008 (Twice), 2009 (Twice), 2010, 2012, 2013*
*Fifth Revised Edition 2014,*
*Reprints 2016, 2017, 2018 (Twice), 2019, 2020, 2021*

***Low Priced Students' Paperback Edition (LPSPE), 2022***

*Reprint 2022*
***Reprint 2024***

**ISBN:** 978-93-550-1053-7 **Product Code:** H6BTC68BTCH10ENAA21L

PRINTED IN INDIA
By Vikas Publishing House Private Limited, Plot 20/4, Site-IV, Industrial Area Sahibabad, Ghaziabad – 201 010 and Published by S Chand And Company Limited, A-27, 2nd Floor, Mohan Co-operative Industrial Estate, New Delhi – 110 044.

# PREFACE TO THE FIFTH EDITION

I feel pleasure in bringing out the thoroughly revised, corrected and condensed edition of **A Textbook of Biotechnology**. Hopefully this edition will be useful for all University/College students where biotechnology is being taught at Undergraduate levels. It will also be useful to all those students who are preparing for civil and various competitive examinations.

Due to expanding horizon of biotechnology, it was difficult to accommodate the current information of biotechnology in detail. Therefore, a separate book entitled **Advanced Biotechnology** has been written for the Postgraduate students of Indian University and Colleges. Therefore, the present form of **A Textbook of Biotechnology** is totally useful for undergraduate students. A separate section of Probiotics has been added in Chapter 18.

Chapter 27 on Experiments on Biotechnology has been deleted from the book because most of the experiments have been written in 'Practical Microbiology' by R.C. Dubey and D.K. Maheshwari. Bibliography has been added to help the students for further consultation of resource materials.

I am thankful to the management and the editorial team of S. Chand & Company Pvt. Ltd. for all help and support in the publication of this book.

Reader's perception in the form of constructive suggestions and valuable comments for improvement of this book is highly appreciated.

**Dr. R.C. Dubey**
Email: profrcdubey@gmail.com

# PREFACE TO THE FIRST EDITION

Recently, biotechnology has been introduced in syllabi of most of the universities, at the undergraduate and/or postgraduate level(s), either as a separate paper or a part of it. The author has been teaching biotechnology for the last two years and realises the problems of students. Therefore, this book is compiled to provide the students, the current informations on different areas of biotechnology. The author does not claim this as his original research work. It is merely an outcome of the compilation of researches done by the scientists in different areas of biotechnology in India and abroad. Help has been taken from a number of standard books, journals, research and review papers to update the knowledge.

This book embodies five different parts, divided in seventeen chapters. The language of the text is simple and the subject matter is fully illustrated. Some of the figures have been drawn by the author himself. Sources of such figures and tables that are modified and presented in the text, are duly acknowledged in their legends. At the end of each chapter, some relevant problems are given which would help the students to grasp the matter.

A short reference of the source materials and the subject- index are listed at the end of the book. In the references, the title of each paper is omitted just to save the space in this elementary book.

The author thanks the researchers, authors of books, publishers, editors whose publications have helped him in making out the text.

I express my indebtedness to Prof. R.S. Dwivedi (my supervisor, C.A.S. in Botany. B.H.U., Varanasi), Prof. J.S. Singh, F.N.A. (B.H.U., Varanasi) and Prof. Bharat Rai, my teacher (B.H.U., Varanasi) who have ever been a source of inspiration.

I must thank my colleagues Prof. S.P. Singh (Head of the Department), Dr. R.D. Khulbe, Dr. Y.P.S. Pangtey, Dr. Sudhir Chandra, Dr. Giribala Pant, Dr. Uma Palni, Dr. Neerja Pandey, Dr. S.C. Sati and Dr. Y.S. Rawat for their help and critical suggestions. The author gratefully acknowledges the help and encouragement given by Dr. R.P. Singh (Head of Forestry Department), Dr. B.K. Singh (Lecturer in Zoology Department) (K.U., Nainital), Dr. C.B. Lal (Gyanpur, Varanasi), Dr. D.B. Singh (Gyanpur, Varanasi), Drs. G.S. Mer, S.D. Tewari, Jeet Ram, and my research students (Mr. H.S. Ginwal and Mr. P.S. Rawat) (K.U., Nainital), Mr. Laxman Pandey and Mr. Rajeev Gaur (B.H.U., Varanasi).

I am grateful to Dr. L.M.S. Palni (Head, Division of Biotechnology, Palampur, H.P.) and Dr. A. Sood (Palampur) for providing a photograph of *in vitro* corm formation in *Gladiolus* sp. *for front cover of the book.*

Thanks are also due to Mr. P.B. Ghansyal for typing the manuscript and to Mr. R.S. Adhikari for drawing the diagrams. I wish to thank my wife for her endurance during the compilation work of the text. I thank my publishers, especially Sri. R.K. Gupta, S. Chand & Company Ltd. and Sri Navin Joshi, Branch Manager, Lucknow for taking pains in bringing out the book. Constructive suggestions, if any, are welcome.

**Dr. R.C. Dubey**

# CONTENTS

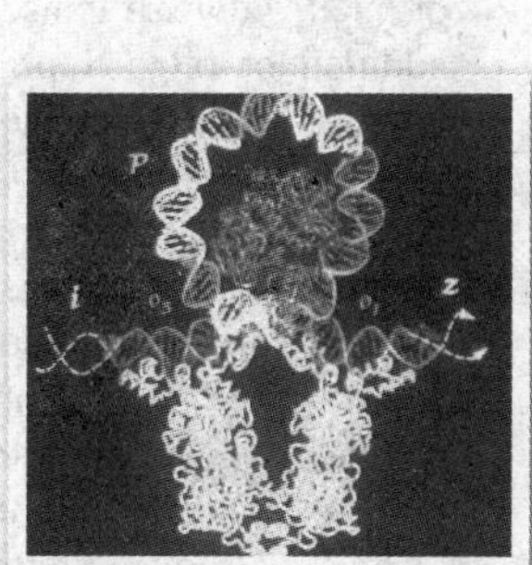

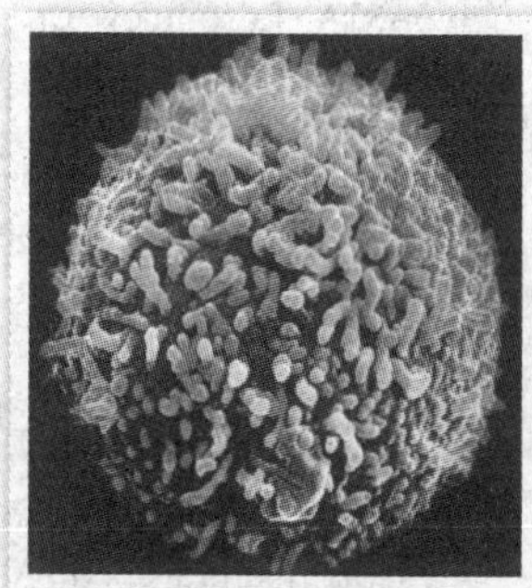



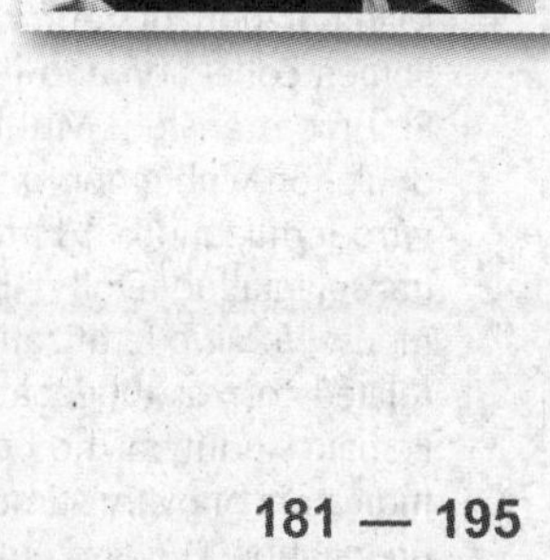

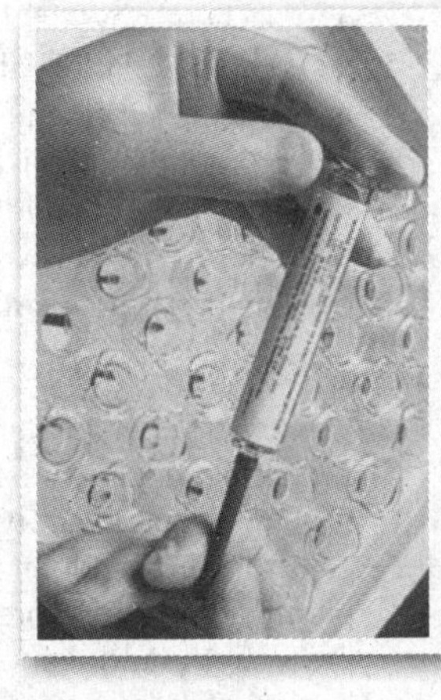

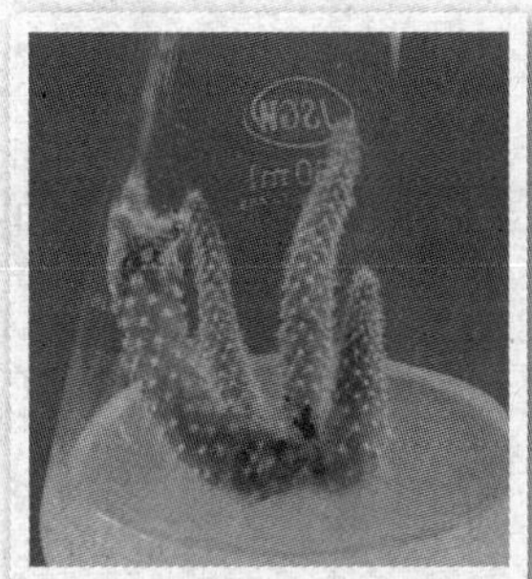

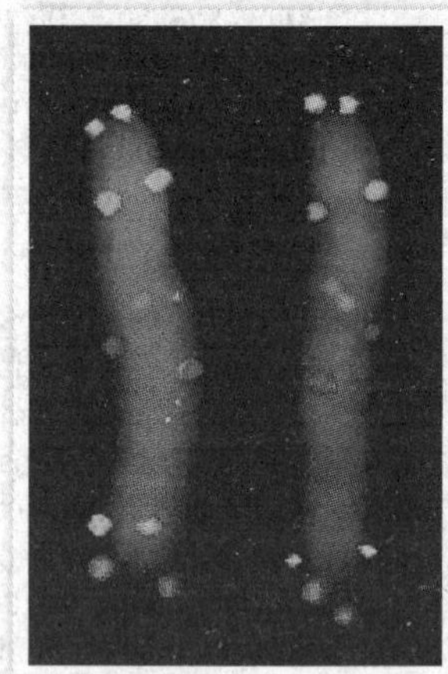

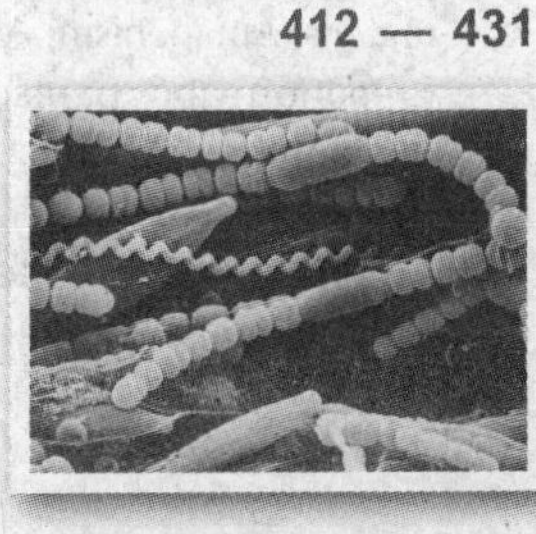

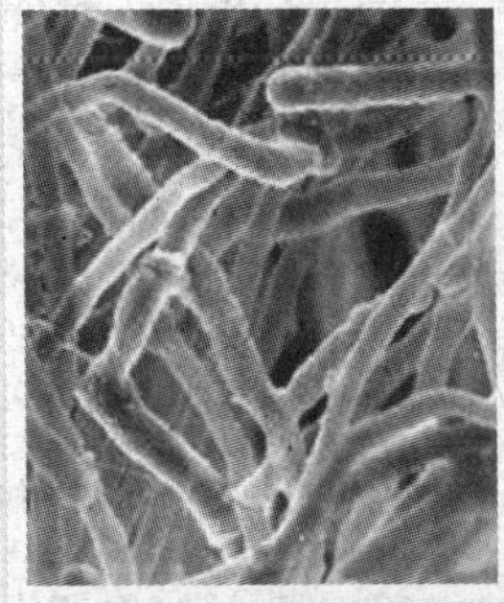

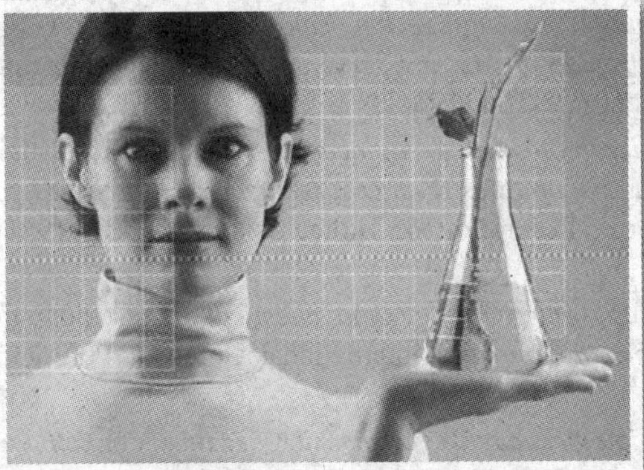

CHAPTER

# 1

# Biotechnology : Scope and Importance

## A. WHAT IS BIOTECHNOLOGY ?

In 1917 a Hungarian Engineer, **Karl Ereky** coined the term biotechnology to describe a process for large scale production of pigs. According to him all types of work are biotechnology by which products are produced from raw materials using living organisms. During the end of 20th century biotechnology emerged as a new discipline of biology integrating with technology; but the route of biotechnology lies in biology. There was no sudden sprout of this discipline, but some of the methods for production of products were developed centuries back. Therefore, biotechnology is concerned with exploitation of biological components for production of useful products. Biotechnology is defined by different organisations in different ways. It has been broadly defined as, "the development and utilization of biological processes, forms and systems for obtaining maximum benefits to man and other forms of life". Biotechnology is "the science of applied biological process" (*Biotechnology* : A Dutch Perspective, 1981). Following are some of the definitions given by other organisations :

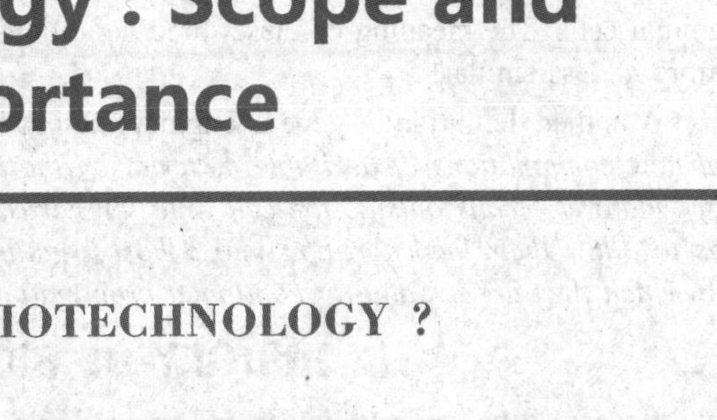

barley and a top-fermenting yeast (*Saccharomyces cerevisiae*) with water at about 60-75°F

- *Biotechnology is the application of scientific and engineering principles to the processing of materials by biological agents to provide goods and service* [The Organisation for Economic Cooperation and Development (OECD), 1981].
- *The integrated use of biochemistry, microbiology and engineering sciences in order to achieve technological application of the capabilities of microorganisms, cultured tissue, cells, and parts their of* [The European Federation of Biotechnology (EFB), 1981; O'Sullivan, 1981].
- *The application of biochemistry, biology, microbiology and chemical engineering to industrial process and products and on environment* [International Union of Pure and Applied Chemistry (IUPAC), 1981.]
- Biotechnology is the "*controlled use of biological agents such as microorganisms or cellular components for beneficial use*" (U.S. National Science Federation).

In the definition given by OECD, "scientific and engineering principles" refer to microbiology, genetics, biochemistry, etc. and "biological agents" means microorganisms, enzymes, plant and animal cells. The meaning of these three definitions and others given by many organisations are more or less similar.

A unified definition of genetic engineering has been given by Smith (1996) as "*the formation of new combinations of heritable material by the insertion of nucleic acid molecules produced by whatever means outside the cell, into any virus, bacterial plasmid or other vector system so as to allow their incorporation into a host organism in which they do not naturally occur but in which they are capable of continued propagation*".

## B. HISTORY OF BIOTECHNOLOGY

If we trace the origin of biotechnology, it is as old as human civilization. Development of biotechnology can be studied considering its growth that occurred in two phases: (*i*) the traditional (old) biotechnology, and (*ii*) the new (modern) biotechnology.

### 1. Traditional Biotechnology

Really the traditional biotechnology is the kitchen technology developed by our ancestors using the fermenting bacteria. Kitchen technology is as old as human civilisation. During *vedic* period (5000–7000 BC), *Aryans* had been performing daily *Agnihotra* or *Yajna*. One of the materials used in Yajna is animal fat (i.e. *ghee*) which is a fermented product of milk. Similarly, the divine '*soma*' (a fermented microbial product used as beverage) had been offered to God. Summarians and Babylonians (6000 BC) were drinking the beer. Egyptians were baking leavened bread by 4000 BC. Preparation of curd, *ghee*, wine beer, vinegar, etc. was the kitchen technology. In spite of all development, preparation of curd, *ghee*, vinegar, alcoholic beverages, jalebi, idli, dosa, have become an art of the kitchen of all Indians (Table 1.1).

Churning of yogurt in earthen pot makes cultured butter in India

> **Table 1.1** Traditional fermented foods prepared in different parts of India

| *Types of foods* | *Substrate* | *Regions* | *Quality and uses* |
|---|---|---|---|
| Ambali | Millet and rice | South | Steamed cake used in snack |
| Bhatura | Wheat flour (*maida*) | North | Flat fried, lavened bread used with chhola |
| Chhurpi | Milk | Himalaya | Cheese like, mild sour, soft mass, used as curry |
| Dosa | Rice and black gram | South | Spongy, shallow fried, used as staple food |
| Dahi | Milk | North | Sour, thick gel with whey |
| Dhokla | Bengal-gram | West | Spongy cake used as snack |
| Gundrum | Leafy vegetable | Himalaya | Sun-dried, sour taste, used as-soup or pickles |
| Idli | Rice and black gram | South | Steamed spongy cake used in breakfast |
| Jalebi | Wheat flour (*maida*) | North | Crispy, deep-fried used as sweet confectionery |
| Khaman | Bengal gram | West | Spongy cake, used in breakfast |
| Khalpi | Cucumber | Himalaya | Sour pickles |
| Mesu | Bamboo shot | Himalaya | Sour pickles |
| Mishti dahi | Milk | East | Thick sweet gel |
| Nan | Wheat flour (*maida*) | North | Leavened flat baked bread used as staple food |
| Papad | Black gram flour (*besan*) | North | Circular wafers used as snack |
| Paneer | Milk | North | Soft cheese used as fried curry |
| Rabadi | Mixture of butter-milk-wheat/pearlmillet/barley | West | Cooked paste used as staple food |
| Srikhand | Milk | West | Concentrated sweetened preparation |
| Sinki | Radish tap root | Himalaya | Sun-dried sour soup pickles |
| Tari | Date palm | East | Sweet alcoholic beverage |
| Tharra | Mahua | North distillation | Sweet alcoholic beverage obtained through |
| Vadai (wada) | Black gram | North | Deep fired cake used as snack |
| Wari | Black gram | North | Spongy cake used as snack |

The traditional biotechnology refers to the conventional technology which have been used for many centuries. Beer, wine, cheese and many foods have been produced using traditional biotechnology. Thus, the traditional biotechnology includes the process that are based on the natural capabilities of microorganisms. The traditional biotechnology has established a huge and expanding world market. In monetary term, it represents a major part of all biotechnology financial profits.

In India who can forget the story of '*Makhan Chori*' (butter stealing) by Lord Sri Krishna during Mahabharat period ? Butter would have been produced following the same kitchen art. Besides breeding of strong productive animals, selection of desirable seeds for enhanced crop production has been the part of human activity since the time immemorial. The Egyptians (about 2000 BC) used to prepare vinegar from crushed dates by keeping for longer time. But the crushed dates produce intoxicants at first. In Egypt, Mesopotamia and Palestine (about 1500 BC) the art of production of wine from crushed grapes, and beer from germinated cereals (malt) using a bread leaven (a mass of yeast) was established. In Indian *Ayurved*, production of '*Asava*' and '*Arista*' using different substrates, and flowers of **mahua** (*Madhuca indica*) or **dhataki** (*Wodfordia fructicosa*) has been well characterised till today since vedic period. In these methods various substrates are transformed into a number of products. Hence, the odour, colour and taste of final products are changed. Moreover, use of salt for preservation of various food has been earlier developed even in Europe. Still preservation methodology of the mummies of Egypt are noteworthy. Possibly, mummification would have been done through dehydration of dead body followed by use of mixture of salts largely sodium carbonate. Thus, the traditional biotechnology was an art rather than a science.

Antony van microscope.

**Role of Microorganisms in Fermentation:** The causes of fermentation could be discovered after observing microorganisms using a microscope by Antony van Leeuwenhoek (1673–1723) at Delft (Holland). During 18th century a significant contribution was made by the chemists on the process and products of fermentation. In 1757, it was demonstrated that a milky precipitate could be obtained when the gas evolved from fermentation was passed through the lime water. It was called *lime water test*. A similar gas comes out from burning of charcoal. Henry Cavendish demonstrated that the gas evolved from brown sugar in water treated with yeast was absorbed by sodium hydroxide solution. Schele and J Priestley (1772–1774) confirmed the identity of oxygen gas present in air. Using analytical technique for carbon estimation, Antoine Lavoisier gave the chemical basis of alcoholic fermentation.

A French man, Nichola Appert (1810) described the method of food preservation. In the same year Peter Durand also gave the use of tin container for food preservation. It was done by putting an air-tight vessel containing food material in the boiling water. It increased the importance of canning industry. Lack of oxygen in such a closed and heated vessel was reported by Gay Lussac. He concluded that oxygen was required for initiation of alcoholic fermentation, but not for further progress of fermentation. After 1830, Charles B. Astier gave the concept that *air is the carrier of all kinds of germs*.

In 1837, Theodore Schwann after a series of experimentation demonstrated that 'the development of the fungus (sugar fungus) on fruit juice causes fermentation'. He was the first to observe and describe the yeast in growing process. Charles Cagniard-Latour (1838) observed yeast budding using a microscope allowing 300–400 power magnification.

Justus von Liebig (1839), a well known chemist, proclaimed that all the activity of yeast cells was the result of chemical and physical reactions going on in the medium.

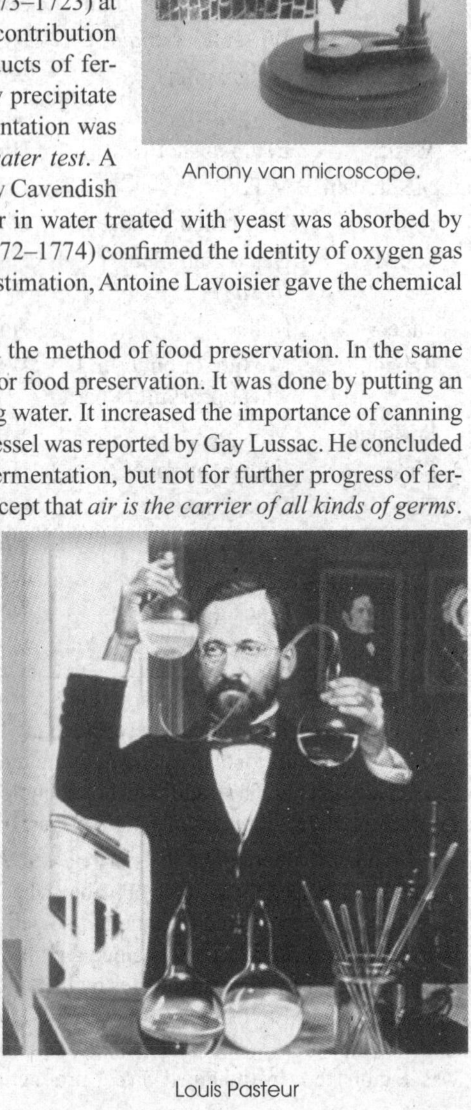

Louis Pasteur

The study of microbiology was started since the first report of Louis Pasteur (1857) on lactic acid fermentation from sugar. He isolated the microorganisms (lactic yeast) that were associated with lactic acid and formed curd. The cells of lactic yeast were smaller than that of beer yeast. Lactic acid production got increased when he added chalk powder to fermentation medium. Pasteur showed the presence of lactic acid in curd by using a *polarimeter*.

In 1860, Pasteur provided a detailed report on the use of synthetic medium for microbiological studies. He concluded that:

(*i*) fermentation is carried out anaerobically by the living cells.

(*ii*) the yeast increased the weight with increase in C and N contents of overall batch during alcoholic fermentation in synthetic medium. The increase in yeast protein in synthetic medium was accompanied by a related decrease in the ammonium nitrogen in the medium.

(*iii*) fermentation of sugar was required for multiplication of yeast cells.

(*iv*) similar phenomenon occurred in fermentation of lactic acid, tartaric acid, butyric acid, etc.

(*v*) because of not using pure culture, some of the fermentation processes were stopped.

(*vi*) growth and physiology of yeast differ when they are grown under aerobic and anaerobic conditions (it was later on called as '*Pasteur effect*'). Under anaerobic conditions a large amount of sugar was converted into alcohol, while under aerobic conditions large amount of sugar was converted into yeast cell mass.

Pasteur suggested that high percentage of microbial population is killed by heating the juice at 62.8° C (145° F) which is now called as *Pasteurization.*

Robert Koch (1881) gave the method for establishing relationship of a pathogen with a disease which is known as 'Koch's postulates' or 'pathogenicity test'. Following this technique he proved that the anthrax disease is caused by *Bacillus anthracis*. Edward Buchner (1897) was the first to demonstrate enzymatically-mediated fermentation reactions. He showed that cell-free yeast juice mixed with concentrated sugar solution evolved upon incubation the carbon dioxide and produced ethyl alcohol. Similar products were also produced in aqueous solution of the other sugars such as glucose, sucrose, fructose and maltose. Buchner called the dissolved substance in juice responsible for sugar fermentation as 'zymase'. This work showed the improved techniques of fermentation. Fermentation could also be demonstrated in test tubes without a living organism. The work of chemistry had more role in understanding the fermentation phenomenon than the microbiology.

Edward Jenner

During 1890s, Hans Buchner and Martin Hahn developed more effective method of getting cell-free extracts after disrupting the microbial cells. They ground the microbial cells with quartz and adding Kieselguhr (to get sufficient consistency in the resulting paste). Thus, sucrose fermentation using cell-free extracts from yeast cells (obtained by the above method) was demonstrated.

Discovery of viruses and their role in disease was possible when Charles Chamberland (1884) constructed a porcelain bacterial filter. Edward Jenner (1798) used vaccination by taking out liquid material from cowpox lesions and introducing into people having smallpox. But Pateur and Chamberland developed an attenuated anthrax vaccine against anthrax disease. After the discovery of toxins produced by *Corynebacterium diphtheriae* (causes diphtheria in human) its antitoxin was developed by Emil von Behring (1890) and Shibasaburo Kitasato. He injected the inactivated toxin into rabbit that induced to produce antitoxin in the blood. The antitoxin inactivated the toxin and protected against the disease. Similarly, tetanus antitoxin was developed by Behring.

## 2. Modern Biotechnology

The two major features of technology differentiates the modern biotechnology from the classical biotechnology: (*i*) capability of science to change the genetic material for getting new products for specific requirement through recombinant DNA technology, and (*ii*) ownership of technology and its socio-political impact. Now the conventional industries, pharmaceutical industries, agro-industries, etc. are focusing their attention to produce biotechnology-based products.

**(*a*) Emergence of Modern Biotechnology:** The new or modern biotechnology embraces all methods of genetic modification by recombinant DNA and cell fusion technologies. It also includes the modern developments of traditional biotechnological processes. The new aspects of biotechnology founded in recent advancement of modern biology, genetic engineering and fermentation process technology are now increasingly find wide industrial application. But the rate of application will depend on: (*i*) adequate investment by the industries, (*ii*) improved system of biological patenting, (*iii*) marketing skill, (*iv*) economics of the new methods, and (*v*) public perception about the biotechnology products.

Canning process.

With the end of 19th century, the traditional biotechnology associated with fermentation was gradually industrialised. This resulted in gradual growth of industries producing beer, whisky, wine, rum, canned food, etc.

In 1920, for the first time, the Leeds City Council (U. K.) established the Institute of Biotechnology. In the late 1960s, OECD was set up to promote policies for sound economic growth of the member countries. In 1978, the European Federation of Biotechnology was set up.

In 20th century, biotechnology brought industries and agriculture together. During World war I fermentation processes were developed which produced the acetone from starch and paint solvent from automobile industries. During World war II the antibiotic penicillin was discovered. Manufacture of penicillin shifted the biotechnological focus towards pharmaceuticals. Linking the fermentation with biochemistry, bioprocess, chemical engineering and instrument designing helped substantially in the progress of industries. During Gulf War (1991) the work on microorganisms dominated for the preparation of biological warfare, antibiotics and fermentation process. Suspected preparation of biological and chemical warfare led to US attack on Iraq in 2003.

A research scientist preparing plasmid DNA

After the discovery of double helix DNA by Watson and Crick (1953), Werner Arber (1971) discovered a special enzyme in bacteria which he called the *restriction enzymes*. These enzymes can cut the DNA strand and generate fragments. The cut ends of two fragments are single stranded sticky ends because the single stranded ends having identical base pairs can re-join. In 1973, S. Cohen and H. Boyer removed a specific gene from a bacterium and inserted into another bacterium using restriction enzymes. This discovery marked the start of *recombinant DNA technology* or *genetic engineering*. In 1976, Baltimore and Baltimore successfully transferred human growth hormone gene into a rabbit. In 1978, the European Federation of Biotechnology was created.

In 1978, a U.S. company 'Genetech' used genetic engineering technique to produce human insulin in *E. coli*. In 1980, trials of new hormone was conducted in the U.S.A., France, Japan, and the United Kingdom. The US Food and Drug Administration gave marketing approval to '*Humulin*' i.e. human insulin made by Eli Lilly (U.S.A) by the end of 1982. Another hormone 'somatotropin' was produced on industrial scale.

In 1993, the first genetically engineered tomatoes, FlavrSavr, were sold in market. In 1996, the first clone lamb 'Dolly' was borne successfully by the efforts of scientists of Scotland. Thereafter, several cloned animals were produced.

In 2001, the sequence of the Human Genome was published in *Nature* and *Science*. Human Genome Project was completed by March 2003. In 2002, a designer baby was born to cure the genetic disease of her elder sister. On December 27, 2002, a claim was made for the birth of a clone baby **'Eve'** by the scientists of 'Human Cloning Society', the **Clonaid** of France. In February 2005, in the meeting of United Nations many countries opposed gene cloning in humans, while a few countries supported with request to allow for the sake of research only.

In May 2005, scientists in South Korea have used a method called **therapeutic cloning** to produce stem cell lines. These are genetic matches to patients. Such stem cell lines could be used for disease research. The U.S.A. condemned this approach. In this method human embryos were produced through cloning (as done for Dolly) and stems cells were obtained from blastocyst. The excised stem cells could be grown *in vitro* and used further.

It is, therefore, interesting that the scientists are engaged in doing the most challenging task of mass production of growth hormones, insulin, vaccines, immunogenic proteins and polypeptides, gene therapy, biofertilizers, biopesticides, producing disease and stress-resistant plants, biomass, enzymes, antibiotics, acids, fuels, etc. Many biochemical companies such as National Pituitary Agency (U.S.A.), E. Lilly (U.S.A.), Kabi Vitram A B (Sweden), Genetech Co. (U.S.A.), Biogen (Switzerland), etc. are producing/trying to produce some of the above products by using genetic engineering techniques. Many Nobel Laureates including Dr. Har Govind Khorana, are associated with these companies.

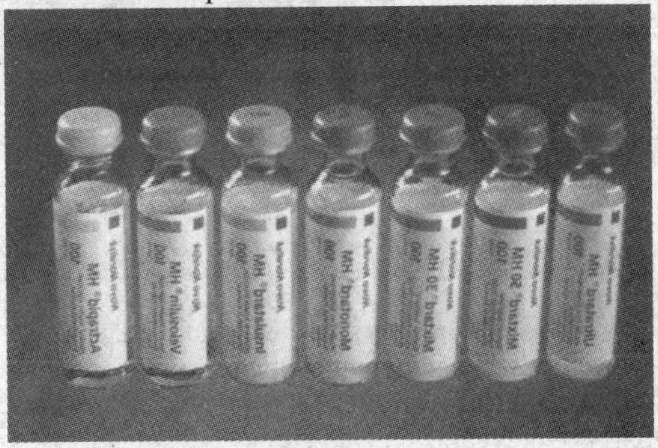

Genetically engineered human insulin available in large scale.

The area of main interest, at present, where such works are being done throughout the world are given in Table 1.2. Many international agencies are making efforts to solve the national and international problems via. collaborative works. Various programmes including molecular and environmental engineering have been taken up by the European Communities with frame work of their collaborative activities. For training purposes, under several biotechnological programmes, centres have been established, for example, the International Centre for Cooperative Research and Training in Microbial Engineering (Japan) and Institute of Biotechnological Studies (U.K.).

Many countries have developed collaborative networks on several aspects such as "Regional Microbiology Network for South-East Asia" (supported by Japan and UNESCO). "Microbiological Resource Centres" (MICRCENS) [supported by UNESCO]; United Nations Environmental Programmes (UNEP), International Cell Research Organisation (ICRO). These two organisations have participations of groups in many countries.

> **Table 1.2** Area of biotechnology.

| *Area of interest* | *Products* |
|---|---|
| 1. Recombinant DNA technology (genetic engineering) | Fine chemicals, enzymes, vaccines, growth hormones, antibiotics, interferon. |
| 2. Treatment and utilization of bio-materials (biomass) | Single cell protein, mycoprotein, alcohol and biofuels. |
| 3. Plant and animal cell culture | Fine chemicals (alkaloids, essential oils, dyes, steroids), somatic embryos, encapsulated seeds, interferon, monoclonal antibodies. |
| 4. Nitrogen fixation | Microbial inoculants (biofertilizers) |
| 5. Biofuels (bioenergy) | Hydrogen (via photolysis), alcohols (from biomass), methane (biogas produced from wastes and aquatic weeds). |
| 6. Enzymes (biocatalysts) chemotherapy. | Fine chemicals, food processing, biosensor, |
| 7. Fermentation | Acids, enzymes, alcohols, antibiotics, fine chemicals, vitamins, toxins (biopesticides). |
| 8. Process engineering reactor, harvesting. | Effluent, water recycling, product extraction, novel |

***(b)* Biotechnology as an Interdisciplinary Area:** Biotechnology is not a sudden discovery, rather a coming of age of a technology that was initiated several decades ago. By the middle of 20th century there had been a tremendous growth in the area of chemistry, physics and biology. Further knowledge of each branch has been advanced substantially. Multidisciplinary strategies were made for the solution of various problems. A novel spectrum of investigation occurred through the true interdisciplinary synthesis. This led to the evolution of biotechnology which is an outcome of integrated effort of biology with technology, the root of which lies in biological science (Fig 1.1).

Biotechnologist Isolating the gene for protein X.

The key difference between biology and biotechnology is their scale of operation. Usually the biologist works in the range of nanograms to milligrams. Biotechnologists working on the production of vaccine may be satisfied with milligram yields, but many other projects aims at kilograms or tonnes. Thus, the main objective of biotechnologists consists of scaling-up the biological processes (Smith, 1996).

The main discipline of biology are microbiology, biochemistry, genetics, molecular biology, immunology, cell and tissue culture. However, on the engineering side it includes chemical and biochemical engineering such as large scale cultivation of microbes and cells, their upstream and down stream process, etc. These processes can be separated into five major operations: (*i*) strain selection and improvement, (*ii*) mass culture, (*iii*) optimisation of cell responses, (*iv*) process operation, and (*v*) down stream processing (*i.e.* product recovery).

Many areas of biotechnology have arisen through the interaction between various parts of biology and engineering, biochemistry, biophysics, cell biology, colloid chemistry, embryology, ecology, genetics, immunology, molecular biology, medical chemistry, pharmacology, polymer chemistry,

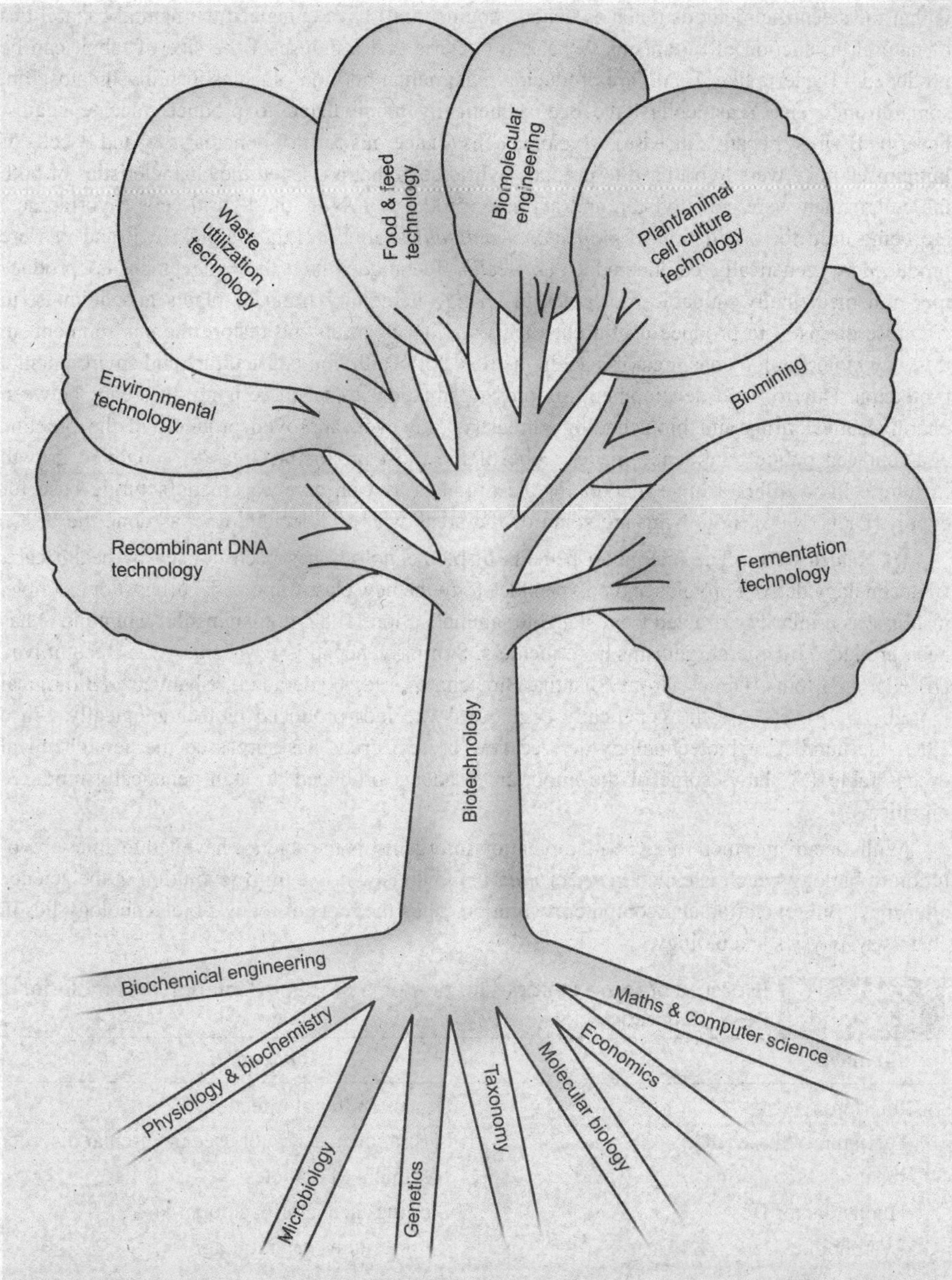


Fig. 1.1. A Biotechnology tree.

thermo-chemistry and virology. The modern biotechnology has developed several technologies extracting the basic knowledge from biology.

**(c) Global Impact and Current Excitement of Biotechnology:** Each and every organism performs its function within its optimum limits. The excitement about the modern biotechnology is that the scientific methods (such as genetic engineering) have enhanced the natural capabilities of natural production of organisms. What a miracle is that a mouse of the size of rabbit can be produced ? Bacteria like *E. coli* are producing mammalian hormons such as insulin, somatostatin, somatotropin, etc. Yeast cells have been genetically manipulated to produce vaccine against hepatitis B virus (hepatitis disease). Myeloma cells (cancerous cells of bonemarrow) and B-cells of immunised mice were hybridised to produce hybrid cells that consisted the characteristics of both the cells which were cell division and antibody production. Now the hybrid cells (hybridomas) are being used for production of *monoclonal antibodies*. In 1982, interferons ($\alpha$, $\beta$ and $\gamma$) were produced by genetically engineered *E. coli* cells. Techniques have been developed to produce rare and medicinally valuable molecules to change hereditary traits of plants and animals, to diagnose diseases, to produce useful chemicals and, to clean up and restore the environment. In this way biotechnology has great impact in the fields of health, food/agriculture and environmental protection. Due to rapid development the present situation is that there is no difference between pharmaceutical firms and biotechnology industry. However, approved products in the pipeline and renewed public confidence made it one of the most promising areas of economic growth in future. India offers a huge market for the products as well as cheap manufacturing base for export (Padh, 1996). Following are some of the areas where biotechnology has done the best.

**(*i*) Health care :** The maximum benefits of biotechnology has been utilised by health care. Biotechnology derived proteins and polypeptides form the new class of potential drugs. For example, insulin was primarily extracted from slaughter animals. Since 1982, human insulin (Humulin®) has been produced by microorganisms in fermenters. Similarly, hepatitis B vaccines *viz.,* Recombivax HB® (from Merk), Guni® (from Shantha Biotechnics Ltd, Hyderabad), Shanvac® (Biological E. Laboratory), etc. are the genetically engineered vaccines produced biotechnologically. Since 1987, the number of biotechnology-derived new protein drugs has surpassed the new chemical drugs. Table 1.3 shows some of the important products produced through genetically modified organisms.

With the advancement of gene manipulation in organisms the science have led to a new revolution in biology which is called *gene revolution*. Obviously, it is a third revolution in the science after industrial revolution and computer revolution. Thus the roots of today's biotechnology lies in chemistry, physics and biology.

**> Table 1.3. Example of some therapeutic products produced through recombinant DNA technology.**

| *Products* | *Application* |
|---|---|
| Interferon | Cancer and viral infection |
| Human urokinase (tPA) | Plasminogen activator used in vascular disorder |
| Insulin | Treatment of diabetes |
| Human factor IV | Clotting factor for haemophilia |
| Lympholines | Auto-immune functioning |
| Serum albumin | In surgery |
| Attenuated pseudorabies virus antigen | Vaccine against rabies |
| Tissue plasminogen activator | In treatment of heart attack |
| Somatostatin | Treatment of human growth disorder |

There are about 35 biotechnology-derived therapeutics and vaccines approved by the USFDA alone for medical use, and more than 500 drugs and vaccines to reach in market (Table 1.4). Similarly, about 600 biotechnology diagnostics are worldwide available in clinical practices (Table 1.5). About 130 gene therapy protocols have been approved by the US authorities. India relies on imports of many immunodiagnostic kits.

**> Table 1.4. Biotechnology-derived drug products** (*Source* : H. Padh, 1996)

| *Products* | *Manufacturers* |
|---|---|
| Human insulin | Eli Lilly, Novo Nordisk, Hoechst |
| Growth factors | Eli Lilly, Novo Nordisk, Genetech, Pharmacia |
| Blood factors | Amgen, J & J, Sankyo, Chugai, Sandoz, Immunex |
| Interferon | Roche, Wellcome, Daiichi, Sumitomo |
| Monoclonal antibodies | J & J, Cytogen |
| Vaccines | Smithkline, Merk, Shionogi |

**> Table 1.5. Approved biotech diagonistics.** (*Source* : H. Padh, 1996)

| *Type* | *Infectious disease* | *Tumour marker* | *Analyte & drug* | *Blood screening* | *Total* |
|---|---|---|---|---|---|
| Monoclonal antibody | 127 | 2 | 433 | 9 | 571 |
| DNA probe | 42 | 0 | 11 | 0 | 55 |
| Recombinant DNA | 11 | 1 | 1 | 0 | 13 |
| Total | 180 | 3 | 445 | 9 | 637 |

***(ii)* Agriculture :** Biotechnology is making new ground in the food/agriculture area. Current public debate about BSTC, bovine somatotropin (a hormone administered to cows to increase milk production) typifies an example of biotechnology product testing public acceptance. Similarly, the FlavrSavr™ tomato (produced by transgenic plants engineered by antisense technology to preserve flavour, texture and quality) is a new breed of value added foods. Food biotechnology offers valuable and viable alternative to food problems, and a solution to nutritionally influenced diseases such as diabetes, hypertension, cancer, heart diseases, arthritis, etc. A transgenic 'Golden rice' has been produced by introducing three genes for the production of vitamin A in 'Taipei' rice. Several insect resistant transgenic *Bt* plants have been produced by incorporating insecticidal toxin producing *Bt* gene of *Bacillus thuringiensis* into the desired plants. A transgenic cotton named 'Ingaurd' was released in Australia which contained *Bt* gene and provided resistance against insects. We can recall the *Bt* cotton prepared for Andhra Pradesh but sown in Gujarat in 2001 which raised debate throughout the country. Biopesticides are coming to the market and their sales are increasing.

Molecular Pharming is a new concept where therapeutic drugs are produced in farm animals, for example, therapeutic proteins secreted in goat milk. There are about a dozen companies that produce lactoferrin, tPA, haemoglobin, melanin and interleukins in cows, goat and pigs. However, it is not surprising that vegetables producing vaccines, insulin, interferon and growth hormones would be available in market in 21st century, beside, human clones and several other miracles.

***(iii)* Human Genome Project (HGP):** The major landmark in human history is the human genome sequence. The HGP is an international research programme. Almost the whole human genome has been sequenced and chromosome map has been developed in various laboratories world-wide through co-ordinated efforts. Human chromosome mapping was completed by March 2003. There are about 33,000 functional genes in human. More than 97% genes are non-functional. They do not encode any polypeptide chain. Objectives of human genome project are to: (*i*) construct the detailed genetic and physical map of human genome, (*ii*) determine the complete nucleotide sequence of human DNA, (*iii*) store information in database, (*iv*) locate the estimated 50,000–100,000 genes within the human genome, (*v*) address the ethical, legal and social issues (ELSI) that may arise from the project, and (*vi*) perform similar analysis on the genomes of several other organisms.

**(*iv*) Environment :** The natural biodegradability of pollutants present in environment has increased with the use of biotechnology. The bioremediation technologies have been found successful to combat the pollution problems (see Chapter 25, *Environmental biotechnology*).

*Bioremediation* is the use of microorganisms to detoxify pollutants, present in the environment usually as soil or water sediments. The pollutants cause several health problems. Microorganisms which show potential to degradation of oil, pesticides and fertilizers belong to the genera of bacteria *Pseudomonas, Micrococcus, Bacillus*, and fungi *Candida, Cladosporium, Torulopsis, Trichoderma*, etc.

**(*v*) Genomics and Proteomics:** Computer-based study and designing of genome is called *genomics*. Genomics deals with sequencing of the complete genome of a particular organism. Similarly, study of proteins present on genome using computer is called *proteomics*. The proteomics can be defined as the study of all the proteins present in genome of an organism. With the help of proteomics and genomics the new molecules that can interact with the other partners could be identified. This gives us deep insight into biological pathway. Now the whole genome is available to biologists for scrutiny. There may be new small molecules as potential drug candidates. Therefore, interaction of some molecules can be studied in greater detail.

The 33,000 genes of human beings are on a microchip. It has helped to design specific drugs for genetic diseases, for which there is no cure so far. For example, a specific gene (*Her*-2 *Neu*) over-expresses in breast cancer patients. A designed drug (Herceptin) is good for treatment of breast cancer. Thus, the field of genomics has helped the growth of pharmacologic, toxicologic and protein studies on animals, therefore, the new areas are called pharmacogenomics, toxicogenomics and proteogenomics, respectively (*see* Chapter 8).

**(*vi*) Bioinformatics:** It is a new field of biotechnology linked with information technology. Bioinformatics may be defined *as application of information sciences (mathematics, statistics, and computer sciences) to increase the understanding of biology, biochemistry and biological data.* The most remarkable success of bioinformatics to date has been its use in the 'shotgun sequencing' (breaking of a large piece of DNA into smaller fragments) of human genome.

## D. BIOTECHNOLOGY IN INDIA AND GLOBAL TRENDS

### 1. Biotechnology in India

The recombinant DNA technology has become the major thrusts in most of the developing countries. In 1982, Government of India set up an official agency, 'the National Biotechnology Board' (NBTB) which started functioning under the Department of Science and Technology (DST). In 1986, NBTB was replaced with a full-fledged department, the Department of Biotechnology (DBT), under the Ministry of Science and Technology for planning, promotion and coordination of various biotechnological programmes.

The DBT is making effort in promoting post graduate education and research. Special M.Sc. courses in Biotechnology in selected group of institution with scholarship are provided by the DBT. The selection of students is done via National Test. In addition, it also provides trained manpower for the rapidly growing biotech industry. It has also raised the level of biology education in certain areas of biotechnology in the country. Moreover, a considerable amount of basic biochemical and molecular biology is imported in these courses.

India has the DBT, DST, CSIR, ICAR, ICMR and IARI, and other agencies which are working under the Government. These agencies and the other National and International Industries are manufacturing Biotech products and marketing them after clinical trials. A Technology Development Board (TDB) has been set up by the Government for the promotion of product development. The TDB works with universities, industries and the National Institutes. The Technology Information, Forecasting and Assessment Council (TIFAC) has prepared a 'Vision 2020' document which consists of biotechnology also.

Since 1980s, India has supported a lot to biotechnology industry and its products. Teaching and research of biotechnology have been included in University's syllabi both at Under Graduate and Post graduate levels. DBT- supported departments are running in several Universities and Institutions. It is hoped that India will play a key role in future as one of the largest market of the world, and as a producer of biotech products.

***(i)* International Centre for Genetic Engineering and Biotechnology (ICGEB).** The United Nations Industrial Development Organisation (UNIDO) recognised the potential of genetic engineering and biotechnology for promoting the economic progress of the developing countries. The initiation taken by UNIDO has led to the foundation of ICGEB. In 1981, in a meeting convened by UNIDO it was proposed to establish an international centre of excellence to foster biotechnology in the developing world. In 1982, this concept was approved by a high level conference of developed and developing nations in Belgrade. The statutes of the centre were signed by 26 countries with the entry into force of statutes on February 3, 1994. The ICGEB has become a fully autonomous international organisation composing of at present 33 member states.

The ICGEB has its two centres, one located in Trieste (Italy), and the other in Jawaharlal Nehru University, New Delhi (India). The Trieste component is currently occupying about 5,700 $m^2$ area, whereas the New Delhi component is occupying about 10,000 $m^2$ area. This centre is functioning in a proper way since 1982.

The organs of ICGEB are the Secretariat, the Board of Governors and the Council of Scientific Advisors. The secretariat component is the Director, two Heads of the components and the scientific and administrative staff operating with the framework of the ICGEB programme. The Board of Governors consists of a representative of each Member State. The Council of Scientific Advisors is composed of eminent scientists and overseas scientific excellence of ICGEB. Funds are provided by the government of Italy and India. From 1999, all Member States have started to finance ICGEB through a scale of assessment adopted by the Board of Governors.

The activities of ICGEB are aimed specifically at strengthening the R & D capability of its Member States by : (*i*) providing the developing countries with a necessary 'critical mass' environment to pursue and advance the research in biotechnology; host research facilities that are technology and capital demanding and, therefore, inaccessible to the great majority of developing countries, (*ii*) training schemes and collaborative research with affiliated centres to ensure that significant members of scientists from Member states are trained in state-of-art technology, in areas of direct relevance to the specific problems of their countries, and (*iii*) acting as the coordinating hub of network of affliated centres that serve as localized nodes for distribution of information and resources located at ICGEB.

Initially a total of six centres were set up at various Universities/Institution namely, Jawaharlal Nehru University (New Delhi), Madurai Kamraj university (Madurai), Tamil Nadu Agriculture University (Coimbatore), National Botanical Institute (Lucknow), and Bose Institute (Kolkata). In addition, the other centres for biotechnology in India are : Indian Agricultural Research Institute (IARI), Jawaharlal Nehru University (JNU), Delhi University, Indian Veterinary Research Institute

(IVRI), Izzatnagar (U.P.); Central Food and Technology Research Institute (CFTRI), Mysore; National Dairy Research Institute (NDRI), Karnal (Haryana); Malaria Research Centre (MRC), Delhi; Regional Research Laboratory (RRL) Jammu; Central Drug Research Institute (CDRI) and Central Institute of Medicinal and Aromatic Plants (CIMAP), Lucknow; Indian Institute of Technology (IIT), Kanpur, Madras, Bombay and New Delhi. Other centres to which DBT has provided infrastructural facilities are Banaras Hindu University, Varanasi; Allahabad University; M.K. University, Madurai; Anna University (Madras); Indian Institute of Science (Bangalore); Pune University, Pune; All India Institute of Medical Sciences, New Delhi; Bhabha Atomic Research Centre (BARC), Bombay, etc. Mr. Rajiv Gandhi, the Late Prime Minister of India, laid a foundation stone on October 4, 1988 at the Centre I.A.R.I. with the name "Lal Bahadur Centre for Biotechnology".

Many public and private institutions working under the Government departments and organisations have advised the DBT to formulate the biotechnology programmes under the following areas : (*i*) plant molecular biology and agricultural biotechnology, (*ii*) biochemical engineering, process optimisation and bioconversion, (*iii*) aquaculture and marine biotechnology, (*iv*) fuel, fodder, biomass and green cover, (*v*) medical biotechnology, (*vi*) microbial and industrial biotechnology, (*vii*) large scale use of biotechnology, (*viii*) integrated systems in biotechnology, (*ix*) Veterinary biotechnology and (*x*) Infrastructural facilities.

In India, the pharmaceutical industry is very strong and vibrant with expertise for chemical drugs. It has little experience in biotech diagnostics and no experience in biotech therapeutics. Moreover, pharma industry is located between Mumbai and Ahmedabad (90% of drug production in India is in Gujarat and Maharashtra). There is no Government institution or university with expertise in this area to help pharma industry. However, for a variety of reasons, the Indian pharmaceutical industry will sooner or later enter in manufacturing of biotechnology-based diagnostics and therapeutics.

***(ii)* Needs for Future Development.** A few developing countries like India, have scientists and technologists related to biotechnology where national strategies of development in biotechnology could be implemented. The scientific and technical manpower has to be properly shifted towards new biotechnology with the aim to produce expertise in biotechnology. In a keynote address, Bachhawat and Banerjee (1985) have described the impact of biotechnology on third world countries. They emphasized "Indian bioscientist must be trained to utilize their knowledge and expertise for application and orientation, for example, a microbiologist must be trained in microbial genetics to be really useful in fermentation technology, or a botanist must be trained in cell culture, protoplast fusion or DNA recombination for practical utility and similarly, people from traditional disciplines in life science may be trained to reorient their knowledge towards application and process of training readjusted according to need."

In India, most of the universities have started teaching biotechonology at under-graduate level. However, at post graduate level teaching and research have been initiated only by a few universities/ institutes on all India entrance test basis. Government of India has selected many thrust areas of national and international relevance as described earlier.

## 2. Global Scenario

Day-by-day biotechnology products are increasing in world market. The high value added biotechnology product to be used in medical field are now in domination for the last few years. The countries which have boosted up biotechnology R & D during the past two decades are the U.S.A., U.K., Japan, France, Australia, Russia, Poland, Germany, as well as India (among the developing countries). The most effective means of promoting international cooperation is through networks. The international networks devoted to applied microbiology/biotechnology are the Regional Microbiology Network for South-East Asia, and the network of Microbiological Resource Centres (MICRENs).

To foster biotechnology inventions, the U.S.A. promotes enterprises through policy development and support. Funding for basic scientific research at the National Institute of Health (NIH) has been supported. The U.S. administration has boosted up the process for improving new medicines so that these may be quickly and safely available in market. The private sector research investment and small business development have also been encouraged through the incentives. Position of therapeutic proteins and vaccines up to June 1998 by the USA has been given in Table 1.6.

**> Table 1.6.** **Position of therapeutic proteins and vaccines up to June 1998 (based on Pharmaceutical Research and Manufacturers of America; Biotechnology Industry Organisation).**

| *Therapeutic proteins/vaccines* | *Status* | |
|---|---|---|
| | *Approved* | *In development* |
| Blood clotting factor | 3 | 3 |
| Gene therapy | 0 | 38 |
| Growth factors | 21 | 8 |
| Human growth hormones | 5 | 8 |
| Inteferons | 7 | 9 |
| Interleukins | 1 | 9 |
| Monoclonal antibodies | 12 | 72 |
| Recombinant human proteins | 6 | 4 |
| Vaccines | 5 | 77 |
| Others | 12 | 52 |

The international markets have been opened for biotechnology research and biotech products. One of the world's best examples of partnership has been established by developing public databases. It enables the scientists to tie up with an enterprise. This has helped in developing partnership among University research-Government and private industries. Science education has been improved. Guidelines have been prepared so that science based regulatory programmes: (*i*) can promote public biosafety, (*ii*) earn public confidence, and (*iii*) can guarantee fair and open international market. The other countries (such as U.K., Japan, Germany, France, Australia, etc) have also prepared similar guidelines for promotion of biotechnology products and bio-business,

## E. POTENTIAL OF MODERN BIOTECHNOLOGY

The modern biotechnology is expected to solve many problems arising at the global level. In 21st century growth and economy of a country will certainly depend on operation of biotechnology. A revolution may occur in some of the areas like medical and health care, agriculture, industry and environment.

DNA fingerprinting has successfully helped the forensic science in the search of criminals making identity of individuals, solving parentage dispute, etc. Human diagnostics and therapeutic drugs have been discovered and commercialised. Gene therapy is hoped to solve the problems of genetic diseases. Biotechnology-based vaccines are the best as compared to conventional vaccines, and so is the insulin. They have no side effects and pose no risk for the presence of live form of viruses in vaccines. Many transgenic plants and animals have been genetically improved. Now they are capable of producing new or improved products. The questions may be raised on complex, ethical, spiritual and philosophical issues.

Hundreds of transgenic plants have been produced and many of them are being sown in field and products are available in market. The plant biotechnology will reduce the dependence of farmers on pesticides and will help to utilise the new technologies. In near future papers and chemicals with less energy and less pollution may be produced through biotechnology.

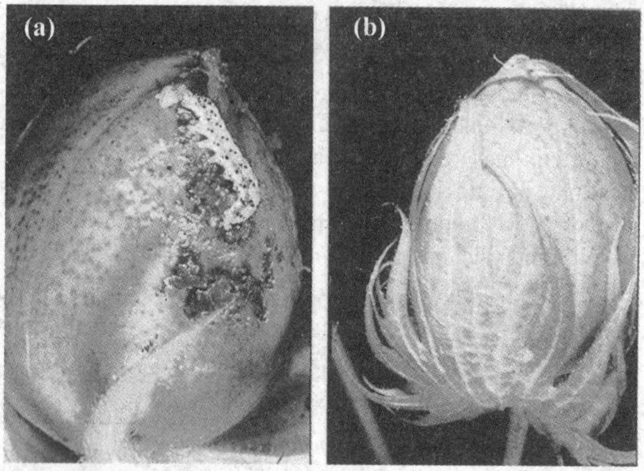

(a) Buds of cotton plants are vulnerable to worm attack
(b) Buds of a modified cotton plant-resist attack.

Abatement of pollution using potential microorganisms (*i.e.* bioremediation) has been used in many countries. Thus the potential microorganisms are looked for better health, better products and better environment of future. Instead of chemical pesticides, biopesticides have been produced and commercialised by some industries in India and abroad. Similarly, biofertilizers (formulation of nitrogen fixing bacteria and blue green algae, *i.e.* cyanobacteria, and phosphate-solubilising bacteria and fungi) have also been made available to farmers in India.

On the other hand, biotechnology has generated new jobs for the youth and stimulated the growth of small business. It has also encouraged the innovations in the industrial and agricultural sectors. In the U.S.A. alone more than 15 lakh youth have been employed in industries.

## E. ACHIEVEMENT OF BIOTECHNOLOGY

In recent years, it has become possible to map the whole genome of an organism to find out the function of the genes, cut and transfer into another organism (*see* Chapters 3 and 4). Owing to the success achieved from gene cloning, many products have been obtained through genetically engineered cells, and hopefully many can be produced during the current decade. Recombinant DNA technology has made it easier to detect the genetic diseases and cure before the birth of a child or suggest accordingly. Gene bank and DNA clone bank have been constructed to make available different types of genes of its known function. Thus, recombinant DNA technology has made it possible to develop vaccines against viral and malarial diseases, growth hormones and interferon (*see* Chapter 7).

Transgenic plant developed by recombinant DNA technology.

Biotechnology has caused revolution in agricultural science. Cell culture and protoplast fusion techniques have resulted in hybrid/cybrid plants through intergeneric crosses which generally are not possible through the conventional hybridization techniques (*see* Chapter 12). It has also helped in the production of encapsulated seeds, somaclonal variants, disease resistant plants, herbicide-and stress-resistant plants, and *nif* gene and *nod* gene transfer as well. Through cell culture techniques, industrial production of essential oils, alkaloids, pigments, etc. have been boosted up. However, many more works are to be done on horticulture and forestry plants as far as micropropagation and establishment of mycorrhizal fungi are concerned (*see* Chapter 13).

For better yield of agricultural crops, use of biofertilizers (seed bacterization, algalization and green manuring) has become an alternative tool for synthetic chemical fertilzers. The biofertilizers arc non-toxic to micro- and macro-biota and to humans as well. This would reduce the constrain on fossil-fuel based industries (*see* Chapters 19 and 20). Moreover, to discourage the use of synthetic pesticides, biocontrol agents have been developed and conditions have been investigated when phenomenon of antagonism takes place (*see* Chapter 21).

For the protection of environment and abatement of pollution, treatment of sewage, transformation of domestic wastes and xenobiotic chemicals have drawn much attention in recent years. To combat these problems such bacterial plasmids have been developed that could be used to degrade the complex polymers into non-toxic forms. Strains of cyanobacteria, green algae and fungi have been developed which could be used for the treatment of municipal and domestic sewage and industrial discharges into nontoxic forms and renew them as source of energy.

Biotechnology has helped the bio-industries in producing the novel compounds and optimization, and scale up of products, for example alcohols, acids, antibiotics and enzymes (*see* Chapters 17 and 22) and single cell protein and mycoprotein (*see* Chapter 18).

Technologies have also been developed to seek an alternative source of energy from biomaterials generated from agricultural, industrial, forestry and municipal sources (*see* Chapter 23). Social forestry and short rotation tree plantation will help to reduce the pressure on forests to meet the demand of fuel in rural sector. In industries, biomass fired system have been developed to meet the energy requirement of engines, such as sugar cane mills. Moreover, urban sewage and plant weeds are used for the production of biogas for cooking and lighting purposes (*see* Chapter 24).

## 1. Ban on Genetic Food

It is a growing concern all over the world that the genetic food may pose risks to human health, ecology and the environment. However, it has forced the government of many countries to re-think on introduction of such designer crop. For the first time the European Commission's Scientific Advisors have recommended that a genetically engineered potato be withheld from the market because they cannot guarantee its safety. Worried at the growing acts of vandalism against the genetically engineered crops in Britain, the environment minister had gone on recod that his government was considering a three year ban on transgenic crops grown for commercial use. The United States, the world's biggest producers of genetically modified foods, has also threatened New Zealand to ban his genetically engineered foods.

In Europe, the boom in the stock market for biotech products also is in wane. British Biotech, Europe's Flagship Biotechnology Company has lost its share value by more than fifty percent. However, it is a general opinion that in solving the global problems of hunger and food shortages biotechnology cannot make food cheaper at any cost.

It is true that cotton pests, especially American bollworm has become resistant to certain pesticides. Moreover, several pests have developed immunity against the Bt gene (*Bacillus thuringiensis* gene, see Chapter 21). If the cotton bollworm too develops resistance against the Bt-cotton, it will force a still large number of farmers to commit suicide as happened in Andhra Pradesh in 1998 and in subsequent years as in Gujrat in 2002.

## F. PREVENTION OF MISUSE OF BIOTECHNOLOGY

In many countries the prevention of misuse of biotechnology is stressed because it may cause a lot of mischief when given a free play in the hands of transnational companies engaged in ruthless persuit of profits. One of these concerns is the rapid pace of genetic erosion. This will lead to a situation where the base genetic material is available to only a few multinational companies in their gene banks. However, this genetic erosion has to be checked by saving the genetic diversity in its own environment, but not gene bank/germplasm bank, by the involvement of people.

We are proceeding from 'green revolution' to 'gene revolution'. New transgenic plants and animals are being produced, and seeds, embryos, sperms are preserved. A day may come when farmers would have to depend on genetically engineered seeds. Still, there is doubt whether such seeds will suit for sustainable agricultural practice free from chemical poison.

The genetically modified organisms (GMOs) should be carefully researched and monitored to ensure that the hazards to users and environment will not occur. In addition, concerted action should be undertaken to ensure that necessary consideration is given to the ethical and social effects of such studies. People should be known about the impacts of GMOs and genetically engineered products.

Such efforts are already being made in some of the countries. The German Green party has called for a 5-year moratorium on commercial release of GMOs. The UK genetics forum is complaining for a partial mentally irresponsible applications of biotechnology. In the USA, a number of groups are strongly opposing the deliberate release of GMOs. In India, gene campaign has been carrying out a public campaign against the patenting of life forms and the misuse of biotechnology. However, such efforts need to be strengthened to check the misuses of biotechnology.

## G. BIODIVERSITY AND ITS CONSERVATIONS

Biodiversity, is a new name for species-richness (of plants, animals and microorganisms) occurring as an interacting system in a given habitat. Biodiversity cannot be replaced because the species becomes adapted in a given habitat after a long course of time. That is why, due to plausticity in their nature and unsustainable resource utilization, over 2.5 lakh species are lost and thousands are threatened to extinction. If a species extincts, it means whole of the gene pool extincts. The real value of biodiversity lies in the informations that are enclosed in the genes. Therefore, there is urgent need, for future, to protect the genes from destruction.

Biodiversity may be defined as *the inherent and externally imposed variability within and among the living organisms present in terrestrial, marine and other ecosystems at the specific time*. Diversity includes the variability in genes, genotype, species, genera, family and ecosystem at a particular time in a specific region. Thus, biodiversity is an expression of both numbers and differences and differences can be seen as a measure of complexity. Commonly biodiversity may be considered at the following three different levels:

**(*i*) Genetic Diversity:** Genes are the functional entities of all organisms because they determine the physical and biological features of organisms. Genes of one organism differ from that of the other organisms. Besides, variations arise due to mutation brought about in genome which is the cause of genetic diversity.

**(*ii*) Species Diversity:** A species is a group of organisms which are genetically identical and interbreed to produce progeny. In contrast horses and zebra are two different species but genetically similar. They can interbreed but produce all infertile offspring. Usually species differ in appearance and thus one can differentiate one species from the other. Thus species diversity is estimated on the basis of total number of species within the discrete geographical boundaries.

**(*iii*) Ecosystem Diversity:** There is diverse ecosystems, and organisms living in such ecosystems are adapted to its respective ecosystems. Therefore, there arises diversity among them. Under ecosystem diversity two phenomenon are frequently referred: the variety of species within different ecosystems *i.e.* more diverse ecosystems contain more species, and the variety of ecosystems found within a certain bio-geographical boundaries.

### 1. Biodiversity in India

India is rich in biodiversity to agriculture, animal husbandry, fisheries and forestry. Much has been described in *Ayurved* and other ancient literature about the indigenous system of medicine, knowledge and wisdom of people. These are supported by a very strong scientific and technological base. In India, over 1,15,000 species of plants and animals have already been identified and described. The country is an important Centre of Diversity and origin of over 167 important cultivated plant spices and domestic animals. A few crops which arose in India and spread throughout the world are : rice, sugarcane, vignas, jute, mango, citrus, banana, millets, spices, medicinal aromatics and ornamentals. No country in the world is as rich in biodiversity as ours. Himalaya itself has the natural wealth of plants, many of which are still unknown and many endangered.

### 2. Conservation of Biodiversity

There is an urgent need for biodiversity rich countries to save it against destruction. However, in most of the developing countries, biodiverisity is attached to environment and forest agencies which have no idea about it. If such countries are not aware of conserving it for sustainable utilization, they would be compelled to export biodiversity and import products for well being of their people.

The agreement between Institute of Biodiversity (INBio) in Costa Rica and Merck (USA) is hailed throughout the industrial world. Under this agreement, extracts, from the wild plants, insects and microorganisms from Costa Rica are supplied to Merck. In return, INBio receives from Merck over 1.35 million US dollars, and expects royalty on the commercial products. INBio has to contribute 50% of royalty to the Government of Costa Rica for National Park Service. The Government has given to INBio rights to bioprospects and share conservation work. Thus, the INBio represents an alliance between biologists/ biochemists and businessmen.

In India, a large number of institutions are involved in conservation and utilization of biodiversity which come under Ministry of Environment and Forest, Agriculture, and Science and Technology. They deal conservation of biosphere reserve, national parks, wild life, sanctuaries, field gene banks, etc. The country needs more expertise and methodologies besides tiger-bird-wild life syndome. India is predominantly an agricultural country. Therefore, the policy makers have to realize that conservation and sustainable utilization of biodiversity must be placed on the top of all developmental plannings.

## H. GENE BANK (GERMPLASM BANK) AND PLANT CONSERVATION

Out of 2,50,000 different plant species, some are already lost and nearly 20,000 seed plants are threatened to extinction. Due to disappearing of natural resources at fast rate, genetic variabilities are being lost for ever, which will result in a dangerous future. Therefore, to save the threatened and gradually vanishing species, and to meet the world demand of food, we would have to conserve the natural heritage. The gene bank (germplasm bank) has taken the challenge of conserving the gene pool of economically, medicinally, socially and ecologically important plant species.

In the first decade of the 20th century a Russian scientist, N.I. Vavilov, was the first to realize the need of conservation of plant genetic resources. Later, in 1920, he established the first genetic resource centre (GRC) of the world at Leningrad. Thereafter, a global concern was raised after the International Biological Programme in 1964.

### 1. World Genetic Resource Centres

During the last two decades, many regional and international GRCs have been set up in different countries. International Rice Research Institute, Manila (Philippines) has rice germplasm bank where 25,000 varieties of rice germplasm has been collected throughout the world. Similarly, maize germplasm have been stored at Maize and Wheat Improvement Centre, Mexico (with more than 12,000 varieties), Institute Colombiano Agropecuario, Colombia (more than 2000 varieties), Instituto National de Investigations Agricolas, Mexico (with more than 7000 varieties), and potato germplasm at International Potato Centre, Lima (Peru). A seed bank has been set up at the National Bureau of Plant Genetic Resources (NBPGR) (New Delhi) which is associated with a world network of gene resource centres coordinated by the International Bureau of Plant Genetic Resources of FAO.

### 2. Conservation and Storage

For the conservation of genetic materials, seeds, pollen grains, vegetatively propagating parts, plant tissues or genome of plant cells are collected. These depend on: (*i*) ease to handle, (*ii*) nature of crops, (*iii*) longevity, (*iv*) expertise, (*v*) methodologies available, (*vi*) space required, (*vii*) problems of genetic erosion of stocks, and (*viii*) expenses. The seeds are compact and easy to handle; hence seed collection and preservation are commonly adopted. Seeds are stored in sealed hermetic containers, dried to a desired moisture level and maintained at desired low temperature.

Viability of pollen grains differs in different species; the species having maximum viability are selected. The families showing the highest longevity of pollen grains are: Pinaceae, Primulaceae, Rosaceae and Saxifragaceae. The longest period of pollen storage recorded is 9 years at –20°C for apples. The vegetatively propagating parts (corn, bulb, tubers, etc) of some plants such as *Dioscorea,* potato, sweet potato, etc. are used for storage purposes.

Techniques have been developed to culture the plant cell, tissue and organs. *In vitro* grown cultures (*e.g.* plantlet, apical meristem culture, etc.) are stored (*see* Chapter 10). Storage of *in vitro* grown cultures has many advantages over the others such as: (*i*) requirement of less space, (*ii*) cheap in maintenance, (*iii*) high propagation potential, (*iv*) least problem of genetic erosion of stock, and (*v*) maintenance of pathogen free stock.

## PROBLEMS

1. What is biotechnology? Discuss in brief the different areas of interest.
2. Why can microbiology, genetics and biochemistry, without combining with technology, not be called as biotechnology?
3. What are the achievements of biotechnology?
4. Write an essay on global impact of biotechnology.
5. Write an essay on History of biotechnology.
6. Discuss in detail how did traditional biotechnology help in the evolution of biotechnology?
7. Give a brief account of emergence of modern biotechnology
8. Give a brief note on potential of modern biotechnology
9. Write short notes on the following :

   (*i*) NBTB (*ii*) DBT (*iii*) OECD (*iv*) ICGEB

   (*v*) Lal Bahadur Shastri Centre for Biotechnology

   (*vi*) Biotechnology in India,

   (*vii*) Biodiversity

   (*viii*) Ban on genetic food.

   (*ix*) Modern biotechnology

   (*x*) Traditional biotechnology

CHAPTER

2

# Genes : Nature, Concept And Synthesis

Richard Dawkins in his book "The Selfish Gene" (1989) has stated about the gene *"they are in you, and in me; they created us, body and minds and their preservation is the ultimate rationale for our existence,......... they go by the name of genes, and we as their survival machines*. The clear concept of gene came out after the work of T.H. Morgan who gave 'the theory of genes' for which he was awarded Nobel Prize in 1933. The recent work on structure of chromosomes and DNA clearly defines the gene as 'a small segement of polynucleotide chain consisting of hundreds or thousands of nucleotides.' Nucleotides are the subunits of a gene. The total complement of genes in an organism is known as its genome. The genes are stored on one or more chromosomes and instruct the cells to make proteins. The size of genes in different organism or within one organism (that consists of more than one thromosome) varies. In this chapter physical and chemical nature of genes, classical and modern concepts of genes, their expression and regulation, and artificial synthesis have been discussed.

## A. CHEMICAL NATURE OF DNA

The DNA is found in all plants, animals, prokaryotes and some viruses. In eukaryotes it is present inside the nucleus, chloroplast and mitochondria, whereas in prokaryotes it is dispersed in cytoplasm. In plants, animals and some viruses the genetic material is double stranded (ds) DNA molecule except some viruses such as ϕX174. In TMV, influenza virus, poliomyelitis virus

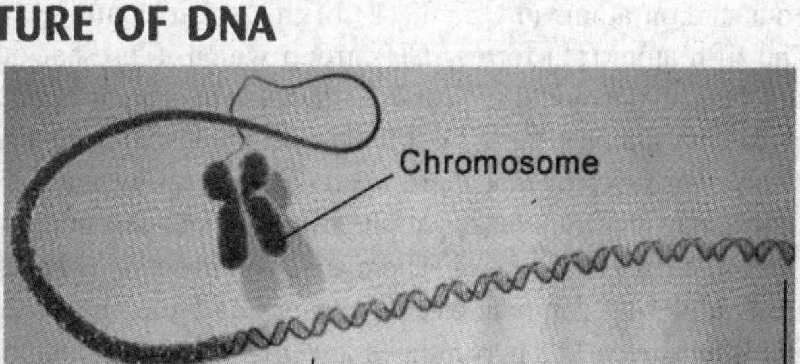

Each chromosome consists of many genes that are made up of DNA.

and bacteriophages the genetic material is single stranded (ss) RNA molecule (Table 2.1). The entire genetic message that controls the chemistry of every cell of the body acting in a specific way is actually written in the language of four nitrogenous bases of DNA *i.e.* purines and pyrimidines. The defined sequence of the four bases constitutes a 'gene' which may be a few or several hundred base pairs long. Genes are the structures of the blue prints called proteins which control the infinite variety of life.

**> Table 2.1. Nature of genetic material.**

| *DNA/RNA* | *Examples* |
|---|---|
| Double stranded DNA (dsDNA) | Higher plants, animals, bacteria, animal viruses (polyoma virus, small pox, herpes virus), Bacteriophages (T-even) |
| Single stranded DNA (ssDNA) | Bacteriophages (ϕ X174 and other bacteriophages), Animal viruses (parvovirus) |
| Double stranded RNA (dsRNA) | Retrovirus, Reovirus, Hepatitis-B virus, animal virus |
| Single stranded RNA (ssRNA) | Plant viruses ( tobacco mosaic virus) |
| | Animal viruses (influenza virus, poliomyelitis virus) |
| | Bacteriophages ( $F_2$, ~ $R_{17}$) |

## 1. Chemical Composition

Purified DNA isolated from a variety of plants, animals, bacteria and viruses has shown a complex form of polymeric compounds containing four monomers known as deoxyribonucleotide monomers or deoxyribotids (Fig 2.1). Each deoxyribonucleotide consists of pentose sugar (deoxyribose), a phosphate group and a nitrogenous base (either purine or pyrimidine). Purines bases (adenine and guanine) are heterocyclic and two ringed bases and the pyrimidines (thymine and cytosine) are one ringed bases. The following components of deoxyribonucleotide have been described:

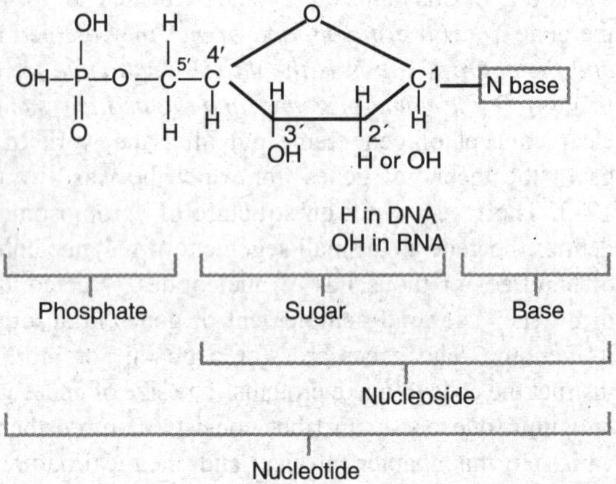


**Fig. 2.1.** A typical nucleotide showing its components: base, sugar and phosphate.

***(i)* A five carbon ring.** Deoxyribose is a pentose sugar consisting of five carbon atoms. Four carbon atoms (1′, 2′, 3′, 4′) of this sugar combine with one oxygen atom and form a ring. The fifth atom (5′) forms -$CH_2$ group which is present outside this ring. Three -OH groups are attached at position 1′, 3′ and 5′ and the hydrogen atoms combine at position 1′, 2′, 3′ and 4′ of carbon atoms (Fig 2.1). In ribonucleotides, the pentose sugar is ribose which is similar to deoxyribose except that there is an -OH group instead of -H at 2′ carbon atom. The absence of -OH group in DNA makes it chemically more stable than the RNA.

***(ii)* Nitrogenous base.** There are two nitrogenous bases, purines and pyrimidines. The purines are double ring compounds that consists of 5-membered imidazole ring with nitrogen at 1′, 3′, 7′ and 9′ position. The pyrimidines are single ring compounds, the nitrogen being at position 1′ and 3′ in 6- membered benzene ring. A single base is attached to 1′-carbon atom of pentose sugar by N-glycosidic bond. Purines are of two types, adenine (A) and guanine (G), and pyrimidines are also of two types, thymine (T) and cytosine (C). Uracil (U) is a third pyrimidine (Fig. 2.2). A, G and C are common in both DNA and RNA. U is found only in RNA.

***(iii)* A phosphate group.** In DNA a phosphate group ($PO_4^{3-}$) is attached to the 3′-carbon of deoxyribose sugar and 5′-carbon of another sugar. Therefore , each strand contains 3′ end and 5′ end arranged in an alternate manner. Strong negative charges of nucleic acid are due to the presence of phosphate group. A nucleotide is a nucleoside phosphate which contains its bond to 3′ and 5′ carbon atoms of pentose sugar that is called phosphodiester.

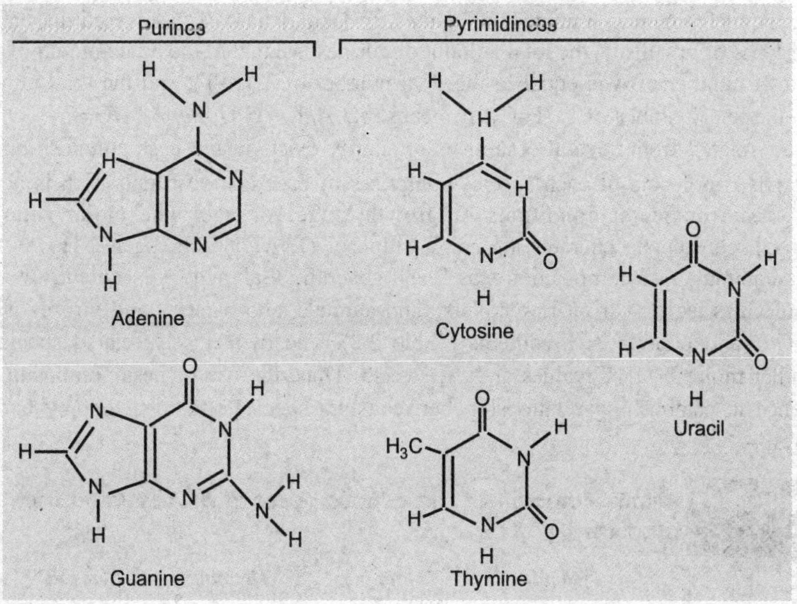


Fig. 2.2. Nitrogenous bases of nucleic acids.

## 2. Nucleosides and Nucleotides

The nitrogenous bases combined with pentose sugar are called nucleosides. A nucleoside linked with phosphate forms a nucleotide (Fig 2.1).

Nucleoside = pentose sugar + nitrogenous base

Nucleotide = nucleoside + phosphate

On the basis of different nitrogenous bases the deoxynucleotides are of following types :

(*i*) Adenine (A) = deoxyadenosine-3′/5′-monophosphate (3′/5′-d AMP)

(*ii*) Guanine (G) = deoxyguanosine -5′-monophosphate (5′-d GMP)

(*iii*) Thymine (T) = deoxythymidine -5′-monophosphate (5′-d TMP)

(*iv*) Cytosine (C) = deoxycytidine -5′-monophosphate (5′-d CMP)

In addition to the presence of nucleosides in DNA helix, these are also present in nucleoplasm and cytoplasm in the form of deoxyribonucleotide phosphates *e.g.* deoxyadenosine triphosphate (dATP), deoxyguanosine triphosphate (dGTP), deoxycytidine triphosphate (dCTP), deoxythymidine triphosphate (dTTP). The advantage of these four deoxyribonucleotide in triphosphate form is that the DNA polymerase acts only on triphosphates of nucleotides during DNA replication.

Similarly, the ribonucleotides contain ribose sugar, nitrogenous bases and phosphate. Except sugar, the other components are similar. However, uracil (U) is found in RNA instead of thymine. Generally, RNA molecule is single stranded besides some exceptions.

## 3. Polynucleotide

The nucleotides undergo the process of polymerization to form a long chain of polynucleotide. The nucleotides are designated by prefixing 'poly' to each repeating unit such as poly A (polyadenylic acid), poly T (polythymidilic acid), poly G (polyguanidylic acid), poly C (polycytidilic acid) and poly U (poly uridylic acid). The polynucleotides that consists of the same repeating unit are called homopolynucleotides such as poly A, poly T, poly G, poly C and poly U.

### 4. Chargaff-equivalence Rule

By 1948, a chemist Erwin Chargaff started using paper chromatography to analyse the base composition of DNA from a number of studies. In 1950, Chargaff discovered that in the DNA of different types of organisms the total amount of purines is equal to the total amount of pyrimidines, *i.e.* the total number of A is equal to the total number of T (A-T), and the total number of G is equal to the total number of C (G-C). It means that A/T = G/C, *i.e.* A+T/G+C = 1. In the DNA molecules isolated from several organisms regularity exists in the base composition.

The DNA molecule of each species comprises of base composition which is not influenced either by environmental conditions or growth stages or age. The molar ratio *i.e.* [A] + [T]/[G]+[C] represents a charateristic composition of DNA of each species. However, in higher plants and animals A-T composition was found generally high and G-C content low, whereas the DNA molecules isolated from lower plants and animals, and bacteria and viruses was generally rich in G-C and poor in A-T contents (Table 2.2). The two closely related species will have very similar molar % G+C values and vice versa. Thus, the use of base composition has much significance in establishing relationship between two species and in taxonomy and phylogeny of species.

**> Table 2.2. Relative amount of nitrogenous bases in DNA isolated from different organisms.**

| *Source* | *Adenine* | *Guanine* | *Thymine* | *Cytosine* | $\frac{A+T}{G+C}$ |
|---|---|---|---|---|---|
| Human sperm | 30.9 | 19.1 | 31.6 | 18.4 | 1.62 |
| Human thymus | 30.9 | 19.9 | 29.4 | 19.8 | 1.52 |
| Sea urchin sperm | 32.8 | 17.7 | 32.1 | 18.4 | 1.85 |
| Wheat germ | 26.5 | 23.5 | 27.0 | 23.0 | 1.19 |
| Yeast | 31.3 | 18.7 | 32.9 | 17.1 | 1.79 |
| *Escherichia coli* | 26.0 | 24.9 | 23.9 | 25.2 | 1.00 |
| *Diplococcus pneumoniae* | 29.8 | 20.5 | 31.6 | 18.0 | 1.59 |
| Bacteriophage $T_2$ | 32.5 | 18.2 | 32.6 | 16.7 | 1.86 |

## B. PHYSICAL NATURE OF DNA

### 1. Watson and Crick's Model of DNA

In 1953, J.D.Watson and F.H.C. Crick combined the physical and chemical data generated by earlier workers, and proposed a double helix model for DNA molecule. This model is widely accepted. According to this model, the DNA molecule consists of two strands which are connected together by hydrogen bonds and helically twisted. Each step on one strand consists of a nucleotide of purine base which alternate with that of pyrimidine base. Thus, a strand of DNA molecule is a polymer of four nucleotides *i.e.* A, G, T, C. The two strands join together to form a double helix. Bases of two nucleotides form hydrogen bonds *i.e.* A combines with T by two hydrogen bonds (A = T) and G combines with C by three hydrogen bonds (G ≡ C) (Fig.2.3A). However, the sequence of bonding is such that for every A.T.G.C. on one strand there would be T.A.C.G. on the other strand. Therefore, the two chains are complementary to each other, *i.e.* sequence of nucleotides on one chain is the photocopy of sequence of nucleotides on the other chain. The two strands of double helix run in antiparallel direction *i.e.* they have opposite polarity. In Fig. 2.3A, the left hand strand has $5' \rightarrow 3'$ polarity, whereas the right hand has $3' \rightarrow 5'$ polarity as compared to the first one. The polarity is due to the direction of phosphodiester linkage.

The hydrogen bonds between the two strands are such that maintain a distance of 20 Å. The double helix coils in right hand direction, *i.e.* clockwise direction and completes a turn at every 34 Å distance (Fig. 2.3B). The turning of double helix results in the appearance of a deep and wide groove called major groove. The major groove is the site of bonding of specific protein. The distance between two strands forms a minor groove, one turn of double helix at every 34 Å. Sugar-phosphate (nucleoside) makes the backbone of double helix of DNA molecule (Fig. 2.3B).

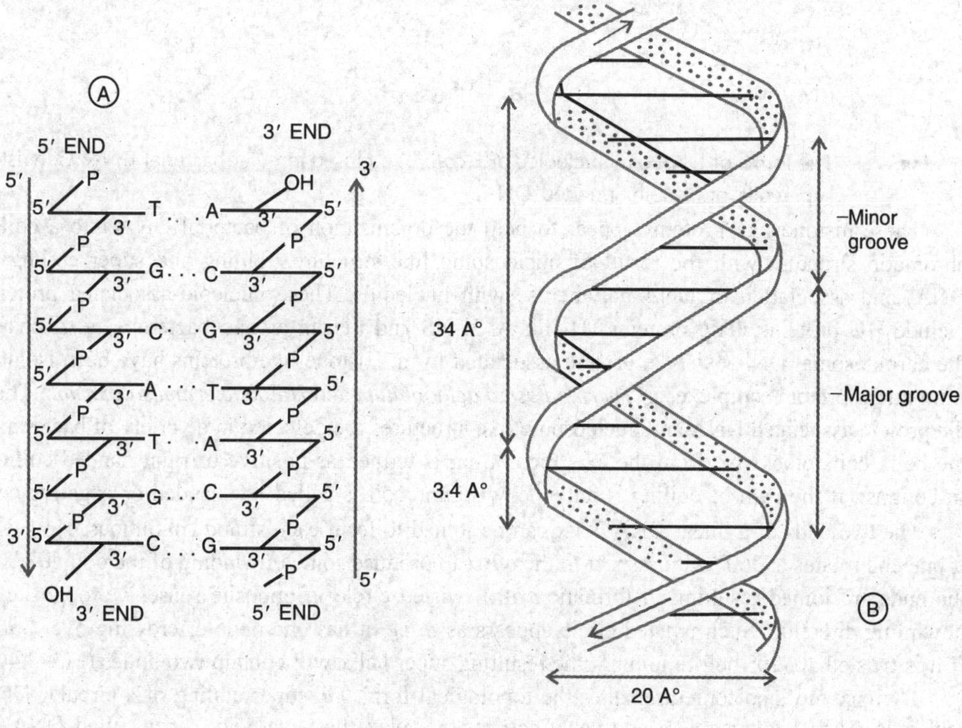


**Fig. 2.3.** Antiparallel orientation of the complementary chains (A), and Watson and Crick's model of DNA double helix (B).

The DNA model also suggested a copying mechanism of the genetic material. DNA replication is the fundamental and unique event underlying growth and reproduction in all living organisms ranging from the smallest viruses to the most complex of all creatures including man. DNA replicates by semiconservative mechanism which was experimentally proved by Mathew, Meselson and Frank W. Stahlin in 1958. If changes occur in sequence or composition of base pairs of DNA, mutation takes place. Though the presence of adenine, guanine, thymine and cytosine is a universal phenomenon, yet unusual bases in DNA molecule also occur. In some bacteriophages, 5′ hydroxymethyl cytosine (HMC) replaces cytosine of the DNA molecule when methylation of adenine, guanine and cytosine occurs. This results in changes of these bases.

### 2. Circular and Super Helical DNA.

Almost in all the prokaryotes and a few viruses, the DNA is organised in the form of closed circle. The two ends of the double helix get covalently sealed to form a closed circle. Thus, a closed circle contains two unbroken complementary strands. Some times one or more nicks or breaks may be present on one or both strands, for example, DNA of phage $PM_2$ (Fig 2.4A). Besides some exceptions, the covalently closed circles are twisted into super helix or super coils (Fig. 2.4B) and is associated with basic proteins but not with histones found complexed with all eukaryotic DNA.

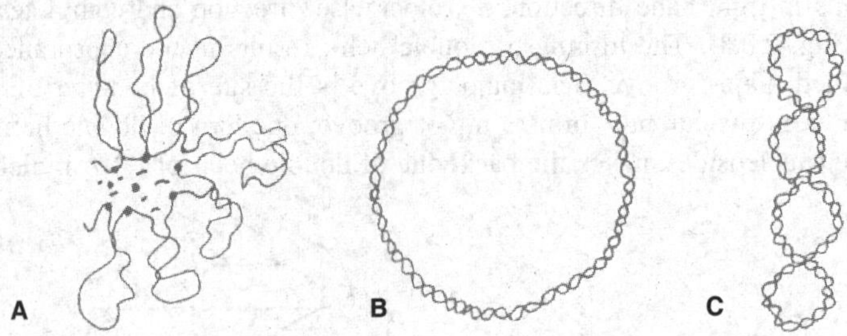


Fig. 2.4. The forms of DNA. A, Nucleoids of *E.coli*; B, a closed circular bacterial DNA; C, twisted supercoils of double stranded DNA.

These histone like proteins appear to help the organization of bacterial DNA into a coiled chromatin structure with the result of nucleosome like structure, folding and super coiling of DNA, and association of DNA polymerase with nucleoids. These nucleoid-associated proteins include HU proteins, IHF, proteins H1, Fir A, H-NS and Fis. In archaeobacteria (e.g. Archaea) the chromosomal DNA exists in protein associated form. Histone like proteins have been isolated from nucleoprotein complexes in *Thermoplasma acidophilum* and *Halobacterium salinarum*.. Thus, the protein associated DNA and nucleosome like structures are defected in a variety of bacteria. If the helix coils clockwise from the axis the coiling is termed as positive or right handed coiling. In contrast, if the path of coiling is anticlockwise, the coil is called left handed or negative coil.

The two ends of a linear DNA helix can be joined to form each strand continuous. However, if one end rotates at 360° with respect to the other to produce some unwinding of the double helix, the ends are joined resulting in formation of a twisted circle in opposite sense, *i.e.* opposite to unwinling direction. Such twisted circle appears as 8, *i.e.* it has one node or crossing over point. If it is twisted at 720° before joining, the resulting super helix will contain two nodes (Fig. 2.4C).

The enzyme topoisomerases alter the topological form, *i.e.* super coiling of a circular DNA molecule. Type I topoisomerases (*e.g. E.coli* top A) relax the negatively super coiled DNA by breaking one of the phosphodiester bonds in dsDNA allowing the 3′-OH end to swivel around the 5′-phosphoryl end, and then resealing the nicked phosphodiester backbone. Type II Topoisomerases need energy to unwind the DNA molecules resulting in the introduction of super coils. One of type II isomerases, the DNA gyrase is apparently responsible for the negatively super coiled state of the bacterial chromosome. Super coiling is essential for efficient replication and transcription of prokaryotic DNA. The bacterial chromosomes is believed to contain about 50 negatively super coiled loops or domains. Each domain represents a separate topological unit, the boundaries of which may be defined by the sites on DNA that limit its rotation.

### 3. Organization of DNA in Eukaryotic Cell

In addition to organization of DNA in prokaryotes and lower eukaryotes as discussed earlier, in eukaryotes the DNA helix is highly organised into the well defined DNA-protein complex termed as nucleosomes. Among the proteins the most prominent are the histones. The histones are small and basic proteins rich in amino acids such as lysine and/or arginine. Almost in all eukaryotic cells there are five types of histones *e.g.* $H_1$, $H_2A$, $H_2B$, $H_3$ and $H_4$. Eight histone molecules (two each of $H_2A$, $H_2B$, $H_3$ and $H_4$) form an octomer ellipsoidal structure of about

11 nm long and 6.5-7 nm in diameter. DNA coils around the surface of ellipsoidal structure of histones 166 base pairs (about 7/4 turns) before proceeding onto the next and form a complex structure, the nucleosome (Fig. 2.5A-B). Thus a nucleosome is an octomer or four histone proteins complexed with DNA.

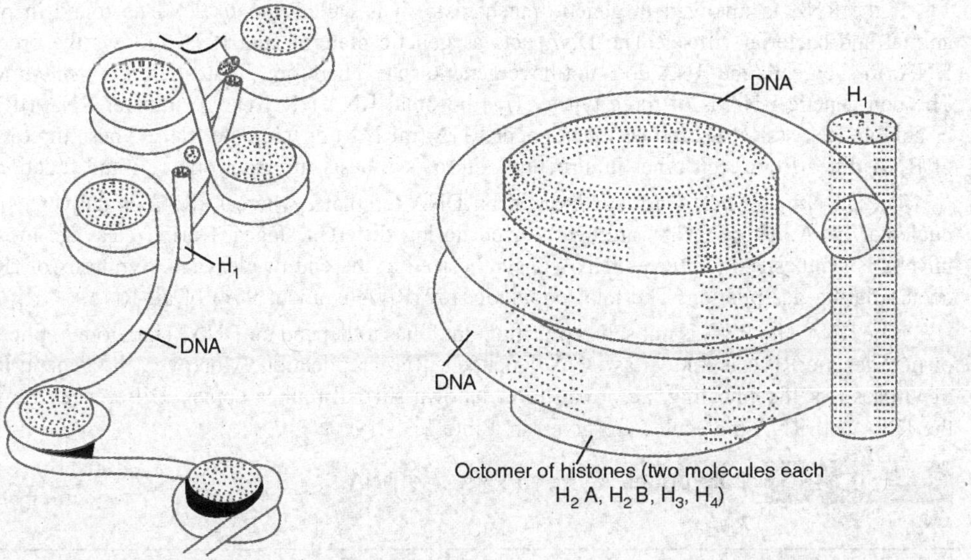


**Fig. 2.5.** Internal organisation of nucleosomes. A highly super coiled chromatin fibre; B, a single nucleosome.

The histones play an important role in determining eukaryotic chromosomes by determining the conformation known as chromatin. The nucleosomes are the repeating units of DNA organization which are often termed as beads. The DNA isolated from chromatin looks like string or beads. The 146 base pairs of DNA lie in the helical path and the histone-DNA assembly is known as the nucleosome core particle. The stretch of DNA between the nucleosome is known as 'linker' which varies in length from 14 to over 100 base pairs. The $H_1$ is associated with the linker region and helps the folding of DNA into complex structure called chromatin fibres which in turn gets coiled to form chromatin. As a result of maximum folding of DNA, chromatin becomes visible as chromosomes during cell division.

## C. STRUCTURE OF RIBONUCLEIC ACID (RNA)

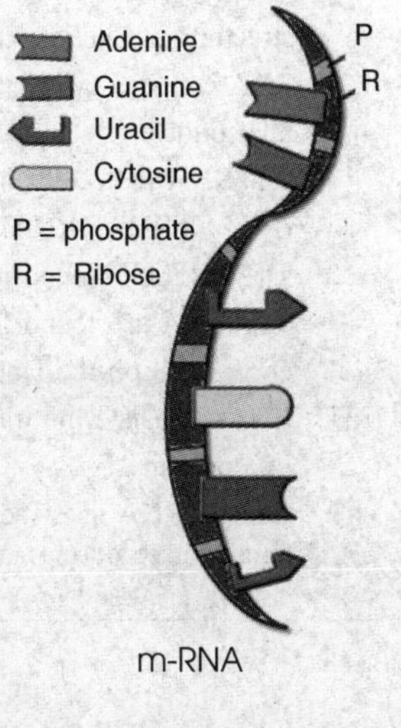


The RNA is usually single stranded except some viruses such as TMV, yellow mosaic virus, influenza virus, foot and mouth disease virus, reovirus, wound tumour virus, etc. which have dsDNA. The single strand of the RNA is folded either at certain regions or entirely to form hairpin shaped structure. In the hairpin shaped structures the complementary bases are linked by hydrogen bonds which give stability to the molecules. However, no complementary bases are found in the unfolded region. The RNA does not possess equal purine-pyrimidine ratio, as it is found in the DNA.

Like DNA, the RNA is also the polymer of four nucleotides—each one contains D-ribose, phosphoric acid and a nitrogenous base. The bases are two purines (A,G) and two pyrimidines (C, U). Thyamine is not found in RNA. Pairing between bases occurs as A-U and G-C.

The nucleotides formed by the four bases are adenosine monophosphate (AMP), guanosine monophosphate (GMP), cytosine monophosphate (CMP), and uridine monophosphate (UMP). These are found freely in nucleoplasm but in the form of triphosphates *e.g.* ATP, GTP, UTP and CTP. As a result of polymerization the ribonucleotides form a polynucleotide chain of RNA.

If the RNA is involved in genetic mechanism, it is called *genetic RNA* as found in plant, animal and bacterial viruses. The DNA acts as genetic material and RNA follows the order of DNA. In such cells the RNA does not have genetic role. Therefore, it is called *non-genetic RNA*. The non-genetic RNA is of three types : (*i*) ribosomal RNA (rRNA), (*ii*) transfer RNA (tRNA) or soluble RNA (sRNA), and (*iii*) messenger RNA (mRNA) or template RNA. These three types of RNA differ from each other in structure, site of synthesis in eukaryotic cell and function.

The mRNA and tRNA are synthesized on DNA template, whereas rRNA is delivered from nucleolar DNA. These RNAs are synthesized during different stages. During cleavage most of mRNA is synthesized , whereas tRNA is synthesized at the end of cleavage. Synthesis of rRNA occurs during gastrulation. The total population of rRNA is about 90% of all RNAs.

Like DNA, the RNA is not self-replicating but it has to depend on DNA. Therefore, replication of non-genetic RNA is known as DNA dependent RNA replication. Moreover, the genetic RNA of viruses is self-replicating, *i.e.* it can form its own several replica copies. Differences between the DNA and RNA molecules are given in Table 2.3.

**> Table 2.3. Differences between RNA and DNA.**

| *RNA* | *DNA* |
|---|---|
| 1. RNA is more primitive than DNA | 1. DNA originated after RNA |
| 2. RNA is the genetic material of some plant, animal and bacterial viruses. | 2. DNA is the genetic material of almost all living organisms |
| 3. Except some viruses (*e.g.* reovirus) most cellular RNA is single stranded. | 3. Except a few viruses (*e.g.* $\phi$X174), most DNA is double stranded. |
| 4. Pentose sugar is ribose. | 4. Pentose sugar is deoxyribose. |
| 5. The bases are adenine, guanine, cytosine and uracil. | 5. The bases are adenine, guanine, cytosine and thymine. |
| 6. Base pairing occurs between adenine and uracil (A-U), and guanine and cytosine (G-C) | 6. Base pairs are A-T and G-C. |
| 7. Base pairing is seen only in hairpin structure and helical region. | 7. Base pairing occurs throughout the length of DNA molecule. |
| 8. RNA contains a few (about 12,000) nucleotides | 8. DNA contains millions of nucleotides *e.g.* over 4 millions. |
| 9. The RNA molecules are of three types: rRNA, mRNA, tRNA. | 9. DNA is only of one type |
| 10. The mRNA is found in nucleolus, tRNA and rRNA are found in cytoplasm. They are formed on the DNA. | 10. DNA is found in chromosomes. DNA is also found in mitochondria and chloroplasts. |
| 11. RNAs translate the transcripts of DNA into proteins. | 11. DNA encodes the genetic masses in a form that transcripts. |
| 12. Genetic RNA uses the enzyme reverse transcriptase during replication. | 12. This enzyme is not required by DNA. DNA after replication forms DNA and after transcription forms RNA. |

## D. GENE CONCEPT

Although the role of herėditary units (factors) in transfer of genetic characters over several generations in organisms was advocated by Gregor John Mendel, yet the mystery of the 'hereditary units′ was unravelled during early 1900s. In 1909, W. Johanson coined the term 'gene' that acts as hereditary units. However, early work done by several workers proposes various hypotheses to explain the exact nature of genes. In 1906, W. Bateson and R.C. Punnet reported the first case of linkage in sweet pea and proposed the *presence or absence theory*. According to them the dominant character has a determiner, and the recessive character lacks determiner. In 1926, T.H. Morgan discarded all the previous existing theories and put forth the *particulate gene* theory. He thought that genes are arranged in a linear order on the chromosome and look like beads on a string. In 1928, Belling proposed that the chromosome that appeared as granules would be the gene. This theory of gene was well accepted by the cytologists. In 1933, Morgan was awarded Nobel prize for advocating the theory of genes. After the discovery of DNA as carrier of genetic informations, the Morgan's theory was discarded. Therefore, it is necessary to understand both, the classical and modern concepts of gene.

T.H. Morgan

According to the classical concepts a gene is a unit of: (*i*) physiological functions, (*ii*) transmission or segregation of characters, and (*iii*) mutation. In 1969, Shapiro and co-workers published the first picture of isolated genes. They purified the *lac* operon of DNA and took photographs through electron microscope.

In 1908, the British physician Sir E.R. Garrod first proposed one-gene-one product hypothesis. In 1941, G.W. Beadle and E.L. Tatum working at St Standford university clearly demonstrated one-gene-one enzyme hypothesis, based on experiments on *Neurospora crassa*. They made it clear that genes are the functional units and transmitted to progenies over generations; also they undergo mutations. They treated *N. crassa* with X-rays and selected for X-ray induced mutations that would have been lethal. Their selection would have been possible when *N. crassa* was allowed to grow on nutrient medium containing vitamin B6. This explains that X-rays mutated vitamin B6 synthesising genes. They concluded that a gene codes for the synthesis of one enzyme. In 1958, Beadle and Tatum with Lederberg received a Nobel prize for their contribution to physiological genetics.

## E. UNITS OF A GENE

After much extensive work done by the molecular biologists the nature of gene became clear. A gene can be defined as a polynucleotide chain that consists of segments each controlling a particular trait. Now, genes are considered as a unit of function (*cistron*), a unit of recombination (*recon*) and a unit of mutation (*mutan*).

### 1. Cistron

One-gene-one enzyme hypothesis of Beadle and Tatum was redefined by several workers in coming years. A single mRNA is transcribed by a single gene. Therefore, one-gene-one mRNA hypothesis was put forth. Exceptionally, a single mRNA is also transcribed by more than one gene and it is said to be polycistronic. Therefore, the concept has been given as one-gene-one protein hypothesis. The proteins are the polypeptide chain of amino acids translated by mRNA. Therefore, it has been correctly used as one-gene-one polypeptide hypothesis.

Moreover, genes are present within the chromosome and their *cis-trans* effect governs the chromosome function. Therefore, S. Benzer termed the functional gene as *cistron* (Fig. 2.6A). Crossing over within the functional genes or cistron is possible. The *cis* and *trans* arrangement of alleles may be written as below:

$$\frac{++}{ab} \qquad \frac{a+}{+b}$$

*Cis* (wild) *Trans* (mutant)

### 2. Recon

Earlier, it was thought that crossing over occurs between two genes. In 1962, S. Benzer demonstrated that the crossing over or recombination occurs within a functional gene or cistron. In a cistron the recombinational units may be more than one. Thus, the smallest unit capable of undergoing recombination is known as *recon* (Fig. 2.6B).

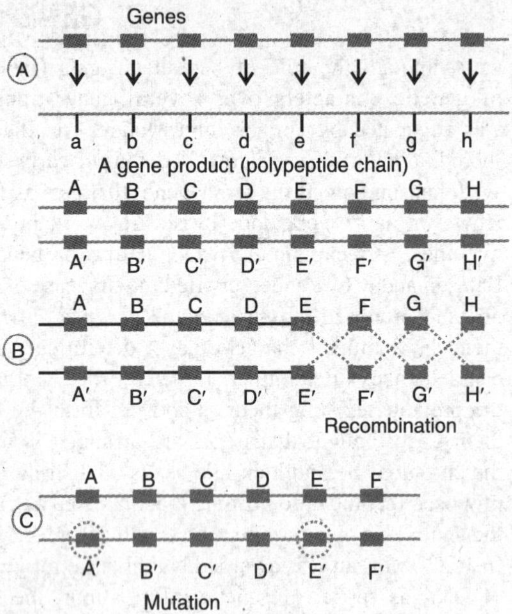


**Fig. 2.6.** The genes as a unit of function *i.e.* cistron (A), recombination *i.e.* recon (B) and mutation *i.e.* muton (C)

S. Benzer (1955) found that the cultures of T4 bacteriophage formed plaques on agar plates of *Eshcherchia coli.* Normal T4 formed small plaques of smooth edges, whereas the mutant T4 phage formed the larger plaques of rough edges. The DNA molecule of T4 phage consists of several genes one of which is called rII region. Formation of rough edged plaques was governed by two adjacent genes (cistrons rIIA and rIIB) in mutant bacteriophage. Both the regions function independently and consist of 2,500 and 1,500 nucleotides, respectively. In rIIA gene over 500 mutational sites are present where crossing over may occur. Through crossing over exchange of two segments of DNA occurs. If crossing over takes place within the gene, by mating two rII mutant of T4 phage a normal wild type phage can be produced. Thus, the work of Benzer lends support that crossing over within a gene occurs, which explains that the recombinational unit (*recon*) is much less smaller than the functional unit *i.e.* cistron.

### 3. Muton

S. Benzer (1962) coined the term *mutan* to denote the smallest unit of chromosome that undergoes mutational changes. Hence, *mutan* may be defined as 'the smallest unit of DNA which may be changed in the nudeotide. Thus, changes at nucleotide level are possible (Fig. 2.6C). The smallest unit of mutan is the nucleotide. Therefore, cistron is largest unit in size followed by recon and mutan. This can be explained that a gene consists of several cistron, a cistron contains many recon, and a recon a number of mutans. However, if the size of a recon is equal to mutan, there would be no possibility in recon for consisting of several mutans.

## F. SPLIT GENES (OR INTRONS)

During 1970, in some mammalian viruses (*e.g.* adenoviruses) it was found that the DNA sequences coding for a polypeptide were not present continuously but were split into several pieces. Therefore, these genes were variously named as split genes or *introns*, *interrupted genes* or *intervening sequences*, *inserts*, and *Junk DNA*. For the discovery of split genes in adenoviruses and higher organisms, Richards J.Roberts and Phillip Sharp were awarded Nobel prize in 1993.

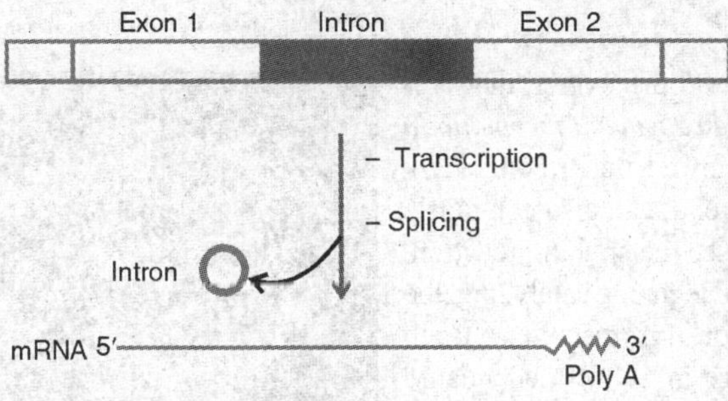


Fig. 2.7. The split genes have exons separated by introns. Removal of introns through RNA splicing.

As shown in Fig 2.7 a DNA sequence codes for mRNA but the complete corresponding sequence of DNA is not found in mRNA. Certain sequences of DNA are missing in mRNA. The sequences present in DNA but missing in mRNA are called intervening sequences or *introns*, and the sequences of DNA found in RNA are known as *exons*. The *exons* code for mRNA.

After transcription a limited RNA transcript has the intron sequences present in the interrupted genes. Genes coding for rRNA and tRNA may also be intervened. The introns are also found in some of the prokaryotes such as cyanobacteria and archaeobacteria (archaea). For some time it was not certain how mRNA is synthesized from a DNA containing introns? Some possible explanations for the mechanism of mRNA synthesis were given: (*i*) DNA rearrangement occurs during transcription with the removal of introns, (*ii*) during transcription RNA polymerase skips the introns and transcribes only exons, (*iii*) individual exon transcribes separately and rejoins to form the complete mRNA, and (*iv*) RNA polymerase may synthesize both introns and exons, and processing of transcripts occurs later on. The transcripts corresponding to introns are removed. Later on it was shown that the fourth mechanism operates in transcription of mRNA.

### 1. RNA Splicing

Initially the RNA transcript introns are synthesised which are later removed by a process called RNA splicing (Fig. 2.7). The junctions of intron-exon have a GU sequences at the intron's 5′-end, and an AG sequence at its 3′OH end. These two sequences are recognised by the special RNA molecules known as small nuclear RNA (snRNA) or *snurps*.

These together with proteins form small nuclear ribonucleoprotein particles called snRNPs. Some of the snRNPs recognize the splice junctions and splice introns accurately. For example, the UI-snRNP recognizes the 5′-splicing junction, and the U5 snRNP recognizes the 3′ splicing junction. Consequently pre-mRNA is spliced in a large complex called a **spliceosome**. The spliceosome consists of pre-mRNA, five types of snRNPs and non-snRNP splicing factors.

Robert and Sharp, the Nobel prize winners in 1993, independently hybridized the mRNA of adenovirus with their progeny or DNA segments of virus. The mRNAs hybridized the ssDNA of virus where the complementary sequences were present. The mRNA-DNA complexes were observed under electron microscope to confirm which part of viral genome had produced the mRNA strand. It was found that mRNA did not hybridize DNA linearly but showed a discontinuous complexes pattern. Huge loops of unpaired DNA between the hybridized complexes clearly revealed the large chunk of DNA strand that carried no genetic information and did not take part in protein synthesis. The adenovirus mRNA contained four different regions of the DNA.

The *B*-globin genes of mice and rabbits, and tRNA genes of yeast tyrosine-tRNA consists of eight genes three of which have been studied in detail. Each gene contains 14 bases (ATTT-AYCAC-TACGA) as intron in the middle. In the same way the pre-tRNA genes contain introns of 18-19 bases. In all the genes introns are present near anticodon. Similarly, a few rRNA genes are also known to contain introns and some of pre-rRNA are self splicing.

### 2. Ribozyme

For the first time, Thomas Cech in 1986 discovered that pre-rRNA isolated from a ciliated protozoa, *Tetrahymena thermophila* is self splicing. Thereafter, S. Altman showed that ribonuclease cleaves a fragment of pre-tRNA from one end, and also contains a piece of RNA. This RNA fragment catalyses the splicing reaction, *i.e.* acts as enzyme. Therefore, this RNA segment catalyzing the splicing reaction is called *ribozyme*. For this discovery Cech and Altman were awarded the Nobel prize in 1989 in chemistry. The best studied ribozyme activity is the self-splicing of RNA. This process is widespread and occurs in *T. thermophila* pre-tRNA, mitochondrial rRNA and mRNA of yeast and other fungi, chloroplast tRNA, rRNA and mRNA, and mRNA of bacteriophage.

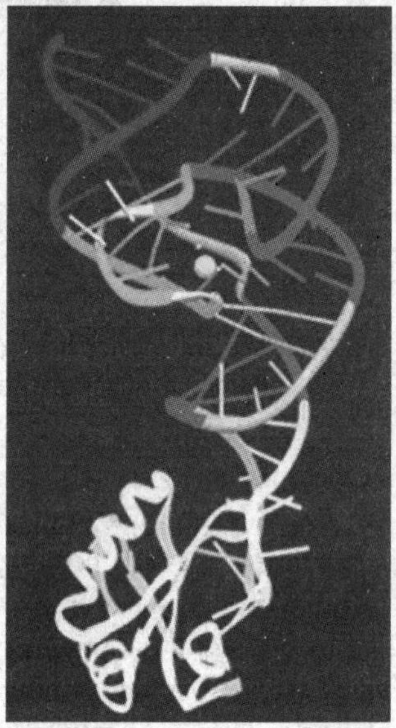

Structure of ribozyme.

The rRNA intron of *T. thermophila* is 413 nucleotide long. The self-splicing reaction needs guanosine and is accomplished in three steps: (*i*) the 3′-G attacks the 5′ group of introns and cleaves the phosphodiester bond, (*ii*) the new 3′-OH group on the left exon attacks the 5′-phosphate on right exon. Consequently two exons join and remove the intron; and (*iii*) the 3′-OH of intron attacks the phosphate bond of nucleotide 15 residues from its end releasing the terminal fragment and cyclizing the intron.

### 3. Evolution of Split Genes

It is supposed that split genes are the ancient condition and bacteria lost their introns only after evolution of most of their proteins. Evidence for the ancient origin of introns has been obtained by the examination of the gene that encodes the ubiquitous enzyme, triosephosphate isomerase (TPI). The TPI is coded by a gene that contains six introns (in vertebrates), five of these are present at the same position as in maize. This shows that five introns were present in the gene before evolution of eukaryotes about $10^9$ years ago.

The TPI plays a key role in cell metabolism that catalyses the interconversion of glyceraldehyde 3-phosphate and dihydroxyacetone phosphate, a central step in glycolysis and glucogenesis. By comparing this enzyme in various organisms it appears that the TPI evolved before the divergence of prokaryotes and eukaryotes from a common ancestor cell, *progenote*. The unicellular organisms under a strong selection pressure minimised the superfluous genome in their cell, whereas there was no such pressure on multicellular organisms. That is why *Aspergillus* has five introns and *Saccharomyces* has none. Precise loss of introns would have occurred by deletion in prokaryotes. The TPI is thought to be evolved to its final three dimensional structure before eubacteria, archaeobacteria and eukaryotic lineage split off from progenote.

## G. OVERLAPPING GENES (GENES WITHIN GENES)

In 1940s, Beadle and Tatum proposed one-gene-one protein hypothesis which explains that one gene encodes for one protein. However, if one gene consists of 1,500 base pairs, a protein of 500 amino acids in length would be synthesized. In addition, if the same sequence read in two different ways, two different amino acids would be synthesized by the same sequence of base pairs. It means, the same DNA sequence can synthesize more than one proteins at different time. It was realized for the first time when the total number of proteins synthesized by ØX174 exceeded from the coding potential of the phage genome. A similar phenomenon is found in the tumour virus SV40 where the total molecular weight of proteins (*i.e.* VP1, VP2 and VP3) synthesized by SV40 genes is much more than the size of the DNA molecule (5,200 base pairs *i.e.* 1,733 codons). From this observation the concept of overlapping genes has emerged.

For the first time B.G. Barrell and coworkers in 1970 gave the evidence for the possibility of the above fact based on the overlapping genes found in bacteriophage ØX174. This virus contains an icosahedral capsid with a knob at each vertex enclosing a single stranded circular DNA (Fig 2.8). F. Sanger and coworkers in 1977 mapped the whole nucleotide sequence of phage ØX174, and phage G4 DNA. The sequences of genes D, E and J, and B, to overlap in the whole sequence of ØX174.

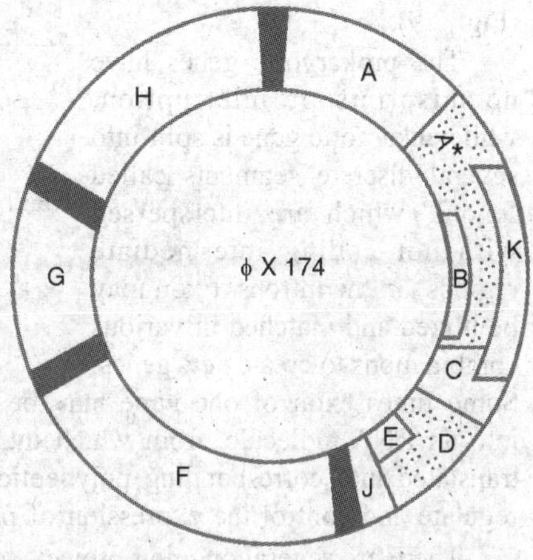


**Fig. 2.8.** Genetic map of ÆX174. The gene B overlaps with gene A; gene K overlaps with A and C; gene E overlaps with gene D.

The ØX174 strand is made up of 5,386 nucleotides of known base sequences. If a single reading frame was used, about 1,795 amino acids would be encoded in the sequence and with and average protein size of about 400 amino acids, only 4-5 proteins could be made. In contrast, ØX174 makes 11 proteins containing a total of more than 2,300 amino acids. The sequence of gene A is now known to contain all of gene B. Gene B is translated in a different reading frame from gene A. Similarly gene E is encoded within gene D. Another translational control mechanism expands the use of gene A. The 37 K Dalton gene A* protein is formed by reinitiating translation at an internal AUG codon within gene A message. The two translational proteins are synthesized by the same translational phase but the functions of the two proteins differ. Protein K initiated near the end of gene A, includes the base sequence of gene B, and terminates in gene C. For example, a reading frame of ....... G,AAG,TTA,ACA...... nucleotides encodes the amino acids lysine, leucine and threonine. However, after reading the frame one nucleotide earlier, the codes become... GAA,GTT,AAC, A... that encode glutamine, valine and asparagine, respectively.

It is obvious that by shifting the reading frame *i.e.* overlapping the code, the same gene can encode two different proteins. Similarly, in the nucleotide sequence .... TAATG....., TAA acts as termination codon of D gene, and ATG acts as the initiation codon of gene J. Here the nucleotide 'A' between A and T overlaps between the two codes. Therefore, the amino acid sequence of A* is similar to a segment of protein A. In addition, overlapping genes have also been detected in animal virus SV40, and tryptophan mRNA of *E.coli.*

## H. GENE ORGANIZATION

The DNA molecules that make up the hereditary elements are called the genome. The functional region of the genome is called *genes*. In a complex genome only a small part is functional, in that it is coded into a protein with the amino acid sequence determined by the DNA sequence. Another small part performs regulatory role by determining the time and extent of decoding in the life of an organism. Protein-coding DNA, along with associated regulatory sequences, makes it sense. A major part of genome is composed of highly repetitive sequences of unknown function which were previously termed 'junk' or 'selfish' DNA.

The DNA is a linear string of symbols, A,T,G and C. Proteins are synthesized by reading a code from DNA sequence, with a triplet of nucleotides (a codon) corresponding to a given

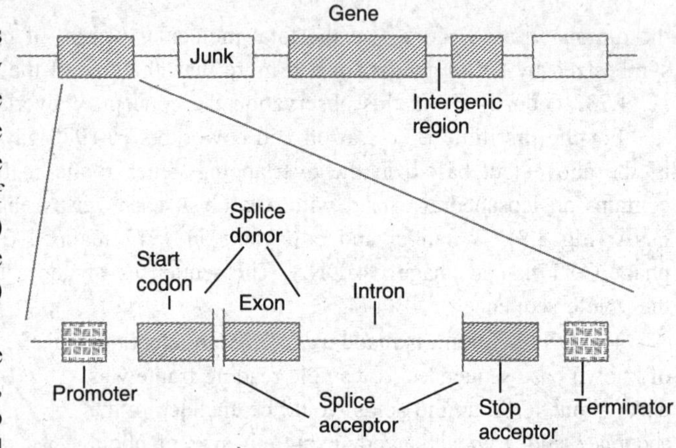


**Fig. 2.9.** Schematic presentation of different regions in and around a gene in a genomic sequence showing the organisation of exons,introns, initiation and termination sites, intergenic spacers and promoters (after S. S.Tewari *et al.*, 1996).

amino acid. Since 20 amino acids are the constituents of naturally occurring proteins, and there are 64 (≡ $4^3$) codons, the genetic code is degenerate. The genetic code also includes a rule for initiation of protein synthesis (the start codon) and a rule to signal the end (the three stop or non-sense codons) (Fig. 2.9).

The prokaryotic genes have no misprints or interruption, while eukaryotic gene is split into several discrete segments called 'exons' which are interspersed with non-coding intermediate regions *i.e.* the introns. Exon may be mixed and matched in various combinations to create new genes. Some times exon of one gene may be intron of another gene. The entire gene is transcribed into an RNA molecule, from which introns are spliced out resulting in mRNA. The mRNA is translated into corresponding polypeptide. There are also ancillary regions or the DNA which regulate and control the expression of proteins at specific time and under specific conditions.

There are several ongoing projects to sequence the entire genome of a number of organisms. The complete genome map of some important organisms may come within a few years, for example, *Drosophila melanogaster* (genome length =165 million bp, consisting of =15000 genes), *E. coli* (4.7 Mbp, 3000 genes), *Saccharomyces cerevisiae* (12.50 Mbp, 6400 genes), *Arabidopsis thaliana* (100 Mbp, 13100 genes), nematode *Caenorhabditis elegans* (100 Mbp, 15000 genes), *Fugu rubripes* (390 Mbp, 8000 genes) and the human genome. Recently, entire genome maps of *Haemophilus influenzae* (1.83 Mbp, 1727 genes) and *Mycoplasms genitalium* (0.58 Mbp, 482 genes) have been sequenced.

## I. GENE EXPRESSION

The DNA has two important roles in the cell, first is replication and the second is expression. Gene expression accomplished by a series of events that contained in DNA is converted into molecule that takes place in the cell. The information contained in DNA is converted into molecules that determines the metabolism of the cell. During the process of gene expression DNA is first copied into an mRNA molecule which determines the amino acid sequence of a molecule of protein. The RNA molecules are synthesized by using a portion of base sequences of single strand of double stranded DNA. This single strand is called template. Hence formation of an mRNA transcript is facilitated by an enzyme RNA polymerase. Therefore, the process of synthesis of an RNA molecule corresponding to a gene is called transcription. By using base sequences and RNA molecule proteins are synthesized in a definite order. Production of an amino acid sequence from an mRNA base sequence is called translation. After completion of translation proteins are synthesized. Therefore gene expression refers to proteins synthesis through two major events, transcription and translation.

**Central Dogma.** DNA itself cannot directly order for the synthesis of amino acids but forms its transcripts first which is then translated into protein. For the first time in 1958, F.H.C.

Crick suggested the unidirection flow of informations from DNA to RNA to protein shown as below:

$$\text{DNA} \xrightarrow{\textit{Transcription}} \text{RNA} \xrightarrow{\textit{Translation}} \text{Protein}$$

This sequencial transfer of information from DNA to protein via RNA is known as *central dogma*. Further more in 1963 H.M. Temin and coworkers reported that Rous sarcoma virus causing cancer contains RNA as genetic material. In 1964, he put forward a hypothesis that RNA tumour viruses synthesise an enzyme reverse transcriptase that synthesized DNA from RNA template. It could also be domonstrated that DNA synthesis prevented after destroying enzyme RNase. It was also shown that RNA specific DNA was synthesized. However, in lymphocytes of leukemia patients RNA-dependent DNA polymerase was discovered. In 1970 Crick again suggested a modified version of central dogma (Fig.2.10).

**Fig. 2.10.** Modified central dogma *i.e.* RNA dependent DNA synthesis.

DNA can undergo replication process to form DNA and transcription to form RNA that in turn undergoes translation. RNA also replicates to form RNA. In special cells only (dot lines) RNA synthesizes DNA, and DNA synthesizes proteins. This process is known as RNA directed DNA synthesis. For detailed discussion of transcription and translation see *A Text Book of Microbiology* by Dubey and Maheshwari (2012).

## J. GENE REGULATION

The DNA of a microbial cell consists of genes a few to thousands which do not express at the same time. At a particular time only a few genes express and synthesize the desired protein. The other genes remain silent at this moment and express when required. Requirement of gene expression is governed by the environment in which they grow. This shows that the genes have a property to switch on and switch off.

Twenty different amino acids constitute different proteins. All are synthesized by codons. Therefore, synthesis of all the amino acids requires energy which is useless because all the amino acids constituting proteins are not needed at a time. Hence, there is need to control the synthesis of those amino acids (proteins) which are not required. By doing this the energy of a living cell is conserved and cells become more competent. Therefore, a control system is operative which is known as *gene regulation* .

There are certain substrates called inducers that induce the enzyme synthesis. For example, if yeast cells are grown in medium containing lactose, an enzyme lactase is formed. Lactase hydrolyses the lactose into glucose and galactose. In the absence of lactase, lactase synthesis does not occur. This shows that lactose induces the enzyme lactase. Therefore, lactase is known as inducible enzyme. In addition, sometimes the end product of metabolism has inhibitory effect on the synthesis of enzyme. This phenomenon is called *feedback* or *end product inhibition*.

From the outgoing discussion it appears that a cell has auto control mediated by the gene itself. For the first time Francois Jacob and Jacques Monod (1961) at the Pasteur Institute (Paris) put forward a hypothesis to explain the induction and repression of enzyme synthesis. They investigated the regulation of activities of genes which control lactase fermentation in *E. coli* through synthesis of an enzyme, β-galactosidase. For this significant contribution in the field of biochemistry they were awarded the Nobel Prize in Physiology or Medicine in 1965.

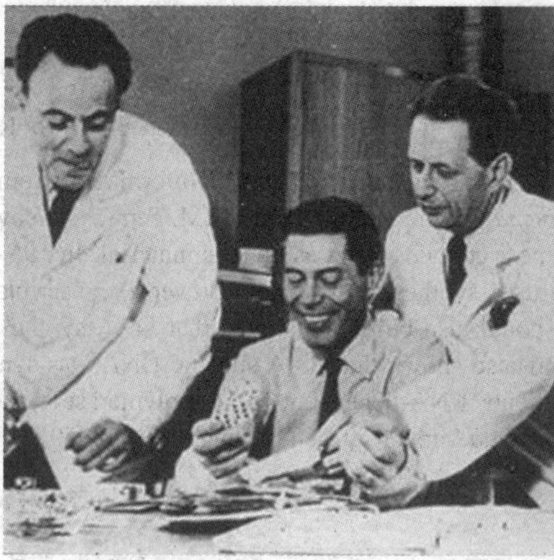

From left to right Francois Jacob (1920–) and Jacques Monod (1910-1976).

Gene expression of prokaryotes is controlled basically at two levels, *i.e.* transcription and translation stages. In addition, mRNA degradation and protein modification also play a role in regulation. Most of the prokaryotic genes being regulated are controlled at transcriptional stage. There are other control measures also that operate at different levels.

## 1. Transcriptional Control

In a living organism it is a general strategy that chemical changes occur by a metabolic pathway through a chain of reactions..Each step is determined by the enzyme. Again synthesis of an enzyme comes under the control of genetic material, *i.e.* DNA in living organisms. Enzyme (protein) synthesis occurs via two steps : transcription and translation. Transcription refers to synthesis of mRNA. Transcription is regulated at or around promoter region of a gene. By controlling the ability of RNA polymerase to the promoter the cell can modulate the amount of message being transcribed through the structural gene. However, if RNA polymerase has bound, it again can modulate transcription. By doing so the amount of gene product synthesized is also modulated. The coding region is also called structural gene. Adjacted to it are regulatory regions that control the structural gene. The regulatory regions are composed of promoter (for the initiation of transcription) and an operator (where a diffusible regulatory protein binds) regions.

The molecular mechanisms for each of regulatory patterns vary widely but usually fall in one of two major groups: negative regulation and positive regulation. In negative regulation an inhibitor is present in the cell and prevents transcription. This inhibitor is called as repressor. An inducer *i.e.* antagonist repressor is required to permit the initiation of transcription. In a positive regulated system an effector molecule (*i.e.* a protein, molecule or molecular complex) activates a promoter. The repressor proteins produce negative control, whereas the activator proteins produce positive control. Since the transcription process is accomplished in three steps (RNA polymerase binding, isomerization of a few nucleotides and release of RNA polymerase from promoter region), the negative regulators usually block the binding, whereas the activators interact with RNA polymerase making one or more steps. In negative regulation an inhibitor is bound to the DNA molecule. It must be removed for efficient transcription. In positive regulation an effector molecule must bind to DNA for transcription.

**(*a*) The *Lac* Operon :** For the first time in 1961, F. Jacob and J. Monod gave the concept of operon model to explain the regulation of gene action. An operon is defined as several distinct genes situated in tadem all controlled by a common regulatory region. Commonly an operon consists of repressor, promoter, operator and structural genes. The message produced by an operon is polycistronic because the information of all the structural genes resides on a single molecule of mRNA.

The regulatory mechanism of operon responsible for utilization of lactose as a carbon source is called the *lac operon*. It was extensively studied for the first time by Jacob and Monod in 1961. Lactose is a disaccharide which is composed of glucose and galactose.

The lactose utilizing system consists of two types of components; the structural genes (*lac*Z, *Lac*Y and *lac*A) the products of which are required for transport and metabolism of lactose and the regulatory genes (the *lac*I, the *lac*O and the *lac*P). These two components together comprises the *lac* operon (Fig. 2.11). One of the most key features is that operon provides a mechanism for the coordinate expression of structural genes controlled by regulatory genes. Secondly, operon shows polarity, *i.e.* the genes Z,Y and A synthesise equal quantities of three enzymes β-galactosidase (by *lac*Z), permease ( by *lac*Y) and acetylase (by *lac*A). These are synthesized in an order, *i.e.* β-galactosidase first and acetylase in the last.

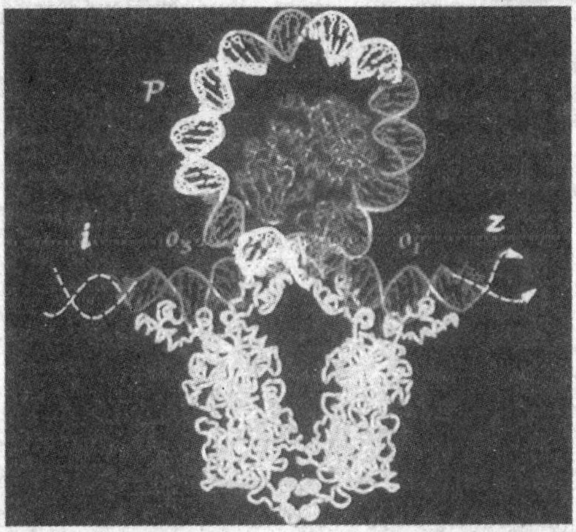


Lactose operon activity is under the control of a repressor protein. The *lac* repressor (violet) and catabolite activator protein (blue) are bound to the lac operon.

(*i*) **The structural genes.** The structural genes form one long polycistronic mRNA molecule. The number of structural gene corresponds to the number of proteins. Each structural gene is controlled independently, and transcribe mRNA molecules separately. This depends on substrates to be utilized. For example, in *lac* operon three structural genes (Z,Y and A) are associated with lactose utilization (Fig. 2.11). β-galactose is the product of *lac*Z that cleaves

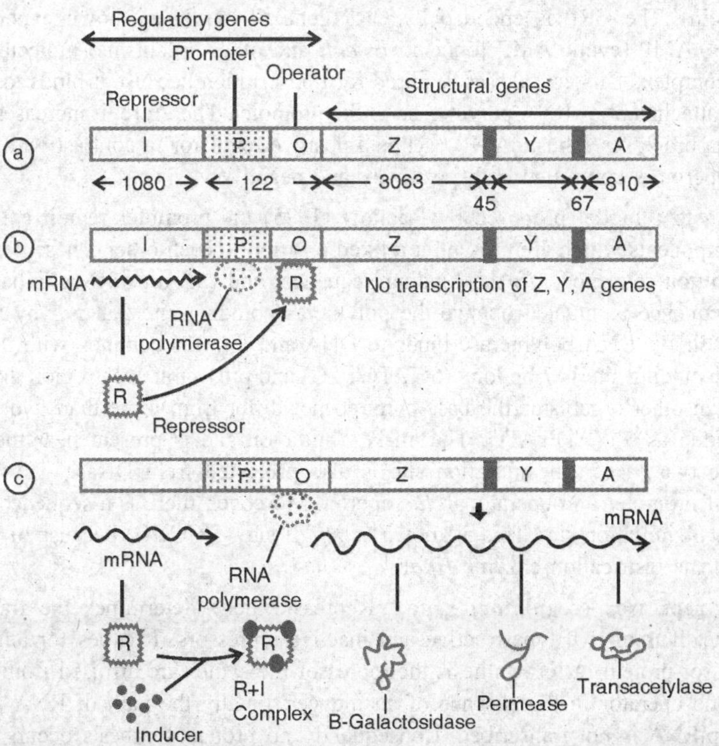


**Fig. 2.11.** The *lac* operon; (*a*) genetic map (the numbers show the number of base pairs that comprise each gene); (*b*) repressed state; (*c*) induced state.

β-1,4 linkage of lactose and releases the free monosaccharides. This enzyme is a tetramer of four identical subunits each with molecular weight of 1,16,400. The enzyme permease (a product of *lac*Y) facilitates the lactose to enter inside the bacterium. Permease has molecular wieght of 46,500. It is hydrophobic. The cells mutant in *lac*Z and *lac*Y are designated as $Lac^-$ *i.e.* the bacteria cannot grow in lactose. The enzyme transacetylase (30,000 MW) is a product of *lac*A whose no definite role has been assigned.

The *lac* operon consists of a promoter (P) and an operator (O) together with the structural genes. The initiation codon of *lac*Z is TAC that corresponds to AUG of mRNA. It is situated 10 bp away from the end of operator gene. However, the *lac* operon cannot function in the presence of sugars other than lactose.

**(*ii*) The operator gene.** The operator gene is about 28 bp in length present adjacent to *lac*Z gene. The base pairs in the operator region are palindrome, *i.e.* show two-fold symmetry from a point (Fig. 2.11). The operator overlaps the promoter region. The *lac* repressor proteins (a tetramer of four subunits) bind to the *lac* operator *in vitro* and protect part of the promoter region from the digestion of DNase. The repressor proteins bind to the operator and form an operator-repressor complex which in turn physically blocks the transcription of Z,Y and A genes by preventing the release of RNA polymerase to begin transcription.

**(*iii*) The promoter gene.** The promoter gene is about 100 nucleotide long and continuous with the operator gene. The promoter gene lies between the operator and regulator genes. Like operators the promoter region consists of palindromic sequence of nucleotides. These palindromic sequences are recognized by such proteins that have symmetrically arranged subunits. This section of two fold symmetry is present on the CRP site that binds to a protein called CRP (cyclic AMP receptor protein). The CRP is encoded by CRP gene. It has been shown experimentaly that CRP binds to cAMP (cyclic AMP found in *E. coli* and other organisms) molecule and form a cAMP-CRP complex. This complex is required for transcription because it binds to promoter and enhances the attachment of RNA polymerase to the promoter. Therefore, it increases transcription and translation processes. Thus, cAMP-CRP is a positive regulator in contrast with the repressor, and the *lac* operon is controlled both positively and negatively.

According to a model proposed by Pribnow (1975) the promoter region consists of three important components which are present at a fixed position to each other. These components are: (*i*) the recognition sequence, (*ii*) the binding sequence, and (*iii*) an mRNA initiation site. The recognition sequence is situated outside the polymerase binding site that is why it is protected from DNase. Firstly, RNA polymerase binds to DNA and forms a complex with the recognition sequence. The binding site is 7 bp long (5'TATGTTG) and present at such region that is protected from DNase. In other organisms the base pairs do not differ from more than two bases. Hence, it can be written as 5' TATPuATG. The mRNA initiation site is present near the binding site on one of the two bases. The initiation site is also protected from DNase. However, there is overlapping of promoter and operator in *lac* operon. Moreover, there is a sequence 5'CCGG, 20 bp left to mRNA initiation site. This is known as *Hpa*II site (5'CCGG) because of being cleaved at this site by the restrication enzyme *Hpa*II.

**(*iv*) The repressor (regulator) gene.** Repressor gene determines the transcription of structural gene. It is of two types: active and inactive repressors. It codes for amino acid of a defined repressor protein. After synthesis the repressor molecules are diffused from the ribosome and bind to the operator in the absence of an inducer. Finally the path of RNA polymerase is blocked and mRNA is not transcribed. Consequently no protein synthesis occurs. This type of mechnism occurs in the inducible system of active repressor.

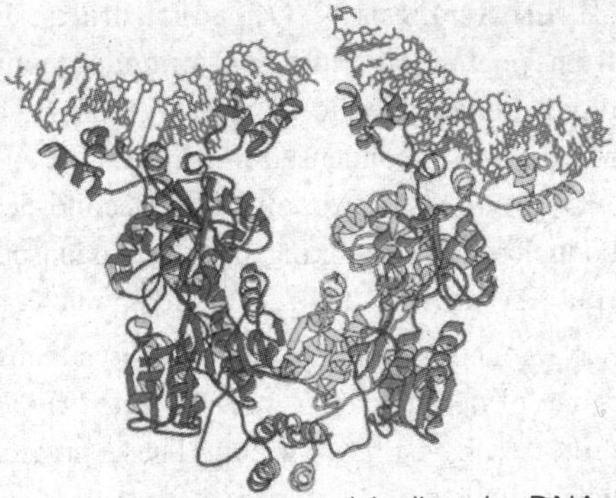

Lactose repressor binding to DNA (shown in blue).

When an inducer (*e.g.* lactose) is present it binds to repressor proteins and forms an inducer-repressor complex. This complex cannot bind to the operator. Due to formation of complex the repressor undergoes changes in conformation of shape and becomes inactive. Consequently the structural genes can synthesise the polycistronic mRNAs and the later synthesizes enzymes (proteins). In contrast, in the reversible system the regulator gene synthesises repressor protein that is inactive and, therefore, fails to bind to operator. Consequently, proteins are synthesized by the structural genes. However, the repressor proteins can be activated in the presence of a corepressor. The corepressor together with repressor proteins forms the repressor-corepressor complex. This complex binds to operator gene and blocks protein synthesis.

In 1961, Jacob and Monod could not identify the repressor protein. In 1966, Gilbert and Muller - Hill succeeded in isolating the *lac* repressor from the *lac* mutant cells of *E. coli* inside which the *lac* repressor was about ten times greater than the normal cells. The *lac* repressor proteins have been crystallized. It has a molecular weight of about 1,50,000 daltons. It consists of four subunits each has 347 amino acid residues and molecular wieght of about 40,000 daltons. The repressor proteins have strong affinity for a segment of 12-15 base pairs of operator gene. This binding of repressor blocks the synthesis of mRNA transcript by RNA polymerase.

## K. ARTIFICIAL SYNTHESIS OF GENES

Har Govind Khorana

For the first time in 1955, Michelson chemically synthesized a dinucleotide in laboratory. Later on in 1970, Har Govind Khorana and K.L. Agarwal for the first time chemically synthesized gene coding for tyrosine tRNA of yeast. For the synthesis of tRNA and rRNA there are specific genes. However, genes of tRNA are the smallest genes containing about 80 nucleotides. In 1965, Robert W. Holley and coworkers worked out first the molecular structure of yeast alanine tRNA. This structure lent support to Khorana in deduction of structure of the gene. A gene is responsible for encoding mRNA, and mRNA for polypeptide chain. If the structure of a polypeptide chain is known, the structure of mRNA from genetic code dictionary and in turn the structure of gene can easily be worked out. There are two approaches for artificial synthesis of the gene, by using chemicals and through mRNAs.

### 1. Synthesis of a Gene for Yeast alanine tRNA

As mentioned earlier that the molecular structure of yeast alanine tRNA was worked out by R.W. Holley and coworkers in 1965 which helped Khorana to deduce the structure of alanine tRNA. They found out that yeast alanine tRNA contains 77 base pairs. It was very difficult to assemble 77 base pairs of nucleotides in ordered form. Therefore, they synthesized chemically the short deoxynucleotide sequences which was joined by hydrogen bonding to form a long complementary strand. By using polynucleotide ligase the double stranded pieces were produced. The complete procedure of synthesizing gene for yeast alanine tRNA is discussed in the following steps.

**(a) Synthesis of Oligonucleotides.** In the first approach, fifteen oligonucleotides ranging from pentanucleotide (*i.e.* oligodeoxynucleotide of five bases) to an icosanucleotide (*i.e.* oligodeoxynucleotide of twenty bases) were synthesized. The chemical synthesis was brought about through condensation between the - OH group at 3′ position of one deoxynucleotide and the -$PO_4$ group at 5′ position of the second deoxynucleotide. All other functional groups of deoxyribonucleotides not taking part in condensation processes were protected so that the condensation could be brought about. The deoxynucleotides were protected as below:

(*i*) The amino group of deoxyadenosine was protected by benzyl group, and the amino group of deoxycytidine was protected by anisoyl group, the amino group of deoxyguanosine was protected by isobutyl group. These protective groups were removed by treating with ammonia when synthesis was over.

(*ii*) The hydroxyl group (-OH) at the 5′ position of receiving deoxynucleotide was protected by cyanoethyl group (HN-$CH_2$-$CH_2$-).

(*iii*) The -OH group at the 3′ position of second or incoming deoxynucleotide was protected by acetyl group.

The different protecting groups used and treatment required to remove the protecting groups are given in Table 2.4. When the groups of deoxynucleotides were protected, the products reacted to form deoxyoligonucleotide. When deoxynucleotides were condensed into oligonucleotides, different protecting groups were removed by treating with ammonia, acid or alkali (Table 2.4). For example, both the cyanoethyl group at the 5′ position and the acetyl group at the 3′ position were removed by alkali treatment.

**> Table 2.4. Different protective groups of nucleotides and their removal.**

| *The groups in base or sugar to be protected* | | *Protected by* | *Protecting groups removed by* |
|---|---|---|---|
| **A.** | **-$NH_2$ group (base)** | | |
| (*i*) | deoxyadenosine | Benzyl group | Ammonia |
| (*ii*) | deoxycytidine | Anisoyl group | Ammonia |
| (*iii*) | deoxyguanosine | Isobutyl group | Ammonia |
| **B.** | **-OH group (sugar)** | | |
| (*i*) | -OH group at 5′ position of first nucleotide | Monomethoxy trityl group | Acid |
| (*ii*) | -OH group at 5′ position of growing chain | Cyanoethyl | Alkali |
| (*iii*) | -OH group at 3′ position | Acetyl group | Alkali |

Finally, condensation between the groups of two, three or four nucleotides was brought about. The receiving segment had a free 3′-OH group and a protected 5′-OH group, whereas the incoming segment had a free 5′-OH group and a protected 3′-OH group. After each addition, the protective group at the 3′ end had removed so that free 3′-OH group could receive another segment.

**(b) Synthesis of three duplex fragments of a gene.** By using 15 single stranded oligonucleotides, three large double stranded DNA fragments were synthesized. These three fragments contained: (*i*) segment of A having the first 20 nucleotides with the nucletides 17-20 as the single stranded, (*ii*) segment B consisting of nucleotides 17-50 with the nucleotides 17-20 and 46-50 as the single stranded, and (*iii*) segment C containing the nucleotides 46-77 with the single stranded region 46-50.

**(*c*) Synthesis of a gene from three duplex fragments of DNA.** The three segments (A,B,C) synthesized as above were joined by using the enzyme polynucleotide ligase to produce the complete gene for alanine tRNA (Fig. 2.12).

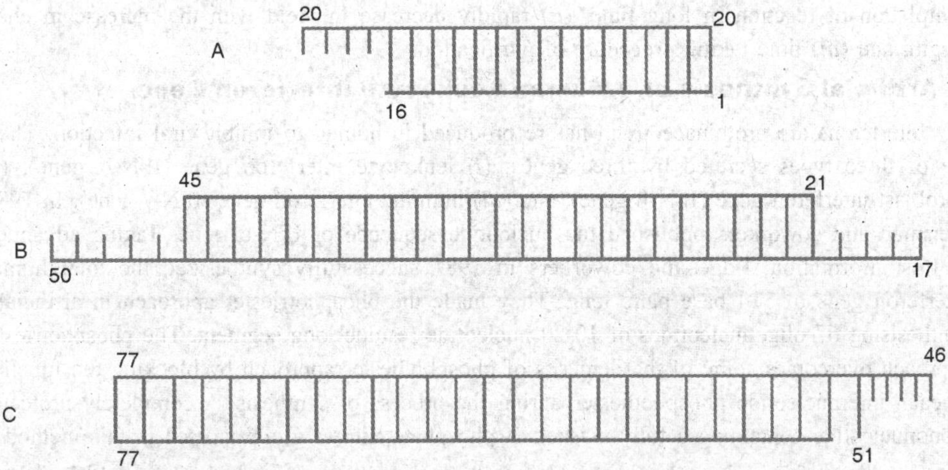


**Fig. 2.12.** The three duplex DNA fragments (AB, and C) were synthesised for the synthesis of the gene for yeast *alanine* tRNA.

The joining of the three fragments was done by any of two following methods:

(*i*) In one approach, fragment A was joined to B by taking advantage of overlapping in nucleotide residues 17-20. Then, the fragment C was added with the overlap in nucleotides 46-50. Thus, a complete double stranded DNA with 77 base pairs was prepared.

(*ii*) In the second approach, the fragment B was joined to C. At the end the fragment A was added to nucleotide residues 17-20 to obtain the complete gene for alanine tRNA.

In 1970, Khorana and co-workers prepared this gene *in vitro* which was used for future work. They found that alanine tRNA gene replicated and transcribed into tRNA just like the natural gene. It is not known whether tRNA prepared from artificially synthesized gene had the molecular organisation similar to alanine tRNA or not.

## 2. Artificial Synthesis of a Gene for Bacterial tyrosine tRNA

In 1975, Khorana and co-workers completed the synthesis of a gene for *E. coli* tyrosine tRNA precursor. *E. coli* tRNA precursors are formed from the larger precursors. The tyrosine tRNA precursor has 126 nucleotides. They synthesized the complete sequence of DNA duplex coding for tyrosine tRNA precursor of *E. coli*, and promoters are terminator genes. Though these segments are not the proper structural gene yet are the regions involved in its regulation. Twenty six small oligonucleotide DNA segments giving rise to tRNA precursor were synthesized which were arranged into six double stranded fragments each containing single stranded ends. These six fragments were joined to give rise complete gene of 126 base pairs for tyrosine tRNA precursor of *E. coli*.

In 1979, Khorana completely synthesised a biologically functional tyrosine tRNA suppressor gene of *E. coli* which was 207 base pairs long and contained: (*i*) a 51 base pairs long DNA corresponding to promoter region, (*ii*) a 126 base pair long DNA corresponding to precursor region of tRNA, (*iii*) a 25 base pair long DNA including 16 base pairs contained restriction site for *Eco*RI. This complete synthetic gene was joined in phage lambda vector which in turn was allowed to transfect *E. coli* cells. After transfection phage containing synthetic gene was successfully multiplied in *E. coli.* Khorana made the phosphodiester approach for synthesizing the oligonucleotides of the biologically active tRNA. The demerits of this approach are: (*i*) the completion of reaction in long time, (*ii*) rapidly decrease in yield with the increase in chain length, and (*iii*) time taking procedure of purification.

### 3. Artificial Synthesis of a Human Leukocyte Interferon Gene

Interferons are proteinaceous in nature produced in human to inhibit viral infection. These are of three types secreted by three genes: (*i*) leukocyte interferon gene (IFN-$\alpha$ gene), (*ii*) fibroblast interferon gene (IFN-$\beta$ gene), and (*iii*) immune interferon gene (IFN-$\gamma$ gene). In 1980, Weisman and coworkers published the nucleotide sequence of IFN-$\alpha$ gene. Taking advantage of this information, Edge and coworkers in 1981 successfully synthesized the total human interferon gene of 514 base pairs long. They made the phosphotriester approach in artificially synthesising 67 oligonucleotides of 10-20 nucleotide residue long segment. The phosphotriester approach overcomes some of the demerits of phosphodiester approach by blocking the function of each internucleotide phosphodiester during the process of synthesis. A completely protected mononucleotide containing a fully masked 3′ phosphate triester group is used in this method.

Coupling of initial nucleotide onto a polyacrylamide resin was done to which further nucleotides in pairs were added. In this way 66 oligonucleotides of 14-21 nucleotide residues were first synthesised. These were arranged in predetermined ways and joined chemically. The 514 base pairs long IFN-$\alpha$ gene contained the initiation and termination signals.

The artificially synthesized gene was incorporated into a plasmid through biotechnological technique. The recombinant plasmid was transferred into *E.coli* cells which expressed $\alpha$-interferon. This technique now-a-days is being adopted to produce interferon commercially.

## L. GENE MACHINE

The fully automated commercial instrument called automated polynucleotide synthesizer or gene machine is available in market which synthesizes predetermined polynucleotide sequence. Therefore, the genes can be synthesized rapidly and in high amount. For example, a gene for tRNA can be synthesized within a few days through gene machine. It automaticaly synthesizes the short segments of single stranded DNA under the control of microprocessor. The working principle of a gene machine includes: (*i*) development of insoluble silica based support in the form of beads which provides support for solid phase synthesis of DNA chain, and (*ii*) development of stable deoxyribonucleoside phosphoramidites as synthons which are stable to oxidation and hydrolysis, and ideal for DNA synthesis.

Fig. 2.13 shows the mechanism of a gene machine. Four separate reservoirs containing nucleotides (A, T, C and G) are connected with a tube to a cylinder (synthesiser column) packed with small silica beads. These beads provide support for assembly of DNA molecules. Reservoirs for reagent and solvent are also attached. The whole procedure of adding or removing the chemicals from the reagent reservoir in time is controlled by microcomputer control system *i.e.* microprocessor.

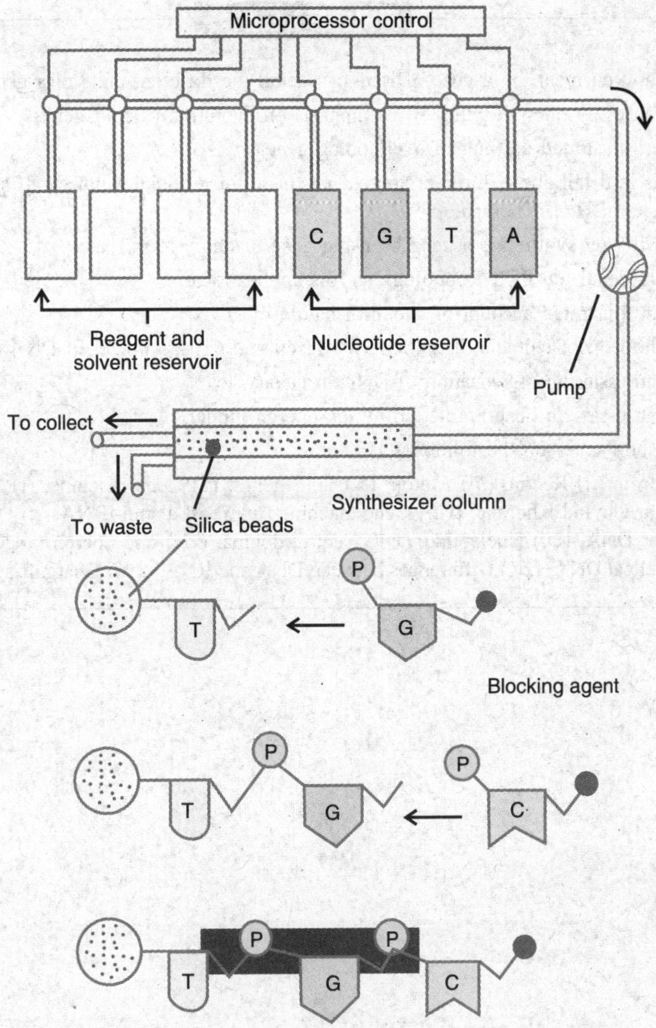


Fig. 2.13. A gene machine and its working mechanism (diagrammatic)

If one desires to synthesise a short polynucleotide with a sequence of nucleotides T, G, C, the cylinder is first filled with beads with a single 'T' attached. Thereafter, it is flooded with 'G' from the reservoir. The right hand side of each G is blocked by using chemicals from the reservoir so that its attachment with any other Gs can be prevented. The remaining Gs which could not join with Ts are flushed from the cylinder. The other chemicals are passed from the reagent and solvent reservoirs so that these can remove the blocks from G which is attached with the T. In the same way this cycle is repeated by flooding with C from reservior into the cylinder. Finally the sequence T.G.C is synthesized on the silica beads which is removed chemically later on.

The desired sequence is entered on a key board and the microprocessor automatically opens the valve of nucleotide reservoir, and chemical and solvent reservoir. In the gene machine the nucleotides are added into a polynucleotide chain at the rate of two nucleotides per hour. By feeding the instructions of human insulin gene in gene machine, human insulin has been synthesized.

## PROBLEMS

1. What do you mean by a gene? Discuss in detail the different units of a gene.
2. Write an essay on split genes with emphasis on evolution of split genes.
3. Give an illustrated account of overlapping gene.
4. Discuss in detail about different approaches made for artificial synthesis of genes with emphasis on works of H.G. Khorana.
5. How will you synthesize a gene by using gene machine or mRNAs.
6. Write in detail the PCR technology and its applications.
7. Give an illustrated account of chemical nature of DNA.
8. With the help of suitable diagram give Watson and Crick's model of DNA double helix.
9. Write an essay on organisation of DNA in eukaryotes.
10. Write an essay on gene regulation by *lac* operon model.
11. Write short notes on the following:

    (*i*) Cistron, (*ii*) Recon, (*iii*) Muton, (*iv*) Split genes, (*v*) Gene machine, (*vi*) Overlapping genes (*vii*) Har Govind Khorana, (*viii*) Gene machine (*ix*) Yeast alanin-tRNA, (*x*) PCR technology, (*xi*) Circular DNA, (*xii*) Nucleotide, (*xiii*) Central dogma, (*xiv*) *Lac* operon model (*xv*) Circular and superhelical DNA, (*xvi*) Differences between DNA and RNA, (*xvii*) Chargaff's rule of equivalence.

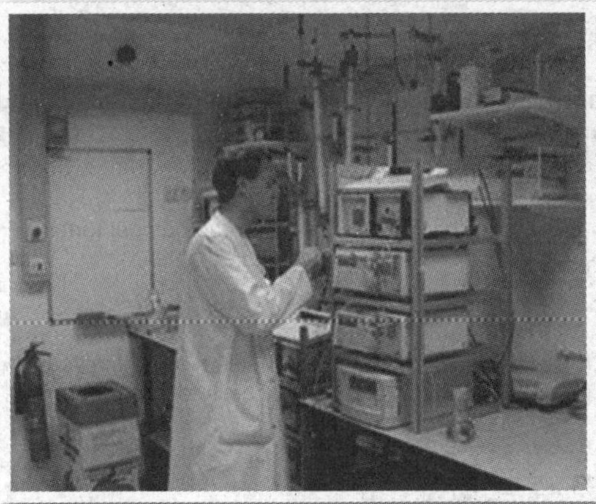

CHAPTER

3

# Tools of Genetic Engineering – I: Basic Requirements

Genetic engineering (= gene cloning, rDNA technology) can be defined as "changing of genes by using *in vitro* processes". A gene of known function can be transferred from its normal location into a cell (that of course, does not contain it) via a suitable vector. The transferred gene replicates normally and is handed over to the next progeny. On confirmation for its presence through biochemical procedures replica of the same cell (*i.e.* clones) can be produced. The derivation of procedures for the reintroduction of the foreign DNA fragment into a bacterium have lead to evolution of new technology, *i.e.* the recombinant DNA technology, gene cloning, gene manipulation or genetic engineering.

The macromolecules such as DNA, RNA, proteins, etc. are synthesised inside the living cells which vary with each other in respect of molecular weight (size), solubility, presence of charges, absorbance of light wavelength (spectrum), etc. There are many techniques that are used to isolate and characterise the macromolecules on the basis of differentiating features. Separation and characterisation of macromolecules are discussed in this chapter.

Size of different types of molecules varies and, therefore, their molecular weight also varies. For example, a macromolecule of small size show less molecular weight and vice versa. Techniques used on the basis of molecular weight are: gel permeation, osmotic pressure, polarity of charges.

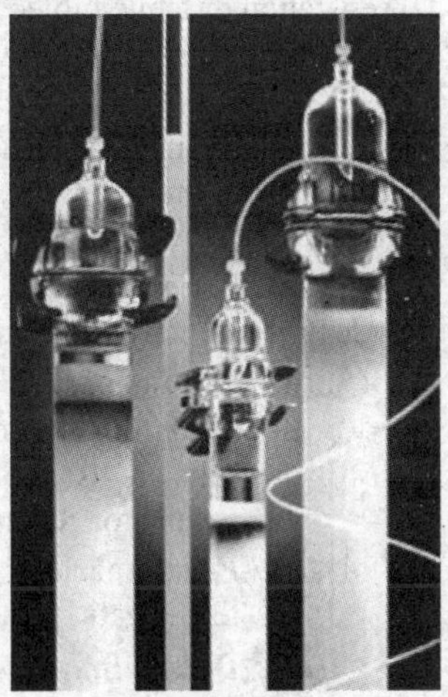

Gel Filtration.

## A. GEL PERMEATION OR GEL FILTRATION

In this technique polymeric organic compound is used to prepare a porous medium. The polymers form a three-dimensional network of pores. The pore size is determined by degree of cross-linking of polymeric chains. Solutes present in the mixture are separated on the basis of their size and shape when they pass through a column consisting of packed gel particles (Fig. 3.1). For such separation many terms like *exclusion chromatography, gel filtration* and *molecular sieve chromatography* are used. The small sized molecules enter the gel beads, whereas the large-sized molecules remain outside (Fig. 3.1). Solvent usually a buffer is used as an eluent.

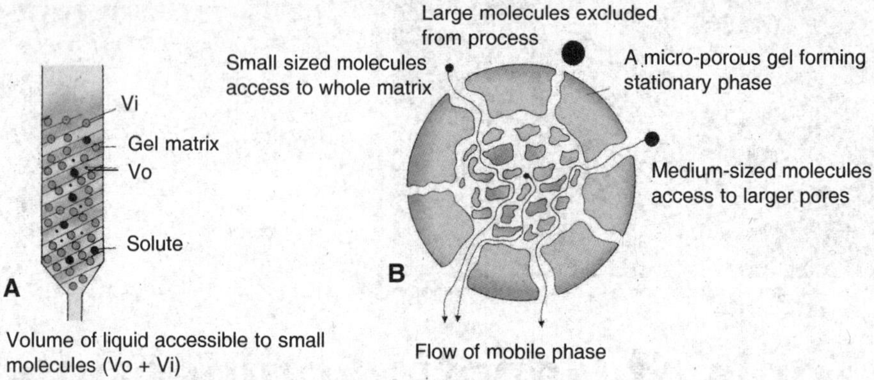


**Fig. 3.1.** Gel permeation chromatography. A- spatial accessibility of molecules during gel filtration; B- movement of molecules of different sizes through microporous gel forming stationary phase.

The small sized molecules are distributed between the mobile phase (liquid) inside and out side of gel particles. $V_0$ refers to the volume of the mobile phase outside the gel particles; where v*i* is the volume of mobile phase present inside the gel particles. Different types of molecules in gel are separated from each other. Distribution coefficient (K*d*) of molecules for a given gel depends upon their size. Very large molecules have no access to mobile phase within the gel, hence their K*d* value is 1. The small molecules have accessible inner mobile phase. Therefore, their K*d* is 1. The K*d* values of molecules govern their separation.

Separation of macromolecules by using gel permeation offers many advantages such as: (*i*) separation of labile molecules, (*ii*) recovery of solutes in maximum quantity, (*iii*) short time and expensive equipment, and (*iv*) high reproducibility.

(*a*) **Characteristics Desired for Gel Filtration Media:** There are certain characteristics desired for gel filtration media such as: (*i*) inertness of gel matrix, (*ii*) presence of low amount of inorganic groups, (*iii*) uniform pore size, (*iv*) wide choice of gel particles and pore sizes, and (*v*) high mechanical rigidity.

These criteria can be fulfilled to varying degree by some of the media such as sephadex polydextran gel, polyacrylamide gel, agarose gel, agarose-acrylamide gel and glass beads.

(*b*) **Application of Gel Permeation:** Gel filtration is used for several purposes as given below:

(*i*) **Desalting or group separation:** In many experiment (e.g. protein separation) inorganic salts are used. Salts should be removed during final preparation by passing through a column of Sephadex G-10.

(*ii*) **Fractionation or purification:** Viral particles and low molecular weight compounds like sugars, proteins and nucleic acids of varying molecular weight can be separated.

(*iii*) **Determination of molecular weight:** The macromolecules are separated on the basis of their relative size.

If molecular weight of a compound is large, it will have lesser elution volume. This method has been used to determine the molecular weight of protein from their elution characteristics (Ve or K*d* value).

## B. ION-EXCHANGE CHROMATOGRAPHY

When molecules are dissolved in solvent, they dissociate into charged ions. Therefore, they develops polarity. On the basis of polarity *i.e.* presence of charges they can be separated. Differences are found in charges, charge densities and distribution of charges on the surface of different molecules. These differences in charge properties of different molecules are taken into account to separate from the other identical molecules. Ion-exchange chromatography is a powerful technique used for separation of two proteins which are very similar to each other, but different from each other in having one charged amino acid. In ion-exchange chromatography solutes are retained on an ion exchanger due to their reversible interaction with the oppositively charged groups on ion exchanger.

Ion exchangers are prepared using some synthetic resins which must be insoluble porous organic molecules. The ion exchanger consists of synthetic or normally occurring organic biopolymers as insoluble matrix to which various groups known as fixed ions are covalently attached. These fixed ions are balanced by equal and oppositively charged mobile ions from the solution called *counter ions*. On the basis of counter ions, the ion exchanger has been grouped into two types: the cation exchangers and anion exchangers.

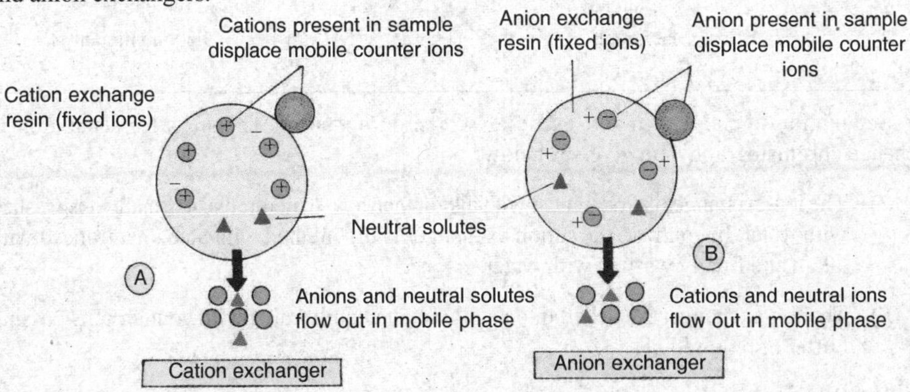


**Fig. 3.2.** Ion-exchange chromatography. A- cation exchanger; B- anion exchanger.

(*a*) **Cation Exchangers:** These are negatively charged exchangers containing positively charged counter ions (cations) (Fig. 3.2A). On the basis of strength of charged functional groups, cation exchangers are divided into three types: *strongly acidic, intermediate acidic* and *weakly acidic* cation exchangers.

(*b*) **Anionic Exchangers:** These are positively charged exchangers containing negatively charged counter ions (anions) (Fig. 3.2B). On the basis of strength of charged functional groups, anions, exchangers are also divided into three types: *strongly basic, intermediate basic* and *weakly basic* anion exchangers.

(*c*) **Matrix Quality:** Inorganic compounds, synthetic resin or polysaccharide based matrix are used in ion exchangers. The chromatographic properties (e.g. efficiency, mechanical strength, chemical stability, flow properties, etc.) of exchangers are governed by the matrix. The matrix also influences the biological activity of molecules. The basic three groups of materials used in construction of matrix for ion exchangers are cellulose, resins (polystyrene or polyphenolic), and polymers or dextran or acrylamide (Table 3.1).

**> Table 3. 1.** **Some commonly used ion exchangers.**

| *Type* | *Matrix* | *Functional groups* | *Name of functional groups* |
|---|---|---|---|
| ***A. Cation exchangers*** | | | |
| Strong cation exchangers | Cellulose | $-SO^{3-}$ | Sulpho |
| | Dextram | $-CH_2SO^{3-}$ | Sulphomethyl |
| | Polystyrene | $-CH_2CH_2CH_2SO^{3-}$ | Sulphopropyl |

| | | | |
|---|---|---|---|
| Weak cation exchangers | Agarose<br>Cellulose<br>Dextram<br>Polyacrylamide | $-COO^-$<br>$-CH_2COO^-$ | Carboxy<br>Carboxymethyl |
| ***B. Anionic exchanger*** | | | |
| Strong anionic exchanger | Cellulose<br>Dextram<br>Polystyrene | $-CH_2N^+(CH_3)_3$<br>$-CH_2CHN^+$ | Trimethylaminomethyl<br>Trimethylaminoethyl |
| Weak anionic exchangers | Agarose<br>Cellulose<br>Dextram<br>Polystyrene | $-CH_2CH_2N^+H3^-$<br>$-CH_2CH_2N^+H(CH_2CH_3)_2$ | Aminoethyl<br>Diethylaminoethyl |

**(*d*) Separation of Solutes in Ion-exchange Chromatography:** The steps for separation in ion-exchange chromatography are given as below:

(*i*) The ion-exchanger is first treated with alkali, then acid to neutralise it. Finally it is washed with water. In contrast, the cation exchanger is first treated with acid, neutralised with alkali and finally washed with water.

(*ii*) The ion-exchanger is packed into a column and equilibrated with counter ions passing buffer of required *p*H.

(*iii*) When samples containing mixture of compounds is applied onto the top of exchanger bed, the solutes having charges similar to that of counter ions (here negative ions) get exchanged with the counter ions and binds reversely but strongly to the ion exchanger (Fig. 3.3A).

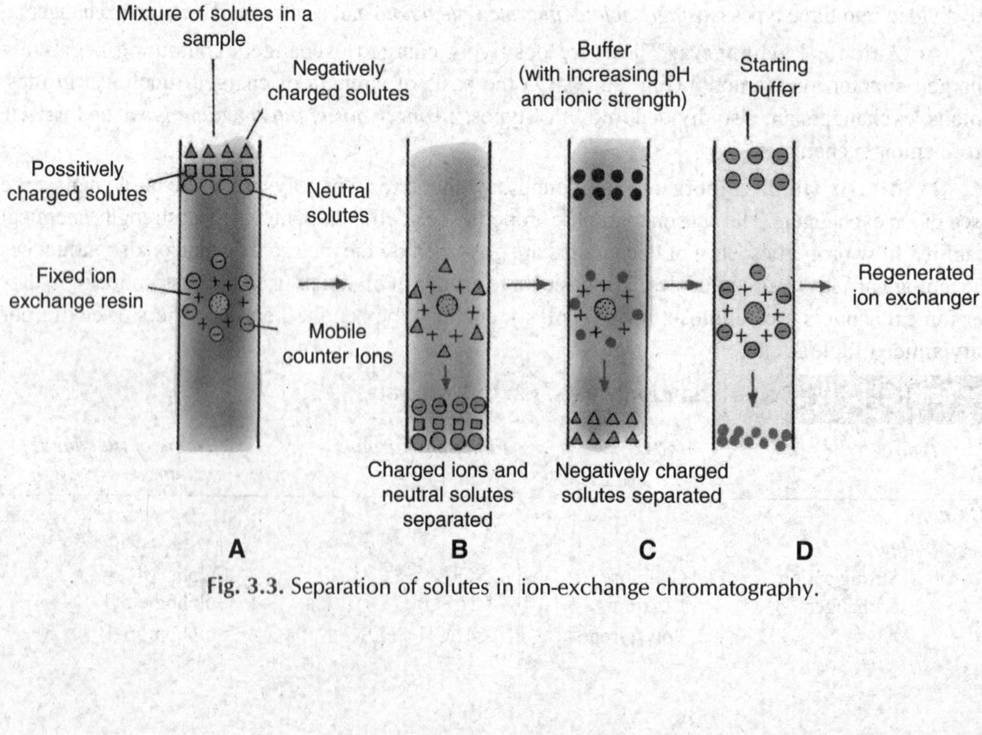


**Fig. 3.3.** Separation of solutes in ion-exchange chromatography.

(*iv*) The other solutes (positively charged) and (neutral ones) have no affinity for the stationary phase and are washed down along with the buffer (B).

(*v*) Thereafter, the bound solutes are released step by step by changing *p*H or ionic strength of the elution buffer (C).

(*vi*) Generally a buffer with increasing *p*H and ionic strength is applied with anionic exchanger, whereas with cation exchanger a buffer with decreasing pH and increasing ionic strength gradient is employed for desorption of bound substances.

(*vii*) The ion exchanger is regenerated after passing the buffer continuously (Fig. 3.3D).

## C. ELECTROPHORESIS

The charged molecules can be separated by applying an electric field which is called electrophoresis. This technique was pioneered by A. Tiselius in 1937. When the charged molecules are placed in an electric field they migrate depending on their net charges, size, shape and applied current. The velocity of movement of molecules can be represented by the following formula:

$$v = \frac{E.q}{f}$$

where, $V$ = velocity of migration of molecules

$E$ = electric field in volts/cm

$q$ = net fractional coefficient which is a function of the mass and shape of molecule

Electrophoresis is applied for separation of RNA, DNA and proteins. DNA molecules have negative charges. Therefore, based on their size DNA molecules migrate to anode, *i.e.* small molecules move faster through the pores of matrix than the larger molecules.

Similarly, protein macromolecules are made up of positively and negatively charged amino acids at specific *p*H. The relative proportion of these two charges govern the net charges of protein macromolecules. If two proteins are of similar size and have identical charges, there shall be a little or no separation of such proteins due to similar charge : mass ratio. Because the electrophoretic mobility (of a molecule depends on charge density (charge/mass ratio).

On the basis of types of support medium electrophoresis is of different types such as: (*i*) *paper electrophoresis* (a strip of Whatman filter paper or cellulose acetate paper is used as support to separate macromolecules on the basis of their varying sizes. Later on the spots developed on filter paper are cut and dissolved in solvent for separation and further work), (*ii*) *starch gel electrophoresis* (starch is partially hydrolysed in buffer to prepare a solution. Then it is heated and cooled to get starch gel. The gel acts as molecular sieve to separate molecules. Molecules of varying sizes migrate in gel and separated), (*iii*) *immuno-electrofocussing* (it works like agar or agarose gel electrophoresis. It is used for separation of proteins based on charge : mass ratio and their antigenicity. First proteins are separated, then allowed to react with antigens through diffusion via gel. Therefore, it is called immuno-electrophoresis), *agarose gel electrophoresis, pulsed field gel electrophoresis, polyacrylamide gel electrophoresis, sodium dodecyl polyacrylamide gel electrophoresis, two dimensional gel electrophoresis*, etc.

## 1. Agarose Gel Electrophoresis

Agarose gels are more porous than polyacrylamide gels and have comparatively larger pore size. Therefore, it is used for separation of large sized macromolecules such as DNA. Agarose is a linear polymer of D-galactose and 3,6-anhydro-L-galactose which is extracted from seaweeds. By changing the concentration of agarose pore size of the gel can be determined. Diagram of an electrophoresis is given in Fig. 3.4.

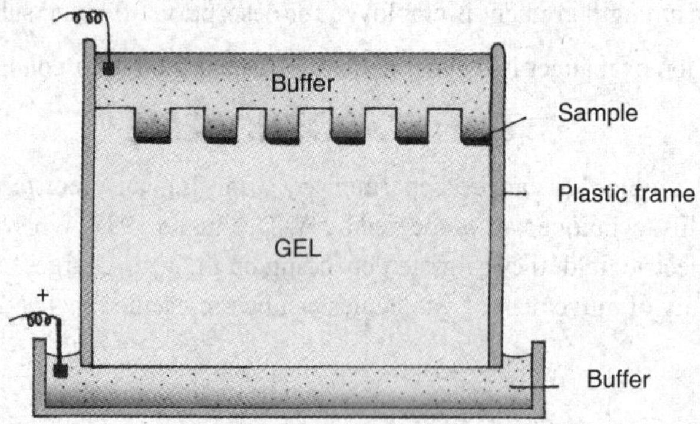


**Fig. 3.4.** Agarose gel electrophoresis - an apparatus for slab-gel electrophoresis.

The negatively charged DNA migrates at neutral pH towards the anode after application of electric field across the gel. It is the standard technique for separation of nucleic acids. Conventional agarose gels cannot resolve fragments larger than 20 kb. Because larger DNA fragments are unable to migrate through pores or sieves of gel as small fragments do.

## 2. Pulsed Field Gel Electrophoresis (PFGE)

Schwartz and Cantor (1984) conceived the PAGE to separate the several megabases long DNA molecules. All linear double stranded DNA molecules, larger than a certain size, migrate through agarose gel almost at similar rate. But agarose gels are unable to resolve linear DNA molecules larger than 750 kb in length.

PFGE varies a little from agarose gel elctrophoresis and can resolve DNA molecules of 100-1000 kb. It changes the direction of electric field in such a way that larger DNA fragments align more slowly with the direction of the new field than do the smaller DNA molecules. The pulsed electric fields applied to a gel force the DNA molecules to reorient before continuing to move like snake through the gels. The larger molecules take much time to re-orient than the smaller molecules (Fig. 3.5). Large DNA molecules are trapped in their tube every time when direction of electric field is changed. Until they are re-oriented along the new axis of electric field, they cannot make further progress through the gel. Hence, longer time is required for larger molecules for such alignment. Such DNA molecules are functional according to their size whose re-orientation time is less than the period of electric pulses. Unlike a single constant electric field in conventional electrophoresis, two perpendicularly oriented alternating electric fields are applied in PFGE for separation of molecules. The smaller molecules change the direction quickly, move fastly and separated from the larger molecules.

Initially, Schwartz and Cantor (1984) electrophoretically separated the individual yeast chromosomes which ranged in size (230-2,000 kb). Since then chromosomal DNA of a wide variety of organisms has been analysed using PFGE.

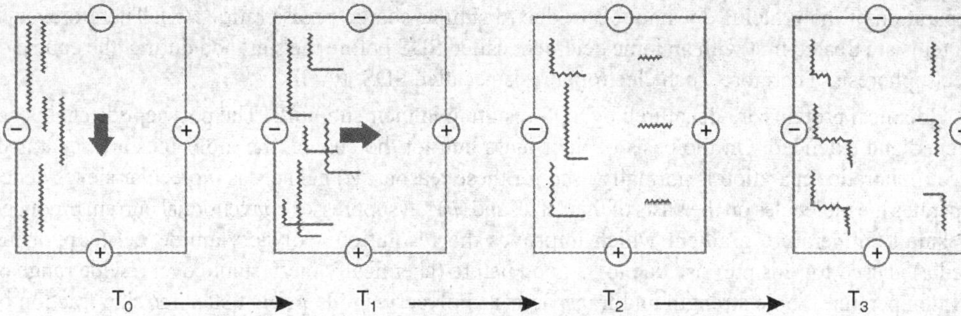


Fig. 3.5. Pulsed field gel electrophoresis.

### 3. Polyacrylamide Gel Electrophoresis (PAGE)

Polyacrylamide gel offers several advantages such as inertness to chemicals, superior resolution, stability over wide range of *p*H, temperature and ionic strength. Polyacrylamide gel is prepared by using the following: (*i*) monomers (e.g. acrylamide, N,N-methylene bisacrylamide, (*ii*) initiator: N,N,N,N,-tetramethylethylene diamine (TEMED), (*iii*) propagators (ammonium persulphate or riboflavin, (*iv*) terminator or inhibitor (oxygen/air).

The pore size of polyacrylamide gel can be changed by varying the concentration of acrylamide and bi-acrylamide monomers in a fixed volume of gelatin solution. Free radicals formed by ammonium sulphate activate the acrylamide gel. Long polymer chain is thus produced after reacting the activated acrylamide with successive acrylamide molecules. Further, bisacrylamide brings about gelation and cross-linking through polymerisation. This results in formation of complex network of acrylamide.

### 4. Sodium Dodecyl Sulphate Polyacrylamide Gel Electrophoresis (SDA-PAGE)

Some of the proteins are not separated due to similar charge : mass ratio. Therefore, such proteins are treated first with an ionic detergent called *sodium dodecylsulphate* (SDS) before start and during the course of electrophoresis (PAGE). Therefore, such electrophoresis is called SDS-PAGE electrophoresis. Identical proteins are denatured by SDS resulting in their sub-units. The polypeptide chains get opened and extended. On the basis of their mass but not the charges, the molecules are separated (Fig. 3.6).

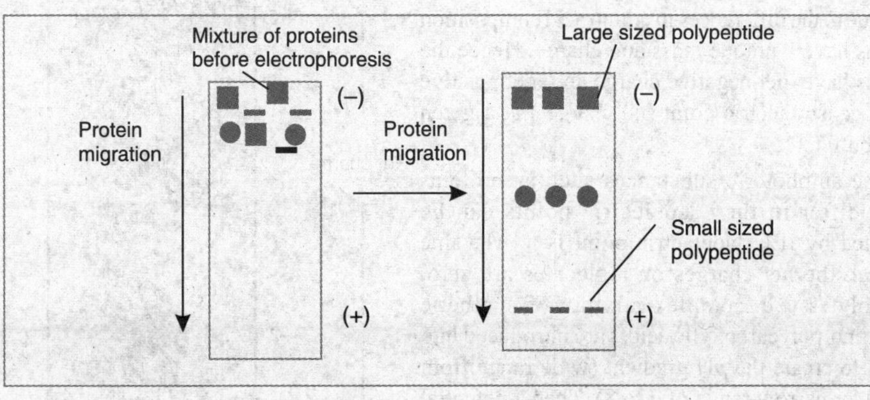


Fig. 3.6. SDS-PAGE analysis of proteins.

SDS–PAGE is used universally for analysing the mixture of proteins according to their respective size. SDS solubilised in soluble proteins makes possible the analysis of the other insoluble mixtures.

Separation of the proteins does not occur due to similar charge : mass ratio (*z/m*). Therefore, such proteins are treated first with an ionic detergent called SDS before the start and during the course of electrophoresis. Therefore, such electrophoresis is called SDS–PAGE.

Identical proteins are denatured by SDS resulting in their sub-units. The polypeptide chains get opened and extended. On the basis of their mass but not the charge, the molecules are separated. Electrophoretic separation is normally used for these reasons: (*i*) gel acts as molecular sieves hence separates the molecules on the basis of their size, and (*ii*) gel suppresses conventional current produced by small temperature gradient which improves the resolution. Polyacrylamide gel (supporting media) is used for this purpose due to its good nature (chemically inert, stable over a wide range of *p*H, temperature, ionic strength and transparent). Polyacrylamide gel is better for size fraction of proteins molecules.

The denatured proteins have a negative charge with a uniform charge to mass ratio (*z/m*) when treated with SDS (anioic detergent). Proteins migrate toward anode at alkaline *p*H through PAGE gel during electrophoresis. The smaller polypeptides moves faster followed by the larger polypeptides. Therefore, the intrinsic charge on proteins is masked in SDS-PAGE. Hence, the separation is based on the size. Mercaptoethanol reduces interpolypeptide disulfide bridge and separates the sub-units of a polymeric protein. Molecular weight of the separated protein can be analysed by comparing the molecular weight of the standard protein and its mobility. In analysis of a complex mixture of proteins the resolution is improved by the initial movement through a stacking gel. The final bands in the separating gel are sharper and focused nicely.

Two-dimensional technique is very useful and effective at separating proteins and can resolve thousands of proteins in a mixture.

## 5. Two-dimensional Electrophoresis

Protein mixture is separated using two-dimensional electrophoresis. The first dimension uses the iso-electric focusing, and the second dimension is sodium dodecyl sulphate polyacrylamide gel electrophoresis (SDS-PAGE).

**Iso-electric Focusing (IEF):** Proteins, DNA and RNA have electric charges which depend on molecule to molecule and the conditions of the medium (*p*H of buffer in which dissolved). Charged molecules can be separated by electrophoresis in gels. Due to the differences in amino acid composition proteins have a unique mass and charge. Hence the proteins have net negative charge and net positive charge or iso-electric point (no charge ) at a given *p*H of buffer.

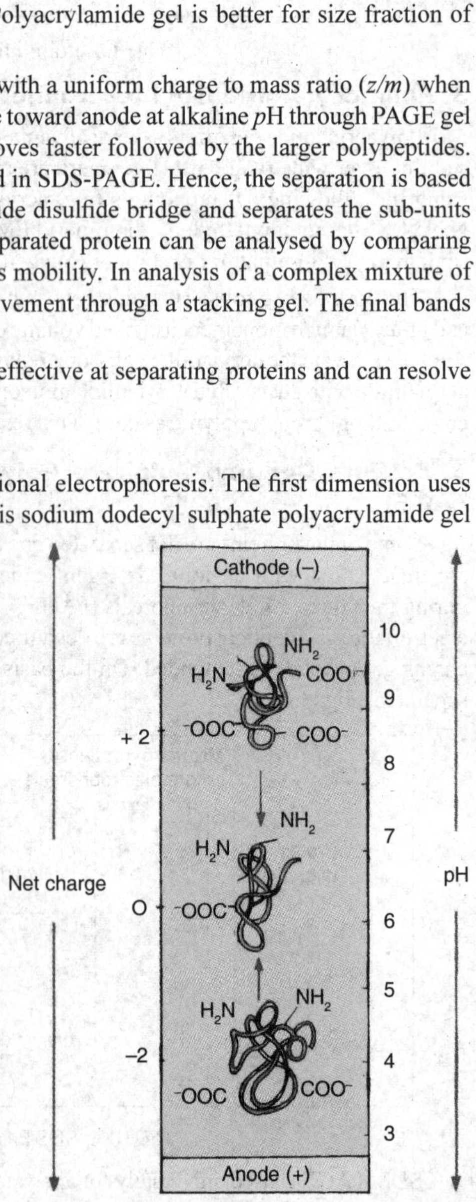

Fig. 3.7. Iso-electric focusing.

The amphoteric substances such as proteins which differ in their isoelectric points can be separated by IEF. Isoelectric point is a *p*H value at which the net charges on molecules are zero. Ampholytes (*i.e.* complex mixture of synthetic polyamino-polycarboxylic acids) are introduced into the gel to create the *p*H gradient (wide range from 3 to 10, or narrow range of 7 to 8). Then a potential difference is applied across the gel. The molecules having differences in isoelectric points by a little as 0.01 *p*H unit can be separated. Proteins migrate depending on their charge until they reach a region which *p*H corresponds to respective iso-electric point at which *p*H proteins possess no net charge and hence got focused (Fig. 3.7).

**(*b*) SDS-PAGE:** SDS-PAGE is carried out as described in earlier section.

## D. MASS SPECTROMETRY

In 1900, for the first time J.J. Thomson introduced mass spectrometer which employed fixed magnetic and electric fields to separate ions of different mass and energy. He recognised that charged particles differing in momentum behaved differently in an electric field, and used this property for separation of ions with different mass.

During 1980s its extensive use for research in various field of biological science was started. During 1990-2000, mass spectrometry became an important technique for genomics and proteomics research leading to the 2002 Nobel Prize in Chemistry to J.B. Fenn and K. Tanaka. Two-dimensional electrophoresis is more powerful when coupled with MS. The unknown protein spot is cut from the gel and cleaved by trypsin digestion into fragments which are then analysed by mass spectrometer and the mass of fragment is plotted. This mass finger print can be used to estimate the probable amino acid composition of each fragment and tentatively identify the protein. The proteome and its changes can be studied very effectively by employing the two techniques together.

The MS can also provide valuable information about covalent modification of proteins which can affect their activity. Mass spectrometry is very useful technique. It is used in identification of unknown compounds, quantification of known compounds and determination of structural and chemical properties of compounds when present in small amount ($10^{-6} - 10^{-8}$ g). This technique involves: (*i*) the production of ions of the materials in sample, (*ii*) their separation on the basis of their mass : charge (*m* : *e*) ratio, and (*iii*) determination of relative abundance of each ion. Therefore, a mass spectrophotometer consists of three components : the *source of ion*, an *analyser*, and a *detector* (Fig. 3.8). It does not directly measure the molecular mass but detects *m* : *e* ratio. Mass is measured in terms of Dalton (Da). One Dalton = 1/12$^{th}$ of mass of a single atom of isotonic carbon ($^{13}C^+$). In recent years, mass spectrometry has become an essential tool for analysis of genome and proteome in its many forms. It is capable of identifying and characterising proteins present even in picomoles ($10^{-12}$). Basically it is an analytical device consisting of the following three main components (Fig. 3.8).

**(*i*) Source of Ignition:** It involves the gaseous ionisation of analyte (the molecule to be analysed) to be examined. When the molecules gain or lose a charge (through electron ejection, protonation or deprotonation), molecular ions are generated.

**(*ii*) Analyser:** The ions generated as above are separated according to their mass-to-charge (m/z) ratio.

**(*iii*) Detector:** The molecular masses of separated ions are determined from the mass spectra obtained after mass spectrometry of analyte. For identification, protein is separated from the crude extract on a two-dimentional gel. The protein spots are excised and then used as search for mass spectrometry or fragmented into small peptides. Protein fragmentation into peptides is done by proteases (*e.g.* trypsin). The peptides are separated by using liquid chromatography (such anion exchange affinity or reverse phase column chromatography). Proteins are easily identified if these have been fragmented into peptides.

The principle of working of mass spectrometry is to create gas phase ions from polar charged biomolecules like peptides, proteins or nucleic acids. The sample (M) undergoes ionisation in the ionisation source upon introduction in the mass spectrometry. Molecular weight more than 1299 Dalton of the sample results in multiple charged ions *e.g.* M+2H. There are many suitable sites of proteins which undergo protonation. All the backbone amide nitrogen atoms and side chain amino group may be protonated. The charged molecules are electrostatically propelled into analyser. On the basis of m/z ratio analyser separates the ions. The detector detects the ions and transfer the received signals to a computer. The information is stored and processed by the computer.

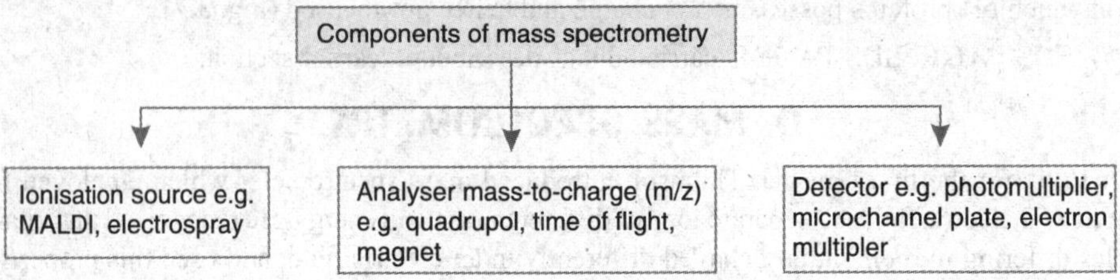


Fig. 3.8. A simplified schematic diagram of mass spectrometry.

The molecules to be examined should be in the form of gaseous ions. There are several methods available for ionisation of biomolecules but MALDI and electrospray (ES) each coupled with TOF-MS are discussed below:

## 1. Matrix-Assisted Layer Desorption-ionisation (MALDI)

MALDI was first developed by Karas and Hillenkamp in 1988. It co-precipitates the large excess of a matrix material with the analyte molecule. In this method a sub-microlitre of the mixture (matrix + analyte) is pipetted onto a metal substrate and allowed to dry. The dried mixture is irradiated by laser pulse (337 nm) which specifically absorbs the selected solid of matrix. The irradiation causes energy transfer and desorption resulting in gas phase matrix ions.

With the help of matrix ions the non-absorbing analyte molecules also get desorbed into the gas phase and ionised. The charged molecular ions of analyte are detected by MALDI-TOF mass spectrometer. Generally, MALDI is used for the study of molecules having mass above 500 Daltons (*i.e.* oligonucleotides and peptides). Proteome analysis will be crucial for understanding different biological processes in the post-genomic era. To achieve this objective mass spectrometry is extensively used for: peptide sequencing, identification of protein, protein expression in different tissues and conditions, identification and post-transcriptional modification (*e.g.* phosphorylation, glycosylation, etc.) of proteins in response to different stimuli, and characterisation of protein interactions that include protein-ligand, protein-protein and protein-DNA interaction

Recent improvement in MS has significantly improved its application in the study of protein structure and function. For the study of macromolecules, the Indian Society for Mass Spectrometry (ISMAS) was established in 1978 with its headquarter at BARC (Bhabha Atomic Research Centre), Mumbai. This society conducts a workshop each year on mass spectrometry. In India, facilities for MALDI-TOF MS have been created at several centres including the Centre for Cellular and Molecular Biology (CCMB), Hyderabad, Indian Institute of Sciences, Bangalore, BARC, Rajiv Gandhi Centre for Biotechnology (RGCB), Thiruvanantpuram, etc.

## 2. Surface Enhanced Laser Desoption-Ionisation (SELDI)

It is a unique combination of affinity chromatography and time of flight (TOF) mass spectrometry on a single platform, a combination designated as SELDI-TOF MS (surface enhanced laser desoption-ionisation time of flight mass spectrometry). This system enables protein capture, purification, ionisation and analysis of complex biological mixtures directly on protein chips array surface and detection of the purified proteins by laser desorption-ionisation tiome of flight (LDI-TOF). MS analysis. SELDI system consists of three parts: *(i)* protein chip arrays (eight spots of 2 mm diameter), *(ii)* protein chip reader (LDI-TOF MS), and *(iii)* specialised software.

## 3. Electrospray Ionisation (ESI)

During the late 1980s, J.B. Fenn developed ESI mass spectrometry and was awarded the Nobel Prize in 2002 in chemistry. In this method the sample is dissolved in liquid with mobile phase (*i.e.* water : acetonitrile : methanol) and pumped through a hypodermic needle at a high voltage (about

4,000 V which causes strong electric field at the top of nozzle of the metal needle). This disperses electrostatically or electrosprays the highly charged small droplets of about one micrometer in size. These droplets are electrostatically attracted to the inlet of MS. Dry gas or heat is applied The solvent around these droplets soon evaporates and impart charge onto the analyte molecules. The electric field on droplets increases with decrease in their size. Hence equal charges repel each other. This ionisation is very gentle which causes no breakdown of analyte ions in the gas phase (Fig.3.9).

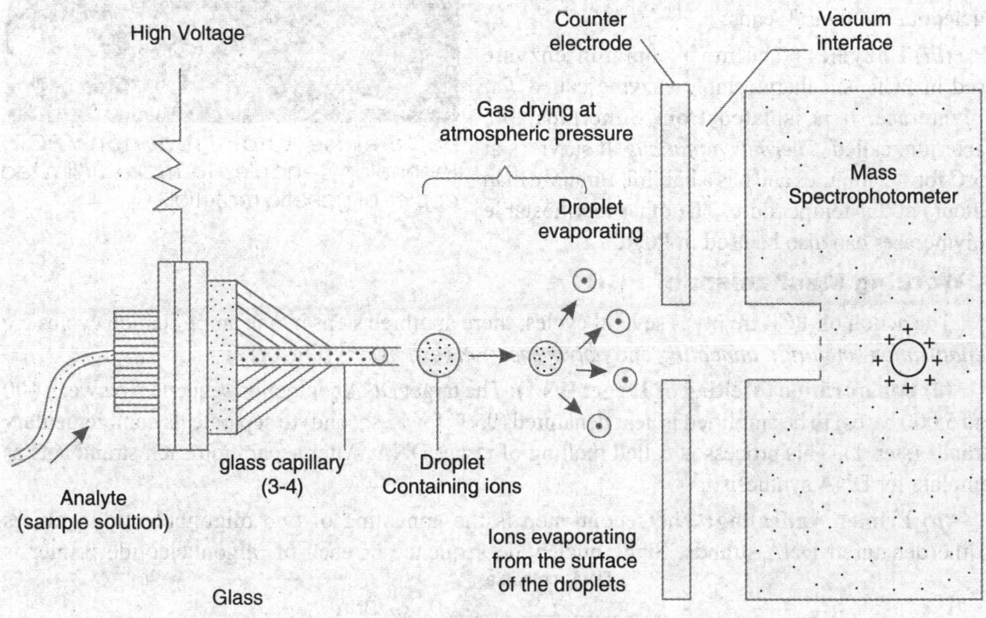


**Fig. 3.9.** Diagram of a standard electrospray ionisation source.

Protons added during passage through the needle give additional charge to the macromolecule. The m/z of the nucleotide can be analysed in the vacuum chamber. ES produces many more ions than MALDI. These ions are transferred into mass spectrometer for analysis with high efficiency. ES MS can analyse a wide range of molecules including proteins, oligonucleotides, sugar and lipids. Since a MS measures m/z, it allows high molecular weight proteins to be observed on a low mass range equipment. Thus ESI generates multiple charged molecules. Peptides or proteins are identified on the basis of m/z ratio. The m/z ratio of ions is analysed after their generation. The new proteins with human genome sequencing project could be identified through the combination of MS and genomics, and also using the tools of 'Bioinformatics' (*see* Chapter 9).

# E. POLYMERASE CHAIN REACTION (PCR)

The polymerase chain reaction (PCR) provides a simple and ingeneous method for exponentially amplification of specific DNA sequences by *in vitro* DNA synthesis. This technique was developed by Kary Mullis at Cetus Corporation in Emery Ville, California in 1985. Kary Mullis shared the Nobel prize for chemistry in 1993. This technique has made it possible to synthesize large quantities of a DNA fragment without its cloning. It is ideally suited where the quantity of biological specimen available is very low such as a single fragment of hair or a tiny blood stain left at the site of a crime. The details of PCR techniques and its mechanism are described by H.A. Erlich in 1989 in his edited book 'PCR Technology'. The PCR technique has now been automated and is carried out by a specially designed machine. The PCR requires the following:

(*i*) **DNA Template:** Any source that contains one or more target DNA molecules to be amplified can be taken as template.

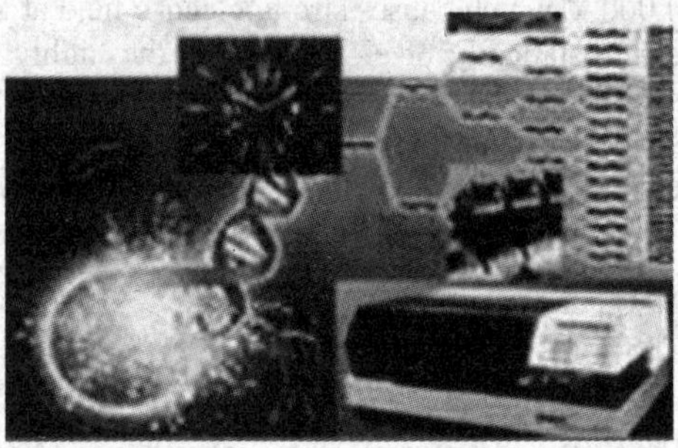

Polymerase chain reaction (PCR) technology enables to make unlimited copies of genetic material.

(*ii*) **Primers:** A pair of oligonucleotides of about 180-30 nucleotides with similar G+C contents act as primers. They direct DNA synthesis towards one another. The primers are designed to anneal on opposite strands of target sequence so that they will be extended towards each other by addition of nucleotides to their 3' ends.

(*iii*) **Enzyme:** The most common enzyme used in PCR is a thermostable enzyme called *Taq* polymerase. It is isolated from a thermostable bacterium called *Thermus aquaticus*. It survives at 95°C for 1-2 minutes and has a half life for more than 2 hours at this temperature. The other thermostable polymerases can also be used in PCR.

## 1. Working Mechanism of PCR

The action of PCR involves several cycles; there are three steps in one amplification cycles *e.g denaturation* (*melting*), *annealing* and *polymerisation* (*extension*) (Fig. 3.12).

(*a*) **Denaturation (Melting of Target DNA):** The target DNA containing sequence (between 100 and 5,000 bases) to be amplified is heat denatured (95°C for 15 second) to separate its complementary strands (step 1). This process is called melting of target DNA. After separation each strand acts as template for DNA synthesis.

(*b*) **Primer Annealing:** The second step is the annealing of two oligonucleotide primers to the denatured DNA strands. Since nucleotide sequence of each of oligonucleotide primer is

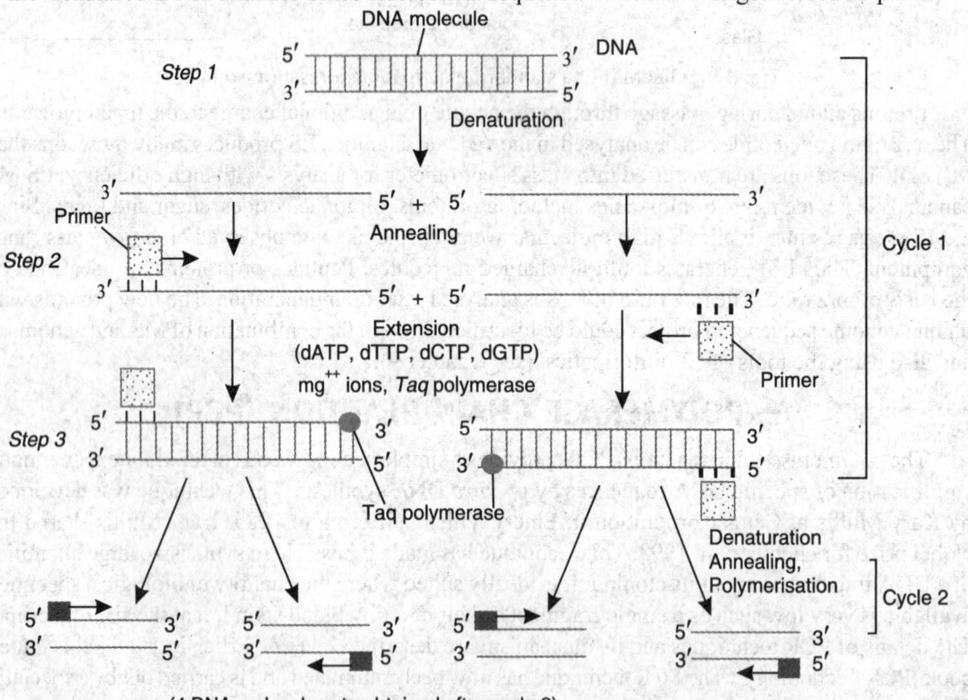


Fig. 3.10. Working of PCR.

complementary to 3' end of single stranded template, the primers anneal (hybridises) the each template. Primers are added in excess and the temperature lowered to about 68°C for 60 seconds. Consequently the primers form hydrogen bonds i.e. anneal to the DNA on both sides of the DNA sequence (step 2). The annealing temperature varies, but the too low temperature favours mispairing. The annealing temperature (in °C) can be calculated using the formula: T = 2 (AT) + 4 (G + C).

**(c) Primer Extension (Polymerization):** Nucleoside triphosphate (dATP, dGTP, dCTP, dTTP) and a thermostable DNA polymerase are added to the reaction mixture. The DNA polymerase accelerates the polymerization process of primers and, therefore, extends the primers (at 68°C) resulting in synthesis of copies of target DNA sequence (step 3). Only those DNA polymerases which are thermostable *i.e.* function at the high temperature are employed in PCR technique. For this purpose the two popular enzymes, *Taq* polymerase (of a thermophilic bacterium, *Thermus aquaticus*) and the *vent* polymerase (from *Thermococcus litoralis*) are used in PCR technology. These enzymes exhibit relative stability at DNA-melting replenishment after each cycle of synthesis. The concentration of $Mg^{++}$ is maintained between 1 and 4 mM. Thus in the first step the target DNA is copied from the primer sites for various distances until the start of second cycle.

After completion of step 3 (of one cycle) the targeted sequences on both strands are copied and four strands are produced. Now, the three step cycle (first cycle) is repeated which yields 8 copies from four strands.

In the second cycle the DNA molecules synthesised in the first cycle get doubled. The second cycle is started with heating of double stranded DNA to result in single stranded DNA. Each single stranded DNA again acts as template i.e. DNA molecules polymerised in first cycle act as template in second cycle. Following the above events all the single stranded DNA molecules of second cycle are converted into the double stranded DNA (Fig. 3.12). Then third and onward cycles are repeated in the same ways to get more DNA products. The $2^n$ molecules of DNA are generated using single stranded DNA as template always after *n* number of cycles.

The third cycle produces 16 strands. This cycle is repeated about 50 times. Theoretically, 20 cycles (each of three steps) will produce about one million copies of the target DNA sequence, and 30 cycles will produce about one billion copies. In each cycle the newly synthesized DNA strands serve as targets for subsequent DNA synthesis as the three steps are repeated upto 50 times. For the working of PCR about 10-100 picomoles of primers are required. The PCR machine can carry out 25 cycles and amplifiy DNA $10^5$ times in 75 minutes. The number of target molecules should be doubled after each cycle resulting in exponential increase in the amplified DNA molecules. But the efficiency of exponential amplification is less than 100% due to sub-optimal DNA polymerase activity, poor primer annealing and incomplete denaturation of the template. PCR efficiency is expressed following the formula as given below:

PCR product yield = Input target amount x $(1+\% \text{ efficiency})^n$; where n is the number of cycles.

You can calculate that by using an amplification efficiency value of 70%, about 26 cycles are required to produce 1 μg of PCR product from 1 pg of a target sequence ($10^6$ amplification [1 μg PCR product = (1 pg target) × $(1 + 0.7)^{26}$].

In recent years the PCR technology has been improved. RNA can also be efficiently used in PCR technology. The *rTth* DNA polymerase is used instead of the *Taq* polymerase. The *rTth* polymerase will transcribe RNA to DNA, thereafter amplify the DNA. Therefore, cellular RNA and RNA viruses may be studied when they are present in small quantities.

## 2. Application of PCR

Some of the areas of application of PCR have briefly been discussed herewith.

**(a) Diagnosis of Pathogens:** Several pathogens are known that grow slowly. Therefore, their cells are found less in number in the infected cells/tissues. It is difficult to culture them on artificial medium. Hence, for their diagnosis, PCR-based assays have been developed. These detect the presence of certain specific sequences of the pathogens present in the infected cells/tissues. Besides, it is useful in detection of viral infection before they cause symptoms or serious diseases.

**(*b*) Diagnosis of Specific Mutation:** Presence of a faulty DNA sequence can be detected before establishment of disease. By using PCR sickle cell anaemia, phenylketonuria and muscular distrophy can also be detected. The other diseases can also be diagnosed by using PCR. For example PCR-based diagnostic tests for AIDS, chlamydia, tuberculosis, hepatitis, human pappiloma virus, and other infectious agents and diseases are being developed.

**(*c*) Prenatal Diagnosis:** It is useful in prenatal diagnosis of several genetic diseases. If the genetic diseases are not curable, it is recommended to go for abortion.

**(*d*) DNA Fingerprinting:** DNA fingerprinting is more successfully used in forensic science to search out criminals, rapists, solving disputed parentage and uniting the lost children to their parents or relatives by confirming their identity. This is done through making link between the DNA recovered from samples of blood, semen, hairs, etc. at the spot of crime and the DNA of suspected individuals or between child and his/her parents/relatives.

**(*e*) In Research:** DNA fingerprinting of new microorganisms isolated from various extreme environment (soil, water, sediments, air, extreme habitats, etc.) is also carried out to confirm their identity by comparing with the DNA sequences of known microorganisms. Their DNA and RNA can be amplified. It is also useful to determine the orientation and location or restriction fragments relative to one another.

**(*f*) In Molecular Archaeology (Palaentology):** PCR has been used to clone the DNA fragments from the mummified remains of humans and extinct animals such as the woolly mammoth and dinosaurs from the remains of ancient animals as recently epitomised in Michel Crighton's Jurassic Park. DNA from buried humans has been amplified and used to trace the human migration that occurred in ancient time.

**(*g*) Disgnosis of Plant Pathogens:** A large number of plant pathogens in various hosts or environmental samples are detected by using PCR, for example, viroids (associated with hops, apple, pear, grape, citrus, etc), viruses (such as TMV, cauliflower mosaic virus, bean yellow mosaic-virus, plum pox virus, potyviruses), mycoplasmas bacteria (*Agrobacterium tumifaciens, Pseudomonas solanacearum, Rhizobium leguminosarum, Xanthomonas compestris,* etc), fungi (*e.g., collectotrichum gloeosporioides, Glomus* spp., *Laccaria* spp., *Phytophthora* spp., *Verticillium* spp), and nematodes (*e.g. Meloidogyne incoginta, M. javanica,* etc).

There are some other PCR-based techniques

## 3. Reverse Transcription-mediated PCR (RT-PCR)

Amplification of double stranded DNA can also be applied to produce double stranded cDNA also, which is synthesised from mRNA using the enzyme reverse transcriptase. Based on this principle RT-PCR was developed. RT-PCR includes a single application combining the process of cDNA synthesis (by reverse transcription) and PCR amplification. Using this principle one can isolate very rare mRNA transcripts from cell samples through RT-PCR. The thermostable enzyme *rTth* uses RNA templates from cDNA synthesis and thus allows single enzyme RT-PCR viral reverse transcriptase (RTase) from avian murine virus (AMV RTase) or mammalian leukemia virus (MLV RTase) can also be used for cDNA synthesis. Both the enzymes i.e. RTase and *rTth* DNA polymerase are used alternatively in the RT-PCR.

RAPD markers have been successfully used for: (*i*) construction of genetic maps e.g. *Arabidopsis,* pine, *Helianthus,* etc., (*ii*) mapping of traits, (*iii*) analysis of genetic structure of population, (*iv*) fingerprinting of individuals, (*v*) identification of somatic hybrids, etc. A detailed account of RAPD is given in Chapter 14.

### 5. Arbitrary Primed PCR (AP-PCR)

In 1990, J. Welsh and M. McCleland developed the AP-PCR and carried out fingerprinting of genomes with arbitrary primers. It is a type of RAPD where single primers of 10-50 bases are used to amplify genomic DNA in PCR. In the first two cycles, annealing in under non-stringent conditions. AP-PCR differs from RAPD in some respects: (*i*) in AP-PCR amplification occurs in three parts, with its own stringency and concentration of constituents, (*ii*) higher concentrations of primers are used in the first cycles of AP-PCR. Primers of different length are arbitrarily selected e.g. M13 universal sequence primer, (*iii*) products of AP-PCR are analysed on acrylamide gels and detected by autoradiography, whereas these of RAPD are analysed on agarose gels, stained with ethidium bromide and visualised under UV light.

## PROBLEMS

1. With the help of suitable diagram describe electrophoresis.
2. Write the principle and working of UV-Vis spectrophotometer.
3. Give a detailed account of mass spectrometry.
4. What is pulsed electric gel elelctrophoresis ? Write in detail its working.
5. What is mass spectrometry. Give a detailed account of it.
6. What is polymerase chain reaction ? Give a detailed account of PCR and its applications.
7. Write brief notes on the following
   (*i*) Iso-electric focusing, (*ii*) Colorimetry, (*iii*) Gel permeation, (*iv*) Ion exchange chromatography, (*v*) MALDI, (*vi*) SELDI, (*vii*) ESI, (*viii*) RT-PCR, (*ix*) AP-PCR, (*x*) Agarose gel electrophoresis.

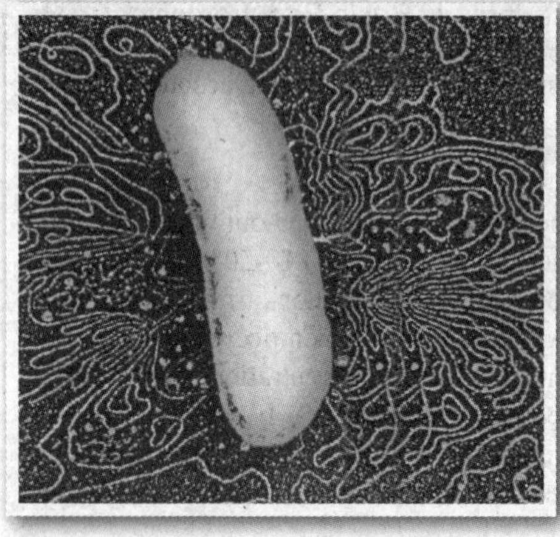

CHAPTER

4

# Tools of Genetic Engineering – II: Cutting and Joining of DNA

Genetic engineering is used for the production of valuable polypeptides, insulin, interferon, growth hormones, and in the transfer of *nif* genes, control of genetic diseases, etc.

There are three basic steps of genetic engineering: (*i*) isolation of DNA fragments from a donor organism, (*ii*) insertion of an isolated donor DNA fragment into a vector genome, and (*iii*) growth of a recombinant vector in an appropriate host. There are various biological tools which are used to carry out manipulation of genetic material and cells as well, for example, enzymes, foreign or passenger DNA, vector or vehicle DNA, gene library. This chapter deals with enzymes used for cutting and joining of DNA molecules.

## A. ENZYMES USED IN GENETIC ENGINEERING

There are many enzymes which are used in genetic engineering as an important biological tool. However, the important enzymes required for genetic engineering are: restriction endonucleases, DNA ligases, exonucleases, DNA polymerases, alkaline phosphatase, polynucleotide kinase, terminal deoxynucletidyl, transferase, reverse transcriptase, etc. Restriction endonucleases are the special types of endonucleases that cut DNA at specific sites.

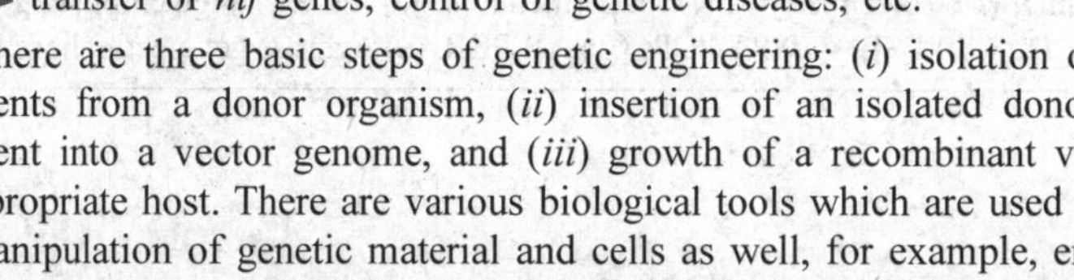


DNA enzymes cut the DNA at specific sites.

Recombinant DNA technology is based totally on different types of enzymes. Some of the enzymes used in gene cloning are given in Table 4.1 and described in this chapter.

**> Table 4.1. Enzymes used in DNA cloning.**

| *Enzyme* | *Use* |
|---|---|
| Alkaline phosphatase | Removes $PO_4$ from 5'end of double stranded or single stranded DNA or RNA |
| DNA ligase | It joins sugar-phosphate backbone of dsDNA with 5'-$PO_4$ and a 3'-OH in an ATP-dependent reaction. The two ends of DNA must be blunt or complementary cohesive ones |
| DNA Pol I | It synthesises DNA complementary to a DNA template in 5'→3' direction beginning with a primer with a free 3'-OH. The Klenow fragment is a truncated version of DNA Pol I lacking 5'→3' exonuclease activity |
| Exonuclease III | It cleaves from the ends of linear DNA and digests dsDNA from 3'end only |
| Mung bean nuclease | It digests ssDNA or RNA but leaves intact double helical region |
| Nuclease S1 | It behaves as mung bean nuclease and cleaves a strand opposite to a nick on the complementary strand |
| Polynucleotide kinase | It adds to 5'-OH end of dsDNA or ssDNA or RNA in an ATP-dependent reaction. DNA becomes radiolabelled if [$\gamma^{32}$P]-ATP is used. |
| Restriction enzymes | These cut both strands of dsDNA within a symmetrical recognition site resulting in blunt or sticky ends (5'- or 3' - ovelhangs) |
| Reverse transcriptase | It is a RNA-dependent DNA polymerase which synthesises DNA complementary to an RNA template in 5'→3' direction by using deoxynucleotide triphosphates (dNTPs) beginning with a primer with a free 3'-OH |
| RNase A | It is nuclease which digests RNA, but not DNA. |
| RNase H | It is also a nuclease which digests the RNA or RNA-DNA heteroduplex |
| T3<br>T7, SP6 RNA polymerases | These are specific RNA polymerases encoded by the bacteriophage T3, T7 and SP6, respectively. These recognise only promoters from its own phage DNA and can be used to transcribe DNA downstream of each promoter |
| *Taq* polymerase | It is a DNA polymerase isolated from *Thermus aquaticus* which operates at 72°C and is stable above 90°C. It is used in PCR |
| Terminal transferase | It adds several nucleotides to 3'-ends of linear ssDNA/dsDNA or RNA molecule |

## 1. Exonucleases

These enzymes act upon genome and digest the base pairs on 5′ or 3′ ends of a single stranded DNA or at single strand nicks or gaps in double stranded DNA (Fig. 4.1A).

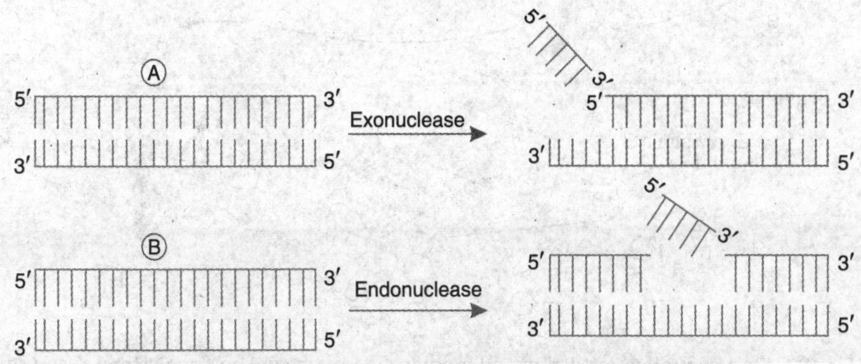


**Fig. 4.1.** Action of exonuclease (A) and endonuclease (B).

On the basis of mode of action the exonuclease are grouped into different types:

### λ Exonuclease:

This enzyme removes nucleotides from 5'end of double stranded DNA. Consequently an improved substrate for terminal transferase is formed. Therefore, this enzyme is used for 5'end modification. The mode of action of this enzyme is given below:

$$\begin{matrix} 5'\text{-GGAATT}_{OH}\text{-}3' & \xrightarrow{\lambda \text{ exonuclease}} & 5'\text{-GAATT}_{OH}\text{-}3' & & \\ 3'\text{-CCTTAAp-}5' & & 3'\text{-Cp-}5' & + & \text{Np-}5' \end{matrix}$$

(Np = nucleotide phosphate)

### Exonuclease III:

This enzyme removes nucleotides from 3'-OH end of the double stranded DNA fragment. Consequently an improved substrate is formed which is used in gene manipulation. This enzyme is used for 3'end modification of DNA fragment as below:

$$\begin{matrix} 5'\text{-GGAATT}_{OH}\text{-}3' & & 5'\text{-G}_{OH}\text{-}3' & +N_P\text{-}5' \\ 3'\text{-CCTTAA}_P\text{-}5' & \xrightarrow{\text{Exonuclease III}} & 3'\text{-CCTTAA}_P\text{-}5' & \end{matrix}$$

## 2. Endonucleases

They act upon genetic material and cleave the double stranded DNA at any point except the ends, but their action involves only one strand of the duplex (Fig. 4.1B).

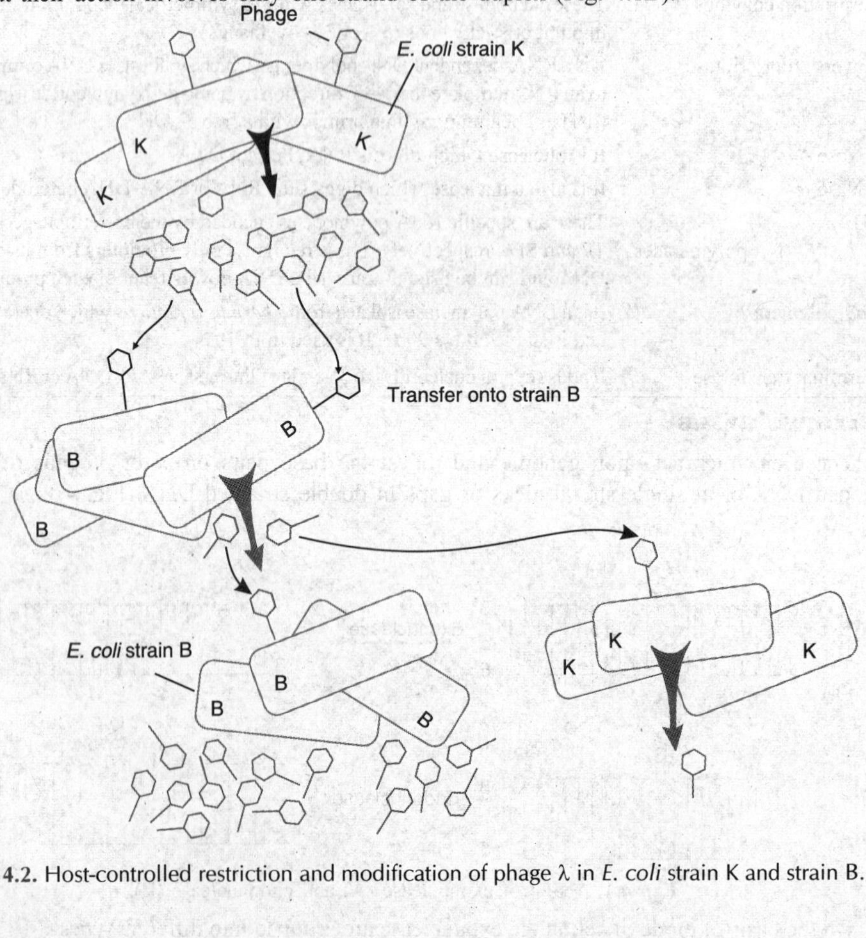


Fig. 4.2. Host-controlled restriction and modification of phage λ in *E. coli* strain K and strain B.

## 3. Restriction Endonucleases

The restriction enzymes are called as '*molecular scissors*'. These act as foundation of recombinant DNA technology. These enzymes are present in bacteria and provide a type of defense mechanism called the '*restriction-modification system*'. Molecular basis of these systems was elucidated first by Werner Arber in 1965. However, there was no understanding about cutting and joining of DNA molecules using DNA enzymes before 1970.

During the early 1950s, it was discovered when phage λ infects *E. coli*, phage DNA is delivered inside the cell. Host-control restriction modification system provides defense to *E.coli*. Bacterium protects itself through: (*i*) *restriction mechanism* (i.e. identifies the introduced foreign DNA and cuts into pieces called restriction endonucleases, and (*ii*) *modification system* (*i.e.* methylation of certain bases of its own DNA by methylase so that phage λ-expressed restriction endonuclease could not cleave its DNA).

Fig. 4.2 shows host's restriction-modification system. If phage λ propagated on *E. coli* strain K (A) is allowed to infect *E. coli* strain B, the efficiency of infection is very low (B). The progeny phage obtained from plaques of B strain can re-infect *E. coli* strain B with high efficiency (C). But infection efficiency of these progeny phage on strain K is also very low (D).

There are three gene loci (*hsd*S, *hsd*M and *hsd*R) that control this system. The *hsd*S encodes a protein which governs the specificity of these system. The *hsd*M gene product acts as modification enzyme (methylase) which interacts with *hsd*R gene product (restriction enzyme) in the cleavage process.

It may be explained that progeny phage particles first grown on strain K and then strain B are restricted when grown on B strain. But the progeny phage particles from strain B are not restricted when re-infect strain B. Again they are restricted when grown on strain K. This shows that the efficiency with which phage grows on a specific host depends on the host's strain on which it was propagated previously. It was confirmed when the restricted phage is adsorbed on the surface of restrictive host and injects its DNA, it is broken down quickly by restriction endonuclease. Hence, the restrictive host protects its own DNA from degradation and modifies its DNA enzymatically. Thus the restriction activity is provided by two enzymes: nuclease and methylase, the genes of which reside on the host chromosome or plasmid.

There are three different types of restriction endonucleases: Type I, Type II and Type III. Each enzyme differs slightly by different mode of action. Type I and Type II are large multi subunit complex which contains both endonuclease and methylase activity. The different bacteria consist of different restriction endonucleases, and in correspondence to these the methylases too are produced. These enzymes occur naturally in bacteria as a chemical weapon against the invading viruses and cut both strands of DNA when certain foreign nucleotides are introduced in the cell. These enzymes cleave a DNA to generate a nick with a 5′ phosphoryl and 3′ hydroxyl termini.

**(*a*) Type I Restriction Enzyme :** The B-K restriction and modification enzyme systems of *E. coli* have been termed as Type I endonucleases such as *Eco*B and *Eco*K. They are complex nucleases which function simultaneously as an endonuclease and a methylase. They move along the DNA in a reaction and require $Mg^{++}$, S-adenosly methionine and ATP as co-factor. They are single multi-functional enzymes consisting of three different sub-units: (*i*) restrictive sub-unit, (*ii*) modification sub-unit, and (*iii*) specificity sub-unit. The third sub-unit specifies the recognition sites. The enzymes show specificity for recognition site but not for cleavage site, *i.e.* they recognise specific sites within the DNA but do not cleave at these sites. This results in production of the heterogeneous population of DNA fragments.

Type I enzyme recognises 15 bp long and cleavage site is about 1000 bp away from the 5′-end of the recognition sequence TCA which is present within 15 bp long recognition site. The adenine position in the restriction site may be methylated. Such DNA molecules are resistant to Type I endonuclease whose recognition sequence is methylated at A residues on both the strands. However, when the recognition sequence is unmethylated, only then DNA is cleaved as below:

*Eco*K 5′- A A A *C N N N N G T C G C -3′
3′- T T T G N'N'N'N' C A *G C G -5′

Here A 'shows methylation at adenine position, N is unspecified bases and N' is complementary bases to N. The unmodified DNA molecules is restricted, *i.e.* cleaved by restriction endonucleases.

**(*b*) Type II Restriction Enzymes:** Type II restriction enzymes were first isolated by Hamilton Smith during 1970s. They are simple consisting of single polypeptide chain and require no ATP for degradation of DNA. They have separate activities for cleaving (endonuclease) and methylation (methylase) of DNA molecules. They are most stable enzymes that require $Mg^{++}$ as co-factor. Daniel Nathans first used these enzymes for mapping and analysing genes and genomes. The first (**I**) Type II restriction enzyme isolated and characterised was from *Escherichia coli* strain RY. Hence, it was named as *Eco*RI. For the discovery of restriction enzymes, W. Arber, H. Smith and D. Nathans were awarded with Nobel Prize in 1978.

**> Table 4.2.** **List of suppliers of restriction enzymes.**

| *Name of Companies* | |
|---|---|
| 1. Benthesda Research Laboratories<br>411 North Stonestreet Avenue<br>PO BOX 6010, Rockvile<br>Maryland, 20850, U.S.A. | 4. Boehringe Mannheim Gmbh Biochema<br>6800 Mannheim 31,<br>West Germany (now Germany) |
| 2. P.L. Biochemical Inc.<br>1037 West McKinley Avenue<br>Milwaukee, Wisconsin, 53205, U.S.A. | 5. Miles Research Products<br>PO Box 37, Stoke Poges<br>Slough, Berkshire SL2 4LY, U.K. |
| 3. Collaborative Research Inc.<br>1365, Main Street Waltham<br>Massachusetts, 02154, U.S.A. | 6. Bangalore Genei Pvt. Ltd<br>8th Main Road, Malleswara<br>Bangalore-560 055, India. |

Thousands of restrictions enzymes have been discovered and some of them are commercially available (Table 4.2). Out of above three types of restriction endonucleases, Type II are used in gene manipulation.Type I enzymes recognise a specific sequence of DNA molecule but cut elsewhere, and Type II make cuts only within the restriction sites and produce two single strand breaks, one break in each strand. Type II enzymes are the most important ones. As a result of their action the broken nucleotides form a DNA duplex which exhibit two fold symmetry around a given point. In some cases, cleavage in two strands are staggered to produce single stranded short projections opposite to each other with blunt or mutually cohesive sticky ends which are identical and complementary to each other. These complementary sequences are also known as *palindrome sequences* or *palindromes*. Therefore, when read from 5′→3′ both strands have the same sequence. They are specific in recognising the nucleotide base pairs on both the strands and cut exactly within the same recognition site. Recognition sites are the palindromes or palimdromic sequences where most of Type II restriction enzymes bind and cut a DNA molecule. Palindromes are the nucleotide-pair sequences that are the same when read forward (left to right) or backward (right to left) from a cerntral axis of symmetry i.e. two strands are identical when both are read in the same polarity *i.e.* in 5′→3′ direction. For example the following phrase is read same in either of the directions (left to right and right to left):

$$\xrightarrow{\qquad\qquad}$$
ANDMADAMDNA
$$\xleftarrow{\qquad\qquad}$$

The length of recognition sites of different enzymes varies which can be four, five, six, eight or more nucleotide pairs (Table 4.3). There are many enzymes that recognise a hexanucleotide (a sequence of 6 nucleotides), whereas the others recognise tetra-, penta or octanucleotide sequences present on DNA molecule. For example, *Eco*RI restriction enzyme isolated from *E. coli* cleaves DNA at the hexanucleotide 5'GAATTC3', whereas *Eco*RII isolated from the same bacterium recognises -$PO_4$ end of pentanucleotide 5'CCTGG3'. Recognition sequences of many type enzymes are palindromic sequences i.e. they exhibit rotational symmetry. The nucleotide sequences are the same when read from 5'→3' in both the strands. For example, nucleotides of recognition sequences of *Eco*RI are the same on both the strands as below:

$\longrightarrow$
5'–GAATTC–3'   Palindromic sequence
5'–CTTAAG–5'
$\longleftarrow$

In addition, a very useful feature of many restriction nucleases is that they make staggered cuts. For example, *Eco*RI enzyme binds to a region of having specific palindromic sequence. The length of this region is 6 base pairs *i.e* hexanucleotide palindrome. It cuts between G and A residues of each strand and produces two single tranded complementary cut ends which are asymmetrical having 5'

**> Table 4.3. Source of type II restriction enzymes, cleavage sites and productions of cleavage**

| *Microorganisms* | *Restriction enzymes* | *Cleavage sites* | *Cleavage products* | |
|---|---|---|---|---|
| *Bacillus amy-loliquefaciens* H | *Bam*HI | ↓<br>5'-GGATCC-3'<br>3'-CCTAGG-5'<br>↑ | 5'-G<br>3'-CCTAG | GATCC-3'<br>G-5' |
| *B. globigii* | *Bgl*II | ↓<br>5'-AGATCT-3'<br>3-TCTAGA-5<br>↑ | 5-A<br>3-TCTAG | GATCT-3<br>A-5 |
| *Escherchia coli* RY13 | *Eco*RI | ↓<br>5'-GAATTC-3'<br>3'-CTTAAG-5'<br>↑ | 5'-G<br>3'-CTTAA | AATTC-3'<br>G-5' |
| *E. coli* $R_2$45 | *Eco*RII | ↓<br>5'-GGATCC-3'<br>3'-CCTAGG-5'<br>↑ | 5'-G-3'<br>3'-CCTAG-5' | 5'-GATCC-3'<br>3'-G-5' |
| *Haemophilus influenzae* Rd | *Hind*II | ↓<br>5'-GTPyPuAC-3'<br>3'-CAPuPyTG-5'<br>↑ | 5'-GTPy-3'<br>3'-CAPu-5' | 5'-PuAC-3'<br>3'-PyTG-5' |
| | *Hind*III | 5'-AAGCTT-3'<br>3'-TTCGAA-5'<br>↓ ↑ | 5'-A<br>3' -TTCGA | AGCTT-3'<br>A-5' |
| *H. parainfluenzae* | *Hpa*I | 5'-GTTAAC-3'<br>3-CAATTG-5'<br>↑ | 5'-GTT<br>3'-CAA | AAC-3'<br>TTG-5' |
| | *Hpa*II | 5'-CCGG-3'<br>3'-GGCC-5'<br>↑ ↓ | 5'-C-3'<br>3'-GGC-5' | 5'-CGG-3'<br>3'-C-5' |
| *Klebsiella pneumoniae* OK 8 | *Kpn*I | 5'-GGTACC-5'<br>3'-CCATGG-3'<br>↑ ↓ | 5'-GGTAC<br>3'-C | C-3'<br>CATGG-5' |
| *Nocardia otitidis-caviarus* | *Not*I | 5'-GCGGCCGC-3'<br>3'-CGCCGGCG-5' | 5'-GC-3'<br>3'-CGCCGG-5' | 5'-GGCCGC-3'<br>3'-CG-5' |

| | | | | |
|---|---|---|---|---|
| | | ↑ ↓ | | |
| *Providencia stuartii* | *Pst*I | 5′-CTGCAG-3′<br>3′-GACGTC-5′<br>↑ | 5′-CTGCA-3′<br>3′-G-5′ | 5′G-3′<br>3′-ACGTC-5′ |
| *Streptomyces albus* G | *Sal*I | 5′-GTCGAC-3′<br>3′-CAGCTG-5′<br>↓ ↑ | 5′-G<br>3′-CAGCT | TCGAC-3′<br>G-5′ |
| *Staphylococcus aureus* 3AI | *Sau*3AI | 5′-GATC-3′<br>3-CTAG-5′<br>↑ | 5′-<br>3′-CTAG | GATC-3′<br>5′ |

Arrows indicate the recognition sites.

overhangs of 4 nucleotides. These ends are called **sticky ends** or *cohesive ends*. Because nucleotide bases of this region can pair and stick the DNA fragments again as given below:

```
Cleavage site
     ↓               EcoRI                            Sticky end
5'-GAATTC-3'     ---------->   5'-G-3'           /       +         5'-AATTC-3'
3'-CTTAAG-5'     <----------   3'-CTTAA-5'                            3'-G-5'
         ↑           Ligation                                       Sticky end
Cleavage site
```

On the other hand some other Type II restriction enzymes cleave both strands of DNA at the same base pairs but in the centre of recognition sequence, and results in DNA fragments with **blunt ends** or *flush ends*. For example *HaeIII* (isolated from *Haemophilus aegypticus*, the order of enzyme III), four nucleotide long palindromic sequence and cuts symmetrically the both DNA strands and forms blunt ends.

```
Cleavage site
     ↓                                         Blunt end
               Cleavage
5'-GGCC-3'   ----->    5'-GG-3'    +    5'-CC-3'
3'-CCGG-5'   <-----    3'-CC-5'         3'-GC-5'
     ↑          Ligation   Blunt end
Cleavage site
```

There are some restriction enzymes whose recognition sites differ. Even then they produce identical sticky ends. For example, recognition sequence of *Bgl*II is 5'AGATCT3', that of *Bam*HI is 5'GGATCC3' and *Sau* is 5'GATC3'. But all the three enzymes cut recognition sequences and produce 5'GATC3' sticky ends. Since, fragments generated by each enzyme have identical sticky ends, they can join complementary sticky ends of one another (Table 4.3).

The properties of Type II restriction enzymes vary with its specificity, *i.e.* its changes with changing the reaction conditions such as lowering NaCl concentration, increasing temperature, replacing $Mg^{++}$ with $Mn^{++}$. For example, the recognition and cleavage sites of *Eco*RI is normally a hexanucleotide sequence (5′ GAATTC3′), but it become a tetranucleotide sequence (5′AATT3′) with the changes in reaction conditions. Moreover, the size of DNA, its base composition and the GC content of recognition sequences govern the number of cleavage sites of Type II enzymes in a DNA molecule. For example, if all the four bases are present in a DNA molecule with 50% AT and 50% GC base pairs, the probability for a particular tetranucleotide is $(X_1)^4$. The tetranucleotide shall be cleaved at a distance of every 256 base pairs. Now-a-days a large number of restriction enzymes are available commercially (Table 4.2). Some of the commonly used restriction endonucleases are given in Table 4.3.

In 1983, R.J. Roberts has given an extensive list of restriction enzymes and the sequences recognised by them. Most of the enzymes share a common central nucleotide in their recognition sequence for example, *Bam*HI, *Bgl*II, *Sau*3A. But these enzymes recognise different sites in DNA and produce identical single stranded 5′ tails which allow the joining of fragments generated by different enzymes within this set. The identical nature of the termini of cleaved DNA fragments from any organism is the very property which permits the annealing and subsequent ligation of the DNA from diverse sources.

***(a)* Nomenclature of Restriction Enzymes :** The restriction enzymes are named based on some of the following principles:

(*i*) Name of the organism is identified by the first letter of genus name and the first two letters of the species name to form a three letter abbreviation in the italic, for example, *E. coli* = *Eco* and *H. influenzae* = *Hin*, etc.

(*ii*) A strain or type identified is written as subscript *e.g. Eco*K for *E. coli* strainK, *Hin*d for *H. influenzae* strain Rd.

(*iii*) In such cases where the restriction and modification systems are genetically specified by a virus or plasmid, the extra chromosomal element is identified by a subscript *e.g. Eco* RI, *Eco*PI, etc.

(*iv*) When a strain has several restriction and modification systems, these are identified by Roman numerals, for example *Hin*dI, *Hin*dII, *Hin*dIII for *H. influenzae* strain Rd. etc. These Roman numerals should not be confused with those in the classification of restriction enzymes into Type I.

***(b)* Type III Restriction Enzymes:** The Type III enzyme is made up of two sub-units, one specifies for site recognition and modification and the other for cleavage. In a reaction, it moves along the DNA and requires ATP as source of energy and $Mg^{++}$ as co-factor. ATPase activity is lacking in these enzymes. Some examples of Type III enzyme are *Hpa*I, *Mbo*II, *Fok*I, etc. They have symmetrical recognition sites and cleave DNA at specific non-palindromic sequences. For cleaving double stranded DNA two sites in opposite orientation must be present. One strand of double stranded DNA is cleaved about 25-27 bp away from the recognition site which is located in its immediate vicinity. Recognition sequence of some of Type III enzymes are given below:

| | |
|---|---|
| *Mbo*II | 5′-GAAGA$(N)_8$-3′ |
| | 3′-CT TCT$(N)_7$-5′ |
| *Fok*I | 5′-GGATGC$(N)_9$-3′ |
| | 3′-CCTACG$(N)_{13}$-5′ |
| *Hgl*I | 5′-GACGC$(N)_5$-3′ |
| | 3′-CTGCG$(N)_{10}$-5′ |

Since the cleavage products of these enzymes are homogeneous population of DNA fragments, they cannot be used for genetic engineering experiments.

***(c)* Other Restriction Enzymes:** There are certain other restriction enzmes which do not fall under the Type I-III. Hence there is need for reclassifying restriction enzymes under the other categories from Type I to Type IV. Examples of the other restriction enzymes are *Eco* 571, *Mcr* (modified cytosine reaction), etc. *Eco*571 is made up of single polypeptide. It shows both nuclease modification and nuclease activity. The S-adenosyl methionine stimulates nuclease activity of this enzyme. It bears some other properties that resemble with Type II enzymes. Therefore, *Eco* 571 should be put under Type II to Type IV.

The DNA of some bacteria and eukaryotes was restricted due to the presence of methyl cytosine in the DNA. This was called modified cytosine reaction. In *E. coli* strain K there are two types of *Mcr* system: McrA and McrBC. McrA is encoded by a prophage-like element, whereas McrBC is encoded by two genes, *mcr*B and *mcr*C. Both the genes are present adjacent to the genes that encode Type I endonucleases in *E. coli* strain K.

Both the strands of DNA are cleaved at multiple sites between the methyl cytosine which is accomplished by GTP. The MerBC system is analogous to the *Rg/B* (restricts glucoseless phage) system which is found in T-even coliphages. T-even bacteriophages consist of glucosylated 5-hydroxymethyl cytosine in their DNA in place of cytosine. The phage DNA is restricted if glucosylation of 5-hydroxymethyl cytosine is inhibited due to mutation of phage or host genes. Hence the host causes the restriction of phage.

### 4. S1 Nuclease

It degrades the single stranded DNA or single strand protrusion of double stranded DNA with cohesive ends. As a result of action of S1 nuclease cohesive ends are converted into blunt ends.

Therefore, S1 nuclease is used to remove the incompatible ends so that overlapping ends may be developed for annealing of two fragments of DNA molecules. The mode of action of S1 nuclease is given below:

$$\begin{matrix} 5'\text{-AG-3}' \\ 3'\text{-TCTTAA-5}' \end{matrix} \xrightarrow{\text{S1 nuclease}} \begin{matrix} 5'\text{-AG-3}' \\ 3'\text{-TC-5}' + 5'\text{-AA-5}' \end{matrix}$$

### Bal 31 Nuclease

This enzyme is very specific single stranded endodeoxyribonuclease. It degrades simultaneously both the 3′ and 5′ strands of DNA molecule. Consequently shortened DNA fragments are formed. These fragments possess blunt ends at both the termini. Its mechanism of action is as below:

$$\begin{matrix} 5'\text{-GGAATT-3}' \\ 3'\text{-CCTTAA-5}' \end{matrix} \xrightarrow{\text{Bal 31 exonuclease}} \begin{matrix} 5'\text{-GGAA-3}' \\ 3'\text{-CCTT-5}' + 3'\text{-AA-5}' \end{matrix}$$

### 5. DNA Ligases

J.E. Mertz and R.W. Davis in 1972 for the first time demonstrated that cohesive termini of cleaved DNA molecules could be covalently sealed with *E. coli* DNA ligase and were able to produce recombinant DNA molecules (Fig. 4.3). DNA ligase seals single strand nicks in DNA which has 5′→3′ -OH (hydroxyl) termini. There are two enzymes which are extensively used for covalently joining restriction fragments : the ligase from *E. coli* and that encoded by T4 phage. The main source of DNA ligase is T4 phage, hence, the enzyme is known as T4 DNA ligase.

For the joining reactions, the *E. coli* DNA ligase uses nicotinamide adenine dinucleotide ($NAD^+$) as a cofactor, while T4 DNA ligase requires ATP for the same. Both the enzymes contain a $-NH_2$ group on lysine residue. In both cases, cofactor breaks into AMP (adenosine monophosphate) (Fig. 4.3.) which in turn adenylate the enzyme (E) to form enzyme -AMP complex (EAC). EAC binds to nick containing 3′ -OH and 5′ $-PO_4$ ends on a double stranded DNA molecule. The 5′ -phosphoryl terminus of the nick is adenylated by the EAC with 3'-OH terminus resulting in formation of phosphodiester and liberation of AMP (Lehman, 1974). After formation of phosphodiester nick is sealed (Fig. 4.3). T4 enzyme has the ability to join the blunt ends of DNA fragments, whereas *E. coli* DNA ligase joins the cohesive ends produced by restriction enzymes. Additional advantage with T4 enzymes is that it can quickly join and produce the full base pairs but it would be difficult to retrieve the inserted DNA from vector. However, cohesive end ligation proceeds about 100 times faster than the blunt end ligation.

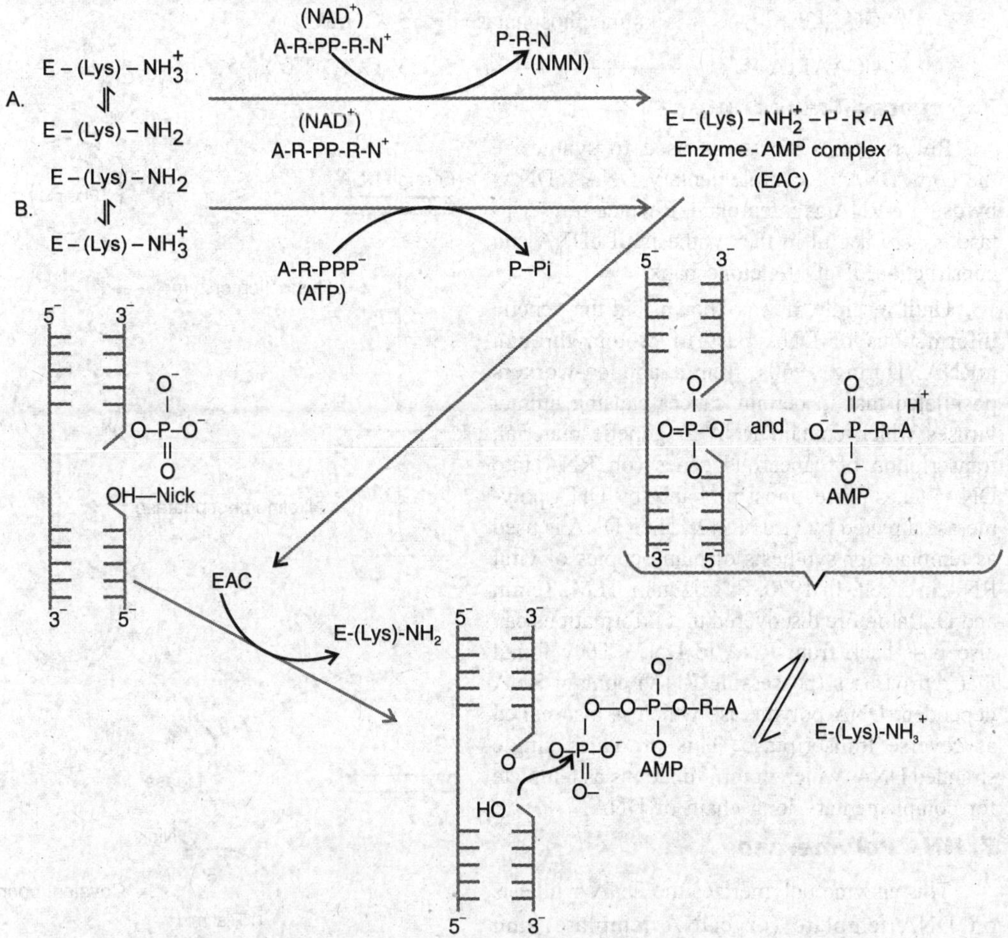


Fig 4.3. Mechanism of DNA ligase enzyme - AMP complex formation and sealing of nick on a double stranded DNA molecule; A, *E. coli* DNA ligase; B, T4 DNA ligase; A, adenine; P, $PO_4$; R, ribose; N, nicotinamide; NMN, nicotinamide mononucleotide

## 6. Alkaline Phosphatase

For joining a foreign DNA fragment when plasmid vector, is treated with restriction enzyme, the major difficulty arises at the same time. Because the cohesive ends of broken plasmids, instead of joining with foreign DNA join the cohesive end of the same DNA molecules and get recircularized. To overcome this problem, the restricted plasmid (*i.e.* plasmid treated with restriction enzymes) is treated with an enzyme, alkaline phosphatase, that digests the terminal 5′ phosphoryl group (Fig. 4.4). The restriction fragments of the foreign DNA to be cloned are not treated with alkaline phosphatase. Therefore, the 5′ end of foreign DNA fragment can covalently join to 3′ end of the plasmid. The hybrid or recombinant DNA obtained has a nick with 3′ and 5′ hydroxy ends. Ligase will only join 3′ and 5′ ends of recombinant DNA together if the 5′ end is phosphorylated. Thus, alkaline phosphatase and ligase prevent recircularization of the vector and increase the frequency of production of recombinant DNA molecules. The nicks between two 3′ ends of DNA fragment and vector DNA are repaired inside the bacterial host cells during the transformation.

The mechanism of action of alkaline phosphatase is as below:

$$\begin{matrix} 5'\text{-GG}_{OH}\text{-}3' \\ 3'\text{-CCAATTA}_{P}\text{-}5' \end{matrix} \xrightarrow{\text{Alkaline phosphatase}} \begin{matrix} 5'\text{-GG-}3' \\ 3'\text{-CCAATTA}_{OH}\text{-}5' \end{matrix}$$

## 7. Reverse Transcriptase

Reverse transcriptase is used to synthesize the copy DNA or complementary DNA (cDNA) by using mRNA as a template. Reverse transcriptase is very useful in the synthesis of cDNA and construction of cDNA clone bank.

Until recently, it was known that the genetic informations of DNA pass to protein through mRNA. During 1960s, Temin and co-workers postulated that in certain cancer causing animal viruses which contain RNA as genetic material, transcription of cancerous genes (on RNA into DNA) takes places most probably by DNA polymerase directed by virus RNA. Then DNA is used as template for synthesis of many copies of viral RNA in a cell. In 1970, S. Mizutani, H.M. Temin and D. Baltimore discovered that informations can also pass back from RNA to DNA. They found that retroviruses (possessing RNA) contain RNA dependent DNA polymerase which is also called as reverse transcriptase. This produces single stranded DNA, which in turn functions as template for complementary long chain of DNA.

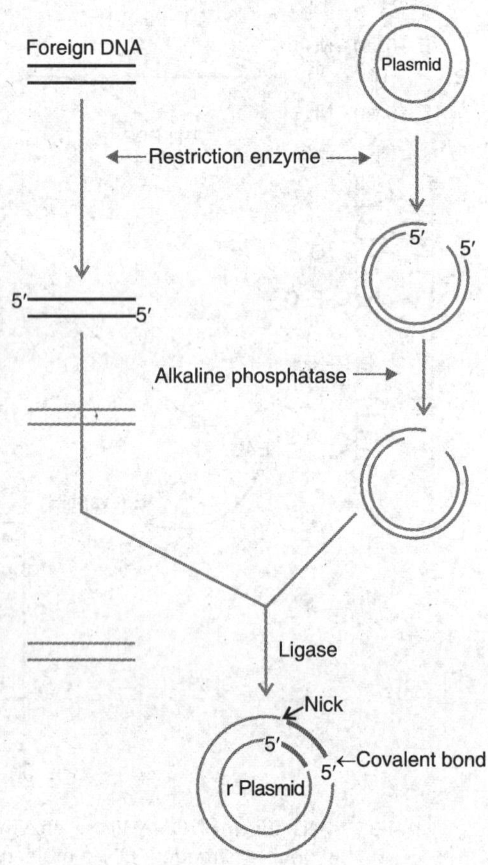


**Fig. 4.4.** Inhibition of recircularization by alkaline phosphates (to increase recombinant plasmids)

## 7. DNA Polymerase

This enzyme polymerizes the DNA synthesis on DNA template (or cDNA template) and also catalyses 5′→3′ and 3′→5′ exonucleolytic degradation of DNA. The DNA polymerase, investigated by A. Kornberg and coworkers in *E. coli* in 1956 is now known as DNA polymerase I (DNA pol I ). The other two enzymes are DNA polymerase II (DNA pol II) and DNA polymerase III (DNA pol III). These have almost similar catalytic activity. DNA pol I (mol wt 109,000) has a single polypeptide chain of about 1,000 amino acid residues. The addition of mononucleotide to the free -OH end of a DNA chain is catalysed by this enzyme. Also it catalyses the other two reaction, *i.e.* 3′→ 5′ exonuclease activity (hydrolysing single nucleotide residues from 5′ - terminus) and 5→3′ exonuclease activity (hydrolysing single nucleotide residues from 5′- terminus). Function of DNA pol II (mol wt 120,000) is little understood. However, it catalyses 3′→5′ exonuclease activity. DNA pol III (mol wt about 140,000) is about several time more active than the other two. It is a dimer of DNA pol III. It requies an auxillary protein DNA copolymerase III and after combination, yields a

DNA pol III - copol III complex. Where there is preformed DNA template it produces a parallel strand in the presence of ATP.

DNA Pol I is susceptible to protease action and can be degraded into fragments. The larger fragments possess polymerase and 3'→5' exonuclease activity but lack 5'→3' exonuclease activity. This portion of DNA Pol I is called **Klenow polymerase**. Instead of intact Pol I, this enzyme is used for polymerase activity as below:

$$\begin{matrix} 5'\text{-G-}3' \\ 3'\text{-CCTTAA-}5' \end{matrix} \xrightarrow[\text{dGTP,dTTP,dATP}]{\text{Klenow polymerase}} \begin{matrix} 5'\text{-GGAATT-}3' \\ 3'\text{-CCTTAA-}5' \end{matrix}$$

### 8. T4 Polynucleotide Kinase

The T4 polynucleotide kinase catalyses the transfer of $\gamma^{32}PO_4$ of ATP to 5′ terminus of a dephosphorylated DNA (i.e. DNA lacking $-PO_4$ residue) or RNA as below:

$$\begin{matrix} 5'\text{-GG}_{OH}\text{-}3' \\ 3'\text{-CCTTAA}_{OH}\text{-}5' \end{matrix} \xrightarrow[\gamma^{22}\text{P–ATP}]{\text{Polynucleotide kinase}} \begin{matrix} 5'\text{GG}_{OH}\text{-}3' \\ 3'\text{-CCTTAA}_{P*}\text{-}5' \end{matrix}$$

This enzyme shows both phosphorylation as well as phosphatase activities; therefore, two types of reaction can be used accordingly. In phosphorylation activity the $\gamma$-$PO_4$ is transferred to the 5′end of dephosphorylated DNA. When ADP is in excess, it causes the T4 polynucleotide kinase to transfer the terminal phosphate (attached to a nucleotide) from phosphorylated DNA to ADP. By using radiolabelled gamma-phosphate ($\gamma^{32}PO_4$), it was found that DNA is rephosphorylated by transfer of $\gamma^{32}P$ from $\gamma^{32}P$-ATP. This shows that the enzyme plays a dual role of phosphorylation activity and phosphatase activity.

Due to the presence of both properties, this enzyme is used in radiolabelling of the 5′ end of duplex during DNA sequencing (*see* DNA sequencing method), constructing the terminally labelled DNA, phosphorylating the synthetic linkers/oligonucleotide linkers/other DNA fragments lacking terminal 5′phosphates, etc.

### 9. Terminal Transferase

Terminal transferase is also known as deoxynucleotidyl transferase. It has been isolated and purified from calf thymus. It has a property to add 'oligodeoxynucleotide tail' to 3′-OH end of double stranded DNA fragments. Thus it extends homopolymer tails and this phenomenon is called homoplymer tailing (for detail see Chapter 5).

## B. USE OF LINKERS AND ADAPTORS

As discussed earlier that treatment of DNA with Type II restriction enxymes results in formation of fragments having sticky ends or blunt ends. Such DNA fragments cannot be ligated with cloning vectors. Therefore, artificially cleavage sites at blunt end can be added as linker or adaptor molecules. Chemically synthesised double stranded oligonucleotides are called **linkers**. They are added to double stranded DNA molecules that act as recognition site for restriction enzyme. For example the following oiligonucleotide possess recognition sequence for *Eco*RI (asterisk nucleotides). It can be ligated to blunt end of any double stranded DNA and cut by *EcoRI* to generate sticky ends to DNA fragments (Fig. 4.5).

5'-C C G A*A *T* T*C *G G-3'
3'-G G C T T A A G C C-3'

The chemically synthesised DNA molecules containing pre-formed cohesive ends are called adaptors. When recognition site is present within the DNA molecules to be coloned, adaptors are used to develop sticky ends. Adaptor is used in the form of 5'-OH so that self polymerisation could not be done (Fig. 4.5). Thereafter, phosphorylation of foreign DNA ligated adaptor is done at 5' end. They joined to restriction enzyme treated fragments.

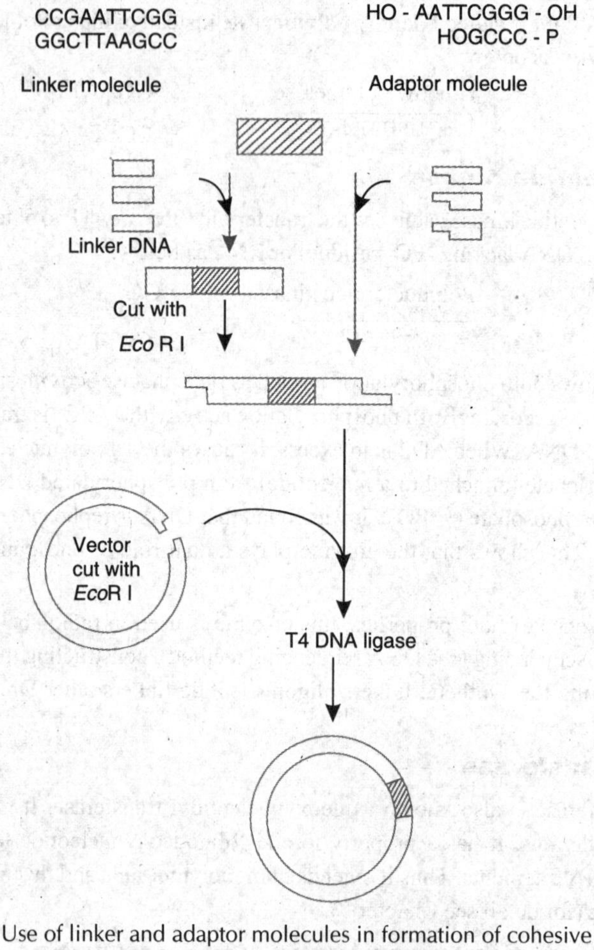


Fig. 4.5. Use of linker and adaptor molecules in formation of cohesive termini.

## C. FOREIGN DNA/PASSENGER DNA

Foreign/passenger DNA is a fragment of DNA molecule which is enzymatically isolated and cloned. The gene is identified on a genome and pulled out from it either before or after cloning. The cloned foreign DNA fragment expresses normally as in parental cell. Thus, the foreign DNA fragments can be procured from a variety of sources depending on the aims and scope of cloning experiments.

Identification and characterization of DNA sequences are rather more difficult on its genome than using mRNA, if it is in pure form. If the gene product translated by mRNA is not well characterized it can be most difficult procedure for cloning. In an average cell or tissue, 1-2% of total cytoplasmic RNA population is mRNA which carry transcripts for coding various proteins. When mRNA is present in low amount it is rather difficult to isolate cDNA clones. The procedures of isolation and purification of mRNA and getting pure cDNA are given in Chapter 6.

## PROBLEMS

1. What is restriction and modification system?
2. What are the DNA enzymes? Discuss in detail their mode of action.
3. Give a detailed account of Restriction endonucleases.
4. Write the mode of action of DNA polymerase and terminal transferase.
5. Write short notes on the following:
   (*i*) Alkaline phosphatase, (*ii*) S1 nuclease, (*iii*) Terminal transferase, (*iv*) Sticky ends, (*v*) Palindomes, (*vi*) Linkers and adaptors, (*vii*) DNA ligases, (*viii*)Type III restriction enzyme.

CHAPTER

# 5

# Tools of Genetic Engineering – III: Cloning Vectors

Similar to DNA enzymes, the other most important requirements for recombinant DNA technology is the cloning and expression vectors. The recombinant DNA is produced by cloning a foreign DNA isolated either from the genome or synthesised chemically or as cDNA using mRNA molecule. However, cloning of this DNA can be done only when 'another DNA molecule' is available that may replicate in the transformed host cell. This 'other DNA molecule' used for joining the foreign DNA is called **vectors**. The vectors are the DNA molecules that can carry a foreign DNA segment and replicate inside the host cell. Vectors may be plasmids, a bacteriophage, cosmids, phagemids, transposons, a virus YAC or BAC. Moreover, a vector must possess the following characteristics:

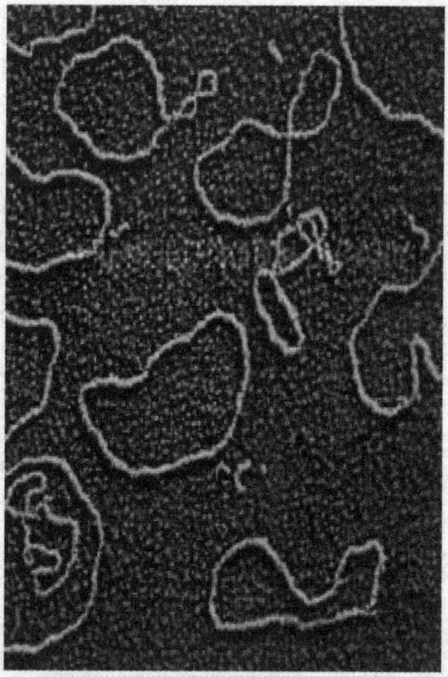

Transmission electron microphaph (TEM) of *E. coli* plasmids.

(*i*) It must replicate (through *ori* gene) in host cells after its introduction.

(*ii*) It must contain marker genes such as $tet^R$ (for tetracycline resistance), $kan^R$ (for kanamycin resistance), $amp^R$ (for ampicillin resistance). These help in selection of transformed cells from untransformed cells.

(*iii*) A unique cleavage site must be present in one of the marker genes. By inactivating it, through restriction enzyme one can detect the recombinant molecules. However, many vectors used commonly contain several recognition and cleavage sites for several restriction enzymes in a small region. These are called *multiple cloning sites* (MCS) or *polylinkers*. Presence of MCS facilitates the use of restriction enzymes of choice.

(*iv*) It should contain specific control systems like promoters, terminators, ribosome binding sites, etc. so that the cloned DNA should express properly.

The best known of these vectors is the plasmid vector. The largest number exists for *E. coli.* During the early 1970s, Stanley Cohen and Herb Boyer selected *E. coli* strain K12 for conducting research work on molecular biology. Since then *E. coli* K12 has been used for various experiments in recombinant DNA technology and production of products. One of the unique features of *E. coli* is that it is found in human intestine and cannot thrive outside the laboratory environment. *E. coli* strain K12 mutant in restriction-modification system is used in gene cloning. *E. coli* strain K12 mutant in *end*H gene (encodes DNA specific endonuclease) increases the yield of plasmid DNA and improves DNA quality. The *rec*A gene is associated with expression of recombination proteins. Therefore, *E. coli* strain K12 mutant in *recA* gene ensures about the biosafety of *E. coli* strain K12 because they can easily be killed by ultra violet light.

## A. BACTERIAL PLASMID VECTOR

The size of plasmids varies from 1 to 200 kb. It depends on host proteins for replication, maintenance and other functions. They are present in a characteristic number of copies in a bacterial cell, yeast cell, in eukaryotic organelles such as mitochondria. The plasmids are maintained as single copy per cell and they are called **single copy plasmid**. There are **multicopy plasmid** also which are maintained as 10–20 copies per cell. Apart from this when the plasmids are under the relaxed replication control, they get accumulated in the cell in very large copies *i.e.* over 1000 copies per cell. Due to increased yield potential, such plasmids are used as cloning vectors. The *ColEl* plasmid replication system of *E. coli* is most extensively studied. It expresses colicins which is an antibacterial protein. The *ColE1* transcribes a DNA sequence near the *ori* gene. Due to increased yield potential these plasmids are used as cloning vectors.

Several plasmids consist of a replicon (a segment of genome that contains an origin of replication) of pMB1, originally isolated from its close relative, *ColE1* replicon. In each bacterial cell about 25-20 plasmids are maintained under normal growth conditions. The *ColE1* or pMB1 replicon relies mainly on long-lived enzyme from host. Replication is primed by an RNA primer and takes place unidirectionally from a specific origin.

*E. coli* also consists of another plasmid called **fertility factor** or **F plasmid** which is associated with conjugation. It is about 95 kb long plasmid and transfer F factor to $F^-$ (F-deficient) cells. One copy of F plasmid is present in each cell and encodes about 100 genes. Some of them are needed during the transfer of F factor. F plasmid forms sex pili, the number of which is governed by the number of plasmids. It also encodes cell surface binding proteins for the filamentous bacteriophage M13.

Plasmids are the extrachromosomal, self-replicating, and double stranded closed and circular DNA molecules present in the bacterial cell. Plasmids contain sufficient genetic informations for their own replication. A number of host properties are specified by plasmids, such as antibiotic and heavy metal resistance, nitrogen fixation, pollutant degradation, bacteriocin and toxin production, colicin factors and phages. Naturally occurring plasmids can be modified by *in vitro* techniques. S.N. Cohen and coworkers in 1973 for the first time reported the cloning DNA by using plasmid as vector.

A plasmid can be considered a suitable cloning vehicle if it possesses the following features:

(*i*) It can be really isolated from the cells,

(*ii*) It possesses a single restriction site for one or more restriction enzyme(s),

(*iii*) Insertion of a linear molecule at one of these sites does not alter its replication properties,

(*iv*) It can be reintroduced into a bacterial cell and cells carrying the plasmid with or without the insert, can be selected or identified (Bernard and Helinski, 1980),

(*v*) They do not occur free in nature but are found in bacterial cells.

In many cases the principal objective of cloning experiment is the insertion of a particular restriction fragment into a suitable plasmid vector and its amplification.

## Amplification of plasmids

Amplification is a process of increasing the number of a plasmid in a bacterial cell. In this process, a cell containing a relaxed plasmid is treated with a drug to inhibit protein synthesis. Consequently, cells stop replicating. The relaxed plasmid pBR322 continues to replicate despite of drug treatment. Replication of relaxed plasmid neither depends on cell replication nor requires protein synthesis. For example, addition of chloramphenicol causes pBR322 to get increased from 3-5 to about 3000 per cell. Finally the ratio of plasmid DNA to chromosomal DNA is increased which makes easy to isolate the plasmid DNA. In order to replicate and clone a fragment of foreign DNA, it is necessary to possess a sequence of nucleotides which are recognised by the host bacterium as an origin of replication. The origin of replication occurs naturally in plasmid and transferred to onward progenies. During cell division these genes are transferred resulting in gradual decrease in number of such genes. However, due to continuous exchange of genetic materials between plasmid and chromosomal DNA, new genes originate with respect to environmental conditions.

There are several plasmid cloning vectors such as pBR320, pSC102, Co1E1, pUC, pRP4, pRK2, pRSFlOlO, pEY, pWWO, Ti- and Ri-DNA plasmids, etc. The plasmid cloning vectors are designated by 'p', some abbreviations and a few numbers. For example, in pBR320, 'p' means *plasmid, BR* refers to the researchers, F. Bolivar and R. Rodriguez who discovered the plasmid, and 322 is the numerical designation. Some of the plasmids are given in Table 5.1.

## 1. pBR322

The plasmids that occur naturally do not possess all the characteristics to be used as cloning vector. Therefore, they are constructed by inserting the genes of relaxed replication and genes for antibiotic resistance.

The pBR322 is the first artificial cloning vector developed in 1977 by Boliver and Rodriguez from *E.coli* plasmid ColEl. It is 4.362 kb long and most widely used cloning vector. In the transformed *E. coli* cells its copy number remains about 15-20 molecules per cell. However, its number may be amplified by incubating the transformed *E. coli* cells on a medium containing chloramphenicol.

**> Table 5.1. Some cloning vectors.**

| *Cloning vectors* | *Natural occurrence* | *Size (Kb)** | *Selective marker*** |
|---|---|---|---|
| **Plasmids** | | | |
| pACYC 177 | *Escherichia coli* | 3.7 | Amp$^r$, Kan$^r$ |
| pBR322 | *E. coli* | 4.0 | Amp$^r$, tet$^r$ |
| pBR324 | *E. coli* | 8.3 | Amp$^r$, tet$^r$, El imm. |
| pMB9 | *R. coli* | 5.8 | Tet$^r$ |
| pRK646 | *E. coli* | 3.4 | Amp$^r$ |
| pC194 | *Staphylococcus aureus* | 3.6 | Ery$^r$ |
| pSA0501 | *S. aureus* | 4.2 | Str$^r$ |
| pBS161-1 | *Bacillus subtilis* | 3.65 | Tet$^r$ |
| pWWO | *Pseudomonas putida* | 117 | Kan$^r$ |

| | | | |
|---|---|---|---|
| **Cosmids** | | | |
| pJC74 | Derived plasmid from Col EL | 16 | Amp$^r$, El imm. |
| pJC720 | do | 24 | El imm., Rif$^r$ |
| pHC79 | Derivative of pBR322 | 6 | Amp$^r$, Tet$^r$ |
| **Viruses** | | | |
| SV40 | Mammalian cells | 5.2 | — |
| Phage M13$^+$ | *E. coli* | 6.4 | — |
| Phage λ | *E. coli* | 4.9 | — |

*, 1 Kb (Kilobase pairs) = 1,000 base pairs = 0.66 mega dalton;

**, Resistance to ampicillin (Amp$^r$), tetracycline (Tet$^r$), erythromycine (Ery$^r$), streptomycine (Strr), Kanamycin (Kan$^r$), rifampicin (Rif$^r$), and colicin production (EL imm.)

A physical map of plasmid pBR322 is shown in ( Fig. 5.1). The pBR322 is constructed from the plasmids of *E. coli,* pBR318 and pBR320. It contains origin of replication (*ori*) that was derived from a plasmid related to naturally occurring plasmid *ColEl.* Therefore, its replication may be more faster than bacterial DNA. It also possesses genes conferring resistance to antibiotics *e.g.* ampicillin (*amp*$^r$) and tetracycline (*tet*$^r$), and unique recognition sites for 20 restriction endonucleases. Six of these 20 sites such as *Eco*RIV, *Bam*HI, *Sph*I, *Sal*I, *Xma*III and *Nru*I are present within the gene coding for tetracycline resistance, two sites (*Hin*dII and *Cla*I) are located within the promoter of the tetracycline resistance gene, and the three sites (*Pst*I, *Pvu*I and *Sca*I) within the beta-lactamase gene that provide resistance to ampicillin. If any foreign DNA is cloned into any of these 11 sites, insertional inactivation of any of the antibiotic resistance markers takes place. For example, if a foreign DNA molecule is inserted at *tet*$^R$ gene cluster, the property of tetracycline resistance will be lost. The recombinant plasmid will allow the cells to grow only in the presence of ampicillin but will not protect them against tetracycline. The presence of an antibiotic resistant gene in a plasmid of bacterium will confer resistance to that antibiotic.

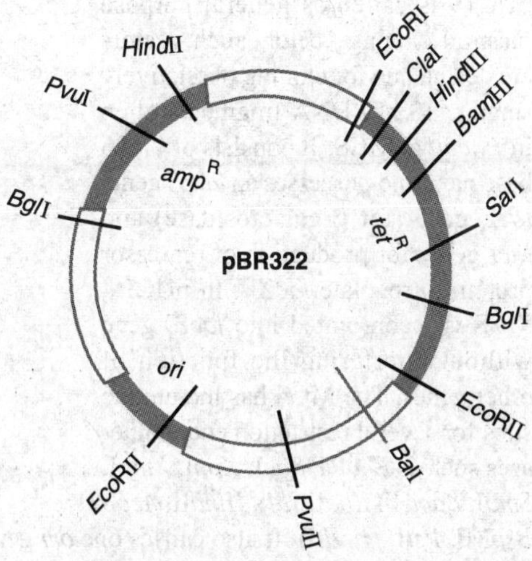


**Fig. 5.1.** A map of pBR322 (4.36 Kb) showing a number of restriction sites and regions encoding for resistance to ampicillin (*amp*$^r$) and tetracycline (*tet*$^r$), and origin of replication (*ori*)

The antibiotic resistant cells can be selected by culturing the cells on nutrient medium supplemented with ampicillin or tetracycline.

## 2. pUC Vectors

In 1983, Messings and co-workers developed the pUC vectors at the *U*niversity of *C*alifornia. These plasmids consist of ColE1 *ori* (origin of replication) gene, *amp*$^R$ gene (for ampicillin resistance), *lacZ*' gene (the shortest derivative of *lacZ* operon for *E. coli* β-lactamase). The pUC19 is one of the popular plasmid vectors of pUC family (Fig. 5.2). The engineered version (*lacZ*') of its *lacZ* gene consists of multiple restriction sites within the first part of the coding region of the gene. This region is known as **multiple cloning site** (MCS). If a target DNA is inserted at any of these sites, the *lacZ* gene is inactivated. Hence, such recombinant vector will form white colony on a suitable plate as compared to nonrecombinant vector. In cloning experiment the choice for application of restriction enzymes is increased after using as MCS. A MCS at the 5'-end of *lacZ* is given below:

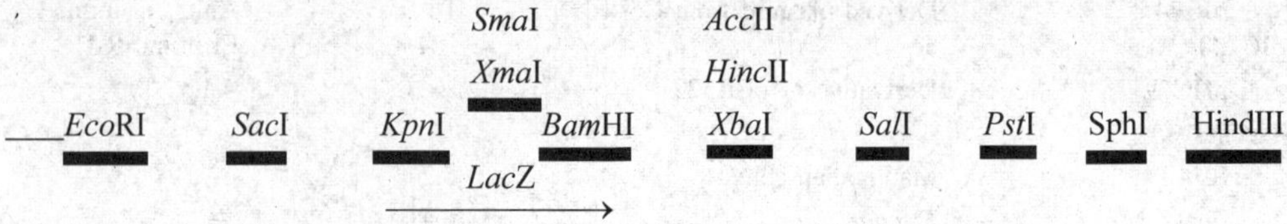


The pUC series consist of several plasmids such as pUC8, pUC9, pUC12, pUC13, pUC18, pUC19, etc. The pUC19 is called as general purpose plasmid cloning vector. Such vectors are designated for cloning of relatively small (<10 kb) DNA fragments often in *E. coli* at MCS. It consists of 2,686 base pairs and possesses an *amp*R gene, *lacZ* gene (for β-galactosidase) and *lacI* gene (for production of repressor protein to regulate *lacZ*'). In pUC19, MCS is incorporated into *lacZ*' gene without interfering the function of other genes. The MCS has the unique sites for several restriction endonucleases such as *Eco*RI, *Sac*I, *Kpn*I, *Xma*I, *Sma*I, *Bam*HI, *Xba*I, *Sal*I, *Hind*II, *Acc*I, *Bsp*MI, *Pst*I, *Hind*III. It also carries one *ori* gene from pBR322.

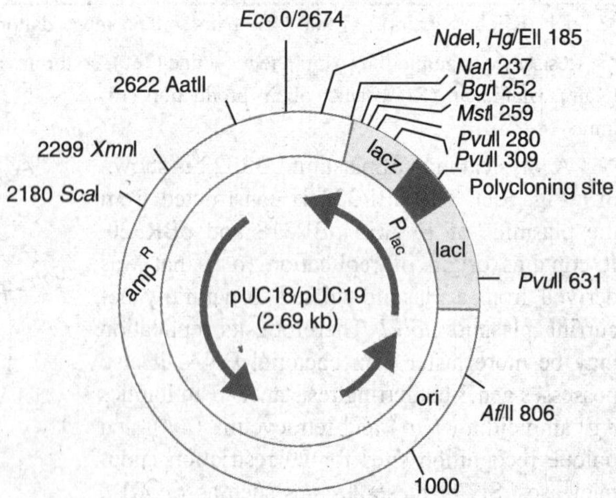


**Fig. 5.2.** Genetic map of the plasmid cloning vector pUC19.

When the cells carrying unchanged pUC19 are grown in the presence of isopropylthiogalactoside (ITG) (an inducer of *lac* operon), the production of *lac*I gene cannot bind to the promoter-operator region of the *lac*Z gene. Therefore, the *lac*Z gene encoded in the plasmid is transcribed. The *lac*Z' protein combines with chromosomal DNA-transcribed protein and forms an hybrid β-galactosidase. The MCS is tagged in the *lac*Z' gene of pUC19 plasmid which does not interfere the production of hybrid β-galactosidase. On the other hand if a substrate 5-bromo-4-chlorionolyl-β-galactosidase (X-gal) is present in the medium, the hybrid β-galactosidase hydrolyses it to produce a blue product. Therefore, colonies containing unmodified pUC19 appear blue in colour. When the DNA insert is cloned in *lacZ*' gene, this gene will be inactivated and will not express β-galactosidase. Consequently, the colonies formed by *E. coli* cells containing recombinant plasmid will be white in colour. This plasmid vector is designed for blue-white screening on X-gal substrate.

### 3. pACYC184

The plasmid pACYC184 was developed by using p15A plasmid. It is a small sized plasmid (4 kb). As described in pBR322, it consists of several restriction sites, *chl*R (for chloramphenicol resis-

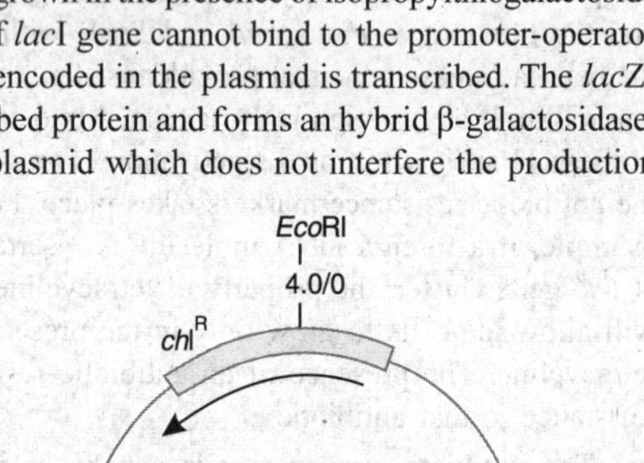


**Fig. 5.3.** Physical map of the pACYC184 vector.

tance) and $tet^R$ (for tetracycline resistance) genes. The plasmid containing cloned DNA fragments can be detected by insertional inactivation of one of the antibiotic resistance genes). Its number can be amplified unlike pBR322 as the latter is not self-transmissible (Fig. 5.3).

## 4. Agrobacterium-based Plasmid Vectors

Several groups of plants have received attention as a means tor DNA cloning and expression of foreign DNA in plant cells. The DNA viruses of plants are of two types: (*i*) double stranded DNA viruses, such as caulimoviruses, and (*ii*) single stranded DNA virus, for example geminivirus. Although plant viruses have major scope for the development as cloning vector, yet this aspect has not been studied so far. A good deal of work is done on soil-borne bacteria, *Agrobacterium tumefaciens* causing crown gall disease and *A. rhizogenes* causing hairy root disease on the stems of numerous plants.

A tumour - inducing plasmid from a common bacterium (*Agro bacterium tumefaciens* causes crown gall tumors which is used by scientists to work in cultures of plant cells to move desirable genes into the plant chromosomes.

**(*a*) Ti-Plasmid.** *A. tumefaciens* consists of a large plasmid which induces tumour in plants, therefore, the plasmid is called as Ti-plasmid. The size of Ti-plasmid ranges between 180 and 250 kb. It contains T-DNA region of about 23 to 25 kb which is transferred into plant cells. Ti-plasmid can be grouped into three on the basis of opine types, for example, octopine, nopaline and agropine. Structural formulae and diagrams of these are given in Figs. 5.4 and 5.5, respectively.

Left side of T-DNA is known as TL-T-DNA and right side of the same as TR-T-DNA. Both the essential (T-DNA) and non-essential regions of Ti-plasmids (both octopine and nopaline types) contain many genes. One or more unusual amino acid derivatives encoded by DNA is known as opines. Opines are neither found normally in plants nor required by plants.

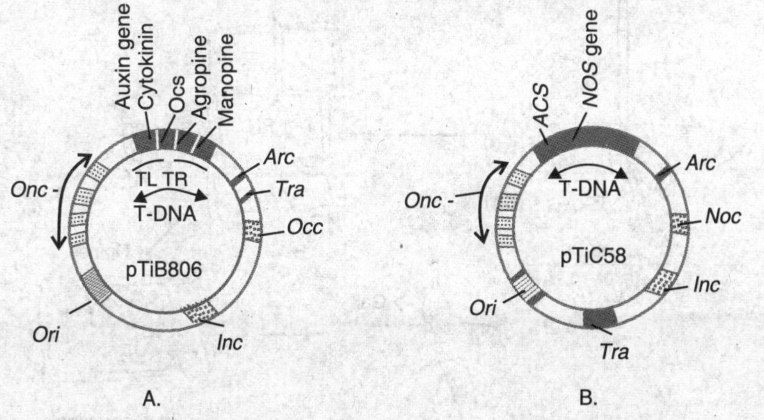


**Fig. 5.4.** A map of octopine plasmid (A) and nopaline plasmid (B).

*A. tumefaciens* uses it as a source of carbon and nitrogen for their growth and multiplication. The genes responsible for the biosynthesis (*ocs* gene for octopine and *nos* for nopaline) and degradation (*occ* for octopine and *noc* for nopaline catabolism) of opines are located at the well defined regions.

*Arc* genes are required for arginin catabolism. The plasmid also carries transfer (*tra*) gene to help the transfer T-DNA from one bacterium to other bacterial or plant cells, *onc* gene for oncogenecity, *ori* gene for origin of replication and *inc* gene for incompatibility.

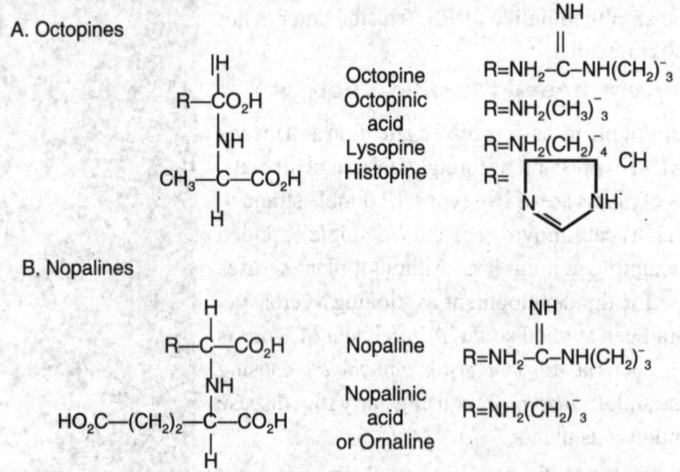


**Fig. 5.5.** Structural formulae of octopines (A) and nopalines (B).

(*i*) **Mechanism of T-DNA transfer.** *A. tumefaciens* lives in soil and attack many dicotyledonous plants (Fig. 5.6 A) most probably at the level of soil surface. Formation of wound is necessary for establishment of bacterial plasmid into plant cells. Lipopolysaccharide, a compound secreted by bacterial cell wall, helps in its attachment with polygalacturonic acid fractions of plant cell wall. From the wounded cell walls of plants, a phenolic compound of low molecular weight (acetosyringone) is secreted which induces the *vir* genes of Ti-plasmids. *Vir* genes encode an enzyme which nick the double stranded T-DNA on the same strand at two points, and produces single stranded DNA molecules. A cut is made on right border of T-DNA and a single stranded T-DNA fragment of 5′→3′ direction is generated. It is carried into plant cells (B).

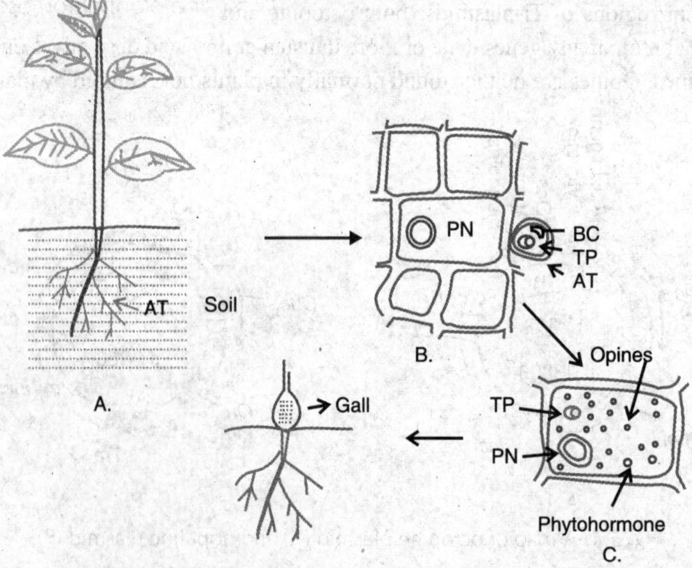


**Fig. 5.6.** Process of infection by *Agrobacterium tumefaciens* (AT) on a dicot plant and induction of tumour or gall (A-D) TP, T-DNA of plasmid, PN, plant nucleus; BC, bacterial chromosome.

T-DNA of Ti-plasmid is stably integrated with plant DNA. The sequencing of nucleic acid at the junction between plant DNA and T-DNA has been done. T-region in both octopine and nopaline plasmids is flanked by a direct repeat of 25 base pairs. *Ops* gene encodes enzymes for the synthesis of opines in transformed cells which are required for poliferation of infecting bacteria. TL-T-DNA encodes two enzymes that are involved in biosynthesis of two phytohormones, auxin and cytokinin (C). This results in disorganised proliferation of cells commonly known as callus, gall or tumour (D). Therefore, the galls are colonized by the bacteria.

***(ii)* Tumour morphology.** Several gene loci for controlling the morphology of tumour have been identified on TL-DNA of two octopine plasmids. These loci are grouped into three categories: *tml* (which causes larger tumour), *tmr* (which induce tumour with large number of roots) and *tms* (which causes tumour with large number of shoots) (Fig. 5.7). These effects are due to the products of loci of TL-DNA on the cytokinin/auxin ratios within the transformed tissues. The function of *tml* is not known so far.

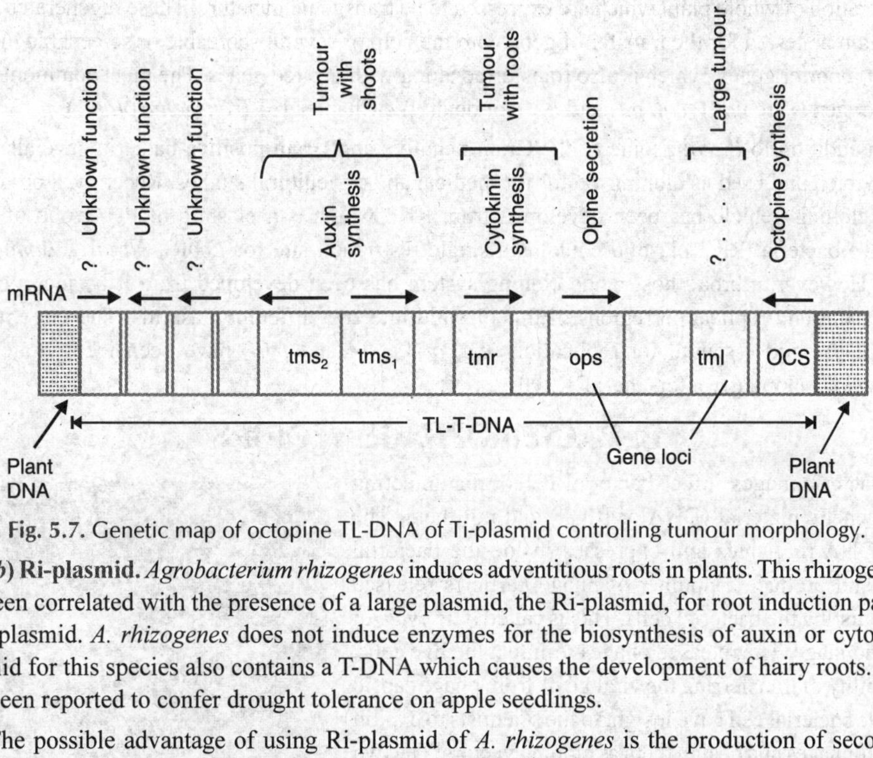


**Fig. 5.7.** Genetic map of octopine TL-DNA of Ti-plasmid controlling tumour morphology.

***(b)* Ri-plasmid.** *Agrobacterium rhizogenes* induces adventitious roots in plants. This rhizogenicity has been correlated with the presence of a large plasmid, the Ri-plasmid, for root induction parallel to Ti-plasmid. *A. rhizogenes* does not induce enzymes for the biosynthesis of auxin or cytokinin. Plasmid for this species also contains a T-DNA which causes the development of hairy roots. It has also been reported to confer drought tolerance on apple seedlings.

The possible advantage of using Ri-plasmid of *A. rhizogenes* is the production of secondary root systems for: (*i*) better anchorage of plants, (*ii*) ability to resist anoxia from flooding of the soil, (*iii*) increased drought resistance due to high root density, and (*iv*) greater chance of interaction with soil-borne mycorrhizal fungi. It may be applied to establish the secondary root systems in certain plants (fruit trees) by inoculation of soil as well as beneficial for those plants which are propagated through cutting and rooting as stimulants for root formation on cuttings.

**(c) Properties of Ti- and Ri- based plasmids.** *Agrobacterium* plasmids have been used for introduction of genes of desirable traits into plants. The properties of Ti- and Ri based plasnids are as below:

(*i*) They do not cause disease because of being disarmed by removing disease causing gene.

(*ii*) They possess sites for insertion of foreign gene which needs to be introduced.

(*iii*) They possess selectable markers *i.e.*, genes which help in selecting the transformed cells.

The selectable marker is a powerful promoter e.g. 35S (for high level of expression of marker) derived from cauliflower mosaic virus (CaMV) and polyadenylation signal (for adding poly A to mRNA). The most commonly used selectable marker is neomycin phosphotrasferase (*npt*II) that confers resistance to neomycin and allows the cells to grow in medium containing neomycin. Thus, after inserting a foreign gene, a vector is designed. The vector is transferred to *A. tumifaciens* or *A. rhizogenes* depending on cases. The transformed bacterium is used to infect the designed host cell where the transformation of gene is to be brought about.

For gene transfer, generally explants are used. The explants are protoplasts, suspension of cultured cells, clumps of callus, thin cell layers (epidermis), leaf discs, sliced tissue, floral tissues, stem and root selection.

The explants are incubated or co-cultured with *Agrobacterium* containing vector with modified foreign gene. The transformed cell or colonies are selected due to the selectable marker and used for regeneration of whole plant which are expressed to be transgenic in nature. These regenerated plants can again be tested for the transfer of gene with the help of several scoreable or screenable markers called 'reporter genes' which is also transferred along with desired genes. The most commonly used reporter genes are *npt*II, *cat, gus* and *lux* (for detail *see* Chapter 13, *Transgenic plants*).

In addition to *E. coli,* some of the Gram-negative and Gram positive-bacteria have also been investigated and used as cloning vector for medical and agricultural studies. Recently, a broad host range cloning vehicle has been developed from RK2 which is a plasmid of P-1 group of Gram negative bacteria. RK2 plasmid contains a single restriction site for *Eco*RI, *Hind*III, *Bam*HI and *Bgl*II. However, a broad host range cloning system has been developed from it by incorporating the transfer and replication regions of this plasmid into two different plasmids. The three regions of RK2 plasmid essential for replication (*e.g. oriV, trfA* and *trfB*) have been incorporated into the plasmid cloning vector, the pRK290.

## B. BACTERIOPHAGE VECTORS

Bacteriophages infect bacterial cells after injecting their genetic material (DNA or RNA) and kill them. The viral DNA replicates and expresses inside the bacterial cells, and produce a number of phage particles released after bursting the bacterial cells. This is called lytic cycle of bacteriophage. The released phages re-infect the live cells. The ability of transferring the viral DNA from phage capsid specific bacterial cell gave insight to the scientists to exploit bacteriophages and design them as cloning vectors. The two bacteriophages *e.g.* phage λ and M13 have been modified, extensively studied and commonly used as cloning vectors.

### 1. Phage λ

The phage vectors are required for cloning of large DNA fragment and, therefore, the gene bank or genomic libraries can be constructed. They are an alternative to bacterial plasmids, and related cosmids. The process of infection and replication of phage in *E.coli* cell are shown in (Fig. 5.8.)

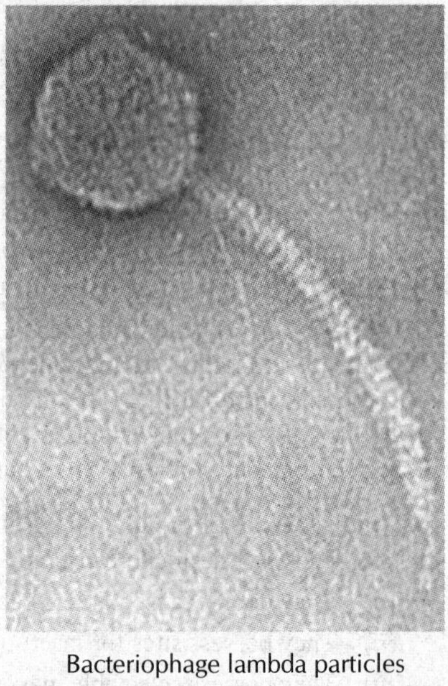

Bacteriophage lambda particles

Phage λ contains a proteinaceous head and a long tail attached to the head. In the head it possesses 50 genes in its 48.514 kb (kilobase pairs) genome of which about half of genes are essential. On attachment with tail to cell wall of *E. coli* it injects its linear DNA into the cell. The linear double stranded DNA molecule cyclizes through the single strand of 12 nucleotides commonly known as *cos* sited at its ends. The *cos* sites are the key feature of the DNA.

5′-CGGGCGGCGACCTCG-3′
3′-GCCCGCCGCTGGAGC-5′

Cleavage (during packaging) ↑ ↓ - Ligation (after infection)

5′C 5′-GGGCGGCGACCTCG-3′
3′-GCCCGCCGCTGGA-5′ GC-5′

The 12 nucleotide long projections (5′GGGCGGCGACCT-3′) show cohesiveness and form the *COS* site (cohesive site). Phage genome has a large non-essential region which is not involved in cell lysis (Fig. 5.8). Taking advantage of it two types of cloning vectors can be produced, either by inserting foreign DNA (insetion vectors) or replacing phage DNA with a foreign DNA (replacement vectors). The upper limit of foreign DNA to be packed is about 23 kb. Replication cycle of phage λ is accomplished into two pathways : the lytic and lysogenic pathways.

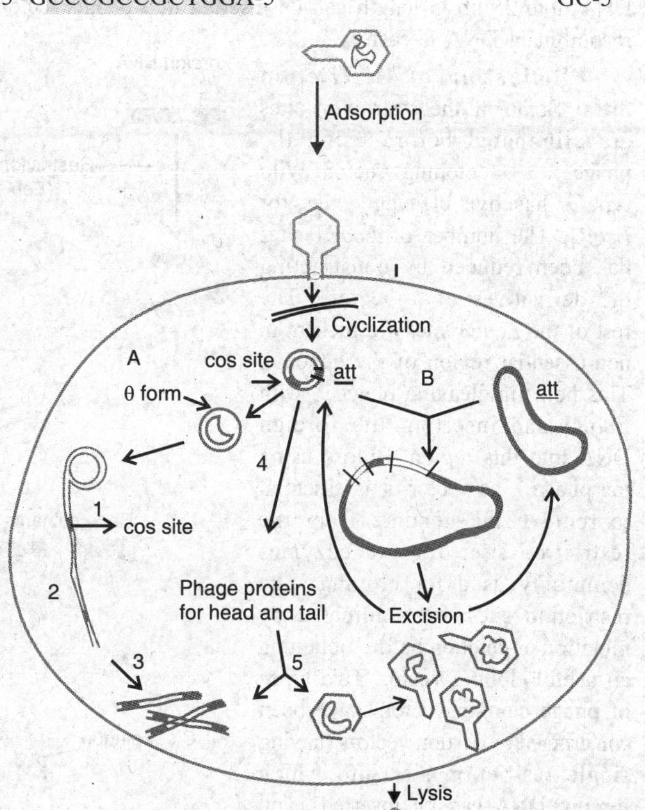


**Fig. 5.8.** Replication of phage lambda in a *Escherichia coli* cell. A - lytic cycle; 1. rolling circle replication; 2. production of concatemers; 3. cleavage at *cos* site; 4. transcription and translation; 5. packaging. B - lysogenic cycle.

In the lytic pathway, early in the infection sites the circular DNA replicates as theta (θ) forms. By a rolling circle mechanism it produces the long concatemeric molecules joined end to end, and composed of several linearly arranged genomes. At the same time phage DNA directs the synthesis of many proteins required to produce empty heads where DNA is packed after the cleavage of concatemeric DNA at its *cos* site to yield fragments of such sizes as to fit in their heads. Eventually, a tail is attached to the head and finally the mature phage particles are released out of the bacterial cell.

The lysogenic pathway for replication is another alternative mode of propagation where it becomes stably integrated into the host chromosome and replicated along with the bacterial chromosome. Phage genome integrates by an attachment site (*att*) with a partially homologous site on the *E. coli* chromosome, where it replicates as a chromosomal DNA segment. In this case a protein is produced by *cI* genes which represses all the genes responsible for lytic pathways. In this pathway no phage structural proteins are synthesized. The interactions of two proteins, the *cI* genes expressed protein (by phage genome) and *cro* gene expressed protein (by *E.coli* chromosomes) decide between the events of the lytic and lysogenic pathways. Phage λ has about 50 genes where only about 50% of these are necessary for growth and plaque formation. Non-essential DNA sequence, therefore, is replaced by donor DNA and the inserted DNA propagated as phage λ recombinant with no detrimental to its replication or packaging. For a detailed description see A *Textbook of Microbiology* by R.C. Dubey and D.K. Maheshwari (2012).

Following are the advantages of phage cloning system over the plasmids: (*i*) DNA can be packed *in vitro* into phage particles and transduced into *E. coli* with high efficiency, (*ii*) foreign DNA upto 25 kb in length can be inserted into phage vector, and (*iii*) screening and storage of recombinant DNA is easier.

**Elimination of Restriction Sites:** Some of the restriction sites are eliminated before using the phage λ as a cloning vector. Wild type λ has five cleavage sites for *Eco*RI. The number of *Eco*RI sites has been reduced by constructing the derivatives of λ vectors. The rest of the *Eco*RI sites are present in non-essential region of the genome. This helps in cleaving phage λ with *Eco*RI and inserting the foreign DNA into this region. Before using the phage λ as vector, it is essential to remove the genome from the restriction sites for the enzymes commonly used for cloning. The restriction sites are eliminated by mutation or deletion before obtaining an useful cloning vector. Two types of phage cloning vectors have been constructed : insertion vectors (having single restriction site into which foreign DNA can be inserted), and replacement vectors.

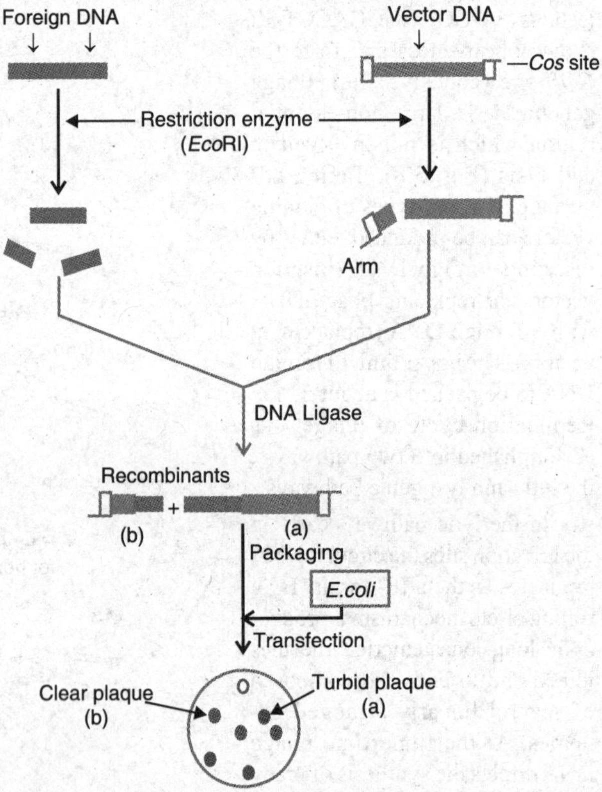


**Fig. 5.9.** Cloning of an insertion vector.

In eukaryotes phage λ and cosmid vectors fulfil the requirements of cloning the whole genome. Phage λ contains about 20–25 kb long segment and cosmid contains 45 kb long segment. Several phage λ-based cloning vectors have been constructed, for example λgt10 and λgt11, EMBL3 and EMBL 4.

**(*a*) Insertion vector :** There are unique cleavage sites in the insertion vectors into which relatively small piece of foreign DNA is inserted. Foreign DNA fragment does not affect the function of phage. The upper and lower limits of size of DNA that may be packed into phage particles is between 35-53 Kb. Therefore, the minimum size of vector must be above 35 kb. It means the maximum size of a foreign DNA to be inserted is about 18 kb. Cloning of an insertion vector is shown in Fig. 5.9.

(*i*) **λgt10:** Phage λ is modified to construct λgt10 that may clone cDNA fragment. The λgt10 is a dsDNA of 43 kb which may clone 7 kb long foreign DNA fragment (Fig. 5.10). After inserting DNA, *cI*$^+$ (repressor) gene is inactivated; therefore, *cI*$^-$ recombinant bacteriophage is formed. The recombinant *cI*$^-$ λgt10 after infecting *E. coli* forms clear plaques which can easily be screened from cloudy plaques formed by non-recombinant λgt10 *cI*$^+$ vector.

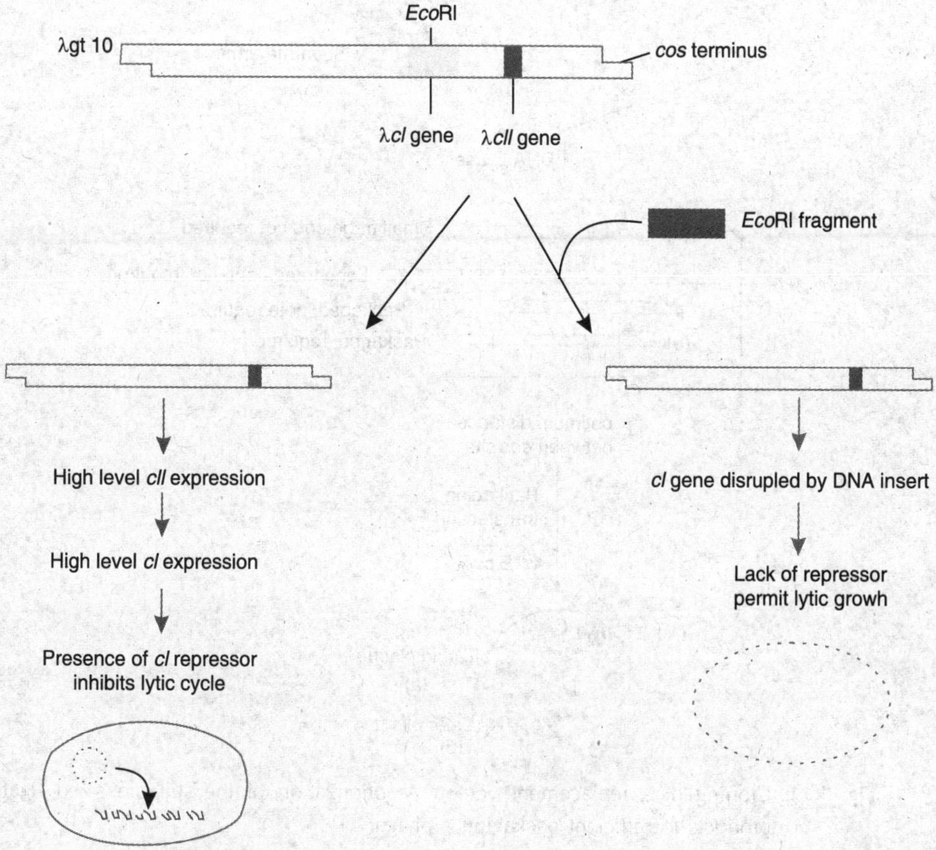


**Fig. 5.10.** Insertional cloning by using phage λgt10 vector.

(*ii*) **λgt11:** It is a 43.7 kb long phage dsDNA constructed for cloning mostly cDNA (less than 6 kb). It is an expression vector where the inserted foreign DNA encodes as β-galactosidase fusion protein. After insertion of a foreign DNA, the recombinant λgt11 is converted into λgt11 *gal*$^-$ and non-recombinant λgt11 remains *gal*$^+$. Therefore, by using IPTG (inducer) and X-gal (substrate) with suitable *E. coli* host cells, recombinant λgt11 can be screened. Recombinant λgt11 *gal*$^-$ forms white (clear) colonies while non-recombinant λgt11 *gal*$^+$ forms blue colonies.

(*b*) **Replacement Vector :** The cleavage site replacement vectors consists of on either side a length of non-essential DNA of phage. As a result of cleavage left and right arms are formed, each arm has a terminal *cos* site and longer a stuffer region, the non-essential region, which can be substituted by foreign DNA fragment Fig. 5.11.

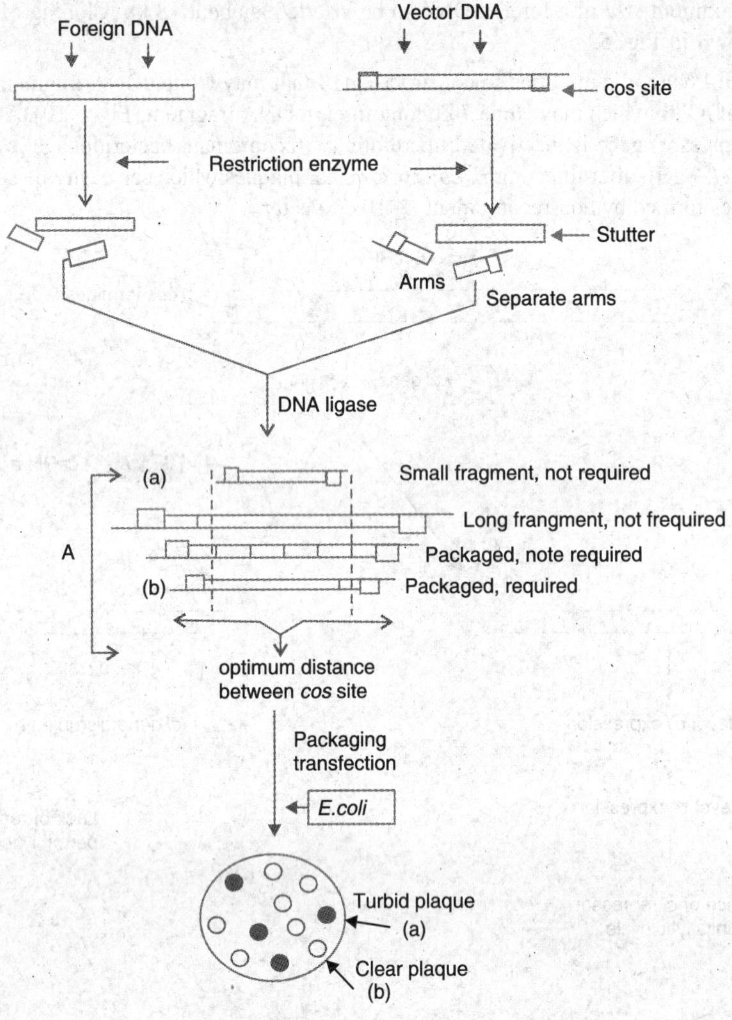


Fig. 5.11. Cloning of a replacement vector. A-optimization of the suitable sized ligation products for efficient packaging a phage.

The maximum size of inserted DNA fragment depends on how much of the phage DNA is non-essential. It has been found that about 25-30 Kb of genome codes for essential products for lytic cycle. The remaining 20-25 Kb of genome could be replaced with the foreign DNA fragments of known essential products. The substituted vectors are gt, *WES* and λ 1059. Non-essential part of the λ genome can be separated from the arms by electrophoresis or velocity gradient ultracentrifugation that facilitates to make size differences. Formation of multiple inserts can be checked by using alkaline phosphatase before ligation with insert fragment. Recombinant DNA formed by multiple inserts has too large genome to be packed in viral head. However, after ligation the recombinant DNA molecules have left arm plus large insert plus right arm linked by their *cos* sites at the arms. Optimum distance from *cos* sites governs efficiency for packaging in *E. coli.* As a result of ligation the size of recombinant DNA fragment may be less or more than the required size or may have more than two *cos* sites or many small fragments of foreign DNA. Those having the range of viral head can be packed *in vitro* using a preparation of head and tail proteins. The viruses thus constructed are allowed to multiply in *E. coli.* Development of plaque turbidity is a useful criteria for the selection of recombinant phages. Plaque turbidity is determined by the presence of nonlysed bacteria. The recombinant phages give clear plaques due to inactivation of *cl* gene.

By using these methods the clones containing recombinant DNA can be isolated from the wild type clones, *i.e.* turbid plaques. The constructed genome has all the informations required for DNA replication and synthesis of the viral protein. The inserts can be identified by colony hybridization technique as described in Chapter 6.

Phage DNA has been most widely used as phage cloning vector because of ease in handling and screening of large number of recombinant DNA containing phages. That is why it is used as a tool in the building of gene bank, for example, rat, yeast, mouse and human gene banks.

Examples of replacement vectors are: EMBL 3, EMBL 4, charon 40, charon 34, charon 35, and λDASH. In these vectors the central non-essential region of about 44 kb long phage DNA is replaced by a foreign DNA of known function. Because the 40% of central region of phage is non-essential for phage propagation.

## 2. Phage M13

M13 is a filamentous phage of *E. coli* which infects only such cells that contain sex pili. It is closely related to the other coliphages such as fd and f1. M13 consists of single stranded circular DNA molecule containing 6,407 base pairs. It lands on sex pilus, binds to receptors specified by F plasmid, transfects and delivers its DNA into the cell through the lumen of pilus. After delivery, the DNA is converted from single stranded molecule to double stranded molecule. It is called *replicative form* (RF) of DNA. The RF of M13 replicates and forms 50-100 RF molecules per cell. Finally, single stranded DNA molecules are produced from RF DNA. Phage M13 encodes single strand binding proteins that bind to the viral single stranded DNA which are synthesised by rolling circle mechanism. Finally, the progeny molecules extrude from the cell as M13 particles. The progeny molecules are packed upon extrusion through the cell wall. The cells are not killed but grow slowly.

The M13 system has two advantages: (*i*) the double stranded RF DNA can be isolated and used as plasmid (Fig. 5.12 A), and (*ii*) the single stranded M13 DNA can be used as a template for sequencing DNA (Fig. 5.12 B). Foreign DNA of about 500 bp can be cloned into a multiple cloning sequence that forms part of a cloned modified *lac*Z gene on the double stranded RF of M13. Different strategies are followed when large sized foreign DNA (< 2,000 bp) is to be cloned.

The competent *E. coli* cells are mixed with the DNA from the ligation reaction. Then the cells are plated on nutrient medium containing X-gal as substrate and the plates are incubated for 24 hours. Thereafter, both white (colourless) and blue colonies (M13 plaques) develop on the medium containing X-gal. X-gal turns blue which is hydrolysed by hybrid β-galactosidase where *lac*Z gene is not disrupted i.e. *lac*Z gene remains functional and lacks any inserted DNA. On the other hand the colourless colonies consist of inserted DNA that has disrupted *lac*Z gene and did not produce β-galactosidase. Therefore, X-gal could not be disintegrated. The M13 particles are isolated from the white colonies. It is used as a source of M13 single stranded DNA which is further employed for DNA sequencing (Fig. 5.12 B). A primer oligonucleotide is added to the M13 single stranded cloned DNA construct that hybridizes to the vector DNA near the insertion site of the cloned DNA. The dideoxynucleotide reaction is carried out; the sequence of the insert is read from the autoradiographs.

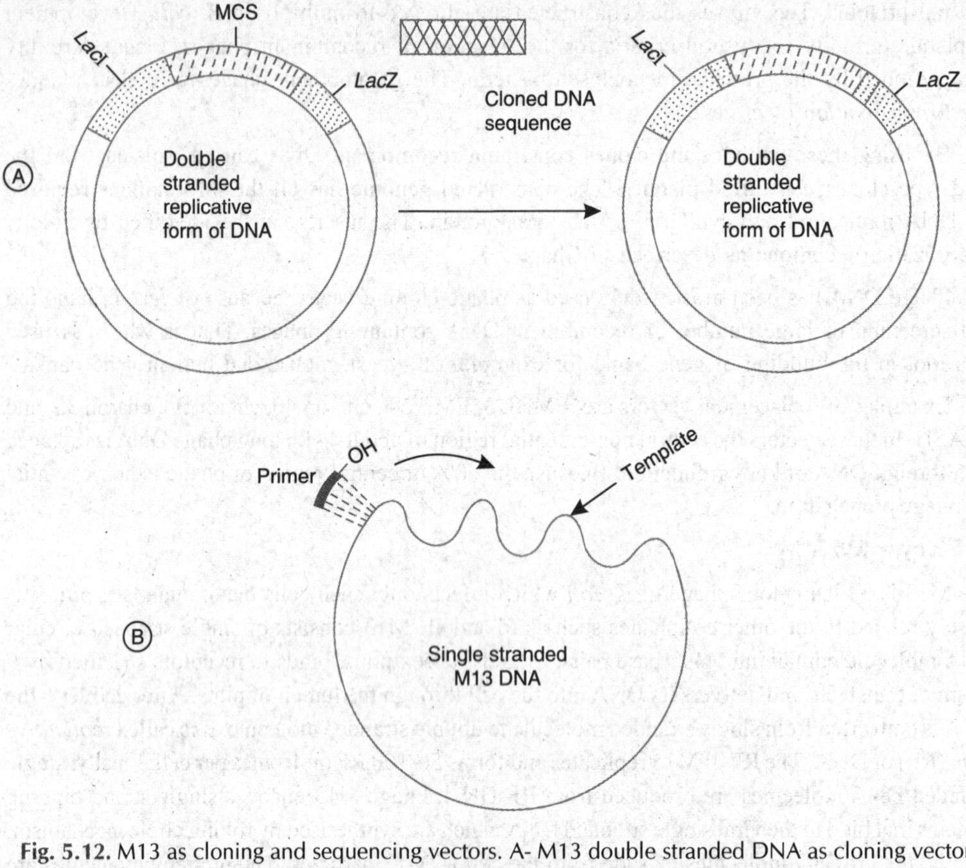


**Fig. 5.12.** M13 as cloning and sequencing vectors. A- M13 double stranded DNA as cloning vector; B- M13 single stranded DNA used for sequencing.

## C. COSMIDS

Based on the properties of DNA and *Çol* El plasmid DNA, a group of Japanese workers showed that the presence of a small segment of phage λ DNA containing cohesive end on the plasmid molecule is a sufficient prerequisite for *in vitro* packaging of this DNA into infectious particles. The cosmids can be defined as the hybrid vectors derived from plasmids which contain *cos* site of phage λ (cosmid = *cos* site + plasmid). For the first time it was developed by H.A. Collins and M. Hohn in 1978.

Cosmids lack genes encoding viral proteins; therefore, neither viral particles are formed with the host cell nor cell lysis occurs. Special features of cosmids similar to plasmids are the presence of: (*i*) origin of replication, (*ii*) a marker gene coding for antibiotic resistance, (*iii*) a special cleavage site for the insertion of foreign DNA, and (*iv*) the small size.

Extra phage DNA of the *cos* site about 12 bases is present. It helps the whole genome in circularization and ligation. The cosmids have a length of about 5 Kb, the upper size limit of the foreign DNA fragment that may be inserted in cosmids and packed into phage particles is, therefore, approximately 45 Kb, much larger than it would be possible to clone in phage λ or plasmid vector (Dahl *et al.,* 1981). According to the size of *cos* sites and upper size limit in the head of phage, the recombinant DNA molecules can be packed into bacteriophage particles in *in vitro* packaging system consisting of packaging enzymes, head and tail proteins. Procedure of DNA cloning by using cosmid vector is shown in Fig. 5.13. Upon transfection of *E. coli* by bacteriophage, the recombinant DNA cyclizes through *cos* sites and then replicates as a plasmid and expresses the drug resistance marker. Recently, based on cosmid vectors, a number of cosmid vectors have been determined from *E. coli*, yeast, and mammalian cells, and gene bank has been constructed.

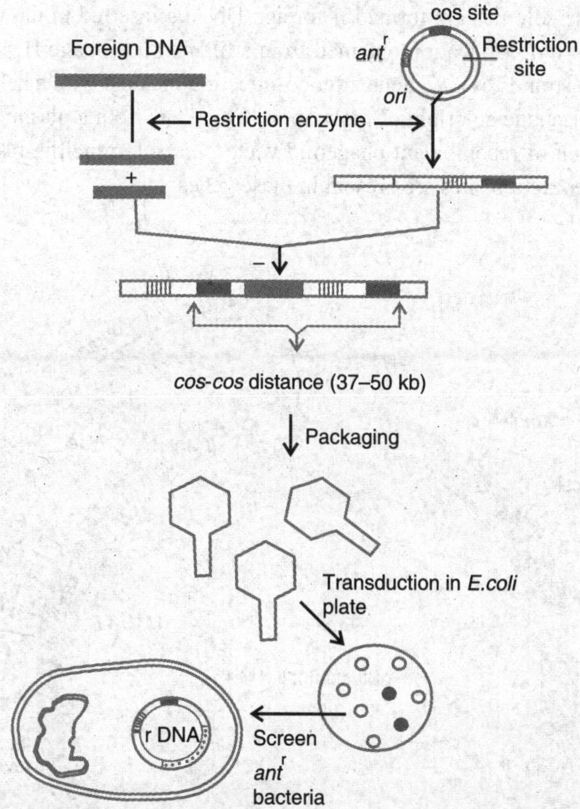


Fig. 5.13. Cloning of a cosmid vector. Transduced bacteria contain a cosmid which show resistance to specific antibiotic. Such bacteria can be screened. *Ant*$^r$, antibiotic resistance; *ori*, origin of replication.

# D. PHAGEMID VECTORS

## 1. Phagemids

Some times generation of a single stranded DNA becomes important for DNA sequencing and site-directed mutagenesis. Using a single stranded DNA containing bacteriophage M13, a series of cloning vectors were developed. The bacteriophage fl is closely related to M13 which also infects *E. coli*. Phagemids are plasmids that contain origin of replication (*ori*) for single strand DNA containing bacteriophage such as fl. *E. coli* maintains a plasmid as double stranded DNA due to plasmid *ori* gene. If *E. coli* cells are infected by the helper fl phage, the *ori* of fl is activated. It switches to a mode of replication and produces single stranded DNA which is packed into phage particles as they are extruded from the host cell. Fig. 5.14 shows a phagemid pBlueScript II KS (+/–) which is used for generation of single stranded DNA molecules.

**(*a*) pBlueScript II KS (+/–):** This vector is 2961 bp long derivative from pUC19. The 'KS' denotes the orientation of polylinker *i.e.* transcription of *lac*Z genes proceeds from the *Kpn*I restriction site towards *Sac*I restriction site. It consists of multiple cloning site (MCS) flanked by T3 and T7 promoters in opposite directions on the two strands (Fig. 15.14). The MCS is set up for *lac*Z blue-white screening. A *lac*I promoter complements with *lac*Z of *E. coli* upstream of *lac*Z region. This feature helps in selection of recombinant vector based on the criteria of development of blue-white colonies (white colonies are formed if foreign DNA is inserted in the vector). It consists of fl (+) and fl (–) origin of replication obtained from a filamentous phage fl. It is used to get sense (+) and antisense (–) strands of *lac*Z gene after co-infecting the host with a helper phage. But in the absence of the helper phage an origin of replication (ColE1 *ori* from a plasmid) is used. Moreover, for antibiotic selection of recombinant phagemid vector, *amp*[R] (ampicillin resistance gene) is also used. A number of restriction sites are present in these genes.

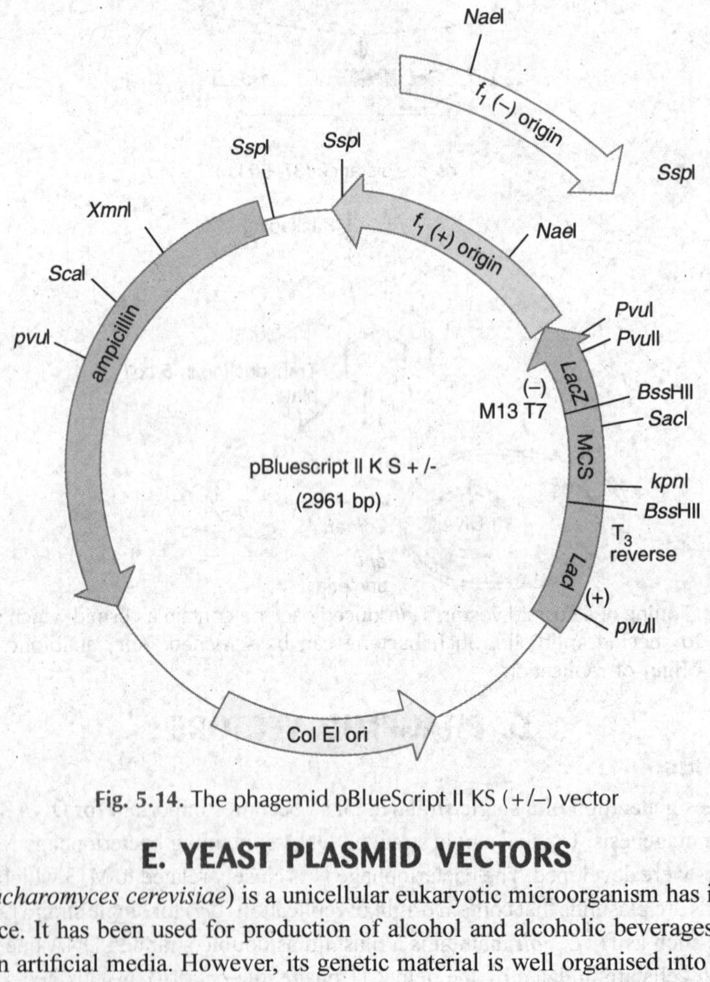


**Fig. 5.14.** The phagemid pBlueScript II KS (+/–) vector

# E. YEAST PLASMID VECTORS

Yeast (*Sacharomyces cerevisiae*) is a unicellular eukaryotic microorganism has its long historical importance. It has been used for production of alcohol and alcoholic beverages. It can easily be cultured on artificial media. However, its genetic material is well organised into chromosomes present inside the nucleus.

Yeast cells contain their own plasmid known as '2 μm circle plasmid' which is used as vector for foreign genes. It is present in many stains in 50–100 copies per cell. The foreign DNA taken up by yeast is integrated by a specific crossing over to yeast DNA containing a homologous sequence on chromosome recognised by the yeast as the origin of replication (*ori*).

In some cases segments of yeast DNA linked with foreign DNA have been demonstrated to transform very efficiently (about $10^4$ cells). The transforming molecules replicate autonomously within the cell. DNA sequences of such property are known as *Ars* (autonomously replicating sequence) fragment. *Ars* fragments have also been isolated from a number of eukaryotic organisms such as *Coenhaorabditis elegans, Dictyostelium discoideum, Drosophila melanogaster, Zea mays*, etc. However, plasmids have been constructed by using *ars* fragments when joined at centromeric sequences (*cen* region of about 6–10 kb DNA) to ensure the replication and stability in yeast. *Cen* is needed for DNA molecule to segregate correctly during meiosis and mitosis. Some times the minichromosomes are lost during meiosis when they are attached by chance to spindle apparatus. Thus, *cen* regions of yeast plasmids are useful fragments because they confer stability.

In 1977, Ratzkin and Carbon inserted a segment of yeast DNA into ColE1 plasmid. They found some of the *E. coli* strain *leu*B as leucine-independent transformants. This showed that the segment of yeast chromosome consisted of the gene for yeast enzyme which was absent in *E. coli* strain *leu*B. After ligation the mutant *E. coli* became able to grow on normal medium. A. Hinnen and co-workers in 1978 found that ColE1 plasmid containing *LEU2* gene (wild type) can transform *leu2* (mutant) yeast sphaeoplast to form leucine-independent phenotype. It is customary with yeast to write wild type allele in capital letter (e.g. *LEU2*) and mutant allele in small letter (e.g. *leu2*). The *LEU2* gene was added into the yeast chromosome and transformants were produced. In this way several yeast genes were isolated and integrated to produce transformants. Such yeast genes are: *LEU2*, *URA3*, *TRP1* and *ARG4*. Thereafter, unique markers were inserted in wild type genes, while introducing the foreign DNA into yeast cells for further selection.

The eukaryotic genes consist of introns; hence, these genes rarely function in bacterial cells. In contrast being eukaryotic microbe, most of yeast genes containing promoters successfully function in bacterial cells. The difficulty associated with yeast was the transformation at low frequency due to crossing over (integrative transformation). Therefore, several plasmids were developed to solve the difficulty such as YIp, YEp, YRp, YCp, YLp, pYAC, shuttle vectors, etc. (Fig. 5.15).

### 1. Yeast Integrative Plasmid (YIp)

The YIp consists of only a yeast marker added to bacterial plasmid. It lacks origin of replication. Hence, it cannot replicate inside the yeast cell without integrating into a chromosome. Therefore, its integration is must for successful transformation. It results in low frequency transformation of yeast cells yielding a transformant per μg of DNA molecule. Integration takes place due to homologous recombination between the inactive gene present on one chromosome and yeast gene carried by YIp as the selectable markers. Crossing over takes place between the chromosomal *LEU2*$^-$ region and *LEU2*$^-$-carrying plasmid. Consequently the flanking region linked to plasmid is integrated to *LEU2*$^-$ region. Autonomously replicating system (*ARS*) is absent in that plasmid; therefore, YIp cannot replicate as a plasmid (Fig. 5.15 A).

### 2. Yeast Episomal Plasmid (YEp)

The YEp is the 2 μm circle-based vector which replicates autonomously without integrating into a yeast chromosome. The 2 μm plasmid is 6 kb in size with so high frequency for transformation. The 2 μm DNA consists of origin of replication and *rep* genes. This plasmid results in high frequency transformation, *i.e.* about $2 \times 10^4$ transformants per μg DNA. The number of restriction sites is limited. It is found in high copy number in the cells. It will lose its stability if 2 μ sequence is lost. It is moderately stable in the cell (Fig. 5.15 B).

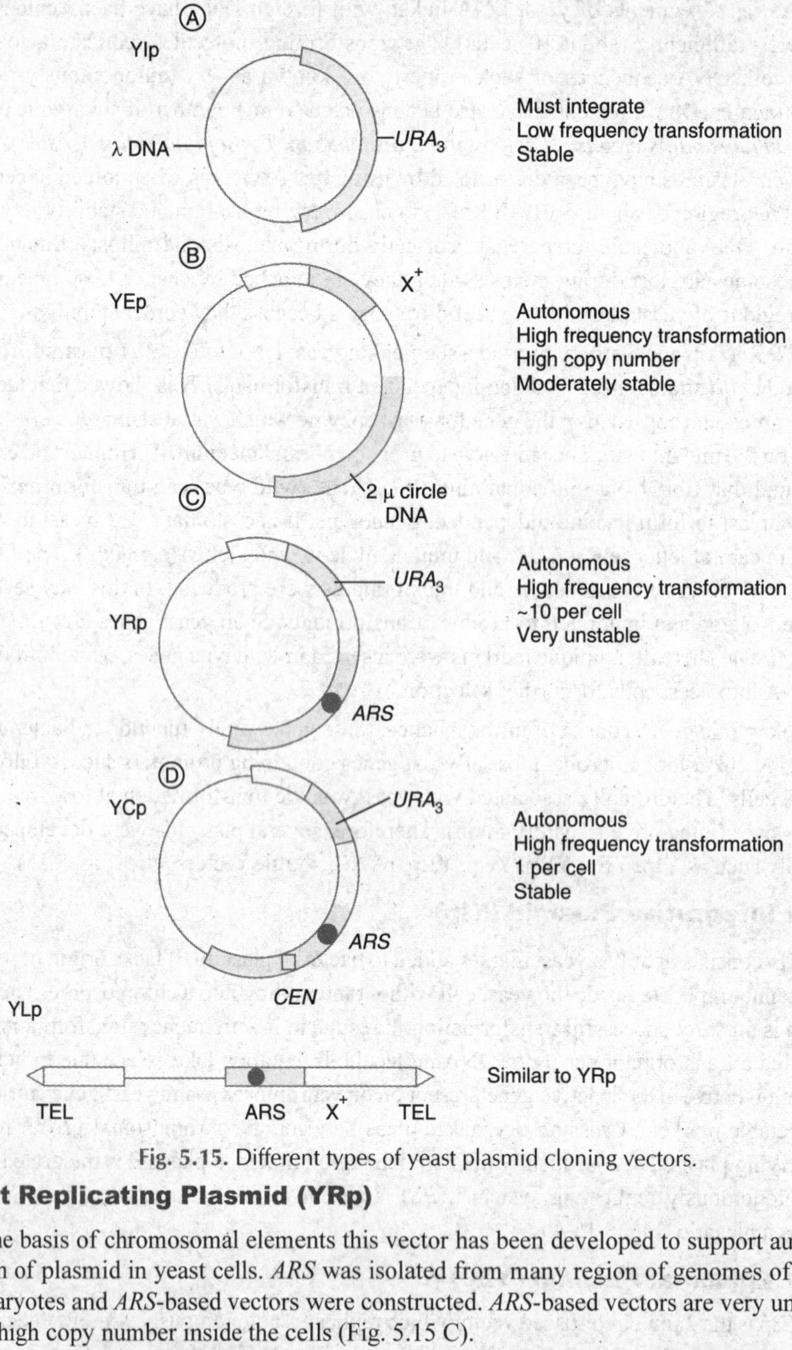


**Fig. 5.15.** Different types of yeast plasmid cloning vectors.

## 3. Yeast Replicating Plasmid (YRp)

On the basis of chromosomal elements this vector has been developed to support autonomous replication of plasmid in yeast cells. *ARS* was isolated from many region of genomes of yeast and other eukaryotes and *ARS*-based vectors were constructed. *ARS*-based vectors are very unstable but can form high copy number inside the cells (Fig. 5.15 C).

## 4. Yeast Centromeric Plasmid (YCp) or Yeast Centromere (CEN)

YCP consists of a sequence of chromosomal centromeric DNA of yeast. It acts as true mini chromosome which becomes circularised. It increases the stability of plasmid and segregates during mitosis and meiosis. Copy number of this plasmid is useful when low copy number of cloned genes are required (Fig. 5.15 D).

### 5. Yeast Linear Plasmid (YLp)

It is a linear plasmid constructed by adding telomeric sequences to both ends of a cleaved *ARS*-based plasmid. Telomere is the end of chromosome. Telomere was isolated from a ciliated protozoan, *Tetrahymena*. This linear plasmid is present in single copy as a chromosome and are very stable and cannot replicate in *E. coli*. Therefore, it has drawn less attention but has raised much possibility to put together the complete set of genes in artificial chromosome of eukaryotes (Fig. 5.15 E).

## F. YEAST ARTIFICIAL CHROMOSOMES (YAC)

The plasmids can accommodate about 10 kb foreign DNA, phage λ can accommodate about 20 kb DNA insert, cosmid can accommodate about 40 kb long four DNA segments. But for stable replication and segregation in eukaryotes, particularly in yeast, a small and well defined sequence should be constructed as recombinant chromosome. With this objective David and co-workers developed, in 1987, **yeast artificial chromosomes** (YAC) by using new approaches where DNA segment of several thousand base pairs (about 1 Mb) can be cloned. Initially it was used for investigation of the maintenance of chromosomes in the cell. Later on it was used as vectors for carrying very large cloned fragments of DNA. It has also been used for physical mapping of human chromosome in 'Human Genome Project'.

YAC consists of centromere element (*CEN*) for chromosomal segregation during cell division, telomere, and origin of replication (*ori*) were isolated and joined on plasmids constructed in *E. coli*. Structure of typical YAC3 vector is shown in Fig. 5.16.

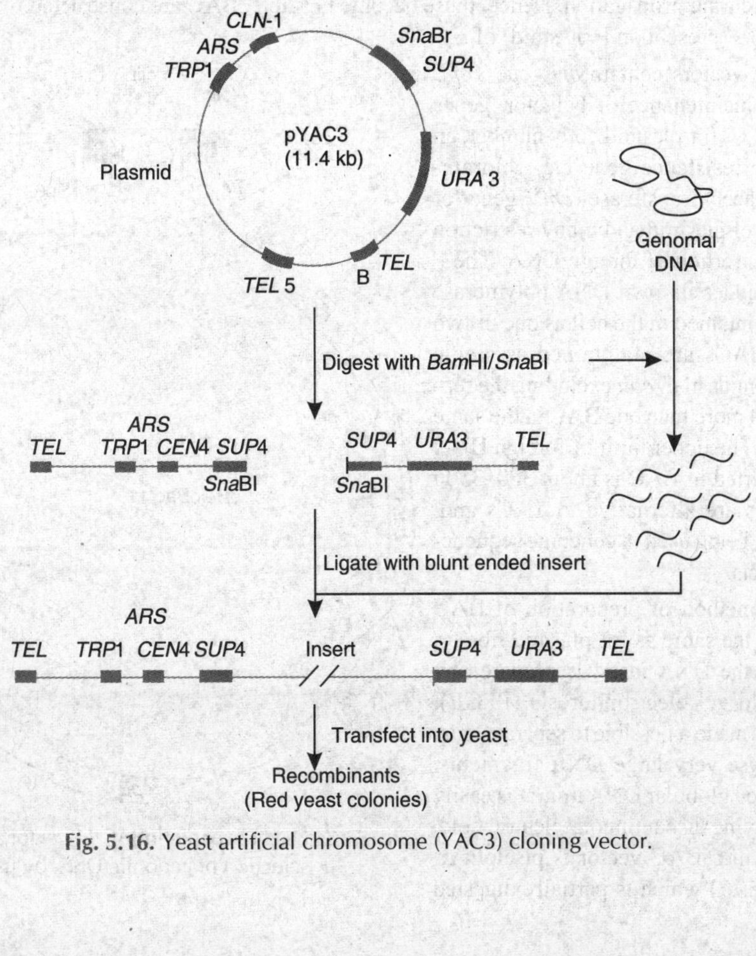


Fig. 5.16. Yeast artificial chromosome (YAC3) cloning vector.

The YAC3 vector contains the *E. coli* origin of replication (*ori*$^{E}$) and a selectable marker (*amp*$^{r}$), a yeast DNA sequence, genes each for uracil biosynthesis pathway (*URA3*) providing centromeric function (*CEN*), autonomously replication sequence (*ARS*) (which acts as yeast *ori*), tryptophan synthesis pathway (*TRP*) and telomeric (T) sequence (which is extended by the enzyme telomerase inside the yeast cell) (Fig. 5.16). The *TRP1* and *URA3* are yeast selectable markers, one for each end, which ensure that properly reconstituted YACs survive in yeast cells, *Sup*4 is inactivated in recombinants and acts as a basis for red-white selection unlike blue-white selection in *E. coli*. There are recognition sites for restriction enzymes such as *Sma*I and *Bam*HI.

Before cloning, YAC is digested with restriction enzyme (*Bam*HIII and *Eco*RI) and recombinants are produced by inserting a large fragment of genomic DNA. This molecule can be maintained in yeast as YAC. The transformants that contain YAC can be identified by red/white colour selection. Non-transformed yeast contain white colonies. Red colonies of yeast contain recombinant YAC molecules. Due to insertion of DNA molecule into *Eco*RI site, *Sup*4 gene is inactivated and no protein will be expressed. This facilitates to develop red colonies.

## G. BACTERIAL ARTIFICIAL CHROMOSOME (BAC)

The BAC was developed by Mel Simon and co-workers as a cloning vector system in *E. coli* and an alternative to YAC vector. BACs are maintained in *E. coli* as a large single copy. These are used as vectors for cloning very large (>50 kb) sequences of DNA. Some times these fragments turn out to comprise non-contiguous (non-adjacent) segments of the genome and frequently lose parts of the DNA during propagation. Hence, these become unstable. BAC are constructed by using fertility or F factors (present on F plasmid) of *E.coli*.

BAC vectors contain *ori* gene, *rep*E gene for maintenance of F factor, genes (*par*A, B, C) for plasmid copy number, an antibiotic resistance gene e.g. chloramphenicol acetyltransferase (*chl*$^{R}$) gene for selection of plasmid and many restriction sites for insertion of foreign DNA. The F factor encodes its own DNA polymerase and is maintained in the cell as one or two copies. BACs are maintained as single copy plasmids in *E. coli* excluding the replication of more than one BAC in the same host cell. The upper limit of foreign DNA to be inserted in BAC is about 300–3500 kb. BACs are alternative to YACs and, therefore, being used in geneome sequencing projects.

The method of preparation of BAC library is the same as for plasmid library. But here the DNA insert is prepared by pulse field gel electrophoresis (PFGE). PFGE has made it possible to separate, map and analyse very large DNA fragments. Large sized globular DNA migrates easily after applying discontinuous electric field. Example of a BAC vector is pBeloBAC 11 (Fig. 5.17) which is partially digested

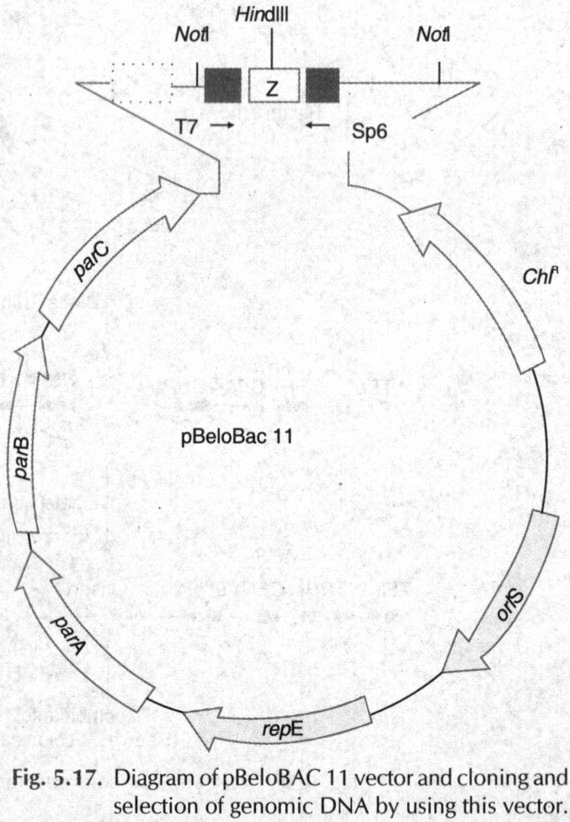


**Fig. 5.17.** Diagram of pBeloBAC 11 vector and cloning and selection of genomic DNA by using this vector.

while cloning a DNA fragment. Genomic DNA is partially digested with *Hin*dIII. Size fractionation of DNA is made by PFGE. Similarly, pBeloBAC 11 is treated with *Hind*III and phosphatase. Such digested vector and fragments of genomic DNA (100–300 kb) are ligated. The recombinant plasmid containing DNA insert is transferred into *E. coli* cells through electroporation (see Chapter 6). *E. coli* can be transformed very efficiently by BAC. Then *E. coli* cells are plated on LB medium containing chloramphenicol, IPTG and X-gal for selection. White colonies of *E. coli* are picked up based on chloramphenicol resistance.

In 1989, by using this vector DNA segments from bithorax gene of *Drosophila* have already been cloned. BAC clones are stable for many generations. In contrast, it lacks positive selection for clones that contain DNA inserts. Therefore, the yield of DNA is very low.

## H. PLANT AND ANIMAL VIRUSES AS CLONING VECTORS

Many eukaryotic genes and their control sequences have been isolated and analysed by using gene cloning techniques based on *E. coli* as host. Many applications of genetic engineering require vectors for the expression of genes on diverse eukaryotic organisms for engineering of new plants, gene therapy and large scale production of eukaryotic proteins. Animal and plant-based vectors have been designed for this purpose.

### 1. Plant Viral Vectors

Virus genome containing DNA insert cause virsuses infection in host plant and amplified naturally. However, there are many viruses that causes systemic infection. Consequently their genes are introduced into all the cells. Since virus infection results in delivery of a new genetic material into host cells, they provide a natural example of genetic engineer. The DNA insert linked with viral DNA that are present in host cell replicates and expresses in host cell along with viral genes.

For insertion of a desired gene into viral genome, it is necessary to study and characterise the virus into detail so that the DNA linked with virus genome may express in the plant. Construction of viral vector that can express in host cell is based on genomic DNA or RNA molecules.

Plant virus vectors are non-integrative and transmitted systematically in plant and assembled in host cell in high copy number. But it never integrates with the host genome. Therefore, to act as viral vector, they should: (*i*) carry extra nucleic acid (DNA or RNA), (*ii*) have broad host range, (*iii*) be easily transmitted, and (*iv*) be preferred to manipulate their genome. There are three groups of viruses (caulimoviruses, geminiviruses and tobamoviruses) which have been worked out in detail.

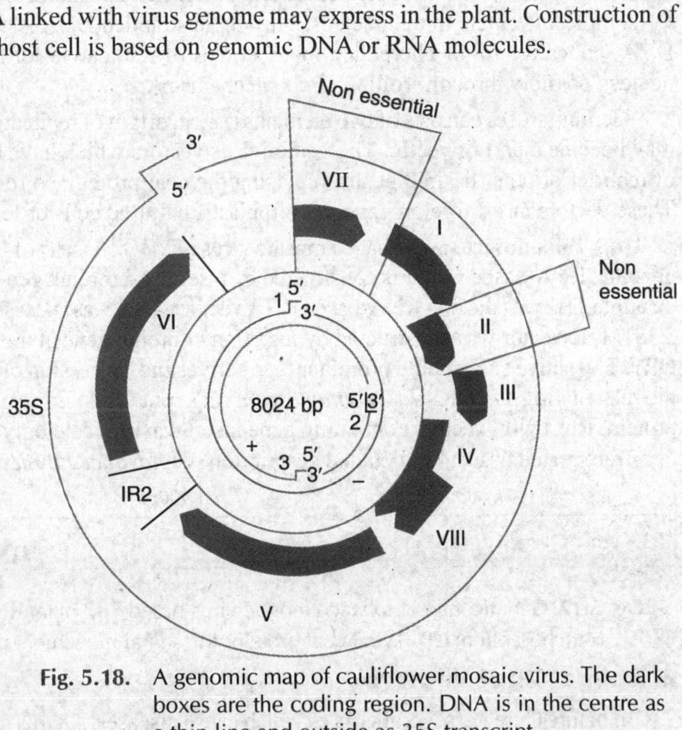


Fig. 5.18. A genomic map of cauliflower mosaic virus. The dark boxes are the coding region. DNA is in the centre as a thin line and outside as 35S transcript.

**(*a*) Caulimoviruses:** Cauliflower mosaic virus (CaMV) comonly known as caulimovirus which has extensively been studied. This virus infects mainly the members of Cruciferae. Genome of CaMV is made up of double stranded DNA which was first manipulated by recombinant DNA technology. The CaMV is transmitted naturally simply by rubbing of leaves or by aphids. Viral particles transmitted are isometric, 50 nm in diameter and isolated from inclusion bodies. Viral particles spread rapidly in 3–4 weeks throughout the plant and found at high copy number (about $10^6$ virions per cell).

Genome of the CaMV is a linear, open, circular molecule of 8 kb twisted or knotted forms (Fig. 5.18). It consists of tightly packed six major and two minor (total I-VIII) open reading frame (ORF) on one coding strand. It replicates through reverse transcription. After infecting the host cell DNA enters inside the nucleus where a supercoiled minichromosome is formed after digestion of a single-stranded overhangs and digestion of gaps. Inside the nucleus of plant cell, the minichromosome acts as a template for nuclear RNA polymerase II forming RNA transcripts. Then the transcripts are transported to cytoplasm where it undergoes either translation or replication into (–)DNA strand using reverse transcriptase. The (+) DNA strand is synthesised from two primer-binding sites that are present near gaps 2 and 3. The RNA synthesis starts from gap 3′ to 5′ end of (–)DNA strand, whereas synthesis started from gap 3 continues to gap 2. Finally, DNA molecules are packed to form viral particles. These particles re-start the similar events after re-entering the cell of a healthy tissue of other plants.

There are two non-essential genomic regions of the virus (ORF I and ORF VII) are non-essential. ORF II synthesises the factors for insect transmission and ORF VII has un-known function. These two regions can be replaced by a foreign DNA and vector constructed. By using this approach, in 1984, Brisson and co-workers transferred CaMV vector containing the bacterial dihydrofolate reductase, into turnip plant cells. This gene is expressed into the plants.

***(b)* Geminiviruses:** This group consists of single-stranded DNA containing plant viruses that cause diseases in several plants. It has geminated (two paired particles) morphology which differentiate it from the other groups of viruses. The capsids are icosahedral, of 10–20×20 nm dimension enclosing covalently closed one or two circular single-stranded DNA molecule(s) (2.6–3.0 kb). After infection DNA molecule enters in the nucleus. The single stranded DNA is converted into a double-stranded DNA replicative form. Thereafter, many copies of replicative forms of DNA is formed inside the nucleus possibly through 'rolling circle mechanism'.

Geminiviruses consists of two elements (*i.e.* an origin of replication and replication protein), each may become a part of vector. The geminivirus vectors which have been constructed so far consists of reporter or other desired genes, replacing the coat protein expressing sequence of viral genome. These vectors insert foreign gene in protoplasts, cultured cells or leaf discs of desired plants.

***(iii)* Tobamoviruses:** Tobacco mosaic virus (TMV) consists of RNA as a linear, large, undivided genome. RNA-based vector is constructed by inserting a foreign gene into an intact viral genome. For example TB2 was the first RNA-based RNA vector which was successfully spread in whole plant (Fig. 5.19). TB2 vector was constructed by inserting a foreign gene at the 3' end of the movement protein ORF. The native subgenome promoter (sgp) drives the expression of the foreign gene and a sgp from the related virus *i.e.* ORSV (*Odontoglossum* ring spot virus) governs the expression of ORF of coat protein. RNA viral vectors containing gene insert can replicate only in protoplast, inoculated leaves or differentiated tissue. For detailed description see *Advanced Biotechnology* by R.C. Dubey (2014).

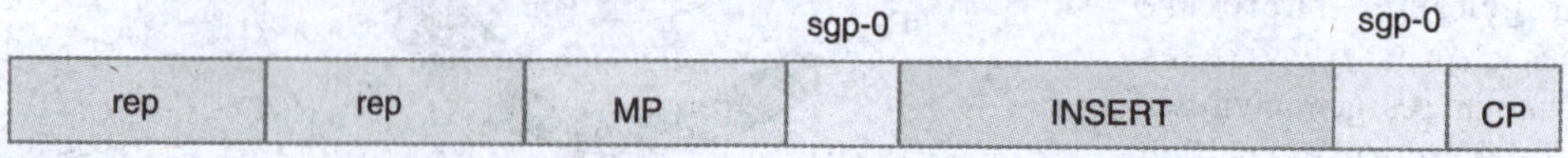


Fig. 5.19. Genetic map of tobacco mosaic virus-based TB2 plant RNA viral vector; insert= *npt*II; α-trichosanthin; CP= viral coat protein; MP=viral movement protein; rep= replication.

## 2. Animal Viral Vectors

In nature there are several viruses which cause diseases. A virus gets adsorbed to the body surface of suitable host and infects the cell. This ability of animal viruses has been exploited and virus-based vectors have been designed to introduce foreign DNA of known function into cultured eukaryotic cell. In 1979, for the first time a simivian virus 40 (SV 40)– based cloning vector was constructed and used in cloning experiment using mammalian cells. Since then several vectors were constructed using adenovirus, papillomaviruses, retroviruses to clone foreign DNA into the mammalian cells, and baculoviruses in insect cells. In recent years, retrovirus-based vectors are commonly used for gene cloning, because they can infect several types of cells. Since they are single stranded, their high concentrations are required for fully differentiated cells like neurons, hepatocytes, etc. The *gag*, *pol* and *env* genes of retroviruses (which are required for replication and assembly of viral particles) can be replaced with foreign DNA. The recombinant DNA is introduced into mammalian cells in tissue culture (Fig. 5.20). For detailed description see *Advanced Biotechnology* by R.C. Dubey (2014).

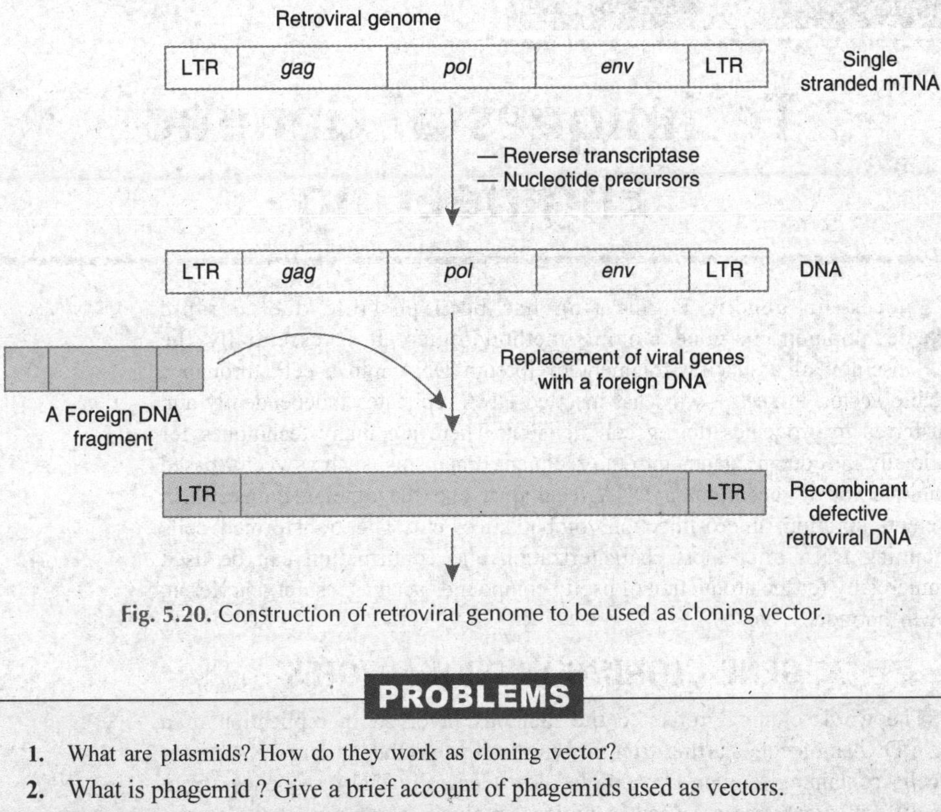


**Fig. 5.20.** Construction of retroviral genome to be used as cloning vector.

## PROBLEMS

1. What are plasmids? How do they work as cloning vector?
2. What is phagemid ? Give a brief account of phagemids used as vectors.
3. Discuss in detail the yeast plasmid vectors.
4. What is a cloning vector ? Why are they necessary ? Describe the cloning vectors studied by you.
5. What is a cDNA bank? How is it constructed?
6. Write short notes on the following:

   (*i*) Cloning vectors, (*ii*) Phage λ , (*iii*) Cosmid and phasmid vectors, (*iv*) Insertion and replacement vectors (*v*) pBR322, (*vi*) *Escherichia coli,* (*vii*) *Agrobacterium.* (*viii*) Restriction-modification system (*ix*) BAC vector (*x*) YAC Vectors (*xi*) P1 vector (*xii*) Plant virus vector (*xiii*) Animal virus vectors.

CHAPTER

6

# Techniques of Genetic Engineering

Success in genetic engineering has been possible due to rapid development in gene cloning methodologies. It is essentially the insertion of a specific fragment of foreign DNA into a cell, through a suitable vector, in such a way that inserted DNA replicates independently and transferred to progenies during cell division. There are many techniques for artificially introducing genes into the recipient organisms, such as vector-based technique called recombinant DNA technique, and the other tecniques such as electroporation, micro-injection, bioballistics, etc. The transformed cells containing DNA after their characterization and confirmation can be used commercially for the production of useful compounds such as insulin, interferon, growth hormones, etc.

## A. GENE CLONING IN PROKARYOTES

The word 'cloning' refers to the methods involved in replication of a single DNA molecule starting from a living cell to produce a large population of cells containing similar DNA molecules. Generally DNA sequences from two diffeent organisms are used in molecular cloning; one that is the source of the DNA to be cloned, and the other that serves as the bost for replication of the recombinant DNA.

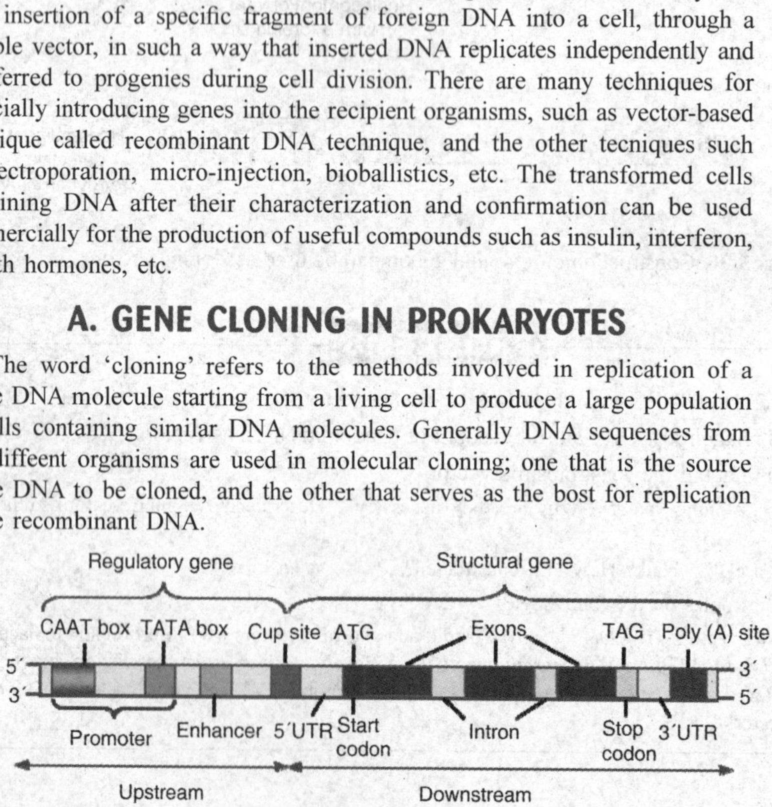


Structural organization of a typical gene construct.

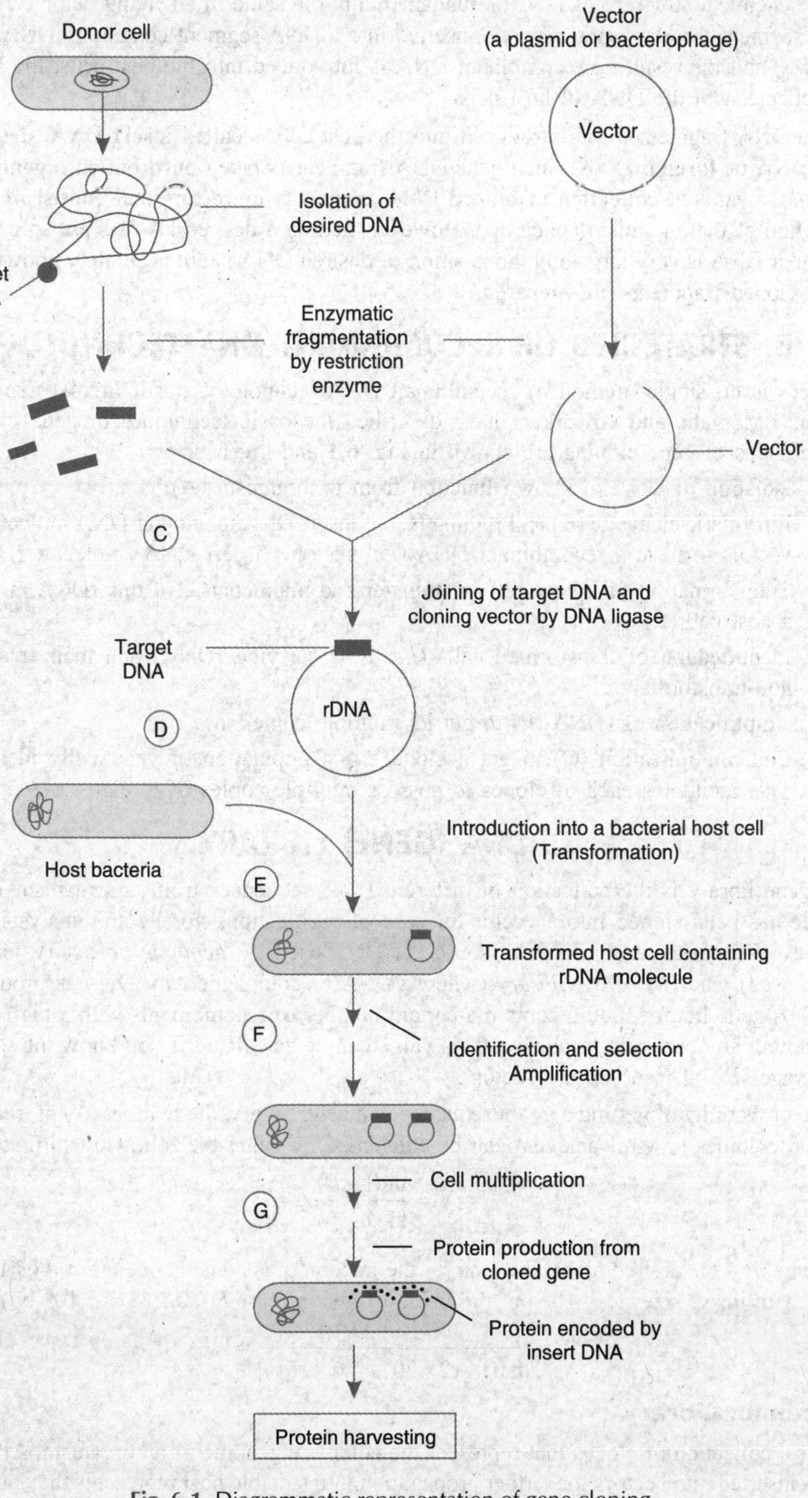


**Fig. 6.1.** Diagrammatic representation of gene cloning.

The chemical structure of DNA is fundamentally the same in all living beings. Therefore, if a DNA segment from any organisms is inserted into a DNA segment containing DNA replication sequences, and the resulting recombinant DNA is introduced into the host cells, the DNA insert will replicate with the DNA of host cells.

The DNA fragment to be introduced into the host cell is called **insert DNA, desired DNA, target DNA** or **foreign DNA**. Such gene/DNA fragment is taken out from an organism or gene library *i.e.* a random collection of cloned DNA fragments in vectors that consist of all genetic information of that organism concerned. However, getting a desired DNA is not so easy because the cellular DNA is very large and the position of desired DNA is not accurately known. Libraries of such cloned fragments are prepared.

## B. STRATEGIES OF RECOMBINANT DNA TECHNOLOGY

There is no single method of recombinant DNA technology, but it involves several steps. In 1976, T.Maniatis and coworkers have described the basic techniques of gene cloning. The principal steps of gene cloning are shown in Fig. 6.1 and given below:

(*i*) Isolation of DNA of known function from an organism (A).

(*ii*) Enzymatic cleavage (B) and joining (C) of insert DNA to another DNA molecule (cloning vector) to form a recombinant DNA (*i.e.* vector + insert DNA) molecule (D).

(*iii*) Transformation of a host cell *i.e.* transfer and maintenance of this rDNA molecule into a host cell (E).

(*iv*) Identification of transformed cells (*i.e.* cells carrying rDNA) and their selection from non-transformants.

(*v*) Amplification of rDNA (F) to get its multiple copies in a cell.

(*vi*) Cell multiplication (G) to get a clone, *i.e.* a population of genetically identical cells. This facilitates each of clones to possess multiple copies of foreign DNA.

## C. DNA (GENE) LIBRARY

A gene library is the collection of different DNA sequences from an organism where each sequence has been cloned into a vector for ease of purification, storage and analysis. There are two types of libraries on the basis of source of DNA used: (*i*) *genomic library* (where genomic DNA is used), and (*ii*) *cDNA library* (where cDNA or complementary DNA is produced using mRNA). A gene library should contain a certain number of recombinants with a high probability of containing any particular sequence. This value can be calculated if you know the genome size and average size of insert in the vector.

The probability of getting a given sequence in a gene library, the number ($N$) of recombinants (bacterial colonies or viral plaques) can be calculated by using the following formula:

$$N = \frac{\ln(1-P)}{\ln(1-f)}$$

where, $P$ = the desired probability and $f$ = the fraction of genome in one insert. For a probability of 0.99 with insert sizes $2 \times 10^4$ bp, the value of genome for *E. coli* ($4.6 \times 10^6$ bp) will be:

$$N^{\text{E.coli}} = \frac{\ln(1-0.99)}{\ln[(1-(2\times 20^4/4.6\times 10^6)]} = 1.1\times 10^3$$

### 1. Genomic Library

It is a collection of clones that represent the complete genome of an organism. All fragments of DNA inserted into vectors for further propagation into suitable host represents the entire genome of an organism. For construction of a genomic library the entire genomic DNA is isolated from host cells/tissues, purified and broken randomly into fragments of correct size for cloning into a suitable vector. There are two basic ways of fragmenting genomic DNA randomly: *physical shearing* (*e.g.* pipetting, mixing or sonication) and partial *restriction enzyme digestion* (by using limiting amount of restriction enzyme the DNA is not digested at every recognition sequence). Using these methods genomic DNA is broken randomly into smaller fragments.

Fig. 6.2 shows a total of 6 random DNA fragments obtained after physical shearing. The vector isolated from bacterium is also digested with the restriction enzyme which digests genomic DNA. The fragments of genomic DNA is inserted into the vector. Each vector consists of different fragments of DNA. The recombinant DNA molecules are transferred into bacterial cells or bacteriophage particles are assembled.

The genomic library for organisms with smaller genome size (*e.g. E. coli*) can be constructed in plasmid vector. Only 5,000 clones (of the average size of 5,000 bp) results in 99% chance of cloning the entire genome of $4.6 \times 10^6$ bp. In addition, libraries from organisms with larger genomes are constructed using phage λ, cosmid, BAC or YAC vectors.

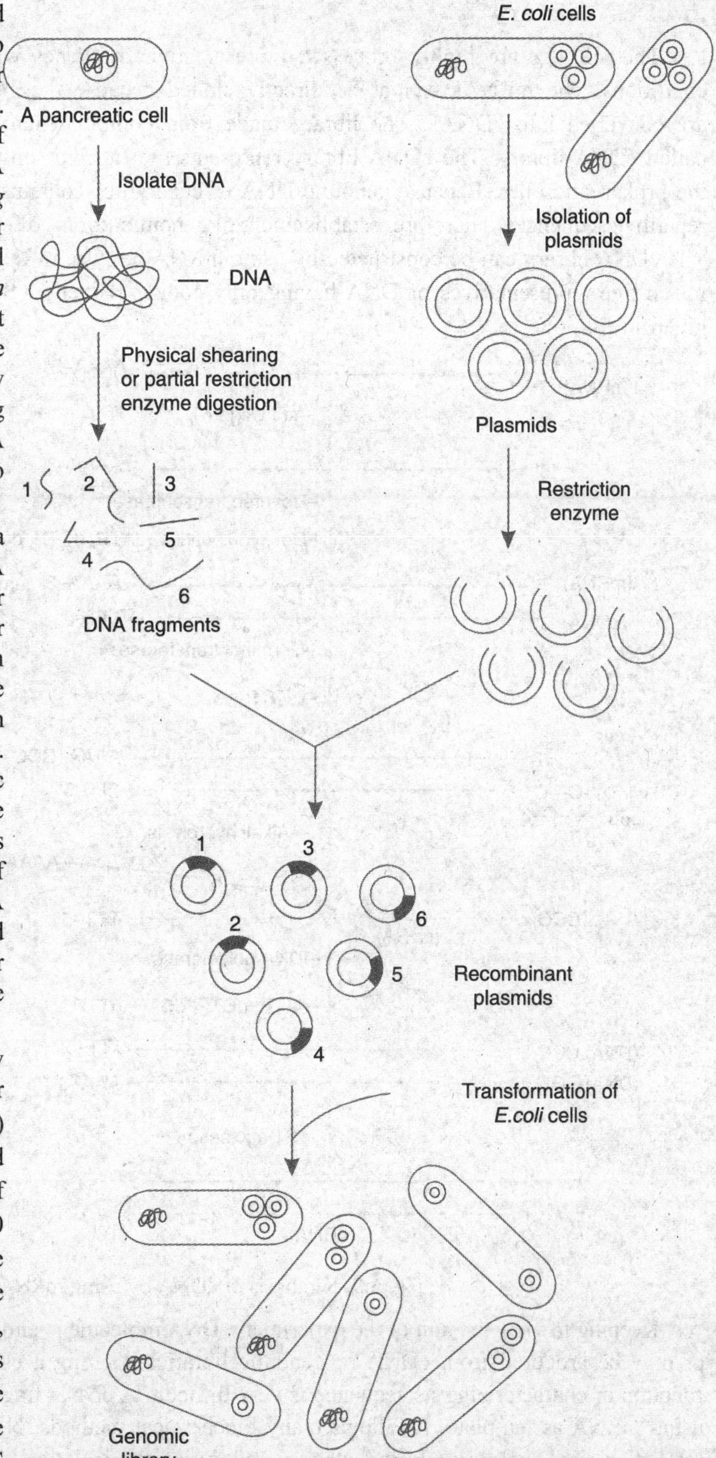


**Fig. 6.2.** Construction of a genomic library.

## 2. cDNA Library

The mRNAs are highly processed representatives of genes which express under specific conditions. The mRNAs cannot be directly cloned because they are unstable hence mRNAs are converted into cDNAs. The library made from complementary or copy DNA (cDNA) is called cDNA library. The cDNA library represents the DNA of only eukaryotic organisms, not the prokaryotic ones. Because genomic DNA of eukaryotes contains introns, regulatory regions, repetitive sequences. Therefore, establishment of genomic library of eukaryotes is not meaningful. The cDNA library can be constructed by using mRNA because mRNAs are the highly processed, intron-free representatives of DNA having only coding sequence. Synthesis of cDNA has been given in Fig. 6.3

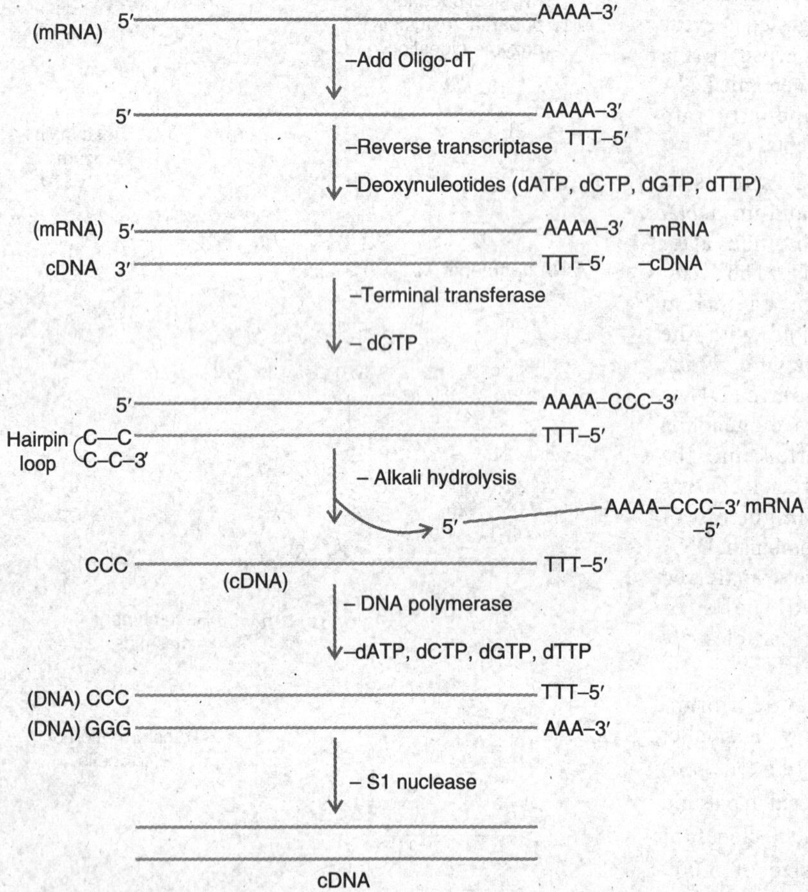


**Fig. 6.3.** Synthesis of cDNA by using mRNA

Keeping in view the aim of the experiment, DNA molecule is either enzymatically fragmented or may be procured from cDNA or genomic libraries. Getting a cDNA fragment of its known function or characterizing its sequence is very difficult. A cDNA fragment is prepared directly by using mRNA as template. It follows many biochemical methods. Necessary enzymes with their restriction sites and cleaved products required during the isolation of DNA have been described earlier, and the important enzymes are given in Table 4.1.

**(*a*) Isolation of mRNA:** It is much difficult to isolate a specific mRNA if it is in low amount in the cell. The majority of mRNA sequence in eukaryotic cells contains a long polyadenylated tract (*i.e.* about 100Å residues) at their 3′ termini (Kates, 1970). By this virtue they can be separated from other types of RNAs. Therefore, mRNA binds with an oligo-dT cellulose affinity column or poly-U sepharose from which it can be eluted. Its integrity can be studied using gel electrophoresis which allows mRNAs to be size-fractionated by recovering chosen regions of the gel lane. Commercial preparations of both these substrates are now available. The mRNA preparations can be enriched for the desired molecules by fractionation according to their size by using sucrose density gradient centrifugation. Preparation of cDNA is shown in Fig. 6.3.

**(*b*) Reverse Transcription:** Reverse transcriptase is required for the synthesis of DNA copy of an mRNA molecule. The mRNAs are treated with oligo-dT primer, reverse transcriptase enzyme and four deoxynucleotide triphosphates (dNTPs), *i.e.* dATP, dGTP, dCTP and dTTP. The oligo-dT primer binds to polyadenylated tail that contains –AAAA(n) residues and provides free -OH site for reverse transcription. Reverse transcriptase adds complementary dNTPs one by one to free 3′ -OH site of primer, forms single strand of DNA resulting in RNA-DNA hybrid. Cellular DNA and total RNA inhibit reverse transcriptase activity. Therefore, it is necessary that mRNA must be in pure form before cloning.

The oligo-dT contains 10-20 nucleotides that hybridize the poly-A tract on mRNA. The oligo-dT primer ensures the initiation of cDNA synthesis at 3′ terminus of RNA in poly-A tract.

**(*c*) Oligo-dC Tailing:** The RNA-DNA hybrid is treated with the enzyme terminal transferase and a nucleotide precursor, dCTP. This enzyme adds dCTP one by one to 3′ -OH group of RNA and DNA strands. Consequently a short oligo-dC tail is produced at 3′ end of both the strands. The residues of cDNA become curved. The curved tail forming a loop like structure is known as hairpin loop.

**(*d*) Alkali Hydrolysis:** Upon hydrolysis, by using alkaline sucrose solution the mRNA-cDNA strands are separated into single strand.

**(*e*) Addition of Oligo-dG Primer:** Oligo-dG primer is added in the reaction mixture and temperature is maintained to about 55°C. This facilitates the binding of dG to oligo-dC tails formed on cDNA.

Now cDNA acts as a template for the synthesis of double stranded cDNA in the presence of DNA polymerase I or AMV (avian myeloblastosis virus) reverse transcriptase.

**(*f*) Synthesis of Second Strand of cDNA:** Klenow DNA polymerase I and dNTP precursors are added to the reaction mixture. This enzyme adds complementary nucleotides to the cDNA one by one to 3′ -OH site of primer. Consequently, from one cDNA, double stranded DNA clone is synthesised the length of which is similar to that of the mRNA molecule initially used.

This is used for cDNA cloning and construction of cDNA library and in gene cloning experiments. The double stranded cDNA is identical to the gene which codes for mRNA( only in prokaryotic genes). However, the majority of eukaryotic genes contain introns (the non coding regions) which interrupt their coding sequences. Methods of removing cDNA in eukaryotes have been described earlier. cDNA can be radioactively labelled by nick translation and used as a hybridization probe for the identification of DNA fragments containing the gene. By using these procedures a cDNA clone bank is built up (*see* preceding section D).

### 3. Screening of cDNA Clone

Since, cDNA clone needs prepared by using a specific mRNA, there is no need of screening the cDNA clones. However, often cDNA clones are the mixture of cDNA molecules. The cDNA clone of desired function may be low or large in number. Therefore, all these clones must be ligated to a suitable vector such as M13 vector, phage λ vectors and transferred into host bacteria. Each bacterial cell possesses a single cDNA clone; hence collection of all recombinant bacteria is called cDNA library.

When gene cloning needs to be done, the insert DNA is isolated from recombinant bacteria and cut with restriction enzyme. Then the cDNA is separated through electrophoresis, amplified and used for gene manipulation.

## D. INSERTION OF A FOREIGN DNA FRAGMENT INTO A VECTOR

The DNA thus isolated as above or procured from DNA or gene bank is fragmented by using the specific restriction enzyme to develop specific cohesive ends. The cloning vector (*e.g.* pBR322) is also treated with the same restriction enzymes so that the cohesive ends generated may have complementary residues similar to foreign DNA. The plasmid contains $amp^R$ and $tet^R$ marker genes to confer resistance against antibiotics, ampicillin and tetracycline respectively. Some time after cleavage by restriction enzymes and ligation of foreign DNA fragment with the plasmid a few important genes are destroyed in the resultant product. For example, *Bam*HI destroys $ter^r$ genes. The sticky ends can permanently join if T4 DNA ligase (which forms a phosphodiester linkage between free 5' phosphoryl and 3′ OH group) is added along with ATP at 4-10$^o$C for a long incubation time. A number of products are formed from this reaction which are highly heterogeneous mixture of recombinant DNA molecules together with parental plasmids.

For insertion of double stranded cDNA into a cloning vector it is necessary to add to both termini single stranded DNA sequence which should be complementary to a tract of DNA at the termini of linearlized vector. In order to get efficient formation of recombinant DNA molecules, addition of sticky ends on both termini is necessary. There are two methods for generation of cohesive ends on the double stranded cDNA, use of linkers and homopolymer tails.

**(*a*) Use of Restriction Enzyme Linkers :** Linkers are the chemically synthesized double stranded DNA oligonucleotides containing on it one or more restriction sites for cleavage by restriction enzymes, *e.g. Eco* RI, *Hind* III, *Bam* HI, etc. Linkers are ligated to blunt end DNA by using T4 DNA ligase (Fig. 6.4). Both the vector and DNA are treated with restriction enzyme to develop sticky ends. The staggered cuts *i.e.* sticky ends are then ligated with T4 DNA ligase with very high efficiency to the termini of the vector and recombinant plasmid DNA (chimeric DNA) molecules are produced. For details see Chapter 4.

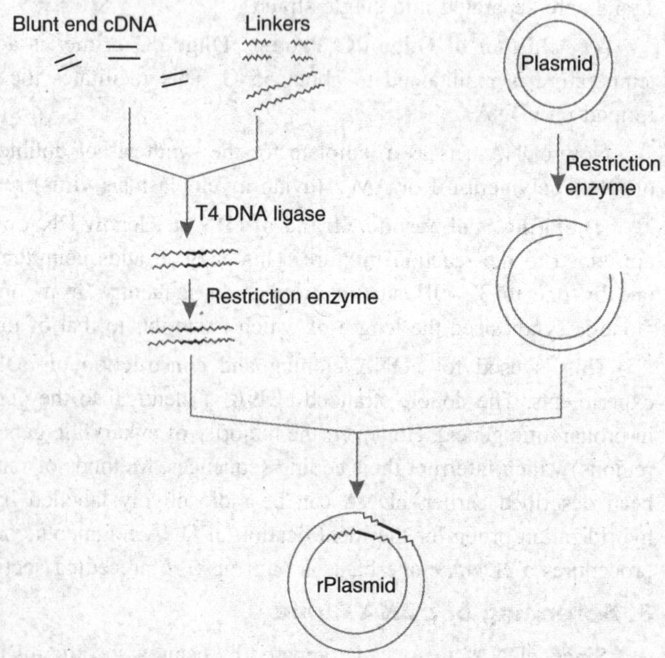


**Fig. 6.4.** Insertion of a foreign DNA fragment into a plasmid by using restriction enzyme linkers.

***(b)* Use of Homopolymer Tails :** D.A. Jackson and coworkers for the first time in 1972 applied this method to construct a recombinant DNA by inserting phage λ DNA into SV40 DNA by using dA-dT homopolymers. Using terminal transferase the synthesis of homopolymer tails of the defined length at both 3′ termini of double stranded DNA and vector can be done. In the presence of precursor dATP, terminal transferase helps to add poly-dA at 3′ termini of vector. Likewise the same enzyme adds poly-T at 3′ termini of DNA molecule, when precursor TTP is present. The linearized vector having tails is incapable of recirculari-zation, unless ligated to a double stranded DNA fragment. The vector and DNA tails are annealed. The poly dA-dT tails are then ligated by T4 DNA ligase (Fig. 6.5). If poly dG-dC tails are used, instead of poly dA-dT tails, a high initiation temperature (upto 37°C) is required for the annealing reaction. The major advantage in using poly dG-dC tails is that the hybrids are more stable than poly dA-dT hybrids It has also been described in chapter 4.

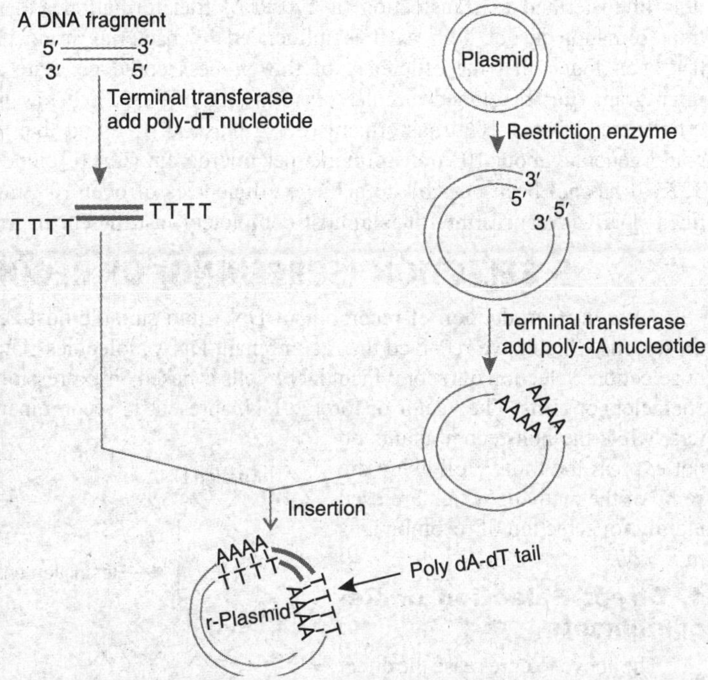

Fig. 6.5. Insertion of a foreign DNA fragment into a plasmid by using homopolymer tail.

In the first method, the inserted DNA molecule can be retrieved by using restriction enzymes (*e.g. Eco*RI) on the cleavage sites of linker, whereas in the second method, however, it is very difficult to retrieve the inserted DNA fragment due to the loss of recognition sites when poly dA-dT tails are added. In the second case, recovery of the inserted DNA fragment is possible only when poly dG-dC tails are used, because it gives recognition site for restriction enzymes.

## E. TRANSFER OF RECOMBINANT DNA INTO BACTERIAL CELL

Several methods have been developed for introduction of recombinant DNA molecule into host cells. According to types of vectors and host cells, the methods are adopted. Some of the methods of gene transfer into host cells are briefly discussed below :

### 1. Transformation

Once a mixture of recombinant DNA is obtained it is allowed to be taken up by the suitable bacterial cells. Originally the transformation procedure was developed in 1979 by M. Mandell and A. Higa. The strains of *E.coli* usually do not have restriction systems, hence it degrades foreign DNA. To escape from degradation exponentially growing cells are pretreated with $CaCl_2$ at low temperature; thereafter DNA is mixed up. The event of entering the plasmid containing foreign DNA fragment into a bacterial cell is known as 'transformation'.

## 2. Transfection

Transfection is the transfer of foreign DNA into cultured host cells mediated through chemicals or viral vectors. The charged chemical substances such as cationic liposomes, calcium phosphate or DEAE dextran are taken and mixed with DNA molecules. The recipient host cells are overlayed by this mixture. Consequently the foreign DNA is taken up by the host cells. For detail see *liposome-mediated gene transfer* in preceding section.

For the first time phage λ was used to transfer the foreign DNA into *E. coli* cell, therefore, it is often termed as transfection (a hybrid of transformation and infection). The efficiency of transformation is not high as it is influenced by bacterial strain and size of foreign DNA. It has been found that the efficiency of this process could generate about $10^5$ transformants per microgram (μg) of cloned circular plasmid (*e.g.* pBR322). It has not been possible to achieve efficiencies of over $10^8$ transformants per g plasmid. If found that the efficiency of this process could generate about $10^5$ transformants per microgram (μg) of cloned circular plasmid (*e.g.* pBR 322). It has not been possible to achieve efficiencies of over $10^8$ transformants per g plasmid. If linear DNA is transformed it is almost completely insufficient in transformation.

# F. SELECTION (SCREENING) OF RECOMBINANTS

After the introduction of recombinant DNA into suitable host cells, it is essential to identify those cells which have received the recombinant DNA molecules. This process is called screening or selection. Selection of recombinant DNA cells is based on expression or non-expression of certain characters or traits. The vector or foreign DNA present in recombinant cells expresses the characters, while the non-recombinants do not express the traits. Following are some of the methods which are used mainly for selection of recombinants in *E. coli.*

## 1. Direct Selection of Recombinants

The above features ease the direct selection of recombinant cells. If the cloned DNA itself codes for resistance to the antibiotic ampicillin (*amp*$^R$) the recombinants can be allowed to grow on minimal medium containing ampicillin. Only such recombinants will grow and form colony on medium that contain *amp*$^R$ gene on its plasmid vector. Following this procedure you cannot say whether the recombinants growing on such medium contain religated plasmid vector or contain recombinant plasmid plus foreign DNA fragment. Because *amp*$^R$ gene is present in both the recombinants.

## 2. Insertional Selection Inactivation Method

This is more efficient method than the direct selection. In this ap-

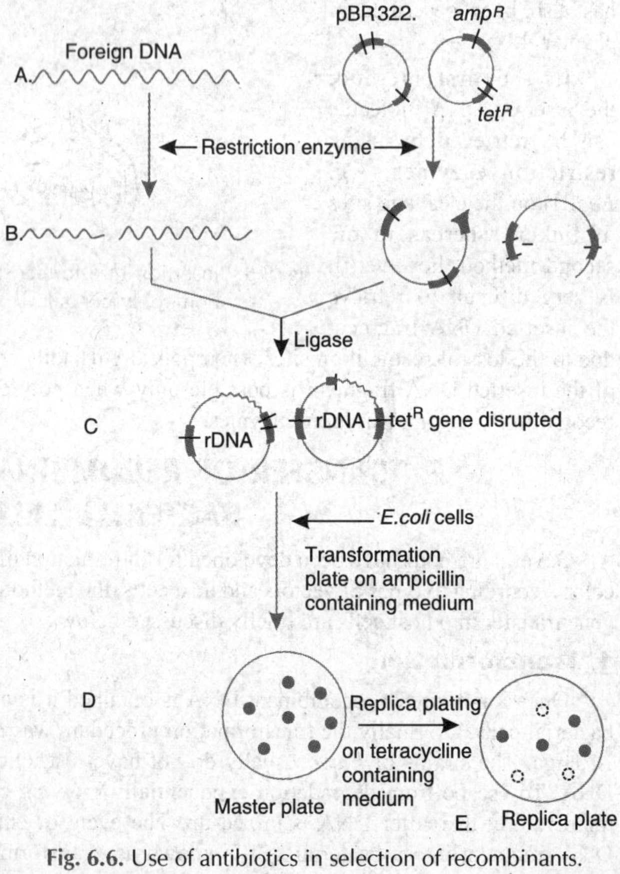


Fig. 6.6. Use of antibiotics in selection of recombinants.

proach, one of the genetic traits is disrupted by inserting foreign DNA. Antibiotic resistance genes acts as a good insertion inactivation system. As discussed in earlier section that plasmid pBR322 contains two antibiotic resistance genes, one for ampicillin (*amp*$^R$ gene) and the other for tetracycline (*tet*$^R$ gene). If the target DNA is inserted into *tet*$^R$ gene using *Bam*HI, the property of resistance to tertracycline will be lost. Such recombinants would be test-sensitive. When such recombinants (containing target DNA in *tet*$^R$ gene) are grown onto medium containing tetracycline, they will not grow because their *tet*$^R$ gene has been inactivated. But they are resistant to ampicillin because *amp*$^R$ gene is functional. On the other hand the self-ligated recombinants will show resistance to tetracycline and ampicillin. Therefore, they will grow on medium containing both the antibiotics.

To ascertain the presence or absence of *tet*$^R$ gene in the inserted DNA fragment in plasmid a replica plating is done from the master plate. Bacterial colonies of master plate are gently pressed with sterile velvet so that a few cells of each colony may adhere on it which is then pressed on other plate containing the nutrient medium amended with tetracycline. Plates are incubated for the growth of bacterial colonies. The appearance of colonies is compared with the master plate and those colonies that fail to grow on replica plate (Fig. 6.6E) can be said to have a plasmid which had insert DNA in the *tet*$^R$ gene of plasmid and had destroyed *tet*$^R$ gene. Hence, *tet*$^R$ gene becomes non-functional.

### 3. Blue-White Selection Method

This method is used for screening of recombinant plasmid. In this method a reporter gene *lacZ* is inserted in the vector. The *lacZ* encodes the enzyme β-galactosidase powerful contains several recognition sites for restriction enzymes. Beta-galactosidase breaks a synthetic substrate, X-gal (5-bromo-4-chloro-indolyl-β-D-galacto-pyranoside) into an insoluble blue coloured product. If a foreign gene is inserted into *lacZ*, this gene will be inactivated. Therefore, no blue colour will develop, because β-galactosidase is not synthesised due to inactivation of *lacZ*. Therefore, the host cells containing rDNA will form white coloured colonies on the medium containing X-gal, whereas the other cells containing non-recombinant DNA will develop the blue coloured colonies. On the basis of colony colour the recombinants can be selected from the agar plates (Fig. 6.7).

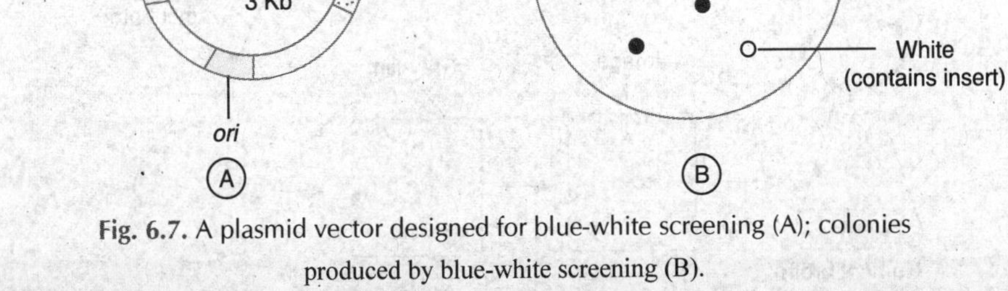


Fig. 6.7. A plasmid vector designed for blue-white screening (A); colonies produced by blue-white screening (B).

### 4. Colony Hybridisation (Nucleic Acid Hybridisation) Technique

The transformed cells grow on nutrient medium, but it is likely that a few cells may have specific DNA as desired among several thousand cells. How is it possible to pick up those cells which have the desired DNA? To make it easy, the "colony hybridization technique" has been developed in 1975 by M. Grustein and S. Hogness. It is suitable for use with plasmids. An analogous method, plaque hybridization technique, has been given for use with phages.

(*a*) **DNA Probes:** The colony hybridisation technique is based on the availability of a radiolabelled DNA probe. A DNA probe is the radio-labelled ($^{32}PO_4$) small fragment of DNA molecule (20-40 bp) which is complementary at least to one part of desired DNA. Generally probe DNA is labelled with $^{32}PO_4$ or some times with $^{125}I$. The $^{32}P$ liberates β-particles, while $^{125}I$ emits γ-rays. In addition to using radioactive isotopic elements, isotopic sulphur or fluorescent molecules may also be used. Therefore, a DNA probe recognises A, T, G and C nucleotides and combines with the complementary sequences of the target DNA. A and T of the probe combine with T and A of target DNA and *vice versa*. Similarly, G and C of probe combine with C and G of target DNA and *vice versa*. The use of DNA probe in searching the functional target DNA is given in nucleic acid hybridization technique. The small sized DNA probe easily catches the target DNA for hybridisation than the long sized probe. Besides, a high rate of hybridisation is achieved on maintaining higher concentration of $^{32}P$-DNA than the target DNA molecules. Moreover, the DNA probe is very specific to its target DNA; this specificity is called **stringency.**

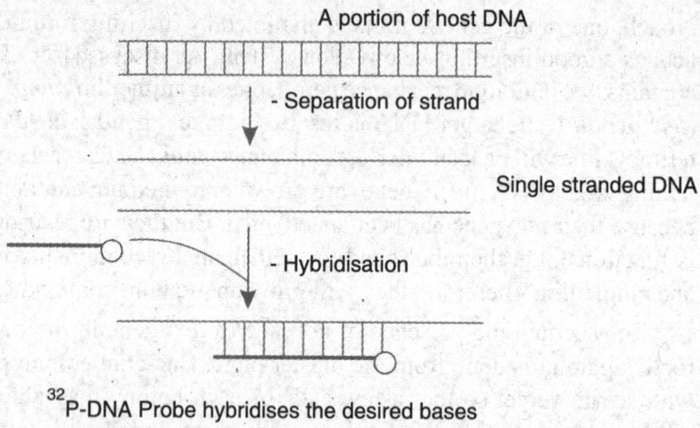


**Fig. 6.8.** Hybridisation of desired DNA with a DNA probe.

The probe may be partially pure mRNA, a chemically synthesized oligonucleotide or a related gene, which identifies the corresponding recombinant DNA. DNA probes have commercial significance. They diagnose specific DNA sequences (genes) and are used in the diagnosis of diseases and microbiological tests.

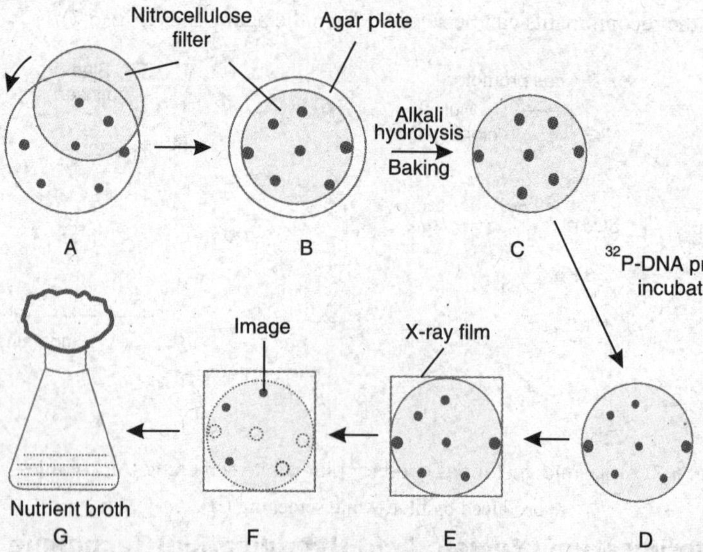


**Fig. 6.9.** Colony hybridization technique.

(*b*) **Colony (or Plaque) Hybridisation Technique:** Colony hybridisation technique was developed by M. Grunstein and D.S. Hogness in 1975. In this technique master plates are prepared as described earlier. Replica plating of colonies is done onto a nitrocellulose filter disc (Fig. 6.9A) which is then placed on the surface of a gelled nutrient medium and both master plate and disc are incubated to develop colonies. Cells growing on nitrocellulose filter disc are nourished through the diffusion of nutrients from gelled nutrient medium (Fig. 6.6B). The filter disc is removed and put on blotting paper soaked with 0.5 N NaOH solution. The alkali diffuses into nitrocellulose, lyses bacterial cells and denatures their DNA. Thereafter, the filter disc is neutralised by tris (hydroxymethyl) aminomethane-HCl buffer by keeping high salt concentrations. Then the filter is dipped in proteinase K solution that removes proteins from the filter. This results in binding the DNA with nitrocellulose disc in the same pattern as the bacterial colonies; to fix the cDNA properly the filter disc is baked at 80°C (Fig. 6.9C).

Then it is incubated with a solution containing radioactive chemical labelled probe ($P^{32}$ DNA) at suitable conditions. The probe will hybridize any bound DNA which contains sequences complementary to the probe (Fig. 6.9D). By thorough washing unhybridized (unbound) probes are removed from the hybridized probe (colonies containing sequences complementary to probe) and is identified by autoradiography of the nitrocellulose filter disc (Fig. 6.8E-F). Colonies which develop positive X-ray image (Fig. 6.9G) are compared with master plate and picked up, and multiplied on the nutrient medium.

**(*c*) Plaque Lifting Method:** In 1977, W.D. Benton and R.W. Davis devised this technique for screening of bacteriophages from plaque forming units. The recombinant phage particles are isolated and used to construct DNA library. In this method lawn of *E. coli* is prepared on agar medium which is allowed to get infected by recombinant phage particles. They infect *E. coli*, multiply inside cells and lyse them forming plaques. As described in Southern blotting technique, replica plates are prepared from the master plate containing plaques by using nitrocellulose filter paper. This filter is treated with alkali solution so that the phage DNA may be denatured. Then filter is baked at 80°C to fix denatured phage DNA and put in a solution containing $^{32}P$-DNA probe. After some times the probe hybridises the desired DNA if complementary bases are present. The filter paper is washed with standard saline citrate solution to remove unhybridised probe and passed through autoradiography. Dark spots are formed where probes have hybridised the desired DNA. Then the recombinant phage particles can be isolated from the plaques.

## 5. Selection of Recombinants (Transgenic) Tissues

For detailed discussion see Chapter 13.

## 6. In Vitro Translation

*In vitro* translation is now used as a method to confirm the identification of recombinant clones. S. Nagata and coworkers in 1980 have used primary translation screened by isolating the interferon gene from total leukocyte poly A+RNA. Translation of poly A+RNA is done when it is microinjected into oocytes from the toad, *Xenopus laevis* with the result of secretion of interferon into culture medium. Interferon is detected by its antiviral activity. It is supposed that the number of interferon mRNA may be between $10^3$ and $10^4$, therefore, about $10^4$ transformants must be repeated to get at least single clone. Thereafter, DNA is immobilized on nitrocellulose or diazobenzyloxymethyl cellulose paper to isolate complementary mRNA by nucleic acid hybridization as described earlier. The RNA

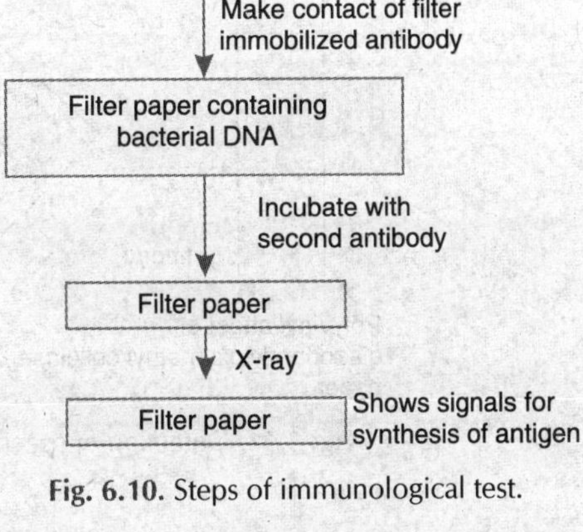


**Fig. 6.10.** Steps of immunological test.

can be eluted from the cDNA-RNA hybrid and microinjected into oocytes and the polypeptides as the translation products are detected by immunoprecipitation techniques. It confirms *in vitro* translation of foreign gene.

## 7. Immunological Tests

The immunological techniques are the final test analogous to colony hybridization technique as described earlier. It is an alternative screening procedure which relies on expression, and generally applicable approach to identify a clone synthesizing a particular polypeptide. This is potentially a very powerful method since the only absolute requirement is that the required mRNA encodes a protein for which a suitable antibody is available. Two *in situ* techniques given by S. Broome and W.A. Gilbert (1978), and Erlich *et al.* (1978) are in current use.

Instead of radio-labelling of DNA molecules, antibodies (immunoglobulins) are used to identify the colonies or plaques developed on master plates that synthesize antigens encoded by the foreign DNA present in plasmids of the bacterial clones. For this purpose a special vector, known as expression vector, is designed where the foreign DNA is transcribed and translated within the bacterial cell. The growth medium containing specific anti-serum may help in detection of viable immunoprecipitate (precipitin) around the colonies or plaques. The method follows : (*i*) the replica plating of bacterial colonies of master plate on nutrient agar, (*ii*) lysis of cells after their growth by exposure to chloroform vapour, or treatment with high temperature, (*iii*) making gentle contact of a solid support, for example, a cellulose filter containing immobilized antibody to solid support with the lysed cells within the colonies to allow absorption of antigen to antibody, and (*iv*) detection of antigen-antibody complex by incubating the cellulose filter with a radio-labelled second antibody. The antibodies that do not react are washed off and position of the antigen-antibody complex is determined by passing the filter through X-ray. It gives the signal of those bacterial cells which synthesize antigen on the master plate (Fig. 6.10).

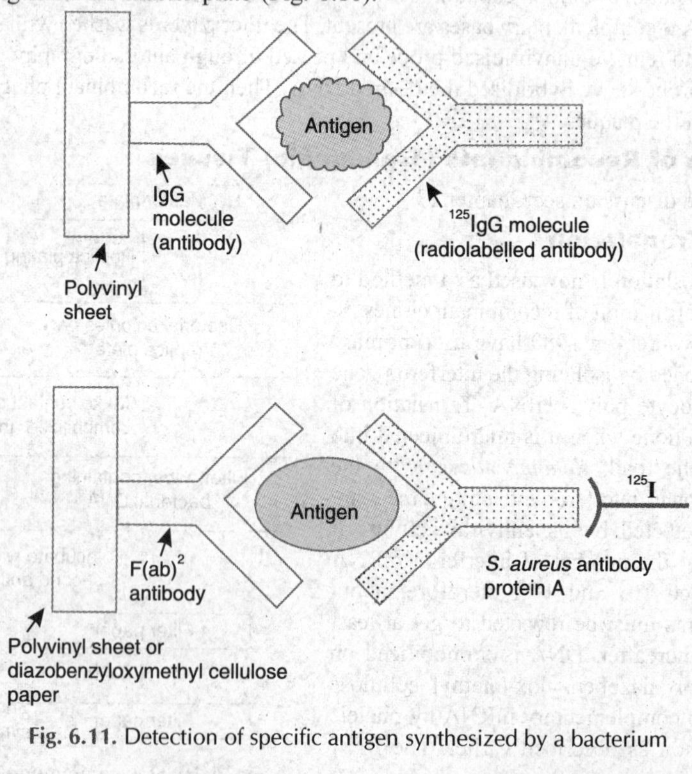


Fig. 6.11. Detection of specific antigen synthesized by a bacterium

S. Broome and W. Gilbert in 1978 purified immunoglobulin G (IgG) and bound it to an agar plate so that the antigens released through *in situ* lysis of bacterial colonies can bound to the fixed antibody (Fig. 6.11). The antigen is then detected by using the same IgG preparation which 4.6 is radioactively labelled. This recognizes determinants on the bound antigen. Erlich *et al.* (1978) used F(ab)½ fragments derived by digestion of pepsin of the immunoglobulin, bound to either polyvinyl cellulose or diazobenzyl oxymethyl cellulose (DBMC) paper. This is incubated first with the antigen and then with undigested antiserum, Fc portion of which will bind radiolabelled *Staphylococcus aureus* A protein.

## G. BLOTTING TECHNIQUES

DNA, RNA and proteins are separated by blotting techniques. These are described below:

### 1. Southern Blotting Techniques

A method, developed by a molecular biologist E.M. Southern (1975) for analysing the related genes in a DNA restriction fragment is called Southern blotting technique. Southern blots can easily provide a physical map of restriction sites within a gene located normally on a chromosome, and reveal the number of copies of the gene in the genome, and the degree of similarity of the gene when compared with the other complementary genes.

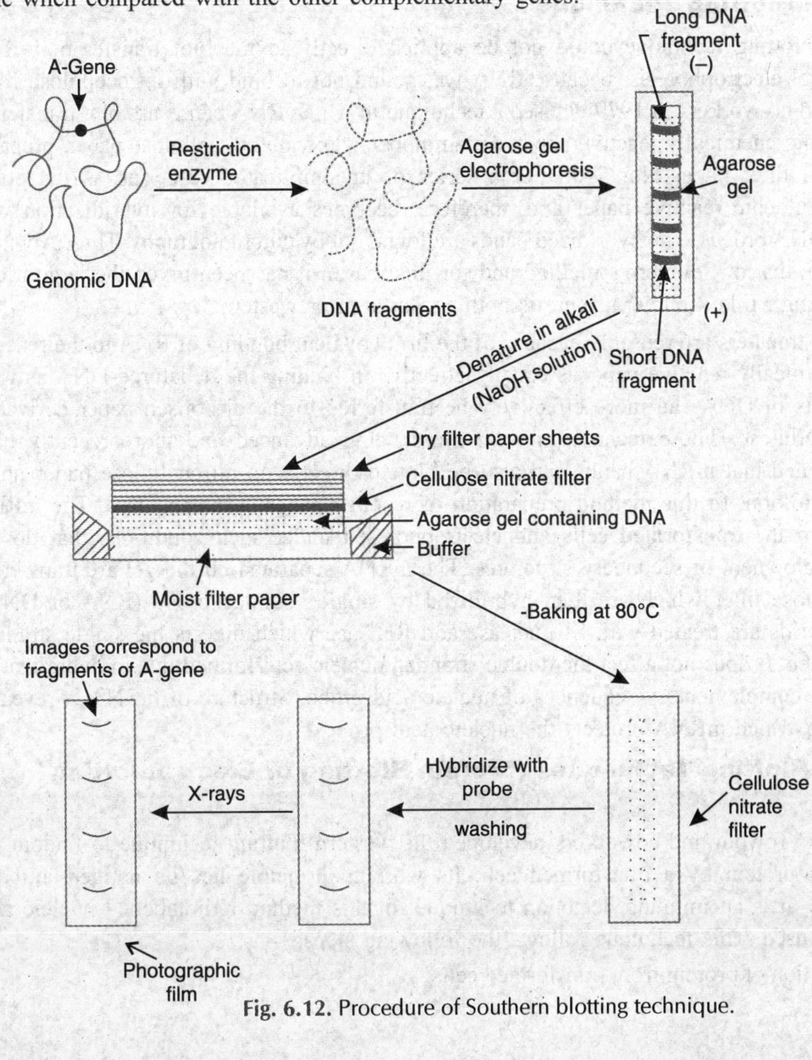


Fig. 6.12. Procedure of Southern blotting technique.

DNA isolated from a given sample is diagested with one or many restriction enzymes (Fig. 6.11). Consequently, DNA fragments of unequal length are produced. This preparation is passed through agarose gel electrophoresis which results in separation of DNA molecules based on their size. DNA restriction fragments present in gel are denatured by alkali treatment. Gel is then put on the top of the buffer saturated filter paper. Upper surface of the gel is covered with nitrocellulose filter and overlaid with dry filter paper. The dry filter paper draws the buffer through the gel. Buffer contains single stranded DNA. Nitrocellulose filter binds DNA fragments strongly when come in contact of it. After baking at 80°C, DNA fragments are permanently fixed to the nitrocellulose filter. Then the filter is placed in a solution containing radio-labelled RNA or denatured DNA probe of known sequences. These are complementary in sequence to the blot-transferred DNA.

The radiolabelled nucleic acid probe hybridizes the complementary DNA on nitrocellulose filter. The filter is thoroughly washed to remove the probe. The hybridized regions are detected autoradiographically by placing the nitrocellulose filter in contact with a photographic film. The images show the hybridized DNA molecules. Thus, the sequences of DNA are recognized following the sequences of probe.

## 2. Northern Blotting Technique

Southern blotting technique could not be applied directly to the blot transfer of mRNA separated by gel electrophoresis, because RNA was found not to bind with nitrocellulose filter. J.C. Alwine and co-workers in 1979 devised a technique in which RNA bands are blot transferred from the gel onto chemically reactive paper. An aminobenzyloxymethyl cellulose paper, prepared from Whatman filter paper No. 540 after a series of uncomplicated reactions, is diazotised and rendered into the reactive paper and, therefore, becomes available for hybridization with radiolabelled DNA probes. The hybridized bands are found out by autoradiography. Thus, Alwine's method extends that of Southern's method and for this reason it has been given the jargon term 'northern blotting'. Like Southern, there is nothing northern or western.

These blot transfers are reusable because of the firm covalent bonding of RNA to the reactive paper. The chemically reactive paper is equally effective in binding the denatured DNA as well. Small fragments of DNA can more effectively be transferred to the diazotised paper derivative than to nitrocellulose. These techniques were more being advanced and more recently have been demonstrated that mRNA bands can also be blotted directly on nitrocellulose paper under appropriate condition. In this method preparation of reactive paper is not required. The mRNA is isolated from the transformed cells and electrophoresed under such conditions that do not permit the development of secondary structures. The mRNA separated on the gel are transferred onto nitrocellulose filter which are then hybridized by single stranded probe (RNA or DNA). Thereafter, hybrids are treated with SI nuclease and RNAase which digests the single stranded RNA/DNA probe. It does not affect the double stranded nucleic acid formed due to hybridization of RNA by the complementary sequences of nucleic acid probe. Structure of mRNA is revealed to the extent to which mRNA protects the nucleic acid probe.

## 3. Western Blotting Techniques (Protein Blotting or Electroblotting Technique)

In 1979, H. Towbin and coworkers developed the western blotting technique to findout the newly encoded protein by a transformed cell. Its working principle lies on antigen-antibody reaction; hence, it is an immuno detection technique. In this method radiolabelled nucleic acid probes are not used. This technique follows the following steps:

(*i*) Extraction of protein from transformed cells.

(*ii*) Separation of protein by using SDS-PAGE (sodium dodecyl sulphate polyacrylamide gel electrophoresis) where SDS acts as solvent for electrophoresis.

(*iii*) Transfer of electrophoresed gel in a buffer at low temperature (40°C) for half an hour.

(*iv*) Blotting of proteins onto nitrocellulose filter paper.

(*v*) Soaking of nitrocellulose filter, Whatman filter and coarse filter in transfer buffer.

(*vi*) Placing of Whatman filter paper on a cathode plate followed by stack of coarse filter, Whatman filter, electrophoresed gel, nitrocellulose filter, Whatman filter paper, coarse filter stack, Whatman filter and anode plate (Fig. 6.13).

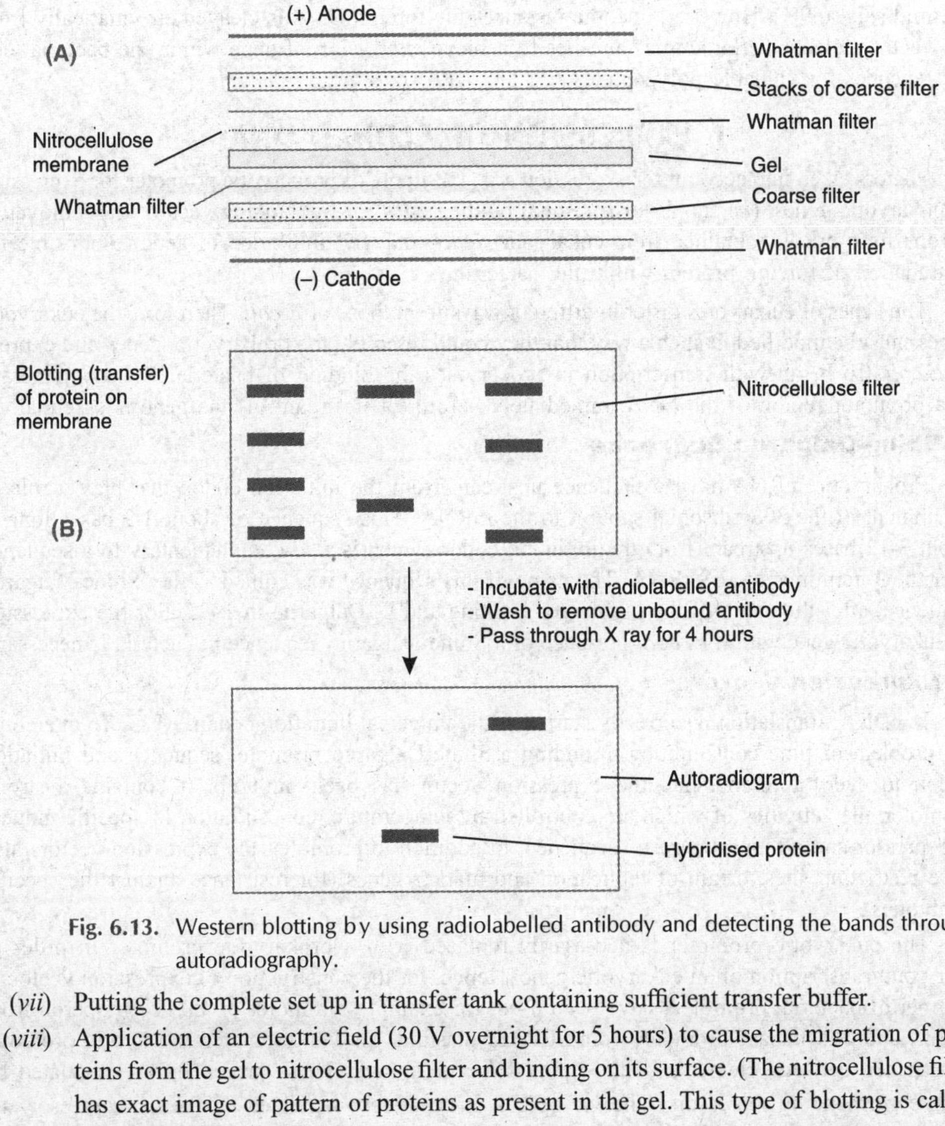


**Fig. 6.13.** Western blotting by using radiolabelled antibody and detecting the bands through autoradiography.

(*vii*) Putting the complete set up in transfer tank containing sufficient transfer buffer.

(*viii*) Application of an electric field (30 V overnight for 5 hours) to cause the migration of proteins from the gel to nitrocellulose filter and binding on its surface. (The nitrocellulose filter has exact image of pattern of proteins as present in the gel. This type of blotting is called western blotting.)

(*ix*) Hybridization of proteins by using radiolabelled antibodies ($I^{123}$- antibodies) of known structure, isolated from the rabbit.

(*x*) Washing of nitrocellulose filter with a wash solution (Tris-buffered saline + Tween 20) to facilitate the removal of unhybridised antibodies.

(*xi*) Detection of hybridized sequences by autoradiography. The dots of diagram shows the presence of desired protein.

## H. RECOVERY OF CELLS CONTAINING RECOMBINANT PLASMID

Colonies complementary to nitrocellulose signal are recovered from the master plate and cultured in nutrient broth in order to get the sufficient bacterial clones containing recombinant plasmid (Fig. 6.9G). However, in some instances the foreign DNA is cleaved enzymatically from the vector molecules. Sometimes aim lies to achieve expression of gene within the bacterial cell if it encodes for valuable polypeptides.

## I. EXPRESSION OF CLONED DNA

Factors taken into account for expression are : (*i*) supply of prokaryotic promoter for expression of eukaryotic genes, (*ii*) supply of ribosonal binding sites for the cloning vector, (*iii*) removal of introns from mRNA obtained from eukaryotic genes and, (*iv*) inhibition of gene responsible for degradation of foreign protein within the bacterium.

The genes of eukaryotes differ in different ways from those of *E.coli.* Therefore, the eukaryotic genes must be modified in such a way that they could resemble the prokaryotic, genes and express in *E.coli.* To begin with transcription in prokaryotes the binding of bacterial RNA polymerase to a promotor region of the DNA immediately before (or upstream of) the gene is essential.

### 1. Shine-Dalgarno Sequence

Prokaryotic mRNA have a sequence upstream from the initiation codon that play a role in attachment of the 30S ribsomal subunit to the mRNA. This sequence of about 3-9 bases long is about 3-12 bases upstream from the initiation codon which is also complementary to a sequence near the 3′ terminus of 16S rRNA. The name of this sequence was coined as the 'Shine-Dalgarno sequence' after the name of the discoverer J. Shine and L. Dalgarno in 1975. For the expression of eukaryotic gene within *E. coli,* presence of a Shine-Dalgarno sequence in bacteria is necessary.

### 2. Expression Vectors

In cDNA, translation is correctly started at the universal initiation codon, AUG. To overcome the problem of time consumption in adding a Shine-Dalgarno promoter sequence and initiation codon to every foreign gene, the expression vector has been built up. It contains a strong promoter, the activities of which are controlled by temperature, concentration of specific inducer or repressor and a Shine-Dalgarno sequence. In addition to promoter, the expression vectors also have restriction sites, origin of replication and marker genes (for resistance against the specific antibiotics).

The eukaryotic promoter is necessarily replaced with a prokaryotic promoter in order to carry out transcription of an eukaryotic gene. Hence, for the construction of expression vectors a number of bacterial promoters have been used, for example, promoter or *lac* or *trp* operon from *E. coli* and ß-lactamase promoter from pBR322 plasmid. The *tac* promoter is a hybrid promoter constructed between elements of the *trp* and *lac* promoters. These promoters are regulated by their corresponding repressors. The *lac* operon is negatively regulated by the *lac* repressor and can be depressed by addition of an inducer, *i.e.* isopropyl ß-D-thiogalactosidase (IPTG). It is positively regulated by the catabolic activator protein (CAP) complexed with cyclic AMP pathway. Similarly, the *trp* operon encodes 5 enzymes for the biosynthesis of tryptophan. The complex formed between the *trp* repressor and tryptophan negatively control the *trp* operon. When the amino acid is in abundant supply the genes responsible for its biosynthesis are repressed and do not work.

But, the eukaryotic proteins are not always stable in prokaryotic cells; it is disintegrated by the bacterium. Protection of protein is necessary. When a foreign gene is inserted in a prokaryotic gene, a hybrid polypeptide is produced. It contains a part of the prokaryotic product attached to *N*-terminus of the foreign polypeptide. The extra sequence of *N*-terminus may stabilize the polypeptide and may also interfere in the right functioning of the foreign protein. Therefore, it is necessary to separate the hybrid protein by chemical cleavage at methionine.

## J. DETECTION OF NUCLEIC ACIDS

There are two approaches for labelling the nucleic acids to be used as DNA probe: radioactive labelling and non-radioactive labelling.

### 1. Radioactive Labelling

Methods for *in vitro* introduction of radioactivity in nucleic acids was developed during 1970s through metabolic labelling. In this method large amount of radioactivity was required. However, this method was time consuming and laborious. The methods developed later on utilised phage T4 polynucleotide kinase (which transfers $\gamma$-$PO_4$ of ATP to 5′-OH terminus in DNA or RNA) or *E. coli* DNA polymerase I (which replaces a normal nucleotide of double stranded DNA with $\alpha$ -$^{32}PO_4$labelled nucleotides). The nucleic acids are labelled *in vitro* by using the following methods:

**(*a*) Nick Translation:** This method is used to label the double stranded DNA probes. If there is no nick, a nick is made on double stranded DNA by using DNase (in a buffered solution containing dNTPs) which exposes 3′-OH group on one strand of DNA. All the four nucleotides are radiolabelled ($\alpha$-$^{32}$P-dNTPs) at alpha position of phosphate group.

Only one nucleotide *i.e.* $\alpha$-$^{32}$P-dCTP is generally radiolabelled, whereas the other nucleotides are unlabelled. *E. coli* DNA polymerase I is used in this method. This enzyme shows 5′ $\rightarrow$ 3′ exocatalytic activity and either removes or hydrolyses the nucleotides from 5′ side of the nick. Nucleotides are eliminated from 5′ end and simultaneously added at 3′ side resulting in movement of the nick along the DNA. Therefore, this method is called nick translation (Fig. 6.14).

The reaction mixture includes probe DNA to be labelled, $\alpha$-$^{32}$P-dCTP, nick translation buffer, the other three dNTPs, DNase 1 and *E. coli* DNA polymerase I. It is incubated at 14°C for 2-4 hours. The hybridised DNA is separated from the rest of dNTPs and used to hybridise the DNA.

**(*b*) Random Primed Radiolabelling of Probes:** A.P. Feinberg and Vogelstein (1983, 1984) developed this technique. The restriction fragments purified by gel electrophoresis from denatured linear or circular DNA are used to prepare probes. In this method primers are used for DNA synthesis but DNA nicking is not done. The purified DNA to be labelled is mixed in a buffer and heated at 100°C so that it may be denatured. Then it is put on ice to lower down the temperature. The reaction mixture (containing Klenow fragment of *E. coli*, DNA Pol I, one radiolabelled nucleotide *e.g.* $\alpha$-$^{32}$P-dCTP and three unlabelled nucleotide triphosphates) is incubated at 37°C for 30 minutes to 16 hours. Since, Klenow DNA Pol I lacks 5′ $\rightarrow$ 3′ exonuclease activity, the radiolabelled probe is synthesised by primer extension but not nick translation as described earlier. The hybridised DNA is separated from unincorporated nucleotides and used as probe for hybridisation.

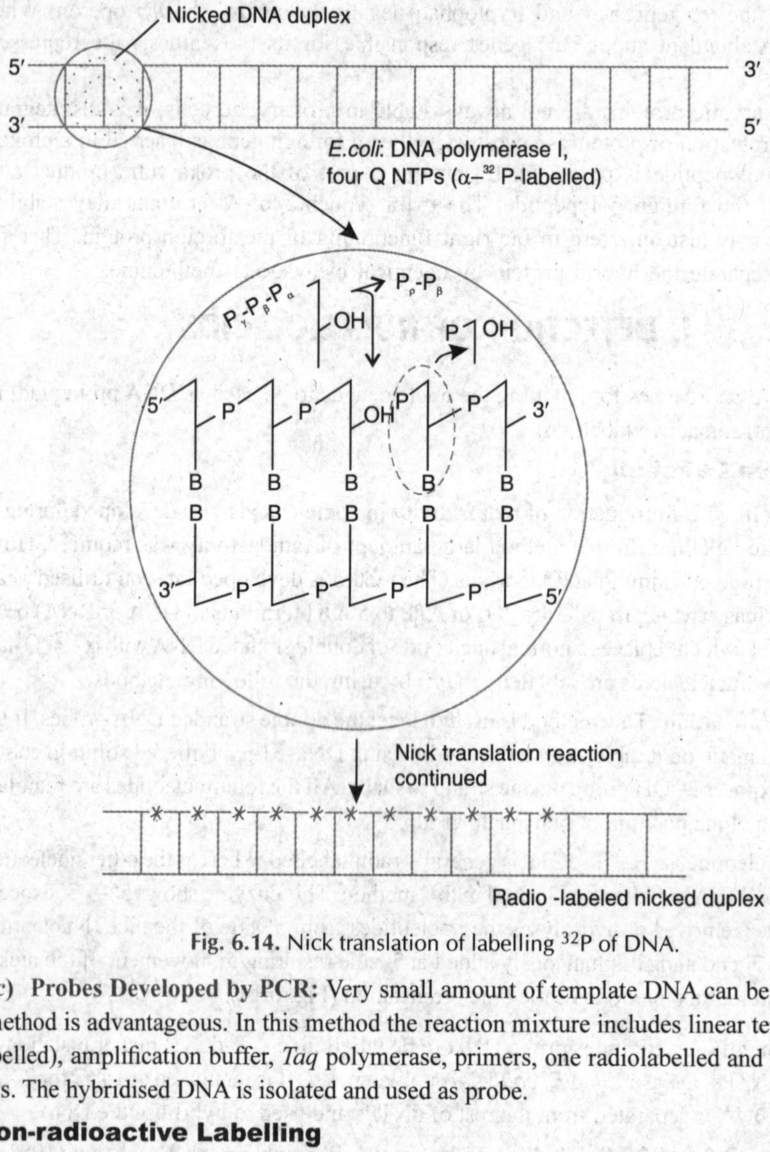


Fig. 6.14. Nick translation of labelling $^{32}P$ of DNA.

(*c*) **Probes Developed by PCR:** Very small amount of template DNA can be used; therefore, this method is advantageous. In this method the reaction mixture includes linear template DNA (to be labelled), amplification buffer, *Taq* polymerase, primers, one radiolabelled and three unlabelled dNTPs. The hybridised DNA is isolated and used as probe.

## 2. Non-radioactive Labelling

The non-radiolabelled DNA probes are also used in detection of desired DNA. There are three systems of non-radioactive routes such as using horseradish peroxidase, biotin probe and steptavidin probes as described below :

(*a*) **Horseradish Peroxidase (HRP) Method:** The HRP is a plant-derived enzyme which is covalently linked to DNA. Chromogenic substrates (such as chloronaphthol) or chemiluminescent substrates (that emit light after reacting with enzyme *e.g.* luminol) are used for detection of HRP-linked DNA. Chloro-naphthol yields an insoluble purple product in the presence of peroxide and HRP. Consequently, presence of DNA is detected. Similarly, HRP oxidises luminol and emits luminescence by using X ray film; hence luminescence can be detected.

(*b*) **Digoxigenin (DIG) Labelling System:** Digoxigenin is a cardenolide steroid derived from a plant *Digitaria* as a hapten. The presence of digoxigenin can be detected by an antibody associated with an enzyme (anti-digoxigenin-alkaline phosphatase conjugate). The probe is labelled with digoxigenin (11)-dUTP which is a nucleotide triphosphate analog. This analog containing digoxigenin moiety is incorporated into DNA by nick translation or random prime labelling. The DNA acts as probe. Anti-DIG (antibody to digoxigenin) conjugated with alkaline phosphatase detects the DIG-labelled probe through enzyme-linked immuno assay. A chromogenic substrate [such as 5-bromo-4-chloro-3-indolyl phosphate (BCIP) and nitrobutane tetrazolium chloride (NBT)] is added which reacts with alkaline phosphatase and produce purple/blue colour (Fig. 6.15).

A chemiluminescent substrate such as dioxetane is also used with digoxigenin or biotin-streptavidin systems. There are several derivatives of digoxigenin which emit light after enzymatic activation by alkaline phosphatase. The denatured DNA present on Southern blots is hybridised by labelled probe. The membrane is placed in detection buffer containing antidigoxigenin (20 μg/ml) and bovine serum albumin (BSA) (5% w/v) and incubated at 37°C for 1 hour. After incubation membrane is washed thrice. Enzyme activity of alkaline phosphatase is detected by using BCIP (0.17 mg/ml) and nitrobutane tetrazolium chloride (NBT) (0.33 mg/ml) as dye. Purple/blue colour develops after enzymatic activation.

(*c*) **Biotin-Streptavidin Labelling System:** Biotin is vitamin H and avidin is found in egg white. Principle of this system is based on the interaction between biotin and the glycoprotein avidin. A biotin-containing nucleotide analog (biotin-14dATP) or probe is prepared through nick translation or random priming methods. Streptavidin is conjugated to alkaline phosphatase. The biotinylated

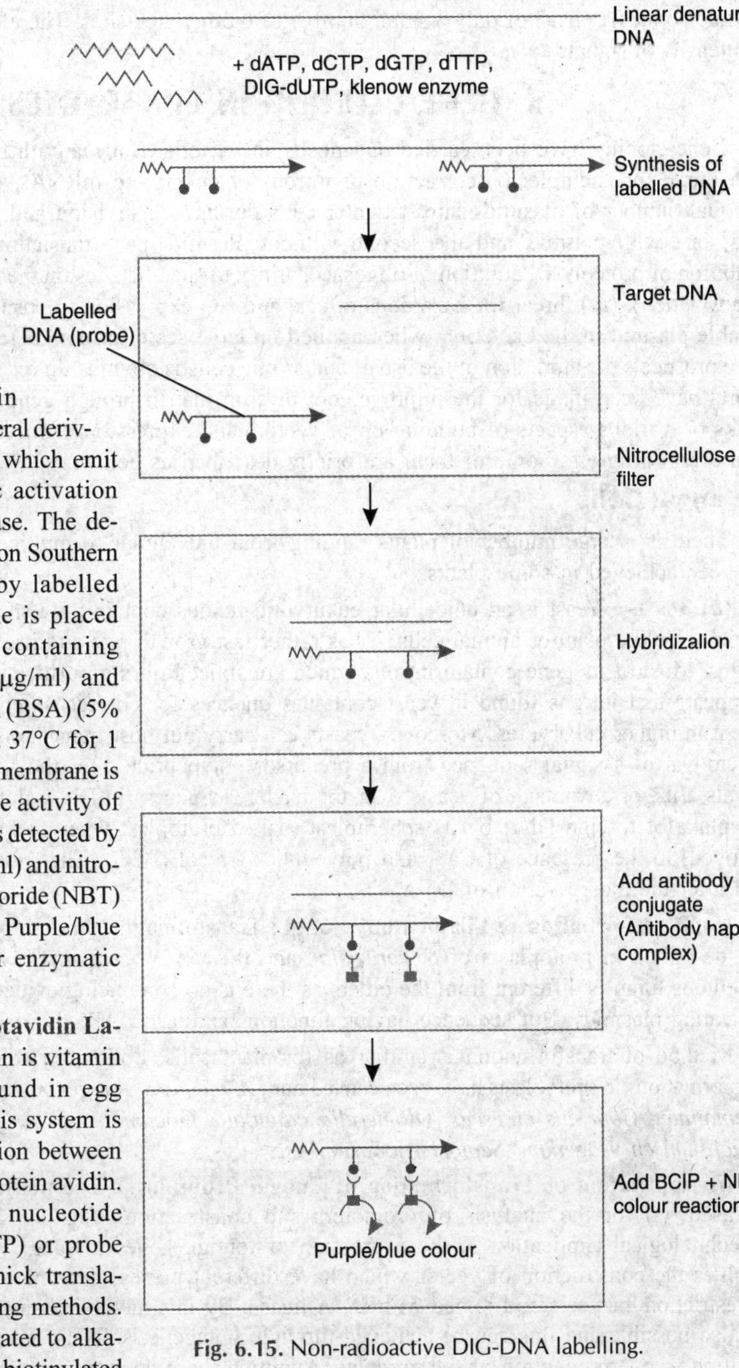


**Fig. 6.15.** Non-radioactive DIG-DNA labelling.

probes are used to hybridise the specific sequence of DNA. The hybridised biotinylated DNA is detected by specific binding of streptavidin-conjugated alkaline phosphatase. Thereafter, a chromogenic or chemiluminescent substrate is added that reacts with alkaline phosphatase and detects the hybrids. The intensity of colour is proportional to the amount of biotin present in hybrid DNA. This method detects the hybrids much faster than the radioactive probes. But very small number of probes (20 nucleotides) consist of only a small number of biotinylated sites. Therefore, there is limitation of intensity of signals.

## K. GENE CLONING IN EUKARYOTES

Gene cloning have been carried out mostly in bacteria. But many difficulties are associated with them, for example: (*i*) correction of introns of eukaryotic mRNAs, (*ii*) failure of transfer of equal number of plasmids into daughter cells during cell division and yield of two types of cells, one with plasmid and the second without plasmid, post-translational modification (*e.g.* inhibition of proteolysis, addition of oligosaccharides to specific sites on the polypeptide chain, the glycosylation), (*iii*) threat for hazardous effects and (*iv*) expression to ensure for the presence of a stable plasmid in the bacterium when applied on large scale for commercial applications. With these prospects the attraction of the use of eukaryotic cells is obvious. In recent years, considerable efforts have been made for the improvement of crop plants through genetic engineering. Many works on various aspects of building up of vectors and expression of insert genes in eukaryotic cells are in progress. Some of them are briefly described as below:

### 1. Fungal Cells

There is a large number of plants ranging from unicellular to multicellular forms. Success has been achieved in some plants.

**(*a*) Yeasts :** Yeast is an unicellular eukaryotic fungus containing a small well characterized genome. Unlike plant or animal cells, it has rather fast growth rate and itself is a non-pathogenic fungus. Most of its gene contain introns which are spliced during purification of mature mRNA. It appears that introns found in yeast contain sequences for correct splicing as they are totally absent in higher eukaryotes. Moreover, yeasts can carry out post-translational modifications such as removal of a signal sequence from a precursor polypeptide after the secretion of cells. This reveals a major advantage of yeasts over the bacteria. Success in DNA cloning in yeast depends on uptake of foreign DNA by its spheroplast (naked protoplast when cell wall is enzymatically removed) in the presence of $Ca^{++}$ and polyethyleneglycol (PEG). The spheroplasts develop cell wall after the incorporation of DNA.

**(*b*) Transformation in Filamentous Fungi :** Transformation through plasmid of *E. coli* was first described in protoplasts of *S. cerevisiae* and then in *N. crassa*. However, the situation in filmentous fungi is different from the others as there is no convincing evidence for autonomously replicating plasmids. But sequence having functions similar to ARS have been isolated.

Method of transformation is similar as for plant cells. Following are the examples where transformations in protoplasts have been carried out : *A. nidulans, A.niger, A. oryzae, Cephalosporium acremonium, Coprinus cinereus, Glomerella cingulata, Gaeumannomyces graminis, Mucor* sp., *Penicillium chrysogenum, Septoriu nodorum, etc.*

**(*i*) Application of Transformation of Fungal Protoplasts.** The transformation approaches are used: (*i*) for the analysis of promoters and construction of expression vectors, and (*ii*) biotechnological application in the expression of foreign genes in fungi. Analysis of promoter involves the construction of vector which have different pieces of promoters. These can be cut by restriction enzymes and joined with DNA ligase. By this method changes in promoter region are also possible and this can be achieved through mutagenesis. Thus, a strain can be improved resulting in overproduction of the products. Adopting these approaches many mammalian genes have now been expressed in *S. cerevisiae* and aspergilli through expression vector (*i.e.* vector which include sequences promoting high level of transcription of the gene and secretion of the gene products).

In the U.S.A. and Europe several companies have demonstrated the expression of foreign DNA introduced in *Aspergillus* sp. through transformation. A. Upshall *et al.* (1987) have constructed a plasmid pM159 for the secretion of human tissue plasminogen activater by *A. nidulans*. Yield of this protein was about 100 mg/litre culture filtrate. Similarly, D.M. Gwynne and collaborators in 1987 found the secretion of human interferon by transformed *A. nidulans* mycelia to about 1 mg/litre.

## 2. Plant Cells

There are large number of plants which have been selected for transformation of cells by using different approaches.

***(a)* Gene Transfer in Dicots by using *Agrobacterium* Ti-DNA as Vector:** Transfer of Ti-plasmid into plant cell is not possible because of its large size, most probably upto 235 kb. Therefore, T-DNA is excised and used as foreign DNA for expression in plant cells. From bacteriological point of view opine production and development of gall are not required for the successful integration and expression of DNA. Hence, replacement of *nos* and *ops* genes with a foreign DNA fragment retaining the opine synthase promoter for expression of inserted DNA fragment has no harm. In 1983, J. Schell and M. Van Montague undertook some experiments where genes for resistance to kanamycin and methotrexate expressed in cultured callus cells. Generally it is easy to use opine promoters for the expression of foreign DNA. It has become possible that the inserted genes, like other plant genes, will express only at certain stages or specific tissues of plants when they are with specific regulated promoters. There are reports of replacement of opine synthase promoters by regulatory promoters obtained from plants.

N. Murai and coworkers in 1983 have introduced gene for bean plant storage protein, the phaseolin (which is supplemented with methionine codon), into the cells of sunflower plant after transformation with Ti-plasmid where phaseolin gene was under the control of its own promoters, the octopine synthase. For detail see Chapter 13.

***(b)* Gene Transfer in Monocots:** The Ti-plasmid is specific only to the dicot plants. *A. tumefaciens* does not induce tumour in monocot plants, however, most of the major crops including cereals are monocots. GM.S. Hooykaas Van and coworkers in 1984 discovered that *A. tumefaciens* could transfer T-DNA into certain monocots resulting in expression of opine gene within the plant cell without inducing the tumour. This discovery made it possible that T-DNA can be expressed into cells of monocots.

***(c)* Plant Cell Transformation :** Plant cell suspension culture coupled with DNA transfer techniques has become a new field of gene manipulation in plants. Cell suspension cultures have been the traditional source of plant material for biochemical selection of plant cell mutants, such as numerous cell lines resistant to various amino analogs. Suspension of protoplasts constitutes an identical plant material from one laboratory to other. The approaches made for plant protoplast transformation are :

(*i*) Introduction of DNA sequence via integration into T-DNA of Ti-plasmid of *A. tumefaciens*,

(*ii*) Transfer of naked DNA into protoplast including bacterial spheroplasts and microinjection, and

(*iii*) Transformation with plant viruses carrying inserts *e.g.* gemini and caulimoviruses.

The plant protoplast treated with PEG, used for protoplast fusion, resulted in efficient delivery of DNA to protoplasts. The PEG method has been employed for the successful transformation of protoplasts with Ti-plasmid.

Once DNA fragment is introduced into plant cells, it integrates with plant chromosomes and expresses from its signal. Transformation of plant cells with a gene for drug and antibiotic resistance linked to plant promoter has been possible. The gene could show expression in plant cell. Therefore, the transformed cells could be selected from the culture medium amended with drugs and antibiotics. The plants regenerated from cells show drug/antibiotics resistance.

**(*i*) Plant cell transformation by ultrasonication :** Ultrasonication has been done for successful gene delivery. The Biotechnology Research Centre, Beijing (China) has used this technique for gene transfer in plants like wheat, tobacco and sugarbeet. When the cultured explants were sonicated with plasmid DNA carrying marker genes such as *cat, npt*II, and *gus,* and sonicated calli transferred to selective medium, shoots were grown successfully. The calli in control set of experiment which were not sonicated with plasmid did not grow on selective medium but they died. Transgenic tobacco plants were obtained at 22% frequency.

**(*ii*) Liposome-mediated gene transfer :** Liposomes are small sized particles containing a phospholipid bilayers which are concentric in nature. Liposomes enclose aqeuous chamber and can entrap water soluble molecules. Therefore, they are called lipid bags. Many plasmids are enclosed in them. By using polyethylene glycol (PEG) they may be stimulated to fuse with protoplast. In several plants like carrot, tobacco, petunia, etc. this technique has been used for successful transfer of genes. Due to endocyclosis of liposomes, DNA enters the protoplast. It gets adhered first to protoplast surface and get fused with protoplast at the site of surface. Consequently plasmids are released inside the cell. There are several advantages in using this technique such as: (*i*) low toxicity, (*ii*) protection of DNA/RNA from nucleases that lyse them, (*iii*) long stable storage of DNA fragments in liposome, (*iv*) applicability in various cell types, and (*v*) high level of reproductivity.

**Sunbean plant.** Scientists at Agrigenetics Corporation (USA) are concentrating on phaseolin gene which is a major protein of french bean. It has low methionin content owing to which food value of the bean is reduced. *Phaseolin* gene was successfully transferred to cultured cells of sunflower. The callus producing bean protein is called as sunbean. The phaseolin gene can be put back into bean plant. By doing so, the nutritional value of the crop can be enhanced.

**(*iii*) Electroporation :** Electroporation (using electric field) is used to transfer the foreign DNA into the fragile cells. Brief pulses of high voltage electricity (about 350 V) is applied to protoplasts suspension containing naked or recombinant plasmids. The electric pulses induce the formation of large pores in the cell membrane. These pores give a passage through which the foreign DNA can enter into the protoplasts and thus, increase the transformation frequency. The transformed protoplasts are cultured for about one month. These develop microcalli which are plated on solid medium containing selective marker (*e.g.* kanamycin). After 37-45 days, the calli are analysed for the presence of differences in transformed cells.

By using electroporation method, genes of choice have been successfully transferred in protoplasts of wheat, rice, petunia, sorghum, maize and tobacco. Moreover, transformation frequency can be improved by using linear DNA instead of circular one, by giving heat sock to protoplast at 45°C for 5 minutes, by adding PEG and using 1.25 kV/cm voltage.

**(*iv*) Particle bombardment gun :** It was developed by Stanford and coworkers of Cornell University (USA) in 1987. As the term denotes, it shoots foreign DNA into plant cells or tissue at a very high speed. This technique is also known as particle bombardment, particle gun method, biolistic process, microprojectile bombardment or particle acceleration. This technique is most suitable for those plants which hardly regenerate and do not show sufficient response to gene transfer through *Agrobacterium* for example, rice, wheat corn, sorghum, chickpea and pigeon-pea etc.

The apparatus consists of a chamber connected to an outlet to create vacuum (Fig. 6.16). At the top, a cylinder is temporarily sealed off from the rest of chamber with a plastic rupture disk. Helium gas flows into the cylinder. A plastic microcarrier is placed close to rupture disk. It contains DNA coated tungsten particle, the microscopic pellets (*i.e.* coated microprojectiles). When to work the apparatus is placed in Laminar flow just to maintain sterile conditions. The target cells/tissue are placed in the apparatus. A stopping screen is put between the target cells and microcarrier assembly. Helium gas is flown in the cylinder at high velocity. When pressure of cylinder exceeds the bursting point of plastic disk, it gets ruptured. Helium shock waves propel the plastic mircrocarrier containing DNA coated micropellets. The stopping screen allows the micropellets to pass through and deliver DNA into target cells. The transformed cells are regenerated onto nutrient medium. The regenerated plant tissues are selected over culture media containing either antibiotics or herbicide. The selected plants are then analysed for expression of foreign DNA. that has been inserted into it. Scientists have got success in delivering foreign DNA into epidermal tissues of *Allium cepa*, scutellar tissues of maize and leaf and cell culture of many crops (Peters, 1993). In addition to bacterial cells, algae, fungi plant organelles (*e.g.* chloroplast and mitochondria), an animal and human cells and fruitfly embryos have been successfully transformed.

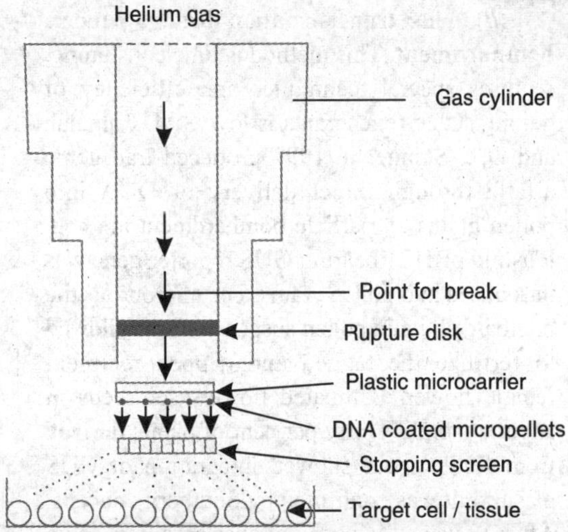


**Fig. 6.16.** Working system of particle bombardment gun.

**Conditions for bombardment.** It requires several conditions required for introduction of foreign gene into the desired cell/tissue/pollen, etc. The conditions required for introduction of β-glucuronidase (GUS) reporter gene (for detail *see* Chapter 13; Section, *Transgenic Plants*) for transformation of Basmati Indica rice have been given in Table 6.1. D. Minhas *et al.* (1996) has reported for the first time the transient expression of GUS reporter gene through biolistic delivery in mature embryo-derived embryonic callus cultures of *Oryza sative* cv Basmati 370. In 1998, scientists of Plant Transformation Group at ICGEB (New Delhi) have got success in transferring human interferon gamma gene into chloroplasts of tobacco, maize, etc. using a particle gum mediated gene transformation.

**>Table 6.1. Bombardment conditions employed on Basmati Indica rice callus.**

| | |
|---|---|
| 1. Distance between rupture disc and microcarrier | 0.5 inch + 15 mm |
| 2. Size of gold particles | 1 μm |
| 3. Distance between microcarrier and target | 6 and 9 cm |
| 4. Density of particles/shot | 9 mg/shot |
| 5. Bombardment medium | MS medium supplemented with 2, |
| 4-D at 2 mg/l (rice callus medium) | |
| 6. DNA concentration | 1 mg/ml |
| 7. Physical parameters | |
| (*i*) helium pressure | 450-1300 pounds/inch$^2$ |
| (*ii*) vacuum | 26 mg Hg |
| 8. Post bombardment culture | On callus medium till GUS assay (48 h) |

*Source* : D. Minhas *et al.* (1996).

**(*ii*) Pollen transformation through particle bombardment.** This method is time consuming, requires special techniques and efficiency of getting stable regenerants is low. S.M. Ramaiah and D.Z. Skinner in 1997 produced transgenic alfalfa through direct delivery of DNA into pollen grains by particle bombardment method. Plasmid pBI121 bearing GUS reporter gene was introduced into pollen grains. The microprojectile bombarded liquid pollen suspension was allowed to fertilize the tagged and tripped recipient female flower. Pollinated flowers set seeds in about a month. Thirty per cent of plants derived from fertile seeds showed integration of GUS plasmid. It was confirmed by Southern analysis. After 10 vegetative generations some transgenic plants lost the integrated GUS plasmid, whereas in few others number, copies of GUS gene decreased due to unknown reasons. An outline of production of transgenic alfalfa plant through pollen transformation is given in Fig. 6.17.

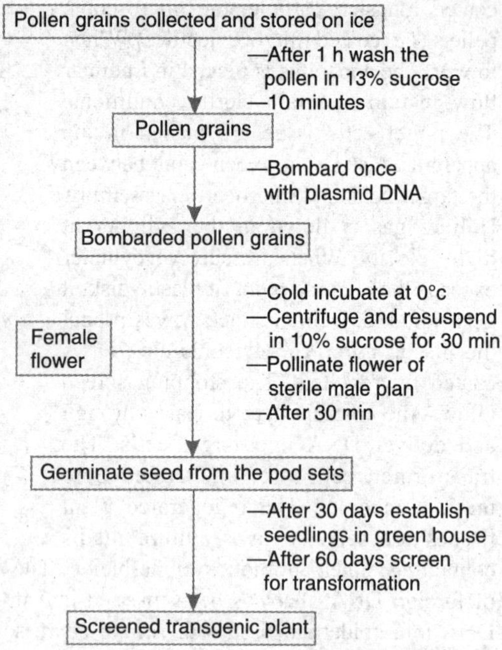


**Fig. 6.17.** Outline of pollen transformation to obtain transgenic alfalfa plants.

## 3. Animal Cells

Animal cell culture on a large scale is more difficult and expensive than plant cells, the bacteria, yeast, etc. This is why the commercialization of natural products of animal origin is limited. However, most of the animal products have been produced by using bacteria as biological tool.

In India, K. Chandrashekharan, Centre for Cellular and Molecular Biology (CCMB), Hyderabad emphasized the use of recombinant DNA technology. Introduction of foreign DNA into mammalian cells. This will help in curing certain genetic defects. Some of the examples of gene cloning in animals are described in Chapter 7.

**(*a*) Animal Viruses :** Several animal viruses are known, for example, simian virus 40 (SV40), adinovirus, retrovirus (ssRNA), vaccinia virus, etc. They can increase the efficiency of animal cell transformation if used as vector. A special feature of viruses is that they contain the strong promoters which can possibly bring about expression of inserted DNA fragment of the viruses. The most commonly used virus is SV40 which contains a circular DNA of about 5.2 Kb. In addition to *ori* region, the DNA contains early genes and late genes. Early genes are required for DNA replication, whereas late genes encode coat protein.

The amount of DNA to be packed into virus capsid is limited. The genome has three identified non-essential regions; hence the use of non-defective viral vectors is limited. The defective virus vectors have missing genes whose infection process has to be helped by viral DNA constituting functional copies of missing genes. If the *cos* cells (the host monkey cells) are used as host, they can be transformed by a plasmid carrying the early region of SV40 with a defective replication origin. *cos* cells express wild type large T antigen and can support the growth of SV40. Recombinants containing foreign DNA inserted into early region can be propagated without the need of helper virus. Thus the recombinant and helper can be separately recovered from the host.

**(*b*) Microinjection :** Foreign DNA can be delivered into a living cell (a cell, egg, oocyte, embryos of animals) through a glass micropipette. This techniqaue is called microinjection. One end of a glass micropipette is heated until the glass becomes some what liquified. It is quickly stretched which forms a very fine tip at the heated end. The tip of the pipette attains to about 0.5 mm diameter which resembles an injection needle. The process of delivering foreign DNA is done under a powerful microscope. Cells to be microinjected are placed in a container (Fig. 6.18). A holding pipette is placed in the field of view of the microscope. The holding pipette holds a target cell at the tip when gently sucked. The tip of the micropipette is injected through the membrane of the cell and contents of the needle are delivered into the cytoplasm.

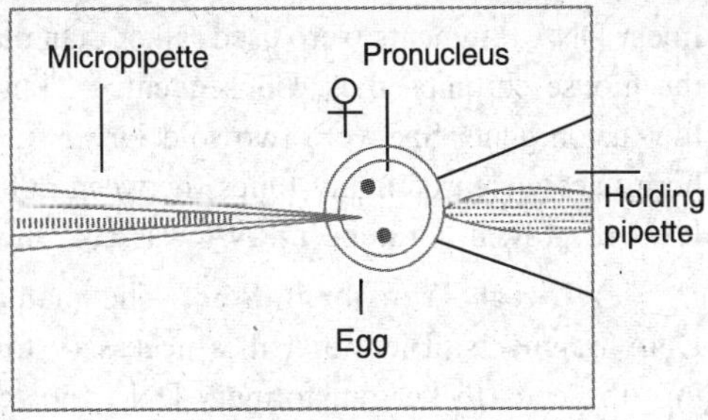


**Fig. 6.18.** A method of microinjection of DNA preparation in egg.

For the study of transcription by microinjection *Xenopus* oocytes have been widely used because oocytes contain between 6,000 and 100,000 or more RNA polymerase molecules than somatic cells. Microinjection is technically easy because of large size of oocytes. Some of the endogenous pattern of gene regulation during development has been characterized. The injected DNA integrates randomly with nuclear DNA and its expression could be possible only when the foreign DNA is attached to a suitable promoter sequence. There are many examples where different types of animal cells have been microinjected successfully.

G. Rubin and A. Spradling in 1982 for the first time introduced *Drosophila* gene for xanthine dehydrogenase into a P-element (parental element) which was microinjected with an intact helper P-element into embryo deficient for this gene. Such embryos later on developed flies with rosy coloured eyes but not mosaic eyes as in the first parental generation.

**Production of Transgenic Animals.** In 1982, R.D. Palmiter of Washington University and R.L.Brinter of Pennsylvanian University isolated the rabbit growth hormone (β–globin) gene, human growth hormone (β-globin) gene as well as thymidine kinase gene and linked separately to the promoter region of mouse associated with the metallothionein I gene (a gene which encodes a metal binding protein). This was joined to pBR322 plasmid to produce the recombinant plasmids. Mature eggs from adult mouse were recovered surgically and fertilized with sperms *in vitro*. Immediately the fertilized eggs were microinjected with recombinant plasmids before the sperm and egg nuclei have fused to form a diploid zygote. The plasmids generally combine homologously with each other within the egg forming a long repeated concatemer which then integrates randomly to give repeated genes at a single chromosomal site. The engineered embryos were then implanted into the uterus of a host mouse mother for further development. The resulting mice are called 'transgenic mice' since part of genome comes from another genetically unrelated organism. Due to introduction of foreign gene before nuclear fusion, chromosomal integration takes place early and progeny contains new genes. Size and body weight of progenies were extremely larger than the normal ones (Fig. 6.19).

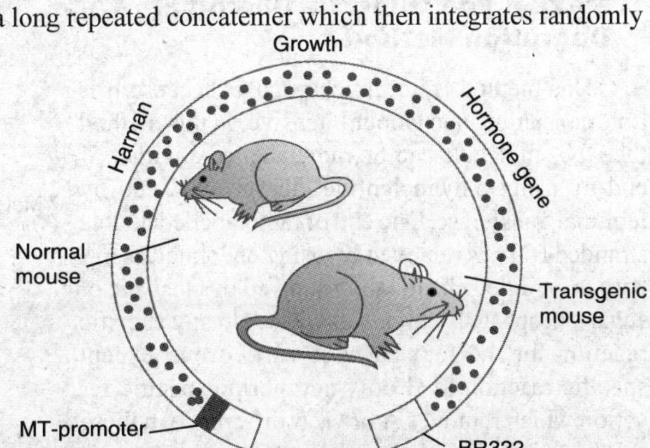


**Fig. 6.19.** Production of transgenic mice: MT, methionine.

Further Palmiter injected mouse embryos with a DNA fragment containing the rat growth hormone gene fused to the promoter region of the mouse metallothionein I gene. In this experiment, linear DNA fragments were used rather than plasmids because these integrate more efficiently into the mouse chromosomes. Consequently, 21 mice were produced, among them seven contained the fusion gene. Six were two fold larger in body weight than the others. The level of growth hormones increased many times (between 200-800 times) more than the control. In addition, the level of growth hormone mRNA was also increased in liver cells.

**(*e*) Direct Transformation :** The mammalian cells can be transformed by the foreign DNA fragments. Therefore, it is necessary to precipitate the DNA with calcium phosphate and mix the cells to be transformed. DNA molecule passes through cell membrane and integrates randomly with mammalian chromosome. Using this technique selective marker can be linked up with DNA fragment to be cloned and can be introduced into mammalian cells. The transformed cells are, thereafter, separated from cell line after plating them on selective medium. Many techniques are used to recover the selectable marker gene from the biochemically transformed cell line. Following these biochemical processes. M. Perucho *et al.* (1980) successfully isolated chicken thyamidine gene (*tk* gene) by ligating with pBR322 and transforming the $tk^-$ (deficient of thyamidine gene) mouse.

## L. DNA (GENE) SEQUENCING

In 1965, Robert Holley and his research group at Cornell University completely sequenced nucleotides of $tRNA^{ala}$ (tRNA for yeast alanine). In 1977, the following two methods were developed. Allan Maxam and Walter Gilbert developed a chemical method of DNA sequencing. In this method, end-labelled DNA is subjected to base specific cleavage reaction before gel separation. In routine sequencing of DNA this method is not commonly followed. In the same year. Frederick Sanger and co-workers developed an enzymatic method of DNA sequencing. It is also called *dideoxynucleotide chain termination method* because dideoxynucleotides are used as chain terminator to produce a ladder of molecules. There are four types of deoxynucleotide triphosphates (dNTPs) such as dATP, dCTP, dGTP and dTTP. Similarly, there are four types of dideoxynucleotide triphosphates (ddNTPs) for example, ddATP, ddTTP, ddGTP and ddCTP.

### 1. Maxam and Gilbert's Chemical Degration Method

This method is not much popular because it is time consuming and labour intensive. In this method the DNA molecule can be radiolabelled at either 5' end by using polynucleotide kinase, or 3' end by terminal transferase. One end of radiolabelled double stranded DNA is removed by using endonuclease. A base is modified chemically followed by cleavage of sugar-phosphate backbone of DNA. No any specific reaction for the four bases is carried out, except specific reaction to G only and purine specific reaction which removes A or G. A difference in these reactions indicates the presence of A.

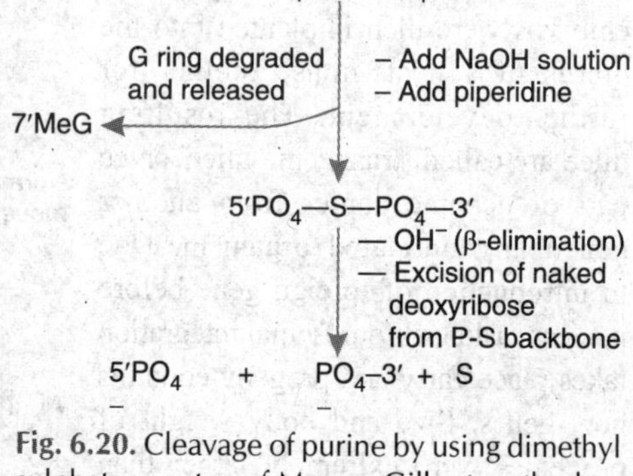


Fig. 6.20. Cleavage of purine by using dimethyl sulphate— a step of Maxam-Gilbert method.

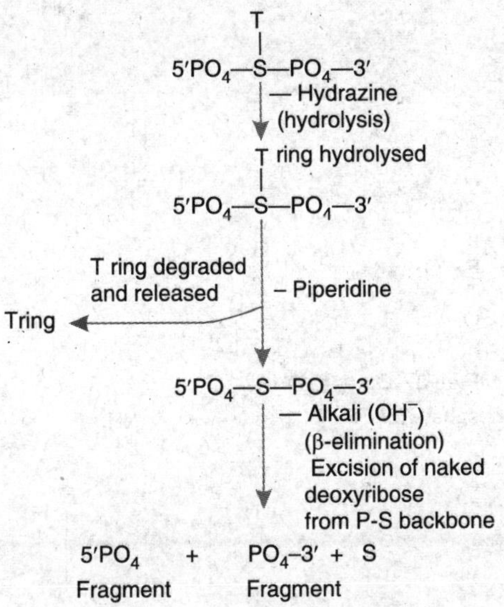


**Fig. 6.21.** Cleavage of pyrimidine by using hydrazine— a step of Maxam-Gilbert method.

*(a)* **Cleavage of Purine:** The mixture is separated in four sets, each treated with a different reagents which degrade only G or C or A and G or C and T. In one set, DNA is treated with acid followed by dimethyl sulphate. This causes methylation of A (at 3' position and G (at 7' position). Subsequently, addition of alkali ( OH) and piperidine results in cleavage of DNA and removal of purines (A or G) (Fig. 6.20).

*(b)* **Cleavage of Pyrimidine:** Similar to the cleavage of purine, pyrimidine (C or T) is also cleaved in the presence of 1-2 M NaCl solution. It works only with C. Differences between these two indicate the presence of T in the DNA sequence. Cleavage of pyrimidines (C or T) through hydrazine hydrolysis is given in Fig. 6.21.

Partial chemical cleavage of DNA fragments as done above generates the populations of radiolabelled molecules extending from radiolabelled terminus to the site of chemical cleavage (Fig. 6.22). These fragments are of different sizes that represent unique pairs of 5' and 3' cleavage products in the random collection. A complete set is formed by these products the length of each number is short by one nucleotide. These can be separated by gel electrophoresis. The fragments containing labelled terminus can be observed by autoradiography of the gel. Following the order of fragments obtained from different digestions the sequence of nucleotides is deduced and interpreted (Fig. 6.22).

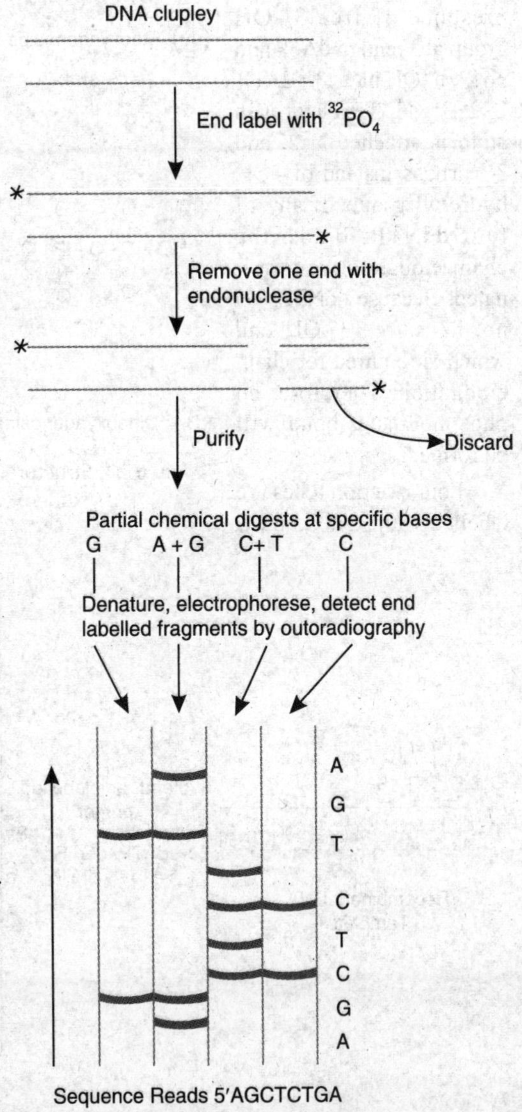


**Fig. 6.22.** DNA sequencing by Maxam and Gilbert method.

## 2. Sanger and Coulson Method (Dideoxynucleotide Chain Termination Method)

Twice Nobel Prize winner, Frederick Sanger and coworkers in 1977 developed a powerful method for DNA sequencing that utilises single stranded DNA as template. This method is also called *dideoxynucleotide cain termination method.* The requirements are : a primer with free 3' -OH ends to start DNA synthesis, DNA polymerase and dNTPs.

Fig. 6.23 shows the presence of free 3'-OH group at 3' end in dATP and no 2',3' -OH in 2',3'ddATP. In 2',3', ddATP a hydrogen atom is attached at 2' and 3' carbons instead of –OH hydroxyl group. If any of four ddNTPs binds, the chain elongation is terminated. Because ddNTPs do not have free 3'-OH end which is required for chain elongation. Therefore, no phosphodiester bond will be formed.

Four reaction tubes are labelled with A,T, G and C

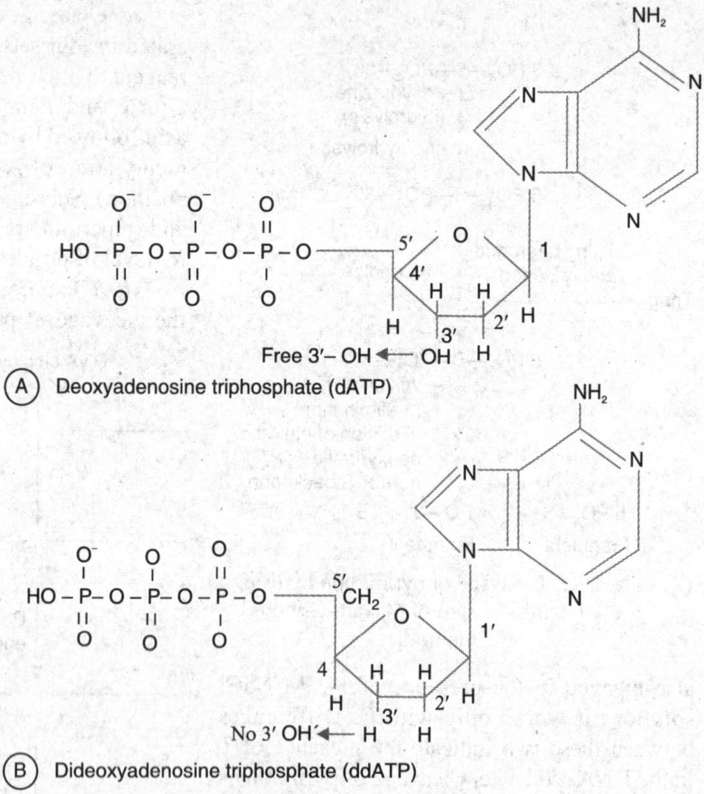


Fig. 6.23. Structure of deoxyadenosine triphosphate (dATP) (A) and dideoxyadenosine triphosphate (ddATP) (B)

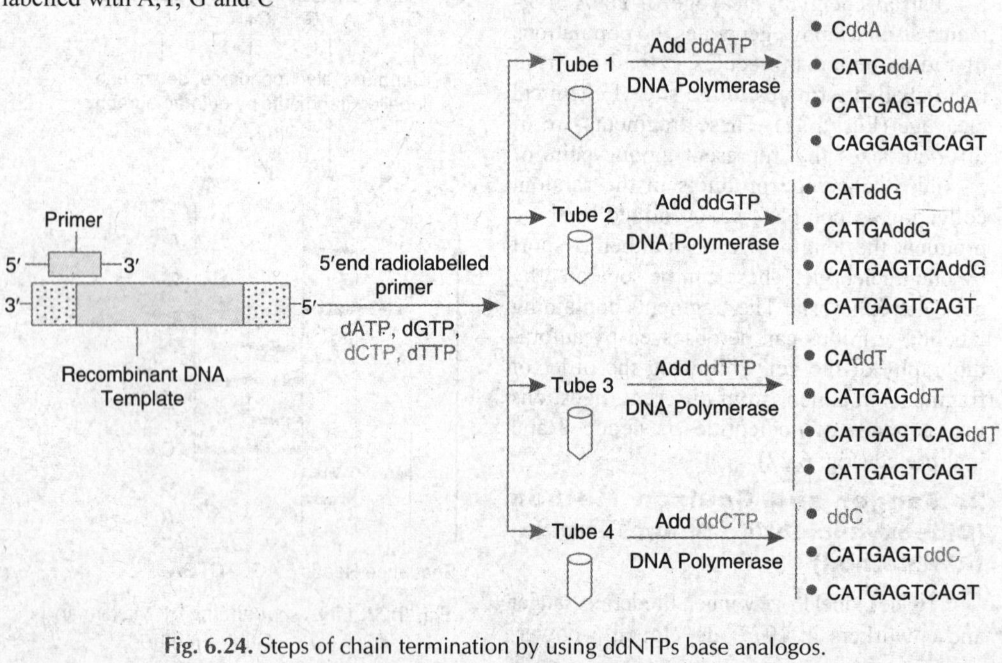


Fig. 6.24. Steps of chain termination by using ddNTPs base analogos.

each containing single stranded DNA termplate (obtained by NaOH hydrolysis), 5'-radiolabelled DNA primer, and all four radiolabelled dNTPs (dATP, dGTP, dCTP and dTTP) (Fig. 6.24). A small amount of ddATP is added to tube 1, ddGTP to tube 2, ddTTP to tube 3 and ddCTP to tube 4. The concentration of ddNTPs should be maintained to about 1% of the concentration of dNTPs. DNA polymerase is added to each tube, DNA synthesis starts and chain elongates. In each tube ddNTP is randomly incorporated and fragments are terminated. The length of each fragment depends on the position of incorporation of ddNTPs.

After completion of reaction, the fragments of each tube are seprated by electrophoresis in four different lanes of high resolution polyacrylamide gel. Then the gel is dried and autoradiography is done so that position of different bands (having radiolabelled 5' end) in each lane is observed (Fig. 6.25). In each lane the ends of fragments contain the base in correspondence to the ddNTPs used. DNA sequence is obtained by reading (from bottom to top of gel) the bands on autoradiogram of four lanes.

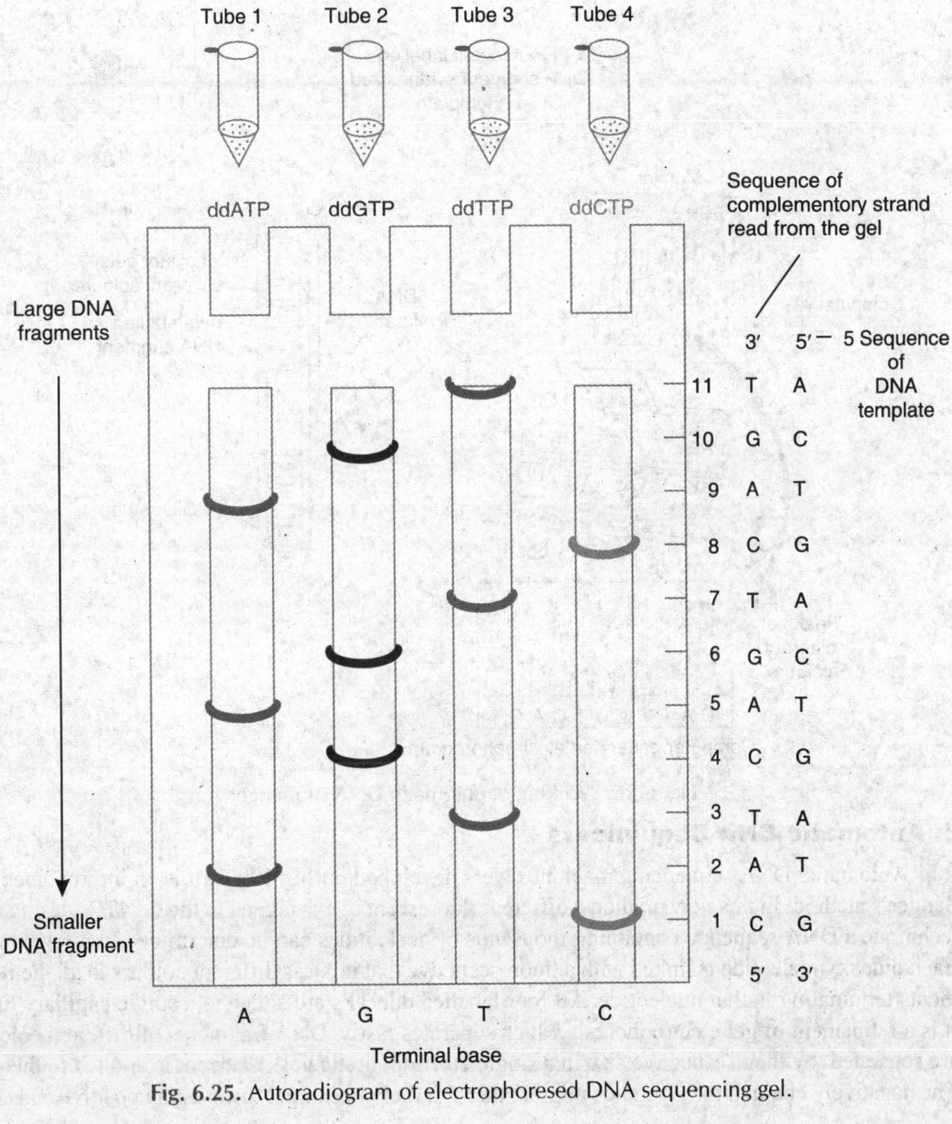


**Fig. 6.25.** Autoradiogram of electrophoresed DNA sequencing gel.

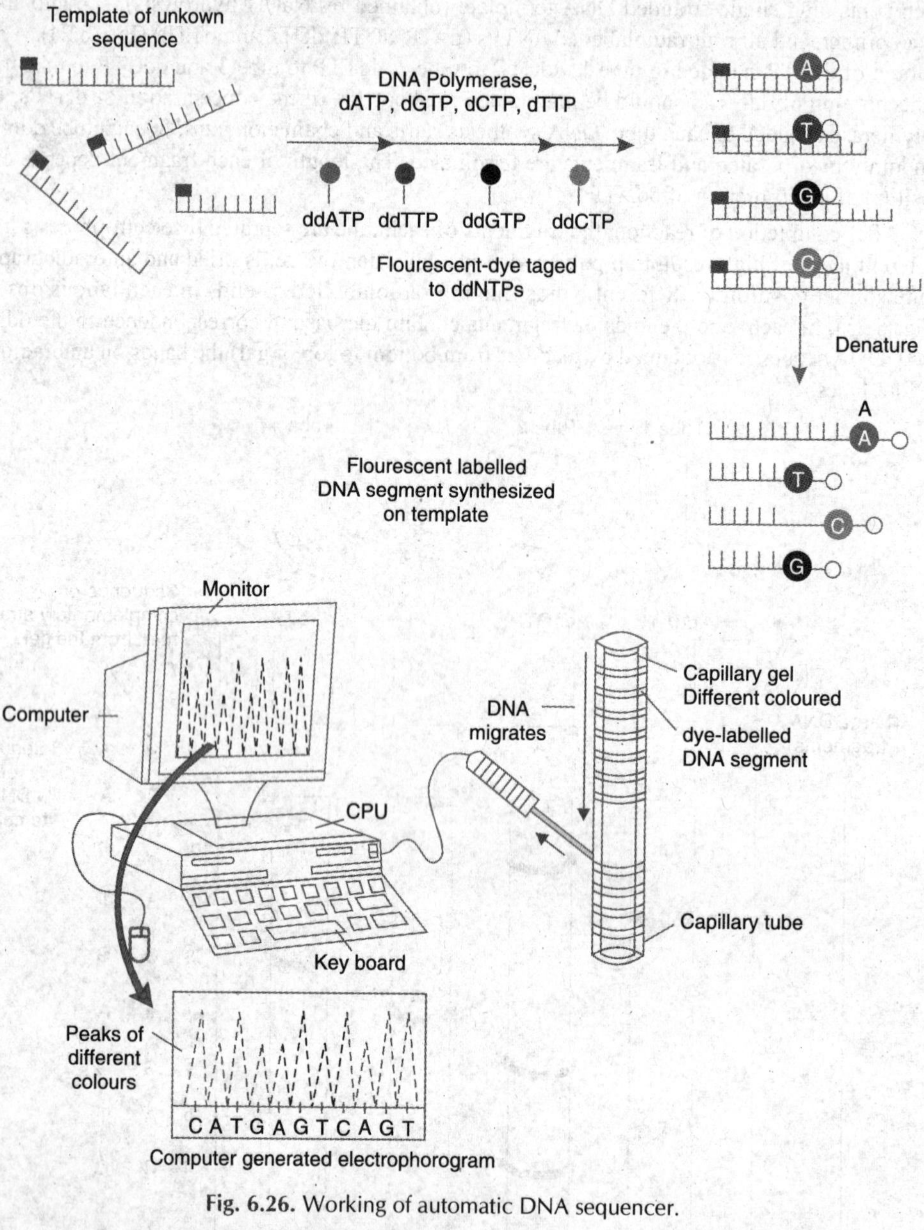


**Fig. 6.26.** Working of automatic DNA sequencer.

## 3. Automatic DNA Sequencers

Automatic DNA sequencing machines were developed during 1990s. It is an improvement of Sanger's method. In this new method a different fluorescent dye is tagged to the ddNTPs. Using this technique a DNA sequence containing thousands of necleotides can be determined in a few hours. Each dideoxynucleotide is linked with a fluorescent dye that imparts different colours to all the fragments terminating in that nucleotide. All four labelled ddNTPs are added to a single capillary tube. It is a refinement of gel electrophoresis which separates fastly. DNA fragments of different colours are separated by their respective size in a single electrophoretic gel. A current is applied to the gel. The negatively charged DNA strands migrate through the pores of gel towards the positive end. The small sized DNA fragments migrate faster and vice versa. All fragments of a given length migrate in a single peak. The DNA fragments are illuminated with a laser beam. Then the fluorescent dyes are excited and emit light of specific wavelengths which is recorded by a special 'recorder'. The DNA sequences are read by determining the sequence of the colours emitted from specific peaks as they pass the detector. This information is fed directly to a computer which determines the sequence. A tracing electrogram of emitted light of the four dyes is generated by the computer (Fig. 6.26). Colour of each dye represents the different nucleotides. Computer converts the data of emitted light in the nucleotide sequences.

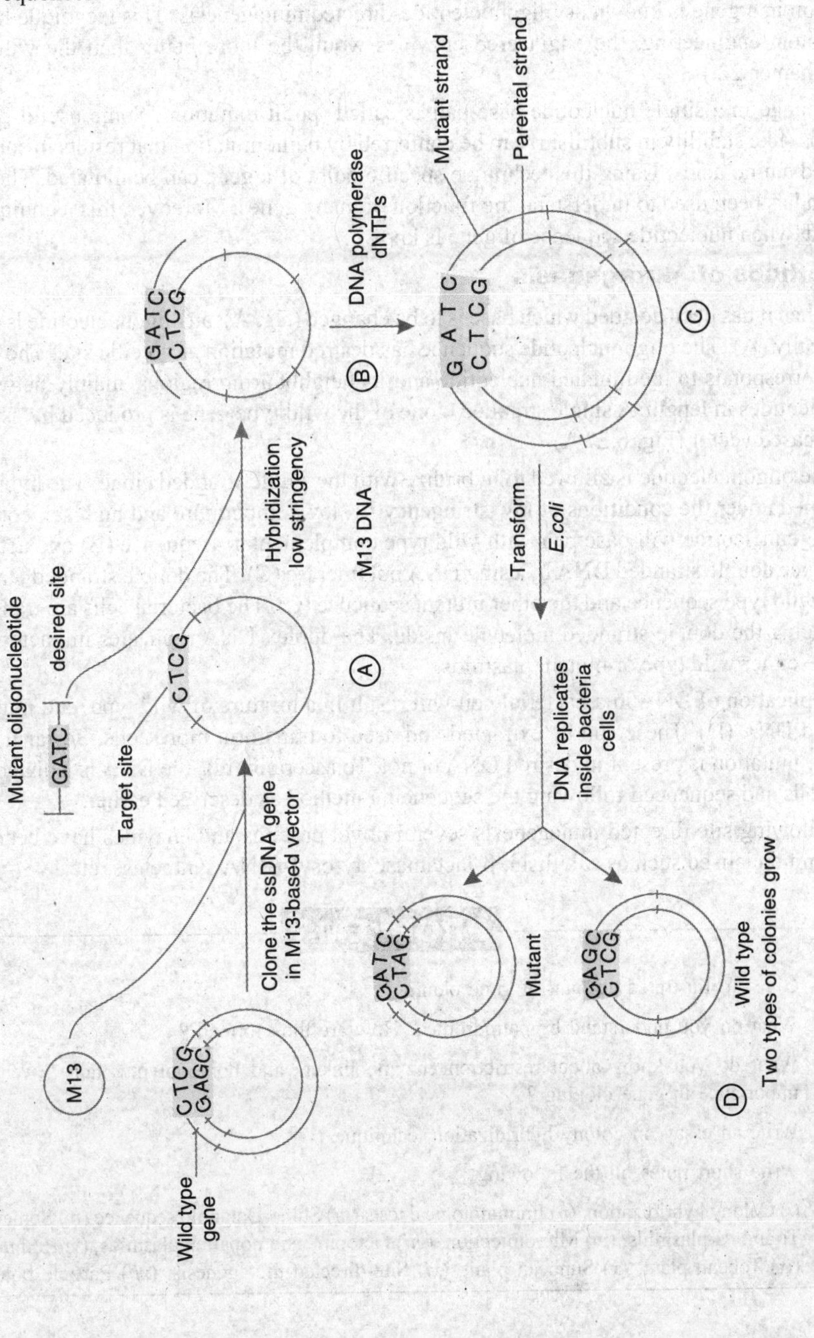


**Fig. 6.27.** Steps of site-directed mutagenesis.

## M. SITE-DIRECTED MUTAGENESIS

Techniques have been developed to mutate the specific portion in the genome in a view to get some novel products of enormous value. Mutations can be done by directing insertions or deletions to a specific site on the DNA but they are unlikely to be of use. These minor changes may have some importance if a single amino acid is altered in a protein in order to improve its properties. For such modifications point-mutations *i.e.* changing of single nucleotide, is done on specific portion in the gene. Thus, the most suitable method that could bring about point-mutation in a gene is known as oligonucleotide-directed mutagenesis. This technique has potential for protein engineering, the engineered enzymes would be more better than the wild type ones (*see* Chapter 22.).

Change in a single nucleotide base pair is called 'point mutation'. Some useful properties in proteins (like stability in subtilisin) can be conferred by point mutation that results in substitution of selected amino acids. Using this technique specific point of a gene can be mutated. Therefore, this method has been used to understand the function of many genes. Moreover, this technique can only be used when nucleotide sequence of gene is known.

### 1. Methods of Mutagenesis

When it has been decided which base is to be changed (*e.g.* A), an oligonucleotide is synthesised chemically (A). The oligonucleotide sequence has desired mutation at specific site. The oligonucleotide corresponds to the mutated nucleotide and its neighbouring regions, mainly between 15 and 20 nucleotides in length. A single stranded clone of the wild type gene is produced by using an M13 phage-based vector (Fig. 6.27 A).

The oligonucleotide is allowed to hybridize with the single stranded clone. The hybridisation is performed under the conditions of low stringency *i.e.* low temperature and high salt concentration. The oligonucleotide will base pair with wild type complementary sequence (B) and acts as primer to produce double stranded DNA by using DNA polymerase (C). The double stranded DNA consists of one wild type sequence and the other mutant sequence (C). The bacterial cells are transformed by introducing the double stranded molecule inside. The duplex DNA replicates in bacterial cell and produce either wild type or mutant plasmids.

Replication of DNA in a bacterial cell will result in a mixture of wild type and mutant double stranded DNA (D). These can be extracted and used to transform more cells. So far it is not sure whether mutation is present in desired DNA or not. To ascertain this, the DNA is isolated from bacterial cells and sequenced following the sequencing method as described earlier.

Following site-directed mutagenesis several novel proteins and enzymes have been produced and commercialised such as subtilisin, β-lactamase, tyrosyl-rRNA synthetase, etc.

## PROBLEMS

1. Give an illustrated account of gene cloning.
2. What do you understand by palindrome ? How are they formed ?
3. What do you know about restriction enzyme linkers and homopolymer tails ? What is their importance in gene cloning ?
4. Write an essay on colony hybridization technique.
5. Write short notes on the following :

   (*i*) Colony hybridization, (*ii*) Immunological tests (*iii*) Shine-Dalgarno sequence (*iv*) Somatostatin (*v*) Ti-and Ri-plasmids, (*vi*) Microinjection, (*vii*) Octopine and nopaline plasmids, (*viii*) Shuttle vector, (*ix*) Tobean plant, (*x*) Sunbean plant, (*xi*) Site-directed mutagenesis, (*xii*) Particle bombardment gun, (*xiii*) Transgenic animals, (*xiv*) Transgenic plants, (*xv*) *In vitro* fertilization technology, (*xvi*) Pollen transformation through particle bombardment.

**6.** What is DNA sequencing ? Give a detailed account of Maxam-Gilbert method of DNA sequencing.

**7.** Discuss in detail about gene library and its construction.

**8.** Give an illustrated account of different approaches made for detection of nucleic acids.

**9.** Write short notes on the following :

(*i*) Western blotting, (*ii*) Sanger and Coulson's method of gene sequencing, (*iii*) *automatic DNA sequencer.*

CHAPTER

7

# Genetic Engineering for Human Welfare

Following the recombinant DNA techniques thousands of DNA fragments of prokaryotes and eukaryotes (plant and animals) have been cloned during the last two decades. According to the utility the cloned DNAs can be used in basic research or commercial production of useful products. Basic research includes DNA probe, expression of cloned genes, colony/ plaque hybridization, genome mapping and many more related to genetic engineering.

Scientists have successfully cloned mammals and crop plants; the new areas of application of genetic engineering are now being explored. It has been used in medicine, pharmacology, reproductive technology and several other research areas. Obviously genetic engineering holds the key in curing many serious diseases, which have claimed many lives. You know that the production of insulin was one of the foremost endeavors for human welfare. Now-a-days, Insulin is being produced through genetic engineering. Interferon secreted by human cells combats virus attack can also be produced commercially in recent years. Gene therapy is also one of the major benefits of genetic engineering, where diseases are cured by repairing or replacing the abnormal non-functional human genes or introducing therapeutic genes, such as Huntington's disease, cystic fibrosis, etc.

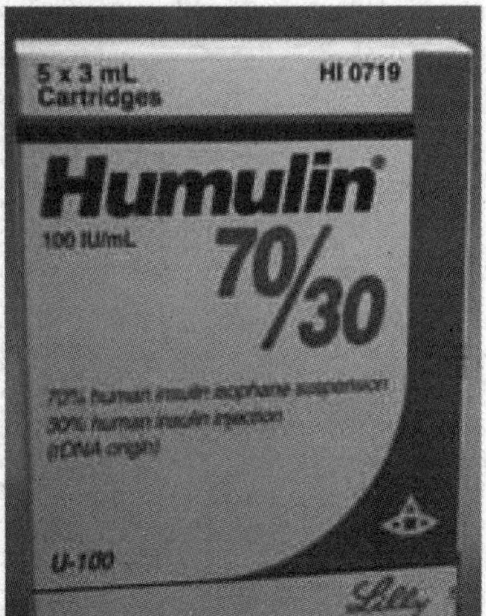


Humulin ® is the first genetically engineered insulin manufactured by Lilly (U.S.A.).

Genetic engineering has also played a key role in the production of new pharmaceutical products. These new pharmaceuticals are created by cloning certain genes. The genetically manufactured human growth hormones are extremely effective in the treatment of skin burns, bone fractures, ulcers in the digestive tract, etc.

Production of genetically modified crops (GMPs) has been possible through genetic engineering; some of the most important GMO crops are corn, soybean, seed oil, cotton, etc. These crops are resistant to insects or deleterious herbicides.

The field of genetic engineering is also involved in production of novel items such as blue roses and glow fish, etc. The scientists can use this tool for the welfare of the entire human race.

Moreover, cloned genes are utilized commercially in medicals, agriculture and industry, and the production of valuable products. Some of the cloned genes and application of genetic engineering techniques are described here.

# A. CLONED GENES AND PRODUCTION OF GROWTH HORMONES, VACCINES AND COMMERCIAL CHEMICALS

## 1. Human Peptide Hormone Genes

In the human body, peptide hormones are secreted after encoding the peptide hormone genes in the specialized cells, for instance insulin and other Human Growth Hormones (hGH).

***(a)* Insulin:** This peptide hormone *i.e.* insulin is secreted by the Islets of Langerhans of pancreas which catabolizes glucose in blood. Insulin is a boon for the diabetics whose normal function for sugar metabolism generally fails.

Insulin consists of two polypeptide chains, chain A (21 amino acid long) and B (30 amino acid long). Its precursor is proinsulin which also contains two polypeptide chains, A and B, and is connected with a third peptide chain-C (35 amino acid long). However, the recent discoveries reveal that precursor of insulin is pre- pro-insulin which is about 109 amino acids long. The pre- pro- insulin is synthesized in beta cells of pancreas, the structure of which is given below :

$NH_2$-(Peptide)-ß-Chain- (peptide-c)- A chain-COOH

In the beginning, efforts were made to isolate mRNA for pre- pro-insulin from rats Islets of Langerhans of pancreas and to synthesize cDNA. Thereafter, it was inserted into a plasmid. The recombinant plasmids were transferred into the *E. coli* cells which secreted pro-insulin.

K. Itakura and collaborators in 1977 chemically synthesized DNA sequence for two chains, A and B, of insulin and separately inserted into two pBR322 plasmids by the side of ß-galactosidase gene. The recombinant plasmids were separately transferred into *E coli* cells which secreted fused ß-galactosidase - A chain and ß-galactosidase - B chain separately. These chains were isolated by detaching from ß-glactosidase in pure form in a amount of about 10 mg/24 g of healthy and transformed cells. Production of recombinant insulin is shown in Fig. 7.1.

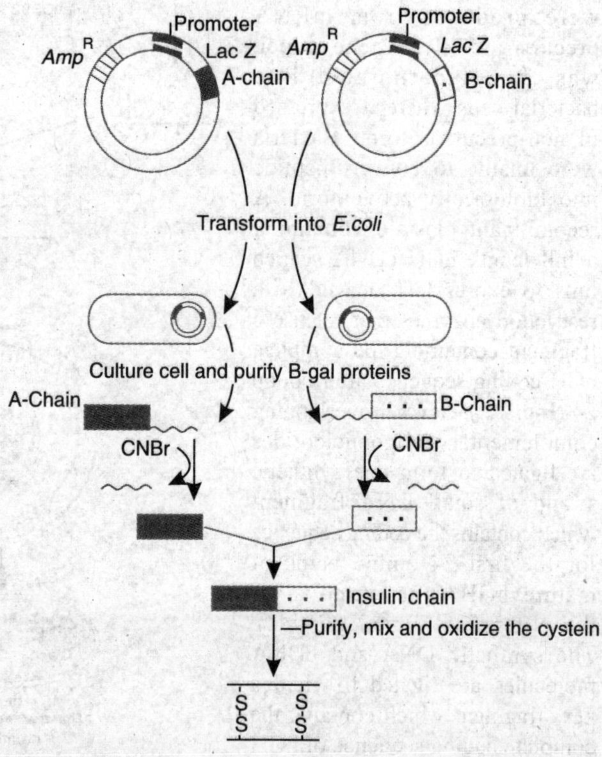


Fig. 7.1. Production of recombinant insulin in *E.coli.*

Detachment of proinsulin could be

possible when an extra methionine codon was added at the *N*-terminus of each gene for A and B chains. The two chains (A and B) were joined *in vitro* to reconstitute the native insulin by sulphonating the two peptides with sodium disulphonate and sodium sulphite. In 1980, W. Gilbert and L. Villakomaroff isolated mRNA for insulin from β cells of rat's pancreas and inserted into pBR322 plasmid in the middle of a gene normally coding the penicillinase, and incorporated it into *E. coli* cells. *E. coli* cells produced a hybrid protein (penicillinase+proinsulin) from which the proinsulin was separated by using trypsin. It is estimated that clones of *E. coli* are capable of producing about one million molecules of insulin per bacterial cell. Human insulin (humulin) is the first therapeutic product produced by means of recombinant technology by Eli Lilly & Co. in 1980. Shreya Life Sciences Pune (India) Co., has started producing the second generation rDNA-based insulin without using DNBr with the name 'Recosulin'.

***(b)* Somatotropin :** Somatotropin is secreted by the anterior lobe of pituitary glands which consists of 191 amino acid units. Its secretion is regulated by two other hormones (somatostatin and growth hormone releasing hormone) produced by hypothalamus. Deficiency of somatotropin in about 3% cases is of hereditary. It has been estimated to about 1 child in 5,000.
Turner's syndrome is one of the most common chromosome disorders in girls and it is characterized by short stature and non-functioning of ovaries affecting approximately 1 in 2,500 live female birth. The extraction of somatotropin pharmaceutically from the pituitary glands could not meet annual demand of this hormone. Biosynthesis of somatotropin was achieved through gene cloning procedures.

Double stranded cDNAs were produced from mRNA precursor of hGH gene which was then incorporated into bacterial cells where it expressed in non-precursor form. Bacteria were unable to convert peptide into biologically active form. A recombinant plasmid containing a full length hGH cDNA (which fails to express) is cleaved with restriction enzymes that release a fragment containing the complete hGH coding sequence after codon 24 (Fig. 7.2). Several overlapping complementary oligonucleotides are ligated to form one synthetic strand of small DNA fragment which contains the coding sequence for the first 24 amino acids of mature hGH (after removal of the N-terminal signal peptide). The synthetic DNA and cDNA molecules are ligated to yield a new fragment which contains the complete coding sequence of hGH DNA. It is ligated into a restriction site just down stream of the *lac*

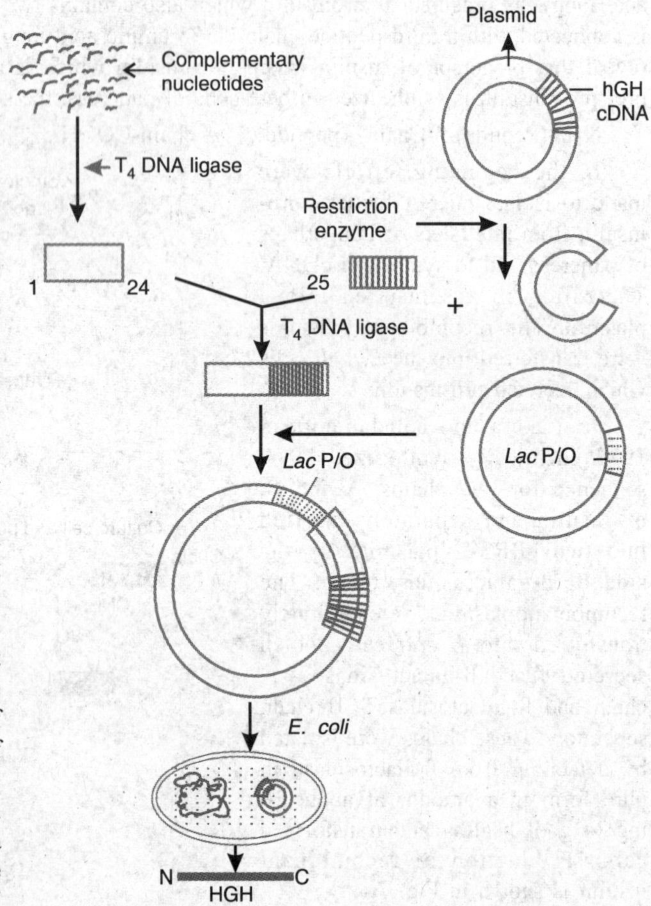


Fig. 7.2. Expression of Human Growth Hormone (hGH) in *E.coli*; *Lac* P/O, lactose promoter/operator.

promoter/operater region cloned on a plasmid. The rDNA plasmids are allowed to transform *E.coli.* Synthesis of hGH is induced by an inducer of *lac* operon (*i.e.* isopropylthiogalactoside (IPTG)). The hGH is subsequently purified. hGH is being produced commercially by this method.

About 100,000 molecules of hormone per cell of *E. coli* have been produced (Newmark, 1979). One of the major difficulties was that at the N-end of polypeptide an extra methionine, the *met*-hGH, was attached. Methods are developed to remove methionine from the *met*-hGH molecules.

(*c*) **Somatostatin :** Somatostatin, a 14 residue polypeptide hormone, is synthesised in the hypothalamus. It is the first polypeptide which was expressed in *E. coli* as part of the fusion peptide which inhibited the secretion of growth hormone, glucagon and insulin. It does not contain any internal methionine. Eight single stranded DNA segments were synthesised chemically which were annealed in an overlapping manner to form a double stranded DNA (the synthetic gene). It had single stranded projections at the each end as the same are formed by *Eco* RI. The synthesised gene contained 51 base pairs which were terminated by two non-sense (stop) codon and preceded by a methioine codon as below:

ATG - (42 basepairs encoding somatostatin) - TGATAG.

Two plasmids, pSOM I and pSOM II-3 , were constructed. The synthetic gene was introduced into *E. coli* β- galactosidase gene at different sites. The chemically synthesised gene was inserted down stream from the *lac* promoter in such a way that the gene fusion should have specified

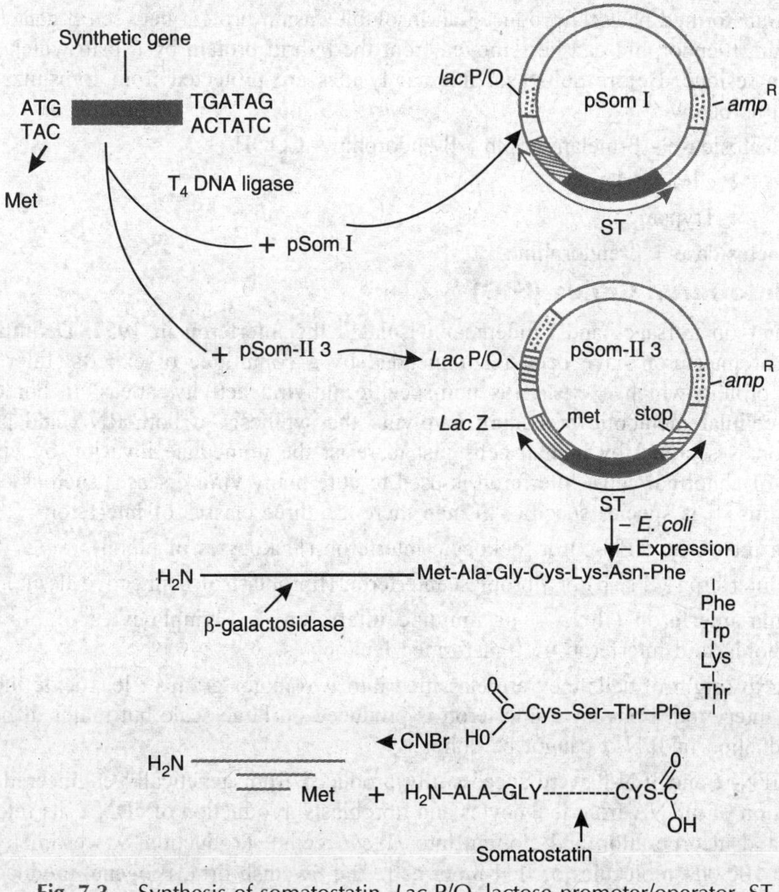


**Fig. 7.3.** Synthesis of somatostatin, *Lac* P/O, lactose promoter/operator, ST, somatostatin; Amp, ampicillin resistance.

a polypeptide in which the first 7 amino acids of β-galactosidase were fused to somatostatin. But the somatostatin was not detected in transformed bacteria; possibly it was degraded in *E. coli* (Glover, 1994). An alternative plasmid, pSOM II-3, was constructed which had the both *lac* promoter /operater and *lac*Z regions. The *lac*Z gene encodes peptide of β-galactosidase. If the resulting frame of *lac*Z is maintained after inserting a DNA, a fusion peptide is produced. The plasmid was cleaved in the *lac*Z segment by restriction enzyme (Fig. 7.3). The synthetic gene was inserted into the plasmid at *Eco* RI site near C-terminus of β-galactosidase. The plasmid in the transformed bacterium directs the synthesis of a fused protein consisting of $NH_2$- termined segment of β-galactosidase fragment coupled by methionine to somatostatin. It was stabilized from proteolytic degradation by β-galactosidase moiety (Glover, 1984). This fused protein was purified and treated with cyanogen bromide (CNBr) which cleaves only protein at the carboxyl side of the methionine. Thus, the methionine linker remains attached to β-galactosidase fragment, and somatostatin was released. Now, it has become possible to inhibit the degradation of foreign protein in *E. coli* by introduction of protease inhibition (PIN) gene of T4 phage.

**(*d*) β-endorphin :** ß-endorphin (30 amino acid long neuropeptide with opiate activity) is another growth hormone which was expressed in genetically engineered *E. coli* cells. In 1980, J.Shine and coworkers integrated DNA sequences of ß-endorphin, obtained from mRNA, adjacent to ß-glactosidase genes on plasmid. The mRNA contained large precursor of protein that consisted of, besides ß-endorphin, the hormones α-melanotropin, corticotropin, ß-lipotropin and ß-melanotropin. The ß-endorphin is cleaved from C-terminus of the precursor peptide. In this way, the transformed bacteria produced an insoluble fusion protein between ß-galactosidase and ß-endorphin. ß-endorphin can be removed from the hybrid protein by tripsin which cleaves only at arginin residue. Before doing so, internal lysines are protected from trypsinization by citraconylation as below :

$NH_3$ -β-glactosidase - β-melanotropin - β-endorphin - COOH

↓ citraconylation

↓ Trypsin

β-galactosidase + β-endorphin

## 2. Human Interferon Genes (HIG)

For the first time, Isaacs and Lindenmann isolated the interferon in 1957. Definition and nomenclature of interferon have been recommended by a committee of experts. Interferon is defined as "a protein which exerts virus non-specific antiviral activity, atleast in homologous cells through cellular metabolic procedure involving the synthesis of both RNA and protein." Thus, interferon is secreted by human cells just to resist the immediate invasion by virus and multiplication of abnormal cells. Interferon is used to cure many viral diseases such as common cold and hepatitis. It is species specific. In man there are three classes of interferon :

(*i*) Alpha interferon (IFN-α) or leukocyte interferon (leukocytes of blood)

(*ii*) Beta interferon (IFN-β) or fibroblast interferon (fibroblast of connective tissue).

(*iii*) Gamma interferon (IFN-γ ) or immune interferon (by lymphocytes of blood) and lymphoblastoid interferon by transformed leukocytes.

According to origin of cells they are classified into two major groups : leukocyte interferon and fibroblast interferon. Leukocyte interferon is produced on large scale but major difficulty is that mass production of IFN-α cannot be done.

In 1980, IFN-α and IFN-β were successfully produced from genetically engineered *E. coli* cells (by isolation of mRNA from leukocytes and fibroblasts, production of cDNA, its integration into pBR322 and incorporation and cloning into *E. coli* cells). Production was estimated to be about 1,000 to 100,000 molecules of IFN-ß per cell. The Swedish firm, Biogene, produced IFN-α and IFN-β through recombinant DNA techniques. It was found that genes responsible for the production of IFN-α and IFN-ß had 865 and 836 nucleotides, respectively.

Later on hybrid plasmid containing cDNA of IFN-ß genes was built up which needed a promoter site on plasmid to express in *E. coli* cells. Similarly, hybrid plasmids were also prepared that contained IFN- genes with *trap* promoter between the leader and ribosome binding sites, so that expression of interferon could be done. Expression of both the interferon could be optimised by varying the spacing sequence between *trap* Shine- Dalgarno sequence and the initiator condon.

IFN-β produced by genetically engineered microoganism showed lower specific activity and decreased stability than natural one. Enhanced specific activity and stability was obtained when the cysteine at position 17 was replaced by a series of site specific mutagenesis resulting in ' IFN-β-Ser' molecule. It was stable for two years and well tolerant in cancer patients. Moreover, genetically engineered *E. coli* is reported to yield 5-10 million units/ml of IFN-β-Ser in a 200 litre batch reactor within 2-3 days of fermentation.

## 3. Genes for Vaccines and Immunogenic Substances

Vaccines are chemical substances prepared from the proteins (antigen) of other animals which confer immunity to a particular virus. Some of the vaccines synthesized biologically through genetic engineering are briefly described as below:

**(*a*) Vaccines for Hepatitis B Virus :** Hepatitis B virus (HBV) is wide spread in man and produces several chronic liver disorders such as Fulminant chronic hepatitis, cirrhosis and primary liver cancer. HBV DNA is a double stranded circular molecule of about 3Kb size and has a large single stranded gap which must be required with an endogenous polymerase before digestion with restriction enzyme for DNA cloning. After infection in human being, HBV fails to multiply and infect a large number of cells and even does not grow in cultured cells. This property has been explained to be due to hinderance of its molecular characterization and development of vaccines. Plasma of human has been detected to have varying amount of antigens. Three types of viral proteins are recognised to be antigenic : (*i*) viral surface antigen (HBsAg), (*ii*) viral core antigen (HBcAg), and (*iii*) the e-antigen (HBeAg).

Although the whole viral genome has been cloned and sequenced, yet there is limited information about amino acid sequence of surface and core antigens. Recently, HBV DNA has been successfully cloned in *E. coli* and mammalian cells, and synthesis of HBsAg and HBcAg particles has been done in the cells. In 1979, C.J. Burrell and coworkers. inserted HBcAg genes in PBR322 near β-glactosidase gene. Production of these genes is needed in order to get production of vaccines on a large scale. In yeast or mammalian systems, these antigens are synthesized more efficiently than in prokaryotes.

**(*i*) Recombinant vaccine for Hepatitis B virus.** After infection, HBV fails to grow and even in cultured cells it does not grow. This property has been explained to be due to inhibition of its molecular expression and development of vaccines. Recombinant vaccine for HBV was produced by cloning HBsAg gene of the virus in yeast cells. The yeast system has its complex membrane and ability of secreting glycosylate protein. This have made it possible to build an autonomously replicating plasmid containing HBsAg gene near the yeast alcohol dehydrogenase (ADH) I promoter (Fig. 7.4). The HBsAg gene contains 6 bp long sequence preceding the AUG that synthesises N-terminal methionine. This is joined to ADH promoter cloned in the yeast vector PMA-56. The recombinant plasmid is inserted into yeast cells. The transformed yeast cells are multiplied in trytophan-free medium. The transformed cells are selected. The cloned yeast cells are culture for expression of HBsAg gene. This inserted gene sequence expresses and produces particles similar to the 22 μm particle of HBV as these particles are produced in serum of HBV patients. The expresed HBsAg particles have similarity in structure and immunogenicity with those isolated from HBV-infected cells of patients. Its high immunogenicity has made it possible to market the recombinant product as vaccine against HBV infection.

**(*ii*) Indigenous Hepatitis-B vaccine.** In India first genetically engineered vaccine (*Guni*) against HBV was developed by a Hyderabad based laboratory (Shantha Biotechnics Pvt. Ltd.). India is the fourth country (after the U.S.A., France and Belgium) to develop this highly

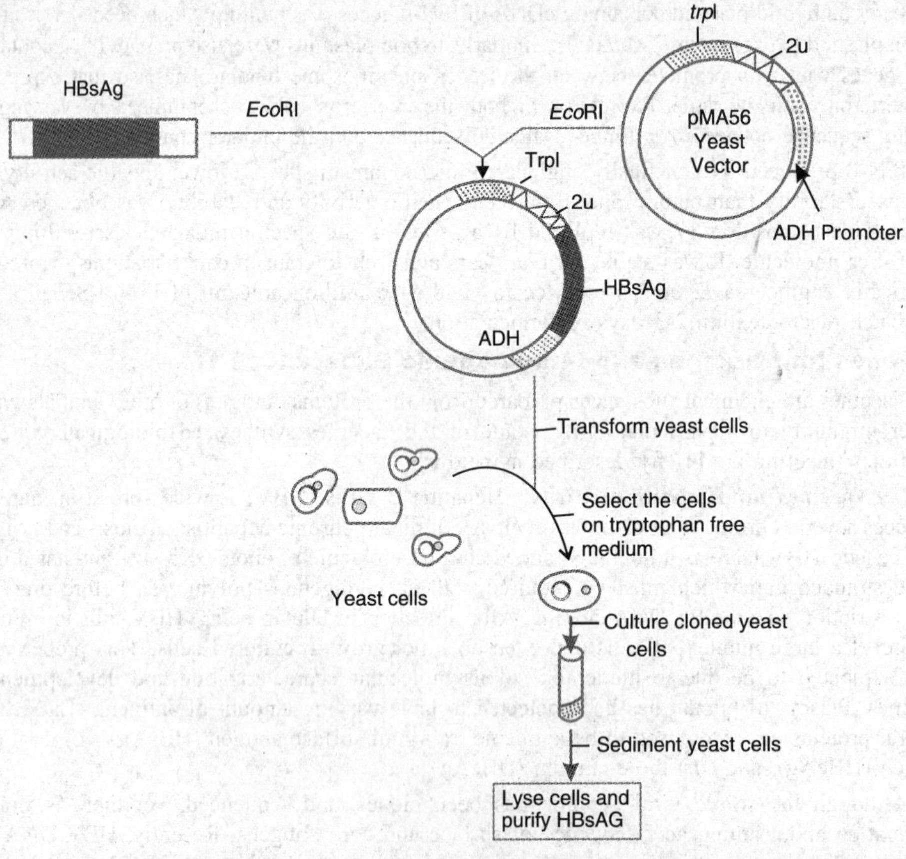


**Fig. 7.4.** Expression of HBsAg gene in yeast.

advanced vaccine. The indigeneous yeast-desired HBV vaccine is one third the cost of the imported vaccine. This new vaccine had undergone human clinical trials at Nizam's Institute of Medical Sciences, Hyderabad and K.E.M. Hospital, Mumbai. The clinical trials clearly proved that the seroprotection is about 98%. It was found more effective than the imported vaccine. The Drug Controller General of India has permitted it for commercial manufacture.

**(*b*) Vaccines for Rabies Virus (RV):** RV causes hydrophobia in animals and humans in many countries like South America, Africa, Asia (including India and Pakistan). Researches are being done to synthesize vaccines by inducing genetically engineered *E. coli* cells. However, attempt has been made to isolate mRNA encoding viral protein from RV infected cells. The genes coding for the production of rabies virus glycoprotein coat has been successfully transferred to *E. coli.* This is the first step towards the production of antirabies virus vaccines as this glycoprotein stimulates antibody production in diseased animals. In 1984, Wistar Institute in Philadelphia developed a new genetically engineered vaccinia virus by inserting a small piece of foreign DNA. The genetically engineered virus synthesised antirabies vaccines for animals. The recombinant vaccinia virus did not cause rabies in those animals that received rabies genome but encoded antigenic molecules and activated the immune system against rabies infection. This new vaccine can be administered orally in animals, and could be decreased the risk of human death due to bites of animals receiving rabies virus.

**(*c*) Vaccines for Poliovirus:** Poliovirus is a causal agent of poliomyelitis in human beings. First attempt to describe the complete structure of poliovirus RNA was made by N. Kitamura and coworkers in 1981. Since then it has become an easy task to have the expression of genetic information of poliovirus. Antipoliomyelitis vaccines through genetically engineered bacterial cells have not been synthesized so far.

**(*d*) Vaccines for Foot and Mouth Disease Virus (FMDV) :** Foot and mouth disease (FMD) is a serious disease caused by *Aphthovirus.* The primary control measure of the disease has been the slaughter of FMDV - infected animals. Chemotherapeutic way of FMD control is vaccination. Vaccines are produced by inactivation of virus grown in bovine tongue epithelium. A detailed study of FMDV reveals that it contains a single stranded RNA covered in a capsid of four polypeptides, for example, VP1, VP2, VP3 and VP4 where only VP1 has a little immunogenic activity. However, the nucleotide sequence encoding for VP1 was identified on the single stranded RNA genome and cloned on double stranded pBR322 in *E. coli.* About 1,000 molecules of VP1 per bacterial cell were synthesized.

Vaccines produced from genetically engineered *E. coli* cells cannot compete with those extracted from virus particle. Therefore, much more work is to be done to make the vaccines available for FMDV on a large scale at low price. An outline for production of FMD vaccine is given in Fig. 7.5.

Synthetic peptides have also been produced and used for immunization against bacterial (*Diphtheria, Streptococcus pyogenes*) and viral (Hepatitis virus, FMDV) diseases. For the production of synthetic polypeptide to be used as vaccines, it is necessary to have the knowledge of structure and function of proteins and the regions involved in immunogenic response. For example, synthetic polypeptides having immunogenic affects against HBV contain disulphide bond in the region between amino acids 117 and 137 corresponding to the viral surface antigen. After injection into mice the polypeptides elicited antibodies against HBsAg and protected half of mice.

Similarly, a synthetic polypeptide has been identified that corresponds several regions of FMDV protein VP1. The region between amino acids 141 and 160 elicited and production of antibodies against FMDV in guinea pigs, rabbits and swine. Likewise, another polypeptide was synthesized and the region between amino acids 200 to 213 VPI also elicited antibodies against FMDV. There are other examples also for the synthesis of vaccines against viral diseases; such vaccines too have a great promise for the immunization purposes.

**(*e*) Vaccines for Smallpox Virus :** Small pox is a very serious disease of humans in most of the countries. The vaccinia virus (the cowpox virus) can be used as the basis of the smallpox vaccine by using the recombinant DNA technology. The genome of vaccinia virus is altered by inserting the foreign genes which encodes the pathogenic antigens. When a person is immunised with a vaccine in preparations, the foreign genes enter the body cells of the patients receiving vaccinia virus and direct the synthesis of the foreign antigenic proteins. The genes replicate

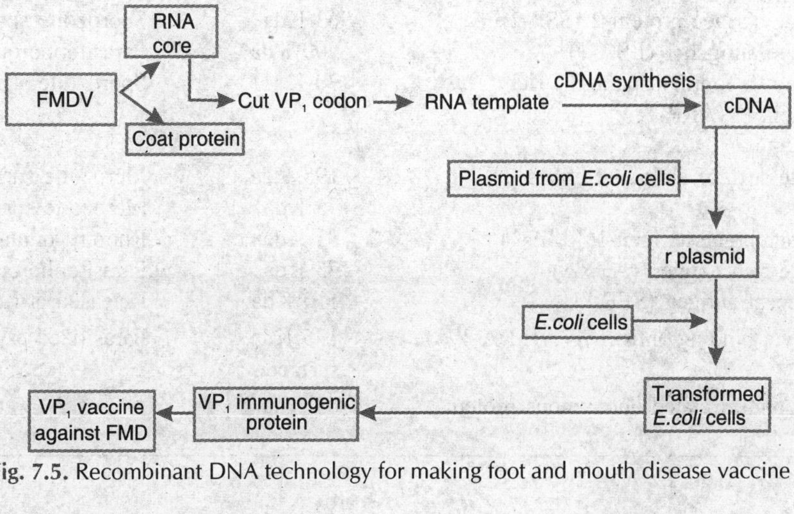


Fig. 7.5. Recombinant DNA technology for making foot and mouth disease vaccine

like vaccinia genes and induce immunity to the smallpox, However, it resemble the other types of vaccine. The vaccinia virus can be modified in such a way that a single preparation confers protection against many viral disease.

(*f*) **Malaria Vaccines:** According to WHO estimates 4 billion people are at the risk of developing malaria and about 500 million cases occur each year resulting in one million death each year mainly of children of 5 years age and pregnant women. In addition, development of resistance against drugs by the species of *Plasmodium,* and insecticides by mosquitoes have been reported. Therefore, threat of malaria disease is still increasing for humans. Therefore, for control of malaria use of vaccines and vector control programmes would be successful. Much work is going on at Indian Institute of Immunology (New Delhi) and ICGEB on development of malaria vaccine by using modern methodologies. All kind of vaccine development through recombinant antigens, synthetic peptides and direct use of DNA are being attempted. All these attempts indicate that development of malaria vaccine is largely complex process. However, progress towards the development of a malaria vaccine has been slow due to several reasons, one of which has been the lack of *in vitro* correlates and the suitable animal models for malaria vaccine trials. *Plasmodium*-rhesus monkey is one of the models for malaria vaccine development.

Malaria vaccines are being developed at three distinct developmental stages of the parasite : (*i*) pre-erythrocytic stage (to eliminate infection by blocking the sporozoites from entering hepatocytes or by destroying the infected hepatocytes), (*ii*) blood stage of parasite (to prevent disease or reduce parasitic load), and (*iii*) sexual stage parasite (to limit transmission of disease).

In India 60-70% malaria is due to *P. vivex* which do not kill host but results discomfort and morbidity. It is more prevalent throughout Asia; but less is known about immune response of host against *P. vivex* as it resisted all attempts of culturing the parasite. *P. cynomolgi* is a simian malaria which is closely related to *P. vivex* in taxonomy and morphology. Hence it is regarded as a good model to study *P. vivex* infection as both share a similar clinical course of infection. At present vaccines are being developed at ICGEB against all stages of life cycle of the parasite but it is believed that an asexual blood stage vaccine is most likely to have the greatest impact on the disease.

**Table 7.1. Asexual stages vaccine target antigens.**

| *Antigens* | *Appropriate Size* | *Location* |
|---|---|---|
| **Sporozoite stage** | | |
| Circum sporozoite surface protein (CSP) | 60 Kda | Sporozoite surface |
| Sporozoite surface protein-2 (SSP-2) | 63 Kda | Sporozoite surface |
| Liver stage antigen-1 (LSA-1) | 200 Kda | Parasitophorous vacuole |
| Sporozoite threonine asparagine rich protein (STARP) | 70 Kda | Sporozoite surface |
| **Blood stage** | | |
| Merozoite surface antigen-1 (MSA-1) | 195 Kda | Merozoite surface |
| MSA-2 | 45 Kda | Merozoite surface |
| Apical membrane antigen-1(AMA-1) | 83 Kda | Rhoptry organelle |
| Acid base rich antigen (ABRA) | 75 Kda | Parasitophorous vacuole |
| Serine repeat antigen (SERA) | 110 Kda | Released at rupture |
| Erythrocyte binding antigen-175 (EBA-175) | 175 Kda | Parasitized erythrocyte surface |
| Throbospondin related anonymous protein (TRAP) | 63 Kda | ? |

***(i)* Expression of vaccine target antigens.** The most successful vaccines have been based on attenuated or killed pathogens. Malaria vaccine is limited to well defined molecules which can induce protective immune responses and easily produced by recombinant DNA technology in various systems. Three important vaccine target antigens such as thrombospondin related adhesive protein (TRAP), apical membrane antigen (AMA) and erythrocyte binding protein (EBP) from *P. cynomolgi* have been cloned and sequenced (Table 7.1). To evaluate their vaccine potential, these antigens have been expressed in *E.coli* using pQE expression system. Each of these proteins was expressed at high levels as insoluble inclusion bodies. Protocols for the large scale production of the correctly folded PcTRAP have been developed. TRAP has a multidomain structure and localised on the cell surface of *P. falciparum* sporozoites.

***(ii)* Animal trials of malaria vaccines.** The rhesus monkeys were immunized with recombinant PcTRAP, parasite lysate or adjuvant to study the protective efficacy. The antigens were delivered intramuscularly in three doses (500 µg each) on 0, 42, and 62 days. Blood from immunised and unimmunised monkeys was collected on days 1, 14, 29, 52 and 70. High antibody titres (8-16 $\times 10^5$ and above) were detected against PcTRAP and parasite lysate as measured by ELISA technique. Then immunised monkeys after injection with *P. cynomolgi* sporozoites ($3 \times 10^4$) were protected from malaria.

Merozoite surface protein-1 (MSP-$1_{19}$), the cysteine rich C-terminal domain of MSP-1 on the surface of *P. falciparum* is a leading malaria vaccine candidate. This is the only part of protein which remains bound to the merozoite membrane after invasion. Similarly the expression and purification of *P. falciparum* acidic basic repeat antigen (ABRA) and its fragments from *E.coli* has also been done. The purified recombinant proteins have been used to assess the antibody responses in human populations living in malaria endemic areas from Kalka village of Rourkela (Orissa), and from Nigeria. It has also showed protective efficacy in immunised rabbits.

Spf66 was the first recognised DNA vaccine for malaria developed by joining three merozoite derived proteins with repeatitive sequences derived from the circumsporozoite protein of *P. falciparum.* This vaccine has given equivocal results on human trials in more than one locations. Through Indo-US collaboration a recombinant multistage *P. falciparum* candidate vaccine has been developed.

***(g)* DNA vaccines :** For the first time J.A. Wolf and coworkers in 1990 injected naked DNA into the muscles of mice which led to expression of encoded marker protein. Thereafter, there has been a surge to use this approach to generate DNA vaccines against a variety of infectious diseases. Thus DNA vaccines are giving hope of a third vaccine evolution.

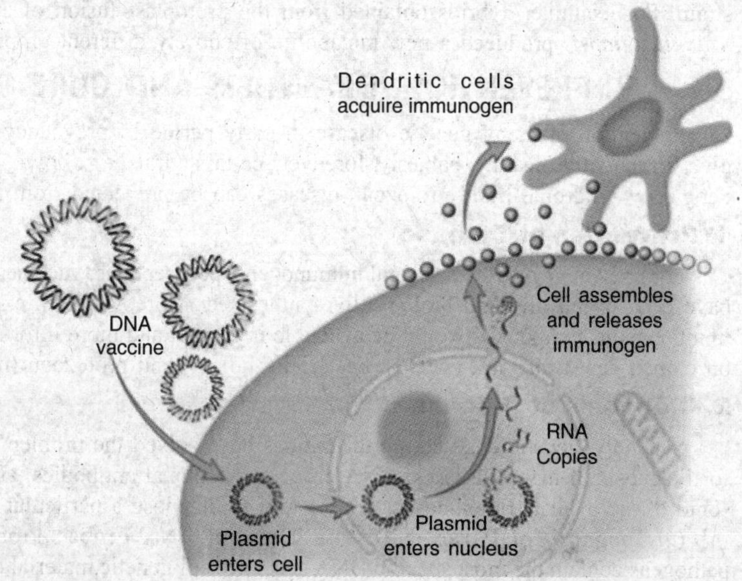

DNA vaccines can be used against a variety of infectious diseases.

The first published report from India indicates modest success in the development of DNA vaccines against rabies and Japanese encephalitis virus (JEV) in experimental animals. The efficacy of DNA vaccine (G protein) against rabies is correlated to levels of neutralising antibodies, whereas in the case of JEV envelop protein, cell mediated immunity appears to be the major mechanism of protection. There is great hope that DNA vaccines can protect against serious infectious diseases. Most likely it is to become a tool to benefit mankind in 21st century as such or in combination with recombinant/cell culture vaccines or as an adjunct to chemotherapy.

In case of malaria, DNA vaccines have distinct advantage, where plasmid DNA encoding different antigens and prepared by the same genetic procedure can be mixed and administered. A mixture of 4 plasmid DNA (pfCSP, pfSSP2, pfEXP-1 and pfLSA-1) has been injected into *Rhesus* monkeys and found to elicit multiple antigen specific cytotoxic T-lymphocytes.

D.H. Lowrie and coworkers in 1999 reported that a DNA vaccine coding for a mycobacterial heat sock protein of Mr65000 (Hsp 65) when administered in 4 doses to mice, 8 weeks after intravenous injection of virulent *M. tuberculosis* H37RV, leads to a dramatic decrease in number of live bacteria in spleen and lungs 2 months and 5 months after the first dose of DNA. Certain other mycobacterial antigens and BCG did not have this effect.

### 4. Production of Commercial Chemicals

There are several chemicals which are produced by using the recombinant DNA technologies. A few of them are as below :

**(*a*) Vitamins :** *See* Chapter 17,

**(*b*) Organic acids :** *See* Chapter 17,

**(*c*) Alcohols :** *See* Chapter 17,

**(*d*) Antibiotics :** *See* Chapter 17,

Recombinant DNA technology has helped in increased production of antibiotics; for example, the rate of penicillin produced at present is about 1,50,000 units/ml against about 10 unit/ml in 1950s. However, protoplast fusion technology of microbes for antibiotic production holds promise for microbes increasing the rate of production. The hybrid cell manifests genetic features of both the species. The hybrid species can produce new antibiotic or increase the productivity of the strain. For example, hybrids obtained from the protoplast fusion of *Streptomyces grieseus* and *S. taniimatiansis* produced a new metabolite of entirely different properties.

## B. PREVENTION, DIAGNOSIS AND CURE OF DISEASE

Prevention of many genetic disease in early period of pregnancy will help the mothers to give birth to the healthy babies. Moreover, certain viral [*e.g.* polyo, AIDS (acquired immune deficiency syndrome)] and protozoan diseases can be prevented from infection in humans.

### 1. Prevention of Disease

For preventive measures several immunogenic polypeptides (vaccines) and proteins (antibodies) have been chemically and biologically synthesized. Moreover, it is expected that in near future immunogenic compounds would be available in market and more informations would be gathered on cloned genes for viral (AIDS), bacterial (cholera) and protozoan (malaria) diseases.

### 2. Diagnosis of Disease

Secondly, genetic engineering techniques have solved the problem of conventional methods for diagnosis of many diseases. DNA probe, monoclonal antibodies, and antenatal diagnosis are some of the available methods used as a tool to diagnose a particular disease.

**(*a*) Diagnosis of Parasitic Disease Through DNA Probe :** Probes used for diagnosis of pathogens contain the most specific DNA sequences of genetic material of parasite. The specificity lies in such a way that the other related species or strain do not contain those sequences. These unrelated unspecific sequences of parasite are first recognised by using DNA hybridization technique. Then a DNA sequence, not present in any species, is identified, cleaved by using restriction enzymes, and inserted into a cloning vector (plasmid). The bacterial cells are transformed by the recombinant vector. The transformants are multiplied. Finally, the foreign DNA fragments are retrieved from the host cells. The DNA sequences of the parasite, thus obtained, are labelled with radioisotope and used as a probe. The probe can also be chemically synthesised. Following are the steps for diagnosis of a particular disease :

(*i*) Isolate the parasite from the infected tissue of the patient. Extract the DNA from parasite and purify it,

(*ii*) Break the DNA by using restriction enzymes to get the DNA fragments of varying size,

(*iii*) Electrophorise the DNA solution of different length by using agarose gel just to get a smear of DNA,

(*iv*) Attach the DNA smear to more firmer support by Southern blotting techniques. Thus the filter paper carries the exact replica of the DNA adhered to it,

(*v*) Hybridize the immobilised DNA on filter paper by incubating it with radiolabelled probe. Probe DNA complementary to certain DNA sequences of parasite DNA sticks to it and forms the hybrid, and

(*vi*) Wash the filter paper to remove unhybridized probe. Keep the filter in contact of x-ray film. Dark bands appear where DNA probe had hybridized the specific target sequence. Thus, a parasitic disease is diagonised positive. If dark bands on x-ray film do not appear, the absence of parasite is noted.

Specific DNA probes have been designed which have shown good results in diagnosing the infection caused by viruses, bacteria or protozoa. Tuberculosis (T.B.) caused by *Mycobacterium tuberculosis* is one of such diseases which can be diagonised by this method. Moreover, a complete testing system for T.B. is marketed by Genprobe Inc, California. Similar efforts have also been made for diagnosis of leprosy, *Kala azar* (caused by *Leishmania donovani*), malaria, etc.

**(*b*) Monoclonal Antibodies :** Antibodies are proteins synthesized in blood against specific antigens just to combat and give immunity in blood. They can be collected from the blood serum of an animal. Such antibodies are heterogeneous and contain a mixture of antibodies (*i.e.*polyclonal antibodies). Therefore, they do not have characteristics of specificity. If a specific lymphocyte, after isolation and culture *in vitro,* becomes capable of producing a single type of antibody which bears specificity against specific antigen,it is known as 'monoclonal antibody'. Due to the presence of desired immunity, monoclonal antibodies are used in the diagnosis of diseases.

Dr. Caeser Milstein received the Nobel Prize for medicine for developing hybridoma cells.

Major difficulties with antibody secreting cells are that they cannot be maintained in culture. But myeloma cells (bone marrow tumour cells due to cancer) grow indefinitely to produce a huge quantity of identical cells (clones) and also produce immunoglobulins in the same amount. The immunoglobulins are infact monoclonal antibodies.

(*i*) **Hybridoma.** Cesar Milstein of Argentina tried to culture myeloma cells. In 1973, in collaboration with Cotton, he succeeded to fuse rat and mouse myeloma cells with the result of production of hybrid cells. The hybrid cells secreted the immunoglobulins which consisted of several types of polypeptides. In 1974, George Kohler and Milstein successfully isolated clones of cells from the fusion of two parental cell lines *i.e.* between lymphocyte from spleen of mice immunized with red blood cells from sheep and myeloma cells. These cells were maintained *in vitro* and produced antibodies. The antibodies for specific antigens immunized the myeloma cells. Moreover, the hybrid cells maintained the character of lymphocyte to secrete antibodies, and of myeloma cells to multiply in culture. These hybrid cell lines (cell clones) are known as hybridomas which are capable of producing unlimited supply of antibodies. For the discovery of production of monoclonal antibodies Kohler and Milstein along with Neils Jerne were awarded Nobel Prize in 1984, in physiology and medicine. Now, techniques are developed to obtain antibodies from hybridoma culture (Fig. 7.6).

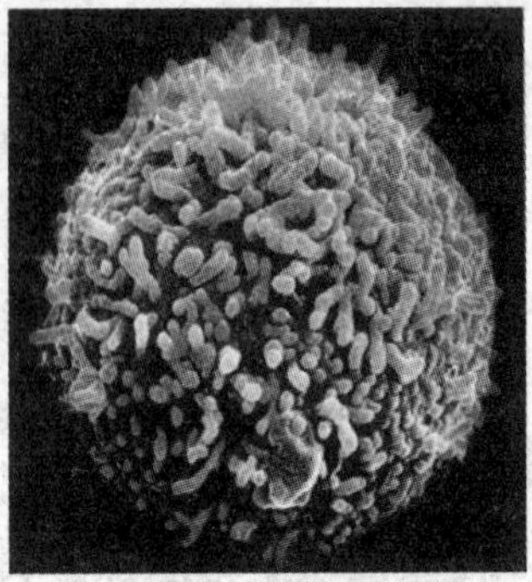

Photograph of a hybridoma cell magnified 2,000 times.

Monoclonal antibodies are produced against a variety of proteins, glycoproteins, glycolipids, nucleic acids and chemically defined groups linked to protein carriers. Possibility of success is related to the number of antibody forming cells in spleen at the time of fusion. This, in turn, depends on immunogenicity of the antigen used. For example, a mouse spleen contains about $10^8$ nucleated cells. Sheep red blood cells (SRBC) act as immunogens and stimulate the replication and differentiation of antibody forming cells. At the time of secondary response each spleen may contain $10^6$ cells producing antibody against SRBC. Among $2 \times 10^5$ spleen cells only one hybrid cell is generated. About 500 hybrid cells per spleen can be obtained where five of these produce specific antibody.

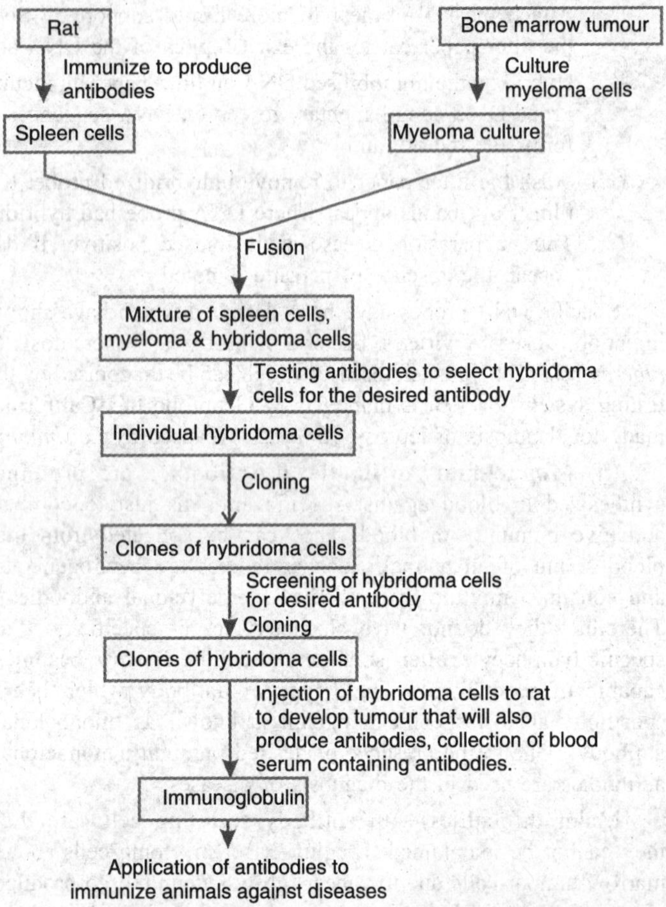


**Fig. 7.6.** Outline of production of monoclonal antibodies.

The cells growing on culture and producing sufficient amount of monoclonal antibodies can be frozen. The cells are collected through centrifugation and resuspended at a density of $10^7$ cells per ml freezing medium in cold at O°C. Samples (0.7-1.0 ml) poured in sterile glass vials are sealed and stored at 80°C.

In India, a hybridoma laboratory has been set up with indigenous material at the Cancer Research Institute (CRI), Bombay. The first experiment conducted at C.R.I. was an attempt to develop leukemia cell reactive monoclonal antibody.

**Uses:** Monoclonal antibodies are useful in the following cases:

(*i*) In diagnosis of ABO blood groups, cancer, pregnancy, allergies, viral diseases and certain hormones.

## C. ANTENATAL DIAGNOSIS OF HAEMOGLOBINOPATHIES

There are as many as 2,500 genetic defects known to occur in human beings. Some of them (*e.g.* phenylketonuria and haemophilia) have been cured but most of them are not understood. If a pregnant women bears a child with a genetic defect, often she is advised to abort it but not to give birth to genetically defected child. Thus, the technique of diagnosis and thereby suggestion for abortion of a genetically defected child is known as antenatal diagnosis. The genetically determined defects of function or synthesis of human haemoglobin is known as haemoglobinopathies.

The inherited haemoglobinopathies are : (*i*) the structural variants where till now 300 amino acid variants are identified, and (*ii*) the thalassaemias, a condition characterized by reduced or absent synthesis of α or ß- globin chain, α and β-thalassaemia respectively.

**(*a*) Methods of Antenatal Diagnosis.** Antenatal diagnosis is done by taking a few millilitres of amniotic fluid from the foetus of about 16-18 week pregnant women. Foetal blood samples are analysed by testing for the ability to synthesize globin chains (alpha and beta). Difficulties with the antenatal diagnosis are that the foetus of less than 3 months of age synthesize very little α or ß- globin chains. At this stage the foetal blood contains normal level of α and low level of β-chain. Synthesis of ß-chain takes place in low amount in foetus of 8 weeks age, and therefore, it is possible to analyse the α globin in a foetus but difficult to analyse ß-chains.

The foetus fluid also contains foetally derived fibroblasts which serve, with or without additional culture, as a source of tissue for preparation of foetal DNA. Structural analysis of DNA is done by: (*i*) isolation of all the globin genes as recombinant DNA molecules, and cloning in bacteria, or (*ii*) without cloning the recombinant DNA in bacteria directly by "Southern blotting technique" and (*iii*) linked polymorphism. There are about 35 diseases which have been identified by antenatal diagnosis. This method helps in suggesting the abortion of defected baby, and if possible, to cure . It can be recommended for gene therapy as well. Thus, the antenatal diagnosis is applied for both genetic counselling and gene therapy.

**(*b*) Genetic Counselling.** Genetic counselling is a technique which is carried out in detail based on antenatal diagnosis and suggested for the possibilities of future progeny. A genetic disease is first identified by a genetic counsellor who understands the family history of genetic disease, and then suggests for the possibilities of giving birth to the child or aborting the child. The child may be normal, genetically defected or carrier of genetic disease. Take the case of thalassaemia. Alpha thalassaemia is caused by the presence of two chromosomes each of them carries one dysfunctional globin gene or one normal chromosome and other dysfunctional genes. Undoubtedly, a person carrying two chromosomes each with single functional α-genes donates one normal genes to their progenies. It is, therefore, impossible to suffer their progenies from 'hydrops fetalis' as it occurs due to complete dysfunctioning of all α-genes. DNA analysis of thalassaemia has revealed that non-function of both α genes on a single chromosome is only due to deletion of both genes from the chromosome.

Recent techniques have made it possible to detect α-thalassaemic couple for the presence of no gene chromosome. However, there is risk for giving birth to hydropic foetuses if both of the couple have α-thalassaemia due to no gene chromosome. If there lies any possibility for gene therapy the patients are also suggested for the same. It has also been suggested that one should go to genetic counsellor before marriage for safe and normal progeny.

### 3. Gene Therapy

There are many diseases which can be cured by using specific medicine synthesised biochemically. Now-a-days techniques have been developed to produce recombinant therapeutic biochemicals, for example insulin, interferon, somatotropin, somatostatin, endorphin, human blood clotting factor VIII:C, immunogenic proteins, etc.

After 1975, a remarkable advancement in recombinant DNA technology has occurred and accumulated such knowledge that has made possible to transfer genes for treatment of human diseases. Several protocols have been developed for the introduction and expression of genes in humans, but the clinical efficiency has to be demonstrated conclusively. Success of gene therapy depends on the development of better gene transfer vectors for sustained, long term expression of foreign gene as well as a better understanding of gene physiology of human diseases.

Genes are the ultimate molecular switches that control various cellular process. The abnormal gene expression can manifest in the form of specific genetic disorders. Until the last decade, delivering genes into humans to correct diseases has been accepted as scientifically viable and recognised as an independent discipline and christened 'gene therapy'.

The ultimate goal of gene therapy is the gene replacement therapy. Gene replacement therapy permits physiological regulation of the transgenes and elimination of the possibility of insertional activation of other cellular genes which occur at the time of random integration of the foreign gene. At present the current strategy for gene therapy largely centres around gene augmentation therapy, where the foreign gene replaces the defective or missing gene.

Overall, there are two strategies for gene transfer: (*i*) the *in vivo* approach which involves introduction of genes directly into the target organs of an individual (it is done in patients therefore, also called patient therapy), (*ii*) *ex vivo* approach where cells are isolated for gene transfer *in vitro* followed by transplantation of genetically modified cells back into the patients.

**(*a*) Types of Gene Therapy** : All the gene therapies that can be done in humans and classified into the following four types:

**(*i*) Somatic gene therapy.** The genetic defects are corrected in somatic cells of the body. It was initially formulated for the treatment of monogenetic defects, but now holds promises for a wide range of disorders such as cancer, neurological disorders, heart diseases and infectious diseases (Table 7.2). Sufficient expertise in performing successful gene transfer in somatic cells is required before carrying out gene manipulation in humans.

**Table 7.2. Genetic disorders and acquired diseases amenable to gene therapy.**

| *Diseases* | *Therapeutic agent* | *Strategy* | *Vector* | *Target cell/tissue* |
|---|---|---|---|---|
| **Genetic Disorders** | | | | |
| Cystic fibrosis | CFTR | *In vivo* | Adenovirus | Nasal epithelium |
| Familial hyper-cholesterolaemia | LDL | *In vivo* | Cationic lipid | Nasal epithelium |
| SCID | ADA | *Ex vivo* | Retrovirus | T cells |
| Haemophilia | Factor VIII/IX | *In vivo* | Retrovirus | Hepatocytes, skin, muscles |
| DMD | Dystrophin | *In vivo* | Retrovirus | Skeletal muscles |
| | | *Ex vivo* | Retrovirus | Myoblasts |
| **Acquired Disorders** | | | | |
| Alzheimer's disease | NGF | *Ex vivo* | Retrovirus | Tumour cells |
| AIDS | HIV antigen | *Ex vivo* | Retrovirus | T cells |
| | RevM10 | *Ex vivo* | Retrovirus | Hepatocyte |

| | | | | |
|---|---|---|---|---|
| | Cytokine | *Ex vivo* | Retrovirus | Hematopoietic stem cells |
| Cancer | Interleukins, | *Ex vivo* | Retrovirus | Tumour cells |
| | HSV-TK | " | " | " |
| | HLA-B4 | *In vivo* | Catonic lipid | Tumour cells |
| | Tumour suppressor gene | " | " | " |
| Cardiovascular disease | tPA | *In vivo* | Adenovirus | Tumour cells |
| Parkinson's disease | TH | *Ex vivo* | Retrovirus | Fibroblast |

CFTR, cystis fibrosis transmembrane regulator; SCID, severe combined immunodeficiency syndrome; DMD, Duchenne muscular systrophy, ADA, adenosine deaminase; HSV-TK, herpes simplex virus thymidine virus, NGF, nerve growth factor TH, tyrosine hydroxylase, tPA, tissue plasminogen activator.

**(*ii*) Germ-line gene therapy.** The functional genes are introduced into the germ cells for correction of genetic defects in the offspring. This therapy is being carried out in laboratory and farm animals. However, it has not been attempted in humans due to technical and ethical problems. One of its types is the *embryo therapy* where embryos are diagnosed for genetic defects. If any such disease is present the patients are advised for embryo therapy or abortion. In young embryo a functional gene is transferred through microinjection technique.

**(*iii*) Enhancement genetic engineering.** This type of gene transfer is done for the improvement of a specific trait in animals; for example introduction of growth hormone gene to increase height. It is being carried out in laboratory and farm animals.

**(*iv*) Eugenic genetic engineering.** Novel genes can be introduced in humans to alter or improve complex traits such as intelligence and personality. This type of therapy is not being attempted in humans because it is far beyond our technical capabilities, and ethical problems.

In 1990, for the first time, Michaele Blease and W. French Andresco of National Institute of Health, Bethesda, U.S.A. attempted gene therapy on a human patient. A four year old girl was suffering from 'severe combined immunodeficiency' (SCID). This disease is caused by a faulty gene which expresses the enzymes adenosine deaminase (ADA). Deficiency of ADA results in the production of a chemical which selectively destroys the T- and β-cells of the immune systems. Finally, the patients die. The scientists introduced a healthy ADA gene into the body of the girl who protected her immune system from damage. This successful trial has given the signal for the dawn of a new era in the field of medical sciences.

**(*b*) Methods of Gene Transfer :** A variety of gene transfer strategies have been developed during the last decade for the treatment of human diseases which can be grouped into the two major categories: the viral and non-viral methods.

**(*i*) Virus vectors.** After 1980, much work has been done on retroviruses as gene transfer vectors, more specifically on murine-leukemia virus (MLV) for gene therapy. Efforts are being made to develop HIV-based vectors so that even non-dividing cells can be injected. The outline of retroviral gene transfer has been shown in Fig. 7.7. The steps of developing a replication-defective recombinant retroviral vectors are : (*i*) the replacement of viral structural genes *e.g. gag, pol* and *env* by the therapeutic foreign genes of interest, (*ii*) transfection of this vector into packaging cell line (*i.e.* producer cells) that provide the viral structural proteins *in trans* so that the recombinant retroviral genome is packed and replication defective retroviruses are generated, (*iii*) transfection of host cells by such viruses, and reverse transcription of recombinant retroviral RNA and random integration into the host genome. In the absence of viral genes, the foreign gene (therapeutic in nature) is transcribed from the viral LTRs, (the long terminal repeats) and desired protein is synthesised (Fig. 7.7). The retroviral vectors are used in *ex vivo* gene transfer experiment, although it has been shown that they can infect a regenerating liver when administered intravenously into hepatectomized animal.

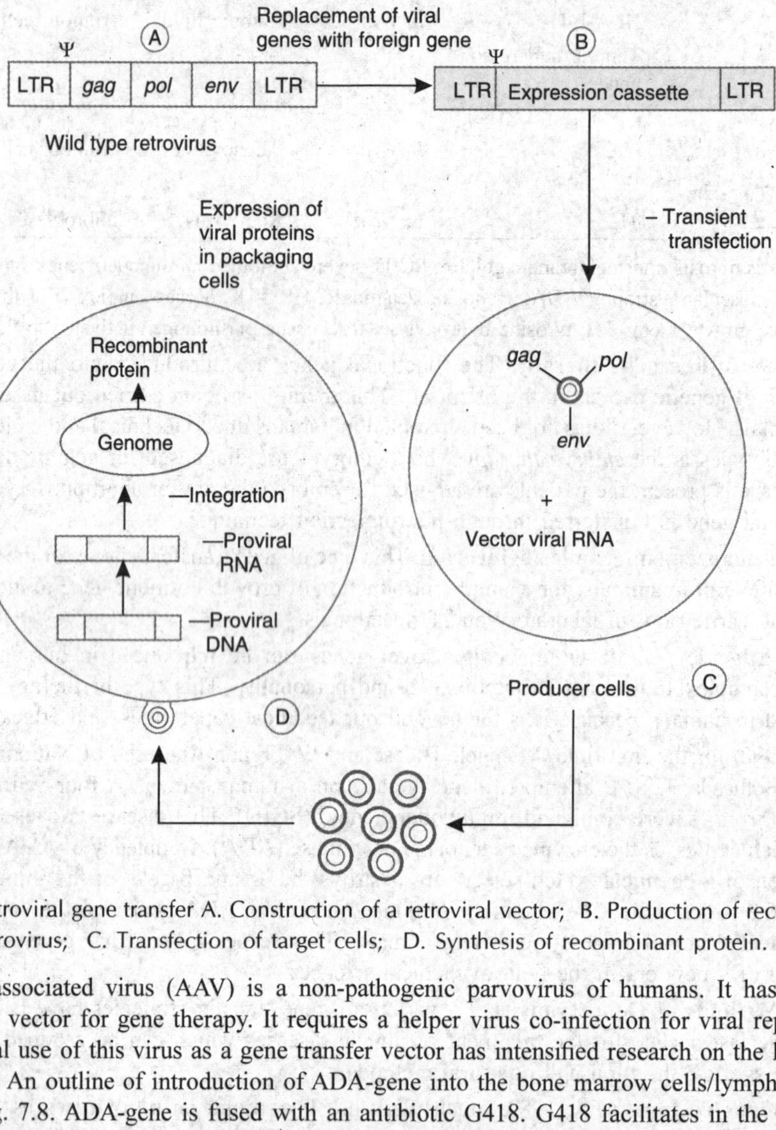


**Fig. 7.7.** Retroviral gene transfer A. Construction of a retroviral vector; B. Production of recombinant retrovirus; C. Transfection of target cells; D. Synthesis of recombinant protein.

Adino-associated virus (AAV) is a non-pathogenic parvovirus of humans. It has aroused interest as a vector for gene therapy. It requires a helper virus co-infection for viral replication. The potential use of this virus as a gene transfer vector has intensified research on the life cycle of the virus. An outline of introduction of ADA-gene into the bone marrow cells/lymphocytes is given in Fig. 7.8. ADA-gene is fused with an antibiotic G418. G418 facilitates in the selection of ADA gene. Therefore, ADA-G418 is cloned into a plasmid containing proviral DNA from a retrovirus. The retroviral genome is defective. Some of the proviral DNA is replaced with the ADA gene. The recombinant DNA is then introduced into fibroblast cell line. A fibroblast cell line harbouring the missing gene is used to produce the mature viral particles upon transfection with the recombinant DNA. When the mature viral particles are released, bone marrow cells are added to growth medium. In turn the viral particles infect bone marrow cells/lymphocytes. The bone marrow cells/lymphoctyes are then selected by using G418 antibiotic in growth medium. Thereafter, the cells are transferred to a second flask containing G418 in growth medium. In this way the selected bone marrow cell/lymphocytes are then re-injected back into experimental mice or for trials in children.

(*ii*) **Non-viral approaches.** Most of approaches in gene therapy has been focused on development of viral vectors rather than clinical efficacy and biosafety. Furthermore, techniques of virus-based gene therapy is very costly and complex. Therefore, this led to the development of non-viral gene therapy. Majority of non-viral methods follow the *in vivo* approaches. Therefore a gene can be used successfully as a drug. In general, the non-viral methods for gene transfer can be grouped into two methods, physical and chemical methods.

**Physical methods.** The various forms of physical methods are given below:

- **Microinjection** : It is a very tedious method. However, it is used in oocytes, eggs, and embryos. In 1993 Jeffrey S. Chamberlain and coworkers from of Human Genome Centre, Michigan University U.S.A. have cured mice that inherited a neuro-muscular disease which is similar to muscular dystrophy of humans. The mutant mice lacking 427 K gene in brain and muscle cells had a weak diaphragm which gradually degenerates like that in dystrophy affected humans. Chamberlain and coworkers microinjected the DNA containing 427 K gene into the zygote of mutant mice. The transplanted genes worked properly and produced necessary protein and thus, prevented the diaphragm.
- **Direct DNA injection.** Direct injection of DNA into skeletal muscle has created a lot of excitement and led to the possibility of using gene as vaccines. Direct gene transfer was attempted since early 1960s but these studies were not done seriously due to inefficient gene transfer and low level of expression. Due to low level of expression therapeutic benefits for the treatment of genetic disorder could not be derived. Thereafter, these studies were further carried out. The low levels of foreign proteins synthesised in the host cells were processed along the class I MHC pathway leading to presentation of peptides on cell surface resulting in a potent immune response against the foreign antigens. This gave the birth to the concept of DNA vaccine or genetic immunization (Rangarajan and Padmanaban, 1996) (see earlier section *DNA Vaccines*).
- **Receptor-meditated gene delivery.** It is an attractive strategy. It takes advantage of normal physiological pathway. The receptors are exclusively present on hepatocytes. They bind to certain glycoproteins lacking the terminal asialic acid. The concept of delivering genes into specific tissue was

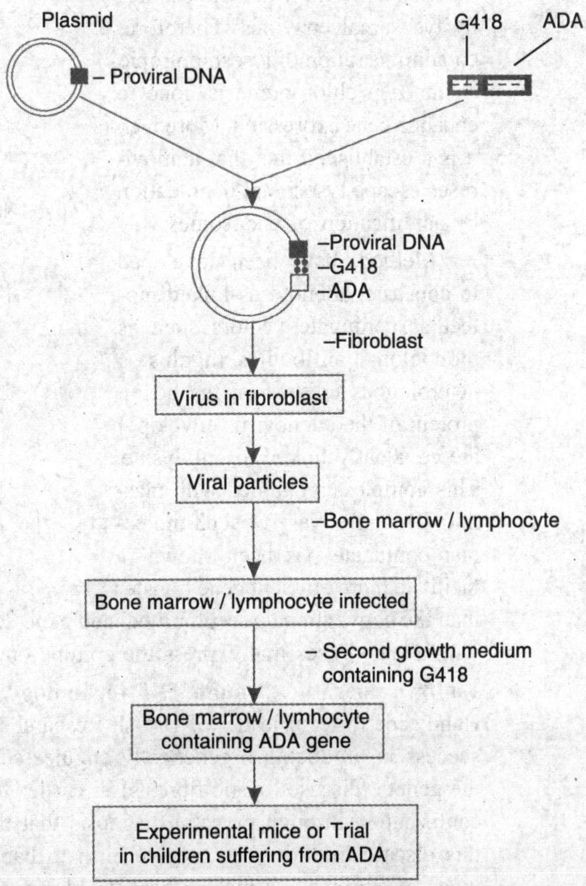


**Fig. 7.8.** An outline for introduction of ADA gene into the bonemarrow cells (diagrammatic).

examined extensively with the liver asialoglycoprotein receptors. An asialoglycoprotein (orosomucoid) is coupled with poly-L-lysine. This conjugate is then condensed with plasmid DNA via electrostatic interactions (Fig. 7.9). This soluble complex *i.e.* asialoglycoprotein-poly-L-lysin-DNA complex (ASGP-PL-DNA) is injected intravenously/intraperitoneally, which gets entered into hepatocytes via the asialoglycoprotein receptors (ASGPR). DNA is delivered into the nucleus. The transgene expression occurs, therefore, recombination protein is detected for several weeks.

In 1991 G.Y. Wu and coworkers later extended the other cell surface receptors such as transferrin receptor. The major problem related to receptor mediated gene delivery system has been the degradation of DNA in the endosomes by lysosomal enzymes. Therefore, co-administration of lysosomotropic agents *e.g.* chloroquine is done to enhance gene expression. Moreover, it is a established fact that adinoviruses escape lysosomal degradation by acidification of endosomes.

Methods have been developed to construct adenovirus-linked molecular conjugate vectors such as monoclonal antibodies (against a heterologous epitope on the hexon protein of the adenovirus envelope) are covalently linked to polylysine. This complex is coupled with plasmid DNA. Delivery of such molecular conjugate has been shown to facilitate more efficient gene transfer than the conventional receptor mediated gene delivery strategy. Such a conjugate interacts with adenoviruses that express the epitope on the envelope.

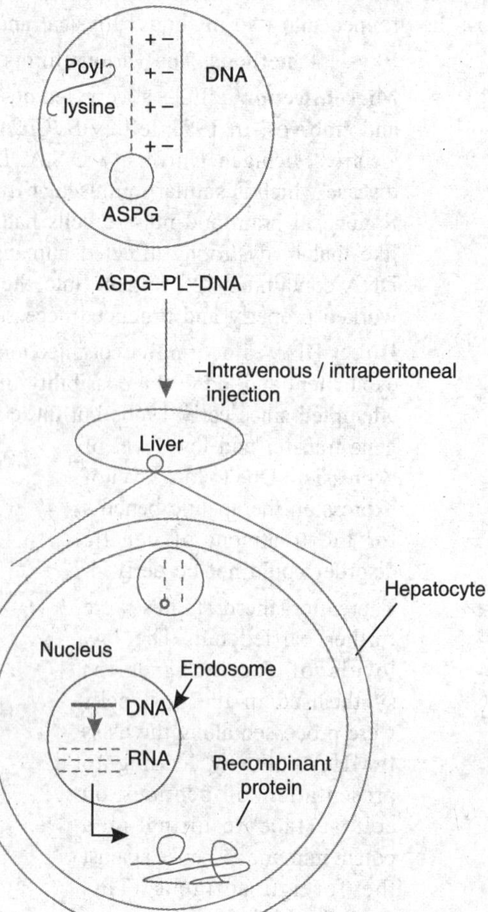


**Fig. 7.9.** Receptor-mediated gene delivery (diagrammatic).

- **Embryo therapy through IVF-technology.** In 1993, adopting IVF-technology, A. Handyside (a team leader of British Researcher at Hammersmith Hospital, London) got success in producing a genetically engineered female baby, whose parents transmitted the genetic disease (cystic fibrosis) in earlier four babies, who died later on. Even after confirmation through prenatal diagnosis that the babies would be sufferer or carrier of this disease they decided to give birth to babies, which failed. This disease clogs the lungs of sufferer and makes them unable to digest the food properly.

Handyside admitted seven couples. Their eggs were collected and fertilized with their respective husbands' sperm in test tubes. After 3 days incubation *in vitro*, when embryo reached 8 cell stage, a single cell was removed for analysis. The cell was magnified one million times by using a magnifying Thermal Cycler. The individual defective gene which causes cystic fibrosis was identified and screened out. Thereafter, the embryo was implanted in the womb of mother. Consequently, a female baby (C. O'Brien) was born who is neither sufferer nor carrier of this diseases, but a normal one. The birth of this genetically engineered baby has increased the hope of millions of couples who transmit the genetic disease to their offsprings, and also of researchers to look upon many more genetic diseases to cure.

(*c*) **Success of Gene Therapy** : The success of gene therapy depends on gene delivery mechanism as well as on the choice of target tissue. In 1996, P.N. Rangarajan and Padmanaban have discussed different conditions leading to success of gene therapy:

(*i*) Cell types capable of dividing *in vitro* (*e.g* myeloblasts, hepatocytes, keratinocytes, endothelial cells, etc.) are amenable for *in vitro* and *in vivo* gene therapy, both the *in vivo* methods are preferred for cell types such as neuronal cells.

(*ii*) The function of gene products also govern the selection of tissues, for example in case of haemophilia a gene can be delivered in any tissue provided the gene product is released into blood stream. In addition, in case of cystic fibrosis the gene should be delivered to specific cell types where introduction of correct gene is required.

(*iii*) Another attractive strategy for the treatment of several disorders is antisense gene transfer, for example in β-thalassemia, α-globin chains are accumulated in RBCs that result in their premature decay. Such types of destruction can be prevented by infection of K562 erythroleukemia cells with AAV expressing human α-globin gene in antisense orientation.

(*iv*) The potential of a gene delivery system is first found out in cultured cells or laboratory animals by using reporter genes (for luciferase, growth hormone, β-galactosidase, etc. For detailed discussion on reporter genes *see* Chapter 9). On the basis of result of these studies again pretrials are conducted in laboratory animals to test the level and duration of expression of the introduced genes. Finally clinical efficacy is evaluated in human patients. But before conducting clinical trials in humans permission is necessary from the Regulatory Agencies such as recombinant DNA Advisory Committee (RAC) and the Food and Drug Administration (FDA) in the USA, the Gene Therapy Advisory Committee (GTAC) in the UK, etc. In India also guidelines have been formulated (see Chapter 22, *Biosafety*). Clinical trials on gene therapy are given in Table 7.3.

**> Table 7.3.** **Clinical trials on gene therapy.**

| *Disease* | *Gene inserted* | *Cell types* | *Remark* |
|---|---|---|---|
| ADA deficient SCID | *ADA*-gene | Peripheral T-lymphocytes and bone marrow cells | Repeated injections to be given |
| Duchenne's muscular dystrophy | Dystrophin gene | Myoblasts | Trials in USA |
| Haemophilia B | Factor IX fibroblast | Autologous skin | Trials in China |
| Cystic fibrosis gene | *CFTR* | – cells via liposomes | Gene delivery to lung |

**(*d*) Future Needs of Gene Therapy in India :** Gene therapy has tremendous scope for treatment of several disorders such as cancer, cardiovascular and neurological disorders. To initiate gene therapy the functional studies should be focussed on (*i*) design of newer vectors for gene delivery, newer approaches to systemic delivery, (*ii*) targeting to specific tissues and cells, (*iii*) stability and duration of expression of gene introduced, (*iv*) the statutes of the introduced genes *in vivo,* (*v*) design of appropriate animal models, (*vi*) assessment of risk-benefit states, and (vii) an understanding of molecular basis of cellular humoral immune response in case of DNA vaccines.

The need of hours is to start more studies in different systems on various aspects realising the significance of approaches of gene therapy. Department of Biotechnology, Govt of India has constituted a group of experts to activate this area of research. At present there are only 3-4 laboratories in this country that have started work on this area. At South Campus of Delhi University, in 1994 Bagai and Sarkar have demonstrated the delivery of macromolecules such as toxins, drugs and nucleic acids into viable cultured cells through the use of reconstituted Sendai virus envelop (RSVE). This approach needs to be extrapolated under *in vivo* conditions. At the Cancer Research Institute (Mumbai) treatment of oral cancer through gene targeting to oral mucosal cells has been demonstrated. In 1995, at the Indian Institute of Science (Bangalore), Prabhu and co-workers have demonstrated the reporter mediated targeting and expression of growth hormone gene in rat liver by using a regulatable promoter (cytochrome P-450 promoter induced by phenobarbitone). In 1996, Rangarajan and co-workers have demonstrated a newer approach to delivering DNA to hepatocytes by perfusion of liver with DNA-lipofectin complex.

## C. DNA PROFILING (FINGERPRINTING)

DNA is the master molecule of all life forms. It constitutes the blue prints of every living organism through which characters are passed generation to generations. However, every living organism differs only due to nucleotide sequences of chromosome. The nucleotides form codes; therefore, this coded genetic informations can be profiled to produce most authentic identity card of any organism. The complicated technology that facilitates the identification of individuals at genetic level is known as *DNA fingerprinting* or more specifically *DNA profiling.* This genetic analysis is based on identifying tiny segments of the hereditary material which testify the unique molecular signature which cannot be altered. With the help of DNA profiling technology identity of burnt or unidentifiable dead body can be done.

The basic requirements for DNA profiling is the availability of biological samples which can be anything including blood stains, a piece of hair with its root, few drops of semen, skin cells, a mouth swab, cells of bone marrow of any body tissue.

In 1984, for the first time, a British geneticist, Alec Jeffreys working at the University of Leicester developed the technique of DNA fingerprinting. This method has been patented, and being used in Europe and America. Its success is based on identification of small segment of DNA which testifies the unique molecular signature that every person on earth possesses and which cannot be changed in one's life time.

**VNTRs:** It is important to note that 99% of base sequences is the same in the DNA of all human beings. Only a short stretches of sequences which is about 3 billion base pairs, differ person to person. Out of the total DNA, about one in 1000 base pairs is a site of variation in the population. This is an unusual sequence of bases of 10-15 base pairs long which is repeated several times. These bases were first isolated and identified by A. Wyman and R. White in 1980 at the University of Utah as hypervariable segments of DNA and termed as 'variable number of tandem repeats' (VNTRs). On the basis of number of repeats present in VNTRs, the length of segment is measured. In different person the length of VNTR differs and, therefore, VNTR is the key of DNA fingerprinting.

It was Jeffreys who first used VNTRs in forensic science. The VNTRs act as genetic markers. The scheme developed by Jeffreys for measuring the length of fragments of DNA containing the repeated sequences is also known as 'restriction fragment length polymorphism' (RFLP) testing.

### 1. Method of DNA Fingerprinting

In this technique DNA is isolated from blood stains, semen stains or hair roots of disputed children or any suspected person and the same from parents, close associates or relatives of suspected criminals (based on the cases). Since hairs contain less amount of DNA, it can be produced in a large amount by using polymerase chain reaction (PCR). RBCs do not contain DNA, therefore, WBCs are the source of DNA. Thus the DNA isolated is cut with restriction enzyme and subjected to Southern blotting. The DNA bands appearing on membrane are hybridized with $^{32}$P-DNA probe, washed in water to remove unhybridized DNA, and passed through X-ray. The hybridized complementary DNA sequences develop images (prints). Identical prints that contain specific DNA sequences appearing on two X-ray films are identified and thus identity is confirmed. Probability of two persons having similar sets of base pairs in the same sequence of VNTR of DNA is one in 30-300 million individuals.

In India, facilities of DNA fingerprinting at international level are available at the Centre for DNA Fingerprinting and Diagnostics (CDFD) (Hyderabad), the then Centre for Cellular and Molecular Biology (CCMB) (Hyderabad). At this centre the Jeffreys technique is not followed. Dr. Lalji Singh has developed an indigenous technique where 'BKM'-DNA probe is used for hybridization of DNA sequences. While he was working on sex determination in snakes in his Ph.D. programme (in BHU, Varanasi), he found a contrast result. Unlike human and other organisms, the female snake contains XY and the male YY sex chromosomes. A segment of DNA was isolated from sex determining Y chromosome of female banded krait (*Bungarus fasciatus*), an Indian poisonous snake. This unique segment was named as 'banded krait minor satellite' (BKM). It is similar to sex determining chromosome in humans. The probe which is used for this purpose is, therefore, called BKM-DNA probe.

### 2. Application of DNA Profiling (DNA Fingerprinting) Technology

DNA profiling technology has applications in many more areas as given in Fig. 7.10 and discussed below:

**(*a*) Setting up of Genetic Databank :** Realising the potential of DNA profiling genetic databank had been/is being set up thoughout the world. Databases of the genetic fingerprints of criminals have been set up in many parts of the USA and Britain. Forensic Science Services Birmingham (Britain) has set up the world's first National DNA Database, where about 5 million record would be available. It has collected samples of tissues and DNA of suspected individuals has been profiled.

The Armed Forces Institute of Pathology (Gaithersberg, USA) has collected DNA samples from every member of the armed services. The samples are stored in freezers in DNA repositories. Each sample consists of blood blots on cards, sealed in individual envelopes. A computer database of all samples has been prepared.

**(*b*) Reuniting the Lost Children :** DNA profiling technology helped in reuniting the lost children with their respective parents or *vice versa*

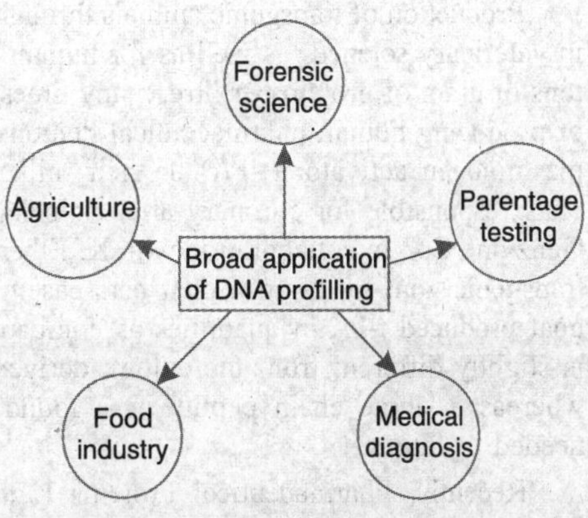


Fig. 7.10. Application of DNA profiling in many areas (based on Chawla, 1998)

who were separated during war, violence or natural disasters. For example, in Argentina many people were abducted during military rule (1976-1983). They were presumed to be died. But an Arzentinian Human Right Group (the Grandmother's of Plaza de Mayo) was of the opinion that the children are still alive and adopted by military personnel after killing their natural parents. In 1987, a genetic databank storing blood samples of grand parents and relatives of lost children was set up. By matching the DNA profile of one or more grand parents with that of lost children about 40 missing children have been reunited with their families.

**(*c*) Solving Disputed Problem of Parentage, Identity of Criminals, Rapists, etc :** To carry out the test a small portion of blotted blood is punched out from which blood cells are isolated and information is taken from computer databank. The first case from preserved sample was done to identify a soldier who was burnt to death in a car accident.

In India, a disputed parentage has been solved. In June 1988, a four year female baby named Laxmi was stolen from Chennai by a missionary and renamed as Merry by the so called parents. After FIR and taking police help the matter was not solved. Therefore, the real parents went to High Court. The High Court sought help from CCMB. The scientist performed DNA fingerprinting tests, of baby and parents (real and false), confirmed the identity of the baby, and submitted the report to the court. On the basis of report of DNA fingerprinting decision was given by the court, and Laxmi was handed over to her real parents. This is the first case in India where a court has given decision based on the report of DNA fingerprinting.

The second DNA fingerprinting–based decision of identity confirmation of a brutally killed lady was done in Delhi, in 1995. During those days the name of unidentifiable dead body was in light. The lady was killed in her home by unknown murderers. The dead body was dumped in a car and taken to a restaurant where it was consigned to the hot flames in a tandoor, and limbs had been brutally severed to unravel the identity. After this incidence no body came forward to receive the dead body neither parents nor relatives. Thanks to DNA fingerprinting technology of CDFD that confirmed the identity of brutally killed and roasted lady as Smt. Naina Sahni. Her husband has been sentenced to life imprisonment by the honourable supreme court of India.

**(*d*) Immigrant Dispute :** More-or-less, in every country, there is the problem of immigration. Ultimately, the identity is not confirmed whether they have crossed the borders or are the resident of that country. Therefore, the problem of immigrants can be solved through DNA databank.

## D. ANIMAL AND PLANT IMPROVEMENT

### 1. Transgenic Farm Animals

Production of transgenic animals through microinjection techniques will play a significant role in veterinary sciences as well as for human welfare. A farm animal can produce milk containing tons of gram of protein per litre, many times more than a litre of bacterial culture. The possibility of producing human pharmaceutical demonstrated when lactating mice secreted the active tissue plasminogen activator (t-PA) in their milk. The t-PA is an enzyme which dissolves the blood clots responsible for coronary artery blockage that results in heart failure. John Mc Pherson at Genzyme Co., in collaboration with K. Elbert and colleagues of Tuft University produced the first transgenic goat by using a goat beta-casein promoter gene linked to t-PA gene. The transgenic goat produced t-PA in quantities as compared to cell culture derived material. But goat peptide is slightly different from the culture derived t-PA. The goat peptide has two chains of peptide, whereas a single chain peptide was found in culture derived t-PA. Therefore, much work is needed in future.

Recently, Pharmaceutical Proteins Ltd. has made contract with Bayer Co. (Germany) to produce alpha-1 antitrypsin by using transgenic sheep. Alpha-1 antitrypsin, a glycoprotein, has now been approved in the U.S.A. as a replacement therapy. The scientists are of the opinion that about 12 kg sheep/lactation alpha-1 antitrypsin can be produced.

Transgenic cow.

Production of transgenic cow has also been demonstrated in the Netherlands but unsuccessful. The eggs of cow collected from ovaries from slaughter houses were fertilized *in vitro* and microinjected with transgenes containing lactiferous gene. The embryos were implanted in the uterus of surrogate mothers. Out of 102 injected embryos, 21 animals were pregnant and 10 calves were borne. Two of these, one male and the other female, carried transgene. The male showed the presence of lactoferrin gene but not the female. A male never gives out milk. Till the animal secretes milk, the experiment is incomplete. But this experiment has given a hope for future.

Caseins are the major proteins of the milk. By transferring genetically manipulated casein genes to farm animals, the texture of cheese and heat stable dairy products can be improved. Casein genes have already been cloned and introduced into bovine genome.

Recently, T.J. Pandian and colleagues of Madurai Kamraj University have succeed in producing transgenic fishes by microinjecting growth hormone gene (human or bovine). A fish produces a large number of eggs at a time. It is much easier to work on eggs for gene manipulation. In developing countries like India, where majority of people suffer from malnutrition, the aquaculture programmes can be intensified for production on a large scale of good quality fishes. For details of animal biotechnology and transgenic animals, see Chapters 10 and 11.

## 2. Crop Improvements

Population of the country is rapidly increasing day by day. However, the most urgent need to the nation is to meet the food demand of the people. To remain self-dependent and to make them available with quality food, the biotechnologists have to boost up the gene revolution programmes for crop improvement so that the improved crops should have high yield, high amount of digestible and quality proteins, and vitamins, disease/pest resistance and drought / herbicide/salt tolerance, etc. These can be done through the transfer of beneficial genes of a prokaryotic microorganism/ eukaryotic incompatible plant in a given crop plant, alteration in metabolic pathways, and making the plants resistant to invasion of pathogens/pests and herbicides/ salt stress. Methods of gene transfer in a desired eucaryotic cell are : through *Agrobacterium,* through virus, co-cultivation of cells, leaf disc transformation, electroporation, etc. These are discussed in the following sub-sections.

(*a*) ***Transgenic plants*** : *See* Chapter 13.

(*b*) ***Nif gene transfer*** : *See* Chapter 13.

(*c*) ***Phaseolin gene transfer*** : *See* Chapter 6; *See* sunbean plant.

(*d*) ***Conversion of $C_3$ plants to $C_4$ plants*** : *See* Chapter 13.

(*e*) ***Herbicide resistant plants*** : *See* Chapter 13.

(*f*) ***Insect pest resistant plants*** : *See* Chapter 21.

(*g*) ***Plant Improvement Through Genetic Transformation*** : See chapter 6

## 3. Crop Protection

See Chapter 21.

# E. ABATEMENT OF POLLUTION THROUGH GENETICALLY ENGINEERED MICROORGANISMS

For detoxification and degradation of toxic chemicals, enzymes are encoded by specific genes present on plasmids. A.M. Chakraborty and co-workers in 1977 succeeded in isolating the microbial culture which could utilize a number of organic chemicals, toxic in nature, such as salicylate, 2,4-D, 3 chlorobenzenes, ethylene, biphenyls, 1,2,4-trimethylbenzene, 2, 4, 5-trichlorophenoxyacetic acid, *etc*.

Genes responsible for degradation of environmental pollutants, for example, toluene, chlorobenzene acids, and other halogenated pesticides and toxic wastes have been identified. For every compound, one separate plasmid is required. It is not like that one plasmid can degrade all the toxic compounds of different groups. The plasmids are grouped into four categories:

(*i*) OCT plasmid which degrades, octane, hexane and decane,

(*ii*) XYL plasmid which degrades xylene and toluenes,

(*iii*) CAM plasmid that decompose camphor, and

(*iv*) NAH plasmid which degrades naphthalene.

Anand Mohan Chakrabarty (an Indian-born American scientist) produced a new product of genetic engineering called as superbug (oil eating bug) by introducing plasmids from different strains into a single cell of *P. putida.* This superbug is such that can degrade all the four types of substrates for which four separate plasmids were required (Fig. 7.11).

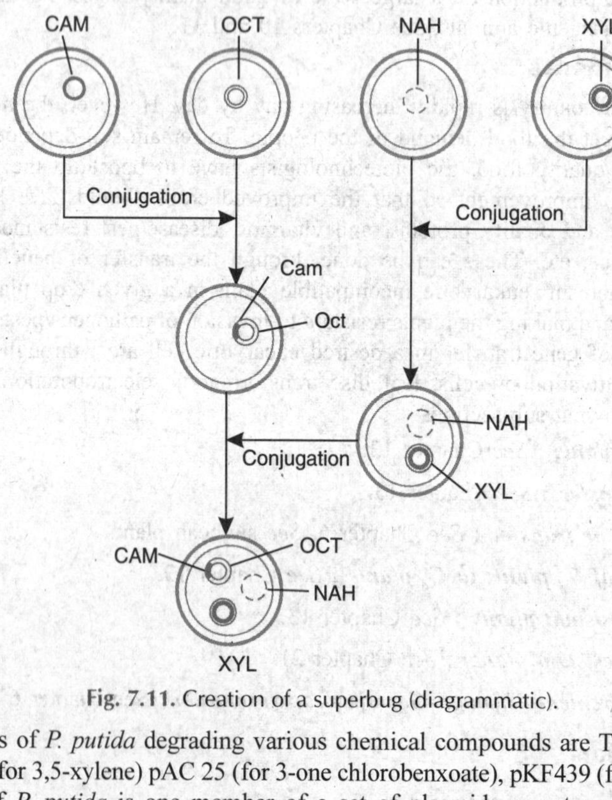


**Fig. 7.11.** Creation of a superbug (diagrammatic).

The plasmids of *P. putida* degrading various chemical compounds are TOL (for toluene and xylene), RA500 (for 3,5-xylene) pAC 25 (for 3-one chlorobenxoate), pKF439 (for salicylatetoluene). Plamid WWO of *P. putida* is one member of a set of plasmids now termed as TOL plasmid. WWO is propagated in *E. coli*.

For a detail discussion *see* Chapter 25.

## PROBLEMS

1. Write an essay on application of genetic engineering principles in medical sciences.
2. How are genetic engineering techniques helpful in crop improvement ? Discuss with examples.
3. What do you know about monoclonal antibodies ? How are they formed ?
4. What is antenatal diagnosis ? What will you suggest to a patient after this test and why ?
5. At what stage genetic counselling can help the persons ?
6. Give a brief account of applications of genetic engineering techniques ?
7. Write an extended note on recombinant malaria vaccine and its future prospects.
8. What do you know about DNA profiling (fingerprinting)? Write in brief application of DNA profiling.
9. Discuss in detail about gene therapy and different methods used in gene therapy.
10. Write short notes on the following :

    (*i*) Humulin, (*ii*) Somatotropin, (*iii*) *Nif* gene, (*iv*) Biologically synthesized vaccines, (*v*) Insulin, (*vi*) Genetic diseases (*vii*) Monoclonal antibodies, (*viii*) Gene Therapy, (*ix*) Antenatal diagnosis of haemoglobinopathies, (*x*) Genetic Counselling, (*xi*) Superbug, (*xii*) Dr Anand Chakrabarty, (*xiii*) DNA vaccine, (*xiv*) DNA profiling, (*xv*) Genetic databank, (*xvi*) Hurdles to DNA fingerprinting, (*xvii*) Methods of gene therapy, (*xviii*) *Recombinant vaccine for hepatitis B virus.*

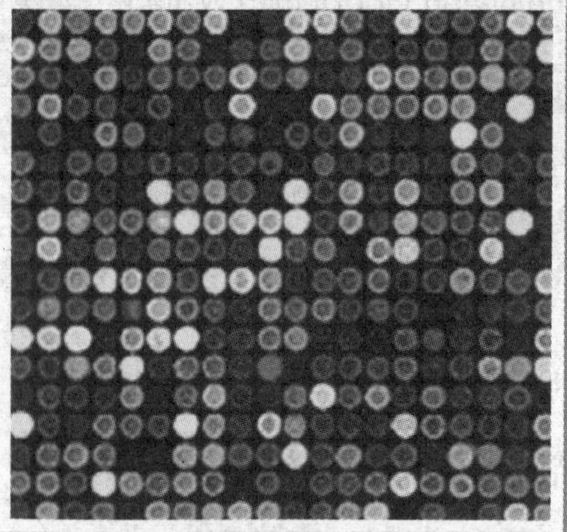

CHAPTER

8

# Genomics and Proteomics

The knowledge of biology is developing rapidly with the development of methodologies. In the beginning of $20^{th}$ century the Mendel's laws were rediscovered. Biologists could know that hereditary traits are transferred from generation to generations through genetic factors. Later on, the genetic factors were discovered as the chromosomes. Thereafter, it could be isolated and manipulated artificially. The end of $20^{th}$ century witnessed that the laboratory-based biology can be used as a tool for earning money as well as storing, analysing and making the use of all genetic information for further uses in multifarious areas.

In future biotechnology will not have to spend much time for cloning genes because they would be able to generate new questions and hypotheses from computer analysis of genome data. Their hypotheses can be tested in laboratory.

## A. GENOMICS

Thomas Roderick

In 1920, H. Winkler coined the term *genome* to describe 'the complete set of chromosomal and extra-chromosomal genes of an organism or virus'. In eukaryotes DNA is also present inside the mitochondria and chloroplasts (in plants only). After a gap of 66 years, Thomas Roderick in 1986 first used the term *genomics* and described it as a scientific discipline of mapping, sequencing and analysing the genome. This definition is unclear but emphasises the systematic exploitation of genome information to answer several queries arising from biology or its related areas. Now the number, location, size and organisation of all genes require to make up an organism can be known. Thus genomics is the study of molecular organisation of genome, their information contents and the gene products they encode.

Significance of genomics was substantially increased when the Human Genome Project was conceived of in 1987. Officially the **Human Genome Project** was started on October, 1990 in the U.S.A.

### 1. Genome Sequencing Projects

Throughout the world, scientists were trying to sequence the genome completely of several organisms of important groups. The reasons of sequencing the genome are given below:

(*i*) It provides knowledge of total number of all genes.

(*ii*) It shows relationships between genes.

(*iii*) It provides opportunities to exploit the sequence for desired experimentation.

(*iv*) It provides all genetic information about the organisms.

(*v*) Genome sequences act as an archive of all genetic information.

Now, the number, location, size, organisation of all genes required to make up an organism can be known.

During the mid 1980s, the U.S. Department of Energy (U.S.D.O.E.) started several projects to: *(i)* construct the detailed genetic and physical maps of the human genome, *(ii)* determine its complete nucleotide sequence, and *(iii)* localise its estimated 1,00,000 genes. The new computational methods were developed for the analysis of genetic map and DNA sequence data. Design of new techniques and instrumentation for DNA analysis was demanded. In order to make the results available rapidly to the scientists, the projects used the advanced means of collaborative work resulting in the development of Human Genome Project.

It was planned to sequence human genome within 15 years and 5 years goal was published. J.D. Watson pledged to encode the genome by 2005 by expending $ 3 billion. Research work on Human Genome Project was successfully completed only due to international collaboration involving 60 countries, 20 genome research centres and more than 1,000 scientists. Up to 1990, some laboratories sequenced 1,00,000 nucleotides. Human Genome Project got assisted internationally and significant informations were generated. A two step process was used to sequence the genome by analysing the mapped clones covering the human chromosomes. This step was called 'shotgun' sequencing (*see* preceding section) and assembly of random fragments of each clone. It was followed by an expensive and labour-intensive 'finishing process' involving the closing of gaps and resolution of ambiguities. Much of the human genome sequencing effort used BAC clones but not YAC clones (*see* Chapter 5) because the YAC clones are less stable during propagation and contain non-contiguous inserts.

By May 2002, sequencing of human chromosome 21 was completed. On June 26, 2000 the 'working draft' of human genome was completed. Human Genome Project and Celera Genomics, Inc. (U.S.A.) jointly announced the completion of a rough draft of the human genome. The problem known as *completion* refers to filling the gaps (such as the 11 small gaps in human chromosomes 22 sequence) which have been difficult to clone and/or sequence. Announcement of completion of this draft captured the imagination of people across the world as man had walked on the moon. 86% human chromosome was evaluated and classified to be highly accurate. The final draft was compiled from 22 billion bases covering the total human DNA sequences. The genome was sequenced in 6 months at the rate of 1,000 bases per second. Human genome is large consisting of $\sim 3 \times 10^9$ base pairs and a lot of repeated sequences.

The 3 billion characters in the DNA sequence that make up the human genome were translated into biologically meaningful information by using computer. This act gave birth to a new field called

'Bioinformatics' (*see* Chapter 9). The field of genomics relies upon bioinformatics. It holds the key to unlock these data for the next generation of innovations.

Besides, complete whole genomic sequences of over 1,000 viruses, over 100 microbes and 9 eukaryotes are known. Hundreds of genome sequencing projects are going on in different parts of the world and many are being started. Some of the completely sequenced genomes include phage $\lambda$, HIV, *E. coli* ($4.639 \times 10^6$ bp), *Helicobacter pylori* causing stomach cancer ($1.67 \times 10^6$ bp), yeast *Saccharomyces cerevisiae* ($12.1 \times 10^6$ bp), nematode worm *Caenorhabditis elegans* ($100 \times 10^6$ bp) and the fruitfly *Drosophila melanogaster* ($160 \times 10^6$ bp). Many other genomic sequencing projects are underway including plants, fish, etc. By the end 1999, 97% chromosome 22 ($34.5 \times 10^6$ bp) was sequenced.

In 1995, the completely sequenced genomes of the first two smallest bacteria, *Haemophilus influenzae* Rd (1,830 kb) and *Mycoplasma genitalicum* (580 kb), were reported. In 1996, the first yeast *Saccharomyces cerevisiae* genome sequencing was completed. In 1997, sequencing of genomes of the two best studied bacteria, *Escherichia coli* and *Bacillus subtilis* was completed. The sequence data of *M. genitalicum* are important because these help to establish the minimal set of gene required for free-living existence. There appear to be 517 genes, out of which 140 genes code for membrane proteins and 5 genes for regulatory function.

Several computer programmes were adopted for data analysis. To carry out sequence assembly, several new computer programmes could also be written. Genome sequencing of many bacteria including those of medical and industrial importance, and from extremes of environment (*e.g.* volcano and deep seavents) were determined at 'The Institute of Genomics Research, (TIGR),' Maryland (U.S.A.). This institute was established by Craig Venter. Complete genome sequencing of *Mycobacterium tuberculosis* was done by several European groups and that of *Bacillus subtilis* at Pasteur Institute and Sanger Centre.

With the establishment of Celera Genomics, Inc. (U.S.A) the commercial era of genomics was started. Celera Genomics sequenced human and mouse genomes. The human genome having over $3 \times 10^6$ bp also possesses repeated sequence. The presence of repeated sequence creates difficulty in sequence assembly. But it may be a little easier to assemble if the parts contain the genes.

With the start of 2000, genomes of 23 different unicellular microorganisms (5 archaeal, 17 bacterial, and 1 eukaryotic organism(s) had been completely sequenced.

## B. METHODS OF GENE SEQUENCING

The strategy of Human Genome Project was to prepare a series of maps of human chromosome with finer resolutions. For this the chromosomes were divided into small pieces which should be cloned. Then the chromosome pieces were arranged in such a way that could correspond to their locations on the chromosome. Each fragment arranged in an order was sequenced after preparation of map.

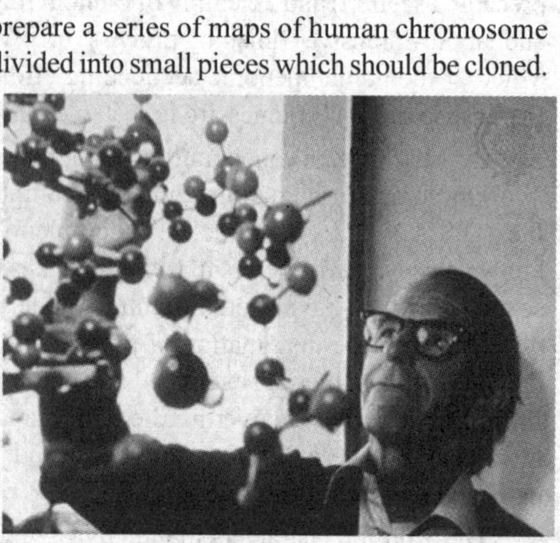

Fredereck Sanger

In 1975 Fredereck Sanger developed a technique which was most widely used for DNA sequencing. He used dideoxynucleotide triphosphates (ddNTPs) in DNA synthesis (*see* Chapter 6). Recently, automated system for DNA sequencing has been developed where ddNTPs are labelled with fluorescent dyes, each of different colours. Each ddNTP fluoresces and imparts different colours which are scanned by a detector and sequence is determined from the order of colours formed in band of gel.

There are several methods used for small scale sequencing of genome. But these methods do not sequence the entire genome. The two direct methods used for genome sequencing and one indirect method (using mRNA but not DNA) have been discussed in this section.

### 1. Direct Sequencing of Bacterial Artificial Chromosome (BAC)

The BAC vectors have already been discussed in Chapter 5. BAC vectors are stable and introduce a complex foreign DNA of 80-100 kb in *E. coli* cells. Therefore, BAC is used in construction of genomic library. Screening of genomic library is done through searching of common restriction fragments. Then BAC clone mapping is done just to determine the arrays of contigs (*i.e.* contiguous clones) which overlap. The large DNA fragments are broken into small pieces and the mapped contigs are sequenced. Thus the direct sequencing procedure involves the sequencing of small pieces of DNA taken from adjacent stretches of a chromosome.

### 2. Random Shotgun Sequencing

It has earlier been discussed in Chapter 6. In this approach different types of cloning vectors are used based on genome size of organisms. For example, shotgun libraries of genomic DNA are constructed in small inserts (10 kb) plasmid vector, and medium insert (10 kb plasmid vectors). However, most libraries from organisms with larger genome are constructed using phage λ, cosmid, BAC or YAC vectors; therefore, DNA inserts of about 23, 45, 350, and 100 kb, respectively can be inserted. Details of steps of insertion and use of enzymes have been described in Chapter 4. The cleaved genomic DNA has several small fragments (Fig. 8.1). These are randomly inserted into plasmids. DNA sequencing of both the small and medium sized insert plasmid libraries is carried out. It is done at both the ends of inserts of randomly selected clones so that it should cover the genome atleast three times.

If you sample a few random genomic DNA containing plasmid, you will find that: (*i*) some plasmids contain inserts which will be different from the others, and (*ii*) some plasmids will have inserts which may contain some regions present in one insert and a few region present in other insert, *i.e.* overlapping inserts. The overlapping inserts come from different regions of the same genomic location. The different regions lie either left or right sides to each other. The both ends of each insert (whether overlapping or non-overlapping) is sequenced. The information of sequence is put in computer database. The overlapping sequences are identified through a computer programme which joins the all sequences into one contiguous stretch. In spite of such effort it is likely that all the inserts would have not been included by a particular sample which were required to provide complete information of the sequence. If there exists such possibility, the specific regions of the genome are cleaved, cloned and sequenced separately.

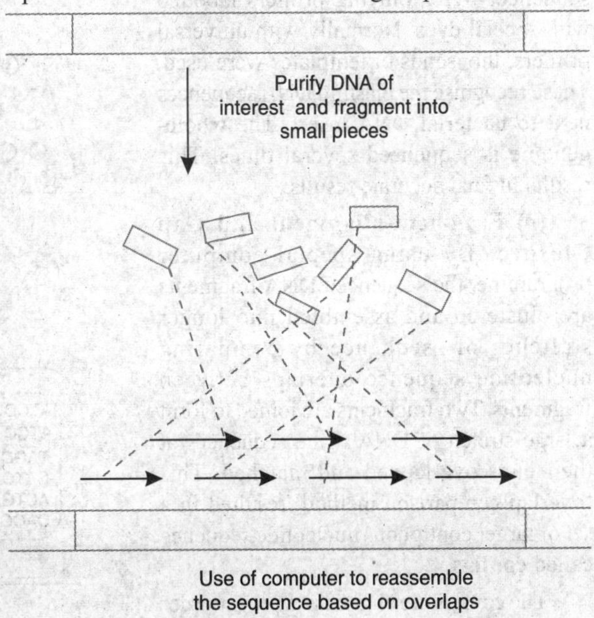


Fig. 8.1. Steps of shotgun sequencing approach.

Thus the complete information of the sequence is generated. This approach rapidly reveals 90% of the desired sequence information. The remaining few gaps are filled by custom oligonucleotide primers. The shotgun sequencing strategy relies on enormous computing power to assemble the randomly generated sequences.

### 3. Whole-Genome Shotgun Sequencing

Before 1995, whole genome sequencing was not possible because available computational power was not sufficient to assemble a genome from thousands of DNA fragments. J. Craig Venter and H Smith developed whole-genome shotgun sequencing, and sequenced the genome of two bacteria, *H. influenzae* and *M. genitalium*. This approach may be categorised into the following four stages:

**(*a*) Library Construction:** The chromosome is isolated from the desired cells following the methods of molecular biology, and randomly fragmented into small pieces using ultrasonic waves. Then the fragments are purified and attached to plasmid vectors (Fig. 8.2). Plasmids with single insert are isolated. A library of plasmid clones are prepared by transforming *E. coli* strains with plasmid that lacked restriction enzymes.

**(*b*) Random Sequencing:** The DNA is purified from plasmid. Thousands of DNA fragments are sequenced using automated sequencer by employing primers labelled with special dyes. Normally with universal primers, thousands of templates were used. These recognise the plasmid DNA sequences next to bacterial DNA insert. The whole-genome is sequenced several times. This results in final accurate results.

**(*c*) Fragment-alignment and Gap Closure:** By using special computer programme, the sequenced DNA fragments are clustered and assembled into longer stretches of sequence by comparing nucleotide sequence overlaps between fragments. Two fragments are joined to form a large stretch of DNA if the sequences at their ends overlapped and matched. This 'overlap comparison method' resulted in a set of larger contiguous nucleotide sequence called **contigs**.

The contigs are aligned in a proper order to form the completed genome sequence. If there exists gaps between the two contigs, these could be analysed and gap filled in with their sequences. Several other methods were

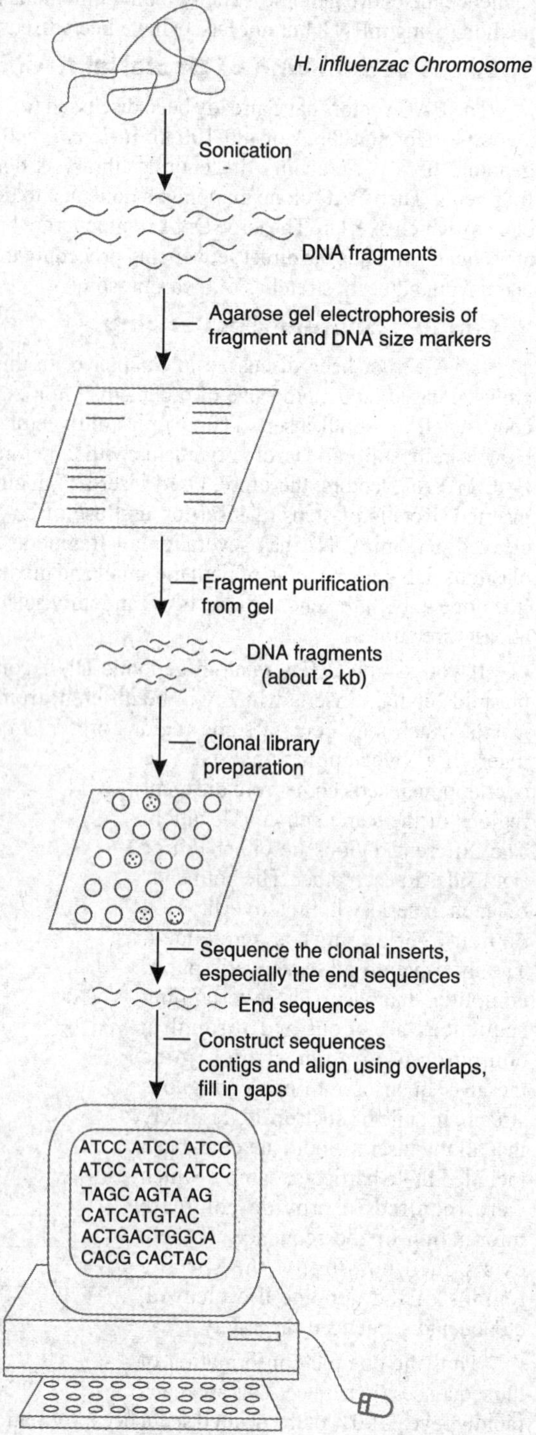


**Fig. 8.2.** Whole-genome shotgun sequencing method.

also used to align contigs and filling the gaps. The large fragments overlaps the previously sequenced contings. Using oligonucleotide probes, the fragments having overlaps with two contigs allow them to place side-by-side and fill in the gap between them.

**(*d*) Proof Reading:** Then the proof reading of sequences is done carefully so that any ambiguities in the sequence could be resolved. The sequence is also checked for the presence of any frameshift mutation; if so, the mutation is corrected.

This approach required too less than four months to sequence the genome (5,00,000 bp) of *M. genitalium*.

## 4. Expressed Sequence Tag (EST) Approach

The EST approach was pioneered by J. Craig Venter and co-workers at the National Institute of Health (NIH) (U.S.A.) in the early 1990s. He developed a new method of investing the genes by focusing the attention of the *active* portion of the genome as mRNA. Venter and co-workers isolated mRNA molecules (instead of fragments of genomic DNA) and constructed cDNA molecules. They treated DNA as a part of chromosomal DNA and sequenced to create 'expressed sequence tags' (ESTs). The ESTs were used as handles for isolating the complete genes. Following EST strategy, plenty of databases of nucleotide sequences were generated. Consequently it helped to prepare the transcript map of human genome at preliminary level. The EST technique demonstrated the possibility of sequencing all genes to the highest levels. This attempt boosted up the growth of genomic industry.

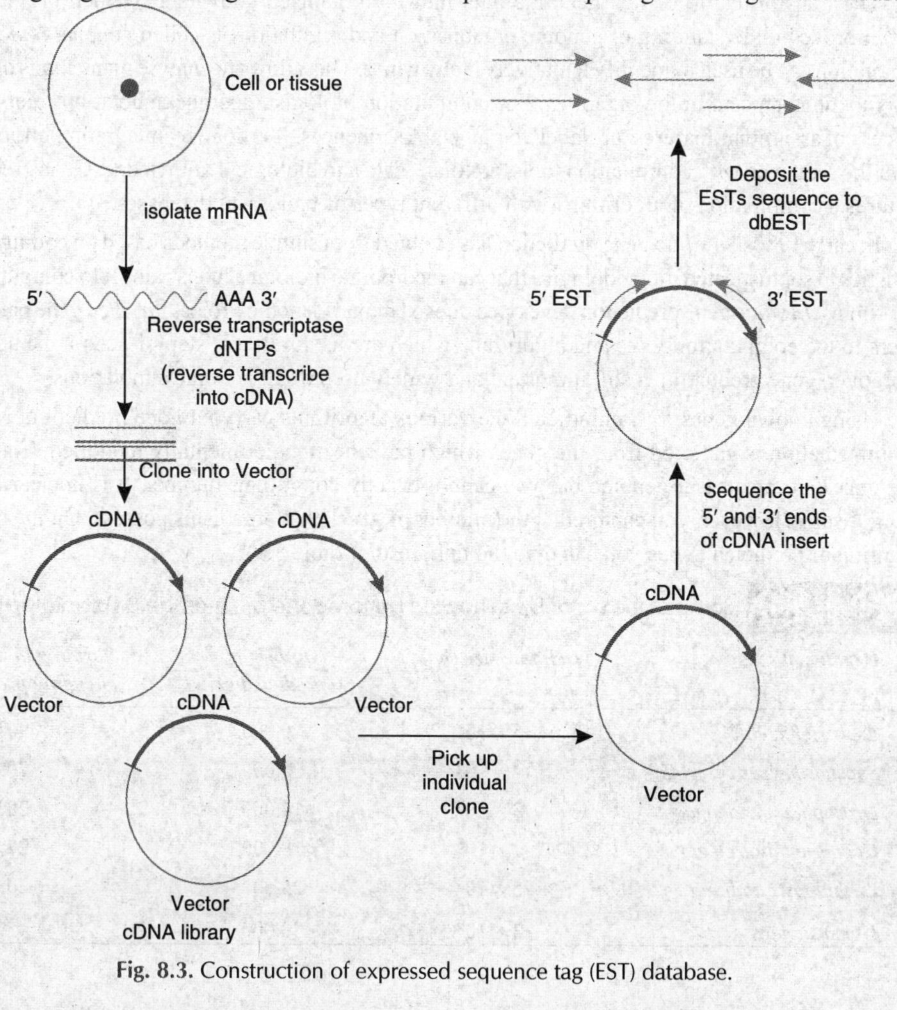


Fig. 8.3. Construction of expressed sequence tag (EST) database.

Thereafter, Venter switched towards sequencing the entire human genome through whole genome shotgun strategy. In this approach he matched the overlapping ends of DNA fragments, fitted contiguous pieces and attempted for genome sequencing by a 'genome assembly programme'. By sequencing the entire genome of some microorganisms, the validity of this method was also proved.

For preparation of expressed sequence tags (ESTs) individual clones are picked up from cDNA library and one sequence is generated from each end of cDNA insert (Fig. 8.3). The contaminating vectors, mitochondria and bacterial sequences are removed before depositing the ESTs into the public database (dbEST). In database, the ESTs are identified by their clone number and presence of 5′ or 3′– orientation. So far the ESTs that have been submitted to the public sequence databases were created from thousands of different cDNA libraries representing over 250 organisms.

The I.M.A.G.E. Consortium has picked up individual clones from many libraries and used for EST sequencing and arranged them for easy distribution. Upto 2002, more than 3.8 million cDNA clones have been arrayed from 360 human and 108 mouse cDNA libraries. The I.M.A.G.E. Consortium at present consists of more than half of ESTs in GenBank.

## C. GENE PREDICTION AND GENE COUNTING

The transformation of raw genomic data into the organised knowledge (which provide new and improved understanding of genome organisation and regulation) is called *genome annotation*. Annotation can be used to identify tentatively many genes. This allows to analyse many kinds of genes and functions present in the organism. For computation biologist, genome annotation refers to the process of assigning 'features' or 'label' to raw DNA sequences. It is done by integrating information from the sequence with computation tools, auxiliary data and biological knowledge. Gene prediction requires a combination of algorithms with different types of biological databases.

In early 1980s, *in silico* gene prediction has evolved from simple methods based on coding region statistics to sophisticated methodologies that can incorporate biological constrains into computational algorithms. *In silico* gene prediction developed due to Human Genome Project. *In silico* gene prediction refers to the computation tools and algorithms which are useful in this step of genome annotation. Moreover, gene prediction is still important and widely used of all genome annotations.

Using known genes as training data the various algorithms carryout gene prediction. Most of the information is gathered from the genes which have been experimentally identified. You know that genes are present in genome but you cannot exactly count their number. It is unclear how to count them? However, you can predict the number of genes that organisms possess. On the basis of counting of predicted genes you can give the final result (Table 8.1).

**> Table 8.1: Some of the organisms showing genome size and number of predicted genes.**

| *Organisms* | *Genome size (bp)* | *Number of predicted genes* | *Part of genome encoding proteins* |
|---|---|---|---|
| *Escherichia coli* | $5 \times 10^5$ | 5,000 | 90% |
| *Saccharomyces cerevisiae* | $14 \times 10^6$ | 6,000 | 70% |
| *Drosophila melanogaster* | $3 \times 10^8$ | 14,000 | 20% |
| *Caenorhabditis elegans* | $1 \times 10^8$ | 18,000 | 27% |
| *Arabidopsis thaliana* | $1.25 \times 10^8$ | 25,000 | 20% |
| *Homo sapiens* | $\sim 3 \times 10^9$ | ~ 30,000 | <5% |

Table 8.1 shows that human genome consists of less number of genes (~30,000) in spite of having largest genome size of ~$3x10^9$ bp, whereas the worm *C. elegans* consists of 18,000 genes in $1x10^7$ bp long genome. The functional genes present in human are <3% of total genome, while in *C. elegans* 27% geneome is functional. Scientists are of the opinion that the number of genes in human should be around ~ 40,000 - 50,000.

In case of microbial genome 40-50% of genes may code for proteins of unknown function. 20-30% genes may encode unknown proteins that are unique to the species.

## 1. Gene Prediction Algorithms

There are several algorithms for gene prediction as given below:

**(*a*) Homology-based Gene Prediction:** It is traditionally the first and most commonly used tool to discover new genes. Homology-based gene prediction falls into two categories as below:

**(*i*) Gene Prediction through Detection of Homology to know the Proteins:** This method uses sequence alignment of the translated DNA sequence (using 6 possible reading frames) with databases (*see* Chapter 9) of known proteins.

**(*ii*) Gene Prediction through Comparison with Expressed Sequence Tags (EST) Database:** The EST has been described earlier. With the appropriate use of sequence alignment parameters about 90% of the genes annotated on human genomic DNA are detected by ESTs.

**(*b*) *Ab Initio* Gene Prediction:** It includes the class of 'statistical learning' algorithms which are used for *in silico* gene recognition. There are several strategies of *ab initio* gene prediction based on: oligonucleotide usage, marker models, statistical pattern recognition and classification, neural networks.

**(*c*) Systenic Gene Prediction:** Systenic gene prediction is gene recognition by using cross-species sequence comparisons to identify and align relevant regions. The presence of exonic features at corresponding positions is searched out in both species simultaneously. The reason behind systenic gene prediction is simple. During evolution the exons (*i.e.* functional regions of DNA sequence) tend to more highly conserved than non-functional regions. Hence local conservation identified through comparisons of genomes of related species indicates biological function. Fig. 8.4 shows the genes in human chromosome that are *systenic* to mouse chromosome.

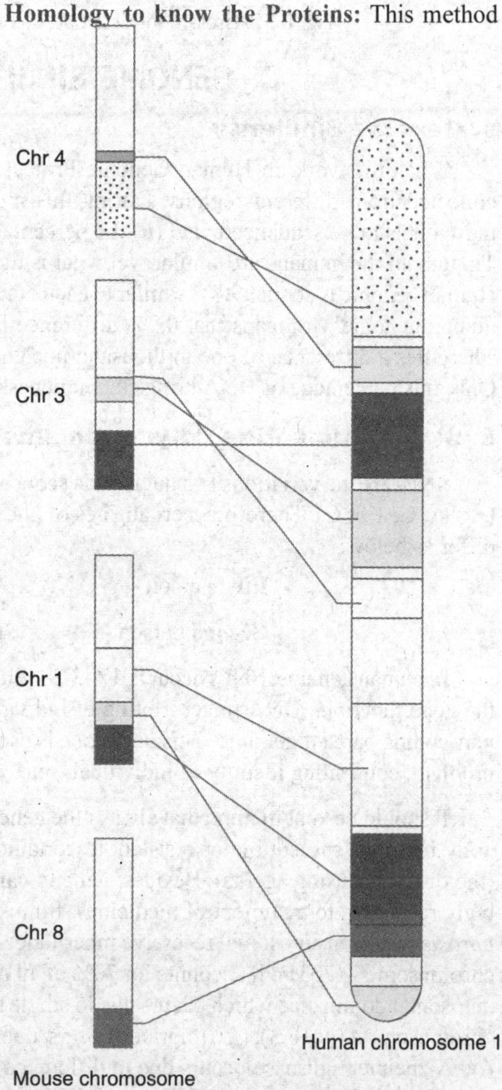


**Fig. 8.4.** Systeny between human chromosome and mouse genome.

## 2. Accuracy and Validity of Gene Prediction Algorithms

Inaccuracies of *in silico* gene prediction algorithms will travel down the line. This results in errors at the transcriptional level (the proteome level) and could ultimately affect or at least hinder our understanding of biology of species. Assessment of accuracy of gene prediction at exon level is given in Fig. 8.5.

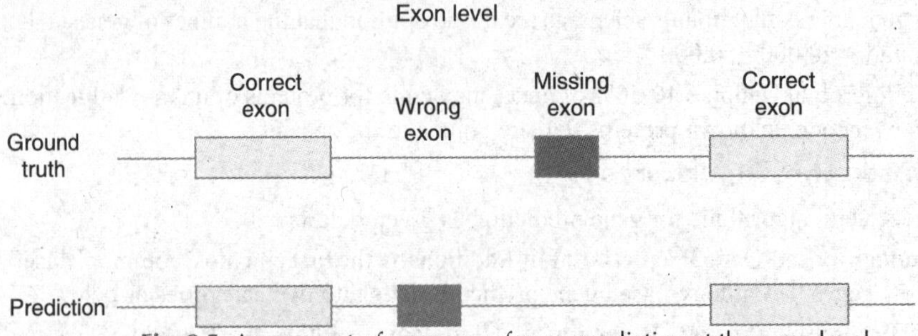


Fig. 8.5. Assessment of accuracy of gene prediction at the exon level.

## D. GENOME SIMILARITY AND SNPS

### 1. Genome Similarity

When the work on Human Genome Project was in progress, several unknown samples were collected from different regions. During this moment two questions were raised: (*i*) genome of individuals to be sequenced, and (*ii*) the percentage of genome similarity between two individuals. Though all the humans are similar, yet what is the percentage of genome similarity between them? Human genome is about 99.8% similar to each other. The percentage of similarity refers to 'consensus human genome'. It means that 0.2% difference in genome makes them different from each other. There lies a difference of one in thousand nucleotides in the genomes of two different individuals. Only this percentage of DNA brings a uniqueness in them.

### 2. Single Nucleotide Polymorphisms (SNPs)

SNPs are the variations in a nucleotide sequence which occur due to change even in a single base (*e.g.* A, G, T or C). Therefore, certain sites of the sequence nucleotide bases of different individuals differ as below:

First person - ...AT**G**CTACG....

Second person - ...TA**T**CTACG.....

In human genome, SNPs occur at 1.6-3.2 million sites. According to changes in bases SNPs affect the gene function. DNA fingerprinting of individuals is possible due to these genetic variations in non-coding parts of genome. This technique is used in search of criminals, rapists, solving parentage problem, confirming identity of individuals, etc.

It should be kept in mind that always the genetic variations are some time harmful. Because our body becomes susceptible or resistant to certain diseases *i.e.* protects from all kinds of pathogens (the disease causing agents). Besides, genetic variations also govern the severity of illness and the body responses to treatment of medicines. It must be noted that: (*i*) single gene mutation in *cystic fibrosis* results in autosomal recessive inheritance and occurs due to deletion of 3 bp at codon 508 on chromosome 7 (7q31). It accounts for 70% of all mutations, (*ii*) Huntington disease is also caused by autosomal dominance which occurs due to single mutation *i.e.* adenine/guanine/cytosine mutation at chromosome 4 (4q16.3), (*iii*) migraine occurs due to polymorphisms which has complex inheritance, (*iv*) Alzheimer's disease occurs due to difference of a single base in *apoE* gene on chromosome 19 (19q13), and (*v*) resistance to HIV and AIDS occurs due to a simple deletion within the chemokine receptor gene *CCR5*.

Fig. 8.6 shows the effect of medicine on patients as decided by a physician on the basis of SNPs present on patient's genome.

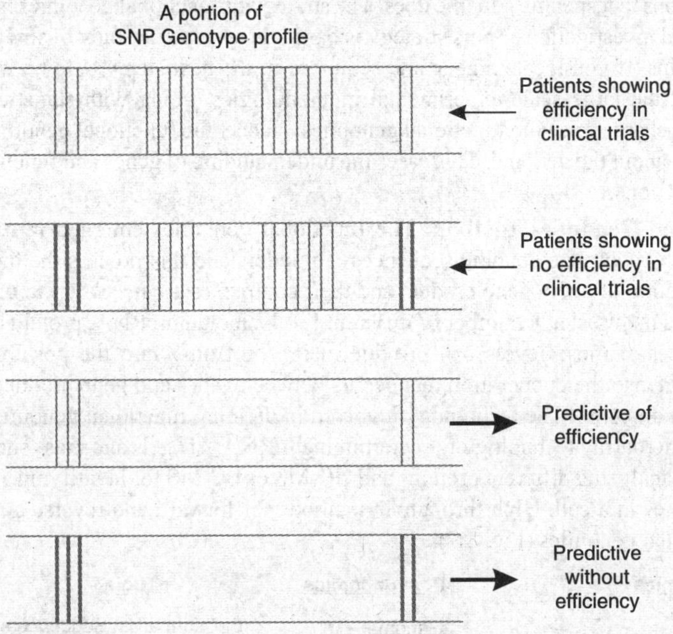


**Fig. 8.6.** Use of SNP map in prediction of medicine response on patients (vertical bars represent different SNPs on patient's genome.

On average, SNPs occur at every 500-1,000 nucleotides in human DNA. SNPs can help to: (*i*) associate sequence variation with heritable phenotypes, (*ii*) facilitate studies in population, and (*iii*) evolutionary biology, and add in positional cloning and physical mapping. SNPs tend themselves to highly automated fluidic or DNA chip-based analyses and have quickly become the focus of several large scale development and mapping in humans and other organisms.

# E. TYPES OF GENOMICS

In the last few years some interesting findings have been recorded and several new branches have emerged. Consequently, the area of genomics has quitely widened. However, the genomics is broadly categorised into three, *structural genomics, functional genomics* and *comparative genomics.*

## 1. Structural Genomics

The structural genomics deals with DNA sequencing, sequence assembly, sequence organisation and management. Basically it is the starting stage of genome analysis *i.e.* construction of genetic, physical or sequence maps of high resolution of the organism. The complete DNA sequence of an organism is its ultimate physical map. Due to rapid advancement in DNA technology and completion of several genome sequencing projects for the last few years, the concept of structural genomics has come to a stage of transition. Now it also includes systematics and determination of 3D structure of proteins found in living cells. Because proteins in every group of individuals vary and so there would also be variations in genome sequences.

## 2. Functional Genomics

Functional genomics deals with the study of function of all gene sequences and their expression in an organism. If you locate a gene and sequence it, the next step would be what function it does?

Based on the information of structural genomics the next step is to reconstruct genome sequences and to find out the function that the genes do. This information also lends support to design experiment to find out the functions that specific genome does. The strategy of functional genomics has widened the scope of biological investigations. This strategy is based on systematic study of single gene/protein to all genes/proteins. If you locate a gene and sequence it, the next step would be what function it does ? Therefore, the large scale experimental methodologies (along with statistically analysed/ computed results) characterise the functional genomics. Hence, the functional genomics provide the novel information about the genome. This eases the understanding of genes and function of proteins, and protein interactions.

(*a*) **Functional Genomics Toolbox:** The functional genomics emerged in response to the challenges posed by complete genome sequences. To understand this process the biochemical and physiological function of every gene product and their complexes are necessary to understand. The activity of genes is manifest at a number of different levels, including RNA, protein and metabolite levels and analyses at these levels can provide insight not only into the possible function of individual gene but also the cooperation that occurs between genes and gene products to produce a defined biological outcome. The technology involved in defining functional genomics are DNA or oligonucleotide microarray technology for determining mRNA; 2D gels and mass spectroscopy and other methods for analyzing different proteins and GC-MS or LC-MS for identifying and quantifying different metabolites in a cell. High throughput methods for forward and reverse genetics are also integral to functional genomics (Fig. 8.7).

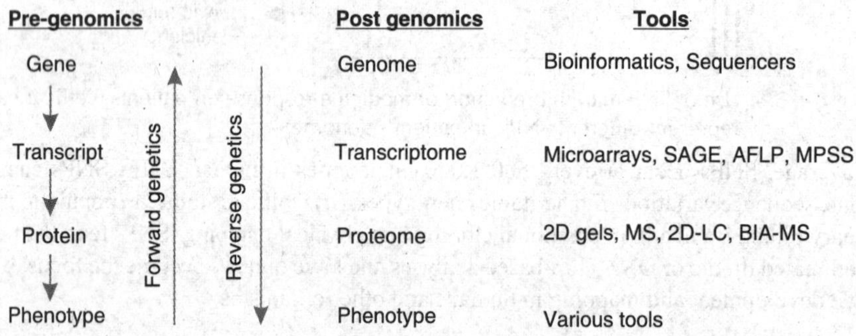


Fig. 8.7. Tools of functional genomics.

(*b*) **Determination of Function of Unknown Genes:** The following approaches are made for determining the function of unknown genes:

(*i*) **Computer analysis:** Computer plays a significant role to locate the genes in the DNA sequence. The most powerful tools used for this purpose is the *homology searching*. All the related genes have similar sequences. Hence, a new gene is discovered when it is similar to an equivalent already sequenced genes procured from different organisms. Those genes are called homologous genes and they share a common evolutionary sequence similarity. Usually identical sequences lack in a pair of homologous genes due to random changes by mutation. But they possess similar sequences because of operation of random changes on the same starting sequences of the common ancestral gene. Homology searching uses this sequence similarity. The analysis is based on the newly sequenced gene must be similar to a second previously sequenced gene. This shows that both the genes have evolutionary relationship and function of the new gene will be similar to the second gene. If the nucleotide sequences are identical by 80%, they show similarity. Such genes are referred to as showing 80% similarity rather than 80% homology. Thus homology analysis gives the information on function of all the genes or its segments. But best results are obtained if its amino acid homology is carried out. There are several software programmes such as BLAST which are used for such analysis. A homology search is done with a nucleotide or an amino acid sequence.

The homology domains have evolved by a single nucleotide change or complex rearrangements resulting in new genes inside which the domains are present. For example, *tudor* domain was first identified in *D. melanogaster*. It is 120 amino acid long motif. A homology search using *tudor* domain has revealed that this domain is present in several proteins. This is the transport protein expressed during oogenesis in *Drosophila*, RNA metabolism in human, and some other functions involving RNA.

Homology analysis of *S. cerevisiae* genome was done to assign functions to new genes. Out of 6,000 genes known, 30% were identified by conventional method and 70% by homology analysis. Out of these 30% genes of the genome were assigned function and about half of genes were the homologous of earlier known genes and about half shows similarity only to domains i.e. showed less similarity. About 10% genes had homologous in databases and 30% of the total did not have homologous. Of the 30%, about 7% were found not to be the real genes because of having unusual **codon bias**.

**(*ii*) Experimental analysis:** Genes express the visible characters called phenotype. In conventional experiments, the functions of such genes have been analysed by searching the mutant organisms where gene for a specific phenotype has been changed. The position of genes can be located through genetic crosses. Then such genes are identified and sequenced following the methods of molecular biology. Thus the method of assigning functions to unknown genes is based on mutating the gene and identifying the phenotypic changes occurring through mutation.

**(*c*) Patterns of Gene Expression:** Always all the genes do not express. Only those genes express whose products are required by the cell. The following approaches are made to understand the active genes in a specific tissue at a specific time which may also help to assess the relative degree of activity of the genes.

**(*i*) Gene expression array by measuring levels of RNA transcripts:** When a gene is switched on, it expresses through formation of an RNA transcript. The gene transcript may be identified to determine the gene. To approach this goal, specific RNA is extracted from the cell. By using this RNA, cDNA is synthesised and used as 'hybridisation probe'. The cDNA copy of the gene being studied is immobilised on a solid support. Then this DNA is hybridised with the hybridisation probe synthesised from RNA transcript. A hybridisation signal is observed when the gene transcripts are present in the extract of RNA. For every coding gene in the genome, this method is repeated and the whole mRNA molecules of a cell is analysed. This is called **transcriptome**. The DNA chip strategy is employed in such conditions.

**(*ii*) Serial analysis of gene expression (SAGE):** It is a method for gene analysis that depends on gene sequencing. SAGE is based on PCR which is used to identify differentially expressed mRNA levels in two different samples. In a cDNA pool the relative proportion of gene, specific ESTs shows the relative abundance of the corresponding mRNA transcripts. Therefore, differentially expressed gene transcripts can be identified by determining the sequence of a significant number of ESTs in one cDNA pool.

By using many mRNA samples, very short 3′ cDNA sequence tags can be generated. Generation of cDNA tags from one mRNA population is shown in Fig. 8.8.

An mRNA is isolated and used to synthesise cDNA. A biotin-labelled oligo-dT primer starts the synthesis of cDNA. It physically allows to separate 3' ESTs by using the magnetic beads-coated with

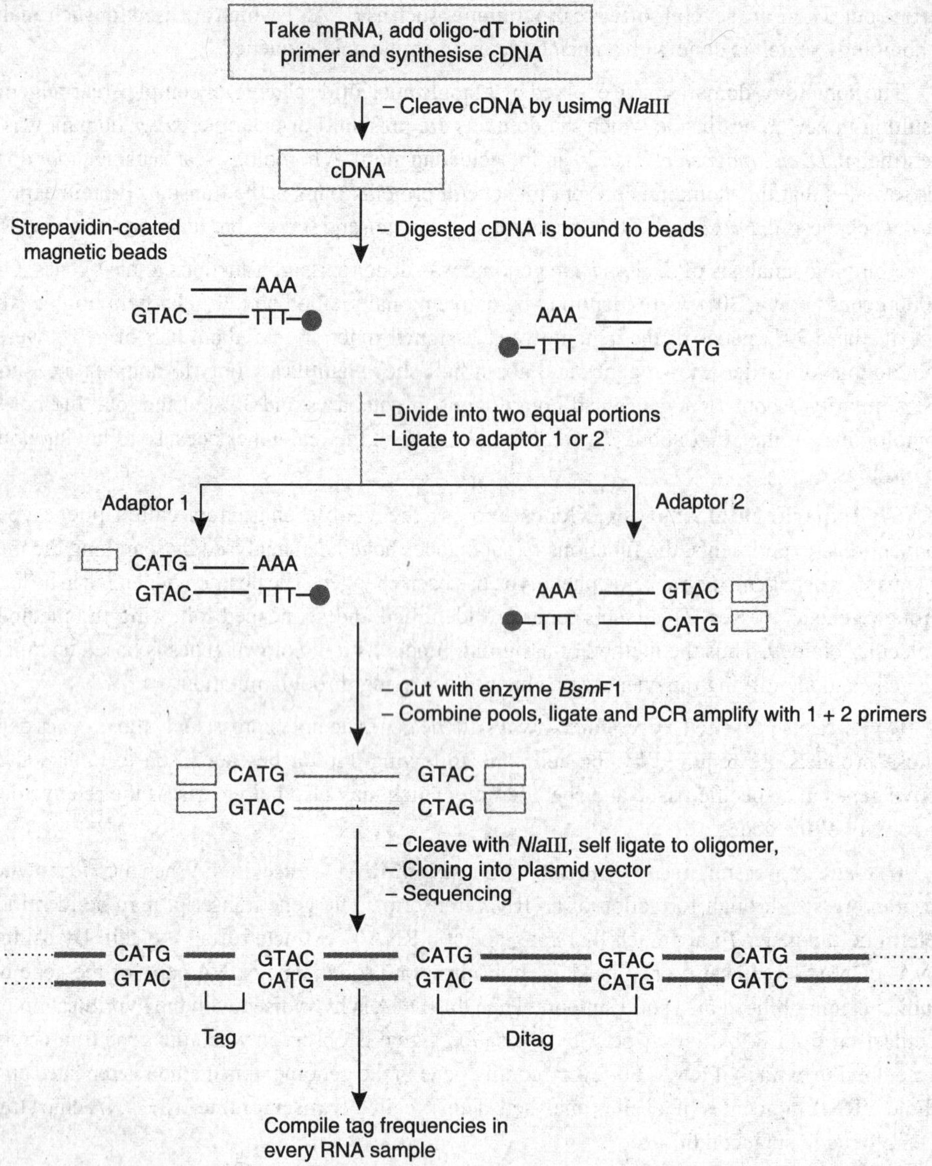


**Fig. 8.8.** Generation of cDNA tags in a SAGE (diagrammatic).

streptavidin. *Nla*III is used to delineate the tag. It recognises CTAG. Similarly, *Bsm*F1 also delineates the tag and cleaves 10 bp downstream of GGGAC sequence. The restriction fragments are ligated to adaptor and PCR amplified so that a large number of short ESTs (10-15 bp long) may be linked into a single array. This array is subjected to automated DNA sequencing. SAGE technique has been studied in yeast and humans, but has not been applied in plants so far.

Moreover, the wealth of knowledge about these untold story is being unravelled by the scientists after the development of **microarray technology** and **proteomics**. These two technologies helped to explore the instantaneous events of all the genes expressed in a cell/tissue present at varying environmental conditions like temperature, pH, etc.

***(iii)* DNA Chip (DNA Microarray) Technology:** A major technological advancement was made in the field of molecular biology during the mid 1990s when DNA chips were produced. It attracted the interest among the biologists throughout the world. There are several synonyms of microarray such as *DNA chips, gene chips, DNA arrays, gene arrays* and *biochips*. One can never differentiate each name. In future, some new techniques *(e.g.* nucleotide arrays, carbohydrate arrays, protein arrays, etc.) are expected to be available. DNA chips are high density miniaturised microarrays of large number of DNA sequences which are attached in a fixed (spotted) location in a systematic order on a solid support *e.g.* glass plates, slides or nylon. DNA fragments from unknown genes are placed using robotic device that accurately deposit nanolitre quantity of DNA solution on a surface measuring only a few square centimetres. The UV light is used to cross link the DNA to the glass slides. The DNA fragments get attached to the surface.

The principle of DNA microarrays lies on the base pairing or hybridization between the nucleotides. Using this technology the presence of one genomic or cDNA sequence in 1,00,000 or more sequences can be screened in a single hybridization. PCR and hybridization have come together in the development of DNA *microarrays* which rapidly screens many thousands of genes. The DNA chips contain known oligonucleotides (20-mers) sequences or cDNA of known function. Thus a single DNA chip can give the 'complete picture' of whole genome of an organism. For application in DNA sequencing, DNA chips will have to possess every possible oligonucleotide sequence. Because the maximum sequence 'read' possible is the square root of the number of oligonucleotide sequences on the chip.

Moreover, microarray technology is useful in many areas such as (*i*) identification of tissue specific genes (*e.g.* insulin gene in pancreas), (*ii*) discovery of drugs (which interact at DNA level), (*iii*) variation in cell cycle, (*iv*) searching out of defects in regulatory genes, (*v*) find out cell response to environmental fluctuation, etc.

There are two types of DNA chips: cDNA-based chips and oligonucleotide-based chips.

- *cDNA-based chips:* Since this type of chips are prepared by using cDNA, it is called cDNA chips or cDNA microarray. It was commercialised by Synteni (Freemont, CA, U.S.A.) which are called probe DNA. The cDNAs are amplified by using PCR. Then these are immobilised on a solid support made up of nylon filter of glass slide ($1 \times 3$ inch). The probe DNA samples are loaded into a spotting pin by capillary action. Small volume of this DNA preparation is spotted on solid surface making physical contact between these two. DNA is delivered mechanically or in a robotic manner. When one DNA spotting cycle is completed, the pin is washed and loaded with fresh DNA to start the second cycle. The cDNA microarrays prepared by Synteni consists of about 10,000 groups of cDNA in an area of about 3.6 $cm^2$.
- *Oligonucleotide-based chips:* The second approach involves the synthesis of oligonucleotides directly on the chip using light-directed solid phase, combinatorial chemistry. It was developed first by Affymetrix (Santa Clara, U.S.A.). A high density of short oligonucleotides (10-20 mers) are found on the chip. Microarrays are prepared in the following ways (Fig.8.9.).
  (*i*) The glass support is coated with light-sensitive protecting groups, *i.e.* photoprotecting DNA base that prevent random nucleotide attachment.
  (*ii*) The surface is covered with a photomask that holds corresponding to the sites for attachment of the desired nucleotides and deprotects the defined regions of the slide.
  (*iii*) To remove the unexposed protecting group, shine the laser light through the photomask holes.
  (*iv*) The chip is bathed in a solution containing the first nucleoside to be attached. The nucleoside has a light-removable protecting group to prevent addition of another nucleotide until the appropriate time.

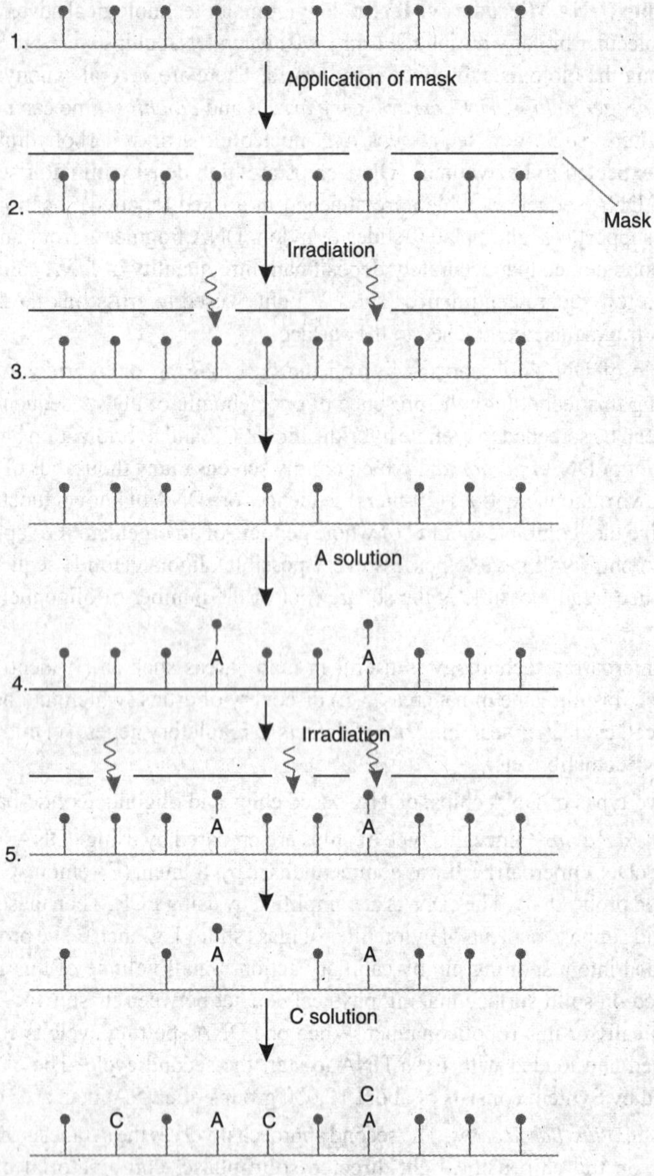


Fig. 8.9. Construction of a DNA chip with attached oligonucleotide sequences, two cycles are shown (based on Prescott *et al.*, 2003)

(*v*) The step 2 is repeated through step 4 with a new mask each time to add nucleoside until all sequences on the chip have been completed.

This method is used for construction of any sequence. The DNA chip available commercially consists of 25 bases long oligonucleotide probes. It is about 1.3 cm on a side and have more than 2,00,000 positions. The probes are often the expressed sequence tags (EST)s derived from cDNA molecules. Now the chips have probes for every expressed genes or open reading frame in the genome (for example, *E. coli* and the yeast).

The nucleic acids to be hybridised, often called **targets** (which may be mRNA or cDNA), are isolated and labelled with fluorescent reporter dyes. The chip is incubated with target-mixture to ensure proper binding to probe with complementary sequences. The unbound target is washed off and chip is scanned with laser beam. Fluorescence indicates that the probe is bound to that particular sequence. The hybridisation pattern shows the genes being expressed. Target samples are labelled with different fluorescent dyes and compared using the same DNA chip. On the basis of gene expression, gene function can be tentatively assigned. Only mRNA can be detected that are currently expressed. If a gene is transiently expressed, its activity may be missed by a DNA chip analysis.

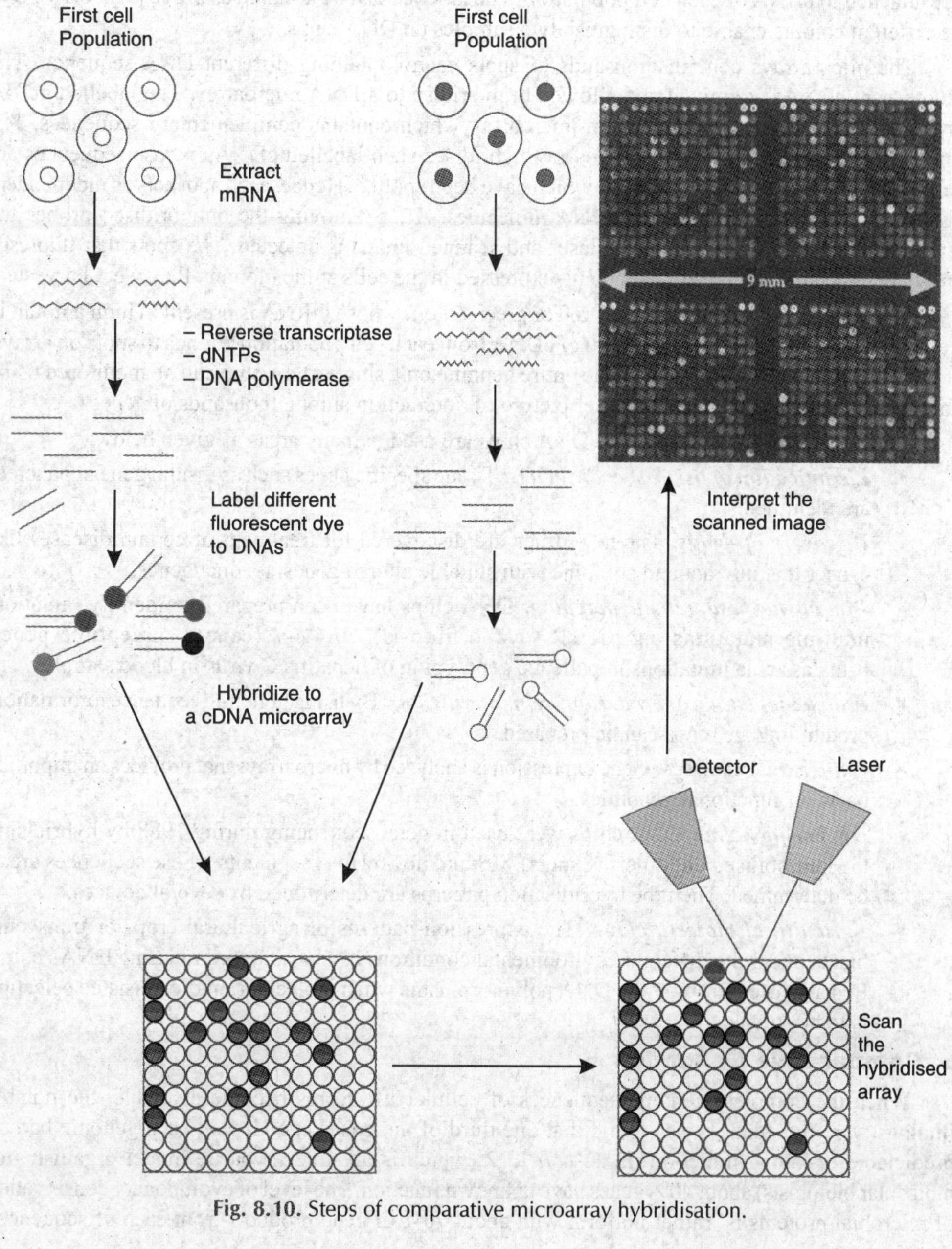


Fig. 8.10. Steps of comparative microarray hybridisation.

***Comparative Microarray Hybridisation:*** The major steps of comparative microarray hybridisation are shown in Fig. 8.10. A gene expressed at a given developmental stage can be detected using microarray technology. Different mRNAs produced by two different cell populations can be compared. For example, if you are interested to compare a normal cell with an abnormal cell (*e.g.* mutant or cancerous cells), then separately isolate sufficient amount (1-2 ml) of mRNA from total cell population and purify them. Soon synthesise cDNA using mRNA and reverse transcriptase following the method as given in Fig. 4.8. The base pair sequence of cDNA is complementary to that of mRNA. Fluorescent dyes or fluors (*e.g.* Texas red, Rhodamine, etc.) of different colours such as red and green are attached to cDNA of each cell population. The labelled cDNA is called as probe (Fig. 8.10). Dye of different colours enable to distinguish two samples on DNA chips.

The microarrays contain thousands of spots each containing different DNA sequence. The fluorescent cDNAs are mixed and allowed to hybridise to a DNA microarray. The labelled cDNA hybridises the DNA of such spots on microarray which contains complementary sequences. The spots of nucleotides present on microarray hybridises such labelled cDNAs whose sequences are complementary to that of DNA arrays which have been spotted. Hence, each spot acts as independent assay which determines different cDNA molecules. After removing the unhybridised probes the hybridised arrays are scanned using laser and scanned image is detected. The spots that fluoresce provide a snapshot of all the genes being expressed in the cells at the moment they were harvested.

Those spots show predominantly red or green colour whose mRNA is present in high amount in cell populations. If about equal amount of cDNA from each cell population bounds to spots on arrays, it impart yellow colour. In this way the entire genome on a single DNA chip can be monitored using this technology which can give a better picture of interaction among thousands of genes.

**Application of DNA chips:** The DNA chips are used in many areas as given below:

- ***Identification of tissue specific genes:*** Tissue specific genes such as insulin gene in pancreas are identified.
- ***Discovery of drugs:*** The new drugs are discovered for treatment of certain diseases like cancer that interact and combine with suitable altered genes i.e. oncogenes.
- ***Diagnostics and genetic mapping:*** DNA chips have been prepared to identify mutation/ multiple mutations e.g. CFTR (cyctic fibrosis), *BRCA-1* (cancer susceptible gene), β-thalassemia mutations in patients, genotyping of hepatitis C virus in blood samples.
- ***Proteomics (study of protein-protein interaction):*** By using genomic sequence information, protein linkage maps can be prepared.
- ***Functional genomics:*** Gene expression is analysed by microarrays that provides an important basis for functional genomics.
- ***DNA sequencing:*** DNA chips were used in gene sequencing during 1988 by hybridising the immobilised oligomers (8 mers) with unknown DNA segments whose sequences are to be determined. Then the hybridisation patterns are determined by several devices.
- ***Agricultural biotechnology:*** The expression patterns of agricultural crops or transgenic plants under the different environmental conditions can be analysed by using DNAs chips. It is also used to study the DNA polymorphisms when molecular marker-assisted selection is to be studied.

### 3. Comparative Genomics

When the complete genome sequences of cellular life forms become available, the notable findings were recorded. It was found that one third of the genes encoded on each genome had no predictable or known function. In *E. coli* K12 (which is all time favourite model organisms of molecular biologists) about 40% genes have unknown function. The level of evolutionary conservation of microbial proteins is rather uniform with about 70% of gene products from each of sequenced genomes having homologous in distant genomes. The function of these genes can be predicted by comparing different genomes and by transferring functional annotations of protein from better studies organisms to their **orthologs** (direct descendants of a sequence in the common ancestor i.e. the 'same' genes in different species that connect vertical evolutionary descedent) as opposed to **paralogs** i.e. genes related by duplication within the genome from less-studied organisms. For better understanding of genomes, biology and organisms, this makes comparative genomics a powerful approach. Comparative genomics includes several distinct aspects; analysis of protein sets from completely sequenced genomes is one of them. There are several databases (*e.g.* general purpose databases and organisms specific databases) used for comparative genomics.

Genome sequencing projects have made very clear that genomes of different organisms (*e.g.* mouse and man) may be very similar. Systeny between human chromosome 1 and mouse chromosome 1, 3, 4 and 8 has been shown in Fig. 8.11. It has been found that the genomes of human and chimpanzee differ by 1-3%. The working DNA of mouse and human is shared by about 97.5%. About 12% of ~18,000 genes of the worm *C. elegans* show sequence similarity to yeast gene. This datum is based on proteins encoded by this much number of genes.

About 2,000 genes of yeast (one third of ~6,000 genes) are functionally similar to the genes of *C. elegans*. Such similarity between the organisms suggests that in spite of evolution of organisms about 100 million years ago, their genomes have not changed much. That is why some percentage of similarity exists between genes of organisms.

Based on the above facts, study of comparative genomics proved a powerful approach for achieving a better understanding of the genomes and, subsequently of the biology of the respective organisms. Recently, some of the genome of the microorganisms viz., *Haemophilus influnenzae, Mycoplasma genitalum, Methanococcus jannaschii, Saccaromyces cerevisae, Escherichia coli, Bacillus subtilis* have been fully sequenced. Computational analysis of complete genomes requires a database (a repository of gene structure of organisns) that store genomic informations and bioinformatics tools. To study completely sequenced genomes, analysis of nucleic acids, proteins, etc. are required.

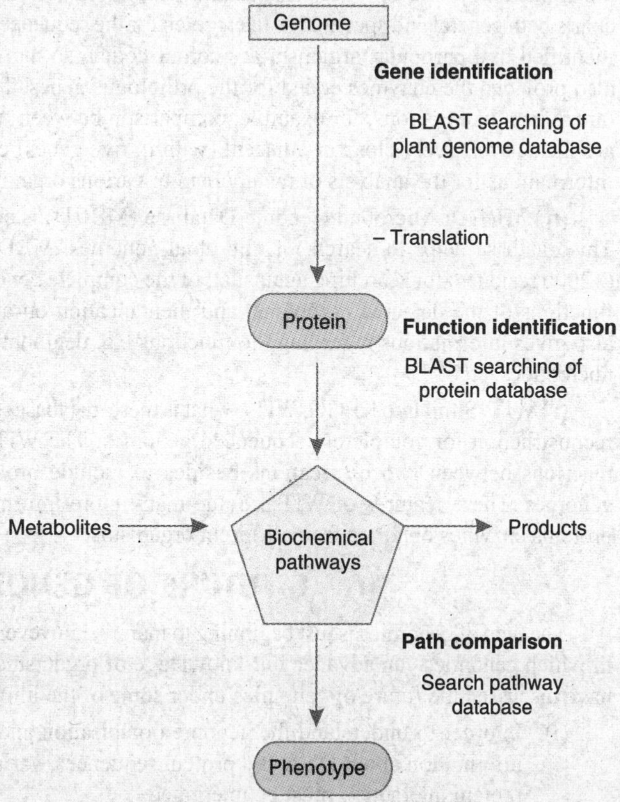


**Fig. 8.11.** From mango genome to phenotype (colour- yellow and flavour- sweet).

**(*a*) Example of Comparative Genomics:** Comparative genomics may be exemplified by the genetic basis of fruit development. The researcher is interested in identifying the genes that are involved in ripening green mangoes into yellow mangoes. The biochemical pathway involved in ripening process is also to be determined. Fig. 8.5 shows a comparative genomics approach that can be used in purely *in silico* effort to determine the genes involved in ripening of mangoes. In this approach the genome of mango fruit is compared to the annotated genomes of similar species to identify the genes and the functions that they do.

**(*b*) Databases for Comparative Genomics:** World Wide Web (www) is accessible to anyone by using Internet.

**(*i*) PEDANT:** This database gives information about the proteins, their three-dimensional structures, enzyme patterns, PROSITE patterns, Pfam domains, BLOCKS and SCOP domains as well as PIR keywords and PIR superfamilies.

**(*ii*) COGs:** Clusters of Orthologs Groups (COGs), is applicable to simplify evolutionary studies of complete genomes and improve functional assignments of individual proteins. It comprises of ~2,800 conserved families of proteins from each of the sequenced genomes. It contains orthologus sets of proteins from at least three phylogenetic lineages, which are assumed to have evolved from an individual ancestral protein. The functions of orthologs are same in all organisms. The protein families in the COGs database are separated into 17 functional groups that include a group of uncharacterized yet conserved proteins, as well as a group of proteins for which only a general functional assignment appeared appropriate. In COGs database due to storage of diverse nature of data on proteins, the similarity searches also give some information for those proteins which has no clear informations in databases. The databases also act as a tool for a comparative analysis of complete genomes.

**(*iii*) KEGG:** Kyoto Encyclopaedia of Genes and Genomes (KEGG) Centers on cellular metabolism as proposed by Kaneshisa and Goto (2000). A comprehensive set of metabolic pathway charts both general and specific has been given for the sequenced (genome) organism. In this, enzymes identified in a particular organism are colour-coded, so that one can easily trace the pathways. It also provided the enzymes coded for the orthologus genes. These genes if located adjacent to each other, form like operons, for example, comparison between two complete genomes in which genes are located relatively close or adjacent (with in five genes) can be made. This site is useful to get informations for the analysis of metabolism in various organisms.

**(*iv*) MBGD:** Microbial Genome Database (MBGD) is situated in University of Tokyo, Japan. This database helps to search for microbial genomes. MBDG accepts several sequences at once (~2000 residues) for searching against all of the complete genomics available displays colour–coded functions of the detected homologs, and their location on a circular genome map. This database also gives informations regarding the functions e.g. degradation of hydrocarbon or biosynthesis of nucleotides, etc.

**(*v*) WIT:** Similar to KEGG, WIT ('what is there' database) gives informations regarding metabolic reconstruction for completely sequenced genomes. The WIT features are to provide sequence of reactions between two bifurcations besides to include proteins from many partially-sequenced genomes. These features of WIT provide many more informations on the sequences of the same proteins/enzymes obtained from different organisms.

## F. FUTURE OF GENOMICS

The field of genomics is just beginning to mature. However, there are challenges and opportunities in which genomics can advance our knowledge of organisms and their uses. Prescott *et al.* (2003) have discussed the future of genomics under some of the important points:

(*i*) In order to understand the genome organisation and working of a living cell, all the new information about DNA and protein sequences, variation in mRNA and protein levels and protein interaction must be integrated.

(*ii*) Genomics can provide insights into pathogenicity and suggest treatments for infectious diseases. Possible virulence gene can be identified and its expression during infection along with host response can be studied. More sensitive diagnostic tests, new antibodies and vaccines may come from genomic studies of pathogens.

(*iii*) The field of pharmacogenomics should produce many drugs for disease treatment. Databases for human gene sequences can be searched for proteins that might have therapeutic value and for new drug targets. Genomics can be used to study variation in drug metabolising enzymes and individual response to medication.

(*iv*) By comparing a wide variety of genomes, the nature of horizontal gene transfer and microbial evolution can be studied.

(*v*) Genomics can be used in industry to identify novel enzymes, with industrial potential, enhance the bioremediation of hazardous wastes and improve techniques for microbial production of methane and other fuels.

(*vi*) Genomics will positively affect the agriculture. It may be used to find out new biopesticides and improve sustainable agricultural practices through enhancement in processes such as $N_2$-fixation, phosphate solubilization, and production of indole acetic acid, HCN, siderophores, etc.

## G. PROTEOMICS

The term '*transcriptome*' has been coined to describe the complement of mRNA transcribed from the genome of a cell. The term *'proteome'* is used to describe the total set of proteins expressed from the transcriptome of a cell. But the term proteome has been devised to describe the proteins specified by genome of an organism. Proteome is a wide term which includes all the variants of a single gene product that are produced from alternative splicing of transcribed RNA and from post-transcriptional modification of a single protein product.

The genome is constant within the cell, whereas transcriptome and proteome vary: (*i*) within a single cell during cell cycle, and (*ii*) between cell types (*e.g.* brain and liver) or when exposed to various external stimuli that exert changes in gene expression (*e.g.* growth factors, hormones, etc.). However, biochemical machinery of the cell can be modulated in response to internal and external factors. This affects synthesis or degradation of proteins, post-transcriptional modification of protein, alteration in cellular localisation, etc.

Proteomics is the identification, analysis and large scale characterisation of proteome (*i.e.* the total protein components) expressed by any given cells, tissues and organs under the defined conditions. The major objectives of proteomics are: (*i*) to characterise post-transcriptional modifications in protein, and (*ii*) to prepare 3D map of a cell indicating the exact location of protein.

Proteomics is the direct outcome of advancement made for nucleotide sequencing of different genomes in large scale. This helps to identify various proteins. Generation of information about protein is necessary. Because, protein governs the phenotypic characters of the cells. Merely genome study cannot provide the understanding of mechanism of disease development and various developmental changes occurring in organisms including humans. Moreover, target drugs for many kinds of diseases can be prepared only after understanding the protein modification and protein functions.

Proteomics provides information about genome function that mRNA studies cannot. Always mRNA and protein cannot be correlated because of the post-translational modification of proteins and protein turnover. Measurement of RNA levels can show the changes of gene expression that occurs inside the cell, whereas proteomics discovers what is actually happening.

A dream of the proteomicist may be to obtain a dynamic map of an organism's entire proteome that reveals how it changes during development and in response to biotic and abiotic challenges to survive. Such a map could have more intrinsic value than the corresponding transcriptome map, which is a more remote indicator of the biochemical activity of the organism. However, proteome map will be more complex than the transcriptome maps, and probably more difficult to achieve because post-transcriptional modification of protein can produce many variants of a single gene product and because proteome map will ultimately need to incorporate information about the location of each protein in a cell.

There are many areas of modern proteomics such as protein expression, protein structure, protein localisation, protein-protein interaction, etc. (Fig. 8.12).

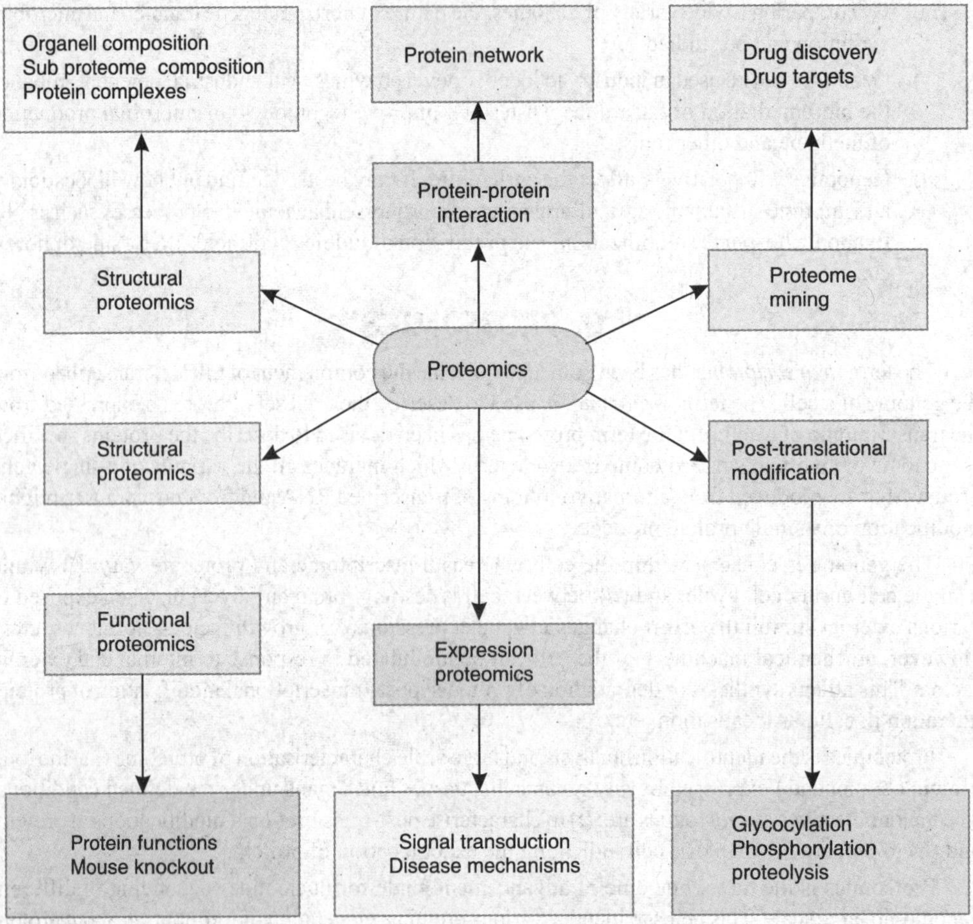


**Fig. 8.12.** Areas of proteomics.

About one third of gene sequences of organisms (of which genome sequence is known) do not carry out any function. The structural genomic projects can be assisted significantly only identifying the proteins completely. One of the objectives of structural genomics is to prepare 3D structure of all proteins expressed by the genome of the cells. Based on the protein structures possible functions may be assigned to respective proteins.

## 1. Relation Between Gene and Protein

In eukaryotes transcription occurs inside the nucleus and translation in cytoplasm. DNA is transcribed under transcriptional regulation into pre-mRNAs by RNA polymerase II (Fig. 8.13). Inside the nucleus, pre-mRNAs undergo various post-transcriptional modifications to increase their stability such as capping at 5'-end, polyadenylation (*i.e.* addition of poly (A) at 3'-OH end) and mRNA additing.

The introns are spliced out by splisosomes through a process called *splicing*. Then the mature mRNAs are transported from nucleus to cytoplasm for translation into protein on the ribosomes. Translational regulation of protein occurs in cytoplasm. Then protein undergoes post-translational modifications to about 200 types. From a single gene, various forms of proteins are generated. This shows that there is no correlation between the number of genes and number of proteins *i.e. the number of proteins are more than the number of genes*.

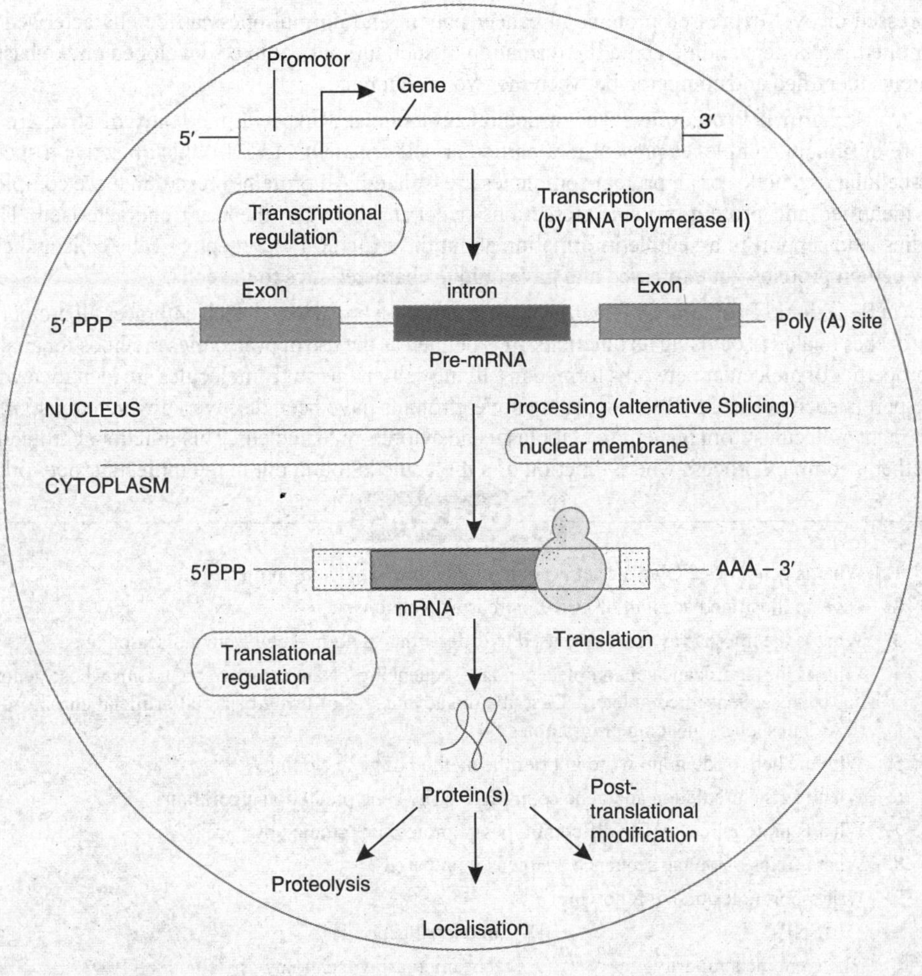


Fig. 8.13. Mechanism of production of multiple gene-products from a single gene.

## 2. Approaches for Study of Proteomics

Although new techniques in proteomics are being developed, the traditional approaches are: two-dimensional electrophoresis, and mass spectrometry. The first dimension uses the iso-electric focusing and the second dimension is SDS-PAGE. These had been discussed in Chapter 3.

## 3. Types of Proteomics

There are many types of proteomics as shown in Fig. 8.12, but expression proteomics, structural proteomics and functional proteomics have been dealt herewith.

***(a)* Expression Proteomics:** Expression proteomics is the quantitative study of protein expression between the samples which differ by some variables. A comparative study of the whole proteome between the samples can be done using this approach. For example, a patient suffering from mouth cancer develops a small tumour. Thus tumour of cancer patient and similar tissues from a normal person may be taken out and analysed for protein expression following different routes. Using techniques of high resolution protein separation and identification (*e.g.* two-dimensional gel electrophoresis, isoelectric focusing, MALDI mass spectrometry, microarray technique, etc.) under-expressed or over-expressed proteins in cancer patient and normal ones can be characterised and identified. An understanding about the formation of such tumour could be developed on the basis of proteins identified and compared between the two individuals.

***(b)* Structural Proteomics:** The structural proteomics deals with the study of structure and nature of protein complexes present in a particular cell organelle. To fulfil this objective a specific sub-cellular organelles or all protein complexes are isolated. All proteins present in these complexes are identified and protein-protein interactions occurring between them are characterised. These studies lend support to assemble information about the structural topography of the cells and clues how certain proteins got expressed and gave unique characteristics to the cells.

***(c)* Functional Proteomics:** Functional proteomics is a broad term which embraces all proteomics approaches related to devising its functions. It is defined as the use of proteomics methods for analysis of properties of molecular networks formed in a living cell. In this study molecules are identified which take part in such networks. Recently, some novel proteins have been discovered which transport the important molecules from nucleus to cytoplasm and cytoplasm to nucleus. This functional proteomics is rather a complex process where function of a molecule is found out in the molecular networks.

## PROBLEMS

1. What is genomics ? Give a brief account of genome sequencing project.
2. Give an illustrated account of DNA microarray technology.
3. What is the microarray technique used to differentiate a normal cell from a diseased one ?
4. What is the significance of complete genome sequencing ? Name any two organisms whose genome has been sequenced completely. Describe the advantages of using bacterial artificial chromosome (BAC) in such sequencing programmes.
5. With the help of diagram write in brief the method of gene cloning.
6. What is gene prediction and gene counting ? Write gene prediction algorithms.
7. What is proteomics ? Write different types of proteomics studied by you.
8. What is expression tag sequence ? How is it prepared ?
9. Write short notes on the following:

   A. SNPs
   B. Genome similarity
   C. Comparative genomics
   D. Functional genomics
   E. DNA microarray technology

CHAPTER

9

# Bioinformatics

In the 21st century biology is being transformed from a purely laboratory-based science to an information science too. The information refers to comprehensive views of DNA sequences, RNA expression and protein interactions. Due to explosion of sequence and structure information available to researchers, people have become optimistic to get answer of fundamental biomedical problems. Similar progress has also been made in computer-based technology also.

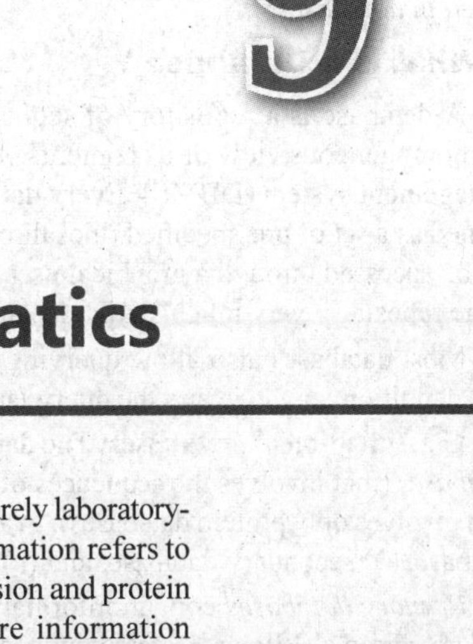


All kinds of 'omics' data are the integral parts of bioinformatics.

In 1987, when the Human Genome Project was conceived of, the field of bioinformatics was in its infancy. Today, bioinformatics has become a recognised discipline on its own, born out of the necessity to bring together the information sciences and the biological sciences in understanding the wealth of data that has been created through various projects around the world.

## A. WHAT IS BIOINFORMATICS ?

Translation of billions of characters in DNA sequences that make the genome into biologically meaningful information has given birth to a new field of science called 'bioinformatics'. The term bioinformatics has been derived by combining ***biology*** and ***informatics***. The key to biotechnology discoveries is locked in the genomes of organisms. The bioinformatics holds the key to unlock these data for the next generation of innovations.

A more precise definition of bioinformatics is *the application of information sciences (mathematics, statistics and computer sciences) to increase outr understanding of biology*. In a broadest sense this term can be considered as information technology applied to management and analysis of biological data. But in view of the recent rapid accumulation of available protein structure, this term is also used to embrace the manipulation and analysis of three-dimensional structure of data. Thus bioinformatics is a multidisciplinary science which aims to use the benefits of computer technologies in understanding the biology of life. Now, as a subject bioinformatics consists of three core areas: (*i*) molecular biology database, (*ii*) sequence comparison and sequence analysis, and (*iii*) the emerging technology of microarrays. In brief, bioinformatics is *the management and analysis of biological information stored in databases*.

### 1. What is a Database ?

A database is a repository of sequences (DNA or amino acids) which provide a centralised and homogeneous view of its contents. The repository is created and modified through a database management system (DBMS). Every data item in the database is structured according to a *scheme*, defined as a set of pre-specified rules through the data definition language. The contents of database can be accessed through a graphical user interface (GUI) that allows browsing through the contents of the repository very much similar as one may browse through the books in library.

Most databases also allow querying of its contents through a specialised query language. The data definition language and the query language form the *data model*.

**Classification of Databases:** The databases are broadly classified into two categories: *sequence databases* (that involves the sequences of both proteins and nucleic acids), and s*tructural databases* (that involves only protein databases). In addition, it is also classified into three categories (*a*) primary database, (*b*) secondary database, and (*c*) composite database.

*Primary databases* contain information of the sequence or structure alone of either protein or nucleic acid e.g. PIR or protein sequences, GenBank and DDBJ for genome sequences. Primary database tools are effective for identifying the sequence similarities, but analysis of output is sometimes difficult and cannot always answer some of the more sophisticated questions of sequence analysis. In 1998, GenBank obtained more than a million of sequences from more than 18,000 organisms. *Secondary databases* contain derived-information from the primary databases, for example, information on conserved sequence, signature sequence and active site residues of protein families by using SCOP, eMOTIF, etc. It is more useful than the primary databases. Orthology provides an important layer of information when considering phylogenetic relationships between the genes. Depending on the type of analysis method used, relationship may be elucidated in considerable detail including superfamily, family, sub-family and species-specific sequence levels. The *composite database* is obviating the need to search multiple resources. The SCOP is structural classification of proteins in which the proteins are classified into hierarchial levels such as classes, folds, super-families.

A moderate database pertaining to protein sequence and structural correlations on the 'Net' was established by A. Bairoch in 1991. This database was called PROSIT which later on was strengthened with a database on sequence analysis and comparison of protein sequences known as SEQUANALREE.

## B. HISTORICAL BACKGROUND

The protein databases were prepared first, then nucleotide databases. In 1959, V.M. Ingram first made attempt to compare sickle cell haemoglobin and normal haemoglobin, and demonstrated their homology. In due course of time the other proteins associated with similar biological function were also compared. This resulted in more protein sequencing and accumulation of vast information. Hence, it is realised to have databases so that using computation software the proteins can be quickly compared.

In 1962, using sequence variability, Zuckerkandl and Pauling proposed a new strategy to study evolutionary relations between the organisms which is called 'molecular evolution'. This theory was based on the facts that similarity exists among the functionally related (homologous) protein sequences. Margaret O. Dayhoff found that during evolution protein sequences undergo changes according to certain patterns such as: (*i*) preferential alteration (replacement) in amino acids with amino acids of similar physico-chemical characteristics (but not randomly), (*ii*) no replacement of some amino acids (*e.g.* tryptophan) by any other amino acids, and (*iii*) development of a point accepted mutation (PAM) on the basis of several homologous sequences.

Further work on sequence comparison on the basis of quantitative strategy was carried out. In 1965, Dayhoff and co-workers collected all the protein sequences known at that time and catalogued them as the *Atlas of Protein Sequence and Structure* which was first published by the National Biomedical Research Foundation (Silver Sring MD). Later on collection of such macromolecular sequences was published under the above title from 1965-1978. The above printed book laid the foundation for the resources that the entire biotechnology community now depends for day-to-day work in computational biology. The development of computer methods pioneered by Dayhoff and her research group is applicable: (*i*) in comparing protein sequences, (*ii*) detecting distantly related sequences and duplication within sequences, and (*iii*) deducing the evolutionary histories from alignment of protein sequences.

In 1980, the advent of the DNA sequence database led to the next phase in database sequence information through establishment of a data library by the European Molecular Biology Laboratory (EMBL). The purpose of establishing data library was to collect, organise and distribute data on nucleotide sequence and other information related to them. The European Bioinformatics Institute (EBI) is its successor that is situated at Hinxton, Cambridge, United Kingdom.

In 1984, the National Biomedical Research Foundation (NBRF) established the protein information resource (PIR). The NBRF helps the scientists in identifying and interpreting the information of protein sequences.

In 1988, the National Institute of Health (NIH), U.S.A. developed the National Centre for Biotechnology Information (NCBI) as a division of the National Library of Medicine (NLM) to develop information system in molecular biology. The DNA Databank of Japan (DDBJ) at Mishima joined the data collecting collaboration a few years later.

The NCBI built the GenBank, the National Institute of Health (NIH) genetic sequence database. GenBank is an annotated collection of all publically available nucleotide and protein sequences. The record within GenBank represents single *contig* (contiguous) selection of DNA or RNA with annotations. In 1988, the three partners (DDBJ, EMBL and GenBank) of the International Nucleotide Sequence Database Collaboration had a meeting and agreed to use a common format. All the three centres provide separate points of data submission, yet exchange this information daily making the same database available at large. All the three centres are collecting, direct submitting and distributing them so that each centre has copies of all the sequences. Hence, they can act as a primary distribution centre for these sequences. Moreover, all the databases have collaboration with each other. They regularly exchange their data.

Now sequence data are accumulating day-by-day. Therefore, there is a need of powerful software so that sequences can be analysed. For the development of algorithms [any sequence of actions (*e.g.* computational steps) that perform a particular task] firm basis of mathematics is needed. Now, mathematicians, biologists and computer scientists are taking much interest in bioinformatics. Moreover, biologists are curious to ask reservoir of all such information because they are widely interconnected through network.

Thus bioinformatics is aimed at: (*i*) the development of powerful software for data analysis, and *(ii)* benefit the researchers through disseminating the scientifically investigated knowledge, etc. The nucleotide and amino acid monomers are represented by limited alphabets. The properties of biopolymers *i.e.* macromolecules (*e.g.* DNA, RNA proteins) are such that they can be transformed into sequences having digital symbols. Genetic data and other biological data are differentiated by these digital data. This resulted in the progress of bioinformatics.

## C. SEQUENCES AND NOMENCLATURE

As mentioned earlier that the sequences of digital symbols are the transformed biopolymers. Indirectly the sequence data means the structure of biopolymer, and structure expresses the function. It shows a reductionist approach. Therefore, the sequence data can be used as context free.

### 1. The IUPAC Symbols

The International Union of Pure and Applied Chemistry (IUPAC) has made certain recommendations. The nomenclature system in bioinformatics is based on these recommendations. Different laboratories of the world follow nomenclature system of IUPAC so that their data set can uniformly and easily be compared. For rapid reproducibility and uniformity, the database institution and editors (who publish journals and research findings) also follow the recommendations of IUPAC.

For routine work, the basic IUPAC nomenclature system of nucleic acids and proteins has been discussed in this section. For detail you should go through the IUPAC web site. Language used in bioinformatics is given below:

| | | |
|---|---|---|
| Alphabet | ⇒ | Nucleotides |
| Words | ⇒ | Gene (prokaryotes)<br>Introns (eukaryotes) |
| Sentence | ⇒ | Operon (prokaryotes)<br>Gene (eukaryotes) |
| Punctuation | ⇒ | Regulatory gene |
| Chapter | ⇒ | Chromosome |

### 2. Nomenclature of DNA Sequences

It is obvious that nucleotides are the building blocks of DNA, and the nucleotides are constituted by four bases (A, G, T and C). Symbols of these four bases and basis of their nomenclature are used as much as they are spelt. Table 9.1 shows the symbols, their meaning and bases of nucleic acid sequences. Often the identity of sequences at specific positions is not clearly identifiable when sequence data are experimentally determined. It happens due to the problems related to the other secondary structures or 'compression' artifacts. In compression secondary structure in DNA fragments causes them to move in the gel so that more than one size of fragments may migrate to the same position.

Generally by repeating the experiment and sequencing the complementary strand, this problem can be solved. However, if ambiguities persist in some cases, the probable possibility can be deduced from the gel reads *i.e.* forward and reverse readings give data from opposite strands of DNA. They provide information about the relative orientations of the **read pairs** (*i.e.* pair of reading) from the same template of fragments.

**> Table 9.1: Recommendations of the single-letter codes.**

| *Symbol* | *Meaning* | *Base* |
|---|---|---|
| G | G | Guanine |
| A | A | Adenine |
| T | T | Thymine |
| R | A or G | PuRine |
| Y | C or T | pYrimidine |
| M | A or C | Amino |

| | | |
|---|---|---|
| K | G or T | Keto |
| S | G or C | Strong (3 hydrogen bonds) |
| W | A or T | Week (two hydrogen bonds) |
| H | A or C or T | not-G, H follows G in the alphabet |
| B | G or T or C | not-A, B follows A in the alphabet |
| V | G or C or A | not-T (not-U), V follows U in the alphabet |
| D | G or A or T | not-C, D follows C in the alphabet |
| N | G or A or T or C | Any |

A new symbol 'S' is used when there is doubt for the presence of G or C but there is surety for absence of A or T. Except a few viruses all the cellular organism consists of double stranded DNA. The two strands are complementary and antiparallel (running from 5'→3' direction) to each other. This is called Watson and Crick base pairing. When one encounters the symbol, the problem arises due to more than one bases at a position. These problems are resolved following IUPAC system of nomenclature. Table 9.2 shows the symbols used in a strand and that of its complementary strand. At certain positions the identical symbols in the strand and its complement are used. This shows that they are the same set of bases (see star, *).

**> Table 9.2: Symbols used in a strand and its complementary strand.**

| Symbol | A | B | C | D | G | H | K | M | S | A | U | W | N | R | Y |
|---|---|---|---|---|---|---|---|---|---|---|---|---|---|---|---|
| Complement | T | V | G | H | C | D | M | K | S* | A | B | W* | N | Y | R |

## 3. Nomenclature of Protein Sequences

You know that there are 20 amino acids which built the protein. But there are a few symbols which represent more than one amino acids. These symbols representing amino acids are given in Table 9.3.

**> Table 9.3: Symbols representing amino acids.**

| *Symbols* | *Amino acids* | *Symbols* | *Amino acids* |
|---|---|---|---|
| A | Ala | M | Met |
| R | Arg | F | Phe |
| N | Asn | P | Pro |
| D | Asp | S | Ser |
| C | Cys | T | Thr |
| Q | Gln | W | Trp |
| E | Glu | Y | Tyr |
| G | Gly | V | Val |
| H | His | B | Asx |
| I | Ile | Z | Glx |
| L | Leu | X | Xaa |
| K | Lys | | |

## 4. Directionality of Sequences

In nucleic acids (DNA and RNA) the nucleotide sequences are synthesised in 5'→3' direction. The 5' primer represents the presence of phosphate group at 5th carbon of sugar, and 3' primer represents the presence of hydroxyl group at 3rd carbon of sugar. Hence, this information is used to collect the data and store it in sequence database. Because data of nucleotide sequences are deposited in the database in the same form as these have been submitted or published.

Always the nucleotide sequences are listed in 5'→3' direction, irrespective of the published order. The nucleotide bases are numbered sequentially starting from 5' end *i.e.* from 5' to 3' direction. A word 'C' is indicated for complementary strand which also shows the orientation of chain in 5'→3' direction. Both the chains run antiparallely *i.e.* one in 5'→3' direction and the other in 3'→5' direction. While depositing sequence data, information on nucleotide sequence of only one strand is submitted in database. The nucleotide sequence of complementary strand is deduced from different web sites or programmes in different packages.

The three letter alphabets of the nucleotide act as codes. Each code represents an amino acid. In nature each cell synthesises proteins from N-terminus to C-terminus (N'→C') where N' represents —$NH_2$ group and C' represents —COOH group of the amino acids. These fundamental phenomenon is universal in all organisms. Hence, this conventional sequence of protein is entered in database. The concept of directionality is a universal fundamental process which is used by different database institutions.

If you have a sequence without any label, how will you search out whether this sequence is DNA or RNA or protein? You will search **sequence search programme** which scans 20 symbol. If the symbols switch between any of 4 bases, the sequence is DNA. If there is U instead of T, it is RNA sequence. If the symbols switch more than 4 bases, it is proteins sequence.

## 5. Types of Sequences used in Bioinformatics

There are different types of sequences which are known to have genetic information. Therefore, such sequences are used in bioinformatics. These sequences have been described in this context.

***(i)* Genomic DNA:** The genomic DNA acts as the reservoir of genetic information of all organisms. In recent years it is routinely sequenced in many laboratories of Molecular Biology. Genomic DNA of prokaryotes differs from that of eukaryotes, as the later differs with respect to location and consists of introns. Construction of genomic library is discussed in Chapter 6.

***(ii)* cDNA:** The double stranded molecules prepared by using mRNA as template and reverse transcriptase are called cDNA. These are expressed genes of the genomic DNA. By using cDNA molecules, substantial number of sequences have been determined and deposited in database. You have to tick at the right position when sequence entry form is to be filled up. This shows that the sequence, which is to be deposited, is the cDNA. Moreover, if you desire to retrieve the sequence this data need to be provided. Details of cDNA production are given in Chapter 6.

***(iii)* Organellar DNA:** Eukaryotic cells consist of different types of organelles *e.g.* chloroplast, mitochondria, Golgi complex, nucleus, etc. In eukaryotes genomic DNA is found in nucleus and organellar DNA molecules are located in mitochondria and chloroplasts.

The DNA molecules of these organelles are usually circular and double stranded of varying size besides a few exceptions. The mitochondrial genomes are relatively small and simple. Human mitochondria genome is made up of 16,569 base pairs. The genome consists of 2 rRNA genes, 22 tRNA genes and 13 protein coding sequences. The size of mitochondrial DNA varies greatly. But the size of chloroplast genome of several organisms has less variation. These are present in many copies. The size of mitochondrial and chloroplast genome of some organisms is given in Table 9.4.

The organeller genomes contain completes genetic system. The chloroplast genome of higher plants contains about 120 genes. The organeller genomes and prokaryotic genomes have several similarities with respect to organization and functions. The organeller genome codes for their own rRNA, tRNA some important proteins. The protein synthesizing machinery of the organelles especially chloroplast resembles that of bacteria but not the eukaryotes. The ribosomes of chloroplasts closely resemble bacterial ribosomes. The mitochondrial ribosomes show both similarity and differences.

**> Table 9.4:** **Mitochondrial and chloroplast genome of some organisms.**

| Types of DNA | Genome size (kilobase pair) | Circular or linear |
|---|---|---|
| **1. Mitochondrial genome** | | |
| Insects and mammals | 16-19 | Circular |
| *Homo sapiens* (humans) | 17 | Circular |
| *Saccharomyces cerevisiae* (yeast) | 33-78 | Circular |
| *Aspergillus nidulens* (fungus) | 32 | Circular |
| *Neurospora crassa* (fungus) | 60 | Circular |
| *Chlamydomonas* (green alga) | 16 | Linear |
| *Paramecium* (protozoa) | 40 | Linear |
| **2. Chloroplast genome** | | |
| *Pisum satium* (pea ) | 120 | Circular |
| *Marchantia polymorpha* (bryophyte) | 121 | Circular |
| *Oryza stiva* (rice) | 134 | Circular |
| *Nicotina tobaccum* (tobacco) | 156 | Circular |
| *Chlamydomonas reinhardii* (alga) | 195 | Circular |
| Higher plants | 120-200 | Circular |

***(iv)* ESTs:** It was Crag Venter who initiated first the sequencing of cDNA molecules using mRNA. The cDNA is cloned into a vector and cDNA library is constructed. Normally each clone has 5' and 3' ESTs associated with it. The average length of sequence is of about 400 bases. While the ESTs are short representing only fragments of genes, but not complete coding sequence. Many sequencing centres have automated the EST production where ESTs are produced rapidly.

Genome Sequencing Centre at Washington University (U.S.A) has been producing over 20,000 ESTs per week. The EST data are used to find out expression pattern using the following formula:

$$\text{Expression pattern} = \frac{\text{No. of ESTs corresponding to each gene}}{\text{Total No. of ESTs}}$$

***(v)* Gene Sequencing Tags (GSTs):** It has been found that the enzyme mungbean nuclease (Mnase) cleaves between the genes of *Plasmodium falciparum*. Therefore, by digesting *P. falciparum* genome a genomic library was established. It helps in identifying the genes of *P. falciparum*. The approach for construction of GSTs is similar to ESTs. It is constructed by isolating one read of sequence from any of the ends 5′ or 3′. The sequences obtained through this approach is called as GSTs.

***(vi)* Other Biomolecules:** The databases also consist of sequences of tRNA and small sized rRNAs. For example, 16S rRNA sequencing is done in tracing phylogenetic relationship among the species. A similar approach can also be made by using other molecule. Like mRNAs, rRNA can be copied into DNA; but this practice is rarely done.

## D. INFORMATION SOURCES

There are several well developed data repositories that have facilitated the dissemination of genome and protein resources of humans and other organisms. Some of the major biological databases are given in Table 9.5. The most comprehensive resources are the NCBI, genome database (GDB) and Mouse Genome Database (MGD). Each database is informative and very useful. Yet it is only after the components become linked to, for a single integrated resource that generate, the information stored in each database that can be analysed as part of the bigger whole. For example, by integrating various forms of information relative to a particular protein, a researcher may be able to elucidate the previously unknown function. Certain steps within a complex biological pathway become clear which were never before understood.

## 1. The National Centre of Biotechnology Information (NCBI)

On November 4, 1988, for the development of information systems in molecular biology, the NCBI was established as a division of the National Library of Medicine (NLM). The NCBI is located in the campus of the National Institute of Health (NIH) in Bethesda, Maryland, (U.S.A.). The NLM was selected to host the NCBI of its experience in biomedical database maintenance. It could establish a research programme in computational biology. Since 1992, the NCBI has been maintaining GenBank, the NIH DNA sequence database. Groups of annotators create sequence database records from the literature obtained from the authors, data exchanged with EMBL and DDBJ. The NCBI is the foremost repository of publicly available genomic and proteomic data. After its establishment the services at NCBI have fully expanded. The NCBI makes available the different kinds of biological data, computational resources for analysis of GenBank data and data retrieval system.

More specifically the NCBI has been charged with creating automated systems for: (*i*) storing and analysing knowledge about molecular biology, biochemistry and genetics, (*ii*) facilitating the use of such databases and software by the research and medical communities, (*iii*) coordinating efforts to gather biotechnology information both nationally and internationally, and (*iv*) performing research into advanced methods of computer-based information processing for analysing the structure and function of biologically important molecules. The NCBI does the following to carry out its multifarious responsibilities:

(*i*) It maintains collaboration with several NIH institutes, academia, industry and other governmental agencies.

(*ii*) It fosters scientific communication by sponsoring meetings, workshops and lecture series.

(*iii*) It supports training on basic and applied research in computational biology for post-doctoral fellows through the NIH intra-mutual research programme.

(*iv*) It engages the members of the international scientific community in informatics research and training through the scientific visitors programmes.

(*v*) It develops, distributes, supports and coordinates access to a variety of databases and software for the scientific and medical communities.

(*vi*) It develops and promotes standards for databases, data deposition and exchange, and biological nomenclature.

The NCBI has developed many useful resources and tools. It may be grouped into different types such as database retrieval tools, BLAST family for search of DNA sequences, ePCR, gene level sequences, chromosomal sequences, genome analysis, analysis of gene expression patterns, molecular structure, LocusLink used in genome catalogue information about gene and gene-based markers, OMIM, UniGene etc.

These tools have their own websites which many be used free of cost. You will learn the practical aspects of some of these tools in your practical. Out of the above resources and tools, three sets of resources are discussed in this regard. Using these resources most of the cases of bioinformatics activity can be carried out. While doing advanced studies, the other resources may be used.

**> Table 9.5: Major biological databases.**

| *Database name* | *Link* | *Information available* |
|---|---|---|
| | ***A. Major Nucleotide Sequence Database*** | |
| NCBI's GenBank | http://www.ncbi.nlm.nih.gov/Genbank | All known nucleotides and proteins sequences<br>International Nucleotide Sequence |

| | | |
|---|---|---|
| Data Collection | | |
| EMBL Nucleotide Sequence Database | http://www.ebi.ac.uk/embl/ | All known nucleotide and protein sequences<br>International Nucleotide Sequence |
| Data Collection | | |
| DNA Data Bank of Japan (DDBJ) | http://www.ddbj.nig.ac.jp | All known nucleotide and protein sequences<br>International Nucleotide Sequence |
| Data Collection | | |
| SWISS-PROT | http://www.expasy.ch/sprot | Annotated protein sequence |
| | **B. Major Mutation Databases** | |
| NCBI's dbSNP polymorphism | http://www.ncbi.nlm.nih.gov/SNP/ | Database of single nucleotide |
| OMIM genomic disorder | http://www.ncbi.nlm.nih.gov/OMIM/ | Catalogue of human genetic and |
| | **C. Major Gene Expression Databases** | |
| GeneExpression expression | http://www.ncbi.nlm.nih.gov/GEO | NCBI's repository for gene |
| | **D. Major Microbial Genomic Databases** | |
| NCBI's Microbial Genome Gateway | http://www.ncbi.nlm.nih.gov/PMGifs/Genomic/micr./html | ? |
| *Escherichia coli* | http://1 ecoli.aist-nara.ac.jp | ? |
| | **E. Major Organism specific Genome Database** | |
| Mouse Genome Database (MGD) | http://www.informatics.jax.org/ | Mouse genetics and genomics |
| *Saccharomyces* genome database (SGD) | http://genome.www.stanford.edu/Saccharomyces/ | *S. cerevisiae* genome information |
| Rice Genome Project (RGP) | http://1 rgp.dna.affrc.go.jp/ | Reporting current data in rice genome project |
| | **F. Major Protein Database** | |
| Protein Database (PDB) | http://www.rcsb.org/pdb | Structure data determined by X-ray crystallography and NMR |

## 2. The GDB

It is the official central repository for genome mapping data created by Human Genome Project. Its central node is located at the Hospital for Sick Children. The GDB holds a vast quantity of data submitted by hundreds of investigators. The GDB has many useful genome resource Web-links on its Resource Page.

## 3. The MGD (Mouse Genome Database)

It is the primary public mouse genomic catalogue resource. The MGD has evolved from a mapping and genetic resource to include sequence and genome information and details on the functions and roles of genes and alleles. The MGD includes information on mouse genetic markers and nomenclature, molecular segments, phenotypes, comparative mapping data, graphical display of linkage, cytogenetic and physical maps.

On the MGD home page, the genes are easily searched through Quick Gene Search box. The MGD contains many different types of maps and mapping data including linkage data, the WICGR mouse physical maps, and cytogenetic band position. The MGD also computes a linkage map that integrates markers mapped on the various panels. It also stores radiation hybrid scores.

## 4. Data Retrieval Tools

GenBank contains 7 millions sequence record covering 9 million nucleotide bases. Unless the databases are easily searched and entries retrieved in a usable and meaningful format, the biological databases serve a little purpose. Moreover, efforts made on sequencing will not be meaningful if biological community as a whole cannot make use of the information hidden within millions of bases and amino acids. There are several database retrieval tools such as ENTREZ, LOCUSLINK, TAXONOMY BROWSER, etc.

**(*a*) ENTREZ:** The integrated information database retrieval system of NCBI is called **Entrez**. It is most utilised of all biological database systems. Using Entrez system you can access literature, prepare genome map, sequences (both protein and nucleotides) and get structural data (3 D). To be very clear, Entrez is not a database, but it is the interface through which all of its component databases can be accessed and traversed. Entrez has ability to retrieve the related sequence structures and references.

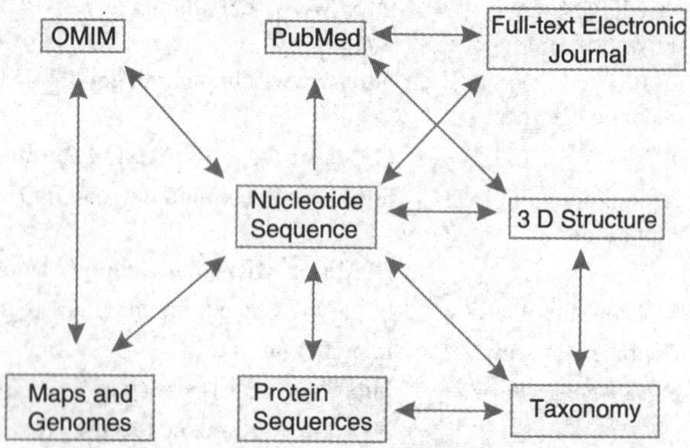


**Fig. 9.1.** Entrez map showing the hard link relationships between databases. Each square show one of the elements that can be assessed through Entrez. OMIM = online Mendelian Inheritance in Man; PubMed = Publishers on Medicine.

The Entrez information includes PubMed records, nucleotide and protein sequence data, 3D structure, information and mapping. The hardlink relationships between databases are shown in Fig. 9.1. All the information can be accessed by issuing only one query. For complete review of the features and complexities you may refer to a tutorial on the Entrez system at http://www.ncbi.nlm.nih.gov:80/entrez/query/static/help/helpdoc.html.

**(*b*) OMIM (Online Mendelian Inheritance in Man):** OMIM is a non-sequence-based information resource that is very much useful in genomics. It is a web-based electronic version of catalogue that contains thousands of entries for human genes and genetic disorders. It serves as a phenotypic companion to Human Genome Project. It was founded by Victor McKusick at the Hohns Hopkins University in 1998. A concise textual information is provided by OMIM from the published literature on the conditions of human having genetic disorders and full citation information. At the NCBI, the online version of OMIM is housed. Also links are provided to Entrez from all references cited within each OMOM entry. Internet resource for OMIM is: http://www.ncbi.nlm.nih.gov/omim.

The OMIM cytogenetic and morbid maps present cytogenetic locations for those genes with published locations and provide an alphabetical list of all the diseases described in OMIM. Therefore, it is necessary to consider the results of web-based biology reported in the scientific literature in order to validate the findings generated through computer-based comparative analysis. Hence, integration of scientific data with the literature is an important step for creating a unified information resource in the life science. For this purpose, individuals are provided with a direct link from OMIM to PubMed, the NCBI literature system. It is very easy to perform OMIM searches. A simple query is performed by search engine on the basis of one or more words typed into a search window. Consequently, a list of documents is returned containing the query words. The users can select one or more disorders from the list so as to see the full text of OMIM entry.

**(*c*) PubMed (Publishers on Medicine):** PubMed is a Web search interface that provides to over 11 million journal citation in MEDLINE and contains links to full text articles at participating publisher's web site. PubMed provides web-based access to over 11 million citations, abstracts and indexing terms for journal articles in the biomedical sciences. It also includes links to full-text journals. At present about 20 million searches are conducted per month and over 1,40,000 users seek information daily through PubMed.

**(*d*) Taxonomy Browsers:** Diversity of organisms is such that millions of species are known. It is hoped that millions of organisms are also unknown. After a species is known, its various features are studied and information is restored in database. So far information of over 79,000 organisms is restored in database. The species are classified according to their characters and similarity with others through the information provided by taxonomy browser. Internet resource of taxonomy browser is http://www.ncbi.nlm.nih.gov/taxonomy/tax.html.

Browsers are used to work at web page. Browsers client-server applications that connect to remote site, download the requested information at the site, and display the information on a user's monitor disconnecting from the remote host.

**(*e*) LocusLink:** LocusLink is an NCBI project to link information applicable to specific genetic loci from several disparate databases. LocusLink provides a single query interface to various types of information regarding to a given genetic locus such as phenotypes, map locations, homologies to other genes and information on official nomenclature. Currently LocusLink search space includes information from human, mice, rats, fruitfly and zebrafish. It carries information on mouse homologue of a given human gene, you cannot get. Beginning with LocusLink query simply by typing the name of the gene into the query box which appears on the top of LocusLink home page, you can select the gene of interest from an alphabetical list. LocusLink links to related web sites including the OMIM. Internet resource for LocusLink is: http://www.ncbi.nlm.nih.gov/LocusLink/.

**(*f*) Sequence Retrieval System (SRS):** The SRS has been created by Swiss Institute of Bioinformatics and the European Bioinformatics Institute, who have also created the Swiss-PROT database. SRS allows retrieval from an extensive catalogue of more than 75 public biological databases. The link button in SRS will allow you to get all the entries in one databank which are linked to an entry (or entries) in another databank. Hyperlinks made links between the entries.

## 5. Database Similarity Searching

**(*a*) Basic Local Alignement Search Tool (BLAST):** Due to genome searching projects on a large scale, the flood of DNA sequence data coming into public databases is staggering. Scientists are relying on deducing the function of putative genes through similarity to well characterised proteins. BLAST is a programme for sequence similarity searching developed at the NCBI. It identifies genes and genetic features. It executes sequences searches against the entire DNA database in less than 15 second.

Additional software tools provided by the NCBI are: Open Reading Frame Finder (ORF Finder), Electronic PCR (ePCR), and sequence submission tools, Sequin and BankIt. All the databases of the NCBI are available from WWW or by FTP. There are several tools to analyse sequence information among the BLAST family of similarity search programme. Sequence similarity searches use alignments to determine a 'match'. The basic operation in database searching is to sequentially align a query sequence to each *subject sequence* in the database. Most users prefer BLAST (http://www.ncbi.nlm.nih.gov/BLAST) or FASTA which rely on heuristic strategies to speed up alignment searches. The theory of BLAST systems is rather complex and out of scope of this book.

Similarity between the two species arises only when they had common ancestors. For example, if two species A and B have two different homologous sequence, it means that they have descended into two different species from a common ancestor. It means that two similar sequences may not be homologous, provided they have common ancestry. If a gene duplicates in a genome, it will also show similarity. Such genes are called **paralogues**, but not homologues. Because paralogues differ in function, whereas homologues do the same function.

There are several variants of BLAST, each is distinguished by the types of sequences (DNA or protein) of the query and database sequences (Table 9.6). The principles of working system of BLAST are: (*i*) using matrices of BLAST a given sequence is compared with database sequence. It specifies scores to either *reward* a match or *penalise* a mismatch, (*ii*) based on the criteria the top scoring matches are ranked. It differentiates the similarity either due to ancestry or random chance. Generally these criteria are not changed. But you can change them as you wish, and (*iii*) further examination of true matches are done using other details which are accessible through Entrez or other tools at NCBI.

**> Table 9.6: BLAST variants for different searches.**

| Programme | Query | Comparison database | Common use |
|---|---|---|---|
| BLASTp | Protein | Protein | Seeks to align an amino acid query |
| BLASTn | Nucleotide | Nucleotide | Align to new DNA sequence to a nucleotide sequence database |
| BLASTx | Nucleotide (translated) | Protein | Analyse new DNA sequence (translated) to find out potential coding regions |
| TBLASTn | Protein | Nucleotide (translated) | Useful for EST analysis |
| TBLASTx | Nucleotide (translated) | Nucleotide (translated) | Useful for EST analysis |

**(*b*) FASTA:** FASTA was the first widely-used programme for database similarity search (Pearson, 2000). FASTS performs optimised search for local alignment using a substitution matrix. It takes sufficient time to apply this strategy exhaustively. This programme uses the observed pattern of word hits to identify potential matches before attempting the more time-consuming optimised search. The *Ktup* parameter controls the trade-off between speed and sensitivity

FASTA format contains a defimilion line and sequence characters. It may be used as input to many analysis programmes. FASTA format is used in a variety of molecular biology software suites. In general BLAST tends to be faster and are more sensitive in detecting more alignments, but FASTA returns fewer false hits.

## 6. Resources for Gene Level Sequences

There are several tools among the resources for gene level sequences *e.g.* UniGene, HomoloGene, RefSeq, etc.

**(*a*) UniGene Database:** The ESTs have been described in previous section. Many redundant ESTs are generated during the course of its production. Because several cDNA clones represent the same gene (Fig. 9.2). Therefore, UniGene (one gene) database was developed at NCBI to control redundancy in EST data. The UniGene clustures ESTs and other mRNA sequences along with coding sequences (CDSs) annotated on genomic DNA into *subsets* of related sequences. The clusters are specific to organisms. At present clustures are available for human, mouse, rat, zebrafish and cattle.

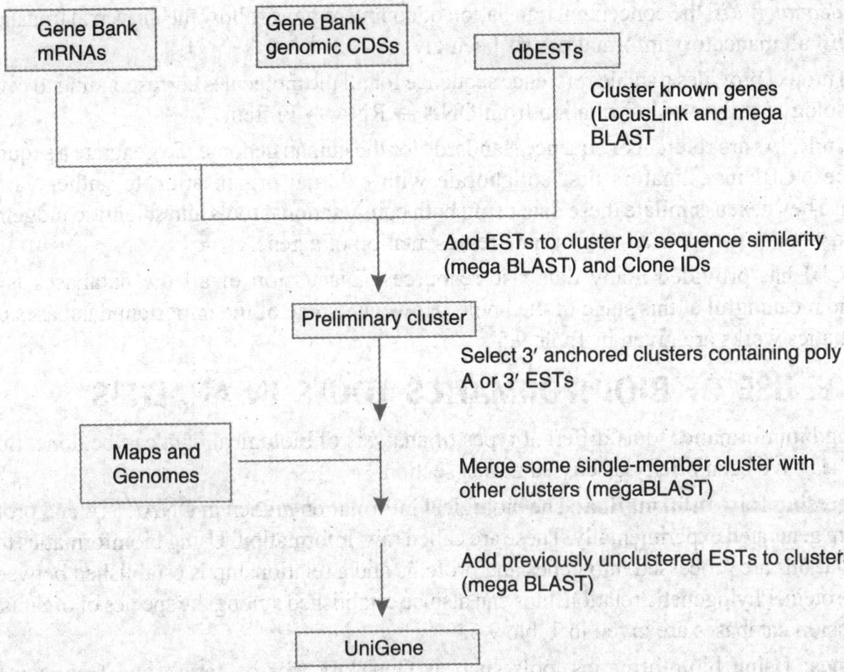


**Fig. 9.2.** Scheme for clustering ESTs by UniGene.

The scheme for clustering ESTs are shown in Fig. 9.3. and steps are given below:

(*i*) First search the sequences for contaminants *e.g.* ribosomal, mitochondrial, repetitive and vector sequences.

(*ii*) Enter the sequences (those contain about 100 bases) into UniGene. The mRNA and genomic DNA are clustered into gene links.

(*iii*) A second sequence comparison links ESTs to each other and to the gene links. All clusters are anchored and contain either a sequence with a polyadenylation site (poly A) or two ESTs labelled as coming from 3' end of a clone.

(*iv*) Clone based edges are added by linking the 5' and 3' ESTs that derive from the same clone.

(*v*) Finally unanchored ESTs and gene clusters of size 1 are compared with UniGene clusters at lower stringency. The UniGene built is updated weekly. Then the sequence cluster may change.

**(*b*) HomoloGene Database:** A new UniGene resource has been created which is called 'HomoloGene' (homologous gene). HomoloGene database includes curated and calculated **orthologs** and **homologs** for genes from human, mouse, rat, zebrafish and cow. This database is also available in LocusLink. Using HomoloGene, homologous relations can easily be inferred. Homologs are identified as the best match between a HomoloGene cluster in one organism and a cluster in a second organism. When two sequences in different organisms are best matches to another the HomoloGene clusters corresponding to the pair of sequences are considered putative orthologs.

**(*c*) RefSeq (Reference Sequence) Database:** It is a database of NCBI which provides a curated nomenclature set of reference sequence standard for naturally occurring biological molecules ranging from chromosomes to transcripts to proteins. For curation process, a curator or annotator is appointed for bioinformatics work. The curators have extensive training in biology. They are very aware of the databases. They ensure that no sequence data is lost during the process of submission. Here, the curator *reviews and checks the newly submitted data and ensures that: (i) the biological features are* *described adequately, (ii)* the conceptual translation of coding regions follow the universal translation rules, and *(iii)* all mandatory information has been given.

RefSeq project provides a stable reference sequence for all the molecules corresponding to central dogma of biology *i.e.* flow of information from DNA → RNA → Protein.

RefSeq mRNAs are reference sequence standards for the human genome. To generate a sequence standard, the NCBI investigators first collaborate with external organisation to gather various information. They next assimilate these data using both computational tools and scientific judgement to determine what sequence is all appropriate representation of a gene.

The NCBI has provided many database resources. Discussion of all the databases is not desirable and meaningful at this stage of the book. However, some of the important databases used in bioinformatics works are given in Table 9.5.

## E. USE OF BIOINFORMATICS TOOLS IN ANALYSIS

By using bioinformatics tools different types of analysis of biological data can be done. Some of the analyses have been briefly discussed in this section.

***(i)* Processing Raw Information:** The biological information hidden in DNA/RNA and protein sequences are generated experimentally. These are called raw information. Using bioinformatics tools these informations are processed into genes and proteins and a relationship is established between a gene and a protein. Phylogenetic relationships can also be established among the species of organisms. Gene expression databases are given in Table 9.5.

***(ii)* Genes:** Using bioinformatics tools such as GenMark (for bacteria), and GenScan (for eukaryotes) gene prediction is carried out in organisms. GenScan can identify introns, exons, promotor sites and poly A signals and other gene identification algorithms.

***(iii)* Proteins:** By using computer programmes protein sequences can be deduced from the predicted genes.

***(iv)* Regulatory Sequences:** Using computer programmes the regulatory sequences can also be identified and analysed.

***(v)* Phylogenetic Relationships:** There are a few web-based applications that will allow you to carryout phylogenetic analysis over the web. By aligning multiple sequences, calculating evolutionary distance and constructing phylogenetic trees, you can establish phylogenetic relationship between two organisms. Internet-based applications that provide phylogenetic analysis capabilities are WEBPHYLIP, PhyloBLAST, GenTree and BLAST2 and Orthologue Search Server. The cluster of orthologous groups (COGs) database simplifies the evolutionary studies of complete genomes and improve functional assignment of individual protein.

A phylogenetic analysis is carried out in four steps: alignment (both building the data model and extraction of phylogenetic data set), determining the substitution model, tree building, and tree evaluation.

***(vi)* Reconstruction of Metabolic Pathways:** Reconstruction of metabolic pathways is one of the indispensable final steps of all genome analyses. It is also a convergence point for the data produced by different methods. The ENZYME database lists name and catalysed reaction for all the enzymes that have been assigned official **Enzyme Commission** (EC) numbers.

***(vii)* Prediction of Function of Unknown Genes:** By using the bioinformatics tools and databases, you can predict the function of unknown genes. Before going for gene prediction you must be aware of the structure of a gene, transcription process, post-transcriptional processing of RNAs and protein folding, etc. There are two approaches used for gene prediction: by using DNA sequences, and using protein sequences.

## QUESTIONS

1. Give a historical account of bioinformatics.
2. List any six resources available from NCBI and their uses.
3. Write the method of nomenclature of protein and DNA sequence.
4. What are the information sources used in bioinformatics ?
5. What are data retrieval tools? Write Entrez map showing the hardlink relationships between databases.
6. Write short notes on the following:

   (*i*) NCBI, (*ii*) Database, (*iii*) BLAST, (*iv*) Directionality of sequence, (*v*) UniGene, (*vi*) ESTs, (*vii*) Entrez, (*vii*) RefSeq database.

CHAPTER

# 10

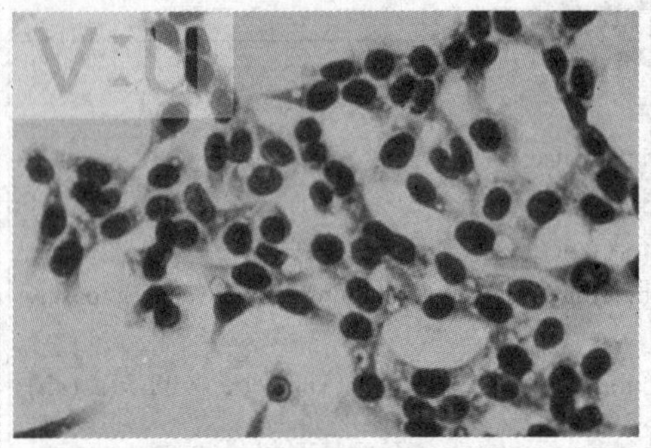

# Animal Cell, Tissue and Organ Culture

The most important areas of today's research which have potential economic value and prospects of commercialization are the cell and tissue culture based production of vaccines, monoclonal antibodies, pharmaceutical drugs, cancer research, genetic manipulation, etc.

Animal cell, tissue and organ can be cultured like plant cell, tissue and organs on artificial media in controlled conditions. A cell differentiates and grows into a large number of cells. Now-a-days several valuable products of medical use have been produced through animal cell culture or genetically engineered cells. However, demand of many others is increasing. Like plant cells, the animal cells of different types for desired colours, sources, etc. may be cultured *in vitro* by providing artificial conditions, and desired products can be obtained from them. One of the most remarkable examples are the monoclonal antibodies, vaccines, etc.

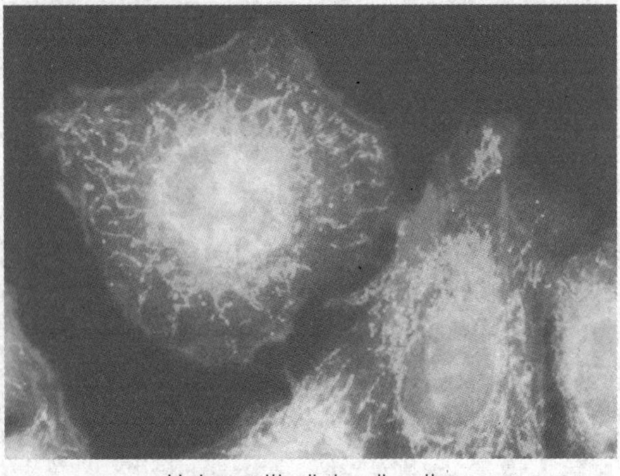

HeLa epithelial cell culture.

In animals tissue regeneration in amphibians and metamorphosis in insects are known since long time but cell and tissue of vertebrates were not cultured for a long time. In recent years, interests arose on cell and tissue culture of birds and mammals. The techniques have been much improved with the development of culture media, creation of microbe-free environment and controlled artificial conditions.

# A. HISTORY OF ANIMAL CELL AND ORGAN CULTURE

Alexis Carrel in 1912 used tissue and embryo extracts as culture media. After using the mixture of chick embryo extract and plasma, cell proliferation was enhanced *in vitro*. The fibrin clot of plasma served as an anchor for cell attachment and the extract provided growth factors and nutrients.

A major innovation to get cell suspension was the use of trypsin for cell disaggregation from tissue explants. This allowed single cell culture. Thus the technique of cell cultures differs from tissue culture. The use of biological fluids and extracts was a major problem because of getting contaminated.

In 1907, Ross Harrison made first attempt to culture animal cells, and cultivated embryonic nerve cells of a frog by using hanging drop method. Thereafter, this method was extended and a wide range of mammalian cells were cultured *in vitro*. However, after supplimenting with embryo extract of chick and plasma, cells proliferated well. These provide nutrients and proliferation factors. Moreover, cell culture method was quite improved after the discovery of antibiotics during 1940s.

Such studies provided an impetus for culture of animal cells in large scale. During this period many human carcinoma cell lines (*e.g.* HeLa cell line) were isolated and grown in culture. Alexis. Carrel, the famous physiologist kept the chick embryo heart alive and its beating continued *in vitro* for about three months. Animal cell culture studies resulted in total stop of using monkeys for multiplication of animal viruses.

In the late 1940s, Enders, Weller and Robbins grew poliomyelitis virus in culture. They provided an impetus to give an easy way to test many chemicals and antibiotics that affect multiplication of virus in living host cells. The significance of animal cell culture was increased when viruses were used to produce vaccines on animal cell cultures in late 1940s. Similarly during 1950s polio vaccine production on cultured cells increased the significance of animal cell culture technology.

In 1966, Alec Issacs infected the cultured cells with viruses and collected filtrate from the infected cells. Again he cultured fresh cells in medium containing filtrate. Such cells were not infected when challenged with viruses. He predicted that the virus-infected cells secreted molecules which coated the surface of fresh cells (uninfected cells). These molecules interfered the entry of viral particles in the uninfected cells. Issac called these molecules as **interferon**. This interpretation could not easily be accepted by the scientific community. It was also said that Issac was mad.

In 1980, the validity of Issac's observation was proved through recombinant DNA technology by cloning and expression of interferon gene in bacterial cells. Now interferon $\alpha$, $\beta$ and $\gamma$ are one of the most successful biotech products which are available in market.

Chinese Hamster ovary (CHO) cell lines were developed during 1980s. Recombinant erythropoietin was produced on CHO cell lines by AMGEN (U.S.A.). Erythropoietin stimulated RBS formation. It is used to prevent anaemia in patients with kidney failure who require dialysis. Within 10 years AMGEN became the first billion biotech company only due to erythropoietin. The Food and Drug Administration (U.S.A.) granted the approval for manufacturing erythropoietin on CHO cell lines. Attempts are being made to culture animal cells for commercial production of recombinant products of animals as the microbial and plant cells are used.

The mammalian cells can be grown in industrial scale cultures in multiple low-productivity roller bottles. In 1982, Thilly and co-workers grew certain mammalian cells to densities as high as $5 \times 10^6$ cells/ml by employing the conventional conditions of medium, serum and $O_2$ and using suitable beads as carriers.

A tremendous excitement has been done in the area of medical science with the development in stem cell technology. It holds a great potential to replace the damaged and dead cells (of blood cancer), neuro-degenerative disease, etc. Since such cells are used as therapeutic cells, hence it is also called cell-based therapy. Attempts are also made in the similar direction for *in vitro* culture of cardiac tissue, neuronal tissue, blood capillaries, etc.

In 1996, Wilmut and co-workers successfully produced a tansgenic sheep named **Dolly** through nuclear transfer technique. Dolly was cloned by transferring nucleus of a mammary (udder) cell into enucleated ovum of an adult sheep. Thereafter, many such animals (like sheep, goat, pigs, fishes, birds, etc.) were produced.

In 2002, **Clonaid**, a human genome society of France claimed to produce a cloned human baby named **Eve**.

The Government of India is promoting cloning of cattle and other animals. For this endeavour research work has been started in certain laboratories of the Department of Biotechnology, Government of India. The National Institute of Immunology (New Delhi) has prepared genetically engineered more than 17 mices for various experiments in the area of health care and genetic improvement in livestock. Similarly Centre for Cellular and Molecular Biology (CCMB) (Hyderabads) has developed transgenic fly by transferring cancer causing genes. Similarly National Dairy Research Institute (Karnal) has developed transgenic cows which can give more milk rich in certain proteins.

## B. REQUIREMENTS FOR ANIMAL CELL AND TISSUE CULTURE

### 1. Some Characteristics of Animal Cell Growth in Culture

Unlike plant and microbial cells, the animal cells can grow only to a limited generations even in the best nutritive media. This growth also depends on the sources of tissue isolated. The special features of different cell cultures are briefly discussed.

(*i*) Neuronal cells constitute the nervous system. In culture the neuronal cells cannot divide and grow.

(*ii*) The cells that form connective tissue (skin) is called fibroblast. Fibroblast can divide and grow in culture to some generations. After completing several generations, they die. It means that all normal animal cells are mortal.

(*iii*) In culture the animal cells divide and grow. Consequently they fill the surface of the container in which they are growing. Thereafter, they stop further growth. This phenomenon is termed as **contact inhibition** *i.e.* inhibition of further cell growth after reaching the wall of container.

(*iv*) The environment of cell growth in culture differs from that of *in vitro*, for example: (*i*) absence of cell-cell interaction and cell matrix interaction, (*ii*) lack of three-dimensional architectural appearance, and (*iii*) changed hormone and nutritional environment. The way of adherence to glass or plastic container in which they grow, cell proliferation and shape of cell results in alterations.

(*v*) In culture the cancer cells apparently differ from the normal cells. Due to uncontrolled growth and more rounded shape, they loose contact inhibition. Therefore, they are piled on each other. These features are exploited by cancer specialists i.e. the cancerologists to identify cancer cells from the normal cells.

Requirements for animal cell and tissue culture are the same as described for plant cell, tissue and organ culture. Desirable requirements are: (*i*) air conditioning of a room, (*ii*) hot room with temperature recorder, (*iii*) microscope room for carrying out microscopic work where different types of microscopes should be installed, (*iv*) dark room, (*v*) service room, (*vi*) sterilization room for sterilization of glassware and culture media, and (*vii*) preparation room for media preparation, etc. In addition, the storage areas should be such where following should be kept properly : (*i*) *liquids*-ambient (4-20°C), (*ii*) *glassware*-shelving, (*iii*) *plastics*-shelving, (*iv*) *small items*-drawers, (*v*) *specialised equipments*-cupboard, slow turnover, (*vi*) *chemicals*-sealed containers.

## 2. Substrates for Cell Growth

There are many types of vertebrate cells that require support for their growth *in vitro* otherwise they will not grow properly. Such cells are called anchorage-dependent cells. Therefore, a large number of substrates which may be adhesive (*e.g.* plastic, glass, palladium, metallic surfaces, etc.) or non-adhesive (*e.g.* agar, agarose, etc.) types may be used as discussed below:

(*i*) **Plastic as a substrate.** Disposable plastics are cheaper substrate as they are commonly made up of polystyrene. After use they should be thrown at proper place. Before use they are treated with gamma radiation or electric arc simply to develop charges on the surface of substrate. After cell growth its rate of proliferation should be measured. In addition, the other plastic materials used as substrate are teflon or polytetrafluoroethylene (PTFE), thermamox (TPX), polyvinylchloride (PVC), polycarbonate, etc. It should be noted that monolayer of cell must be grown. Moreover, plastic beads of polystyrene, sephadex and polyacrylamide are also available for cell growth in suspension culture.

(*ii*) **Glass as a substrate.** Glass is an important substrate used in laboratory in several forms such as test tubes, slides, coverslips, pipettes, flasks, rods, bottles, Petri dishes, several apparatus, etc. These are sterilized by using chemicals, radiations, dry heat (in oven) and moist heat (in autoclave).

(*iii*) **Palladium as a substrate.** For the first time palladium deposited on agarose was used as a substrate for growth of fibroblast and glia.

**Substrate Treatment.** To increase the efficiency of cultured cells through increased cell attachment, the surface of the substrates is treated with purified fibronectin or collagen (or gelatin or poly D-lysine). This chemical is poured onto the surface of dish (substrate), excess amount drained, chemical is dried and sterilized by using UV light. In addition, the surface of substrates is also treated with monolayer of special types of cells which is called feeder layer because it feeds the growing cells. The special cells used as feeder layer are glial cells, normal foetal intestine, mouse embryo fibroblast, etc. The feeder layer is used for the growth of neurons, epithelium of breast and coelome, and production of transgenic animals.

## 3. Culture Media

Culture of animal cells and tissue is rather more difficult than that of microorganisms and plants because the latter synthesise certain chemical constituents from inorganic substances. However, the culture media provide the optimum growth factors (*e.g.* pH, osmotic pressure, etc) and chemical constituents (unlike microbes). There are two types of media used for culture of animal cell and tissue, the *natural media* and the *synthesized media*.

(*a*) **Natural Media :** Natural media are the natural sources of nutrient sufficient for growth and proliferation of animal cells and tissue. These are of three types : (*i*) *coagulans or plasma clots* (it is used since long time but now available in market in the form of liquid plasma kept in silicon ampoules or lyophilized plasma. Plasma may also be prepared in laboratory taking out blood from male fowl and adding heparin to prevent blood coagulation), (*ii*) *biological fluid* ( it is obtained in the form of serum from human adult blood, placental, cord blood, horse blood, calf blood or in the form of biological fluids such as coconut water, amniotic fluid, pleural fluid, insect haemolymph serum, culture filtrate, aquous humour (from eyes), etc. The most commonly used fluids are human placental, cord serum and foetal calf serum. Before use its toxicity should be checked, and (*iii*) *tissue extract* (extract from some tissues such as embryo, liver, spleen, leukocytes, tumour, bone marrow, etc. are also used for culture of animal cells, where embryo extract is of most common use. Tissue extract should be used before a week or stored at 27°C.

**(b) Synthetic Media** : Synthetic media are prepared artificially by adding several nutrients (organic and inorganic), vitamins, salts, $O_2$ and $CO_2$ gas phases, serum proteins, carbohydrates, cofactors, etc. However, different types of synthetic media may be prepared for a variety of cells and tissues to be cultured. It can be prepared for different functions. Basicaly, synthetic media are of two types, *serum-containing media* (*i.e.* the media containing serum) and *serum-free media* (*i.e.* media devoid of serum). Example of some of the media are: minimal essential medium (MEM) CMRL 1066, RPMI 1640 and F12.

When the synthetic media are devoid of serum in culture medium, it is called serum-free media. Example of some serum-free media for certain cells and cell line are given in Table 10.1. By doing so the medium could be made selective for a particular type of cells because each type of cells requires different chemical constituents and physical factors. Serum-free media should not be used commonly until cheap and better serum-free media are available. Serum itself has several disadvantages as given below: (*i*) it deteriorates within a year and differs with batches, (*ii*) a number of batches are required if more than one cell types are used which make difficult for maintaining and co-culturing of cells difficult, (*iii*) supply of serum is less than its demand, therefore, medium becomes several times costly, and (*iv*) undesirable growth stimulation and inhibition may occur. Fiechter (1996) has enlisted the advantages and disadvantages of using the serum in culture media (Table 10.1).

**> Table 10.1. Serum-free medium for certain cell and cell lines.**

| *Serum* | *Serum-free medium* | *Cell or cell lines* |
|---|---|---|
| 1. CS | MCDB 202 | Chick embryofibroblasts |
| | CMRL 1066 | Continuous cell line |
| | MCDB 110, 202 | Fibroblasts, human diploid fibroblasts |
| | MCDB 402 | Fibroblasts, mouse embryofibroblasts, 3T3 cell |
| 2. FB | MCDB 130 | Endothelium |
| | F12 | Skeletal muscles |
| | HoS | Mouse leukemia, mouse erythroleukemia, skeletal muscles |

*Source* : based on R.I. Freshney (2000).

**> Table 10.2. Advantages and disadvantages of serum in culture media.**

**Advantages**

* Serum contains a complete set of essential growth factors, hormones, attachment and spreading factors, binding and transport proteins.
* It binds and neutralises toxins.
* It contains protease inhibitors.
* It increases buffering capacity.
* It provides trace elements and other nutrients.

**Disadvantages**

* It is not chemically defined and, therefore, it is of variable composition lot to lot.
* It may be a source of contamination by viruses, mycoplasma, prions, etc.
* Its components may bind, inactivate, antagonise or mimic the action of added medium ingredients.
* It increases difficulties and cost of down stream processing.
* It is most expensive ingradient of the culture media

*Source* : based on Fiechter (1996).

## 4. Sterilization of Glassware, Equipments and Culture Media

The thoroughly washed glassware and equipments and carefully prepared culture media are sterilized by heat, steam or millipore filter paper (Table 10.3 ). There are different methods used for s sterilization of liquid media as given in Table 10.4. The sterilized materials are removed when temperature cools down and used according to procedure adopted. For detailed description *see* Chapter 12.

**> Table 10.3:** **Sterilization of glassware and equipments.**

| *Sterilization* | *Items* |
|---|---|
| **Autoclaving*** | Apparatus containing glass and silicon tubing, disposable tips for micropipettes, dispensor tubing for pupet, glass bottles with screw caps, magnetic stirrer bars, screw caps, silicon tubing, stopper-rubber, silicon, millipore filters, etc |
| **Dry heat**** | Glassware, glass coverslips, glass slides, instruments, Pasteur pipettes, pipettes, test tubes, etc. |

* 15 lb/in$^2$ (= 121°C) for 20 minutes; ** 160°C/ 1 hour.

**>Table 10.4.** **Sterilization of liquid by different methods.**

| *Sterilization* | *Storage* | *Solution* |
|---|---|---|
| **Autoclave*** | Room temperature | Agar, antibiotics, EDTA, glucose, glycerol, HEPES, lactalbumin hydrolysate, $NaHCO_3$ solution, NaOH solution, phenol red, salt solution (with glucose), tryptose, water |
| **Autoclave** | 4°C | Methocel |
| **Filter**** | Room temperature | Glucose solution, HCl, $NaHCO_3$ solution, NaOH solution, |
| **Filter** | -20°C | Collagenase, antibodies, glutamine, sodium pyruvate solution, transferrin, trypsin, vitamines |
| **Filter-stacked filters** | 4°C | Bovine serum albumin |
| **Filter-stacked filter** | -20°C | Serum |
| **Steam (30 min at 100°C)** | 4°C | Carboxylmethylcellulose |

* 15 lb/in2 (= 121°C) for 30 minutes, ** 0.2 µm pore size filter.

## 5. Equipment Required for Animal Cell Culture

The equipment required for animal cell culture are given below :

(*a*) **Laminar Air Flow (LAF) :** LAF hood acts as aseptic working table for inoculation of animal cells. Culture manipulation in aseptic conditions protects from contamination by any microorganisms such as bacteria and fungi. The contaminated bacterial/fungal cells grow more rapidly than the cultured animal cells. Therefore, the growth of animal cells fails to occur in the presence of contaminants. The working area of LAF hood is first made sterile by using 70% ethanol. Manipulation of any cell is done by keeping the LAF in ON position.

The LAF hood performs two functions : (*i*) it provides a sterile environment for cell manipulation (*i.e.* protects tissue culture from operator), and (*ii*) protects the operator from the potential infection risk from the culture. There are different types of LAF hoods, **airflow** hoods (cabinets) are commercially available in various size and shapes. In this apparatus, sterile air flows inside the space of cabinet which maintains the sterile conditions required for all transfer work (Fig. 10.1). Different types of laminar airflow hoods are available on the basis of nature of the cells and kinds of organisms and tissue culture as below:

(*i*) **Class I hoods:** It protects the operator and to some extent the cell culture too. Air comes from the open front and passes over the cell culture. Then it is released through the top of hood.

Sterile air is sucked into the hood. The users wear special protective clothing within the sterile work area.

*(ii)* **Class II hoods:** It is very common type of hood found in tissue culture. It provides protection to the operator and cell culture. Through the top of hood, filtered air is drawn which passes over the tissue culture. Then air is released through the bottom of work area. Besides, air also enters from the half open front of the operator and released from the front of work area through the grill. The inflow of stream of sterile air into the base of work area protects the culture and operator from contamination.

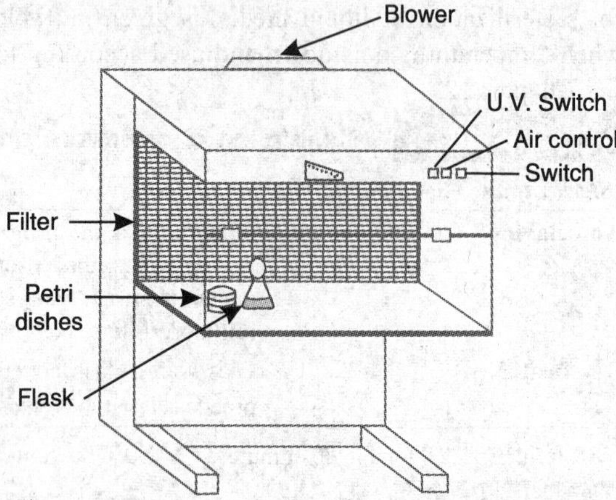


**Fig. 10.1.** Laminar airflow used for aseptic manipulation of cultured tissue.

***(iii)* Class III Hoods:** Highly pathogenic microorganisms are handled in the class III hoods of laminar air flow. A physical barrier separates the workers from the inoculation work. The open front is replaced with glass or perspex. A couple of heavy duty protective gloves are attached to it. The manipulative work is assessed from this glass. The knowledge of special safety consideration is required before the use of class III hood, Help should be sought from the relevant local biosafety officer.

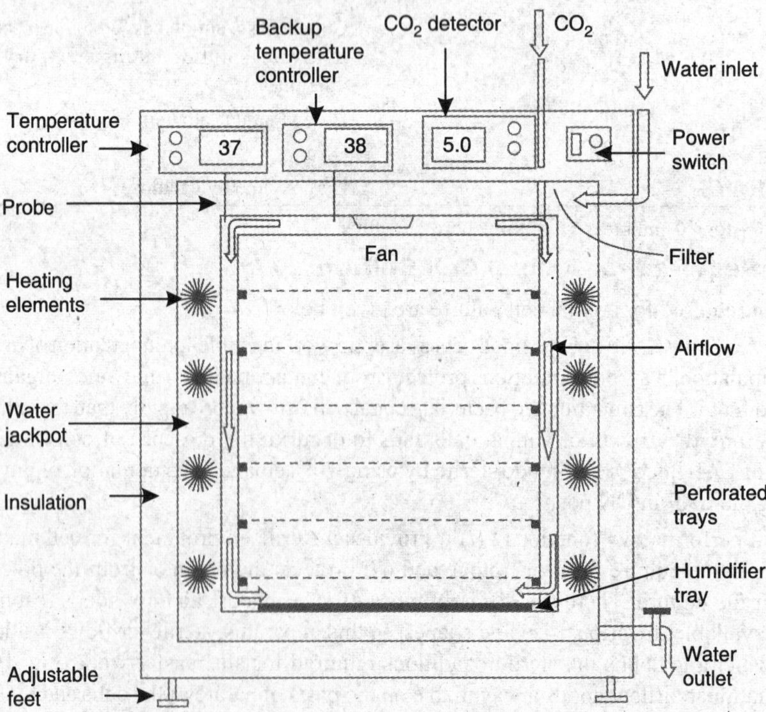


**Fig. 10.2.** Design of a $CO_2$ incubator.

(*b*) **$CO_2$ Incubators :** The $CO_2$ incubators provide the suitable environmental conditions to the growing animal cells. A silicon gasket is used on the inner door which makes the chamber of incubator airtight. It separates the internal environment from the external environment. The external environment possesses the microbial contaminants, while the internal environment remains always contamination-free. The filtered air *i.e.* a high efficiency particulate air (HEPA) is injected inside the chamber which maintains the internal environment of chamber sterile. The relative humidity inside the chamber remains high; therefore, correct osmolarity is maintained and medium is protected from desiccation. The $CO_2$ incubators are designed in such a way that: (*i*) they maintain sterile conditions inside the chamber, (*ii*) they maintain an atmosphere of high relative humidity, (*iii*) they provide an atmosphere with a fixed level of $CO_2$, and (*iv*) they maintain a constant temperature at which they are fixed. A design of $CO_2$ incubator is given in Fig. 10.2.

(*c*) **Centrifuges :** There are different types of centrifuges based on speed. A low speed centrifuge is needed for most of the cell cultures. The separated beads of cells are disrupted simply by a gentle breaking action. Commonly cells are centrifuged at 20° C. The motor evolves heat which rises the temperature. Therefore, use of low temperature for centrifugation is preferred so that the cells should not be exposed to high temperature.

(*d*) **Inverted Microscope :** Use of an inverted microscope is important to observe cell cultures *in situ*. Because, the cells are found on bottom of the tissue culture vessel (*e.g.* Petri plates). The culture medium remains above the growing cells in plates. If such plates are put over the stage of an ordinary microscope, the growing cells at bottom cannot be observed. Therefore, the inverted microscope is used for such purpose.

In inverted microscope the optical system is at the bottom and light source at the top. When a plate is observed in the inverted microscope, the cells of the culture growing at the bottom of the plate can easily be observed. If any one want to count cells using a counting chamber, a standard microscope with movable slide holder is required for this purpose.

(*e*) **Culture Room:** All types of cultured plant tissues are incubated under the conditions of well controlled temperature, humidity, illumination and air circulation. The culture room should have light and temperature control systems. Generally temperature is maintained at 25 ± 2°C and 20-98% relative humidity and uniform air ventilation. Generally cultures are grown in diffused light and darkness each for a period of 12 hours.

(*f*) **Data Collection (Observation):** The cultures are monitored at regular intervals in the culture room for the growth and development of cultured tissues. Observation is also made under aseptic area in laminar airflow.

## 6. Isolation of Animal Material (Tissue)

During the experimental work, attempts should be made that animal materials are not contaminated. When glassware and media are sterilized, animal materials should be handled. Before that a balanced salt solution (BSS) is required. This solution consists of 1000 units of penicillin and 0.5 mg of streptomycin or neomycin per ml. Before culture animal materials are washed in BSS aseptically to avoid contamination. The tissue to be cultured should be properly sterilized with 70 per cent ethanol, and removed surgically under aseptic conditions. Attempts should be made to avoid contamination. Thus, the tissue isolated is either stored in freeze or used immediately.

(*a*) **Disaggregation of Tissue :** Some of tissues consist of cells which are tightly aggregated. Tissue like epithelium is impregnated with $Ca^{2+}$ and $Mg^{2++}$ ions that provide integrity to it.

Therefore, for getting primary culture it is necessary that tissue must be disaggregated either mechanically or using enzymes or chemicals so that cell suspension could be obtained. The cells in suspension grow to produce primary culture.

**(*i*) Physical (mechanical) disaggregation.** The tissue after careful removal from a given spot is aseptically kept in a sieve of 100 μm sieve (Fig. 10.3). It is put in a sterile Petri dish containing buffered medium with balanced salt solution. The cells are alternately passed through sieves of decreasing pore size (50 μm and 20 mm mesh). If desired the process is repeated to get more disaggregation of cells. The debris remaining on sieves are discarded and medium containing cells is collected. The cells are counted using a haemocytometer. If necessary, medium is diluted by serum to raise the level of cells to $10^4$ cells/ml. The other methods of mechanical disaggregation of cells are forcing the cells through syringe and needle or repeated pipetting. Although the physical method is quick and cheap yet it damages many live cells.

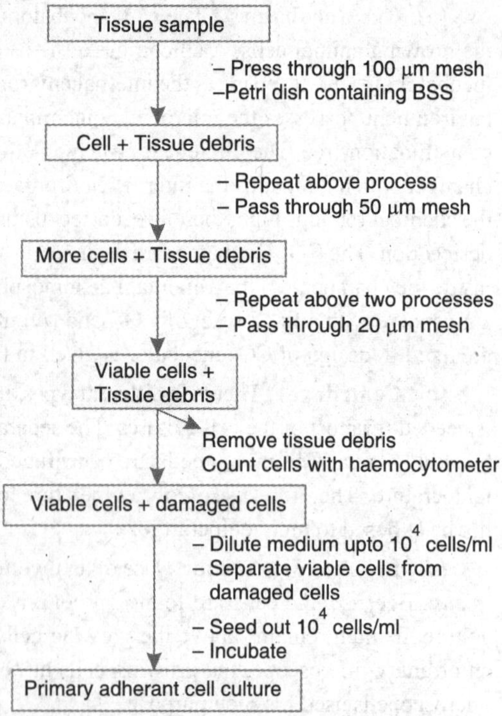


**Fig. 10.3.** Physical disaggregation of tissue by sieving (diagrammatic).

The viable dissociated cells are now termed as 'primary cells'. When the primary cells are seeded on culture medium in high density, they grow well. Thus the primary viable cells of primary culture are called adherant culture and the cells adherant cells. Moreover, at this stage some of the non-viable cells if growing along with adherant cells can be separated out by using the second medium. Similarly, primary culture can also be grown in suspension. In suspension the non-viable cells be removed from primary disaggregates by centrifugation using Ficoll and sodium metrizoate. In this way viable cells are separated from non-viable cells.

**(*ii*) Enzymatic disaggregation.** Enzymes are also used for dislodging the cells of tissue. By using enzymes a high number of cells is obtained. Moreover, in embryonic tissue a high number of undifferentiated cells with least extracellular matrix is found. Therefore, disaggregation of embryonic tissue occurs more readily than that of adult or new borns. In addition, in fragile tissue such as tumours the chances of cell death and cell recovery are more than the normal tissues. There are two important enzymes used in tissue disaggregation, collagenase and trypsin.

• **Use of collagenase.** Collagenase is used for disaggregation of embryonic, normal as well as malignant tissues. The intracellular matrix contains collagen, therefore, collagenase disaggregates normal and malignant tissues. Moreover, the epithelial cells can be damaged by it but the fibrous tissues remain unaffected. The crude collagenase also contains non-specific proteases. First, the biopsy tissues are kept in medium containing antibiotics. Thereafter, the tissue to be disaggregated is dissected into pieces in basal salt solution containing antibiotics (Fig. 10.4). The chopped tissue is properly washed with sterile distilled water and transferred in complete medium containing collagenase. After five days of treatment the mixture is pipetted so that the medium may get dispersed. When the whole treatment is left for some times, the residual clusture of epithelial cells settles on bottom of test tubes. Clustures present in test tubes are washed by settling or the dispersed cell suspension is made free from the enzyme collagenase by centrifugation. Suspension consists of enriched fibroblast fraction which is plated out on medium. Similarly, the clusture which is washed by settling consists of enriched epithelial fraction. It is also plated out on medium.

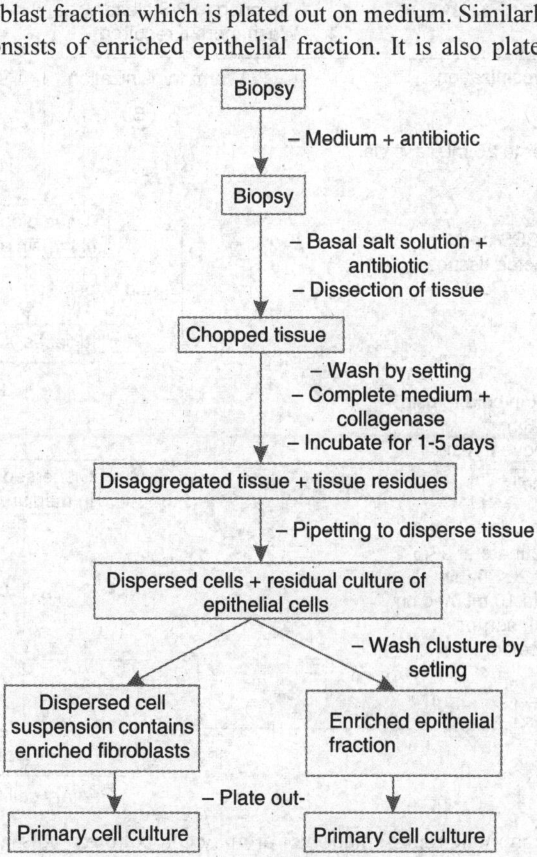


**Fig. 10.4.** Tissue disaggregation by collagenase.

- **Use of trypsin.** Use of trypsin for disaggregation of tissue is called trypsinization. However, the enzyme trypsin in crude form is commonly used for embryonic tissue because many kinds of cell can tolerate it and different types of tissues are significantly affected. Besides, serum or trypsin inhibitors (*e.g.* soybean trypsin inhibitor) can neutralise its residual enzyme activity only in serum-free medium. On the basis of role of temperature on trypsin, activity is of two types, cold trypsinization and warm trypsinization.

  ***Cold trypsinization.*** The tissue sample to be disaggregated is chopped into 2-3 small pieces and kept in small sterile glass vial (Fig. 10.5A). If necessary, these may be washed with sterile distilled water. The pieces are removed from vials, dissected keeping in BSS, The whole content again transferred in glass vial is placed on ice and soaked in cold trypsin for 4-6 hours. This allows penetration of enzymes in tissue. Further, trypsin is removed and tissue is incubated at 36.5ºC for 20-30 minutes. The vials that contain tissue pieces, 10 ml of medium containing serum is added and cells are dispersed by repeated pipetting. The cells are counted by using haemocytometer. Cell density is maintained to $10^4$ cell/ml with dilution in growth medium. These are plated and incubated for 48-72 hours for cell growth.

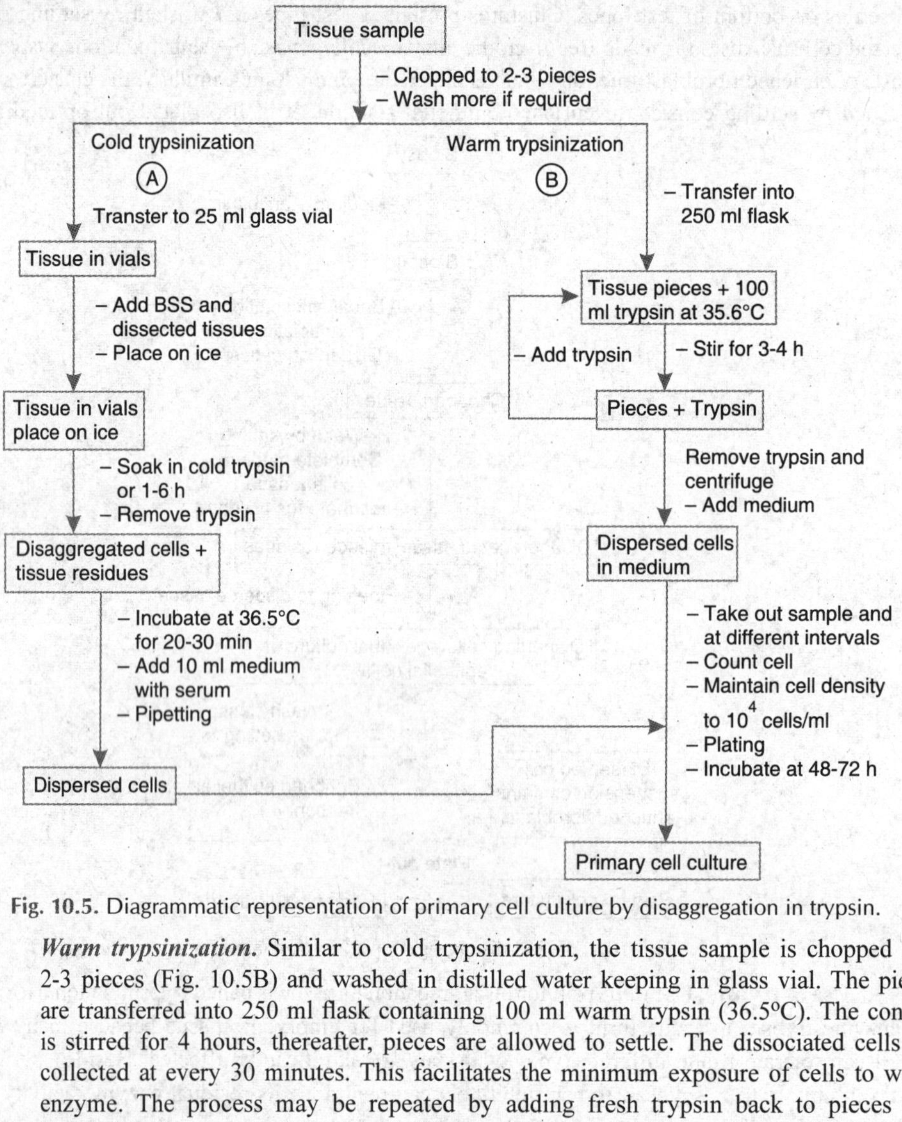


**Fig. 10.5.** Diagrammatic representation of primary cell culture by disaggregation in trypsin.

***Warm trypsinization.*** Similar to cold trypsinization, the tissue sample is chopped into 2-3 pieces (Fig. 10.5B) and washed in distilled water keeping in glass vial. The pieces are transferred into 250 ml flask containing 100 ml warm trypsin (36.5°C). The content is stirred for 4 hours, thereafter, pieces are allowed to settle. The dissociated cells are collected at every 30 minutes. This facilitates the minimum exposure of cells to warm enzyme. The process may be repeated by adding fresh trypsin back to pieces and incubating the contents. The trypsin is removed by centrifugation after 3-4 hours during which complete tissue may be disaggregated. The glass vials containing dispersed cell pellets in medium are placed on ice. After different trypsinization time, samples are pipetted, cells counted using haemocytometer and cell density maintained to $10^4$ cells/ml. The cells are plated on medium and incubated for 48-72 hours for cell growth.

**(*b*) Establishment of Cell Cultures :** There are many type of animal cells that can grow *in vitro* such as tumour cells, pigmented melanoma cells, neuroblastoma cells, steroid producing adrenal cells, growth hormone prolactin secreting cells, teratoma cells capable of differentiating in artificial conditions pigmented or cartilage cells, etc. On the basis of purpose of experiment, a continuous (immortal) cell line can be developed from cultured tissues. Healthy animal tissues capable of dividing are cultured on artificial nutrient media that proliferate and differentiate into heterogeneous mixture of different types of cells (Fig.10.6).

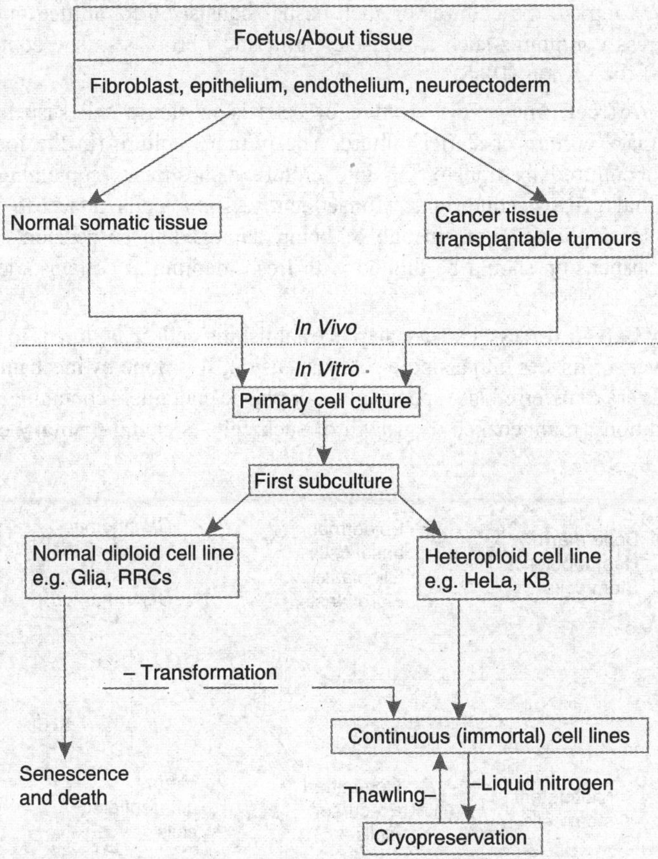


**Fig. 10.6.** Cell culture from somatic tissue.

However, it is not necessary that the cells which are plated on medium will start immediate growth. On the basis of growth responses culture cells are divided into the three types: precursor or master cells or stem cells, undifferentiated but committed cells, and the mature differentiated cells. The precursor cells have the ability to proliferate but they do not differentiate until the proper conditions for induction are applied in the medium. This is done to facilitate some or all cells to mature to differentiated cells. These are totipotent (also called multipotent or pluripotent) cells which have the ability to differentiate into different types of cells. The entire blood and immune systems are produced by the totipotent stem cells. In contrast, the pluripotent cells differ from totipotent cells in being less general and still capable of differentiating into many kinds of cells. Now the cell cultures are at a stage of equilibrium of three type of cells *viz.*, multipotent stem cells, undifferentiated but committed precursor cells and mature differentiated cells. This equilibrium may be shifted by manipulating the growth conditions. For example, in the presence of low serum, suitable hormones, cell matrix interaction and high cell density, cell differentiation is promoted, whereas in the presence of high serum, suitable growth factors and low cell density, cell proliferation is promoted.

The types of cells growing in culture are determined by their respective sources from where these have been derived. For example, a high number of stem cells are found in cell lines derived from embryos because they are capable of frequent cell division as compared to adult cell culture. Also the culture of such adults comprises of stem cells which are capable of undergoing continuous removal *in vivo* such as intestinal epithelium, haemopoietic cells, epidermis, etc. In contrast, the cultures of such tissues consist of committed precursor cells that renew only in stress conditions such as muscles, glia and fibroblasts. The committed precursor cells have limited life (Anon, 1988).

***(i)* Evolution of cell lines.** This mixture of cells is in single cell suspension which may be used as a primary culture or starter culture. The primary culture (in the form of single-cell suspension) is subcultured by transferring into culture dishes/flasks containing special growth nutrients at optimal growth conditions. Consequently, some cells attach to the surface and proliferate to yield single cell line, inspite of being damaged in suspension. Therefore, while subculturing the suspension should be diluted with fresh medium at certain ratio and transferred into a flask (Anon, 1988).

***(ii)* Primary Cell Cultures :** The first step in establishing cells in culture is to dissociate organs (*e.g.* kidney or liver) or tissues into a single cell suspension. It is done by mechanical or enzymatic methods. The cells are transferred into special glass or plastic containers containing culture medium. Under these conditions, maintenance of growth of such cells is called **primary cell culture** (Fig. 10.7).

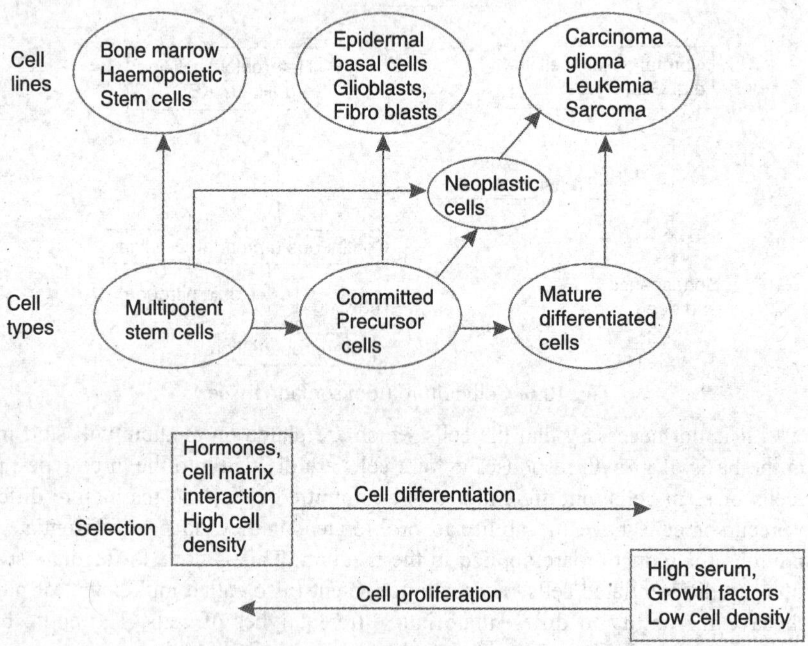


**Fig. 10.7.** Culture conditions affecting cell types and origin of cell lines.

The cells are enumerated together by a proteinaceous material. Therefore, crude preparation of proteolytic enzymes (trypsin and collagenase) are commonly used to break the cementing material and separate cells of a given tissue. The characteristics of the animal cells govern the characteristics of the cells in culture. The growing cells are of two different types : (*i*) anchorage-dependent (adherent) cells, and (*ii*) anchorage-independent cells (suspension culture). Commonly the adherent cells can be obtained from such organs that are fixed at a place (*e.g.* kidney, liver, etc.). The cells too are not mobile but remain fixed in connective tissues.

In contrast the suspension cells grow continuously in liquid medium and do not attach to the surface of the container. The source of cells is the governing factor for suspension non-adherent cells. All suspension cultures are raised by culturing the blood cells. You know that all blood cells are vascular in nature and get dissolved in plasma. The primary culture becomes a cell line only after its first subculture. The subculture is needed when the nutrients present in medium for cell growth diminish. These may be subcultured several times on the fresh medium and propagated accordingly. By doing so a continuous growth of cells is maintained.

***(iii)* Secondary Cell Cultures and Cell Lines :** The primary cell culture (Fig. 10.6) cannot remain viable for a long time because the cell utilise all nutrients of the medium. Therefore, sub-culturing needs to be done on another fresh medium. Hence, the primary culture is removed, adherent cells are dissociated enzymatically or by repeated pipetting. Then the cells are diluted with fresh medium and passed into fresh culture flask.

This results in multiple copies of a single type of cell with a negligible amount of non-proliferating cells. During the course of repeated sub-culture and selection the cell line gets evolved and properly established consisting of rapidly proliferating cells. Thus the unaltered form of cell line (only for a limited number of generations) is called continuous cell line which propagates in logarithmic ways (Fig. 10.7).

Sub-culturing is done on fresh medium at certain intervals. It provides sufficient nutrient and space to growing cell lines. Characteristic features of such cell type governs that how quickly sub-culturing shall be done. However, the cells may exhaust the medium and die if they are not split frequently. During sub-culturing the *in vitro* conditions produce mostly undefined selection pressure. Consequently, a certain cell type (*e.g.* fibroblast) is selected. After some generations the normal diploid cells die.

Some times some cells of secondary cell cultures *in vitro* can be transformed spontaneously or chemically. Such transformed cells are immortal (*i.e.* they will not die) and hence give rise to continuous cell lines *i.e.* cancerous cell lines. These cells proliferate idefinitely with a doubling time of about 10 to 25 hours. Such cultures consist of mixture of cell types where a particular cell type may dominate over the others. For details *see* preceding section. Any change in continuous cell line may discontinue the increase in cell number. This may be brought about by chemicals, spontaneous mutation or viruses (*e.g.* Epstein Barr virus). The phenomenon of alteration in continuous cell line is called '*in vitro* transformation'.

A cell line consists of several similar or dissimilar cell lineages. A defined cell lineage having specific properties is called 'cell strain'. The cell strain is identified when cells of that culture are in bulk. On the basis of life span of culture, the cell line or cell strain may be continuous or finite. The other types of cell line may be immortal *i.e.* which will not die. The life of finite cell line is limited upto 20-80 generations, thereafter, they die. Properties of some continuous and finite cell lines are given in Table 10.5. It should be noted that the cell lines are represented in abbreviated form either derived from the sources of cells, name of institute or association with viruses (for example EB, Epstein Barr; WI, Wistar Institute). They are numbered if more than one cell line are derived from the same cell line (*e.g.* EB1, EB2). Furthermore, a cell line is given the number of population doubling *e.g.* EB1/1, etc.

**> Table 10.5. Properties of finite and continuous cell lines and cell strains.**

| *Cell lines* | *Source* | *Morphology* | *Age* | *Tissue* | *Characteristics* |
|---|---|---|---|---|---|
| **1. Finite Cell Line** | | | | | |
| IMR90 | Human lung | Fibroblasts | Embryonic | Normal | Infected by human virus |
| MRC5 | Human lung | Fibroblasts | Embryonic | Normal | Infected by human virus |
| MRC9 | Human lung | Fibroblasts | Embryonic | Normal | Infected by human virus |
| WI38 | Human lung | Fibroblasts | Embryonic | Normal | Infected by human virus |

| **2. Continuous Cell Line** | | | | | |
|---|---|---|---|---|---|
| EB | Human | Lymphocytic | Juvenile | NP | EB virus, +ve |
| HeLa | Human | Epithelial | Adult | NP | G6PD Type A |
| LS | Mouse | Fibroblastic | Adult | NP | Grow in L929 suspension |
| P388D | Mouse | Lymphocytic | Adult | NP | Grow in suspension |
| S180 | Mouse | Fibroblastic | Adult | NP | Cancer chemotherapy |
| 3T3A | Mouse | Fibroblastic | Embryonic | Normal | Readily transformed |

NP, Neoplastic tissue.

(*c*) **Types of Cell Lines :** There are different types of cell lines produced from different tissues or organs. Broadly they are grouped into two types : finite cell lines and continuous cell lines.

(*i*) **Finite cell lines :** The cell lines which grow through a limited number of cell generations and have a limited life are called finite cell lines. The cells grow slowly and form monolayer. Their doubling time ranges from 24 to 96 hours. The characteristics of cell lines are anchorage dependent, contact inhibition and density limitation.

(*ii*) **Continuous cell lines :** Continuous cell lines are obtained either from transformed cell lines *in vitro* or cancerous cells (Fig. 10.8). They divide rapidly. Their generation time is 12 to 14 hours. They have no or reduced density limitation. The transformed cells *in vitro* bear the following differences:

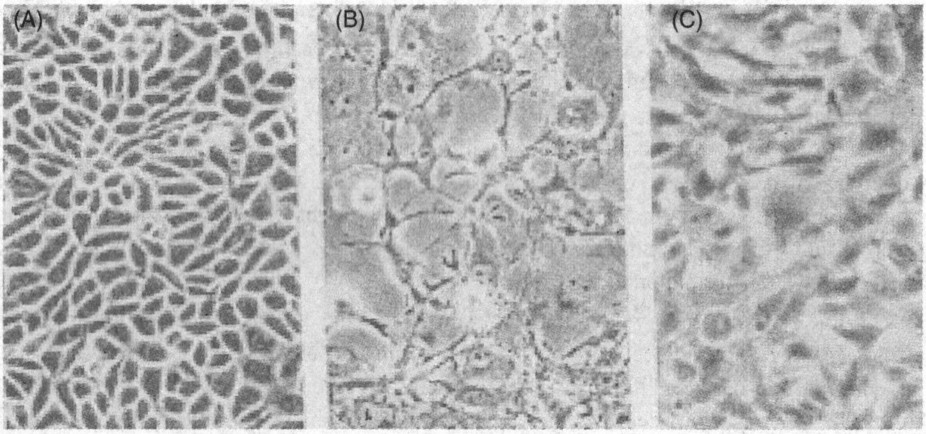


**Fig. 10.8.** Suspension culture of Chinese hamster ovary (CHO) continuous cell line (A). Cos-7 cell line from monkey Kidney (B), HeLa cells from human cervical carcinoma (C).

(*i*) Enhanced growth and proliferation due to rapid growth rate,

(*ii*) Existence of altered ploidy *i.e.* aneuploidy or heteroploidy due to altered chromosome number,

(*iii*) Different cell shape and organisation of microfilaments,

(*iv*) Ability to translocate,

(*v*) Different energy metabolism,

(*vi*) No contact inhibition and no anchorage dependence, and

(*vii*) Different growth factor requirements and responses to regulate molecules.

Different types of continuous cell lines are shown in Fig. 10.8.

**(*d*) Factors Affecting Subculture *In Vitro*.** There are several factors that influence differentiation and proliferation of cells when subcultured *in vitro* as below:

(*a*) The mammalian cells need attachment to a suitable surface. The maximum cell numbers are limited by the surface area available. Therefore, mammalian cells should be grown on microcarriers such as beads of anion exchange resin, etc.

(*b*) Serum is also added in medium. It allows better cell growth in agitated and/or aerated cultures. Serum is a highly complex mixture of several kinds of molecules that provides both growth-promoting and growth-inhibitory factors to the cells. Some constituents which are very useful are hormones, albumin binding proteins, transport protein, growth factors, attachment factors and micronutrients. Low serum or serum-free cultures are more susceptible to fluid-mechanical damage. It protects the cell from physical damage caused by agitation in bioreactor because bubbles cause damage but in the absence of gas, cells are also damaged, therefore, slow agitation rate of medium in bioreactor is required.

(*c*) Various types of additives are also used to protect freely suspended animal cells in culture from agitation or aeration damage. These additives are: cellulose and starch derivatives, protein mixtures pluronic polyals, polyvinyl-pyrrolidones, dextrans polyvinyl glycol (PEG), polyvinyl alcohols (PVA), methylcelluloses (MCs), etc. Effect of these additives on physiological and product expression of cells, cell aggegation should be arrested under both static and bioreactor growth conditions before their use.

(*d*) MCs have been found as reliable protectant, therefore, MCs and other cellulose derivatives have been influenced as media additives in the culture of different types of cells. Tilly *et al*. (1982) got success in growing certain mammalian cells to a level to about $5 \times 10^6$ cell /ml of cell density by employing conventional conditions of medium, serum and oxygen, and suitable bead carriers.

(*e*) The optimum pH of medium should be maintained between 7.0 and 7.6. This is controlled with a dual system provided with acid and base. In animal culture the COD/bicarbonate buffering system is used in culture medium.

(*f*) The animal cells are very sensitive to temperature, therefore, by using a thermostate temperature of bioreactor medium should be maintained at 37°C.

## C. CULTIVATION OF ANIMAL CELLS EN MASSE IN BIOREACTOR

The cells are cultivated in two different phases as freely suspended cells in liquid phase and immobilised cells on solid phase.

### 1. Suspension Culture

In suspension culture the cells are dispersed in liquid medium and grow freely but not attached with any solid. The medium is agitated so that they should not form sediments. While adherent cells are permitted to get attached on solid support and they grow as immobilised culture. Therefore, it depends on choice of cells whether they are capable of growing freely in suspension or as immobilised culture.

A fully automated bioreactor is used that maintains the physicochemical and biological factors to optimum level and maintains freely the suspended cells in an agitated low viscous liquid medium. The most suitable bioreactor used is a compact-loop bioreactor that consists of marine impellers (Fig. 10.9). It also relies on the integrated monitoring and control of physical, chemical, biological and biochemical parameters.

Bioreactor for animal cell culture.

As compared to bacteria, animal cells grow slowly. This results in bringing the cells in unfavourable metabolic state due to even small changes in culture. The main carbon and energy sources are glucose and glutamine. Lactate and ammonia are their metabolic products that affect growth and productivity of cells. Therefore, on-line monitoring of glucose, glutamate and ammonia can be carried out by on-line flow injection analysis (FIA) by using gas chromatography (GC), high performance liquid chromatography (HPLC), etc.

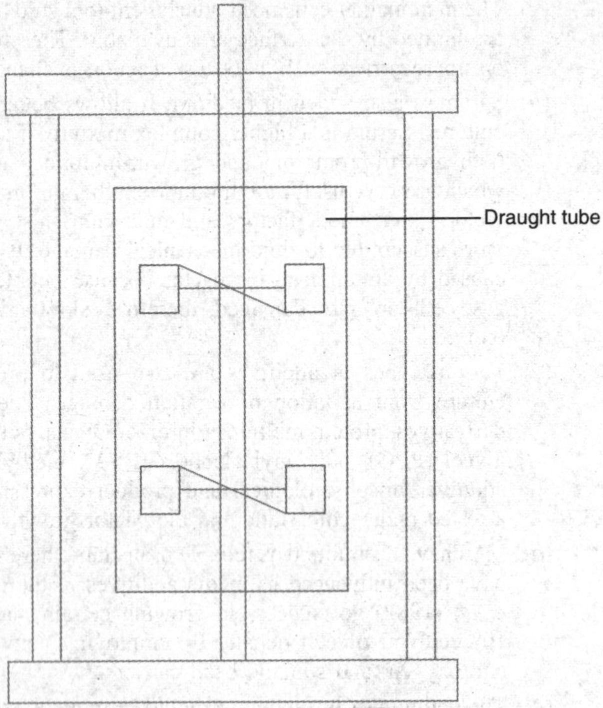


**Fig. 10.9.** A compact loop bioreactor (diagrammatic).

Moreover, cell death also occurs due to certain reasons. Therefore, for measuring differences between viable and dead cells some enzyme activities are measured and correlated with cell density. Lactate dehydrogenase is used to distinguish cell density of viable cells (Fiechter, 1996).

Shear occurs due to stirring and sparging. Cells attached to air bubbles are exposed to enormous forces when bubbles leave the bulk liquid at the surface and burst due to decompression. Cells in sparged cultures can be protected from shear forces by using medium with a high viscosity. This can be achieved through high cell density ($> 10^7$ cells/ml).

Similar to microorganisms, the animal cells can also be cultivated as batch, fed-batch and continuous culture. Continuous process can be carried out as a chemostat or **perfusion culture** (*i.e.* a continuous culture when biomass is retained in a reactor and cell-free culture liquid is removed).

## 2. Methods for Scale-up of Cell Culture Process

In batch culture there are various methods for scaling up of animal cell culture process. These methods include **roller bottles** with microcarrier beads (for adherent cell culture) and **spinner flasks** (for suspension culture).

***(a)* Roller Bottles :** Early commercial production with anchorage-dependent cells was often performed in roller bottles (Fig. 10.10 A-B). Inside the roller bottles animal cells adher to curved surface area of microcarrier beads. Consequently, this increases the total surface area for available spaces for the growth of the cells. Typically, a surface area of 750–1500 $cm^2$ with 200–500 ml medium will yield $1 - 2 \times 10^8$ cells. Moreover, a large surface area could be obtained by the use of microcarriers in stirred tank reactor.

The roller bottles are well attached inside a specialised $CO_2$ incubators. The attachments rotate the bottles along the long axis. The entire cell monolayer is exposed to the medium after each full rotation of the bottle. The volume of the medium superficially covering the monolayer is sufficient for cell growth. This system has three major advantages over static monolayer culture : (*i*) increase in surface area, (*ii*) constant and gentle agitation of the medium, and (*iii*) increased ratio of surface area of medium to its volume, which allows gas exchange at an increased rate through the thin film of the medium over the cells which are not submerged in the medium (Freshney, 2000).

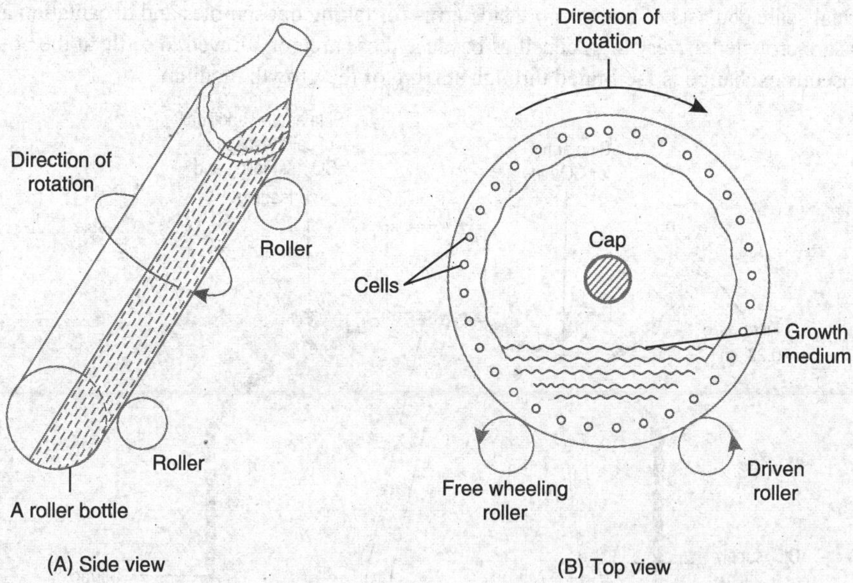


**Fig. 10.10.** A side view of roller bottle (A); cell culture inside roller bottle (B).

***Microcarrier Beads*** : Microcarriers are also called microcarrier beads because these are small spherical particles (having 90–300 micrometer diameter) looking like beads (Fig. 10.11). For animal cell culture microcarrier beads are used to increase the space for cell growth. This results in an increase in number of the adherent cells per flask. These beads are made up of dextran or glass. These beads can also be used with spinner culture flasks as they are buoyant. At the recommended concentration when microcarriers are resuspended they provide 0.24 $m^2$ area for every 100 ml of culture flask. This condition supports the growth of adherent cells to a very high density. Otherwise they shall be crowded and cause problem. The rapid cell growth quickly exhausts growth medium.

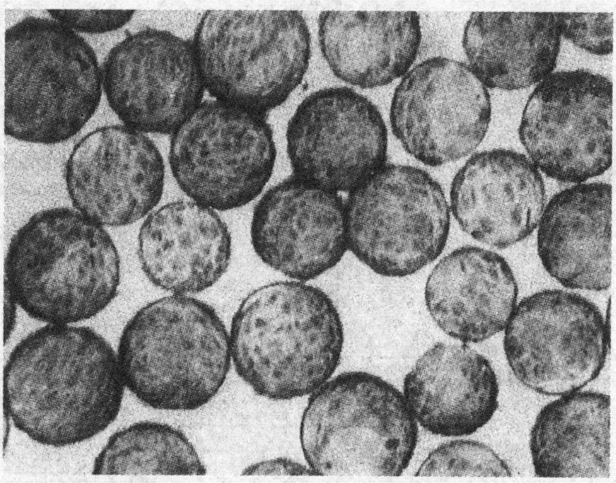

**Fig. 10.11.** Microcarrier beads.

**(*b*) Spinner Cultures :** Scaling-up of process requires several intermediate steps. The first step is the transfer of cells from stationary culture to shake flask or spinner flask. The spinner flask was originally developed to provide gentle stirring of microcarriers. But now it is used for suspension culture also. Merits of spinner culture are : very simple device, and cheaper cost. Its demerits are : (*i*) it is un-suitable for scale up because large number of single units required are time consuming, (*ii*) cultures are inhomogeneously mixed, and (*iii*) it is badly suited for process control.

The spinner flask consists of a flat surface glass flask. A teflon paddle (impeller) hangs down in the flask from the lid and remains suspended without touching the bottom. The impeller rotates and agitates the medium when flask is put in a magnetic stirrer (Fig. 10.12). The spinner flask used at commercial scale consists of one or more side arms for taking out samples and decantation as well. The cells are crowded at very high densities because these are not allowed to settle at the bottom of flask. Gaseous exchange is facilitated through stirring of the growth medium.

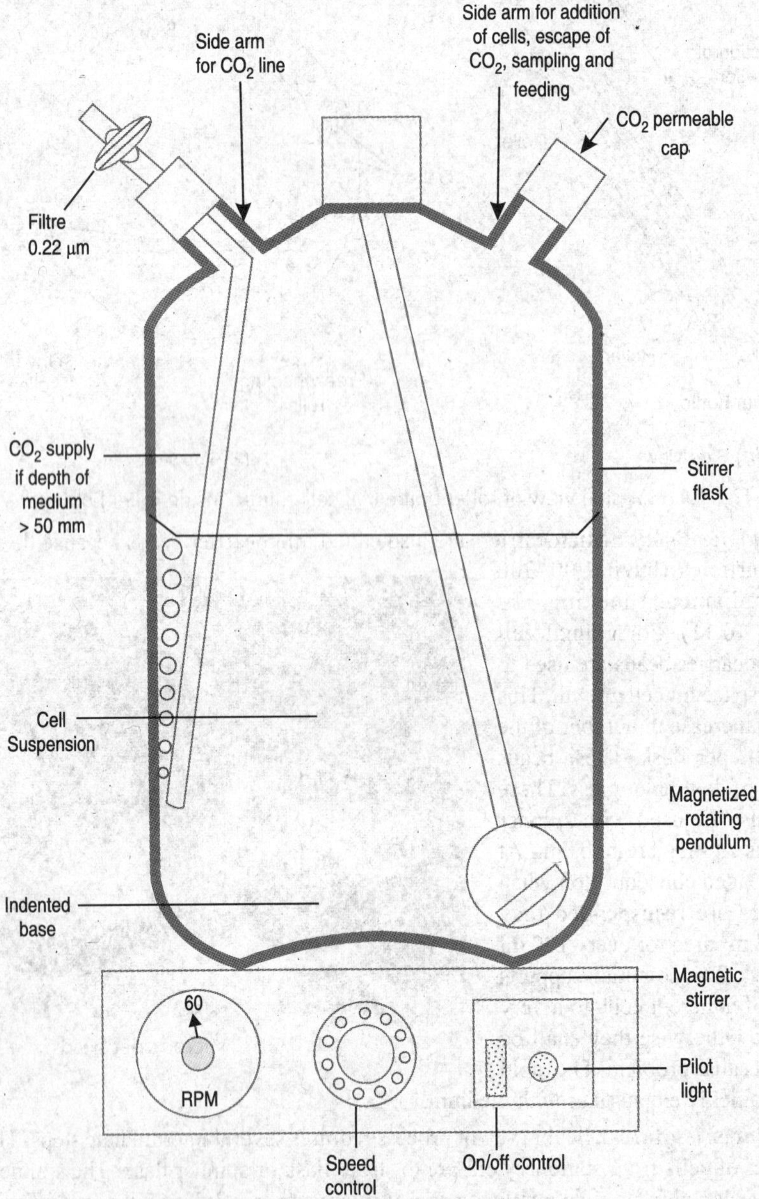


**Fig. 10.12.** Spinner cell culture.

Many animal cell lines can be cultivated in suspension culture similar to microganisms. Some cell types (*e.g.* typical normal diploid cells) are anchorage-dependent (*i.e.* they require a surface to grow on). Such cells are grown on the inside surface of plastic or glass bottles or on the surface of microcarriers suspended in liquid medium. An artificial culture environment maintains its optimum pH (6.7 to 7.9), dissolved $O_2$ (20–80%) and temperature. The animal cells are sensitive to shear forces.

## D. IMMOBILISED CELL CULTURE

The properties of microcarrier beads used for cell immobilisation are that they develop charged surfaces, either positive or negative charges, for example, DEAE sephadex (diethyl - aminoethyl type of cross-linked dextran) developed positive charges, polysysteine negative charges, gelatin beads slightly positive or negative charges. The beads require surface coating for adhesion of sufficient number of cells, because energy is required for cell adhesion. In addition glass and ceramic materials are used as good carriers as they contain high surface energy.

In different types of bioreactors different materials such as glass spheres are packed for cell attachment. The materials increase surface area for cultivation of cells *en masse*. This type of bioreactor is called packed bed bioreactor. For a bioreactor of 100 litre capacity about 50 kg of 3 mm diameter spheres are used. This gives a total surface area of about 20 $m^2$.

## E. INSECT CELL CULTURE

Insects are the vectors for a large number of viruses causing diseases on economic plants as well as animals including humans. Some important diseases are also caused by certain insects such as *Aedes aegypti* for yellow fever, *Anopheles* female for malaria, *Bombax mori* for silk disease, *Heliothis zea* cottonboll, etc.

Insect cells are grown in suspension in suitable bioreactor. The bioprocess technology developed for certain mammalian cell is also applicable for insect cells with certain alterations. The physico-chemical factors affecting cell expression for production of valuable products are cell viability, cell density, insect species, types of tissue, concentration of dissolved $O_2$, composition of culture medium, property of cell attachment and nature of substratum.

Significant achievement gained in the area of health care and agriculture has focussed the attention of biotechnologists in recent years. Cultures of insect cells have become a good tool for expression of heterologous gene products (Table 10.7). Baculoviruses infect only invertebrates, but neither vertebrates nor humans nor plants. Genetically modified vector expression in cultured insect cells and expression of foreign genes if introduced into cells for the production of biotechnological products *viz.*, recombinant proteins and viral insecticides, cell culture of certain insects (such as moths, flies, butterflies, mosquitoes, bollworms, loopers, etc) are being used. Nuclear polyhedrosis viruses (NPVs) are the most widely used baculoviruses. The NPVs consist of double stranded DNA. They are the natural pathogens of the insects of the group Lepidoptera. Therefore, they grow well in cell cultures of members of Lepidoptera.

**Production of Commercial Products from Insect Cell Cultures.** Commercially desired proteins may be produced *in vitro* by using a susceptible continuous cell line of insect in a bioreactor. The fully grown cells are allowed to be infected by genetically engineered baculovirus. Baculovirus infects the cell line and lyse them resulting in release of protein products in the medium. Thereafter, protein is purified.

Adopting the same method bioinsecticides (occlusion bodies) can also be produced by using wild type virus. The occlusion bodies are then isolated from the bioreactor and used for the management of crops against the attack of insects. Biotechnological application of some products derived from insect cell culture are given in Table 10.6.

**Table 10.6. Biotechnological applications of some insect cell culture systems.**

| *Insects* | *Application* |
|---|---|
| *Aedes aegypti* (yellow fever mosquito) | Arbovirus antigens, vaccines, diagnostics |
| *Autographa californica* (alfalfa looper) | Bioinsecticides |
| *Bombax mori* (silkworm) | Bioinsecticides |
| *Heliothes virescens* (tobacco bollworm) | Bioinsecticides |
| *Spodoptera frugiperda* (fall armyworm) | Recombinant proteins |
| *Trichoplusia ni* (cabbage looper) | Recombinant proteins |

## F. SOMATIC CELL FUSION

In animals, fusion of two different cells and production of a hybrid cell have been successfully achieved. These hybrid cells have significant biotechnological applications in many more areas such as: (*i*) study of control of gene expression and differentiation, (*ii*) gene mapping, (*iii*) malignancy, (*iv*) viral replication, and (*v*) antibody production through hybridoma technology. In 1960s, in France, the hybrid cells were successfully produced from mixed cultures of two different cell lines of mouse.

Moreover, within the body fusion of myoblasts and formation of multinucleate fibres may be exemplified. They can also be allowed to fuse *in vitro* and form heterokaryons. Macrophages fuse around the foreign body or bacterial cells in the tissues. Bone cells are also known to undergo somatic cell fusion. Cells growing in culture are induced by some of the viruses such as 'Sendai virus' to fuse and form hybrids. This virus induces two different cells first to form heterokaryon (Fig. 10.13). During mitosis chromosome of heterokaryons are brought towards two poles which later on fuse to form hybrids. Removal of surface carbohydrates is necessary before establishment of cell fusion. Some chemicals such as polyethylene glycol also induce somatic cell fusion. It is interesting to note that the cells of taxonomically different animals can fuse and form hybrids. This suggests that there is no campatibility between membranes, nuclei, organelles of two different groups of animal cells.

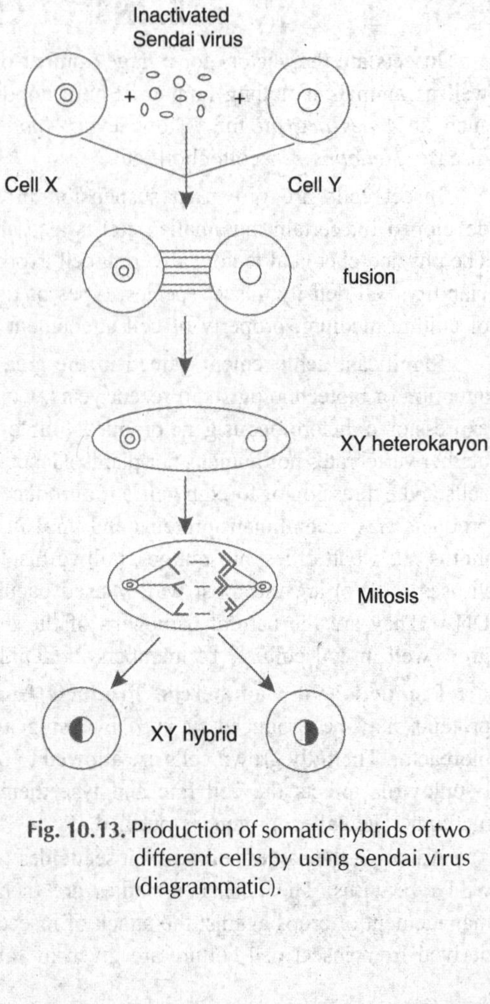


**Fig. 10.13.** Production of somatic hybrids of two different cells by using Sendai virus (diagrammatic).

## G. ORGAN CULTURE

Not whole but pieces of organs can be cultured on artificial medium. For organ culture care should be taken to handle in such a way that tissue should not be damaged. Therefore, organ culture technique demands more tactful manipulation than tissue culture. The culture media on which organ is cultured are the same as described for cell and tissue culture. However, it is more easy to culture embryonic organs than the adult animals. Methods of culturing embryonic organ and adult organs differ. Besides, culture of whole or part of animal organ is difficult because these require high amount of $O_2$ (about 95%). Special serum-free media (*e.g.* T8) and special apparatus (Towell's Type II culture chamber) are used for adult culture. In addition, the embryonic organs can be cultured by applying any of the following three methods:

**(*a*) Organ Culture on Plasma Clots.** A plasma clot is prepared by mixing five drops of embryo extract with 15 drops of plasma in a watch glass placed on a cotton wool pad. The cotton wool pad is put in a Petri dish. Time to time cotton is moistened so that excessive evaporation should not occur. Thereafter, a small piece of organ tissue is placed on the top of plasma clot present in the watch glass. In the modified technique the organ tissue is placed into raft of lens paper or ryon. The raft makes easy to transfer the tissue, excess fluid can also be removed.

**(*b*) Organ Culture on Agar.** Solidified culture medium with agar is also used for organ culture. The nutrient agar media may or may not contain serum. When agar is used in medium, no extra mechanical support is required. Agar does not allow to liquefy the support. The tumours obtained from adults fail to survive on agar media, whereas embryonic organs grow well. The media consist of ingredients: agar (1% in basal salt solution), chick embryo extracts and horse serum in the ratio of 7:3:3.

**(*c*) Organ Culture in Liquid Media.** The liquid media consist of all the ingredients except agar. When liqiud media are used for organ culture, generally perforated metal gauze or cellulose acetate or a raft of lens paper is used. These possibility provides support.

**(*d*) Whole Embryo Culture.** During 1950s, Spratt studied how metabolic inhibitors affect the development of embryo *in vitro*. Old embryo (40 h) was studied upto another 24-48 h *in vitro* until died. For embryo culture a suitable medium prepared is poured into watch glasses which are then placed on moist absorbent cotton wool pad in Petri dishes. For the culture of chick embryo, eggs are incubated at 38°C for 40-42 h so that a dozen of embryos could be produced. The egg shell sterilized with 70 per cent ethanol is broken into pieces and transferred into 50 ml of BSS. The vitelline membrane covering the blastoderm is removed and kept in Petri dish containing BSS. With the help of a forcep the adherant vitelline membrane is removed. The embryo is observed by using a microscope so that the developmental stage of blastoderm could be found out. The blastoderm is placed on the medium in watch glass placed on sterile adsorbent cotton wool pad in Petri dishes. Excess of BSS is removed from medium and embryo culture of chick is incubated at 37.5°C for further development.

## H. VALUABLE PRODUCTS FROM CELL CULTURE

From cultured animal cells several valuable products such as human monoclonal antibodies, and biochemicals can be produced on a large scale. Several million dollors have been earned from this industry in Europe, America, Africa, Japan and India. This industry has better future. More interestingly, the genetically engineered cells have revolutionized the cell culture industry. Several specific promoters of human origin are utilized for high expression of foreign genes.

For large scale production of certain biochemicals, the genetically engineered baculovirus-infected animal cells are also in use in a bioreactor. To fulfil the process several 'perfusion systems' have been developed that retain the cells in the bioreactor at the time of replacement of conditioned medium with fresh medium. This results in increase in cell density and in turn cell productivity. For commercial production of products a large scale cell culture system and scaling up of process are required. Therefore, 'master cell banks' (MCBs) are established to meet out the demand. The MCBs are used to develop master working cell bank (MWCB) which meets the demand of production system. After several subculturing, the MWCB is regularly checked for any kind of changes occurring in cells. Thus, the large scale cultures are the source of all valuable products which are produced in a bioreactor. Diagram of a compact bioreactor is given in Fig. 10.9.

Some of the important products which are produced from animal cell cultures are : (*i*) enzymes (asperagenase, collagenase, urokinase, pepsin, hyaluronidase, rennin, trypsin, tyrosin hydroxylase), (*ii*) hormones (leutinizing hormone, follicle stimulating hormone, chorionic hormone and erythropoietin), (*iii*) vaccines (foot and mouth disease vaccine, vaccines for influenza, measles and mumps, rubella and rabies), (*iv*) monoclonal antibodies, (*v*) interferons, etc. (Table 10.7). Large quantities of human interleukin-2 or T-cell growth factor have been produced by culturing a permanent lymphoblastoid T-cell line in a large batch suspension culture in a bioreactor.

## 1. Tissue Plasminogen Activator (tPA) (e.g. Elastase)

The enzyme tPA is a protease that occurs in mammalian cells naturally. In human body proteins exist in small amount, and converts plasminogen into plasmin. Plasmin is an active enzyme that degrades fibrin of the blood clot.

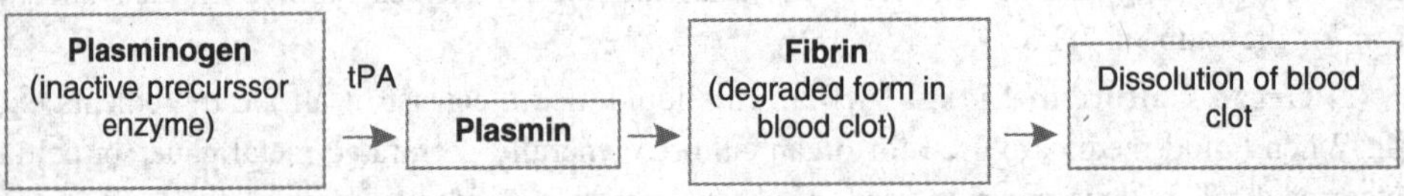


Besides, it is also used medically for the prevention and treatment of coronary thrombosis *i.e.* dissolution of blood clots of coronary veins, cerebral haemorrhage and other cases of bleeding. Initially high concentration of tPA is used to get it circulated in whole body. But at high concentration it causes internal bleeding. Therefore, a long-lived tPA of increased specificity for fibrin in blood clots and non-inducer of internal bleeding is desirable.

The tPA is also used in acute myocardial infarction (AMI). The tPA catalyses the proteolytic processing of inactive pro-enzyme plasminogen (in blood) into the active protein plasmin. Plasmin cleaves insoluble fibrin of blood clots into soluble fibrin fragment peptides. Hence, blood clot is dissolved and the blood vessel is opened.

Using site-directed mutagenesis tPA of increased stability and specificity has been produced through rDNA technology and animal cell culture. The tPA is the first drug which was produced in 1987 by a biotechnology company **Genetech** (U.S.A.). By using recombinant DNA technology the tPA gene was introduced into mammalian cells (CHO or melanoma cells). The cDNA of tPA gene was inserted into the plasmid. The recombinant plasmids were inserted into mammalian cells (Fig. 10.14.).

The transformed cells that secreted high amount of tPA were screened from the mixed population of cells. These were cultivated in a large fermentor to get high amount of tPA.

Now, tPA is commercially produced by a mammalian cell line carrying a high expression vector (*e.g.* pSV2) as a fusion of mouse DHFR (dihydroxy folate reductase) gene with the tPA gene. Recombinant protein is secreted into the medium since the tPA signal sequence is cloned along with the tPA gene.

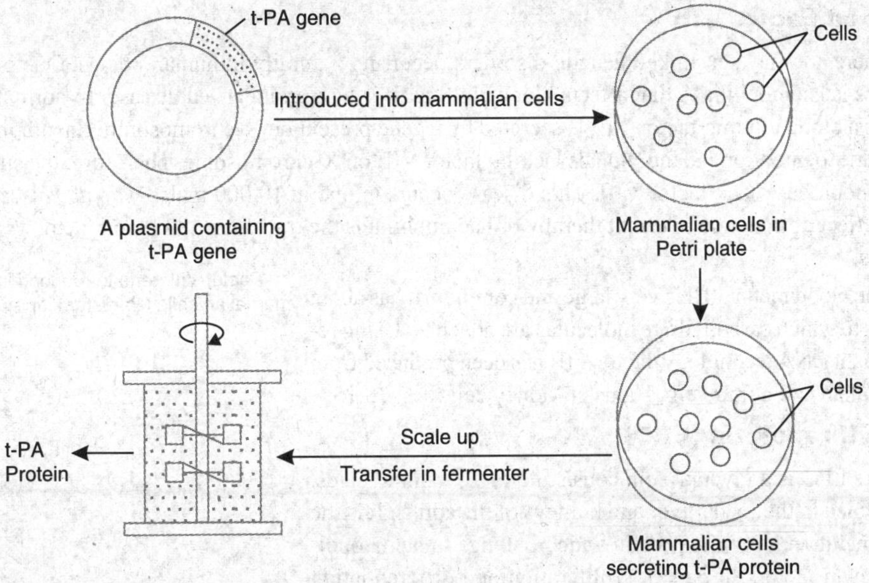


Fig. 10.14. Steps for production of tPA by mammalian cells.

> **Table 10.7. Some products of medical use derived from animal cell cultures.**

| *Products* | *Application* |
|---|---|
| **Erythropoietins** | |
| Erythropoietin-α | Anaemia resulting from cancer and chemotheray |
| Erythropoietin-β | Anaemia secondary to kindney disease |
| **Human growth hormones** | |
| hGH | Human growth deficiency in children, renal cell carcinoma |
| Somatotropin | Chronic renal insufficiency, Turners' syndrome |
| **Monoclonal antibodies (therapeutic)** | |
| Anti-lipopolysacharide | Treatment of sepsis |
| Murine anti-idiotype/human B-cell lymphoma | B-cell lymphoma |
| **Monoclonal antibodies (diagnostics)** | |
| Anti-fibrin 99 | Blood clot |
| Tcm-FAb (breast) | Blood cancer |
| PR-356 CYT-356-in-111 | Prostate adinocarcinoma |
| **Plasminogen activator** | |
| Urokinase type plasminogen activator | Acute myocardial infarction, acute |
| Tissue type plasminogen activator | stroke, pulmonary embolism, |
| Recombinant plasminogen activator | deep vein thrombosis |
| **Vaccines** | |
| HIV vaccines (gp120) | AIDS prophylaxis and treatment |
| Malaria vaccine | Malaria prophylaxis |
| Polio vaccines | Poliomyelitis prophylaxis |

*Source:* based on Feichter (1996).

### 2. Blood Factor VIII

There are may sex-linked genetic disorders occurring naturally in human beings. One of such diseases is haemophilia A that affects blood clotting. It is also an inherited disease. In normal individuals a blood clotting factor VIII is secreted by a gene present on X-chromosome. Haemophilia A arises due to mutation *i.e.* individuals lacking factor VIII on X-chromosome. Therefore, the sufferes of haemophilia A lack factor VIII. This disease occurs in one in 10,000 males who are susceptible to the effect of mutation. Current therapy of haemophilia is the transfusion of blood factor VIII into patients.

The blood factor VIII is very large and complex. It has about 25 sites to which carbohydrate molecules are attached. Using recombinant DNA technology, factor VIII has been produced from mammalian cell culture *e.g.* Hamster kidney cell (Fig. 10.15.).

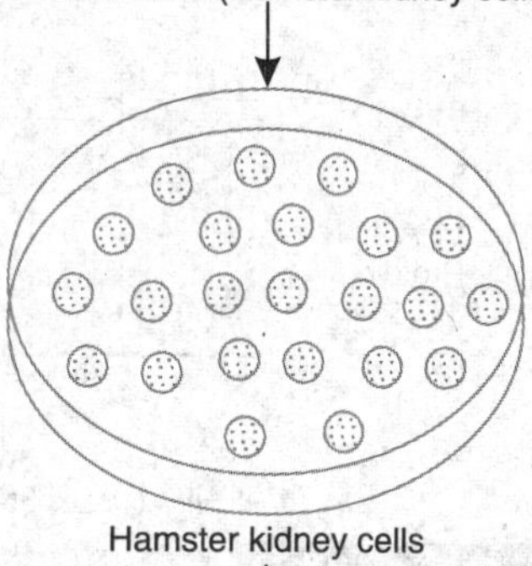


**Fig. 10.15.** Production of blood clotting factor VIII protein from transformed Hamster kidney cell culture.

### 3. Erythropoietin (EPO)

The EPO is a glycoprotein consisting of 165 amino acids. It is formed in the foetal liver and kidneys of the adults. It is the first generation hormone protein which belongs to hematopoietic growth factors. It causes proliferation and differentiation of progenitor cells into the erythrocytes (erythroblasts) in the bone marrow. In the case of anaemia, EPO is secreted by the kidney when oxygen lacks totally (anoxia) or oxygen shortage (hypoxia) condition exists in body.

Recombinant human EPO (r-HuEPO) has been produced in mammalian cell system [*e.g.* Chinese Hamster ovary (CHO) cell lines] and commercialised. Virtually, the r-HuEPO is similar to the hormone produced naturally. The recombinant human EPO is a biological factor active as the natural EPO. It is used to treat anaemia which arises after surgery. It is an effective treatment in the patients suffering from AIDS, cancer and renal failure. The use of r-HuEPO offers several benefits over blood transfusion : (*i*) no donors are required, (*ii*) no transfusion facilities are needed, and (*iii*) there is no risk of spread of any disease like HIV in patients who receive the blood.

## I. HYBRIDOMA TECHNOLOGY

Blood consists of two major components, various kinds of cells and the fluid called **serum**. Serum consists of heterogeneous population of antibodies. Different antibodies are produced by different types of B-lymphocytes. Hence, serum consists of different types of antibodies which are often called **polyclonal antibodies**. The antibodies bind to specific domains of antigens (called antigenic determinants or **epitopes**) and neutralise them.

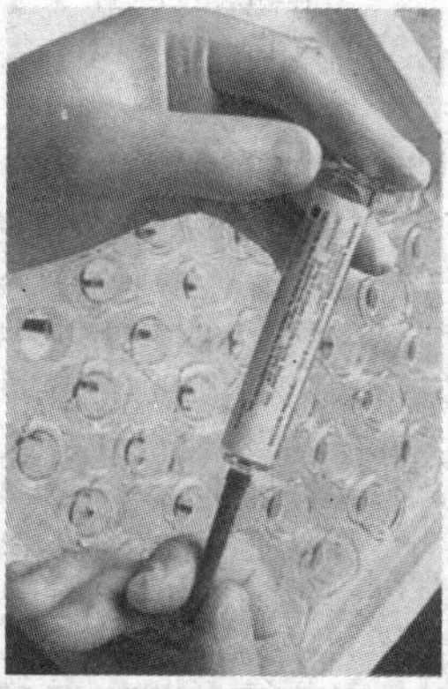

Vials of monoclonal Antibodies.

### 1. Monoclonal Antibodies

Production of monoclonal antibodies (MoAb) and hybridoma technology has been described in Chapter 7. The mouse MoAb have revolutionised the field of biology and more specifically immunology. The mouse MoAb have been used in human patients with varying level of success who were suffering from leukemia, lymphoma, melanoma and colorectal cancer. Clinical trials have indicated several limiting factors such as : (*i*) heterogenecity of tumour cells (not all malignant cells carry relevant antigen), circulating free antigens (they bind Fab on antibody molecules and thus block MoAb from binding to the target cells), (*iii*) antigenic modulation (antigen modulated off the cell surface as a consequence of binding of MoAb to the cancer cell as in leukein. This problem can be overcome by applying monovalent MoAb.

Therefore, several strategies have been developed for the production of human MoAb. These are (*i*) human-human hybrids (sensitised human B-cells from an individual are exposed naturally or by vaccination and then fused with human immortal cells obtained from tumours of lymphocytic cells. The first human MoAb was produced through this technique but the process was slow, (*ii*) interspecific hybrids (human B-cells isolated from bone marrow, periplasmal blood, spleen, lymph node, etc. are fused with non-secretary mouse or rat myeloma cells. From the hybrid cells human chromosomes were eliminated. The interspecific hybridoma produces MoAb which are of human type as expressed by human B-cells, (*iii*) EBV-transformation (The Epstein Barr virus transforms sensitised B-cells and results in production of immortal line of B-cells; the transformed B-cells secrete low amount of antibodies).

## 2. Applications of Monoclonal Antibodies

The monoclonal antibodies are used in the four main ways *i.e.* disease diagnosis, disease treatment, passive immunisation and detection and purification of biomolecules. However, their exploitation for various uses is based on two features : (*i*) exacting the specificity of antibody binding, and (*ii*) presence of similar structure of all antibody molecules.

**(*a*) Disease Diagnosis** : Using MoAb medical conditions and diseases of sufferers can be diagnosed. One of the approaches is the 'antibody-sandwich' strategy which is also called ELISA (enzyme-linked immunosorbent assay). The circulatory antigens in blood can be assayed using ELISA or radio-immuno assay (RIA) technique. By using ELISA you can test HIV, hepatitis, typhoid, herpes, etc. For each test, separate ELISA kit is used.

A preparation of monoclonal antibody is attached to the surface of a plastic dish called **well** (Fig. 10.16.). Then antigen-containing solution is added. The solution is washed to remove unbound antigens. A second monoclonal antibody preparation (attached with alkaline phosphatase or other enzyme) is added to the wells. If the specific antigen is attached to the first MoAb, the second antibody will bind attaching the enzyme to the surface of the dish as an antibody-antigen-enzyme labelled antibody sandwich. If there is no antigen, no enzyme will be attached. When substrate for enzyme is added, coloured product appears which confirms the presence of antigen.

Fig. 10.16. Steps of ELISA testing.

ELISA is used to diagnose HIV, cancer, Epstein-Barr virus as well as pregnancy testing. It is also used to check meat for the presence of pathogenic microbe (or their toxin) such as streptococcal pharyngitis.

**(*b*) Disease Treatment :** The MoAb has highly specific antigen-binding properties. This makes them an ideal candidate for use in disease therapy.

**(*i*) Therapeutic monoclonal antibodies – OKT3 :** Certain metabolic pathways are selectively blocked by the MoAb. If an antigen is present on cell membrane, it is neutralised by the specific antibody. Thus an antibody inactivates a 'specific cell surface antigen' present on T cells.

Cell-mediated immune system (CMI) plays a key role during organ or bone marrow transplantation. T-cells of the CMI system rejects any of the foreign tissues transplanted. Therefore, patients undergo immuno-suppressive therapy following organ transplantation.

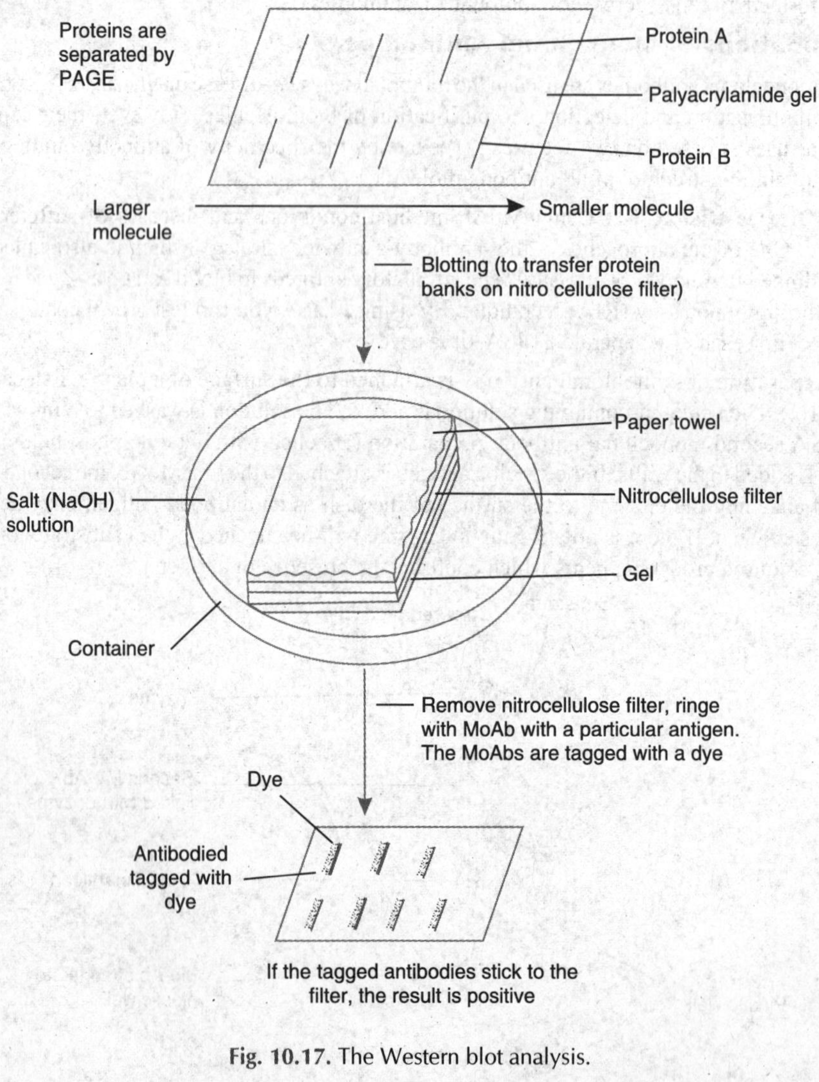


Fig. 10.17. The Western blot analysis.

The patients are given immuno-suppressive drugs to suppress T-cells. Otherwise the transplantation operation would get failed. The MoAb can be used selectively to eliminate T-cell population responsible for transplant rejection. This treatment has been successful in renal (kidney) transplantation.

CD3 is an antigen present on the surface of mature T-cells (lymphocytes). If T-cell population is depleted, the rejection of cellular transplanted organ does not occur.

An antibody that acts against CD3 surface antigen of T-cell is called OKT3 *i.e.* anti-CD3 MoAb. The OKT3 removes antigen-containing T-cells from blood circulation and from **allograft** (allograft is tissue transferred between genetically different members of the same species. In humans, organ grafts from one individual to another are allograft unless the donor and recipient are identical twins). Hence, OKT3 can be used to prevent acute renal allograft rejection in humans. OKT3 has also been used to purge donor bone marrow prior to its use in marrow transplantation.

**(*c*) Passive Immunisation :** Passive immunisation refers to transfer of antibodies from one individual to another. For example, administration of antivenom in the patients against snake-bite comes under this category. Use of MoAb is done for passive immunisation against certain diseases. Approved MoAb-based drugs are already being used to treat septic shock (a set of symptoms appearing due to the presence of a particular toxin in the cell wall of many infectious bacteria).

**(*d*) Detection and Purification of Biomolecules :** MoAb are very useful in determining the presence and quantity of specific proteins through Western Blotting technique (Fig. 10.17). First the proteins from various samples are prepared and separated by polyacrilamide gel electrophoresis (PAGE). The protein bands are transferred onto a nitrocellulose filter. The resulting Western blot is incubated with primary antibody. The antibody will bind if the specific antigen is present. This binding will allow a secondary antibody tagged to specific dye to be used to locate the first one. The Western blot is photographed to show the exact location of antigen as dark band.

## PROBLEMS

1. What are the requirements for animal cell and tissue culture ?
2. Write an essay on animal cell culture medium, its different ingredients and methods of sterilization.
3. Discuss in detail different methods of disaggregation of animal tissue.
4. Give a detailed account of establishment of cell cultures and evolution of cell lines.
5. Write in detail on culture of animal cells *en masse* in a bioreactor.
6. Write an essay on production of commercial products from animal cell culture.
7. Write short notes on the following:

(*i*) Sterilization, (*ii*) Substrates for cell growth, (*iii*) Culture media, (*iv*) Advantages and disadvantages of serum in medium, (*v*) Enzymatic disaggregation, (*vi*) Physical disaggregation, (*vii*) Cold trypsinization, (*viii*) Warm trypsinization (*ix*) Immobilised cell culture, (*x*) Insect cell culture, (*xi*) Somatic cell fusion, (*xii*) Organ culture, (*xiii*) Production of biochemicals from cultured cells, (*xiv*) Type of cell lines, (*xv*) Hybridoma Technology, (*xvi*) tPA, (*xvii*) EPO, (*xviii*) Application of MoAb.

CHAPTER

# 11

# Manipulation of Reproduction and Transgenic Animals

Animals are a source of high protein, food, fibre and leather which men have been using since the time immemorial. In India milk and milk products have become a part of our civilization. Even in any small festival, demand of these two gets increased as several milk-based sweets and food items are prepared. Right from God worship to dining table use of milk and milk products is unavoidable. However, milk production could not keep pace with milk demand. Therefore, to meet milk demand synthetic milk and synthetic khoa are illegally sold in market even after several punishment threats declared by the Government. Due to unaware use of synthetic milk several infants have been reported to lose their lives.

Transgenic mouse.

Production of food from animals is inexpensive and exhausting, and less efficient than plants. The most common farm hoofed animals are cattle, sheep, goats and pigs but unlike plants their management is expensive. Inspite of expenses rendered on nutrition, shelter and management, animals have been our companion since the dawn of civilization. Moreover, there is regional variation on availability of animal protein to human diet due to cost, culture and religion. In biological sense, despite several similarities animals differ from plants in their reproduction systems such as number of gametes (ova), completion of life cycle, production of products, etc. Except pigs females of all common livestocks produce one egg at about monthly intervals. Therefore, through biotechnological methods discovered in recent years, improvement in reproductive systems of animals, their milk proteins and milk-products, and several novel products which normally lack in animals can be carried out.

# A. MANIPULATION OF REPRODUCTION IN ANIMALS

One of the natural laws of life is to be many from one which is possible through reproduction. In animals sexual reproduction occurs through fertilization of female gametes (ova) by male gametes (sperms) followed by fusion of two nuclei coming from two gametes. Thus sexual reproduction maintains the genetic traits of organisms. But offsprings develop and born in natural way. To meet the demand of animal products due to gradual increasing population there is need of increase in number of animals too which is possible biotechnologically because natural process of increase in number is slow.

In nature female mammals produce only one egg in a month except pigs. Secondly, ruminant females can have only one offspring in a year. Therefore, biotechnology can help to meet increasing demand and need of quality products of animals. In this regard principles of artificial breeding system has been discussed that may be useful further in the area of animal biotechnology.

## 1. Artificial Insemination

A male animal produces millions of sperms daily. Theoretically, it can inseminate females regularly and produce several offsprings. This excess capacity of male has been utilized through developing new technologies for artificial insemination which can be said as the first animal biotechnology.

The most effective factor that has increased the productivity of cattle is the artificial insemination. However, the breeder must replace the nature through artficial insemination if he ensures about the ovulation of female in herd together at a time. In contrast, if a breeder awaits for female to ovulate and then inseminate separately, the importance and economic significance of artificial insemination get reduced. Therefore, through artificial insemination technology increased number of females can be inseminated by a male.

***(a)* Semen and its Storage** : In addition methods have been developed to produce semen from male by ejaculation. Semen ejaculate is collected and diluted (extended). Sperm motility and their number per millilitre are examined under the microscope. About 0.2 ml bull semen

These tanks contain the semen of "super cattle" preserved in liquid nitrogen for use in artificial insemination.

contains about 10 million motile sperms. The diluted sperms may be used fresh within a few days or cryopreserved at -196ºC by using liquid nitrogen (for detailed description *see* Chapter 10, Cryobiology). The cryopreserved semen can be stored for a long time and easily transported across the states or countries. Thus, cryopreserved semen of a single male is capable of inseminating thousands of females of a country or other countries. For example, one ejaculate semen of a bull is sufficient to inseminate about 500 cows.

**(*b*) Ovulation Control :** In many animals it is difficult to find out oestrous (sexual heat) in animals because it persists only for a few hours and occurs mostly at night. After ovulation (which is indicated by oestrous) females are inseminated. But in a herd it would be economical, easy and simplified management if females are inseminated at a time. However, it is possible only when all the female ovulate at a time; in practice it is not possible to get synchronisation of oestrous. Moreover, it is possible to bring about ovulation in about 80 per cent of females by using hormones, for example progesteron and/or prostaglandin. These hormones regulate ovulation cycle of female and result in total synchrony of oestrous.

**(*c*) Sperm Sexing :** Sperms are produced in the testes of males and ova in female's ovaries. Sperms and ova contain half of chromosomes as compared to somatic cells. An ovum possesses autosomes and one X chromosome. Similarly, a sperm contains autosomes and one Y chromosome. In animals sex is determined genetically *i.e.* by sex chromosomes. X chromosome determines femaleness and Y maleness. All the ova contain X chromosome, whereas a sperm consists of either X or Y chromosomes. One sperm ejaculate contains half X and half Y chromosomes.

In dairy industry demand of females is more than the males. Secondly, females have more desirable characteristics. The livestock industry prefers animals of one sex. Therefore, through artificial insemination technology X and Y chromosomes can be detected and sex of progenies determined accordingly.

There is a fluorescence dye (Hoechst 33342) that stains X and Y chromosomes with different intensities. Thus, these two chromosomes possibly can be separated by using an instrument, fluorescent activated cell sorter (FACS). Sperms are present in the form of suspension. The FACS converts a suspension of sperms into microdroplets. Each droplet consists of a single sperm cell. Individual microdroplet passes through a laser beam. Microdroplets of different intensities are deflected into separate collection tubes as the fluorescence of dye is measured electronically. The sperms separated by using FACS have recently been used and pre-sexed calves have been produced through *in vitro* fertilization technique (*see* preceding section). Moreover, FACS is very expensive and slow. It takes about 24 h to process one semen ejaculate, whereas the sperm cannot remain viable for a long time. Therefore, more refinement in technique is required for its use on a large scale.

## 2. Embryo Transfer

In 1890, the first case of developing pregnancy in rabbits through embryo transfer is known in literature. During 1930s the same method was used in goat and sheep. In cattles cases of embryo transfer were reported after 1950 (BIOTOL series, 1992). Embryo transfer method cannot be used widely because of its high cost, technical difficulties and limited supply of embryo from superovulated donors. The embryo develops in foster mother (recipient) which simply acts as incubator and does not make any genetic contribution to the offspring.

Secondly, ruminant female carries one pregnancy at a time as only one egg is produced and fertilized with male's sperm. Thus, there is a chance for increasing the number of egg production at a time and transfer of fertilized eggs *i.e.* embryo into uterus of less important foster mothers other than original female in farm animal.

**(*a*) Multiple Ovulation (Superovulation) :** The reproductive cycle of ruminant female is such that the ovarian follicle of a non-pregnant female matures and releases single egg at a time. The time of ovulation differs in different animals, for example, 21 days in cows and horse, 16 days in sheep and goats, etc. Normally ovulation occurs as a result of circulation of gonadotropic hormone. But by increasing the concentration of hormone the number of egg production gets increased. In well managed domestic cattles 8-10 eggs are superovulated; the number may go to 60. However, this depends on health, nutrition, breed of animals and environment in which they live.

Multiple ovulation and superovulation. This techniques have spread from cattle and sheep to goats, horses and deers. He has emphasized that (*i*) general selection for increased litter size is useful only in sheep, (*ii*) the gonadotropic hormone induces superovulation in goats, sheeps and cattle but the response varies so much, and (*iii*) immunisation against ovarian steroid hormones can increase litter size in sheep. Therefore, different molecular forms of follicle stimulation hormones (FSH) should be characterized in the farm animals.

After injecting the gonadotrophin the females are induced for superovulation. During follicular phase (second phase of oestrous cycle) about 20 ovarian follicles are induced. Eventually these grow and filled with fluid. The space of follicle which is filled with fluid is called antrum, and such follicles as antral follicles. In normal course, only one follicle develops and releases one egg after maturation. However, before ovulation the follicles laying against surface of ovary looks large sized (8 mm in sheep and pigs, 15 mm in cattle). Therefore, immature oocytes from follicles of donor females are recovered surgically by using laparoscope.

The females to be superovulated are frequently injected with prostaglandin F2a (PGF2a) so that synchronised oestrous could develop in them. After 10 days of oestrous they are injected with hormone FSH upto 4 days followed by PGF2a treatment so that oestrous may be maintained. The FSH treatment induces superovulation. The females are artificially inseminated. The eggs are fertilized. After fertilization embryos undergo developmental stages (Fig. 11.1).

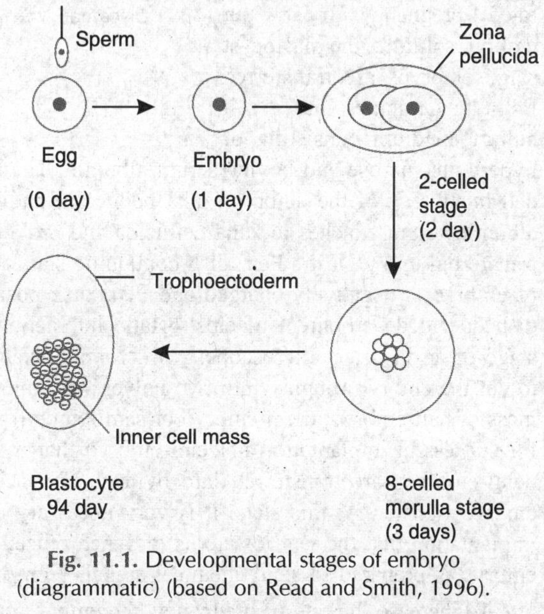

**Fig. 11.1.** Developmental stages of embryo (diagrammatic) (based on Read and Smith, 1996).

**(*b*) Multiple Ovulation with Embryo Transfer (MOET) :** After 6-8 days of fertilization (for sheep and goat) the embryos are recovered. During this stage embryos have come to morula or blastula stage and remain in female's oviduct. In cattle the embryos are recovered without surgery by inserting a catheter into oviduct. A saline solution is drained in oviduct and embryos flushed out through catheter into a storage bottle. The oviduct of sheep is small, therefore, it is exposed surgically and embryos are recovered by syringe adopting the same method for cattle.

In storage bottle embryos settle down and solution decanted. Embryos with small volume of saline are poured into a Petri dish. Then they are examined under microscope. The identified embryos are transferred into synchronised recipient females (as artificial insemination), when they are in 6-8 days of cycle of embryo development, stored and transported or manipulated as desired. In general, one superovulated female results into 5-6 progenies. In cattle 50-60 per cent of pregnancy can be achieved from embryo transfer method.

**(c) Embryo Splitting (Demi-embryo) :** Embryo of blastocyst stage (Fig. 11.2) is differentiated into two regions, trophoectoderm and inner cell mass (ICM). Trophoectoderm is a single layer of trophoblast cells lining the inner side of zone surrounding the blastoceal (a fluid-filled cavity). Zona becomes the foetal membrane of placenta. The ICM is the mass of cells .which develops into foetus.

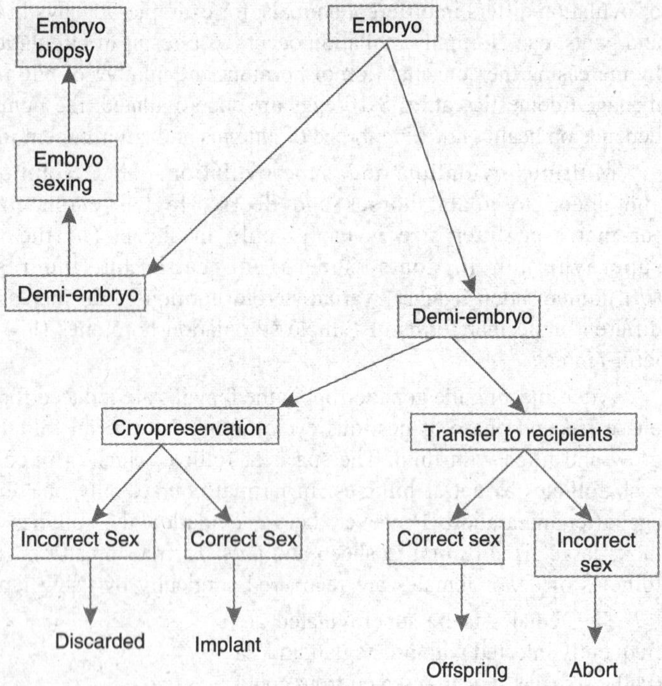


Fig. 11.2. An outline of embryo splitting.

The embryos of blastocyst stage can be split into equal two halves and transferred to females to produce identical twins. Thus, embryo splitting technology has increased the rate of pregnancy. To carry out embryo splitting, the blastocyst stage embryos are transferred for a few minutes into a cell culture medium consisting of hypertonic sucrose and bovine serum albumin. The medium of high osmotic strength enters into cell membrane of the embryo (zona pellucida) and cells contract due to exosmosis. The bovine albumin serum attaches to zona pellucida and provides negative charge to the embryo. Then it is transferred into a plastic Petri dish containing standard cell culture medium. The outer membrane of embryo is negatively charged and Petri dish positively. Embryo sticks to the surface of Petri dish due to development of electrostatic interaction of two charges. The Petri dish is kept on stage of an inverted microscope. A micromanipulator equipped with fine surgical blade is used to cut the embryo roughly into two halves with minimal damage to the cells. The hemispherical mass of half embryo, demi-embryo or semi-embryo reforms spheres. However, it is not necessary for successful implantation of demi-embryo that it should be enclosed in a zona pellucida. The demi-embryos are transferred into oviduct of synchronized recipients as described for normal embryo transfer. At this stage it is very necessary to know the situation for successful embryo implantation that the wall of uterus of synchronized female recipient is waiting for embryo, and embryo is prepared to send chemical signals from the trophoblast.

To increase the rate of pregnancy, the embryo can be cut into equal three or four pieces, and transplanted in synchronized females. But too small trophoblast will not induce the uterus for pregnancy. So far it is unknown how cells of ICM are required for successful pregnancy, but it is clear that increasing in embryo splitting will have less probability of pregnancy.

Embryo biopsy (removal of small number of cells for genetic analysis) should be combined with splitting so that the twins which will be produced should be identical and of known genotype. Biopsy is very necessary in breeding so that sex and genetic diseases could be detected. In case the embryo has any genetic disease, it can be prevented from implantation in recipient females.

**(d) Embryo Sexing :** Before the implantation of embryo its sex should be detected from the biopsy sample. The principle for sexing is very common. The presence of Y chromosome makes the offsprings male and that of X makes female. The second method is the use of polymerase chain reaction ( PCR) machine in sex detection. PCR amplifies DNA sequence of Y chromosome and reaction products can be seen directly. Single blastomere is isolated from early embryo from a womb, amplified DNA sequences of Y chromosome and carried out embryo sexing before implantation into uterus. It is true that the PCR was first commercially implemented for embryo sexing of livestock (for detail description *see* Chaper 2, Genes: Nature, Concept and Synthesis). Now the sexed embryos are available commercially but these are very costly.

## B. IN VITRO FERTILIZATION (IVF) TECHNOLOGY

The term *in vitro* means in glass or in artificial conditions, and IVF refers to the fact that fertilization of egg by sperm had occurred not in uterus but out side the uterus at artificially maintained optimum condition. In recent years the IVF technology has revolutionized the field of animal biotechnology because of production of more and more animals as compared to animal production through normal course. For example, an animal produces about 4-5 offsprings in her life through normal reproduction, whereas through IVF technology the same can produce 50-80 offsprings in her life. Therefore, the IVF technology holds a great promise because a large number of animals may be produced and gene pool of animal population can also be improved. In India M.L. Madan, an animal embryo-biotechnologist at National Dairy Research Institute, Karnal (Haryana) has got success in producing more calves in cows.

Nowadays, exogenous hormones have been developed through naturally and recombinant DNA technology which are used to induce superovulation in farm animals. Superovulation is a phenomenon of producing greater than the normal number of eggs through hormonal treatment by a single female at a time.

The IVF technology is very useful. It involves the procedure : (*i*) taking out the eggs from ovaries of female donor, (*ii*) *in vitro* maturation of egg cultures kept in an incubator, (*iii*) fertilization of the eggs in test tubes by semen obtained from superior male, and (*iv*) implantation of seven days old embryos in reproductive tract of other recipient female which acts as foster mother or surrogate mothers. These are used only to serve as animal incubator and to deliver offsprings after normal gestation period. The surrogate mothers do not contribute any thing in terms of genetic make up since the same comes from the egg of donor mother and semen from artificial insemination.

**(*a*) *In Vitro* Maturation (IVM) of Oocytes** : The immature oocytes are incubated *in vitro* so that they can be mature. However, immature oocytes should be taken out from follicles because they cannot mature in it but degenerate. Therefore, full potential of superovulation and all the oocytes can be utilized by IVF technology. Moreover, metabolic and hormonal requirement for oocytes during IVM should be found out so that the present rate of maturation (20%) could be improved. In majority of cases ovarian follicles never reach maturity and degenerate due to unknown causes. Possibly there may be genetic defects associated with them.

**(*b*) Culture of *In Vitro* Fertilized Embryos** : IVF of eggs is carried out in small droplets (microdroplets) of culture medium. Each microdroplet comprises of about 10 oocytes. The medium should be supplemented with penicillamine, hypotaurin, and epinephrine because they facilitate penetration of sperms into oocytes. Moreover, one dose of sperm is given that consists of about one million sperms per ml of medium. Thus, IVF embryo must be maintained at *in vitro* conditions for a few days so that it may develop into blastocyte. It takes about seven days for sheep and goats and eight days for cattle. There are many laboratories where about 60 per cent IVF embryos of cattles are cultured to blastocyte stage.

The term delivery from cultured embryo is very low due to occurrence of high loss during first two months of pregnancy. This may be due to abortion of foetuses arising from the presence of genetic defects. It should be noted that before birth about 80 per cent genes play a key role in differentiation and development of foetuses. The oocytes which are forced to mature *in vitro* occasionally bears some defects. Some times environmental mutagenesis occurs in eggs, sperms or embryos. Artificial culture media should be improved as oxygen may have toxic effect. Therefore, gas atmosphere should be carefully controlled.

The other most successful method of IVM is to place the fertilized zygotes into agar (so that it may wrap around it) and implant them in oviduct of synchronized sheep or rabbit where the environment for early development of embryo is perfect. For early bovine embryo the oviduct of rabbit and sheep has been used as *in vitro* culture system. Hundreds of cattle eggs can be put into oviduct of a sheep and many of these are recovered after a week. A good quality of embryo develops at the late morula/blastocyte stage of development with a yield of about 40 per cent or more.

The First IVF calf took birth after getting success in fertilizing the eggs recovered from ovulated cow. Thereafter, hundreds of IVF calves have been born in Japan, India, U.K., etc.

## 1. Embryo Cloning

A clone is a population of cells or organisms derived asexually from a single ancestor. They are genetically identical to each other to their common ancestor. Cloning means the production of exact genetic replica copies of an individual. They can not be considered as an offspring but simply the copy of a given individual. Much work has been done on cloning in plants and microorganisms. However, the techniques used in plants can not be applied for animals. Moreover, many animals from a single genetically superior embryo can be produced. Still there is no method of finding out which embryos are capable of cloning. It is useless to clone an embryo if it is not superior.

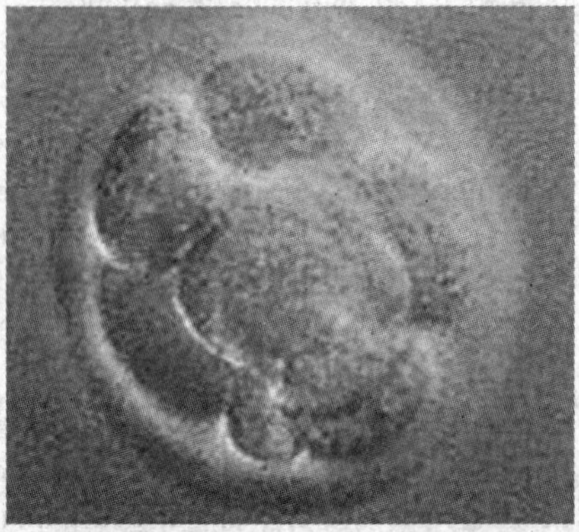

Scientists claim human embryo cloning will bring medical benefits.

**(*a*) Quadriparental Hybrids :** During 1960s, Beatrice Mintz at Cancer Research Institute, Philadelphia (USA) demonstrated the interesting experimentation. She carried out fusion of embryos of two different species of mouse. This resulted in the formation of a single embryo which finally developed into a normal healthy animal having four parents. Embryo A was derived from the cross of male x female of one species, and embryo B derived after cross of male x female of the other species. The embryo A and B were united together and produced a single mass. In this experiment Mintz removed zona pellucida membrane of two early embryos and placed them in a suitable culture medium. The embryonic cells of the two embryos of blastula stage united randomly into a single mass of double sized embryo (blastocyst). A fresh membrane developed around the embryo. Then the embryo was transferred into the uterus of a foster mother. The foster mother was mated with a sterile male to bring her into proper stage for implantation. The first offspring with four parents was born in 1965. Similarly, exciting experiments have been done on another mammal. Following the same technique a hybrid of goat and sheep named geep was produced.

At present two types of techniques for embryo cloning viz., nuclear transplantation (transfer) and embryonic stem cells, are being developed.

**(*b*) Nuclear Transplantation :** Nuclear transplantation (also nuclear transfer) involves removal of a single blastomere from a cleavage stage embryo with a fine micropipette of glass, and placing it under the outer membrane of an unfertilized mature enucleated oocyte (whose haploid nucleus has been removed by using micropipette or destroyed by UV light). For the first time in 1955, Robert Briggs and Tom King at Cancer Research Institute, Philadelphia (USA) carried out nuclear transplantation experiment on embryonic cells of frog. They transferred nucleus of undifferentiated blastula (a stage soon after fertilization of egg) into an enucleated egg cell. They noticed the normal development of the embryo. When they performed serial transplantation of differentiated nucleus from late gastrula (a stage after blastula) into a nucleus-free unfertilized egg, abnormal embryos were formed. This shows that cell nucleus is differentiated with embryo development. In 1960s, J.B. Gurdon at Oxford University, U.K. transferred differentiated intestinal nucleus of a frog into nucleus-free unfertilized egg of different amphibian species (*Xenopus laevis*). The embryo developed into tadpole and matured into frog. This new enucleated cell developed into normal embryo. Any damage to the donor nucleus during transplantation leads to abnormal development (Fig. 11.3).

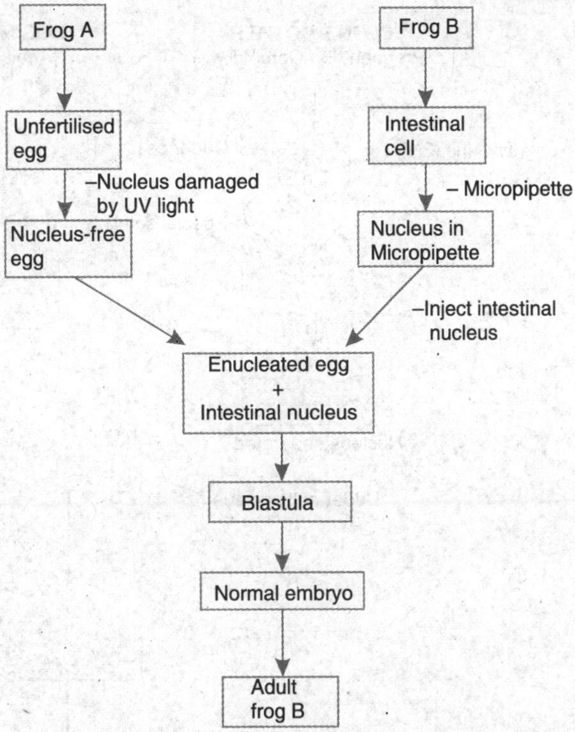


**Fig. 11.3.** Nuclear transplantation in frog (diagrammatic).

Dolly – the first mammalian clone.

**DOLLY - The first mammalian clone.** 'Dolly', the worlds' first mammalian clone has been created from a fully differentiated non-reproductive cell of an adult sheep. It was born in February, 1996. The name Dolly has been given after an American country singer, Dolly Parton. In 1995, Ian Wilmut and his team of researchers at Roslin Institute, Edinburgh, Scotland, took udder (a fully differentiated tissue) from six year old sheep, Fin Dorset Ewe, and placed it in special solution that controlled cell cycle of cell division. The cell was deprived off certain nutrient. At the same time an unfertilized egg was obtained from another adult sheep (Fig. 11.4). Its nucleus was carefully removed leaving the intact cytoplasm in egg. The nucleus of udder cell was taken out and transferred into nucleus-free egg. This was facilitated by applying mild electric shock. The

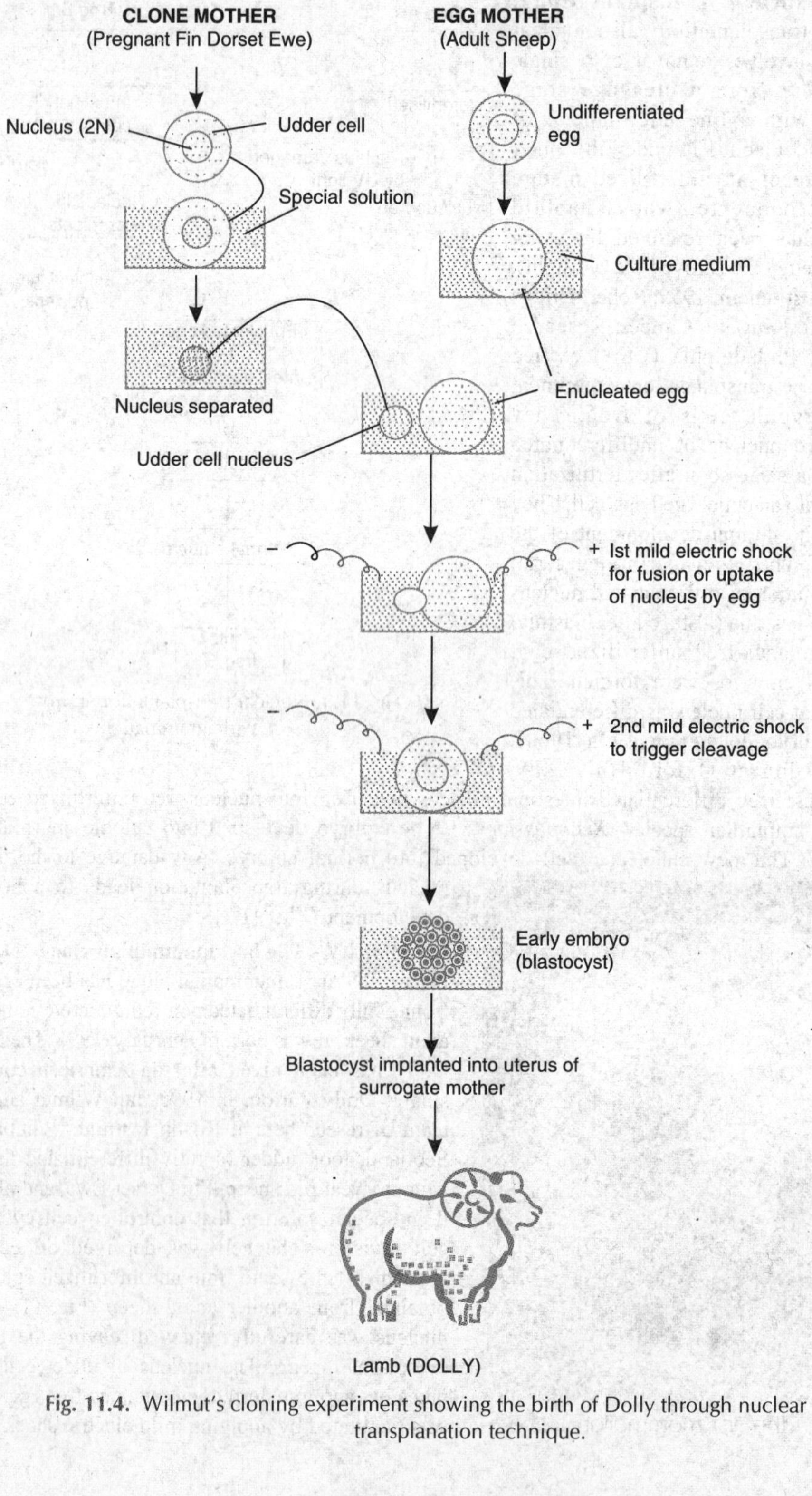


**Fig. 11.4.** Wilmut's cloning experiment showing the birth of Dolly through nuclear transplanation technique.

newly transplanted nucleus soon became functional according to the new cytoplasm in which it had been artificially transferred. This viable combination underwent cleavage like normal zygote. This so called embryo was then transplanted into the uterus of a third adult sheep (surrogate mother/foster mother) for its further development.

Finally, a normal healthy little lamb, Dolly was born in February, 1996 which was genetically similar to the *clone mother* from which nuclear DNA was taken out. It does not have any similarity with that sheep from which egg was taken out or surrogate mother because they did not contribute any genetic character. Thus, Dolly has only a single parent because she has born asexually, a characteristic feature found in lower forms of animal life, not in mammals. Although behind this great success the rate of success is very slow, yet it has given some hope to embryo-biotechnologists to bring about refinement. Out of 277 nuclei transferred singly to enucleated egg, only 29 eggs grew into embryos. Out of these, only 13 embryo could be successfully transplanted into surrogate mothers. Of these only one ewe was successful in giving birth to an offspring, Dolly.

The significant considerations that can be derived from this experiment are that: (*i*) the genes of differentiated cells have inherent totipotency, (*ii*) the interplay between the regulatory system of the genome in nucleus and the cytoplasmic factors of egg may make a cell totipotent, (*iii*) possibly the cytoplasm of enucleated egg makes the transplanted nucleus totipotent like that of normal fertilized egg nucleus, (*iv*) the maternally derived information in egg cytoplasm has an important role in cleavage which usually occurs after fertilization. Hence, it is the egg cytoplasm but not the nucleus that regulates cleavage. Because the udder cell nucleus has limited potential for mitosis; it is the egg cytoplasm that interacted intracellularly with udder cell nucleus and stimulated to undergo repeated mitosis, (*v*) carbon copy of the adult sheep could be produced without involving sperms from male partner, and (*vi*) the cloned animal produced via nuclear transplantation technique will be capable of restoring fertility as in 1998 Dolly gave birth to a little lamb named *Bonny.* Thereafter, several animal clones were developed such as sheep, cat, etc.

## 2. Embryonic Stem (ES) Cells

Cloning of mice could not be done as in sheep via nuclear transplantation. This was due to acceleration of developmental programmes of embryo. However, it is evident that before first embryonic division the cell has started its process of differentiation. Therefore, for cloning of mice an alternative approach has been made *i.e.* the use of ES cells.

The ES cells are the pluripotent cells isolated from inner cell mass of early embryos. ES cells are isolated without injection of immortalising (transforming) agent in mice. For many years it has been possible to grow mouse ES cells as cell lines in laboratory. These ES cells can be induced to generate many different types of cells. Mouse ES cells has been shown to give rise the muscle cells, nerve cells, liver cells, pancreatic cells as well as hematopoietic cells.

The ES cell lines have a great value in the area of production of transgenic animals. Since the cells are plentiful and maintained in culture, they can be transformed very easily. The resultant transformants can be fully characterised *in vitro*. Their genome can be modified. This may be the area of greatest impact on live stock. Using *in vitro* fertilisation (IVF) technique you can fertilise a mouse egg artificially *i.e.* outside the animal body and grow in tissue culture. You can observe that the embryo will undergo several steps of cleavage. The dividing cells accumulate at one corner of embryo. These accumulated cells are called the **inner cell mass** (ICM).

The ES cells can be used for cloning in two ways : (*i*) creation of chimera, and (*ii*) nuclear transplantation. In the presence of irradiated fibroblast cells, the ICM should be maintained in tissue culture. It was interesting to note that these cells: (*i*) retained the characteristics of the embryo founder cells even after prolonged culture, (*ii*) fully irradiated into embryogenesis when returned to early embryo, (*iii*) could be used to produce chimeric mouse, (*iv*) maintained a stable euploid karyotype, and (*v*) self-renewal without differentiation into cell types.

(*a*) **Production of Chimeric Mouse :** An early embryo is isolated from the fertilised mouse of black colour. The ES cells of trophoblast stage of embryo are grown on culture medium. A small number of ES cells can be injected into blastocoel space of an embryo of a white (albino) mouse through microinjection technique. The ES cells of black mouse intermingle with that of albino. The microinjected embryo is transplanted into the uterus of a **surrogate mother** (another mouse of which ova are not used). The progeny born has black and white skin colour. Such mouse was called **chimera** or chimeric mouse (Fig. 11.5).

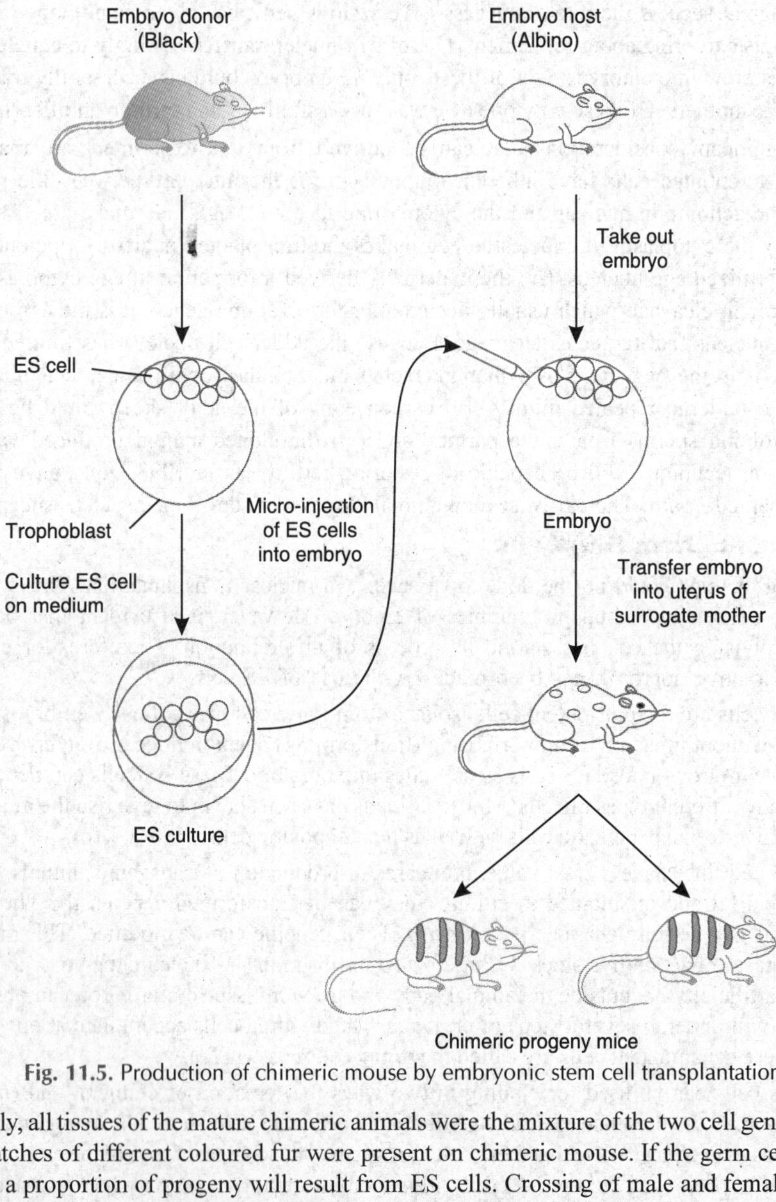


**Fig. 11.5.** Production of chimeric mouse by embryonic stem cell transplantation.

Ideally, all tissues of the mature chimeric animals were the mixture of the two cell genotype. That is why patches of different coloured fur were present on chimeric mouse. If the germ cells are also chimeric, a proportion of progeny will result from ES cells. Crossing of male and female chimeras allows the selection of a homozygous variants of mice derived solely from the ES cells.

# C. *IN VITRO* FERTILIZATION AND EMBRYO TRANSFER IN HUMANS

Due to several complecacies associated with humans, there has been the birth of *invitro* fertilization technology. The IVF technology is a boon to childless couples. Initially this technique was pioneered by Prof. Robert Winston during 1970s which was later applied by P. Steptoe and R. Edwards for the production of first test tube baby, Louse Joy Brown, who was born on July 25, 1978. Since then more than 25,000 test tube babies have been produced so far through out the world.

## 1. Infertilities in Humans

There is a large number of reasons for developing infertilities in humans. By using IVF many more infertilities have been treated.

**Who Benefits IVF.** It is not likely that any patient of any age or any group of the society can go for treatment for IVF. There are several prerequisite conditions applicable for male and female partners. The male should not be azoospermic; however, if he is azoospermic and desires for a baby, he can procure semen either from sperm bank or elsewhere. Besides, the female partner should also be healthy and fit for IVF because it is she who is to bear the pregnancy. Therefore, the female must be such that: (*i*) any surgical procedure should not create any trouble to her, (*ii*) her ovaries should be accessible for oocytes, (*iii*) her uterus should accept and bear pregnancy for the required period of about nine months, and (*iv*) her cervical canal should be easily negotiable for embryo transfer.

Petri dishes containing human ova await the addition of sperm to complete in vitro fertilization.

**(*a*) Male Sterility** : In a fertile human the total count of sperms should be 15-20 million per ml. When the number of sperms decreases, the person is called oligospermic and the condition as *oligospermia*. Similarly, when there is very low number of motile sperms, the patient is called azoospermic and the phenomenon as azoospermia. Generally in *azoospermic* humans the number of non-motile sperms is high. In addition, for IVF, the presence of motile sperms should be around 10-20,000 per ml.

**(*b*) Female Sterility** : There are several reasons and types of female infertility; few of them are : (*i*) *non-functional* (*inaccessible or absent*) *ovaries* (some times ovaries of female are non-functional or inaccessible, therefore, healthy oocytes will not be produced. In such cases oocytes can be procured from donor for IVF), (*ii*) *non-functional* (*or absent*) *uterus* (in this condition the oocytes are obtained from female and subjected to IVF, and embryo transferred to surrogate mother for pregnancy and further development), (*iii*) *idiopathic infertility* (no definite cause is known for idiopathic infertility. Possibly it may be due to abnormal fertilization or failure in fertilization. Therefore, IVF followed by embryo transfer will cure this infertility, and (*iv*) *tubal infertility* (fallopian tubes receive oocytes from follicles of ovaries for fertilization. Any deformity in fallopian tubes results in tubal infertility. IVF followed by embryo transfer in uterus has replaced the function of fallopian tubes. Therefore, the patients will not be subjected to surgical treatment).

## 2. How the Patients for IVF are Treated?

First of all the female aspirant for IVF is suggested to keep the records of her menstrual cycle upto six months. This information supports the doctors to take fruitful decision for the possibility of IVF. The female patient is hospitalized to ensure the date for collection of urine for the presence of luteinising hormone (LH) surge, and to fix date for administrating human chorionic gonadotrophin (hCG). The hCG controls the final stage of formation of follicle and oocytes. The oocytes may be recovered during any menstrual cycles, *i.e.* natural cycle, stimulated cycle or controlled cycle. According to the nature of three cycles, the treatment for IVF is recommended.

During natural cycle, spontaneous LH surge is found out at 3-6 h intervals by taking urine or plasma sample. Under the influence of LH surge fully developed follicles of both the ovaries burst and ova collected in fallopian tubes. In natural cycle, only one ova at a time is formed. Secondly, the stimulated cycle is created by administrating clomiphene and /or human menopausal gonadotrophin (hMG). Clomiphene (150 mg) per day is given for the 5th to 9th day of the cycle. However, the duration may be prolonged for 10 minutes but there will be more risk of uterus receptivity and abortion. The hMG alone or along with Clomiphene is also given that does not allow the high rate of abortion. But several abnormalities are also caused by hMG. Consequently multiple follicles and oocytes are stimulated instead of one. At this time hCG is also given to female patients so that inhibition of LH surge can be checked. The disadvantages associated with these are the ultrasonographic assessment of the hormonal effect and development of some of abnormal oocytes that result in abnormal pregnancy. Thirdly, hCG is administered at the optimum stage of maturation that arrests the development of follicle. However, it is difficult to predict the optimum time for hCG administration that varies for different preovulatory follicles. Patients need not stay in hospital for a long duration because the doctor can (to his own and patient's convenience) recover oocytes through laparoscopy. His success rate in oocyte recovery, embryo transfer and pregnancies are achieved by the stimulated and controlled cycles.

**(*a*) Indicators for Ovary Stimulation :** It is very necessary to monitor the stimulation of ovary so that the oocytes can be collected at right time of development. There are several parameters which are used as indicators for stimulation of ovary, these include: (*i*) prediction of LH surge on the basis of menstrual cycle, (*ii*) rise in temperature of body during preovulatory and postovulatory days, (*iii*) changes in cervical mucus as it is secreted five days before LH surge, goes maximum a day before LH surge and falls sharply thereafter, (*iv*) estimation of progesterone, (*v*) levels of oestrogen in urine, etc.

**(*b*) Oocyte Recovery and Culture :** Oocytes are recovered through recently developed equipment called laparoscopy. Using microscopic equipment the follicles can be observed and aspirated with the aspirating apparatus which consists of 23 cm long needle made up of stainless steel (with internal diameter of 1.6 mm and external diameter 2.1 mm) and 52 cm long tubing. The total space between needle and tubing is 1.2 ml. The ovulating follicles are selected for aspiration. It consists of ova of less than 1.5 cm. From such follicles fluid (4-12 ml) is aspirated by using needle and tubing. After aspiration wall of follicle gets collapsed. Fluid is of straw coloured some times it contains blood also. When there appears blood, heparinised culture medium should be added in tube so that blood clotting may be prevented. However, if fluid contains blood, ova could not be easily identified and isolated.

The aspirate is observed under the microscope for the presence of oocytes. If ova are not present in aspirate, the heparinised culture medium should be reflushed into follicles to get oocytes. The oocytes are identified under the microscope and incubated for 5-10 h depending on maturity of oocytes and follicles. The media used for oocyte culture and IVF are modified Ham's F10 medium, modified Whitten's medium, Earl's solution and Whittingham's T6 medium. Any one of these media may be used.

(*c*) **Semen Preparation** *:* The semen is procured either from husband of female partner or from semen bank or elsewhere. Husband's semen is collected at site when required through masturbation 60-90 minutes prior to insemination. Semen is subjected to centrifugation; sperm pellets are recovered and transferred to culture medium for suspending it again. It is then centrifuged and semen pellets are retransferred to culture medium. The sperms are incubated at 37°C for 30-60 minutes. The most active spermatozoa float at surface. Therefore, samples for IVF should be taken from surface of culture medium containing spermatozoa.

### 3. *In vitro* Fertilization and Embryo Transfer

Insemination is done by mixing about 1 ml culture medium containing oocytes with 1 ml of semen preparation (that contains 10,000-50,000 spermatozoa/ml). Oocytes are incubated for 12-13 h. Thereafter, they are observed keeping them in a Petri dish for the presence of two pronuclei and two polar bodies. Besides, the abnormal fertilized oocyte contains more than two pronuclei and granulation in cytoplasm. First cleavage occurs after 24-30 h of insemination. Delay in cleavage after insemination confirms that the embryo is abnormal. Thus, the abnormal embryo is discarded; only normal embryo is used.

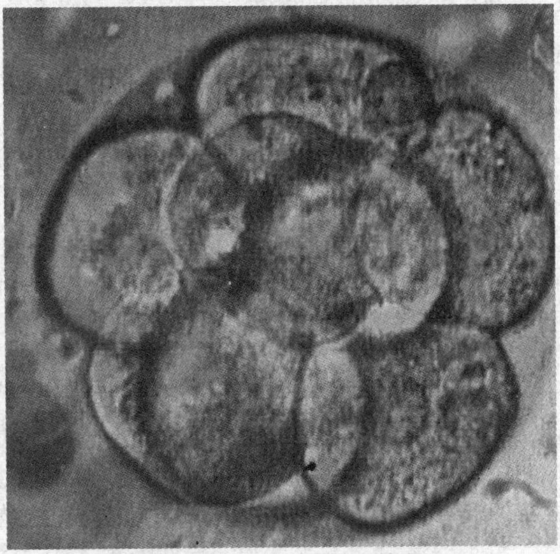
Eight-cell stage of human development.

After fertilization, embryo should not be detained for a long time *in vitro*. It should be transferred to treated recipient female when it is in 2-4 cell stage. However, success has been achieved with transfer of 1-16 celled embryos. However, early embryo fails to survive in uterus. This stage is very critical, therefore, much care should be taken. Some times, multiple embryo transfer is also done. It may increase the chance of embryo survival, but results in multiple pregnancy to which the female does not like. The transferred embryo stays in fallopian tube up to 8-16 cell stage, thereafter moves to uterus. Embryo of prolonged stage *in vitro* should not be transferred because of less chance of survival and pregnancy.

Embryo transfer should be done in an operation theatre. No anaesthesia should be given to female. She is given only Diazepam (Valium) of 10 mg orally before embryo transfer. The embryo is to be transferred through cervical canal; therefore, the patient is placed in lithotomy (*i.e.* in the position of knee and chest with tilting the head down) so that the fundus should be in lower position than cervical canal (Fig. 11.6). The cervix is visualised by inserting a sterile bivalve vaginal speculum.Then uterine cavity is aligned to cervical canal by manipulating the uterus. Distance between fundus and external cervical is earlier measured through ultrasonography. The embryo is drawn into a teflon catheter along with 10 ml of tissue culture medium. The catheter is gently inserted into uterine cavity just short of fundus and embryo is gently injected along with culture medium. It should be made sure that the embryo has been transferred or not. The catheter is observed under microscope. Absence of embryo in catheter ensures for its transfer in uterus. Wrong placement of catheter or high amount of medium will result in tubal pregnancy or expulsion of embryo from uterine cavity. Therefore, care should be taken for proper handling. After embryo transfer, the female is allowed to take rest for about 2-7 h and abstain from intercourse for about 10 days. The patient follows the advice of doctor and undergoes for regular check up.

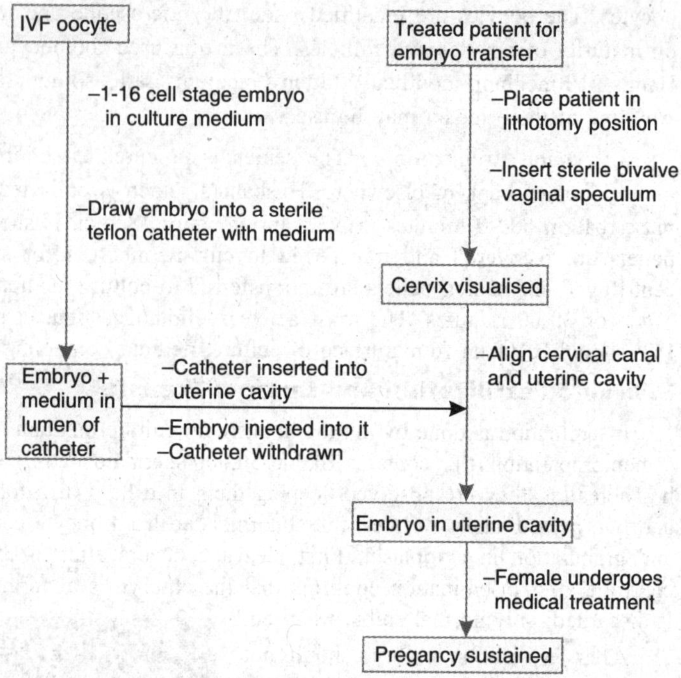


**Fig. 11.6.** Outline of embryo transfer in treated female human.

**(*a*) Bioethical Problems Prelated toTest Tube Babies.** Although the IVF technology is a boon to childless couples, yet there are several problems related to it, if it becomes commercialised. These problems may be religious, ethical, emotional or political. For example, the Catholic Church does not approve IVF techniques as it proclaims that conception should never be taken out of the body. The Gamete intrafallopian transfer (GIFT) is acceptable, while zygote intrafallopian transfer (ZIFT) is not. Muslim countries like Malaysia believe the sperm donation as immoral. Children borne of donated sperms are considered illegitimate. However, there is no controversy in India so far. A sperm bank was set up in New Delhi in January, 1994 to help the childless couples.

One of the emotional problems is the unused extra embryos. Whether they will be thrown away or implanted in surrogate mothers. The first one is the moral question. The surrogate mothers act as animal incubator and deliver baby after the normal gestation. They don't contribute the genetic materials as it come from the donors. Therefore, the surrogate mothers could be commercialised. The child will not be their real one. Will the children borne through donated egg/sperm be given social or religious recognition ? Will the children be known of their biological parents ? Will their parents accept them ?

## D. TRANSGENIC ANIMALS

Much has been described about the methods of transfer of genes/DNA fragments/chromosomes into animal cells in earlier chapters either through microinjection, drug delivery system, electroporation or mediated by viruses. Cell manipulation has been aimed at production of novel chemicals, pharmaceutical drugs or improvement in animal breed. The other purpose has been to study the structure and function of genes through molecular markers and genome mapping (*see* precedent section). Improvement in livestock has already given good results for increased milk production in cattles, increased growth rate of livestock and fish, production of valuable proteins in large quantities in milk, urine and blood of livestock, wool production in transgenic sheep.

Thus the transgenic animals can be used as *bioreactors* for large scale production of valuable recombinant chemicals such as hormones, interferons, proteins, etc. Thus, manufacturing of recombinant drugs through transgenic animals is called 'molecular farming' or 'molecular pharming'. Areas of possible investigation of domestic livestock by using the transgene technology has been given in Fig. 11.7.

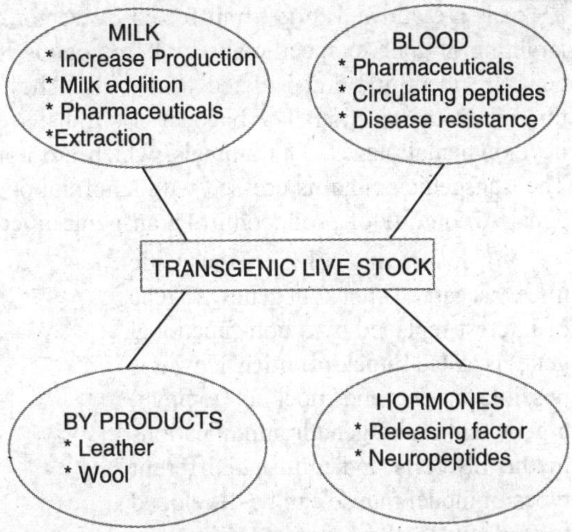


**Fig. 11.7.** Areas of investigation of domestic livestock via transgenic technology.

## 1. Strategies for Gene Transfer

Desired foreign genes are transferred into animal cells/embryos via virus, microinjection, targeted gene transfer methods, etc as discussed below:

**(*a*) Transfection of Animal Cells/ embryos :** During 1970s, attempts were made to produce transgenic animals. The first success was achieved in 1976 when mouse embryos were infected by retrovirus. The *retroviruses* are supposed to be an efficient vector that transfect the animal cells and deliver its genes leading to production of transgenic cells/animals. The other viruses used for this purpose are *vaccinia virus, adeno-associated virus, herpes virus* and *bovine papiloma virus.* For detailed description of retroviruses, their morphological nature and genome, see *A Text Book of Microbiology* by Dubey and Maheshwari (2012).

The transfected cultured mammalian cells have been used for diagnostics of oncogene (*i.e.* cancerous gene) as well as for gene therapy. The steps for detection of cancer gene follows : (*i*) isolation of DNA from tumour cell line, (*ii*) DNA fragementation into 30-50 kb long pieces through mechanical shearing, (*iii*) dissolution in phosphate buffer followed by pre-cipitation by adding $CaCl_2$, (*iv*) pouring of this solution onto a layer of mouse 3T3 cells; foci of cells developed, and (*v*) use of transfected cells in detection of cancer causing genes. The other application of transfected cells is in gene therapy. Genes of desired function are inserted in cultured cells. The latter is placed in patient's body to rectify the malfunctioning gene.

**(*b*) Transfer Through Micro-injection :** Microinjection method has also been developed and variously used in production of transgenic animals. So far gene transfer has been successfully carried out in several classes of animals viz., fish, birds, insects, mammals, etc. (for detail *see* Chapter 7).

**(*c*) Gene Targeting :** The other approach is the targeted gene transfer that involves transfer of genes at homologous sites in the host genome. It is done just to replace the wild type of mutant genes. For the first time it was done in bacteria and yeast. In 1985 success has been achieved in human also where human β-globin gene was transferred into recipient cell through recombination.

Targeted gene transfer is possible because the homologous DNA sequences are present at the targeted site as well as in vector that carries the desired gene of foreign origin. Besides, marker genes are also used to select the cells in which gene has been transferred at targeted site. It is achieved by: (*i*) using marker genes for antibiotic resistance, (*ii*) hypoxanthine phosphoribosyl transferase (HPRT), and (*iii*) polymerase chain reaction.

Gene targeting is also done by using embryonic stem (ES) cells as described earlier. The

ES cells are allowed to get transfected by vector containing desirable genes. In transfected cells targeting of gene to specific site by homologous recombination occur. Thereafter, the transfected cells are identified and isolated from bulk. They are multiplied and introduced into blastocyst through microinjection. The blastocyst is transferred into uterus of a surrogate mother for further developmental stages. The animals, which are born, are checked for the presence of transgene. The transgenic animal is crossed with a normal one to study the inheritance of introduced foreign gene. An outline of production of transgenic mice is shown in Fig. 11.8.

(*d*) **Knockout Mice :** Transgenic mice that carry a knockout gene (*i.e.* gene of interest replaced by a non-functional gene) is called **knockout mice.** Now it is possible to select and knockout (remove) a gene and make genetic modifications in the ES cells and mouse. Different types of model mouse can be developed to understand the function of various genes *e.g.* disease development. For example, knockout mice have helped the immunologists to understand the effect of knockout gene on immune system in animals. Various knockout mice are being used in immunological research. Production of knockout mice (gene targeted) is accomplished in the following steps (Fig. 11.9).

(*i*) Isolation and culture of ES cells from inner cell mass of a mouse embryo.

(*ii*) Induction of a mutant or disrupted gene into the cultured ES cells and selection of homologous recombinant cells in which genes of interest have been knockedout.

(*iii*) Injection of homologous recombinant ES cells into a recipient mouse embryo and transfer of manipulated embryo into uterus of surrogate mother mice.

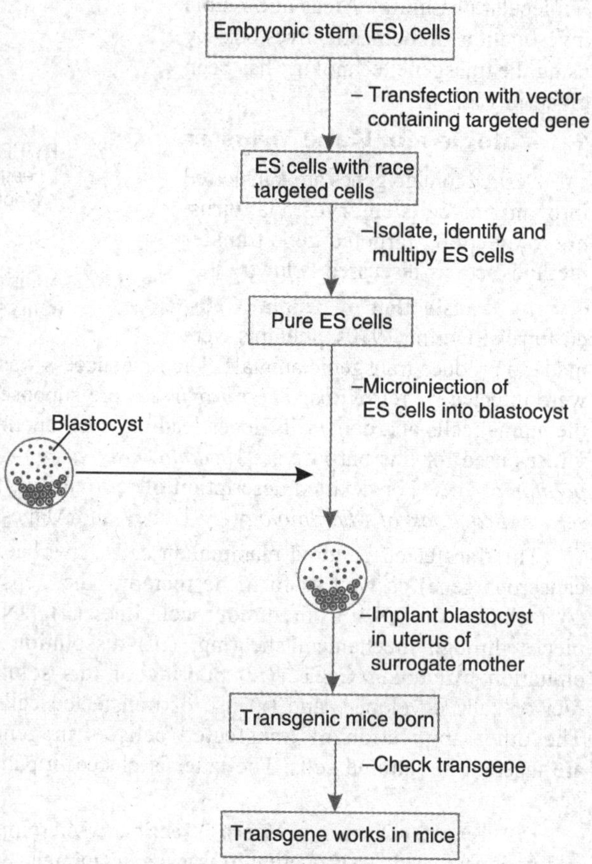


**Fig. 11.8.** Production of transgenic mice by targeted gene transfer through ES cells (diagrammatic).

(*iv*) Mating of chimeric offspring heterologous for disrupted gene to produce homozygous knockout mice.

## 2. Transgenic Mammals

For the first time in 1982, there appeared a report on the transfer of human growth hormone gene of rat fused to the promoter of mouse metallothioneine I gene. It was done by microinjection method. As a result of presence of a novel gene, there has been a drastic increase in body weight of mice (for detail method *see* Chapter 7). Since then a large number of transgenic mammals and other animals have been produced such as cow, pig, rabbit, goats, sheep, fish, etc. The purpose of production of transgenic animals has been to produce more protein in milk and meat, disease

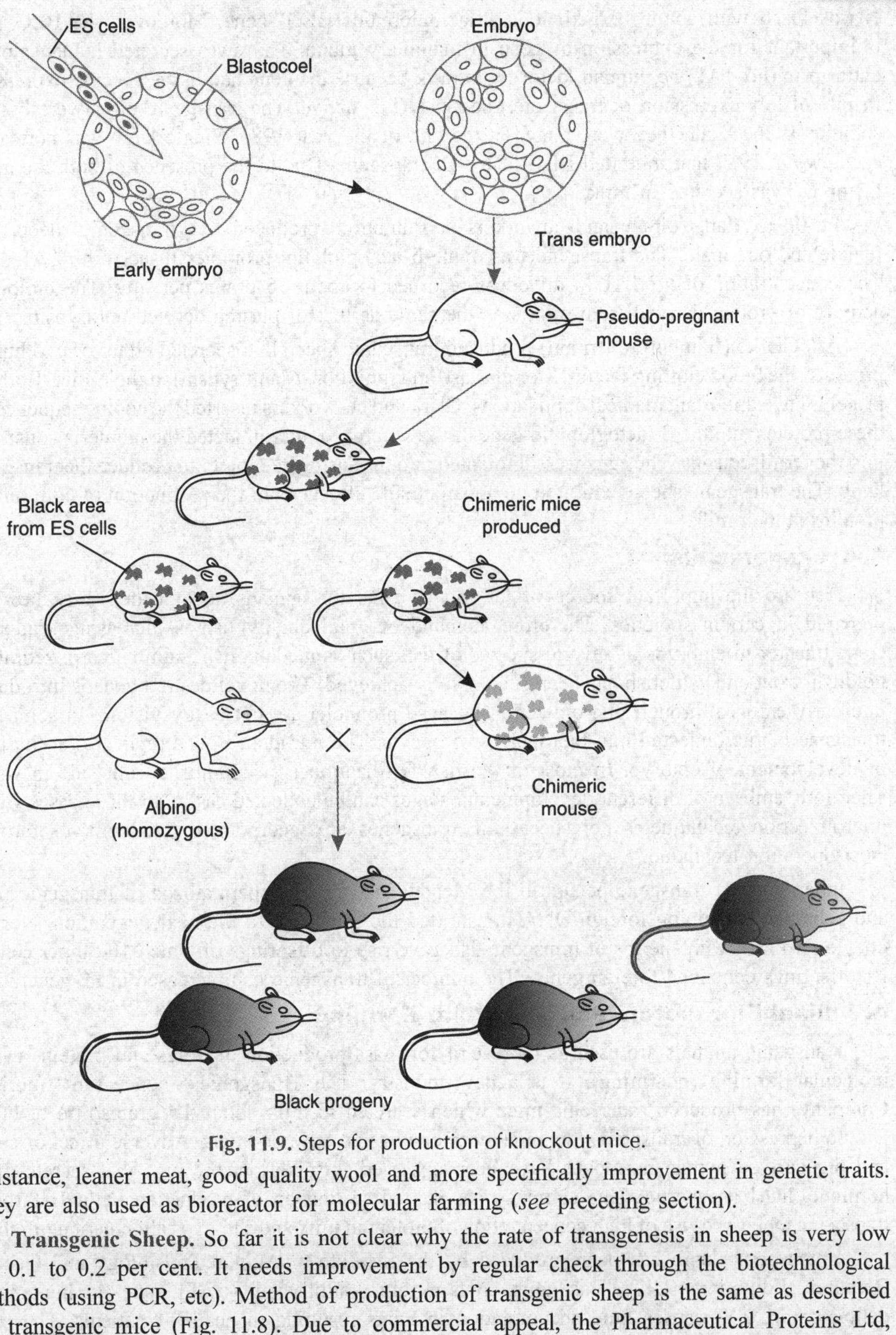


**Fig. 11.9.** Steps for production of knockout mice.

resistance, leaner meat, good quality wool and more specifically improvement in genetic traits. They are also used as bioreactor for molecular farming (*see* preceding section).

**Transgenic Sheep.** So far it is not clear why the rate of transgenesis in sheep is very low *i.e.* 0.1 to 0.2 per cent. It needs improvement by regular check through the biotechnological methods (using PCR, etc). Method of production of transgenic sheep is the same as described for transgenic mice (Fig. 11.8). Due to commercial appeal, the Pharmaceutical Proteins Ltd, Cambridge (UK) provided fund to J.P. Simons for the production of transgenic sheep. In 1988, Simons reported first production of transgenic sheep. He first produced two transgenic ewe that consisted of about 10 copies of human antihaemophilic factor IX gene in the form of cDNA.

It was fused with about 10.5 kb long β-lactoglobulin (BLG) gene. Moreover, the BLG gene is important for the expression of gene in mammary gland. The ewes secreted human alpha-1 antitrypsin (hα-1AT) *i.e.* human factor IX in milk because the gene had tissue specific expression. Inspite of low expression of transgene, the hα-1AT is active. The transgenic ewes were born in summer 1986. Again they were mated at the end of the year 1986. Single lamb was born from each ewe in 1987 that inherited BLG-factor IX transgene. Due to the presence of both the genes factor IX was secreted in milk.

In 1991, Alan Colman and coworkers at Edinburgh produced five transgenic sheep, four female and one male. The transgene was ovine-β-lactoglobulin promoter fused to hα-1AT gene. The concentration of hα-1AT in milk was recorded to about 35 grams per litre. The biological activity of protein derived from milk was the same as that of plasma derived antitrypsin.

A.J. Clarke (Edinburgh, Britain) produced transgenic sheep that secreted either of two human proteins (the blood clotting factor IX or elastase and inhibitor α-lantitrypsin) in their milk. Both the proteins have important medical applications. Clark and co-workers inserted the coding sequence for these proteins into the β-lactoglobulin gene of the sheep and microinjected the chimeric constructs into the fertilised eggs. The eggs were implanted into the surrogate mother to produce the transgenic lamb. The transgenic sheep exhibited no apparent side effects from the production of both human proteins in their milk.

### 3. Transgenic Fish

Fish are the important source of fat and proteins for humans and the delicious diets are prepared in certain societies. Therefore, demand of good quality fish is increasing gradually. Gene transfer in embryos of several species of fish such as medaka fish, salmon, carp, zebrafish, goldfish, trout and cattlefish has been successfully achieved. The novel desired gene is introduced into early embryo through microinjection because pronuclei are not easily visible. In some fish transgene is microinjected into the nucleus of oocyte. The introduced gene replicates at the time of development of embryo. In most of the fish fertilization is accomplished out side in water. Therefore, embryo of different developmental stages can be collected easily. In those cases where microinjection technique is not successful, transgenes are incorporated into embryos through electroporation technique.

Inheritance of transgene occurs in the Mendalian way. The spermatozoa of transgenic trout and zebrafish containing foreign DNA transmitted the transgene to the next generation. Normal oocyte fertilized with sperms of transgene fish gave rise to offsprings of which 10-50 per cent of F1 offsprings contained foreign genes. The number of transgenic fish increased in F2 generation.

### 4. Animal Bioreactors and Molecular Farming

Transgenic animals are used as bioreactor for mass production of drugs and proteins called molecular farming (pharming). It is a new industry. The Transgenic Science (an American Company) has produced transgenic mice which secreted in milk about 0.5 grams/litre of hGH. Due to expression of transgene, this mice hormone has no adverse effect. Adverse effect of cattle and pig derived human hormones have been observed. In addition, the pigs that secrete extra hormone, hGH are sterile. This company has started scaling up of production of hGH through transgenic rabbit because of high concentration of protein in milk and short gestation period of rabbit.

A transgenic lamb has been produced by the Institute of Animal Physiology and Genetic Research, Edinburgh (UK). This lamb contains alpha-antitrypsin (αAT) gene. In human the deficiency of αAT causes fluid accumulation in lungs resulting in death of patients. The αAT inhibits the enzyme elastase that digests foreign particles and clears the lungs. In the absence of αAT elastase is not inhibited and, therefore, it digests lung tissue. In lactating milk of transgenic lamb, presence of αAT can be tested. Transgenic sheep can also be used for production of αAT commercially.

# E. APPLICATION OF MOLECULAR GENETICS

Methods have been developed to isolate animal genes and to characterize them. The first approach is to identify DNA sequence associated with economic trait loci. The second step is to incorporate the identified gene into three maps : (*i*) a physical map of chromosome (where DNA sequences are assigned to specific sites on specific chromosomes), (*ii*) a linkage map (where linkage of different genes and markers is assigned in the same chromosome), (*iii*) a genetic map (where inheritance of economic traits is corrected with the inheritance of genes and markers).

The unkown and novel genes associated with genetic diseases can also be identified through their expression. Then the protein sequences are identified through amino acid sequencing or injecting the protein into suitable animal (*e.g.* mouse, rabbit, chicken, etc) to raise the antibodies.

The animal production industry is gradually reaching towards revolution due to the development of detailed linkage and genetic map, more knowledge of expression and regulation of genes and techniques for large propagation. The application of molecular genetics in animals currently being used are : breeding selected traits into livestock, animal cell culture, and production of transgenic animals.

## 1. Selected Traits and Their Breeding into Livestocks

There are several genetic traits known which can be bred into livestock if required, for example, disease resistant traits. Some traits are determined from the coordinated expression of many genes but not single gene. The discipline of quantitative genetics unravels the total number of genes contributing to a trait, contribution of individual trait and its location on the chromosomes. Altogether they are termed as quantitative trait loci (QTLs), therefore, herds are surveyed for reference animals that have been bred for a broad range of phenotypic variations.

**(*a*) Diagnosis, Elimination and Breeding Strategies of Genetic Diseases** : An increase in homozygosity of recessive genes and the frequency of specific alleles at QTLs have been resulted due to isolation of domestic animals in small breeding group. Consequently most breeds of livestocks comprises of variant genes which are harmful or lethal at homozygous stage (when it contain two variant genes). Such allele may be found in 15 per cent cases with 5 per cent homozygotes. Several variant genes responsible for genetic diseases have been identified. Diagnostic tests are available which permit animals as normal, carrier (*i.e.* heterozygous with one normal allele or one variant allele) or affected (*i.e.* homozygous with two variant alleles). Now PCR is used to amplify the region of genes which has been affected. Thus amplified DNA fragement is used to identify the variants.

It is well known fact that genetic disease occurs when a gene malfunctions and enzyme is not expressed. When both alleles are defective in homozygote these results in phenotypic defects. The carriers (heterozygote) are phenotypically normal. Therefore, the genetic defects can be diagnosed by genetic tests and rectified biotechnologically during pregnancies that affect foetus. The second approach for elimination of defect from a herd is the use of normal (homozygous) animals for mating. Use of MOET with embryo biopsy and splitting on the best animals is the other method of spread of genetic disease and curing them. Pre-identified carrier or affected embryos should not be transferred (this aspect has been discussed earlier).

## 2. Application of Molecular Markers in Improvement of Livestock

During the last two decades the progress in recombinant DNA technology and gene cloning has brought in revolutionary changes in the field of basic as well as applied genetics. Several new approaches for genome analysis have been made. Now it is possible to uncover a large number of genetic polymorphism at the DNA sequence level and to use them as markers for evolution of genetic basis for the observed phenotypic variability.

Usually a marker is considered as a constituent that determines the function of construction. Variations occurring at different levels *i.e.* at the morphological, chromosomal, biochemical or DNA level can serve as genetic markers. Thus genetic markers can be defined as any stable and inheritable variation that can be measured or detected by a suitable method, and that can be used to detect the presence of a specific genotype or phenotype other than itself. The markers revealing variations at DNA level are referred to as 'molecular markers'. (for detail see chapter 14). So far an unlimited number of molecular markers is known since the first demonstration of DNA level polymorphism in 1974, which is called restriction fragment length polymorphism (RFLP). The molecular markers are classified into two broad categories, the hybridization-based markers and the PCR-based markers.

**(*a*) The Hybridization-based Markers :** This includes the traditional RFLP analysis. During RFLP analysis well labelled probes for important genes (*e.g.* cDNA or genomic sequence) are hybridized onto filter membrane containing restriction enzyme digested DNA. Then these are separated by gel electrophoresis and subsequently transferred onto these filters by Southern blotting. Thereafter, the polymorphisms are observed as hybridization bands. The individuals that carry different allelic variants for a locus will show different banding pattern. Hybridization can also be carried out with the probes (*e.g.* genomic or synthetic oligonucleotide) for the different families of hypervariable repetitive DNA sequences such as minisatellite, simple repeats, variable number of tandem repeats (VNTR) and microsatellite to reveal highly polymorphic DNA fingerprinting pattern.

**(*b*) The PCR-based Markers :** There is no need of probe-hybridization step. The PCR-based markers have lead to discovery of several useful methods which are easy to screen. On the basis of types of primers (*i.e.* primers of specific sequences targeted to particular region of genome or primers of arbitrary sequences) used for PCR. These markers further can be subdivided into two groups, the sequence-targeted PCR assay, and the arbitrary PCR assay.

**(*i*) The sequence-targeted PCR assay.** In this assay system, a particular fragment of interest is amplified using a pair of sequence-specific primers. In this category, PCR-RFLP or cleaved amplified polymorphic sequence (CAPS) analysis is a useful technique for screening of sequence variations that give rise to the polymorphic restriction enzyme (RE) sites. A specific region of DNA encompassing the polymorphic RE sites is amplified. The amplified DNA fragment is digested with respective RE. The variations in sequence are screened by several approaches.

**(*ii*) The arbitrary PCR assay.** In this assay randomly designed primer is used to amplify a set of anonymous polymorphic DNA fragments. When primer is short, there is high probability of priming taking place at several sites in genome that are located within amplifiable distance and are in inverted orientation. Polymorphism detected by using this method is called randomly amplified polymorphic DNA (RAPD). Based on this principle several techniques have been developed which differ in number and length of primer used, stringency of PCR conditions and the method of fragment separation and detection.

**(c) Properties of Molecular Markers :** In genetic analysis, many types of markers viz., morphological, chromosomal, biochemical and molecular markers are used. Morphological (*e.g.* pigmentation and other features) and chromosomal (*e.g.* structural and numerical variations) markers usually show low degree of polymorphism, therefore, they are not very useful. Biochemical markers have been tried out extensively but have not been found encouraging as they are sex linked, age dependent and influenced by the environment. The molecular markers capable of detecting the genetic variations at the DNA sequence level, have removed the limitations. They possess unique genetic properties that make them more successful than the genetic markers. They are numerous, distributed on genome, follow typical Mendalian inheritance and are multiallelic giving heterozygosity of more than 70 per cent and unaffected by environmental factors.

For genetic analysis, molecular markers offer several advantages. There are several advantages of molecular markers: (*i*) the DNA samples can easily be isolated from blood, tissues *e.g.* sperms, hair follicle as well as archival preparations, (*ii*) the DNA samples can be stored for a longer time and readily be exchanged between the laboratories, (*iii*) the analysis of DNA can be carried out at an early age or even at the embryonic stage, irrespective of sex, (*iv*) once the DNA is transferred onto a solid support *e.g.* filter membrane, it can be repeatedly hybridized with the different probes and heterologous probes and *in vitro* synthesised oligonucleotide probes can also be used, and (*v*) the PCR-based methods can be subjected to automation.

**(*d*) Application of Molecular Markers :** Polymorphism observed at the DNA sequence level has been playing a major role in human genetics for gene mapping, pre- and post- natal diagnosis of genetic diseases, and anthropological and molecular evolution studies. Similar approach for exploitation of DNA polymorphism as genetic markers in the field of animal genetics and breeding has opened vistas in livestock improvement programmes. Much interest has been generated in determining variability at the DNA sequence level of different livestock species, and in their assessment whether these variations can be exploited efficiently in conventional as well as transgenic breeding programmes.

Molecular markers can play important role in livestock improvement through conventional breeding strategies. The various possible applications of molecular markers are: short-range (immediate) applications and the long-range applications (Table 11.1).

**> Table 11.1. Molecular markers useful in conventional livestock breeding.**

| *Application* | *Marker system* |
|---|---|
| **1. Short-range/immediate Application** | |
| Parentage determination | DFP, microsatellite |
| Genetic distance estimation | DFP, RAPD, microsatellite |
| Determination of zygosity/ Freemartinism | RFLP, PCR-RFLP, microsatellite |
| Sex determination | RFLP, PCR-RFLP, DFP, microsatellite |
| Identification of disease carrier | RFLP, CAPS, microsatellite |
| **2. Long-range Application** | |
| Gene mapping | Type II markers *e.g.* VNTR, minisatellite, microsatellite, RAPD |
| Marker-assisted selection | Any marker having direct/indirect association with the performance traits/QTL under question |

RAPD, randomly amplified polymorphic DNA; CAPS, cleaved amplified polymorphic sequence; DFP, DNA fingerprinting; QTL, quantitative trait loci.

**(*e*) Transgenic Breeding Strategies :** The current breeding strategies of livestock largely rely on the principle of selective breeding. In this method genetic improvement is brought about by increasing the frequency of advantageous alleles of many loci. The actual loci are rarely identified. In these methods genes cannot be moved from distant sources like different species or genera due to reproductive barrier.

The recent development in molecular biology has given rise to new technology called *transgenesis*, which has removed the breeding barriers between different species or genera. Transgenesis has opened up many vistas in understanding behaviour and expression of a gene. It has made possible to alter the gene structure and modify its function. There are many applications of transgenesis, but the most convincing one is the development of transgenic dairy animals for the production of pharmaceutical proteins in milk, and animals with altered milk composition.

The transgenesis first starts with identification of the genes of interest. In this context, molecular markers can serve as reference point for mapping relevant genes. After successful production of transgenic animals, appropriate breeding methods could be followed for multiplication of transgenic herd. Molecular markers can also be used for identification of the animals carrying the transgenes. Most of the QTL are polygenic in nature and transgenesis presently single gene traits are being manipulated. The technology holds future promises in moving polygenic QTL across the breeding barriers of animals. However, it is expected that molecular markers will serve as a potential tool to geneticists and breeders to evaluate the existing germplasm, and to manipulate it to create animals of desired traits as demanded by the society.

## F. BIOETHICS IN ANIMAL GENETIC ENGINEERING

In earlier sections several benefits of animal genetic engineering have been discussed in bio-medical research and diagnostics, production of therapeutic proteins, development of transgenic live stock with beneficial traits (*e.g.* increased milk). Besides all benefits, genetic modification raises many ethical concerns. These refer to moral objections about the genetic modification brought about in animals. Because these modifications result in 'unnaturalness' of this technology; for example,

(*i*) Transfer of human gene (*e.g.* blood factor IX) into the food animals (*e.g.* sheep, etc.).

(*ii*) Transfer of animal genes into food plants (*e.g.* interferon-alpha gene into plants) which may be used by vegetarians.

(*iii*) Transfer of human genes into such organisms that can be used as animal feed (*e.g.* modified yeast producing medically important human protein).

(*iv*) Transfer of genes from such animals (whose flesh is not eaten by some religious groups) into the others that they normally eat.

Considering the religious belief of people, it was recommended that: (*i*) if alternative food is available, gene transfer in food organism should not be carried out, (*ii*) products from transgenic organisms should be properly labelled as 'Transgenic' or 'GM Products' so that some religious group whom these are ethically unacceptable may have the choice open.

If there occurs suffering of animals by genetic engineering, it may face many moral opposition. Earlier such oppositions have been met. For example, there are evidences where animals suffer from severe arthritis that have injected with transgenic hormones for the improvement of quality of their meat.

Therefore, justification must be made for producing and using GM animals that their use will not pose any risk to humans, other animals, and the environment.

Some of the concerns about safety which must be considered seriously are given below :

**(*i*) Concerns about Escape of Transgene :** The transgenic animals may breed with other domestic or wild animals. Therefore, there is risk for transferring the transgene to other populations too.

**(*ii*) Concerns about Risk for Escape of Retroviruses :** Retroviruses are used as vector for the production of genetically modified animals. There is risk for escaping retroviruses from laboratory and transferring transgene to other animals also.

**(*iii*) Concerns about the Risk to Human and Animal Health :** Human and animals consume products from transgenic animals. Still there is fear and risk in the society that the GM products and trangenic food animals may create the health problem like permanent disability, allergy, etc.

***(iv)* Concerns about the Risk from Drug Resistance Gene Markers :** Certain gene markers (for drug resistance *e.g.* kanamycin resistance, *kan*$^r$) are used in certain process of genetic engineering. These marker genes might inadventently get transferred and expressed.

***(v)* Ecological Concerns :** There is a major concern about the wide effects of producing disease resistant animals.

***(vi)* Concerns in Xenotransplantation :** During transplantation of animal organs in human there is risk for the contamination of animal organs by animal viruses. If so, the wide population may be infected by such viruses.

However, most of the people of our society do not know about the scientific discoveries of biotechnology, and its merits and demerits. Hence to them transgenic animals, gene therapy and GM products look threatening. Therefore, there is an urgent need to have mass communication and societal education about the ethical and moral issues raised due to biotech products. Hence, popularisation of biotechnology will certainly remove the fear from possible risks and enhance **bio-business**.

## PROBLEMS

1. What do you know about artificial insemination ? Write in detail about semen storage and sperm sexing in animals.
2. What do you understand with 'multiple ovulation' ? In what ways it can be induced, and embryo transferred in animals ?
3. Write an essay on embryo splitting and implantation in animals.
4. Give an illustrated account of *in vitro* fertilization in animals with suitable example.
5. What do you know about 'embryo cloning'? Write in detail about nuclear transplantation in animals.
6. Write an essay on cloning of embryonic stem cells.
7. Give a detailed account of *in vitro* fertilization in humans.
8. With the help of suitable diagram discuss in detail about embryo transfer in humans.
9. What do you know about transgenic animals ? Write in brief transgenic mammals, sheep and fish.
10. Write an essay on animal bioreactor and molecular farming.
11. Discuss in detail about application of molecular markers in livestock improvement.
12. Write an essay on application of molecular genetics in animals.
13. Write short notes on the following: (*i*) Semen and its storage, (*ii*) Ovulation control, (*iii*) Sperm sexing, (*iv*) Embryo sexing, (*v*) Embryo splitting, (*vi*) Multiple ovulation, (*vii*) *In vitro* fertilization, (*viii*) Dolly—the first cloned mammal, (*ix*) Targeted gene transfer, (*x*) Quadriparent animals, (*xi*) Transgenic animals, (*xii*) Animal bioreactor and molecular farming, (*xiii*) Application of molecular genetics, (*xiv*) Application of molecular markers in livestock improvement, (*xv*) Transgenic breeding strategies and molecular markers, (*xvi*) Chimeric mouse, (*xvii*) Knockout mouce.
14. Give a brief account of bioethics in animal engineering.

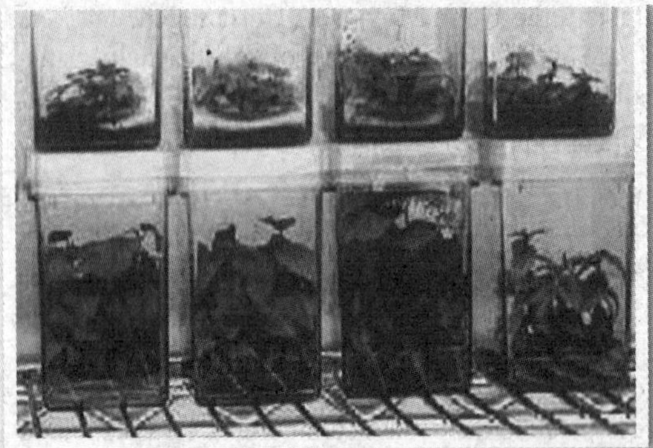

CHAPTER

# 12

# *In Vitro* Culture Techniques of Plant Cells, Tissues and Organs

During the last two decades plant cell, tissue and organ culture have developed rapidly and become a major biotechnological tool in agriculture, horticulture, forestry and industry. Those problems which were not feasible through conventional techniques, now have been solved via these techniques, for example, inter- and intra-specific crosses, micropropagation, somaclonal variation, encapsulated seeds, etc.

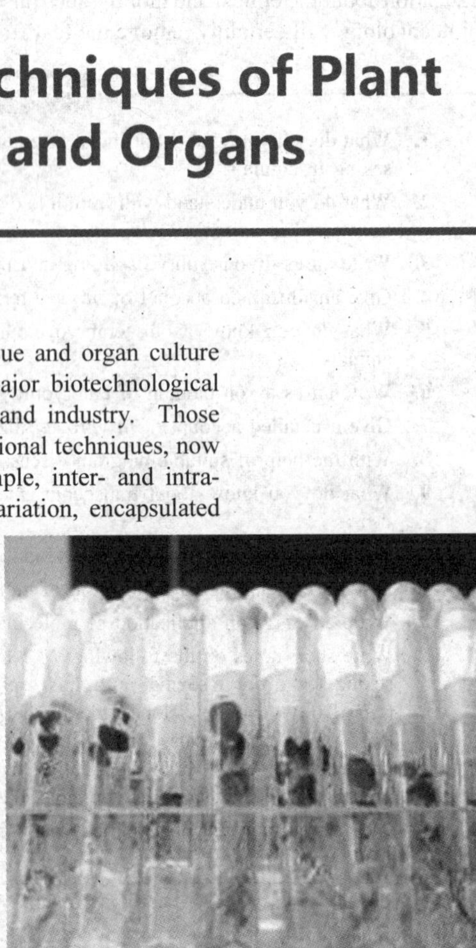

Micropropogation of potato plants.

To boost up these areas ICGEB in a workshop (held at New Delhi from September 18 to 20, 1985), recommended the need of more research in developing countries on plant cell culture, differentiation, regeneration and transformation in tropical grain legumes, woody legumes and cereals. The emphasis was laid to improve growth under stress condition, pest and disease resistance, improved nutritional quality, nitrogen fixation and the control of partitioning within the plants. A detailed account of plant cells, tissue and organ culture is given in this context.

**Totipotency : The basis of plant cell and tissue culture.** Each living cell, of a multicellular organism, is capable of independent development, when provided with suitable conditions. In 1901, T.H. Morgan coined the term 'totipotency' to denote this capacity of cell to develop into an organism by regeneration. However, the concept of totipotency is important in tissue culture. Use of multicellular organisms in research, as biological units, is rather difficult; therefore, attempts to study an organism by reducing to its constituent cells and subsequently the cultured cells as basic organism, are of fundamental importance.

## A. HISTORICAL BACKGROUND

Historically Henri-Louis Duhamel du Monceau (1756) pioneered the experiments on wound healing in plants through spontaneous callus (unorganised mass of cells) formation on decorticated region of elm plants. But the science of cell and tissue culture could be advanced after propounding the cell theory by Schleiden and Schwann (1839). Trecul (1853) observed callus formation in a number of plants. Vochting (1878) suggested the presence of polarity as a key feature that guide the development of plant fragments. He observed that the upper portion of a piece of a stem always produced buds and the basal region produced callus or roots.

During 19th century the idea of development of callus (a disorganized proliferated mass of actively dividing cells) from isolated stem fragments and root apices came into existance. Callus could also be developed from buds, and root and shoot fragments of about 1.5 mm in size without using nutrient medium.

In 1902, a German Botanist Gottlieb Haberlandt (in Berlin) developed the concept of culture of isolated cells of *Tradescantia* in artificial condition. Though his experiment failed to induce the cells to divide, he did not succeed because by that time even auxin was not discovered. But he lent a foundation to plant physiology. He described the cultivation of mesophyll cells of *Lamium purpureum* and *Eichhornia crassipes*, epidermal cells of *Ornithogalum* and hair cells of *Pumonaria*. Cell survived for 3–4 weeks. Due to this endeavour, Hamberlandt is regarded as the **father of tissue culture**.

From 1902 to 1930 attempts were made for organ culture. Hannig (1904) isolated embryos of some crucifers and successfully grew on mineral salts and sugar solutions. Simon (1908) successfully regenerated a bulky callus, buds, roots from a poplar tress on the surface of medium containing IAA which proliferated cell division.

The term 'Tissue Culture' can be applied to any multicellular culture growing on a solid medium (or attached to substratum and nurtured with a liquid medium) that consists of many cells in protoplasmic continuity. But in organ culture (*e.g.* excised roots) the cultured plant material maintains its morphological identity, more or less, with the same anatomy and physiology as *in vitro* of the parent plants (Doods and Roberts, 1985).

Until the early 1930s, R.P. White (USA), Gautheret (France) and Nobercourt (France) independently cultured tissues excised from several plants on the defined nutrient media for a long period. Gautheret (1939) cultured cambium tissue of carrot on Knop's solution supplemented with other chemicals in trace amount. P.R. White in 1939 cultured tobacco tumour tissue from the hybrid *Nicotiana glauca,* and *N. langsdorffii.*

During 1940 to 1970, suitable nutrient media were developed for culture of plant cells, tissue, protoplasts, anthers, roots tips and embryos. *In vitro* morphogenesis( *i.e.* regeneration of complete plant from cultured tissue) of plants was always successfully done. In 1941, van Overbeek and co-workers used cocount milk (embryo sac fluid) for embryo development and callus formation in *Datura.*

In the 1950s several important achievements were made in the field of plant physiology. The understanding of plant growth hormone in rapid multiplication of totipotent cell was developed. In 1958, F.C. Stewart and J.Reinert obtained regeneration in callus tissue culture of *Duccus carrota.* The foundation of commercial plant tissue culture was laid in 1960 with the discovery of G.M. Morel for a million fold increase in clonal multiplication of an orchid, *Cymbidium.* At the university of Wisconsin, Skoog and co-workers found out the role of cytokinins in tissue culture. Consequently, several chemicals were tested which stimulated callus. Adenine in the presence of auxin was found to induce callus growth and bud formation in tobacco cultures (Skoog and Tsui, 1948). Eventually a potent cell division factor from degraded DNA preparations was isolated, identified and named as kinetin. The term cytokinin was given to this group after substituting the aminopurine compounds that stimulate cell division in cultured plant tissue and behave physiologically similar to kinetin. Later on other cytokinins (the naturally occurring plant hormones) such as zeatin, and isopentyl adenin were discovered. In 1957, F. Skoog and C.O. Miller advanced the hypothesis of organogenesis in cultured callus by varying the ratio of auxin and cytokinin in the growth medium. The shoot was formed with keeping the ratio of kinetin higher and root developed when ratio was lower.

When fragments of callus are transferred into a liquid medium and aerated on a shaker, it gives a suspension of single cell and aggregate of cells. A cell can be propagated by subculturing. In 1953, W.H. Muir developed a successful technique for the culture of single isolated cells which is commonly known as paper-raft nurse technique (placing a single cell on filter paper kept on an actively growing nurse tissue). Later on, attempts were also made for single cell culture by hanging drop and agar plate method. During this period phenomenon of totipotency was fully developed by demonstrating that a single isolated cell can divide and regenerate a whole plant.

In 1952, the Pfizer Inc., New York (U.S.A.) got the US Patent and started producing industrially the secondary metabolites of plants. The first commercial production of a natural product shikonin by cell suspension culture was obtained.

Cell wall creates a barrier in plant protoplast culture. In 1960s, the role of enzymes *e.g.* cellulase and pectinase in dissolution of cell wall in buffer solution at suitable pH, and isolation and culture of protoplast was developed. During this period extensive study of plant tissue culture was done by using the explants (excised plant parts ) from different parts of gymnosperms and angiosperms. S. Guha and P. Maheshwari in 1966 developed techniques for the production of vast numbers of embryos from cultures of pollens and sporogenous tissues of anther. In 1974, C. Nitsch gave methods to double the chromosome number in microspors of *Nicotiana* and *Datura,* and collected seeds from the homozygous diploid plants within 5 months. However, the benefits from protoplast culture currently availed are : (*i*) intraspecific, interspecific and intergeneric protoplast fusion and some times between plant and animal, (*ii*) Transfer of mitochondria and plastids into protoplast, (*iii*) Uptake of certain beneficial genes of blue-green algae, bacteria and viruses by protoplasts, and (*iv*) Transfer of genetic informations into isolated protoplasts.

I.K. Vasil in 1982 has emphasized to develop the following techniques of cell culture and somatic genetics of mainly grasses and cereals: (*i*) Rapid clonal propagation, (*ii*) Regeneration from single cell and protoplast (*iii*) Androgenetic haploids, (*iv*) Mutant/variant cell lines and plant regeneration, (*v*) Somatic hybridization, (*vi*) Transformation, and molecular biology.

In India, work on tissue culture was started during mid 1950s at the Department of Botany (University of Delhi) by Panchanan Maheshwari who is regarded as **father of embryology** in India. Different tissue culture methodologies were involved for morphogenic studies involving ovary, embryo, endosperm, ovules, etc. At the University of Delhi, Sipra Guha Mukherjeee and S.C. Maheshwari (1964-67) for the first time developed the haploid through anther and pollen culture. Discovery of haploid production was a land mark in the development of plant tissue culture.

The advancement made in cell and tissue culture technology are due to the development in composition of culture media. Based on the success of plant cell culture techniques many recent advances have been done in the area of micropropagation, production of secondary metabolites and pathogen-free plants, genetic manipulation (e.g. *in vitro* pollination, somatic hybridation/cybridisation, induction of haploid, genetic transformation and production of transgenic plants.

During 1980, recombinatnt DNA technology made possible to transform artificially cultured plant cells by introducing foreign genes. Thus several transgenic plants of special characteristics have been produced (see Chapter 13). The gene revolution has made the second green revolution. Now it is possible to develop plants of desired genetic characters.

# B. REQUIREMENTS FOR IN VITRO CULTURES

## 1. A Tissue Culture Laboratory

Culture of plants cells and tissues *in vitro* is not an easy task. It requires all the nutrients and physico chemical factors in maintained in a laboratory.

A a good laboratory, similar to a microbiological laboratory, is required which must have facilities for: (*i*) nutrient medium preparation, sterilization, cleaning and storage of supplies, (*ii*) aseptic condition for working the living materials, (*iii*) a controlled environmental conditions for growth and development of cultures, (*iv*) observation and evaluation of culture as hoped, and (*v*) recording the observations made during the experiment. But the most basic facility that an individual needs for tissue culture requires the following:

**(*a*) Washing and Storage Facilities:** A separate area is required which should have large sink with provision for hot and cold running water, distillation apparatus, washing machine, pipette washer, drier and cleaning brushes, keeping and weighing the chemicals, and putting glassware.

**(*b*) Media Preparation Room:** An area is required for preparation of media. In such space there should be provision for bench space for chemicals, labware, culture vessels, closures and miscellaneous equipment required for media preparation and dispensing. In this room provision is also made for placing hot plates or stirrers, pH meter, balance, waterbath, burners, oven, autoclave, culture, vessel refrigerator, etc.

Vitamins and growth hormones are carefully weighed. Stock solutions of chemicals are kept in refrigerator to avoid contamination.

**(*c*) Transfer Area:** Earlier transfer of plant tissue was done in open in laboratory bench under clean and dry atomsphere conditions. Later on closed plastic box was constructed which consisted of UV tube. UV light from the tube and use of 95 % ethanol helped to maintain sterile conditions inside the space of plastic box.

For sterilization of glassware, working tables, nutrient media and plant materials different techniques are applied. The glassware (culture tubes or Erlenmeyer conical flasks) containing sterilized nutrient medium after inoculation with plant material, are plugged with non-absorbent sterile cotton, and finally kept in growth chamber. Environmental conditions such as temperature, moisture and light are controlled as needed. For cell suspension culture, orbital shaker or aeration instrument is placed in growth chamber. Generally temperature is set between 25--27°C. Environmental conditions may vary according to plant species and nature of experiment. Light intensity, quality and photoperiod (light and dark cycles) are also regulated. Other laboratory accessories are dissection microscope, compound microscope, centrifuge, first aid kit, etc. which are required during *in vitro* culture of plant material.

## 2. Nutrient Media Composition and Preparation

Vital activity of a cell is the absorption of nutrients through cell membrance and rapid proliferation into inumerable cells. P.R. White in 1934 observed the unlimited growth of isolated root tissues when provided with nutrient medium containing inorganic salts, sucrose, vitamins, growth hormone and a few amino acids.

**(*a*) Inorganic Minerals :** Inorganic nutrients include macronutrients (*e.g.* nitrogen, phosphorus, potassium, calcium, magnesium and sulphur) in the form of salts in large amount and microelements (*e.g.* boron, molybdenum, copper, zinc, manganese, iron and chloride). A concentrated stock solution is prepared in advance and finally added to medium as required. To overcome the problem of solubility, the stock solution of iron is prepared in a chelated form as the sodium salt of ferric ethylenediamine tetra acetate (Fe-EDTA).

***(b)* Growth Hormones :** Several growth hormones are known which stimulate the biological activity in cultured materials.

Cytokinins promote cell division and regulate growth and development similar to kinetin (6-furfuryl aminopurine). Auxin resembles indole acetic acid (IAA) and stimulates shoot elongation. Gibberellins are of less importance, however, $GA_3$ is used in apical meristem. The most widely used cytokinins are adenin, kinetin, zeatin, benzyladenin; and auxins are IAA, NAA ($\alpha$ -Naphthalene acetic acid), 2,4-D. For the induction of callus the amount of kinetin should be 0.1 mg/litre. Structural formulae of cytokinins and auxins are given in Fig. 12.1.

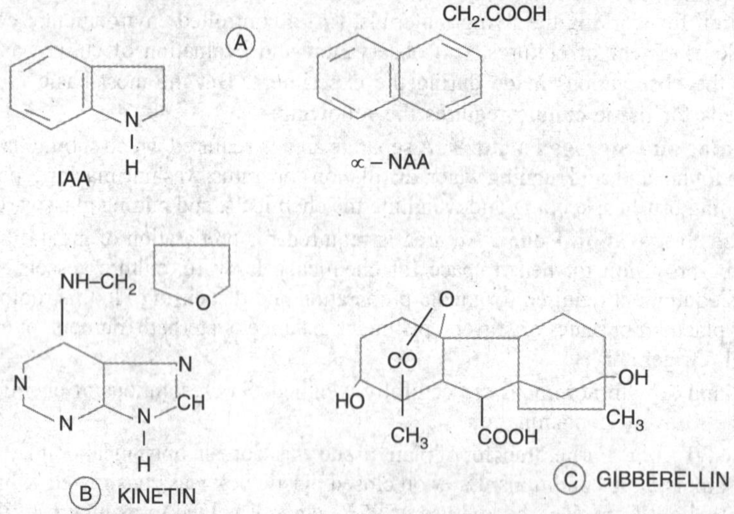


**Fig. 12.1.** Chemical structure of growth hormones. A-auxin; B-Cytokinin; C-Gibberellin $A_3$.

***(c)* Organic Constituents :** The organic compounds serve a source of carbon and energy. They are used in high concentration, *e.g.* 20-30g/litre. Sucrose and D-glucose (carbohydrates) are commonly used; but glycerol and myoinosital are also the principal source of carbon. Other complex organic compounds are peptone, yeast extract, malt extract, coconut water, tomato juice, etc.

***(d)* Vitamins :** Vitamins are required in trace amount as they catalyse the enzyme system of the cells. Vitamin $B_1$ (thiamine) is the most commonly used vitamin for all plant tissue cultures. Other groups of vitamins which stimulate growth are niacine (nicotinic acid) vitamin $B_2$ (riboflavin) vitamin $B_6$ (pyrodoxin), vitamin C (ascorbic acid), vitamin H (biotin) and vitamin $B_{12}$ (cyanocobalamin).

***(e)* Amino Acids :** Although nitrogen sources are present in the inorganic salts, yet various amino acids and amides are used in plant tissue culture media. The most widely used amino acids are L-aspartic acid, L-asparagin, L-glutamic acid, L-glutamine and L-arginin.

***(f)* Solidifying Agents:** Most commonly agar (a polysaccharide obtained from a seaweed *i.e.* a red alga, *Gelidium amansii)* is used as solidifying or gelling agent. Agar gels do not react with constituents of media and not digested by plant enzymes. Generally 0.5 -1% agar is used to form gel. Before use of agar, gelatin (10%) had been used as gelling agent. Demerit of gelatin is that it melts at low temperature (25%C)

***(g)* pH:** The pH affects the uptake of ions. Optimum pH between 5.0 to 6.0 is required for growth and development of cultured tissues. Therefore, optimum pH of the medium should be maintained before sterilisation of the medium.

### 3. Maintenance of Aseptic Environment

During *in vitro* culture maintenance of aseptic environment is the most difficult task. Because the cultures are easily contaminated by fungi and bacteria present in the air. The contaminants produce toxic metabolites which inhibit growth of cultured plant tissues. Therefore, each stem must be handled aseptically and with great care. Following are some of the sterilisation methods for aseptic manipulation of plant tissues. However, detailed account of sterilisation has been given in Chapter 10.

**(*a*) Sterilisation of Glassware:** Glassware (Petri platcs, vials, culture tubes, flasks, pipettes, etc.) , metallic instruments are sterilised in a **hot air oven** at 160-180° C for 2-4 hours.

**(*b*) Sterilisation of Instruments:** The metallic instruments (*e.g.* forceps, scalpels, needles, spatulas, etc.) are flame sterilised i.e. dipping them in 75% ethanol followed by flaming and cooling. It is called **incineration.**

**(*c*) Sterilisation of Culture Room and Transfer Area:** Floor and walls of culture room are washed first with detergent then 2% sodium hypochlorite or 95% ethanol. Larger surface area is sterilised by exposure to UV light. The cabinet of laminar airflow is also sterilised by exposing UV light for 30 minutes and 95% ethanol 15 minutes before beginning of work inside the cabinet of laminar airflow. The UV radiation is harmful to eyes. Therefore, UV light should be exposed when no experiment is in progress and no person is allowed to see UV light.

**(*d*) Sterilisation of Nutrient Media:** Culture media are properly dispensed in glass container, plugged with cotton or sealed with plastic closures and sterilised by autoclaving (steam sterilisation) at 15 psi (that gives 121°C) for 30 minutes. Minimum time required for autocalving of nutrient media is given in Table 12.1.

During autoclaving vitamins, plant extracts, amino acids and hormones are denatured. Therfore, the solution of these compounds are sterilised by using millipore filter paper which has 0.2 µm pore diameter.

**> Table 12.1.** Minimum time required for autoclaving of nutrient media at 15 psi.

| *Volume (ml)* | *Sterilisation time (minutes)* |
|---|---|
| 1-200 | 15 |
| 200-1000 | 30 |
| 1000-2000 | 40 |

**(*e*) Sterilisation of Plant Materials :** Surface of all plant materials have microbial contaminants. Therfore , **disinfectants** (*e.g.* sodium hypochlorite, hydrogen peroxide, mercuric chloride, or ethanol) should be used to make plant materials sterile. Then the chemicals must be washed 6–8 times using sterile distilled water. Then explants are transferred aseptically on nutrient medium inside the cabinet of laminar airflow.

## C. METHODS OF PLANT CELL, TISSUE AND ORGAN CULTURE

There is a little variation in the methods of plant cell, tissue and organ culture, but the basic steps are almost the same.

### 1. Basic Steps

The basic steps for regeneration of whole plants form an explant or cells as shown in Fig. 12.2 and described belwo:

**(*i*) Preparation of Suitable Nutrient Medium :** Suitable nutrient medium as per objective of culture is prepared and transferred into suitable containers (*e.g.* flasks, Petri plates, culture tubes) and autoclaved at 15 psi (pound per inch square ) for 30 miuntes. The hormones and vitamins are sterilised using millipore filter..

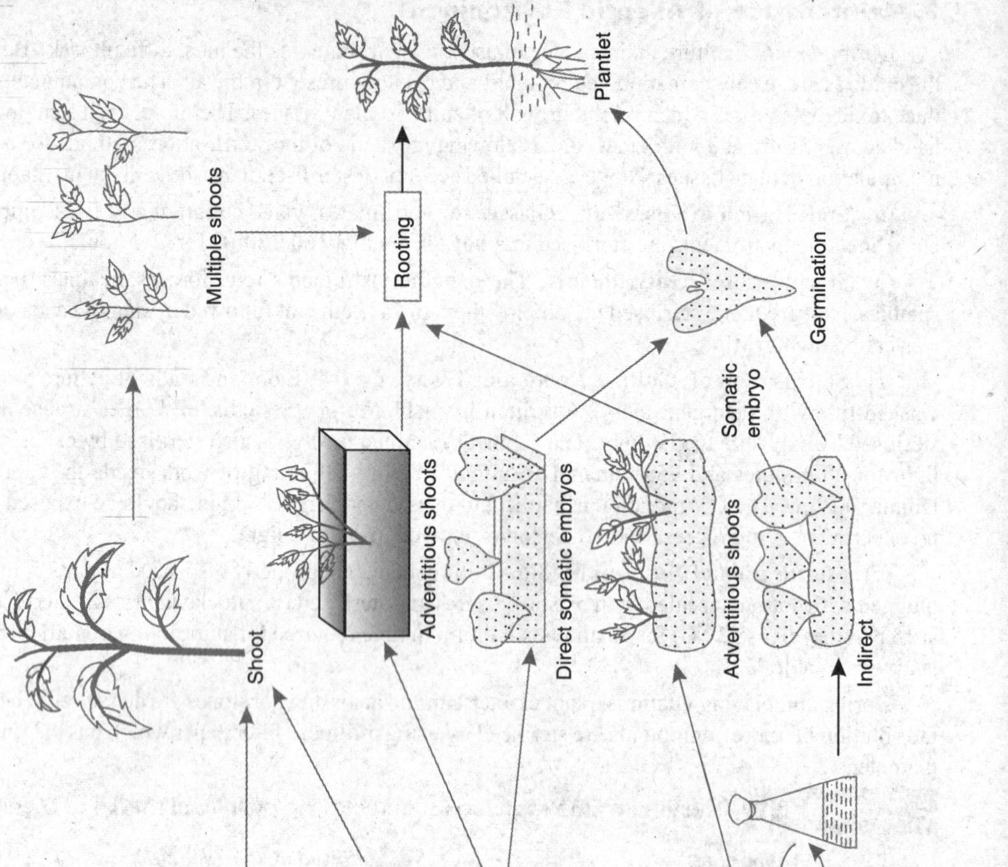


**Fig. 12.2.** Steps for regeneration of a plant by tissue culture method.

***(ii)* Selection of Explants:** Explants are any excised part of plant to be used in tissue culture, for example, axillary bud, leaf and stem segmenats, root tip, shoot tip, anther, ovary, endosperm, etc. Always young and healthy parts of the plants are selected as expalnt.

***(iii)*** **Sterlisation of Explants:** Sterilisation is the complete eradication of microorganisms present on the surface of any material. The explants are sterilised by disinfectants (*e.g.* sodium hypochlorite –NaOCI, mercuric chloride – $H_gC1_2$) and washed aseptically for 6-10 times with sterilised distilled water.

***(iv)*** **Inoculation (Transfer):** The sterile explant is inoculated on the surface of solidified nutrient medium under aseptic conditions. The cabinet of laminar airflow provides the sterile conditions (See Fig.10.1)

***(v)*** **Incubations:** The cultures are incubated in the growth chamber/tissue culture room at 25 ± $2^0$C, 50 – 60% relative humidity and 16 hours of photoperiod (*i.e.* light and dark regime is created artificially in growth chamber). After defined period callus develops on the medium or shoots/ roots develop from explant.

***(vi)*** **Regeneration:** Plantlets regenerate after transferring a portion of callus onto another medium and induction of roots and shoots or directly from expalnts.

***(vii)*** **Hardening :** Hardening is the gradual exposure of plantlets for acclimatisation to environmental conditions.

***(vii)*** **Plantlet Transfer :** After hardening process plantlets are transferred to green house or field conditions.

## D.TYPE OF CULTURES OF PLANT MATERIALS

There are different types of cultures which are produced through cultured plant material *e.g.* explant culture, callus culture, cell or suspension culture, protoplast culture, organ culture.

### 1. Explant Culture

There are a variety of forms of seed plants viz., trees, herbs, grasses, which exhibit the basic morphological units *i.e.* root, stem and leaves. These again vary with differences in cells and tissues, and their topography.

Parenchyma is the most versatile of all types of tissues. They are capable of division and growth. However, development of a tissue is characterized by three types of cell growth: cell division, cell elongation and cell differentiation. For this purpose, the explant from healthy and young part of the plant is used. Presence of parenchyma is first consideration in a particular species. Parenchyma from stems, rhizomes, tubers, roots is easily accessible and will generally respond quickly to culture conditions *in vitro*. However, parenchyma cells have an identical morphogenetic potential.

Explant cultures are the cultures of plant material (Fig.12.3). Any part of plant may be explant such as young and healthy pieces of leaf, stem hypocotyl, cotyledons, etc. Explant cultures are generally used for induction of callus or regeneration of plant.

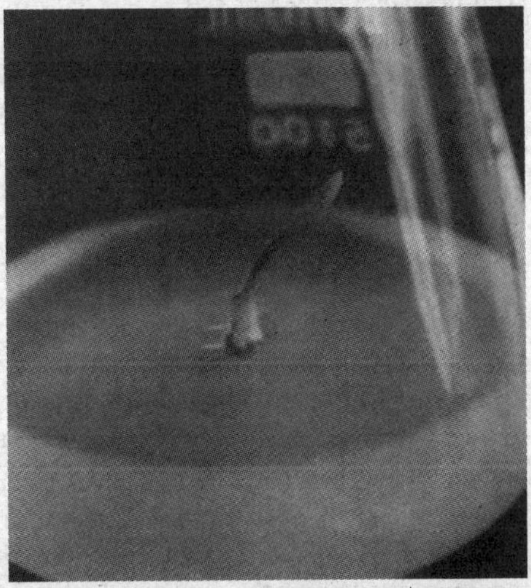

**Fig. 12.3.** Explant culture (Courtesy: Dhar and co-workers, G.B. Pant Institute of Himalayan Research and Development, Almora)

### 2. Callus Formation and its Culture

In nature, callus develops by infection of microorganisms from wounds due to stimulation by endogenous growth hormones, the auxins and cytokinins. However, it has been artificially developed by adopting tissue culture techniques. Explant, a 2-5 mm sterile segment, excised from a stem, tuber or root is transferred into nutrient medium and incubated at 25-28°C in an alternate light and dark regime of 12 h. Nutrient medium supplemented with auxins induces cell division. Soon the upper surface of explant is covered by callus. A callus is an amorphous mass of loosely arranged thin walled parenchyma cells developing from proliferating cells of the parent tissue (Dodds and Roberts, 1985). The unique feature of callus is that the abnormal growth has biological potential to develop normal root, shoots and embryoids ultimately forming a plant. (Fig. 12.4).

**Fig.12.4.** Callus induction (Courtesy : Dhar and co-workers, G.B. Pant Institute of Himalayan Research and Development, Almora)

Callus formation is governed by the source of explant, nutritional composition of medium and environmental factors. Explants of meristematic tissues develop cells more rapidly than thin three walled and lignified cells of tissue. Callus is formed through three developmental stages: induction, cell division and differentiation.

During induction metabolic rate of cells is stimulated, duration of which depends on physiological status, and nutritional and environmental factors. Owing to increased metabolic rate, cells accumulate high contents of nutrients and finally divide to form many number of cells. Cellular differentiation and expression of certain metabolic pathways start in the third phase leading to secondary products. Some times callus appears of different colours, for example, yellow, white, green or red.

Within the cell population of callus, the genetic instability results in variations in phenotypes which may be attributed to developmental (epigenetic) or genetic basis. Epigenetic changes involve selecting gene expression. They are stable and heritable at cellular level.

When callus has grown on nutrient medium after a long time it becomes essential to subculture it within 28 days on a fresh medium. Otherwise there develops nutrient depletion in original medium which results in paucity of water and accumulation of toxic metabolites.

### 3. Organogenesis

Root, shoot and leaves (but not embryo) are the organs that are induced in plant tissue culture. Since embryo is an independent structure and does not have vascular supply, it is not supposed to be the plant organ. Organogenesis (*i.e.* development of organs) starts with stimulation caused by the chemicals of medium, substances carried over from the original explants and endogenous compounds produced by the culture.

F. Skoog in 1944 for the first time indicated that the organogenesis could be chemically controlled. He observed root initiation (rhizogenesis) and shoot inhibition (caulogenesis) after addition of auxin to the medium. In 1957, Skoog and co-workers gave the concept of regulation of organogenesis by a balance between cytokinin and auxin. A high ratio of auxin: cytokinin stimulates the formation of root in tobacco callus, but a low ratio of the same induced shoot formation. The hypothesis of organogenesis was advanced, it explains that organogenesis in callus starts with the development of a group of meristematic cells, *i.e.* meristemoids that can respond to the factors within the system to initiate a primordium which, depending on kinds of factors, induces either root, shoot or embryoid. (Table 12.2). Shot formation in a cactus from callus is shown in Fig. 12.5.

**> Table 12.2.** ***In vitro* control of organogenesis by auxins and cytokinins.**

| *Auxin* (mg/l) (mg/l) | *Cytokinin* (mg/l) | *Organogenesis* |
|---|---|---|
| 0.0 | 0.2 | No growth |
| 0.03 | 1.0 | Shoots |
| 3.0 | 0.02 | Roots |
| 3.0 | 0.2 | Callus |

*Source* : Nandi and Palni (1992)

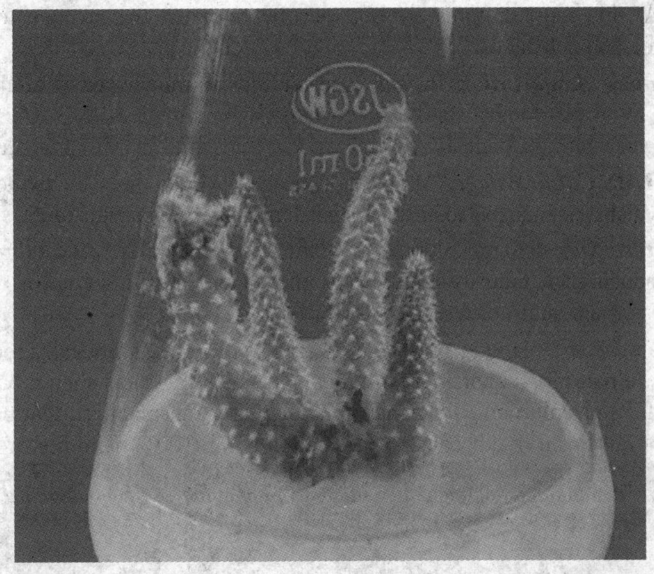

Fig. 12.5. Organogenesis; shoot formation in a cactus from callus.

## 4. Root Culture

Generally excised root is cultured in liquid medium. It has several advantages over solid media. The techniques of root culture give certain important informations such as : (*i*) nutritional requirements, (*ii*) infection by *Rhizobium* and nodulation, and (*iii*) physiological activities, for example, production of alkaloids, nicotine, etc. (Dodds and Roberts, 1985). For the first time, P.R. White in 1934 reported the successful organ culture *e.g.* potentially unlimited growth of excised tomato roots. Subsequently, roots of several species of gymnosperms and angiosperms have been successfully cultured.

## 5. Shoot Culture and Micropropagation

In shoot culture, apical meristem (the region of shoot apex laying distal to leaf primordium) is cultured. It clearly differs from shoot apex by having shoot apex and a few leaf primordia. Because of culture of relatively large tips (5-10 mm long sections), this technique is also known as meristem culture, meristemming and mericlones. Shoot tip culture is extensively applied in horticulture, agriculture and forestry. In 1960, G. Morel pioneered the work of shoot apex culture of orchid, *Cymbidium,* for the clonal multiplication. Among the workers Murashige significantly contributed for establishing micropropagation technique and its further biotechnological application. Because of minute size of the propagules in culture, the *in vitro* propagation technique is known as micropopagation.

In 1947, T. Murashige has described the procedure of development of micropropagation into three different developmental stages: Stage I, establishment of explant aseptically; stage II, multiplication of propagules by repeated subcultures on a specific nutrient medium; and stage III, rooting and hardening of plantlets and planting into soil.

In 1987, R.A. Fossard have described the following four stages of micropropagation:

***Stage I*** : Selection of a suitable explant and inoculation into nutrient medium;

***Stage II*** : Multiplication and growth of culture which take about 2 months, followed by repeated subculturer. At this stage rootless (with apical dominance, AD) cultures are obtained. In contrast, in fast multiplying cultures rootless multishooted (MS) cultures are also developed;

***Stage III*** : At this stage cultures are obtained by changing the medium to planting out in the following three sub-substages:

(*a*) ***Microcutting Stage (MC):*** In this case, culture is transferred to multishoot inducing medium as to get longer shoot and to harvest by MC. Finally MC is subcultured on medium for rooting. Stage II AD cultures also respond to MC procedure;

(*b*) ***Stage-III MS Culture Stage:*** At this stage, stage II MS cultures are divided into individual shoots which are induced to form apical dominant shoots with roots;

(*c*) ***MS Culture Stage:*** Small clumps of culture from stage II MS culture are transferred to AD inducing medium to get a bushy plant. Generally this type of plant is preferred for many ornamental horticulture plants;

***Stage IV*** : This is the planting out stage where plantlets are aseptically removed from test tube environment to natural and harsh environment. At this stage roots should be fully functional in potting mix (the soil environmental where plantlets are transplanted). During this procedure generally plantlets fail to survive because of desication (from 100% humidity of test tubes to low

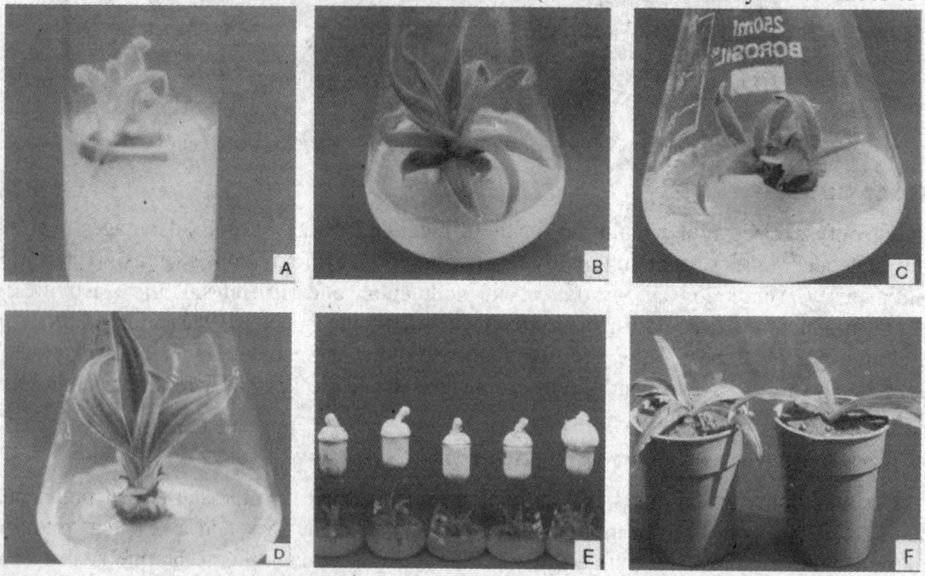


**Fig. 12.6.** Different stages in micropropagation of Ratanjot. **A.** explant; **B.** elongation; **C.** multiplication; **D.** rooting; **E.** rooted plantlets; **F.** acclimatisation (Courtesy: U. Dhar and co-workers; **G.** B. Pant Institute of Himalayan Environment and Development. Almora).

humidity under ambient conditions), harsh environment, invasion of soil microorganisms, unadjustibility from dependent (artificial medium) to independent nutrition (by photosynthesis).

Mature tissue impregnated with high amount of phenolic compounds is difficult to culture. Therefore, oxidation of these phenolics is necessary before culturing them. C.Y. Hu and P.J. Wang in 1983 have suggested for: (*i*) adding antioxidants to the medium, (*ii*) pre-soaking of explants in antioxidant solutions before culture, (*iii*) sub-culturing to a fresh medium, and (*iv*) providing light or no light during the start of culture.

In addition to nutritional composition, light and temperature are equally important in propagation. For morphogenesis and biosynthesis of chlorophylls, fluorescent light is necessary. For stage I and II light intensity of 1,000 lux has been found best, and for stage III it was recorded between 3,000-10,000 lux.

A medium that does not initiate callus formation is thought to be suitable because the genetic instability of callus will lead to a high degree of genetically aberrant plants (Dodds and Roberts, 1985). Chemical constituent that favour callus development should be avoided. The source of auxin is usually IAA, $\alpha$-NAA and IBA (indole-3-butyric acid). The auxin and 2, 4-D should be discarded as they stimulate callus formation and inhibit organogenesis. Exogenous GA3 (0.1 m1/ litre) is required for the development of shoot tips of isolated potato, carnation, *Chrysanthemum* and *Dahlia* and their subsequent micropropagation.

At planting out stage, it is necessary to develop acclimatization capability in plantlets before removing them from the test tubes. This can be done by (*i*) induction to develop some normal and functional leaves, (*ii*) induction of functional roots, and (*iii*) exposing the *in vitro* cultures to harsh environment before first two weeks of planting out. R.A. Fossard in 1987 described the successful potting mix as peat alone, vermiculite alone, mixture of loam-peat, peat-perlite-vermiculite-ash, perlite-pulverized pine bark-peat-river sand, and perlite-vermiculite-sand (see Chapter 13). Different stages in micropropagation of Ratanjot is given in Fig. 12.6

In 1990, Usha Yadav and co-workers successfully cultured shoot tips and nodal explants of two tropical trees (*e.g. Morus nigra* and *Syzygium cuminii*) *in vitro*. From these explants multiple shoots and roots were proliferated by providing different nutrient conditions. Finally the regenerated plants were transferred into the soil. In a separate experiment, inflorescence (in S. *cumminii*) and fruits (in *M.nigra*) were also developed successfully *in vitro*.

## 6. Cell (Suspension) Culture

Cell suspension is prepared by transferring a fragment of callus (about 500 mg) to the liquid medium (500 ml) and agitating them aseptically to make them cells free. It is difficult to have suspension of single cell. However, the suspension includes single cell, cell aggregates (varied number of cells), residual inoculum and dead cells (Dodds and Roberts, 1985). In 1980, P.J. King has described that a good suspension consists of a high proportion of single cells than small cluster of cells. It is more difficult to have a good suspension than to find optimum environmental factors for cell separation.

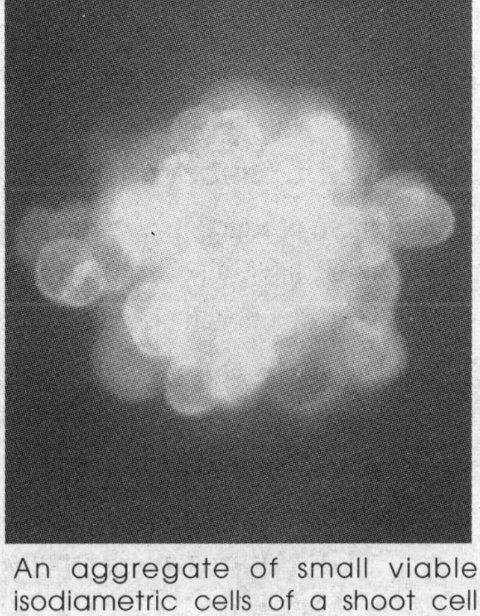

An aggregate of small viable isodiametric cells of a shoot cell suspension culture.

In 1997, P.J. King and H.E. Street described the techniques of cell separation by changing the nutritional composition of medium. No standard technique for separation of cells from callus has been recommended. When cells are transferred into a suitable medium they divide after lag phase (no cell division) and linearly increases their population. After some time, based on nutrient level, the rate of cell division decelerates until it comes to stationary phase (see Fig. 14.2) At this stage, to keep the cells viable, it is essential to subculture the cells. By using plate technique, cell lines can be raised where a mass of cells is spread over medium. Further, growth of plated cells depends on cell density.

In 1977, H.E. Street suggested that cell density should be determined before subculturing. Growth of culture depends on a critical cell density below which culture will not grow. For instance, for a clone of *Acer pseudoplatanus* 9–15x$10^3$ cells/ m1 are required. Generally cultures are agitated on orbital shaker or magnetic stirer for good result. Thus different types of works can be carried out by using cell suspensions.

The cultured cells of a higher plant is inherently dualistic. On one hand it possesses necessary genetic informations for existance on cell level (reproduction, growth, mature state, programmed death). On the other hand, the cultured cells retain supplementary informations that determine the production of substances which are important for integrative functions and biocoenotic interactions of the plants. The informations determining the progression to a programme for the development of a whole plant are also redundant for existance at cell level. Cell culture systems have been employed in numerous morphological analyses by varying the origin of cells and physicochemical factors.

The properties of cultured plant cell suspensions in common with culture of microorganisms are as: (*i*) they grow in sterile environment, (*ii*) they are homogeneous in size, (*iii*) they have a doubling time which is longer than that of microorganisms but considerably shorter than cells *in situ*, and (*iv*) they can be grown on a large scale.

The cell suspension cultures are used: (*i*) induction of somatic embryos and shoots, (*ii*) *in vitro* mutagenesis and selection of mutants, (*iii*) genetic transformation studies, (*iv*) production of secondary metabolites.

**(*a*) Benefits from Cell Culture :** Cell suspension cultures have many advantages over the callus cultures as : (*i*) the suspension can be pipetted, (*ii*) they are less heterogeneous and cell differentiation is less pronounced, (*iii*) They can be cultured in volumes upto 1,500 litres, (*iv*) they can be subjected to more stringent environmental controls, and (*v*) they can be manipulated for production of natural products by feeding precursors.

## 7. Somatic Embryogenesis

Embryo production is a characteristic feature of the flowering plants. However, such structures (embryoids) have also been artificially induced in cultured plant tissues, besides zygote.

H.W. Kohlehbach (1978) proposed the following classification of embryos:

(*a*) **Zygotic embryos-** It is formed by the zygote.

(*b*) **Non-zygotic Embryos-** It is formed from the cells other than zygote. It is of following types:

- (*i*) **Somatic Embryos-** It is formed from somatic cells *in vitro*
- (*ii*) **Parthenogenetic Embryos-** It is formed by unfertilised egg.
- (*iii*) **Androgenic Embryos-** It is formed by pollen grains.

Generally, somatic embryos called **embryoids** are similar to zygotic embryos (or seed embryos) except they originate from somatic cells and are larger in size.

Somatic embryogenesis can be initiated in two ways : (*i*) by inducing embryogenic cells within the preformed callus, and (*ii*) directly from pre-embryonic determined cell, (without callus) which are ready to differentiate into embryoids. In the first case, embryoids are initiated in callus from superficial cell aggregates where cells contain a large vacuole, dense cytoplasm, large starch granules and nucleus.

Two nutritional media of different composition are required to obtain embryoids. First medium contains auxin to initiate embryogenic cells. Second medium lacks auxin or reduced level of auxin is needed for subsequent development of the embryonic cells into embryoids and plantlets. In both the cases reduced amount of nitrogen is required. The embryogenic cells pass through 3 different stages *e.g.* globular, heart shaped and torpedo shaped, to form embryoids (Fig. 12.7).

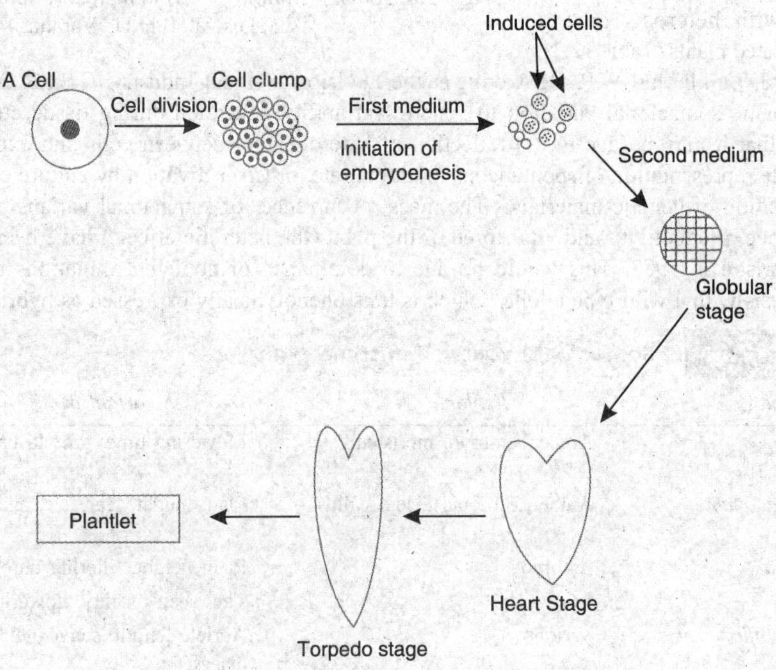


Fig. 12.7. Events of somatic embryogenesis.

These embryoids can be separated and isolated mechanically by using glassbeads. When embryoids reach torpedo stage they are transferred to filter paper bridge (a sterile and pluged culture tube containing about 10 ml MS liquid medium supplemented with Kinetin (0.2 mg/lit) and sucrose (2% W/V) on which Whatman No.1. Filter paper is placed to make a bridge (Dodds and Roberts, 1985). Some plants in which somatic embryogenesis has been induced *in vitro* are A*tropa belladona, Brassica oleracea, Carica papaya, Coffea arabica, Citrus cinensis, Daucus carrota, Nicotiana tabacum, Pinus ponderosa and Saccharum officinarum.* Somatic embryos produced in mango are shown in Fig.12.8.

## 8. Somaclonal Variation

The genetic heterogeneity of cells in a population represents continuity of genotypes, whereas phenotypically the population is represented as a discrete sum of subclones. After cloning a single cell, from the population of a strain of *Dioscorea deltoidea*, there developed a subpopulation that differed in their growth rate and sapogenin (diosgenin and yamogenin) content. The traits varied independently. In one clone of 12 studies a high growth rate was combined with a high diosgenin content.

**Fig. 12.8.** Somatic embryogenesis in mango (Courtesy: Prof. V.S. Jaiswal, B.H.U., Varanasi)

The genetic nature of variability in cultured cells explains the appearance of a large number of somaclonal variants with heterosis among the regenerated plants (Table 12.3). In 1981, P.J. Larkin and W.R. Scowcroft at the Division of Plant Industry, C.S.I.R.O., Australia gave the name somaclonal variation to genetic variability generated during tissue culture. They explained that it may be due to: (*i*) reflection of heterogeneity between cells and explant tissue, (*ii*) a simple representation of spontaneous mutation rate, or (*iii*) activation by culture environment of transposition of genetic materials. The mass occurrence of somaclonal variants, increase in the resistance, productivity and vital force of the plant (the heterotic effects) have been explained by P.S. Carlson in 1983. This would be due to dominance of nonlethal mutations that lead to heterozygosis with a wild type allele which is thus phenotypically expressed as hybrid heterosis.

**Table 12.3. Somaclonal variation in some plants.**

| *Species* | *Explants* | *Variant traits* |
|---|---|---|
| *Brassica* spp. (Brassicaceae) | Anther, embryo, meristem | Flowering time, growth habit, wax |
| *Nicotiana tabacum* (Solanaceae) | Anther, protoplast, leaf callus | Plant height, leaf size, alkaloids |
| *Oryza sativa* (Poaceae) | Embryo | Plant height, tillering number, panicle size, seed fertility, flowering date |
| *Saccharum officinarum* (Poaceae) | Various | Auricle length, sugar yield, pathogenic disease |
| *Solanum tuberosum* (Solanaceae) | Leaf callus, protoplast | Plant habit, disease resistance and shape, yield and maturity date of tubers |
| *Triticum aestivum* (Poaceae) | Immature embryo | Plant height, spike shape, maturity tillering, $\alpha$-amylase, leaf wax |
| *Zea mays* (Poaceae) | Immature embryo | Endosperm and seedling mutant, mtDNA sequence rearrangement |

*Source* : W.R. Scowcroft (1984).

One of the most complex and difficult problems involving the population of cultured cells is the problem of retention and expression of the trait of totipotency *i.e.* the ability to realize the programme of development from cell to plant. The genetic variability of cultured cells is the basis for obtaining somaclonal variants of the plants that have valuable traits. But this variability makes it difficult to realize the results of cell and genetic engineering as the trait may be lost in the row of cell generations leading to the regeneration of the plant. Stability is an essential requirement for clonal micropropagation and stabilizing selection in plant breeding.

Most significant achievement of somaclonal variation was made by Shepard and co-workers in old variety of potatoes. In 1980, J.E. Shepard and coworkers screened about 100 somaclones produced from leaf protoplasts of Russet Burbank and found a significant and stable variation in compactness of growth habit, maturity, date, tuber uniformity, tuber skin colour and photoperiodic requirements. The characters of greater tuber uniformity and early onset of tuberization were agronomic improvements over the parent variety. Moreover, somaclonal variation is applicable for seed propagated plants only *e.g.* rice, wheat, maize, tobacco, *etc*. not for vegetatively propagated species.

Thus, somaclonal variation has proved an alternative tool to plant breeding for generating new varieties that can exhibit disease resistance and improvement in quality and yield in plants such as cereals, legumes, oil seeds, tuber crops, fruit crops, *etc*.

In 1984, P.J. Larkin and co-workers have performed a detailed genetic analysis of wheat somaclonal variation, coupled with qualitative and quantitative assessment for yield and other attributes of commercial importance. The main findings are as below : (*i*) heritable variation is encountered for characters *e.g.* height, awns, tiller number, grain colour, gliadin protein and amylase regulation, (*ii*) besides morphological variation, numerical and structural aberration in chromosome occur, (*iii*) morphological variations may not be correlated with chromosomal variation, (*iv*) the occurrence of translocations has a major scope for exploitation by introgression of genes from wild relatives by culture of immature embryos of wide crosses, and (*v*) the somaclonal variation holds great scope for creating utilizable germplasm for crop improvement.

In Hawaii, tissue culture is integrated in the breeding programme and somaclonal variation has been exploited for the isolation of clones resistant to certain diseases, for example, viruses, downy mildew and eye spot disease. Some of the resistant clones have advantages in having higher sugar content and yield than parental ones.

In India, Sugarcane Breeding Institute, Coimbatore has released varieties produced through the process of somaclonal variation.

**Bio-13.** In India, a somaclonal variant of *Citronella java,* a medicinal plant has been released as 'Bio-13' for commercial cultivation by Central Institute for Medicinal and Aromatic Plants (CIMAP), Lucknow. Bio-13 yields 37% more oil and 39% more citronellon than the control variant.

**Supertomatoes.** Heinz Co. and DNA Plant Technology Laboratories (USA) developed Supertomatoes with high solid component by screening somaclones which reduced shipping and processing costs. According to an estimate made in 1986, in the USA alone, for every percent increase in tomato solid processors would save about 100 million dollars per annum.

**Oxalate removal.** Through selection of somaclonal variants, nutritional stress factors have been estimated in some vegetables. First application of this technique was shown in *Amaranthus gangeticus,* a leafy vegetable.

## 9. Protoplast Culture

Protoplasts (cell minus cell wall) is the biologically active and most significant material of cells. When cell wall is mechanically or enzymatically removed the isolated protoplast is known as "naked plant cell" on which most of recent researches are based.

Plant cell wall acts as physical barrier and protects cytoplasm from microbial invasion and environmental stress. It consists of a complex mixture of cellulose, hemicellulose, pectin, lignin, lipids, protein, etc. For dissolution of different components of the cell wall it is essential to have the respective enzymes.

**(*a*) Isolation of Protoplasts :** Until suitable methods were developed, protoplasts were isolated by cutting the plasmolysed plant tissues and releasing protoplast through deplasmolysis of cells. In 1960, E.C. Cocking for the first time isolated the protoplasts of plant tissues by using cell wall degrading enzymes viz., cellulase, hemicellulase, pectinase, and protease extracted from a saprophytic fungus *Trichoderma viride.* Later on protoplasts were cultured *in vitro.*

Microorganisms are well equipped with a system to produce substrate specific extracellular enzymes, the extent of which depends on the genetic variability of the specific species and strains. However, the basic techniques of isolation and culture of protoplast are given in Figs. 12.9. and 12.11 with a brief description.

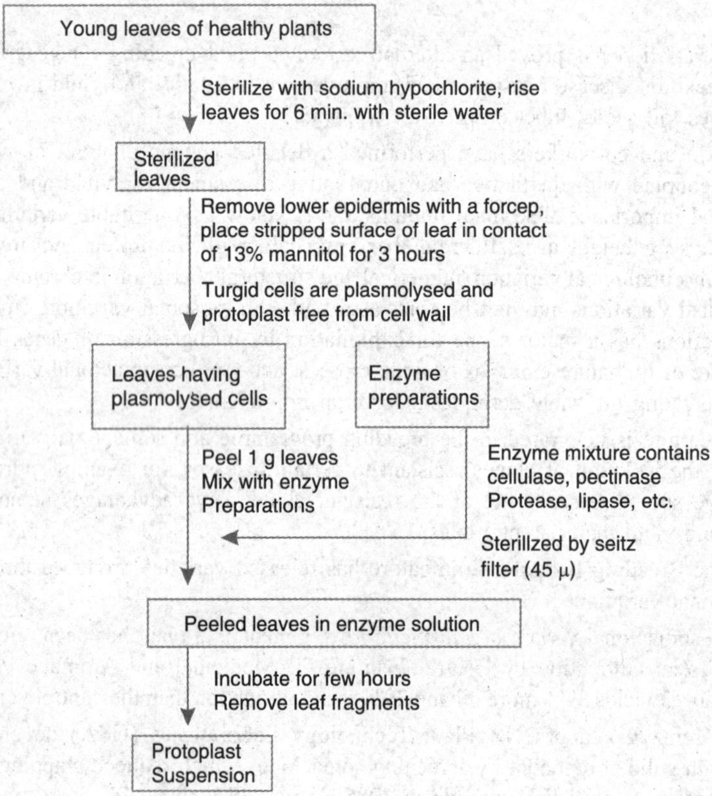


**Fig. 12.9.** Methods of isolation of protoplasts.

**(*i*) Surface sterilization of leaf Samples :** Mature leaves are collected from healthy plants which are washed in tap water to remove adhering soil particles and sterilized with sodium hypochlorite solution.

**(*ii*) Rinsing in suitable osmoticum :** After 10 min, sample is properly washed with sterile distilled water or MS medium adjusted to a suitable pH and buffer to maintain osmotic pressure. Washing should be done for about 6 times to remove the traces of sodium hypochlorite.

**(*iii*) Plasmolysis of cells :** The lower epidermis covered by thin wax cuticle is removed with a forcep. Stripping should be done from midrib to margin of lamina. The stripped surface of leaf is kept in mannitol solution (13% W/V) for 3 hours to allow plasmolysis of cells.

**(*iv*) Peeling of lower epidermis :** Thereafter, about 1 gm leaves are peeled off and transferred into enzyme mixture already sterilized through a Seitz filter (0.45μm). This facilitates the penetration of enzyme into tissue within 12-18 hours at 25ºC.

**(*v*) Isolation and purification of protoplasts :** Leaf debris are removed with forcep, and enzyme solution containing protoplasts are filtered with a nylon mesh (45 μm). Filtrate is centrifuged at 75 X g for 5 min and supernatant is decanted. Again a fresh MS medium plus 13% mannitol is added to centrifuge. Repeated washing with nutrient medium, centrifugation and decantation are done for about three time. Finally specific concentration of protoplast suspension is prepared. Protoplasts isolated from leaf mesophylls are given in Fig. 12.10.

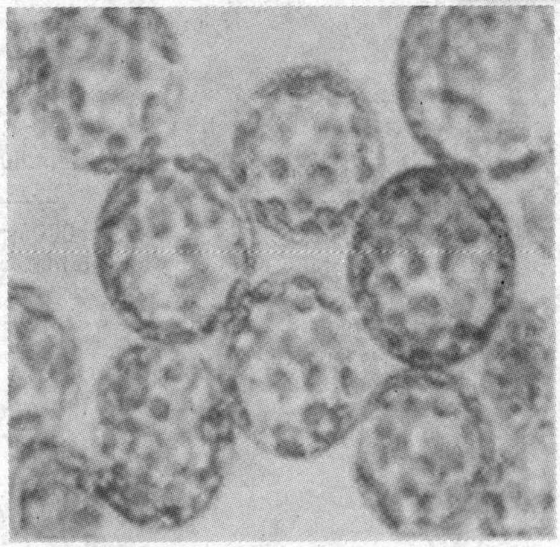

**Fig. 12.10.** Protoplasts isolated from leaf cells.

**(*b*) Protoplast Culture and Regeneration :** From the protoplast solution of known density (about $10^5$ protoplast/ml) about 1 ml suspension is poured on sterile and cooled down nutrient medium in Petri dishes. The plates are incubated at 25ºC in a dim white light.

The protoplasts regenerate a cell wall, undergo cell division and form callus. The callus can also be subcultured. Embryogenesis begins from callus when it is placed on nutrient medium lacking mannitol and auxin. The embryo develops into seedlings and finally mature plants.

Examples of plant species that have regenerated from protoplasts are *Cucumis sativus, Capsicum annum, Ipomoea batata, Beta vulgaris, Helianthus annuus, Glycine max, Rosa sp., Chrysanthemum* sp., etc.

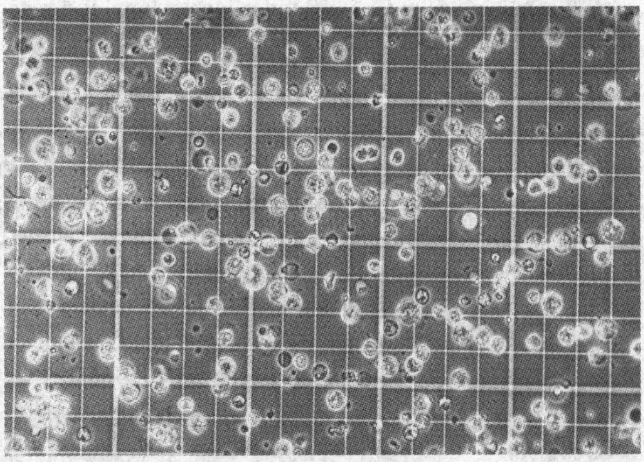

*Brassica* protoplasts on a grid.

The isolated protoplasts are used for various purposes such as: (*i*) biochemical and metabolic studies, (*ii*) fusion of two different somatic cells to get somatic hybrids, (*iii*) fusion of nucleated (containing uncleus) and enucleated (without nucleus) cells to produce **cybrid** (cytoplasmic hybrid), (*iv*) genetic manipulation, (*v*) drug sensitivity, etc.

## 10. Protoplast Fusion and Somatic Hybridization

With the development of techniques for enzymatic isolation of protoplasts and subsequent regeneration, a new tool of genetic manipulation of plants has now become available. Moreover, the fusion of protoplasts of genetically different lines or species has also been possible. For example, some plants that show physical or chemical incompatibility in normal sexual crosses, may be produced by the fusion of protoplasts obtained from two cultures of different species. Somatic hybridization of crop plants represents a new challenge to plant breeding and crop improvement. In the field of pest and disease resistance and transfer of $C_3$ photosystems into $C_4$ crop plants somatic crosses show most interesting promises (Dodds and Roberts, 1985).

The results obtained so far from somatic hybridization represent that it is possible to recover fertile and stable amphidiploid somatic hybrids after protoplast fusion. There are four major aspects of protoplast fusion as: (*i*) production of fertile amphidiploid somatic hybrids of sexually

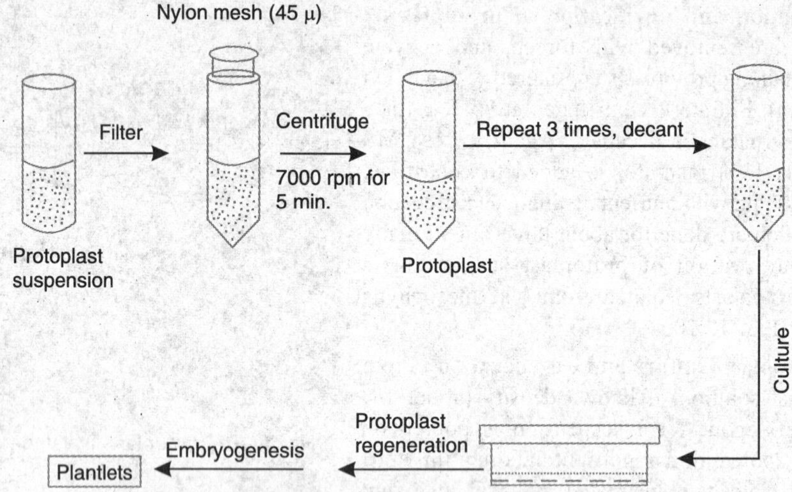


**Fig. 12.11.** Purification, culture and regeneration of protoplasts.

incompatible species is achieved. Induced fusion of protoplasts from two genetically different lines of species must result in a variety of homo-as well as heterokaryotic fusion products (heterokaryon or heterokaryocytes). Selection of a few true somatic hybrid colonies from the mixed population of regeneration protoplast is a key step in successful somatic hybridization technique; (*ii*) production of heterozygous lines within one plant species which normally will be propagated only vegetatively *e.g.* potato; (*iii*) the transfer of only a part of genetic information from one species to another using the phenomenon of chromosome elimination, and (*iv*) the transfer of cytoplasmic genetic information from one to a second line or species.

It has been possible to transfer useful genes (*e.g. nif* genes, disease resistance genes, rapid growth genes) from one species to another, thereby, to widen the genetic base for plant breeding.

**(*a*) Fusion Products—the Hybrids and Cybrids :** Fusion of cytoplasm of two protoplasts results in coalescence of cytoplasms. The nuclei of two protoplasts may or may not fuse together even after fusion of cytoplasms. The binucleate cells are known as heterokaryon or heterocyte (Fig. 12.12.) When nuclei are fused the cells are known as hybrid or synkaryocyte and when only cytoplasms fuse and genetic information from one of the two nuclei is lost is known as cybrid, *i.e.* cytoplasmic hybrid or heteroplast (Doods and Roberts, 1985). Production of cybrid (through fusion of nucleated and enucleated different somatic cells) is called **cybridisation,** while production of hybrid (by fusion of two nucleated cells) is called **hybridisation.** There are some genetic factors which are carried in cytoplasmic inheritance, instead of nuclear genes, for example, male sterility in

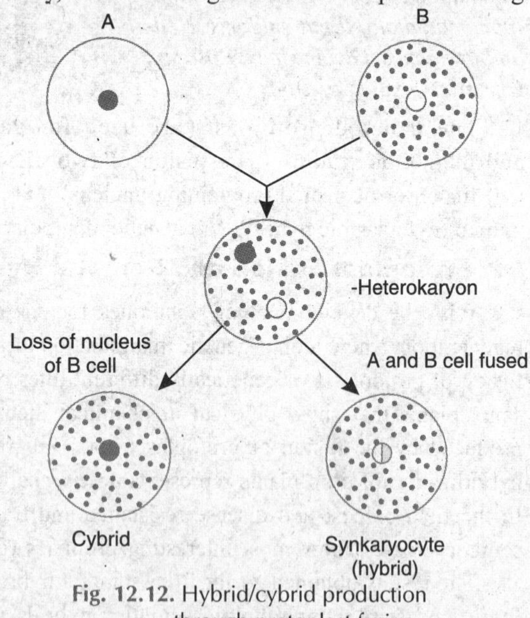


**Fig. 12.12.** Hybrid/cybrid production through protoplast fusion.

some plants. Susceptibility and resistance to some of the pathotoxins and drug are controlled by cytoplasmic genes. Therefore, production of cybrids which contain the mixture of cytoplasms but only one nuclear genome can help in transfer of cyloplasmic genetic information from one plant to another. Thus, informations of cybrid can be applicable in plant breeding experiments.

In China cybrid technology in rice is a great success. Such plants are very useful in producing hybrid seeds without *emasulation*, However, cybrid technology has successfully been applied to carrot, *Brassica* sp, *Citrus,* tobacco and sugar beet.

**(*b*) Methods of Somatic Hybridization:** Procedure for successful somatic hybridization is as below: (*i*) isolation of protoplasts from suitable plants, (*ii*) mixing of protoplasts in centrifuge tube containing fugigenic chemicals *i.e.* chemicals promoting protoplast fusion, such as polyethylene glycol (PEG) (20%, W/V), sodium nitrate ($NaNO_3$), maintenance of high pH 10.5 and temperature 37ºC (as a result of fusion of protoplasts viable heterokaryons are produced. PEG induces fusion of plant protoplasts and animal cells and produces heterokaryon, (*iii*) wall regeneration by heterokaryotic cells, (*iv*) fusion of nuclei of heterokaryon to produce hybrid cells, (*v*) plating and production of colonies of hybrid cells, (*vi*) selection of hybrid, subculture and induction of organogenesis in the hybrid colonies, and (*vii*) transfer of mature plants from the regenerated callus.

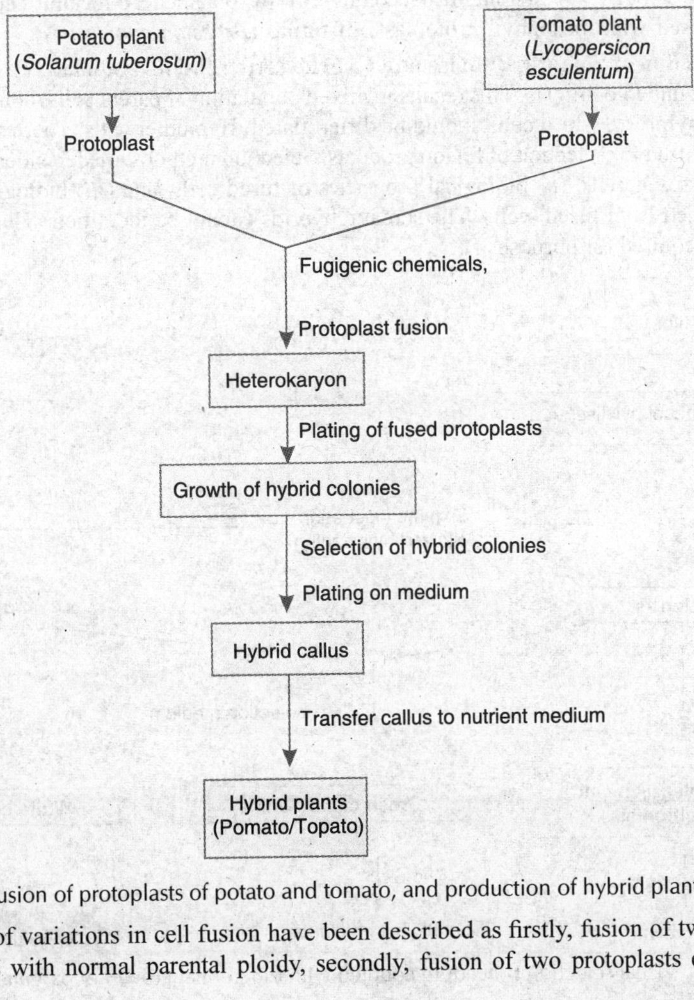


**Fig. 12.13.** Fusion of protoplasts of potato and tomato, and production of hybrid plant (pomato).

A number of variations in cell fusion have been described as firstly, fusion of two protoplasts each from cells with normal parental ploidy, secondly, fusion of two protoplasts each obtained from haploid cell sources (the fusion of two diploid cells of *Solanum* sp. would yield a tetraploid which is the normal ploidy for potato, and thirdly, fusion of a protoplast with an enucleated protoplast. The result of this type of fusion would bring about cybrid with a single nucleus of one parent and the respective ploidy.

Protoplast fusion of *Nicotiana glauca x N. langsdorffii* has been carried out successfully. In 1985, A.A. Kuchko achieved somatic hybrid through inter-specific cross of wild and cultivated variety of potatoes *i.e. Solanum tuberosum x S. chacoense.* Intergeneric hybrids which may be difficult to achieve by sexual crosses have been obtained through protoplast fusion *e.g.* tomato + potato, *Datura* + *Atropa*, barley + wheat, barley+ rice, wheat + oat, and sugarcane + sorghum. Potato+tomato cross is shown in text ( Fig. 12.13.). With regard to inter familial cell hybrids, there have been very few successful experiments to obtain such hybrids, for example, hybrids of *Glycine sp* + *Nicotiana* species. Based on detailed studies on inter familial hybrids, it has been concluded that inter familial hybrids of plant cells are genetically unstable and show species specific elimination/ reconstruction of chromosome belonging to one of the parents; this was usually the parent represented by the mesophyll cells in hybridization process. These hybrids are incapable of regeneration and morphogenesis into plants. Similarly, intertribal hybrids were studied more widely due to their higher genetic stability; the first intertribal cell hybrid was obtained from callus cells of *Arabidopsis thaliana* crossed with mesophyll protoplasts of turnip (*Brassica compestris*).

**(*c*) Selection of Somatic Hybrids and Cybrids:** After fusion protoplasts on media regenerate cell walls and undergo mitosis. This results in mixed population of parent cells, homokaryotic fusion product and hybirds. Hybrid cells should be differntiated from other cells. There are various selection methods used for selection of fusion products. Selection methods are dependent on: (*i*) physical properties of fused cells, (*ii*) biological properties of fused cells, and (*iii*) biological properties of colonies formed from fused cells. The somatic hybrids cannot be identified. Hence, biochemical markers are required for purpose.

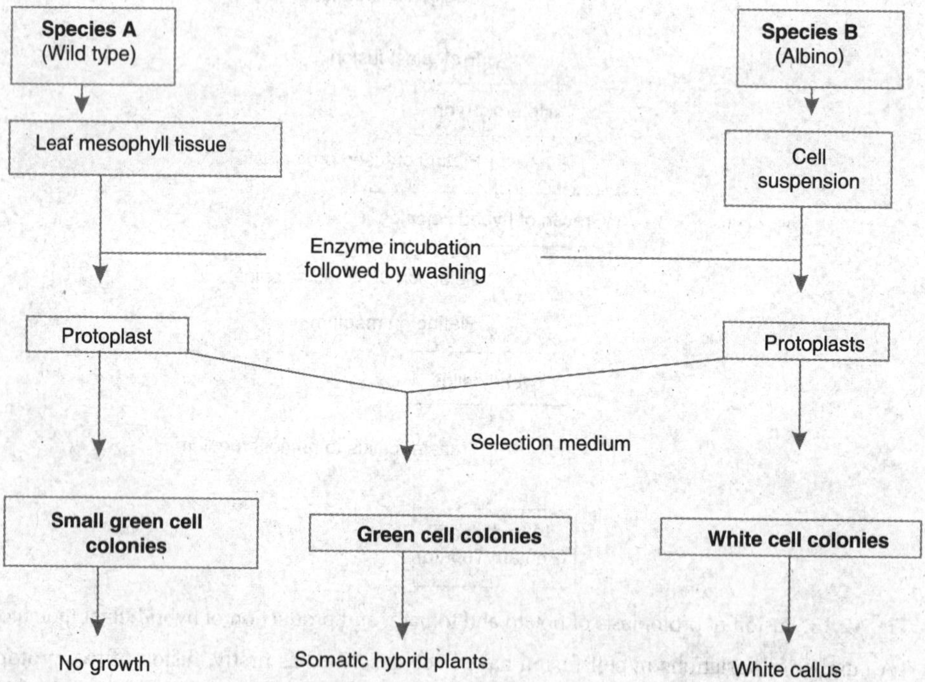


Fig. 12.14. Visual selection procedure coupled with differential growth of parental protoplasts.

In this 1972, for the first time P.S. Carlson and co-workers produced first intersperct somatic hybrids between *Nicotiana glauca and N.langsdorfii.* In 1978, Melchers and workers developed first intergenic somatic hybrids between potato (*Solanum thuberosum*) tomato (*Lycoperscon esculentum*). The somatic hybrids was called **pomato or topato.**

P.S. Carlson and co-worker (1978) used nutrition requirement of mesopto chloroplasts. They found that protoplasts of somatic hybrid grew and formed callus, parental types did not form calli.

(*i*) Besides, drug sensitivity test and auxotrophic mutant-based selection are applied.

(*ii*) In most of somatic hybridisation experiment, selection procedure includes full of chlorophyll-deficient (non-green) chloroplasts of one parent with the protoplasts of the other parents. This help in visual selection of heterokaryon under microscope (Fig.12.14)

(*iii*) Biochemical method is used for selection of hybrids and cybrids by staining some isozymes (*e.g.* esterases and peroxidases) and fluoresecent dyes (FACS). Radioimmunoassay techniques were employed in selection of high yielding lines of *Catharanthus* and *Anchusa* cells. It could be made more efficient by making automatic via using flow-cytometric sorting systems.

(*iv*) Fluorescent antibodies can also be used to detect the desired component present on the cell surface.

## 11. Anther and Pollen Culture (Production of Haploid Plants)

Anther, a male reproductive organ, is diploid in chromosome numbers. As a result of microsporogenesis, tetrads of microspores are formed from a single spore mother cell. They are known as pollen grains after release from tetrads (Bhojwani and Bhatnagar, 1974). The aim of anther and pollen culture is to get haploid plants by induction of embryogenesis. Haploid plants have single complete set of chromosomes that in turn may be useful for the improvement of many crop plants. Moreover, chromosome set of these haploids can be doubled by mutagenic chemicals (*e.g.* colchicine) or regeneration technique to obtain fertile homozygous diploids.

W. Tulecke in 1951 cultured pollen grains of *Ginkgo biloba* (gymnosperm) and succeeded to induce the development of haploid callus. In 1964, S. Guha and S.C. Maheswari made a remarkable discovery by culturing pollen grains of an angiospermic plant, *Datura innoxia* on the nutrient agar medium and also developed torpedo- shaped embryoids that metamorphosed into plantlets through the process.

The anthers to be cultured should be one of the three categories *i.e.* premitotic, mitotic and postmitotic. In premitotic anthers, where the microspores have completed meiosis but not started first pollen division, the best response is achieved *e.g. Hordeum vulgare.* In mitotic anthers where first pollen division has started the optimum responses are achieved *e.g. N. tabacum* and *D. innoxia.* In post mitotic anthers, the early bicellular stage of pollen development is the best time to culture *e.g. Atropa belladona.*

Haploid plants are very useful in: (*i*) direct screening of recessive mutation because in diploid or polyploid screening of recessive mutation is not possible, and (*ii*) development of homozygous diploid plants following chromosome doubling of haploid plant cells (Fig. 12.15).

In China, the most widely grown wheat is a doubled haploid produced through homozygous diploid lines. Anther culture of rice is also successfully grown. Haploid plants have been produced in tobacco, wheat and rice through pollen culture. These are used for the development of disease resistant and superior diploid lines. The Institute of Crop Breeding and Cultivation (China) has developed the high yielding and blast resistant varieties of rice **zhonghua No. 8** and **zhonghua No.9** through transfer of desired alien gene.

At present, more than 247 plant species and hybrids belonging to 38 genera and 34 families of dicots and monocots have been regenerated using anther culture technique. They include economically important crops and trees such as rice, wheat maize, coconut, rubber trees, etc.

**(*a*) Culturing Techniques :** Methods of anther culture are shown in Fig. 12.15. Anthers are superficially sterilized and washed with double distilled sterile water. They are excised from the flower buds and their proper developmental stages are determined under microscope. On confirmation of a stage, (*i*) the anthers are directly transferred on nutrient agar or liquid medium where induction of embryogenesis occurs, or (*ii*) the pollen grains are aseptically removed from the anthers and cultured on liquid medium.

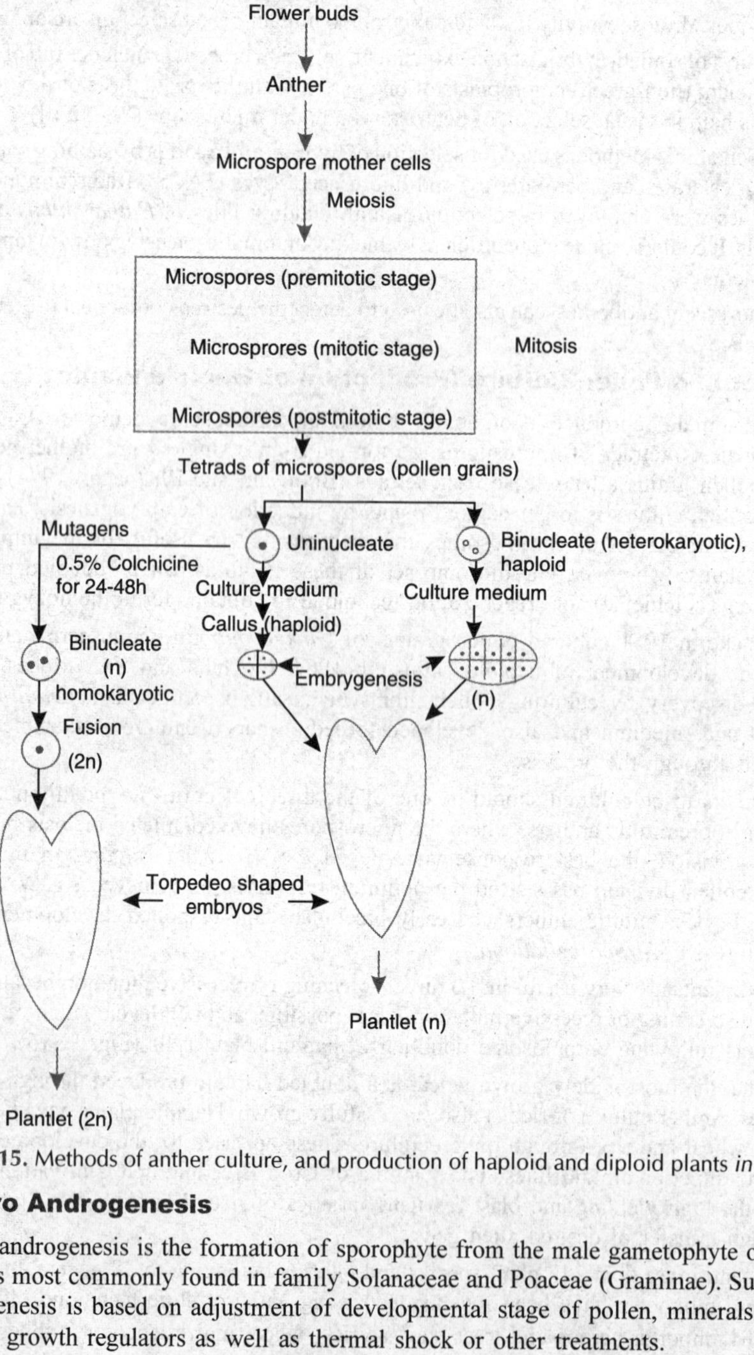


Fig. 12.15. Methods of anther culture, and production of haploid and diploid plants *in vitro*.

## 12. In vitro Androgenesis

*In vitro* androgenesis is the formation of sporophyte from the male gametophyte on artificial medium. It is most commonly found in family Solanaceae and Poaceae (Graminae). Success of *in vitro* androgenesis is based on adjustment of developmental stage of pollen, minerals of culture medium and growth regulators as well as thermal shock or other treatments.

Methods of *in vitro* androgenesis is given in Fig. 12.16. Pollen grains are isolated from excised anther and extracted in a beaker. Well collected pollens are washed properly and centrifuged and decanted. They are inoculated in liquid medium, then subcultured in solid MS medium. In culture pollen grains can be induced to produce callus or embryo from which whole plants are regenerated in one month. This technique is most successful in *Brassica, Datura, Petunia,* etc. Mainly there are two ways of *in vitro* morphogenesis of pollen grains, direct and indirect (Fig. 12.16)

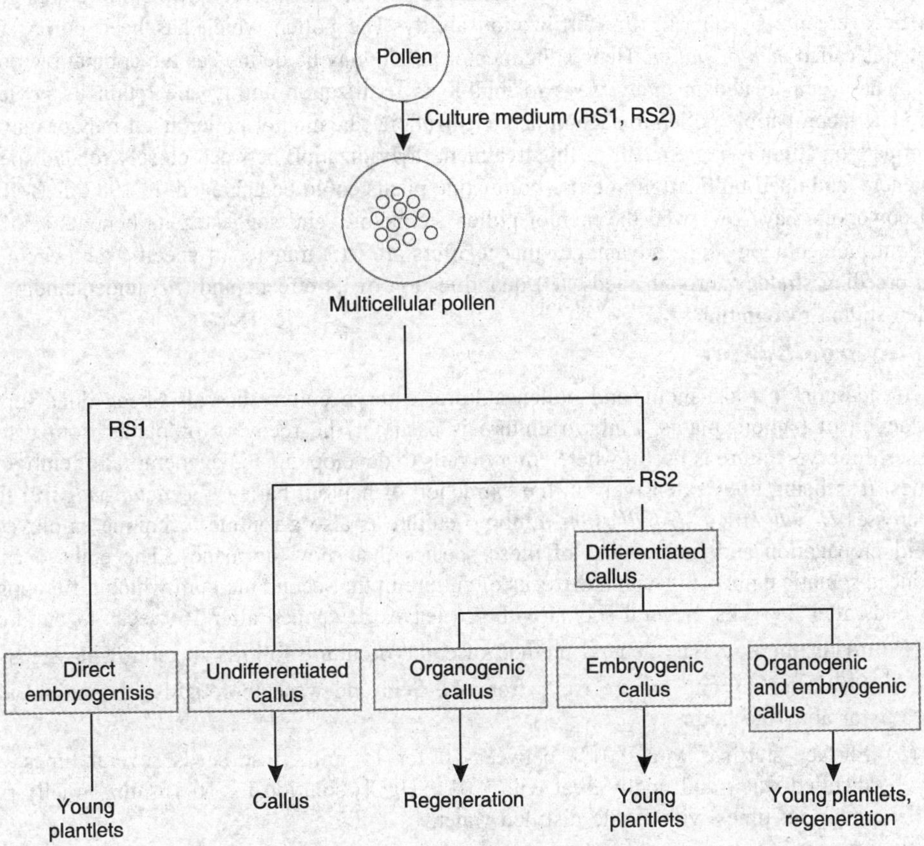


**Fig. 12.16.** Types of *in vitro* pollen morphogenesis in *Datura* (based on S. Sangwan, 1981)

**(*a*) Direct androgenesis.** Direct androgenesis is also called pollen derived embryogenesis. Here pollen directly acts as a zygote and, therefore, passes through various embryogenic stages similar to zygotic embryogenesis. When pollen grains has reached globular stage of embryo, wall of pollen is broken and embryo is released. The released embryo develop cotyledons, then the plants. Direct androgenesis is very common in many plants of the family Solanaceae and Brassicaceae (Fig. 12.16).

**(*b*) Indirect androgenesis.** In indirect androgenesis the pollen grains, instead of normal embryogenesis, divide erratically to develop callus (Fig. 12.16). Indirect androgenesis has been found in barley, wheat, *Vitis,* coffea, etc. Possibility of pollen morphology from the dividing pollen varies. The haploid callus, embryo or plantlets may originate from (*i*) the continued division of vegetative cells of the pollen (the generative cell soon degenerate), (*ii*) multiple division of generative cells (non-vegetative cell), and (*iii*) division products of both generative and vegetative cells.

## 13. Mentor Pollen Technology

Pollination is a natural phenomenon which occurs in many plant species. However, there are certain plants where pollen grains remain viable but unable to set seed on its own pistil *i.e.* they are self incompatible. Similarly, in several other plants, besides being viable they even fail to germinate on stigmas of another species *i.e.* they are interspecific incompatible.

After the birth of recent mentor pollen technology the incompatibility problem of pollen grains can be overcome by altering the fertilization ability. The pollen which has been purposefully treated is called *mentor pollen.* Hence, the mentor pollen may be defined as the compatible pollen which has been treated in many ways to inhibit its fertilization ability and retain its power to stimulate incompatible pollen to accomplish fertilization. The mentor pollen itself may or may not germinate on stigmas. As a result of this treatment, hybridization between closely related species or genera, and itself pollination in cross-pollinating plants could be achieved. In 1987, R.B. Knox and coworkers have reviewed the mentor pollen technology and suggested its benefits as a tool in plant biotechnology. The advantages that it offers are : (*i*) transfer of specific genes required in a breeding strategy through seeds, (*ii*) quantification of its effects, and (*iii*) understanding the pollen-stigma recognition.

## 14. Embryo Culture

In addition to root, shoot, and pollen culture, embryo culture has also been done for the production of haploid plants. Embryo culture is used for the recovery of plants from distinct crosses. Embryo culture is useful where embryo fails to develop due to degeneration of embryonic tissues. It is being used extensively in the extraction of haploid barley (*Hordeum vulgare*) from the crosses *H. vulgare* × *H. bulbosum.* Embryo culture is also a routine technique employed in orchid propagation and in breeding of those species that show dormancy. The embryo callus produced somatic embryoids within 8 weeks of culture in the second medium which differentiated into buds after 2 weeks. Several shoots with 4-6 leaves developed after 16 weeks of culture.

**Culturing method.** The general method of embryo culture follows the following steps:

(*i*) Pluck healthy and mature fruits from the field and wash thoroughly in running water for about an hour.

(*ii*) Surface sterilize with 0.01% between-20 for 15 min, rinse seeds several times with distilled water and finally treat with 0.01% $HgCl_2$ solution for 10-15 min. Finally rinse it for six times with sterile distilled water.

(*iii*) Break seeds aseptically and isolate the embryo.

(*iv*) Culture embryo on callus proliferation medium. Supplement the basal medium of Murashige and Skoog (1962) with different combinations and concentrations of sugar, vitamins, hormones and other growth adjuvants for callus proliferation and shoot regeneration.

(*v*) Incubate the cultures at 22-25°C under a 16 h photoperiod of 2000 lux luminous intensity.

(*vi*) After two weeks of inoculation the embryo begins to swell on callus proliferation medium. Distinct callus growth is observed after 4 weeks.

(*vii*) After 8 weeks of inoculation transfer the callus on shoot regeneration medium. Within 4 weeks of transfer into second medium the callus turns green and produces soft spongy tissue. Some of these tissues are differentiated into embryoids.

(*viii*) The embryoids produce cluster of budlets when subcultured onto shoot regeneration medium. The budlets grow into shoots and produce 2-3 leaf appendages within 12 weeks. Thereafter, they are separated into individual shoots and then subcultured into a fresh medium of the same composition until shoots develop.

## 15. Embryo Rescue

Viable hybrids are produced as a result of sexual crosses between two varieties of the species. However, if sexual crosses are done between the species of the same genus or between two different genera, production of hybrid is rather difficult because of several barriers arising either during pollination, fertilization or embryogenesis. It has been observed that in some cases inspite of successful pollination and fertilization, embryo does not develop. This is due to inherent deficiencies or incompatibility between the developing embryo or endosperms. In such cases, immature embryos are dissected out from the fruit (seed) and grown artificially on medium which differentiate into shoot, root and plantlets. This technique of growing immature embryo is termed as 'embryo rescue'.

Embryo rescue technique is very useful in wide hybridization, complete growth of embryo in plant, breaking dormancy of certain seeds where dormancy period is very long. By using embryo rescue technique wild varieties can be crossed with cultivars. As compared with cultivars, the wild species have greater resistance to pests and pathogens, and produce grains of better quality. At the International Crop Research Institute for Semi Arid Tropics (ICRISAT), Hyderabad, this technique has been used to improve groundnut, pigeon pea and chick-pea. A new hybrid variety of pigeon-pea (ICPH8) has been developed at ICRISAT that matures in 100 days instead of 200 days. This variety is resistant to pathogens and pests and yields 20% higher than original cultivars. However, it can grow in a wide range of conditions.

## 16. Triploid Production

Double fertilisation occurs in a majority of individuals except the familes: Orchidaceae, Podostemaceae and Trapaceae. This results in two fusion products: (*i*) the zygote (through fusion of an egg and one of the male gametes), and (*ii*) the triploid primary endosperm (fusion of the second male gamete with two polar nuclei). The zygote gives rise to embryo and the latter forms endosperm (also called secondary embryo). The endosperm nurses the developing zygotic embryo.

In 1949, LaRue succeeded to produce callus from immature endosperm. From India, B.M. Johri and S.S. Bhojwani (1965) at the University of Delhi reported the endosperm culture. Some examples of triploid plants raised from endosperm cultures are; *Asparagus officinalis,* barley (*Hordeni vulgare*), rice (*Oryza sativa*), maize (*Zea mays*), *Prunus persica, Pyrus malus, Citrus grandis,* sandle palnt (*Santulum album*).

The triploid plants are self-sterile and usually seedless. This characteristic increases edibility of fruits and desirable in plants such as apple, banana, grape, mulberry, mango, watermelon, etc. These are commercially important edible fruits. The triploids of poplar (*Popu's tremuloides*) have better quality pulpwood. Therefore, it is important to the forest industry.

## 17. Protoplast Fusion in Fungi

The essence of protoplast fusion in fungi is the improvement of strains to be used for commercial purposes. By this method compatibility barriers between several species can be overcome. In recent years, several companies have realized to exploit this technique as a breeding tool for strain development in fungi. For example, two titre strains of *Cephalosporium acremonium* were crossed and cephalosporin C was improved. But attempts to improve citric acid production in *A. niger* were unsuccessful.

Method for isolation of protoplasts is the same as described for plant cells. Generally, PEG is used for inducing protoplast fusion. But many of isolated protoplasts die due to PEG treatment. This results in low rate of survival which facilitates to detect them by using selection techniques.

Electrofusion, a two stage process, is another method for inducing protoplast fusion. In this method, the protoplast membranes are brought into close contact by dielectrophoresis and a field pulse is used to cause the fusion event. For the first time, it was applied in *S. cerevisiae.*

The fungal protoplast techniques are applied in the following four major areas to get the crosses (*i*) between isogenic strains which may provide an opportunity for genetic mapping, (*ii*) between different strains or isolates of a species primarily for breeding, (*iii*) between apparently incompatible strain of the same species, and (*iv*) between different species and genera.

**(*a*) Intraspecific Protoplast Fusion.** Intraspecific protoplast fusion is the cross between the same species in an individual which involves the isogenic strains or the non-isogenic ones. The true value of protoplast fusion as a mean for establishing parasexual crosses has been realized so far in a few fungi. For example, in *C. acremonium* this technique offers the only way of carrying out crosses and genetic analysis. Difficulty arises through the conventional methods because of production of heterokaryons and somatic diploids. However, through this technique the fusion products were not the heterokaryons but the haploid recombinants. Protoplast fusion technique made it possible to produce a preliminary genetic map of 8 linkage groups for *C. acremonium*. Genes which enhance the production of antibiotics have been identified and allied to specific linkage groups. Another example is the yeast, *Candida,* of which biotechnological potential would be of great use. The other examples are : *Absidia glauca, Candida maltosa, Pleurotus ostreatus, Aspergillus niger, A. sojae, Fusarium graminearum, F. lycopersici, Penicillium verruculosum, Trichoderma, harzianum, T. reesei,* etc.

**(*b*) Interspecific Protoplast Fusion.** Interspecific protoplast fusion is the crosses between two different species. Interspecific protoplast fusions are of much importance in the area where new products are to be produced. Due to new genetic set up many noval secondary metabolites such as, antibiotics may be produced. Some of the examples where interspecific hybrids were produced through protoplast fusion are: *S. cerevisiae* x *S. fermentali, S. cerevisiae* x *S. lipolytica, S. cerevisiae* x *S. rouxil, P. chrysogenum* x *P. notatum, P. chrysogenum* x *P. citrinum, P. chrysogenum* x *P. baarnense, Aspergillus nidulans* x *A. rugulosus, Ganoderma applanatum* x *G. lucidum, Pleurotus ostreatus* x *P. florida.*

## PROBLEMS

1. Write a brief history of development of *in vitro* techniques of plant cell culture.
2. What are the requirements for establishing a tissue culture laboratory?
3. Give a brief account of methods of sterilization.
4. How will you prepare the culture medium? Write the formulation and nutritional component of culture medium.
5. What is an explant? How will you induce callus from it?
6. Write an eassy on organogenesis and its application.
7. Give a brief account of benefits from root culture.
8. How will you prepare cell suspension? Discuss the benefits of using aqueous medium over the solid one.
9. What do you know about micropropagation? How does it differ from vegetative propagetion?
10. Write an eassy on somatic embryogenesis.
11. How will you isolate protoplasts from the plant cells? Give the applications of protoplast culture.
12. What are the benefits from pollen and anther culture?
13. Why does a cultured anther permit pollens to develop embryos but not the cultured pollen grains?
14. Write short notes on the following ?

    (*i*) Explant, (*ii*) Callus, (*iii*) Totipotency, (*iv*) Organogenesis, (*v*) Cell culture, (*vi*) Somaclonal variants, (*vii*) Hybrid and cybrid, (*viii*) Fugigenic chemicals, (*ix*) Anther and pollen culture, (*x*) Micropropagation, (*xi*) Protoplast fusion in Fungi, (*xii*) *In vitro* androgenesis, (*xiii*) Mentor pollen technology, (*xiv*) Embryo culture, (*xv*) Embryo rescue.

CHAPTER

# 13

# Applications of Plant Cell, Tissue and Organ Cultures

Plant protoplast, cell and tissue cultures have become an important tool for crop improvement, commercial production of natural compounds and many more in the development of forestry. In 1987, M.S. Swaminathan emphasized the significance of biotechnological application of *in vitro* cultured plant protoplasts/cells/tissues as below : (*i*) Tissue culture applications in order to capitalise upon the totipotency of cells, (*ii*) Cell and protoplast cultures coupled with DNA vectors to overcome problems caused by barriers to gene transfer through sexual means; (*iii*) Culture of plant cells for the production of useful compounds, (*iv*) Extension and increase of efficiency of biological nitrogen fixation, and (*v*) Transfer of genes for nitrogen fixation ability to non-fixing species.

Through tissue culture technique there has been an increase in productivity of different crop plantations.

Due to playing with plant tissues in laboratory, this technique has been referred by some researchers as a 'botanical laser' whose numerous uses are yet to be fully understood.

In Gurgaon (Haryana), the tissue culture laboratories (Euro-India Biotech Ltd. in collaboration with Haryana Agro-Industries Cooperation, and Kiwi Callus ( NZ) Ltd. New Zealand) have been set up for: (*i*) micropropagation of plants, (*ii*) development and production of transgenic plants and seeds, (*iii*) development of genetically engineered plants, and (*iv*) production of pure and virus free plants of rare species of high value. Various aspects for plant improvement through tissue culture technology is given in Fig. 13.1.

# A. APPLICATIONS IN AGRICULTURE

## 1. Improvement of Hybrids

Development of cell fusion and hybridization techniques has solved the problem of incompatibility of plants and widened the scope of production of new varieties within a short time. In 1985, A.A. Kuchko obtained somatic hybrid of wild and cultivated potatoes (*S. tuberosum* and *S. chacoense*) and succeeded in the induction of organogenesis. The somatic hybrid plant inherited many characters viz., intermediate leaf morphology, stomata, forms and colour of tubers, prolonged flowering, large and fertile pollen grains, high yield, resistance against Y-virus.

The best example of the application of anther culture in crop breeding and improvement is the production of anther culture derived rice and wheat varieties in China. About 50 varieties in rice and 20 in wheat have been developed by using this technique. The advantages of anther culture as a tool in plant breeding are : (*i*) the availability of rapid method of advancing heterozygous breeding lines to homozygosity, (*ii*) getting gametoclonal variations, and (*iii*) early expression of recessive genes as well as variants and new forms. However, haploid plant materials available as protoplast, cell and tissue culture systems are currently being evaluated for the use in transfer of foreign genetic material to select plant species by protoplast fusion, transformation, transduction and organelle transfer.

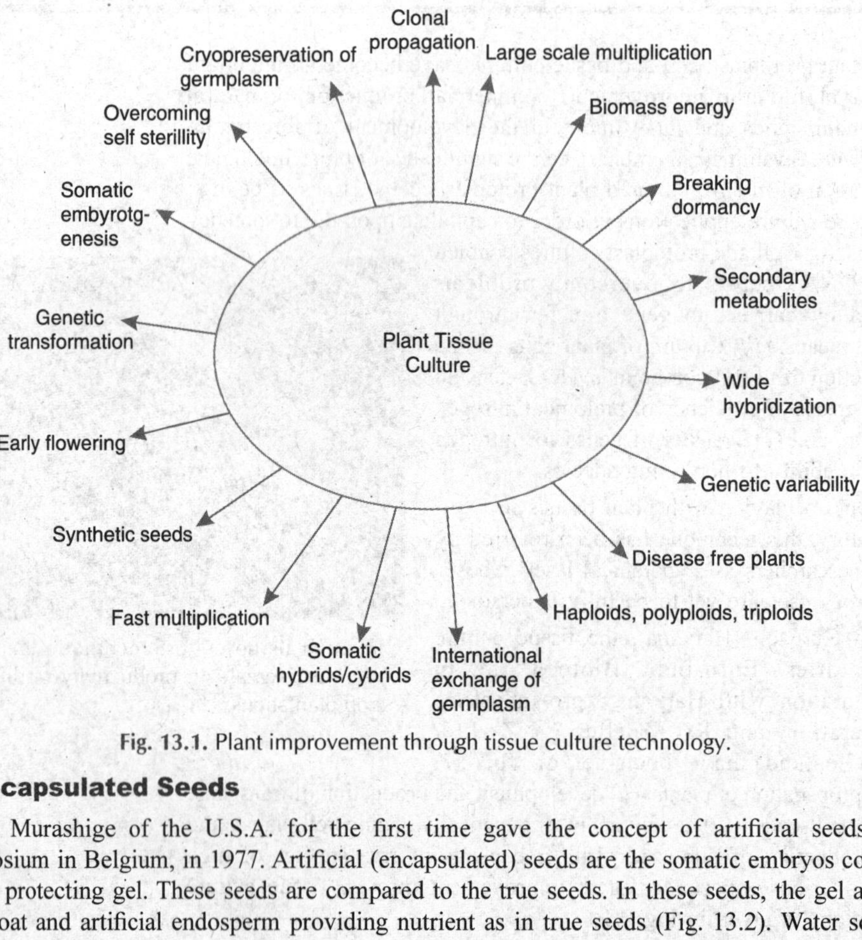


Fig. 13.1. Plant improvement through tissue culture technology.

## 2. Encapsulated Seeds

T. Murashige of the U.S.A. for the first time gave the concept of artificial seeds at a Symposium in Belgium, in 1977. Artificial (encapsulated) seeds are the somatic embryos covered with a protecting gel. These seeds are compared to the true seeds. In these seeds, the gel acts as seed coat and artificial endosperm providing nutrient as in true seeds (Fig. 13.2). Water soluble gels (hydrogel) must be used as the protective gel. Usually Na/Ca alginate (a product of brown algae) is selected for encapsulation purpose because it is less toxic to embryos and easy to handle.

***(a)* Methods of Production** : Following are the steps for the production of encapsulated seeds :

(*i*) Induce pseudoembryos (artificial embryos) from cell suspension culture.

(*ii*) Mix embryos well with 2 per cent Na- alginate,

(*iii*) Drop the embryos in a bath of calcium salt *e.g.* solution of Ca $(NO_3)_2$ for 30 minutes. This results in very quick complex formation at surfaces due to exchange of ions *i.e.* $Na^+$ and $Ca^+$. Consequently, individual embryo is enclosed into a clear and hardened beads of about 4 mm,

(*iv*) Sieve the bead through a nylon mesh; $Ca(NO_3)_2$ solution can be recycled, and

(*v*) Test the growth vigour of beads by plating in sand or soil amended with pesticides.

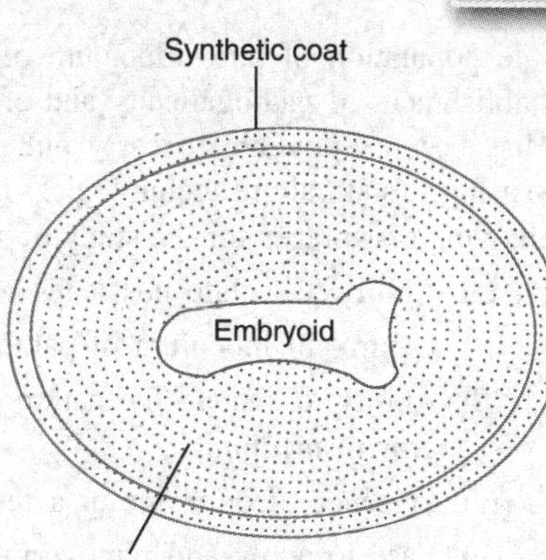


**Fig. 13.2.** Diagram of an encapsulated seed of plant.

In 1985, S.L. Kitto and J. Janick produced *Citrus* embryos *in vitro* and tested 8 compounds for their synthetic coating properties on embryos. Out of the chemicals tested, a polyethyleneoxide (polyox WSR-N75) revealed good encapsulating properties. It was selected for use in further research with *in vitro* produced carrot embryos. Later on Polyox coated embryos were kept in dry condition and then allowed for germination at suitable conditions. The germination percentage of the encapsulated embryos was low. The uncoated embryos did not survive after dry treatment. A number of presumptive hardening treatments are applied such as: (*i*) high inoculum density (0.8 embryo suspension per 25 ml medium), (*ii*) 12 per cent sucrose instead of 2 per cent (*iii*) chilling at 4°C during the last three days of the embryo-induction phase, and (*iv*) amending 1µm abscisic acid in the nutrient medium at the time of embryo induction. These treatments should be combined with polyox coating. All the treatments increased the survivability of the coated embryos. Similarly, production of artificial seeds by encapsulation of somatic embryos in *Eucalyptus* sp. has been reported by E.M. Muralidharan and A.F. Mascarenhaus in 1987.

Synthetic seeds of *Selinum tenuifolium* (bhutkeshi) produced by using somatic embryos are given in Fig. 13.3.

## 3. Production of Disease Resistant Plants

Many plant species, which propagate vegetatively are systematically infected by virus, bacteria, fungi, and nematodes. Their inoculum is carried over several generations resulting in continued adverse effect of productivity and quality of crops. In order to ensure highest possible yield and quality, it is necessary to provide disease free stock plants to growers. Tissue culture techniques have solved the problem and minimized the time of biological testing. Unless large

**Fig. 13.3.** Somatic embryos (Courtesy: U. Dhar and co-workers, Almora).

scale population of pure inoculum of test pathogens are available, it is difficult to persue the establishment of pathogenecity and crop loss - assessment as it is done in field condition. Now it has become possible to carry out such experiments in laboratory within short span of time by using tissue culture technologies. In 1983, S.A. Miller and D.P. Maxwell have discussed the following advantages for the study of several aspects of host-pathogen interactions and responses:

(*i*) Ability to isolate host cells without wounding,

(*ii*) Control of inoculum of pathogen and number of host cells,

(*iii*) Ability to change the nature of host pathogen interaction by altering the constituent of growth medium,

(*iv*) Presence of only one or a few major host cell types, and

(*v*) Easy to apply and remove materials *e.g.* labelled precursors from cultured cells.

**(*a*) Production of Virus-free Plants :** About 10 per cent of viruses transmit through seeds. In some cases, they are confined to seed coat (*e.g.* TMV) or internally seed-borne (in legumes). Moreover, viruses result in great loss, for example, potato leaf roll virus or potato virus X can cause upto 95 per cent reduction in tuber yield and potato virus X between 5 and 75 per cent depending on virus strain and host cultivar.

Tissue culture technique can be utilized for the production of virus-free plants either through meristem culture or chemotherapy or selective chemotherapy of larger explants from donor plants or dormant propagules or a combination of the two. Above plants were made virus-free by meristem culture by 1983.

A. Sood and L.M.S. Palni in 1992 got early flowering in tissue culture raised and virus tested Easter lily plant (*Lilium longoflorum*). They have described various steps for production of virus-free plants of Easter lily (Fig. 13.4) : (i) bulb scale segments from commercially available bulbs were tested positive (virus indexing); segments were subjected to hot water treatment (40ºC for 24 h), (*ii*) surface sterilized segments transferred to MS medium supplemented with 2.5 mg/1 kinetin and 0.5 mg/1 IAA and solidified agar; direct organogenesis (shoot bud formation) was observed, (*iii*) explants transferred to MS medium supplemented with 6-benzylaminopurine (2.5 mg/1 and α-naphthalene acetic acid (0.5 ppm) for extensive rooting, (*iv*) rooted plants transferred after 6 weeks to pots without any growth regulators, (*v*) the plantlets transferred to soil. Normal flowering was observed in transplanted plants within 6 weeks. Virus indexing was carried out at various stages as indicated in diagram. Only those plants were retained which tested negative.

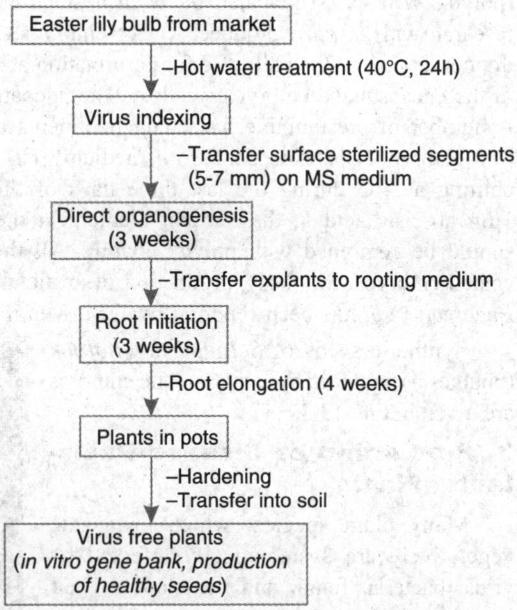


**Fig. 13.4.** Outline for production of virus-free plants of Easter Lily (based on Sood and Palni, 1992).

In 1952, G. Morel and C. Martin for the first time successfully obtained a virus free plant through shoot meristem culture. Later on, several virus-free plants were regenerated *in vitro* such as *Pisum sativum, Trifolium repens* and *Citrus* sp., *Dactylis glomerata* (from mild mosaic mottle viruses) and *Lolium multiflorum* (from ryegrass mosaic virus).

**(*b*) In vitro Selection of Cell Lines for Disease Resistance:** For the study of disease resistance *in vitro,* one of the important considerations is the selection of suitable type of culture *e.g.* callus tissue, suspension culture, isolated cells or protoplasts. However, callus cultures have been widely used for study of expression of race-specific and non-host resistance, and offer several advantages over suspension cultures, isolated cells or protoplasts. The advantages are : (*i*) ease of initiation and maintenance of tissue in culture, (*ii*) ability to add inoculum (spores and zoospores, etc.) directly on callus surface so that the culture medium does not act as direct source of nutrients for the pathogen, (*iii*) the ability to follow the progress of infection and colonisation of callus tissue by the pathogen using histological/cytological methods, and (*iv*) phytoalexin accumulation can be determined in pathogen challenge tissue and related to the extent of colonization.

Subclones regenerated from callus cultures have been found more resistant than the original material against eye spot (*Helminthosporium sacchari*), downy mildew (*Sclerospora sacchari*) and Fizi disease (virus). D.V. Rao and M.S. Palni in 1992 have reviewed *in vitro* selection of cell lines for diseases resistance in plants. D.V. Rao in 1989 raised a cell line of bajra pearlmillet (*Penisetum americanum*) which was resistant to downy mildew caused by *Sclerospora graminicola* (Fig. 13.5). Heavily infected cultures with developing plasmodia lost their capacity to differentiate and finally died. The occasional appearance of green embryos from necrotic tissues was interpreted as possible sign of resistance which is expressed in cultures. Similarly, induction of resistance against *Alternaria solani* has been achieved. The toxins isolated from *A. solani* have been used as stress factor to toxin resistant clones of potato which eventually confer resistance to the pathogens.

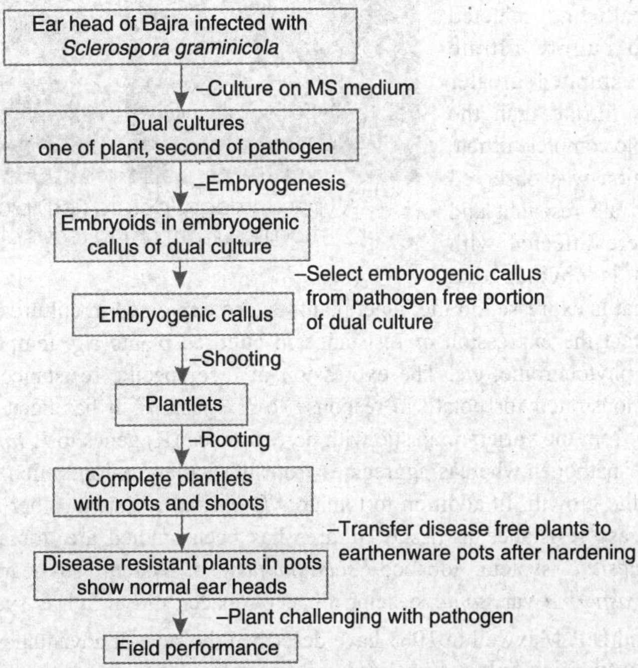


**Fig. 13.5.** *In vitro* selection of a cell line of Bajra resistant to downy midew.

In 1980, J.E. Shepard *and* coworkers have found an increased resistance to leaf blight disease caused by *Phytophthora infestans* and *Alternaria solani* in clonal populations generated from mesophyll cell protoplasts of the potato cultivar *Russet Burbank.* They solved the technical problems of regenerating plants from mesophyll protoplast and then compared a number of protoclones (as they called them). Some were more resistant than others to *P. infestans* and some to *A. solani.* The work of Shepard and coworkers marks a new era *i.e.* the entry of genetic engineering into plant breeding for disease resistance.

These techniques can be extensively applied for mass rearing of nematodes *in vitro* and screening of resistant breeding materials and nematicides (and fungicides for fungal pathogens).

As a result of host cell-pathogen interactions, protoplasts of plant tissues are damaged due to secretion of toxins and enzymes. Therefore, by *in vitro* test in a small flask, millions of protoplasts can be screened for resistance. They are equivalent to thousand acres of growing plants in the field conditions. Protoplast fusion or *in vitro* pollination can, however, produce hybrids between species and genera of plants where normal sexual procedures fail.

A number of investigations show that intact plant resistance to pathogens also manifest itself when their cultivated tissues, cells or protoplasts are treated with the respective toxins. Toxin resistant callus plants have also been produced. P.S. Carlson (1973) for the first time obtained disease resistant plant by treating the tissues to be cultivated with toxins, then by regeneration plants from the stable cell clones. Chlorosis resistant plants were regenerated after treating haploid tobacco cells with an analogue to a bacterial toxin. Disease resistant plants induced by this techniques are corn, sugarcane, cloves, tobacco, potato, etc.

A virus damages potato (left) plant but a modified plant (right) resists attack.

M. Behnke in 1980 has shown that leaves of potato regenerants, produced from calli and selected for resistance to culture filtrate of *P. infestans,* exhibited greater resistance to this filtrate than the leaves of control. No complete filtrate resistance correlation was observed when both the filtrate resistant and control plants were infected with fungal spores. It is obvious that resistance gene, that is expressed in the intact plant, is also expressed in cultured tissue. However, several factors affect the expression of resistance in cultured plants *e.g.* temperature, inoculum density, balanced phytochrome, etc. The expression of race specific resistance in tissue culture has also been demonstrated for potato in response to *P. infestans.* It has been found that tissue culture aggregates from the variety majestic with no resistance (R) genes to *P. infestans* stimulated growth of races of pathogen,whereas aggregates from the variety onion which contain $R_1$ genes did not stimulate the growth. In addition to tomato / *P. infestans* system, other systems in which expression of disease resistance in tissue cultures has been studied are, tomato (*Lycopersicon esculentum*)/*P. infestans* system, tobacco/*Pseudomonas* sp. system, soybean (*Glycine max*)/ *Phytophthora megasperma* var. *sojae* system, tobacco/tobacco mosaic virus system, etc.

S.A. Miller and D.P. Maxwell in 1983 have described the several advantages of tissue culture systems over the suspension cultures, isolated cells or protoplasts for the study of expression of race specific host resistance. These are : (*i*) the ease of initiation and maintenance of cultured tissue, (*ii*) the ability to add inoculum directly to callus so that the tissue culture medium would not be a direct source of nutrition for the pathogens, and (*iii*) the ability to follow cytologically the progress of infection and colonization of callus tissue by the pathogens and the hosts response.

## 4. Production of Stress Resistant Plants

Biochemical mechanisms exist in cultured cells which determine the resistance to biocide chemicals and provide the theoretical promise for selection *in vitro.* In 1974, M.H. Zenk reported cell suspension tolerant to 2, 4-D when the cells were subcultured for 6 months in liquid medium supplemented with increasing amounts of herbicides. The cells were able to grow in 1 mM (milli mol) 2,4-D, while the control suspension was completely inhibited at 0.3 mM 2, 4-D. The mechanism of tolerance involves an enhancement in the metabolism of 2, 4-D. The carrot cells lost the tolerance to the herbicides when transferred to 2, 4-D free medium. It suggested that tolerance was the result of induction of enzymatic systems responsible for the degradation of 2, 4-D.

In 1978, R.S. Chaleff and M.F. Persons reported for the first time that regenerated plants from stable cell lines were resistant to herbicides. They also obtained mutant tobacco plants with monogenetic dominant resistance to picloram. The promising possibility of selecting for herbicide resistance *in vitro* was supported by many workers in many plants *e.g. Citrus,* barley and tomato.

Uses of protoplasts as a system to select cell lines tolerant to herbicides have not been extensively explored. Protoplast technology can be used to increase the possibility to obtain monoclonal lines and offer the opportunity for intraspecific transfer of cytoplasmic factor of resistance to some type of herbicides. A potential application of transfer of herbicide tolerance factor is the transplant of cytoplasmic organelles, for example, chloroplasts.

## 5. Transfer of *nif* Gene into Eukaryotes

Nitrogen fixing ability, a genetic character, exists in prokaryotic diazotrophs. However, one of the major tasks is the transfer of this character to eukaryotes. In recent years, researches are being done to solve this problem through tissue culture techniques coupled with the recombinant DNA technology. Historically, nitrogen fixation by rhizobia was believed to occur through symbiosis. For the first time an excitement was caused in scientific community with the discovery by R.D. Holsten and coworkers in 1971. They obtained active rhizobia in the absence of nodules, leghaemoglobin and bacteroids which were apparently necessary in the intact plants. They established *Rhizobium japonicum* on cell suspension of soybean roots. Callus induced from root explants of soybean on a specific medium was inoculated with *R. japonicum.* Later on it was microscopically observed that infection threads were formed by the bacteria which were present between intracellular spaces. They multiplied inside cells. Moreover, development of nitrogenase in soybean callus- *Rhizobium* system growing on solid medium was also observed by J.J. Child and T.A. La Rue in 1974. It was found that only specific (isomorphic) cells of callus are vulnerable to infection by the bacteria.

In addition to improvement in the bacterial strains and increased nodulation, it is necessary to seek those genotypes with the efficient photosynthesis and improved partitioning of carbohydrates to nodules. The real challenge lies in achieving greater input of biologically fixed nitrogen into nonlegume crops. Improvement of associative $N_2$ fixation by sugarcane, wheat and the crops associated with *Azospirillum* will have the same objective as *Rhizobium* works.

## 6. Future Prospects

In addition to work done successfully on *nif* gene transfer, there are other important genes which have been clones, for example, (*i*) phaseolin and leghaemoglobin genes of soyabean, (*ii*) storage protein genes in soyabean, (*iii*) genes of ribulose bisphosphate carboxylase/oxygenase

(Ru BP case) of pea, maize, wheat etc. Success achieved on these aspects would certainly promote in green revolution.

Moreover, improvement in primary productivity by conversion of $C_3$ plants into $C_4$ ones through genetic engineering techniques hopefully would increase the primary productivity. It is well established fact that photorespiration lowers the capacity of $C_3$ plants in temperate zones which results in the inhibition of net biomass of about 50 per cent. This is caused by RuBPcase due to its reaction with carbon dioxide and oxygen. The same enzyme leads the reaction of both carboxylation and oxygenation. When the concentration of carbon dioxide is low and that of oxygen is high, the latter competes with the former and leads oxygenation by RuBP oxygenase with the production of glycolic acid induced by photorespiration rather than producing phosphoglyceric acid in a normal way through carboxylation by RuBP carboxylase.

By genetic engineering techniques, if the gene encoding RuBPcase is modified which would combine only with carbon dioxide but not with oxygen, only then it would be possible to convert $C_3$ plants into $C_4$ ones.

## B. APPLICATIONS IN HORTICULTURE AND FORESTRY

### 1. Micropropagation

In recent years, the application of micropropagation techniques as an alternative mean of asexual propagation of important plants has increased the interest of workers in various fields. The micropropagation techniques are preferred over the conventional asexual propagation methods because of the following reasons : (*a*) in this method only a small amount of tissue is needed as the initial explant for regeneration of millions of clonal plants in a year; (*b*) this method provides a possible alternative method for developing resistance in many species; (*c*) it provides a means for international exchange of plant materials, hence the problem for introduction of disease can be solved in quarantine; (*d*) *in vitro* stock can be quickly proliferated as it is not season dependent, and (*e*) valuable germplasm can be stored for a long time.

Regeneration of plantlets in cultured plant cell and tissues has been achieved in many trees of high economic value. Many of the studies are aimed at large scale micropropagation of important trees yielding fuel, pulp, timber, oils or fruits. Therefore, clonal forestry and horticulture are gaining an increasing recognition as an alternative for tree improvement. However, strategies for transferring cultured plants from *in vitro* to field conditions are based on relatively higher priced horticultural species rather than agricultural and forestry species. Regeneration of plantlets in cultured tissue has been described to be accomplished into three stages (*see* shoot culture).

At planting out stage the plantlets fail to survive bacause of sudden change in the environment and invasion by soil microbes. The regenerates should be transferred first to green house and then to field. The humidity should be controlled by covering the plants with transparent polyethylene sheets. This acclimatization requires several weeks which should be followed by potting into sterile peat or soil.

Plantlet produced through micropropogation.

In recent years, the interest has aroused in commercializing the *in vitro* propagation of forest trees. This will bring about refinement in the existing procedures to make micropropagation more cost effective. In 1989, Mascarenhas, Muralidharan and co-workers are making efforts to commercialize this biotechnology with respect to forest trees. However, development of automated procedure, plant delivery systems using somatic embryos and artificial seeds are also in progress.

For betterment and improvement of tree plants of high economic value a breakthrough in forestry research has come with production of artificial seeds in *Eucalyptus* and genetic transformation and *in vitro* regeneration in conifers. Moreover, micropropagation has been successfully done in many trees.

In 1989, A.F. Mascarenhas and E.M. Muralidharan reviewed the tissue culture studies carried out on forest trees in India. Some of the important plants are : *Acacia nilotica, Albizia lebbeck, A. procera, Azadirachta indica, Bauhinia purpurea, Butea monosperma, Dalbergia* sp., *Dendrocalmus strictus, Eucalyptus* sp. *Ficus religiosa, Morus* sp., *Populus* sp., *Shorea robusta, Tectona grandis* (*all angiosperms*), *Biota orientalis, Cedrus deodara, Cryptomeria japonica, Picea smithiana, Pinus* sp. (all gymnosperms).

### 2. *In Vitro* Establishment of Mycorrhiza

Mycorrhizal fungi show highest level of specialization of parasitism. But the major problems with them is their failure to grow on an artificial medium in laboratory. Therefore, establishment and multiplication of mycorrhizal fungi on cultured tissue of the same host plant, if successfully developed, may be a good tool for handling mycorrhizal fungi, production of high potential inoculum and their establishment in root systems of nursery plants in horticulture and forestry, and plantation of mycorrhiza-infested seedlings into field. Only one report is available on this work.

Many attempts have been made to establish Vesicular Arbuscular Mycorrhizal (VAM) fungi in axenic culture but unfortunately none of them got success. It was assumed that self inhibition of hyphal growth occurs in the growing germ tubes and the self inhibition compounds were recovered by adding activated charcoal into an agar medium that absorbs inhibitory compounds produced by germ tubes into medium. Cultures of mycorrhizas synthesized aseptically are grouped into two : the whole plant cultures and excised root cultures. Both the types of cultures are known as genetobiotic or monogenic systems. Due to presence of two organisms, it is also known as two member culture.

In 1962, B. Mosse for the first time, reported the establishment of two member cultures. Appressorium formation and root penetration were much more likely to occur if a *Pseudomonas* sp. was present in culture. It is, therefore, suggested that three organisms *i.e.* fungus-plant-bacterium might be necessary for the development of symbiosis.

Moreover, mycorrhizal fungi have been cultured only on cortical tissues of roots which were separated from the whole plant, as in root organ cultures where it acted as food base. VAM fungi have very high degree of specialization for food base on root cortex.

## C. APPLICATIONS IN INDUSTRIES

Production of useful compounds by cultured plant cells has become a field of special interest in various biotechnological programmes. However, much attention has been paid on the production of pharmaceutical and other secondary compounds such as essential oils, food flavorings and colourings which are used by the convenience food, ice cream and confectionary industries.

### 1. Secondary Metabolites from Cell Cultures

Plant cells cultured *in vitro* have been considered to potential source of specific secondary metabolites. Cell cultures may contribute in atleast four major ways to the production of natural products. These are as : (*i*) a new route of synthesis to establish products *e.g.* codeine, quinine, pyrethroids, (*ii*) a route of synthesis to a novel product from plants difficult to grow or establish *e.g.* codeine, quinine, pyrethroids, (*ii*) a route of synthesis to a novel product from plants difficult to grow or establish *e.g.* thebain from *Papaver bracteatum,* (*iii*) a source of novel chemicals in their own right *e.g.* rutacultin from culture of *Ruta,* (*iv*) as biotransformation systems either on their own or as part of a larger chemical process *e.g.* digoxin synthesis (Fowler, 1983). Natural products of plants and their associated industries are given in Table 13.1

**> Table 13.1. Natural products from plants and their associated industries.**

| *Plant products* | *Plant species* | *Industry* | *Industrial uses* |
|---|---|---|---|
| Codeine (alkaloid) | *Papaver sominifera* | Pharmaceutical | Analgesic |
| Diosgenin (steroid) | *Dioscorea deltoidea* | " " | Antifertility agent |
| Quinine (alkaloid) | *Cinchona ledgeriana* | " " | Antimalaria |
| Digoxin (glycoside) | *Digitalis lanata* | " " | Cardiac tonic |
| Scopolamine | *Dhatura stramonium* | " " (alkaloid) | Antihypersensitive |
| Vincristine (alkaloid) | *Catharanthus roseus* | " " | Antileukaemic |
| Pyrethrin | *Chrysanthemum cinerariaefolium* | Agrochemical | Insecticide |
| Quinine (alkaloid) | *C. ledgeriana* | Food & drink | Bittering agent |
| Jasmine | *Jasmium* sp. | Cosmetics | Perfume |
| Saffron | *Crocus sativa* | Food | Flavouring/colouring agent |
| Taxol | *Taxus brevifolia* | Pharmaceutical | Ovarian and breast cancer |

This aim can be achieved by selection of specific cells producing high amount of desired compounds and the development of a suitable medium. In general, secondary metabolites produced by plant cell cultures are rather in small amount but strains of cells producing the same in greater amount than those found in the intact plants have been isolated by clonal selection. Commonly, two methods are employed for the selection of specific cells : single cell cloning and cell aggregate cloning. The difficulties associated with isolation and culture of single cells limit application of this method. The latter may appear to be more time-taking but easier than the first one.

In 1982, T. Yamakawa and coworkers isolated anthocyanin pigment from grape cell cultures. Callus started from white cells always contained a few red cells and red cell culture produced a mixture of both red and white cells.

Employing two stage culture system, anthocyanin at 13 per cent dry weight has been achieved with an yield of 830 mg/lit/15 days. Nowadays, Canadian and Israel biotechnology firms have ventured into commercial exploitation of anthocyanin by plant cell culture. Mitsui Petrochemical Ltd., Japan has commercialized the production of a red pigment (shikonin) from cell culture.

Recently, *in vitro* production of high amount of useful compounds has increased with the success obtained so far in experimental studies. It is hoped that in near future, the industrial production of such compounds by using these techniques would be possible.

**(*a*) Cell Suspension and Biotransformation :** Biotransformation is a process through which functional groups of organic compounds are modified by living cells. Biotransformation done by plant cell culture system can be desirable when a given reaction is unique to a plant cell and the product of reaction has a high market value.

Plant cells are used in a number of ways for biotransformation purposes. The most basic procedure is to supply the cell suspension with the components that is to be transformed, and harvest the products from the culture medium after incubation at suitable conditions. A V-fermenter is used for bio-transformation (Fig. 13.6). Plant cell culture has potential for bioconversion of flavonoids (tannins, anthraquinone), mevalonales, phenyl propanoids (steroids, cardiac glycerides) and alkaloids.

During the period 1950 to the mid 1970s generally low level of product yield was observed with certain exceptions. This was probably our steady improving knowledge of the physiology and biochemistry of cell cultures. In addition, exceeding amount of product yield or those at part was measured in some cell cultures, for example diosgenin, ginseng, saponins, harmin and visnagin.

Therefore, it is necessary that comparison between the yield from the higher plant and those from cell culture must be done (Table 13.2). Improvement in product yield through cell culture should be made. This is particularly important in terms of process development and scale up. Cell suspensions are more amenable to scale up from a biochemical engineering standpoint that requires simple bioreactors as compared with more organised tissue system.

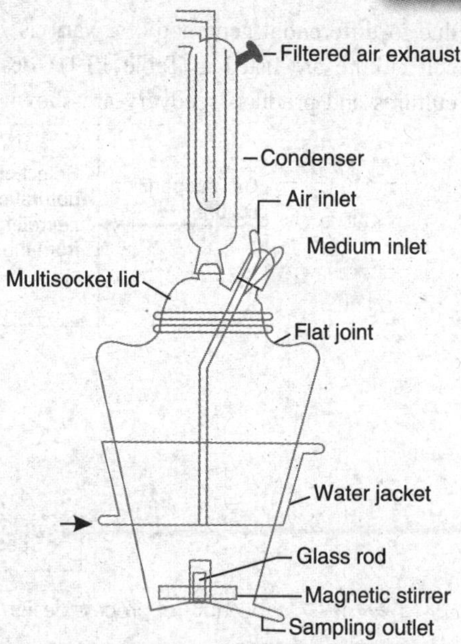


**Fig. 13.6.** Diagram of a V-fermenter used for the production of plant metabolites.

**>Table 13.2. Cell cultures accumulating alkaloids in higher amount than their mother plants.**

| *Plants* | *Alkaloids* | *Yield (% dry weight)* | |
|---|---|---|---|
| | | *Cell culture* | *Whole plant* |
| *Ailanthus* | Canthin 6-ones | 1.27 | 0.01 |
| *Berberis* | Jatrorrhizine | 10.0 | 2.0 |
| *Catharanthus* serpentine | Ajmalicine, | 1.3 | 0.26 |
| *Ephedra* | Pseudoephidrine | 2.25 | 0.6 |
| *Macleaya* | Protopine | 9.4 | 0.32 |
| *Nicotiana* | Nicotine | 3.4 | 2.5 |
| *Stephania* | Biscoclaurines | 2.29 | 0.92 |

*Source :* Based on L.A. Anderson *et al.* (1986).

The useful natural products are synthesized through secondary metabolism, hence they are also known as secondary metabolites. The secondary metabolites include alkaloids, terpenoids, tannins, glycosides (steroids and phenolics) and saponins. Their chief applications are in pharmaceuticals, in food flavouring and perfumery. According to one estimate 4300 different flavour compounds have been identified in food. Certain flavours consist of one or few related compounds such as 2-isobutylthiazole (tomato flavours), methyl-ethyl cinnamates (strawberry), methyl anthranilate (grape), benzaldehyde (cherry), menthol (mint), safranal (saffron).

During metabolism in growing cells, the secondary metabolites are either deposited in vacuoles or excreted from gland cells. Genotype, physiological conditions as much as location within a given plant determine the formation of secondary metabolites. For example, a meristem cell, the nature of which is frequent division, will rarely exhibit terpenoids, phenolics and alkaloids, while a mature parenchyma cell may be loaded with a whole spectrum of products. This differs due to differential gene activity. Various secondary metablites that have been produced through cell culture are listed in Table 13.1 ; design of procedures in biotechnological process of cell cultures and products recovery are shown in Fig. 13.7.

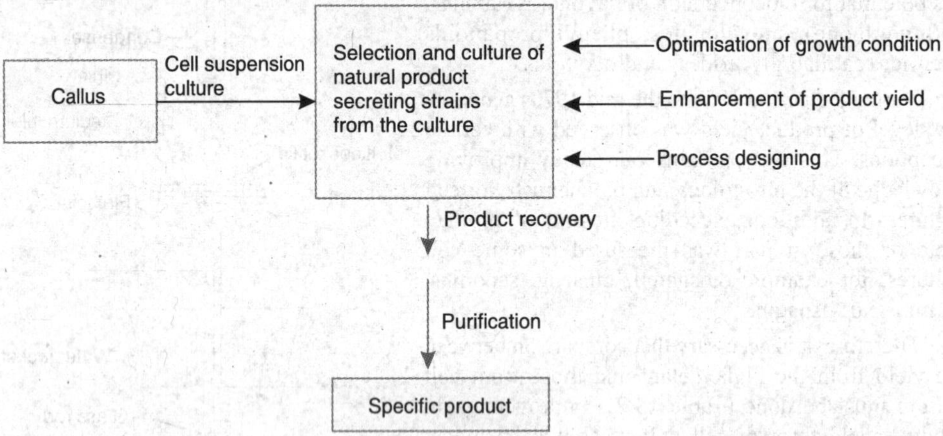


**Fig. 13.7.** Procedure of process design and product recovery from the cultured plant cells.

**> Table 13.3. Secondary metabolites produced through cell suspension culture.**

| *Plant species* | *Secondary metabolites* |
|---|---|
| *Acer pseudoplatanus* | Flavonols and phenolics |
| *Catharanthus roseus* | Serpentine |
| *Daucus carota* | Anthocyanin |
| *Datura stramonium* | Tropane |
| *Lithospermum erythrorhizon* | Shikonin |
| *Mentha canadensis* | Menthol, terpenes |
| *M. piperata* | Menthol |
| *Morinda citrifolia* | Anthraquinones |
| *Nicotiana tabacum* | Tobacco alkaloids, quinone |
| *N. rustica* | Nicotine |
| *Panax ginseng* | Saponines |
| *Populus nigra* | Anthocyanins |
| *Rosa gallica* | Essential oil |
| *Ruta graveolens* | Furnanocoumarins |
| *Scopolia japonica* | Peptides |
| *Solanum lacinalium* | Solasaline |
| *Vinca minor* | Indole alkaloids |

*Source* : Research/review papers.

Similarly, monogenic cultures of nematodes have been used for the study of mechanisms of action of nematicides. This technique can be used extensively in industry to supply nematodes for nematicide screening programmes.

**(*b*) Factors Affecting Product Yield :** There are a number of factors that affect the selecting high-yielding cell lines, some of them are discussed herewith.

**(*i*) Tissue origin genetic character.** Plant cells are genetically totipotent, therefore, proper environmental conditions should be given so that any cell may be induced to produce any substance according to the characteristic of parental plants. However, it has been found that low yielding plants produce high contents of products and *vice versa.* Therefore, such plant parts should be chosen where there is the highest concentration of desired product.

***(ii)* Culture conditions.** Chemical composition of nutrient media influences the potential synthetic machinery (*i.e.* biomass) and synthesis of secondary products. A balance should be maintained between the production of biomass and secondary product. However, excess increase in biomass reduces the yield of desired products. The major chemicals that affect biomass are carbohydrates and their different sources, nitrogen, potassium, phosphorus, trace elements, vitamins, etc. Secondly, the physiological factors such as light, temperature, pH, etc. also affect the product yield.

***(iii)* Selection and screening.** Cell clones from better strains are selected which is rather a difficult task. The more difficult work is to detect the very small amount of desired product present in single cell or small population of cells. To reach the goal mutagenic techniques as a selection pressure are applied to develop high yielding cell lines. Radioimmunoassay (RIA) technique has been applied with much success for the screening of a variety of cultures and products. Moreover, the enzyme linked immunosorbent assay (ELISA) has also been used for screening of products.

## 2. Secondary Metabolites from Immobilised Plant Cells

Large scale yield of secondary metabolites from cultured plant cells can be increased simply by changing the physiological and biochemical conditions from growth medium. But one of the methods of production on increased rate is the use of immobilised plant cells. The method of immobilised plant cells has been found very effective for the production of secondary metabolites as it provides a stable and uniform environment. The plant cells are immobilised in inert matrix and bathed in a medium which does not allow the cell division but keeps the cells in viable conditions for a long time. This obviate the need of further subculturing. There are two commonly used methods for immobilisation : (*i*) immobilisation of cells or subcellular organelles (entrapment in some matrix such as alginate, polyacrylamide, and collagen, or in combination of gels), and (*ii*) adsorption to an inert substrate such as glass beeds. Examples of cell immobilisation are given in Table 13.4.

**> Table 13.4. Plant cells immobilised by two methods.**

| *Source of cells* | *Immobilised substrates* | *Methods* |
|---|---|---|
| *Catharanthus roseus* | Alginate, agarose, | Immobilised in |
| polyacrylamide, carrageenan, | inert substratum | alginate and gelatin |
| *Digitalis lanata* | Alginate | do |
| *Morinda citrifolia* | Alginate | do |

*Source :* K. Lindsey and M.M. Yeoman (1983).

Different support systems and viability and biosynthetic performance of cells within them have been widely worked out. Calcium alginate is frequently used as support. The enclosing matrix can be contained in a more rigid framework to form a column. Beads are packed into a column or maintained in suspension as free beads or the original alginate cell matrix can be added to an inert support *e.g.* nylon mesh. These conditions are sufficient to allow the cells to transform the precursors into secondary metabolites.

The slow growing cells accumulate larger amount of secondary metabolites than the fast growing cultures. When the amount of secondary metabolites is in high concentration, there develops chloroplasts and results in greening of tissue culture.

### 3. Future of Plant Tissue Culture Industry in India

There are over hundred companies globally, each capable of producing more than a million plants per annum through micropropagation. Each year the number of such companies and tissue cultured products is increasing. More than 10 export-oriented units for the mass propagation of tissue cultured plantlets of flowering and ornamental plants have been licensed.

Mass propagation is possible only through tissue culture. However, a commercially viable tissue culture laboratory is that which has capacity of producing plantlets less than a million. During 1990, the global production of tissue cultured plantlets was about 500 million of which India accounted for 5 million.

In India commercialization of plant tissue cultured seeds was started at a small scale by A.V. Thoman & Co. (AVT) at Manalaroo (Kerala). By using small scale technology developed by NCL and AVT perfected the process and created a super variety of cardamom. This variety could be cropped in two years instead of the usual tree. The yield increased from 70 to 250 kg per hectare, and earning also increased accordingly from Rs. 6,000 to Rs. 25,000 per hectare. Another Indian Company, ITC Agro Tech has developed sunflower hybrid named PAC3425 which has been found to have 11% enhancement in seed production and 26% increase in oil yield as compared to the best quality cultivars.

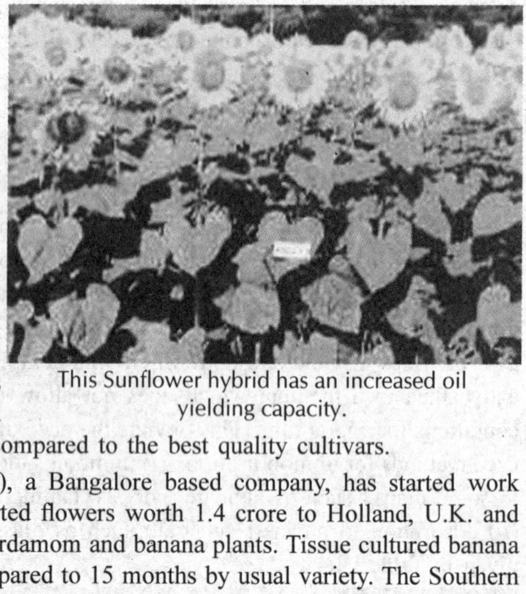

This Sunflower hybrid has an increased oil yielding capacity.

An Indo-American Hybrid Seeds (IAHS), a Bangalore based company, has started work on plant biotechnology. In 1991, IAHS exported flowers worth 1.4 crore to Holland, U.K. and Denmark. It has introduced better strains of cardamom and banana plants. Tissue cultured banana plants begin to bear fruits in 9 months as compared to 15 months by usual variety. The Southern Petroleum Chemical Industries Corporation (SPIC) has planned to increase the export of tissue cultured ornamental plants *viz.*, carnation, lillies, and chrysanthemum to Holland, Australia and Europe.

The major crops multiplied by tissue cultures are ornamental foliage plants, orchids, fruit trees and plantation crops. The Department of Biotechnology (Govt. of India) has identified 14 important forest tree species to be propagated by tissue culture. Two pilot plant units have been set up at NCL and Tata Energy Research Institute (New Delhi) which have the production potential of a few million plant per year. The plant tissue culture industry has recorded a tremendous pace of growth reaching 500 million Mark in 1990 from a production of only 130 million in 1985-86. It has aimed to reach producing 15 billion plants per annum.

## D. TRANSGENIC PLANTS

The plants, in which a functional foreign gene has been incorporated by any biotechnological methods that generally not present in plant, are called transgenic plants. However, a number of transgenic plants carrying genes for traits of economic importance have either been released for commercial cultivation or are under field trials. There are several methods discussed in previous chapters which are used in gene transfer. These includes: (*i*) electroporation, (*ii*) particle bombardment, (*iii*) microinjection, (*iv*) *Agrobacterium*-mediated gene transfer, (*v*) co-cultivation (protoplast transformation) method, (*vi*) leaf disc transformation method, (*vii*) virus-mediated transformation, (*viii*) pollen-mediated transformation, (*ix*) liposome-mediated transformation, etc.

During the last 20 years, considerable progress has been made on isolation, characterisation and introduction of novel genes into plants. As per estimate made in 2002, transgenic crops are cultivated world-wide on about 148 million acres (587 million hectares) lands by about 5.5 million farmers. Transgenic crop plants have many beneficial traits like insect- resistance, herbicide tolerance, delayed fruit ripening, improved oil quality, weed control, etc. Some examples of transgenic crops approved by Food and Drug Administration (FDA) (U.S.A) is given in Table 13.5.

**> Table 13.5.** **Some transgenic crops approved by Food and Drug Administration (U.S.A.).**

| *Producers* | *Transgene inserted* | *Improved traits* | *Crop plants* |
|---|---|---|---|
| AgroEvo | PAT & Barnase/Barstar | Weed control and Hybrid production | Canola |
| Calgene | Thioesterase | High laurate oil | Canola |
| | Nitrilase | Weed control | Cotton |
| | Antisense PG | Delay ripening | Tomato |
| Ciba-Geigy/Northrup King | *Bt cry I* A (*b*) | Insect resistance | Corn |
| Cornell University | Coat protein | Virus resistance | Papaya |
| Dekalb Genetics | *Bt cry I* A (*c*) | Insect resistance | Corn |
| DuPont | Acetolactate synthase | Weed control | Cotton |
| | Gmfad2-1 | Improved oil | Sugarbeet |
| Monsanto | EPSP synthase | Weed control | Canola, corn, soyabean, cotton |
| | *Bt cry I* A (*b*) | Insect resistance | Corn |
| | *Bt cry I* A (*c*) | Insect control | Cotton, tomato |
| | *Bt cry III* A and coat protein | Insect and virus control | Potato |
| Northrup King | *Bt cry I* A (*b*) | Insect resistance | Corn |
| University of Hawai | *Coat protein* | Virus resistance | Papaya |

Initially, some plants were produced by using reporter genes (see preceding section). Later on several genes for known traits of economic importance were incorporated into many crop plants. In others promoter sequences have been used that reduced/enhanced tissue specific expression of the adjacent genes according to requirement. In some cases antisense RNA genes have been introduced to inhibit expression of some existing genes in a desirable manner. All these approaches led to the development of transgenic crop plants of economic importance. More than 1000 field trial tests with transgenic crop plant have been conducted. Some of the commercially grown transgenic crop plants in developed countries are: 'Flavr Savri' and 'Endless Summer' tomatoes, 'Freedom II' squash, 'High-lauric' rapeseed (canola) and 'Roundup Ready' soyabean, etc. During 1995, in the USA full registration was granted to genetically engineered *Bt* gene containing insect resistant 'New Leaf' (potato), 'Maximizer' (corn), 'BollGard' (cotton).

Currently, India is importing both grain legumes and edible oils to meet people's demand. by 2050, India's population is expected to reach about 1.5 billion. It is hoped that 30% India's population will be suffering from malnutrition. Therefore, nutritional security for everyone would require the extensive availability of grain legumes, edible oil fruits and vegetables, milk and poultry products. These challenges can be met by better resource management and producing more nutritious and more productive crops.

To strengthen further research in the area of crop biotechnology, a new institute, the **National Centre for Plant Genome Research** (NCPGR) has been established in New Delhi to strengthen plant biotechnology research in India. Department of Biotechnology (DBT) (Ministry of Science & Technology) has made enough fund for promotion of crop biotechnology. Table 13.6 shows the major Indian developments in transgenic research and application in public sector.

**> Table 13.6. Major Indian developments in transgenic research and application in public sector.**

| *Institute* | *Crop plant* | *Transgene inserted* | *Aims of the project* |
|---|---|---|---|
| Bose Institute of Kolkata | Rice | 5-adenosylmethionine decarboxylase | To generate stress tolerant plants |
| Central Institute for Cotton Research, Nagpur | Cotton | *Bt cry* gene (s) | To generate insect resistant crop plant |
| Central Potato Research Institure, Shimla | Potato | *Bt cry* IA (b) | To generate insect resistant crop plant |
| Central Tobacco Research Institute, Rajahmundry | Tobacco | *Bt cry* IA (b) and *cry* I (c) | To generate insect resistant crop plant |
| CCMB, Hyderabad | Rice | *bar* | To generate herbicide tolerant crop plant |
| Central Rice Research Institute, Cuttack | Rice | *Bt cry* IA(b) Xa21 | To develop plants resistant to pests, bacteria |
| Delhi University (South Campus) | Mustard/rapeseed | *bar, barnase, barstar* | To generate herbicide tolerant crop plant |
| | Brinjal | Chitinase, glucanase genes | To develop plants resistant to diseases |
| | Rice Basmati | Pusa *cod*A, *cor*47 | To develop resistance against biotic and abiotic stress |
| India Agricultural Research Institute, New Delhi | Brinjal, tomato, cauliflower, cabbage | *Bt cry* IA(b) | To develop insect resistant plant |
| International Centre for Genetic Engineering and Biotechnology, New Delhi | Tobacco | *Bt cry* IIa5 | To develop insect resistant plants |
| Madurai Kamraj Univ. | Coffee | Chitinase, β-1, 3-glucanase and osmatin genes | To develop plants resistant to fungal infections |
| National Botanical Research Institute, Lucknow | Rice | *Bt cry* (b) gene | To develop plants resistant to bacteria and fungi |
| Punjab Univ., Ludhiana | Rice | *Bt cry* IA(b), *Bt cry* IA(c) | To develop pest resistant plants |
| Tamil Nadu Agricultural University, Coimbatore | Rice | GNA gene | To develop plants resistant to pest gall midge |
| University of Agricultural Science, Bangalore | Muskmelon | Rabies glycoprotein gene | To develop edible vaccine |

So far more than 60 transgenic dicot plants including herbs, shrubs and trees and several monocots (*e.g.* maize, oat, rice, wheat, etc.) have been produced. In future, the number of these crops certainly will go up. These transgenic plants contain certain selected traits such as herbicide resistance, insect resistance, virus resistance, seed storage protein, modified ripening, modified seedoil, agglutinin, etc. Moreover, in the light of future need, the transgenic plants are being looked up as bioreactor for molecular farming *i.e.* for the production of novel biomedical drugs such as growth hormones, vaccines, antibodies, interferon, etc. (*see* preceding section).

## 1. Selectable Marker Genes and Their Use in Transformed Plants

When plant cells are transformed by any of the transformation methods as given earlier, it is necessary to isolate the transformed cells/tissue. However, it is possible to do now. There are certain selectable marker genes present in vectors that facilitate the selection process. In transformed cells, the selectable marker genes are introduced through vector. The transformed cells are cultured on medium containing high amount of toxic level of substrates such as antibiotic, herbicides, etc. For each marker gene there is one substrate (Table 13.7). For a model transgenic system, tobacco is the most common plant that is found everywhere. The young explants such as leaf discs are aseptically cut into pieces. These pieces are transferred onto tissue regeneration medium supplemented with an antibiotic, kanamycin. From the transformed discs shoots grow directly. The cells which do not undergo transformation will die due to kanamycin. Therefore, antibiotics and herbicides should be used carefully because even in low concentration many cells are damaged. When regeneration has accomplished, selection should be done thereafter. Besides, another difficulty associated with successful selection is the regeneration of shoots from transformed calli because the ex-plants may be heterogeneous and non-transformed cells could not be selected. Therefore, such methods should be used that can ensure escape of only few non-transformed shoots from selection. However, it is ensured by using leaf discs as only the cells which are in direct contact of medium containing antibiotic/herbicide will undergo regeneration.

**Table 13.7.** Selectable marker genes of vectors and their applications.

| *Substrates used for selection* | *Marker genes* |
|---|---|
| **1. Antibiotics** | |
| Bleomycin | Gene *ble* (unknown enzyme) |
| G418, Kanamycin, Neomycin | Neomycin phosphotransferese (*npt*II) |
| Gentamycin | Gentamycin acetyl transferase (*gat*) |
| Hygromycin B | Hygromycin phosphotransferase (*hpt*) |
| Methotrexate trimethoprim | Dihydrofoate reductase (*dfr*) |
| Streptomycin | Streptomycin phosphotransferase (*spt*) |
| **2. Herbicides** | |
| Chlorosulfuron imidazolinones | Mutant form of acetolactase synthase (*als*) |
| Bromoxynil | Bromoxynil nitrilase (*bnl*) |
| Glyphosate synthase (*aro*A) | 5-enolpyruvate shikimate-3-phosphate (EPSP)- |
| PPT (L-phosphinothricin, also called bialaphos) | Phosphinothricin acetyltransferase (*bar*) |

In addition, there is an alternative procedure where there will be no selection pressure imposed on cells/shoot that develop from ex-plants. In this method, samples of tissue from regenerated shoot are taken, the samples are tested for expression of a marker gene. There is a number of marker genes which are commonly described as *reporter genes* or *scoreable genes* or *screenable genes*

(Table 13.8). Some of the reporter genes which are most commonly used in plant transformation are: *cat, gus, lux, npt*II., etc. They are briefly discussed as below :

**(*a*) Chloramphenicol acetyl transferase (CAT) gene :** The *cat* gene is not used as a selectable but as reporter gene. It was first isolated from the bacterium *E.coli* but it is absent in higher plants and mammals. In transformed cells, its presence can be detected by assaying the enzyme CAT on $^{32}$P-chloramphenicol mixed growth medium. Therefore, the enzyme uses acetyl Co-A-chloramphenicol-$P^{32}$ as substrate and transfer acetyl CoA to chloramphenicol converting the latter into acetyl chloramphenicol which is detected autoradiographically.

**>Table 13.8. Examples of some of the reporter genes used as screenable markers.**

| *Reporter genes* | *Enzymes expressed* | *Substrate/ assay* | *Identification* |
|---|---|---|---|
| *cat* acetyl transferase (TLC) separation) | Chloramphenicol +Acetyl CoA | $^{14}$C-chloramphenicol chloramphenicol by autoradiography | Detection of acteyl |
| *gus* | β-glucuronidase X-GLUC, REG, NAG) | Glucuronides (PNPG, colorimetric, photometric | Detection of fluorescence, |
| *lac*Z | β-galactosidase | β-galactosidase + X-gal | Colony colour |
| *lux* | Luciferase | Decan and FMNH$_2$, ATP+O$_2$+luciferin | Bioluminescence on exposure of X ray film |
| *npt*II phosphotransferase | Neomycin | Kan+$^{32}$P-ATP (*in situ* assay) | Detection of radioactivity |
| *nos* | Nopaline synthase acid + NADH | Arginine+ketoglutaric | Electrophoresis |
| *ocs* | Octopine synthase | Arginine+pyruvate+NADH | Electrophoresis |

**(*b*) Neomycin Phosphotransferase (NPTII) Gene (*npt*II Gene) :** The *npt*II gene confer resistance against kanamycin and detoxifies it by phosphorylation. It encodes enzyme NPTII. Presence of *npt*II gene in transformed tissue can be detected by selecting them on kanamycin supplemented medium. Similarly, the presence of this enzyme is also detected in transgenic plants or transformed tissue. Commonly *nos* promoter is linked with *npt*II gene so that synthesis of enzyme NPTII may be started well. However, if *npt*II gene has adverse effect on expression of desirable gene, its expression can be improved by using an alternative approach.

Enzymes in glowworm and fireflies convert chemical energy into light to create spectacular natural displaces.

Firstly, NPTII is fractionated by using non-denaturing polyacrylamide gel electrophoresis (PAGE). In agar layer, radiolabelled $^{32}$P-ATP is used with kanamycin. The gel (that contains the enzyme NPTII) is covered with agar containing both $^{32}$P-ATP and kanamycin. The entire preparation is incubated at 35°C. As a result of phosphorylation of kanamycin $^{32}P$ is incorporated into it, the presence of which is detected autoradiographically.

**(*c*) Luciferase Gene (*lux* Gene) :** The *lux* gene is found in glow-worm, firefly and bacteria that secretes the enzyme luciferase. Due to secretion of this enzymes the glow-worm becomes luminescent in dark. The *lux* gene has been transferred into tobacco through Ti-plasmid of *Agrobacterium*. Consequently *lux* gene containing bioluminescent tobacco plants were produced.

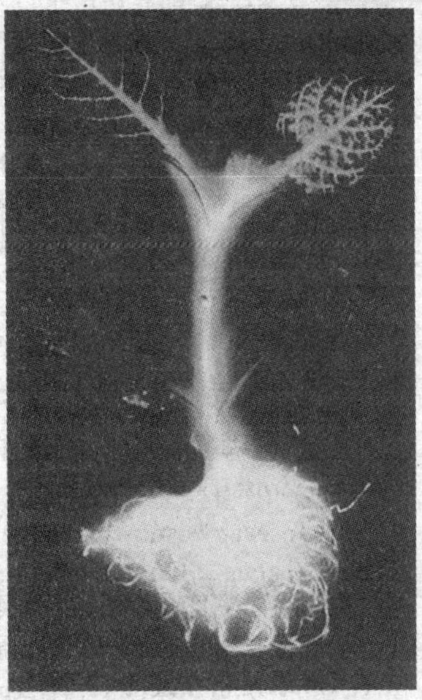

A gene from a firefly has been incorporated into this modified tobacco plant that glows in the dark.

Similarly a green fluorescent protein (GFP) isolated from the jellyfish, *Aequorea victoria* is used as reporter gene or tag in a wide variety of organisms. These act as visible marker for gene expression.

**(*d*) The β-galactosidase Gene (*lac*Z Gene) :** The *lac*Z gene that encodes β-galactosidase is a polylinker as it contains several restriction sites but maintains the proper reading frame. Most DNA fragments cloned into polylinker disrupt *lac*Z gene and abolish β-galactosidase activity. When a foreign gene fused with *lac*Z gene is inserted into a microbial cell, its presence and function can be detected. When the genetically engineered microbial/plant/animal cells contained a reporter gene is allowed to grow on medium containing a chemical X-gal (*i.e.* 5-bromo -4-chloro-3-indolyl-β-D-galacto-pyranoside), β-galactosidase hydrolyses X-gal, and releases an insoluble blue dye. The release of dye shows the presence of foreign gene. If there appears no colour, it means the gene is disrupted.

## 2. Transgenic Plants for Crop Improvement

For crop improvement several genetic traits have been introduced in plants and thereby plant's efficiency has been increased several times as compared to the normal plants. In this regard some of the plants have been described:

**(*a*) Virus Resistant Transgenic Plants :** Plant viruses cause severe disease on crop plants and result in yield loss in several economically important plants. There are two approaches for developing genetically engineered resistance in plants : *pathogen-derived resistance* (PDR) and *non-pathogen-derived resistance* (non-PDR). For PDR, complete or part of viral gene is introduced into the plant which interferes the essential steps in the life cycle of the virus.

For the first time Roger Beachy and co-workers introduced coat protein (CP) gene of tobacco mosaic virus (TMV) into the tobacco. They observed the development of TMV-resistance in transgenic plants. Now, there are several host-virus systems in which it has been fully established. In many crops virus-resistance transgenics have been developed by introducing either CP gene or replicase gene encoding sequences. Coat protein-mediated resistance (CPMR) is the most favoured strategy to make virus-resistant plants.

Several important crops have been engineered for virus resistance using CPMR approach and released for commercial cultivation. Transgenic plants can yield more than non-transgenic plants (Table 13.9). Normal papaya and transgenic papaya resistant to papaya ring-spot virus are shown in Fig. 13.8.

**> Table 13.9. Comparative performance of transgenic virus resistant plants.**

| *Host* | *Transgene* | *Increase in yield (5) over non-transgenic plants* |
|---|---|---|
| Tomato | TMV, CP gene | 40 |
| | CMV, satellite gene | 14 |
| Potato | PVX, PVY CP | 38 |
| Squash | CMV + ZYMV + $WMV_2$ CP | 97 |
| | ZYMV + $WMV_2$ CP | 90 |
| Papaya | PRSV CP | 90 |

TMV-tobacco mosaic virus; CMV-cucumber mosaic virus; PVX-potato virus X; PVY-potato virus Y; ZYMV- Zucchini yellow mosaic virus; WMV2- watermelon mosaic virus 2; PRSV- papaya ring spot virus; CP- coat protein.

**(*b*) Insect-resistant Transgenic Plants** : The *Bt* gene of a bacterium, *Baccilus thuringiensis* has been found to encode the toxins called endotoxin which pose cidal effect on certain insect pests. These toxins are of different types such as beta-endotoxin and delta-endotoxin.

The *cry* genes of *Bacillus thuringiensis* (commonly called *Bt* gene) was found to express proteinaceous toxin inside the bacterial cells. When specific insects (species of Lepidoptera, Diptera, Coleoptera, etc.) ingest the toxin, they are killed. Toxin denatures the epithelium of gut by creating many holes at alkaline pH (7.5 to 8).

Fig. 13.8. Normal papaya (A) and transgenic papaya resistant to PRSV (B).

The insecticidal toxin of *B. thuringiensis* has been classified into the four major classes : *cry I*, *cry II*, *cry III* and *cry vi* based on insecticidal activities against many insects. These toxins affect to the specific group of insects. They do not harm the silkworm and butterflies or other beneficial insects.

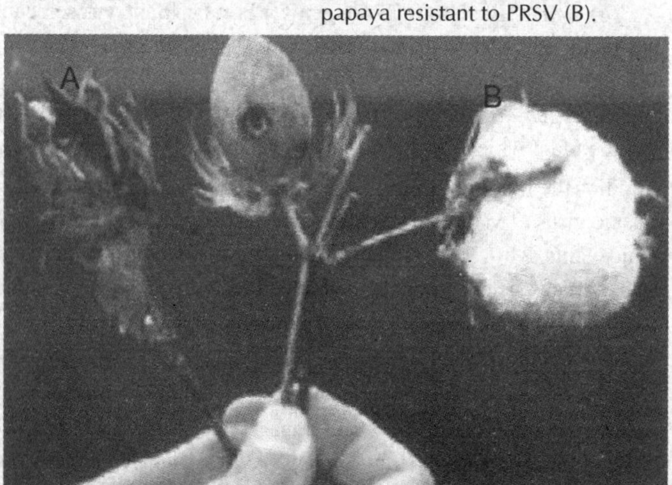


Fig. 13.9. Non-transgenic cotton (A) and transgenic cotton (B).

The approach has been the isolation of the *Bt* gene and its introduction into Ti-DNA plasmid of *Agrobacterium tumifaciens*. The genetically modified *A. tumifaciens* was allowed to infect the desired plant. Thus Ti-plasmid mediated transformation of several plants has been done,

for example tobacco, cotton, tomato, corn etc. Field experiments has also been done with *Manducta sexta* which is a serious pest of tobacco. It has been found that 75-100 per cent larvae of *M. sexta* died when chewed the leaves of transgenic tobacco, whereas the control plants (that were not transgenic) were severely damaged by the insect. Besides, when tobacco plant was crossed with a normal (control) plant, the resistance gene was inherited as per Mendelian principle.

Using biotechnological approaches many transgenic crops having *cry* gene *i.e. Bt*-genes have been developed and commercialised. Some examples of *Bt*- crops are brinjal, cauliflower, cabbage, canola, corn, cotton, eggplant, maize, potato, tabacco, toamato, rice, soyabean, etc.

In India, *Bt*-cotton was permitted to sow at large scale in field. It contains *cry* (*c*) gene that provides resistance against bollworm (*Helicoperpa armi-gera*) which is a notorious pest of cotton (Fig. 13.9.). Several transgenic crop plants have been developed and commercialised at national and International levels (Tables 13.5 and 13.6). However, many transgenic plants are under field trials.

**> Table 13.10. Transgenic plants which have been produced by using recombinant DNA technology by inserting valuable traits.**

| *Traits* | *Plants* |
|---|---|
| Herbicide resistance | Corn, cotton, oilseed rape, potato, tobacco, tomato |
| Insect resistance | Corn, cotton, oilseed rape, potato, tobacco, tomato |
| Virus resistance | Corn, cucumber, melon, papaya, potato, tobacco, tomato |
| Modified seed storage protein | Sunflower, rice, soyabean |
| Modified ripening | Tomato |
| Modified seed oil | Oilseed rape |
| Agglutinin (wheat germ) | Corn |

In India, the private sectors are encouraged to develop and commercialise transgenic plants. Table 13.11 shows transgenic plants which are under field trials in private sectors in India.

**> Table 13.11. Transgenic lines in advanced stage of development for field trials in private sector in India.**

| *Crop plant* | *Transgene introduced* | *Characters* | *Private sector* |
|---|---|---|---|
| Cotton | *cry I* A(*c*) | Lepidoptera pest resistance | Ankur Seeds P. Ltd., (Nagpur), MAHYCO (Mumbai) |
| | *cry x* gene | Lepidoptera pest resistance | MAHYCO (Mumbai) |
| | CP4 EPSPS | Herbicide glyphosate resistant plants | MAHYCO (Mumbai) |
| Corn | *cry* IA (*b*) | Lepidoptera pest resistance | MAHYCO (Mumbai) |
| Rice | Bacterial Blight | Bacterial blight resistance | MAHYCO (Mumbai) |
| | *cry* IA (*b*) | Lepidoptera pest resistance | Hybrid Rice International, Gurgaon |
| Tomato | Alfalfa glucanase | Viral and fungal resistance | Indo-American Hybrid Seeds, Bangalore |
| | *cry* IA (*b*) | Lepidoptera pest resistance | Proagro PGS (India) Ltd, Gurgaon |
| Maize | *cry I* (*A*) | Lepidoptera pest resistance | Syngenta India Ltd., Pune |

*Bt*-cotton was released for commercial cultivation in March 2003, following bio-safety procedures laid down by the Government of India. These cotton hybrids have been granted permission for field sowing and are currently in the fields in the six states namely, Maharastra, Gujarat, Madhya

Pradesh, Andhra Pradesh, Karnataka and Tamil Nadu. Transgenic mustard hybrids after trials have been found suitable and released for cultivation.

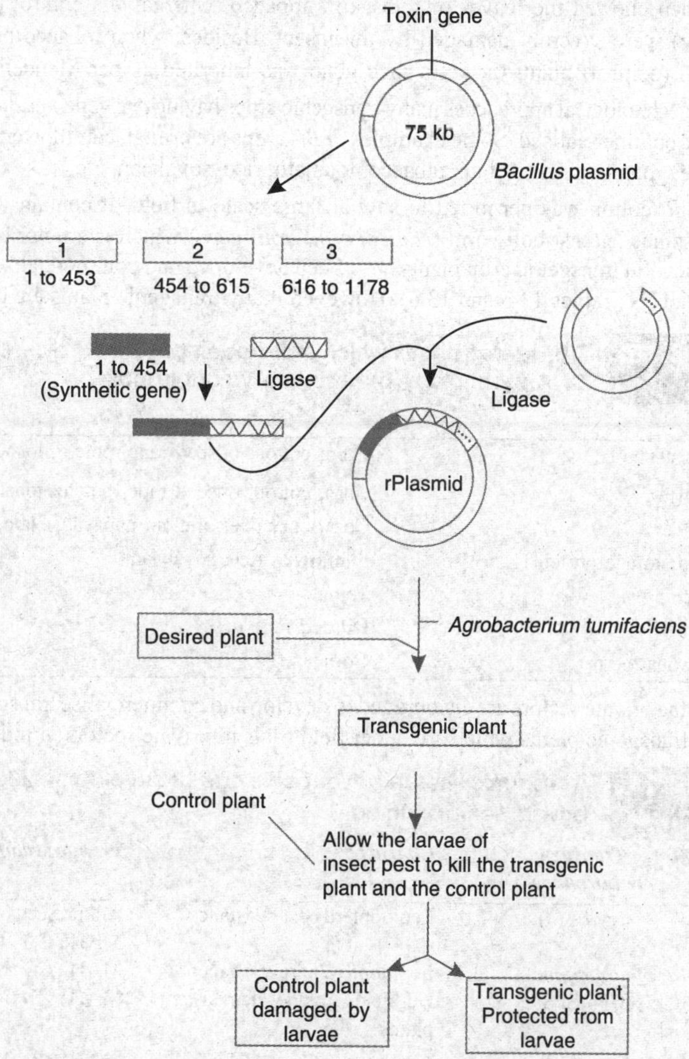


**Fig. 13.10.** Construction of a transgenic plant and expression of Bt gene against insect larvae.

Similarly, transgenic tomato plants has also been produced through cell/tissue culture and transformation techniques. Outline of introduction of *Bt* gene in a crop plant is shown in Fig. 13.10.

In 1996, the first two companies 'Mycogen' and 'Ciba Seeds' commercialized *Bt* insect resistant seeds of corn which has shown very effective protection from the European corn borer.

The second insect resistant gene is cowpea tripsin inhibitor (CpTI) gene. In cowpea (*Vigna unguiculata*) the level of CpTI is high and, therefore, it is resistant to the attack of major storage pest of seeds called bruchid beetle (*Callosobruchus maculatus*). CpTI has been found toxic in nature to many insect pests. Therefore, the *cpti* gene was isolated and joined to CaMV 35S promoter with one marker gene and incorporated into *Agrobacterium*. The genetically engineered *Agrobacterium* infected leaf discs of tobacco and delivered *cpti* gene. Finally from the infected leaf discs transgenic tobacco plants were produced that contained high level of CpTI.

(*c*) **Herbicide-resistant Transgenic Plants** : Herbicides are used in agriculture for killing the weeds (the unwanted plants). However, the herbicides disturb the metabolic activity of photosynthesis or synthesis of amino acids. In addition, due to their use in free hand, environmental pollution occurs and, therefore, biodegradable new herbicides are being developed which will be eco-friendly and environmentally safe. For the development of herbicide resistant plants, two strategies are being applied: (*i*) modification of target molecules that may be insensitive to herbicides, and (*ii*) degradation of herbicides.

The mechanism of action of different herbicides differ. Therefore, attempt must be made to develop resistance against at least three herbicides *e.g.* glyphosate, sulphonylurea and imidazolinones. A herbicide resistant gene for RPSPS (5-enolpyruvate-shikimate-3-phosphate-synthase) was isolated from plants resistant to glyphosate (active ingredient of Roundup herbicide). The resistant gene for EPSPS was transferred to petunia plants and transgenic petunia was developed which was resistant to glyphosate. The other transgenic plants of tomato was developed by introducing a mutant *als* gene (for the enzyme ALS, acetolactate synthase) of tobacco or *Arabidopsis.* The enzyme ALS was inhibited by the herbicides sulphonylurea compounds (active ingredient of Gleen & Qust herbicide) and imidazolinones.

Fig. 13.11. Normal cotton (A) and herbicide tolerant cotton (B).

Herbicide-tolerance is a genetic trait (Fig. 13.11). The herbicide-tolerance gene expresses enzyme which detoxify the herbicide and tolerate the effects.

Using biotechnological approaches transgenic plants have been produced by introducing herbicide tolerant gene in chloroplast degrading enzyme and detoxify herbicides. For example initially Monsanto (U.S.A.) produced glyphosate under the trade name **Roundup**® which is a widely used non-selective herbicide. Transgenic plant **Roundup Ready** has been produced and commercialised. It is tolerant to the herbicide Roundup®.

A gene resistant to PPT (L-phosphinothricin), an active ingredient of herbicide 'Basta', was isolated from *Medicago sativa.* It inhibits the enzyme GS (glutamine synthase) which is involved in ammonia assimilation. This gene resistant to PPT was incorporated into tobacco, as a result of which transgenic tobacco was produced which was resistant to PPT. Hoechst AG (Germany) produced phosphinothricin under the trade name Basta which is also a non-selective herbicide. Similar enzyme has been isolated from *Streptomyces hygroscopicus* by the scientists of Hoechst (Germany) and Plant Genetic System. This enzyme also inactivates the herbicide 'Basta'. Transgenic plants resistant to 'Basta' has been produced by introducing the bacterial gene.

A number of microorganisms are associated with the degradation of herbicides (*see* Chapter 21, *Environmental Biotechnology*). Obviously, degradation is accomplished by genes encoding specific enzymes such as PAT (phosphinothricin acetyl transferase encoded by *bar* gene of *Streptomyces* spp. that degrades herbicide PPT), nitrilase (encoded by *bxn* gene of *Klebsiella bromoxynil*), GST (glutathione-S transferase that degrades the herbicide Atrazine), etc. By using these genes several transgenic crop plants have been produced, for example, transgenic potato, oilseed rape, and sugarbeet (all containing *bar* gene) and transgenic tomato (containing *bxn* gene). The other herbicide resistant transgenic crop plants are corn, cotton, soyabean, etc. (Table 13.8). These have been released for commercial cultivation after field trials.

### 3. Molecular Farming from Transgenic Plants (Transgenic Plants as Bioreactor)

Animal life is possible due to plants. Plants are the natural producers. Plant cells act as the nature's cheapest 'factory'. The cell uses $CO_2$, water, minerals and sunlight to synthesise thousands of valuable and complex products which are the basis of animal's life.

In recent years, transgenic plants are used by biotechnology industries as 'bioreactor' for manufacturing special chemicals and pharmaceutical compounds. Normally, these chemicals are produced in low amount or not produced by the plants.

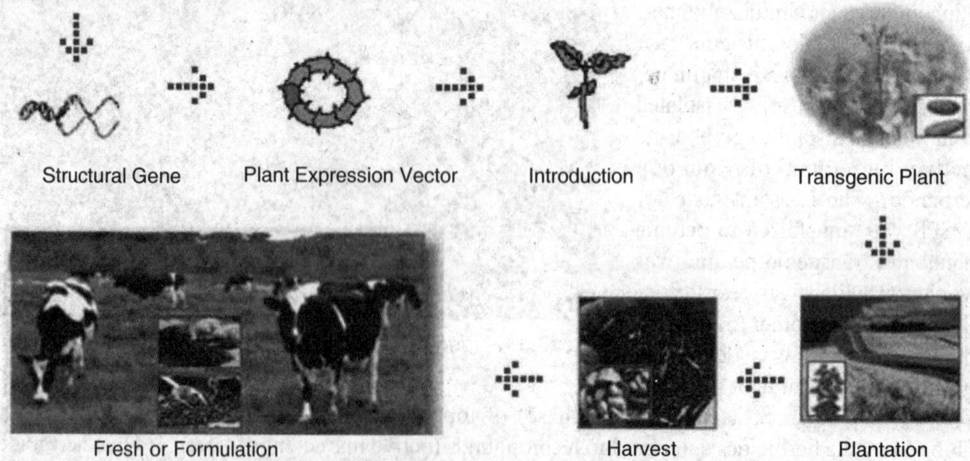


Molecular Farming.

The plants are being looked upon as potential bioreactors or biofactories for the production of immunotherapeutic molecules *i.e.* molecular farming. Transgenic material in the form of seed or fruit can be easily stored and transported from one place to another without fear for its degradation or damage. Besides, transgenic plants capable of producing several different products. can be created at any time by crossing the plants that produce different products.

In successful trials, transgenic plants have been found to produce monoclonal antibodies, functional antibody fragments, proteins, vitamins and the polymer polyhydroxybutyrate (PHB). The PBH can be used to prepare biodegradable plastics. Some of the examples have been discussed in this section.

**(*a*) Nutritional Quality :** Nutritional quality of plants can be improved by introducing genes. Transgenic plants have been produced that are capable of synthesising cyclodextrins, vitamins, amino acids, etc. Consumption of such plant will help in improving the health of malnurished people in poor countries. In this context some examples are given below :

**(*i*) Cyclodextrins :** Cyclodextrins (CD) are cyclic oligosaccharides containing 6, 7 or 8 glucose molecules in $\alpha$, $\beta$ or $\gamma$ linkage respectively. CDs are synthesised from the starch by the action of cyclodextrin glucosyl transferase (CGTase) enzyme. It is used in pharmaceutical delivery system, flavour and odour enhancement and removal of undesirable compounds (*e.g.* coffeine) from food. A CGTase gene isolated from *Klebsiella* was transferred successfully into potato. The transgenic potato tubers produced CDs.

**(*ii*) Vitamin A :** Vitamin A is required by all individuals as it is present in retina in eyes. Deficiency of vitamin A causes skin disorder and night blindness. Throughout the world 124 million children are the sufferers of vitamin A. Each year about 20 million new children are victimised due to deficiency of vitamin A.

You know that rice is used as staple food almost in every country. The contents of vitamin A is very low in rice. Vitamin A is synthesised from carotenoid which is precursor of vitamin A. Carotenoid is synthesised by three genes. Prof. Ingo Potrykus and Peter Beyer produced genetically engineered rice by introducing three genes associated with biosynthesis of carotenoid. The transgenic rice was rich in pro-vitamin A. Since the seeds of transgenic rice is yellow in colour due to pro-vitamin A, the rice is commonly known as **golden rice** (Fig. 13.12.).

Fig. 13.12. A photograph showing seeds of golden rice.

Golden rice is an interesting development which could open the way for improving nutritional standards in rice-eating cultures. Similarly, the work done in India by Ashish Dutta (1992, 2000) on the introduction of *amal* gene (encoding balanced amino acid-protein) from *Amaranthus* into potato holds promise for enhancing nutritional value of low protein food. The transgenic potatoes having *amal* gene are undergoing field trials.

***(iii)* Quality of seed protein :** Seeds are the reservoir of all proteins, amino acids, oils, etc. and used as food throughout the world. However, nutritional quality of legumes and cereals can be affected due to deficiency of certain essential amino acids such as lysine (in cereals like rice, wheat), methionine and tryptophan (in pulses *e.g.* pea). Following recombinant DNA technology improvement in quality of seed protein has been done. The two approaches were done for improvement in nutritional quality of seeds. In the first strategy a gene (encoding protein containing sulphur-rich amino acid) tagged with seed-specific promotor was transferred into cultured tissue of pea plant (rich in lysine but deficient in methionine and cystein). The transgenic pea produced protein containing sulphur rich amino acids. In the second strategy, improvement in endogenous genes is done. The modified gene introduced in cereals produces higher amount of essential amino acids such as lysine.

***(b)* Immunotherapeutic Drugs :** For the first time Hiatt *et al.* (1989) produced antibodies in plants which could produce positive immunization. But the first report on production of edible vaccine appeared in 1990 in the form of a patent application. In 1992, C.J. Arntzen and co-workers expressed hepatitis B surface antigen in tobacco to produce immunologically active ingredients via genetic engineering of plants. During 1980s, great effort has been made to transform plant by foreign genes. Various foreign proteins including serum albumin, human α-interferon, human erythropoietin, and murine IgG and IgA immunoglobulins have been successfully expressed in plants. Antigens and antibodies expressed in plants can be administered orally as any edible part of the plants, or by parental route after purification from the plants. The edible part of the plant to be used as vaccine is fed as raw material to experimental animal or humans. After cooking they are denatured. Therefore, for the production of edible vaccines or antibodies, it is desired to select a plant whose products are consumed raw to avoid degradation during cooking. The plants of choice are tomato, banana and cucumber.

In 1999, the Indian scientists at ICGEB, New Delhi have successfully produced transgenic maize, tobacco, rice, etc. capable of producing interferon gamma (INF-γ).

***(i)* Edible vaccines.** The plants are capable of producing vaccines in large quantities at low cost but the purification may require more cost. Therefore, attention has been paid to produce such antigens that stimulate mucosal immune system to produce secretary IgA (S-IgA) at mucosal surface such as gut and respiratory epithelia because of their effectiveness on sites as most of the pathogens invade these regions. For example, bacteria and viruses are transmitted via contaminated food or water and cause diseases such as diarrhoea, whooping cough, etc. In 1990, the first report of the production of edible vaccine (a surface protein from *Streptococcus*) in tobacco at 0.02 per cent of total leaf protein level was published in the form of a patent application under the International Patent Cooperation Treaty. Thereafter, expression of a number of antigens in plants was successfully made and reported (Table 13.12).

**> Table 13.12. Antigens produced in transgenic plants.**

| *Proteins* | *Plants* |
|---|---|
| Hepatitis B surface antigen | Tobacco |
| Rabies virus glycoprotein | Tomato |
| Norwalk virus capsid protein | Tobacco |
| *E.coli* heat-labile enterotoxin β-subunit | Potato |
| Cholera toxin β-subunit | Potato, tobacco |
| Mouse glutamate dehydrogenase | Potato |
| VP1 protein of Foot & mouth disease virus | *Arabdiopsis* |
| Insulin | Potato |
| Glycoprotein of swine-transmissible gastroenteritis coronavirus | *Arabdiopsis* |

Acute watering diarrhoea is caused by enteroxigenic *Escherichia coli* and *Vibrio cholerae* that colonize the small intestine and produce enterotoxin. Cholera toxin (CT) is very similar to *E.coli* toxin. The CT has two subunits, A and B. Attempt was made to produce edible vaccine by expressing heat-labile enterotoxin (CT-B) in tobacco and potato.

A tobacco plant was produced that expressed CT-A or CT-B subunits of the toxin CT-A. CT-A produced in plant was not cleaved into CT-A1 and CT-A2 subunits which generally happens in epithelial cells. Similarly, CT-B subunit when expressed in potato was processed in natural way, the pentameric form (the naturally occurring form) being the abundant form. Even after boiling transgenic potato tubers till they became soft, about 50 per cent of the CT-B was present in the pentameric GM1 ganglioside-binding form.

Similarly, a rabies virus coat glycoprotein gene has been expressed in tomato plants. Orally administered protein elicited protective immunity in animals. It was hoped that after further effort an edible oral vaccine against rabies can be developed. The value of vaccine can be improved by providing other adjuvants which either enhance the immunogenic potential or reduce degradation of the active ingredients by the micro-organism of gut.

In transgenic tobacco plants the hepatitis B surface antigen (HBsAg) accumulates to 0.01 per cent of soluble protein level. The HBsAg was recovered in virus like particles of 22 nm diameter (similar to yeast-derived HBsAg-based vaccine) which is known to be a prerequisite for better immunogenicity (Fig. 13.13). A crude extract from plant was used for immunization in mice. Immune response included all IgG subclasses and IgM against hepatitis B.

One of the alternative strategies of producing a plant-based vaccine is to infect the plants with recombinant virus carrying the desired antigen that is fused to viral coat protein. The infected plants have been reported to produce the desired fusion protein in large amounts in a short duration. The technique involved either placing the gene downstream a subgenomic promoter, or fusing the gene with capsid protein that coats the virus (Fig. 13.14; Table 13.13).

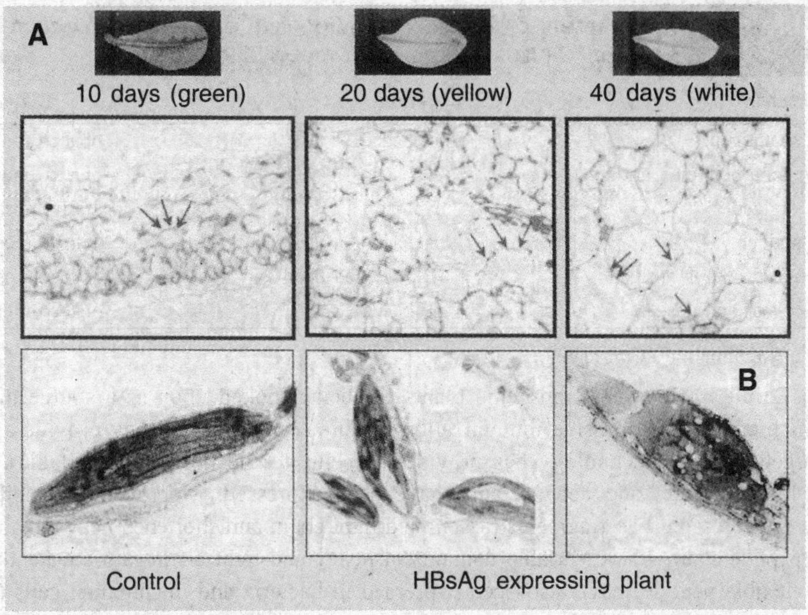


**Fig. 13.13.** Expression of HBsAg by transgenic tobacco leaves; A- enlarged cell size (see arrow); B- loss of stacked stroma and grana in transgenic chloroplast.

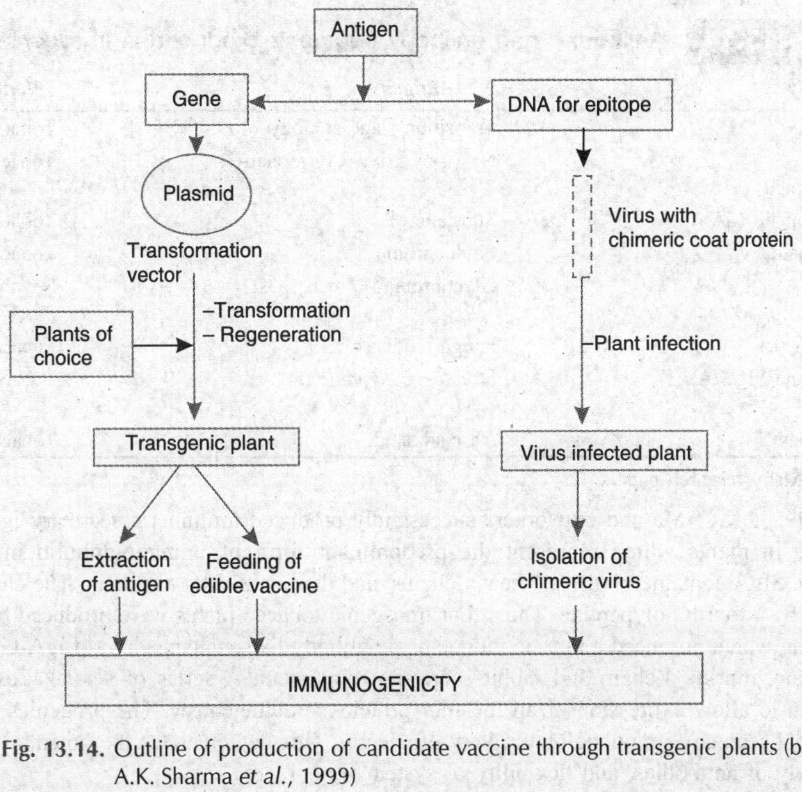


**Fig. 13.14.** Outline of production of candidate vaccine through transgenic plants (based on A.K. Sharma *et al.*, 1999)

**> Table 13.13. Transient production of antigens in plants after infection with plant viruses expressing a recombinant gene.**

| *Protein* | *Plant* | *Carrier* |
|---|---|---|
| Influenza antigen | Tobacco | Tobacco mosaic virus |
| Murine zona pellucida antigen | Tobacco | Tobacco mosaic virus |
| Rabies antigen | Spinach | Alfalfa mosaic virus |
| HIV-1 antigen | Tobacco | Alafalfa mosaic virus |
| Mink enteritis virus antigen | Black eyed bean | Cowpea mosaic virus |
| Colon cancer antigen | Tobacco | Tobacco mosaic virus |

*Source* : A.K. Sharma *et al.* (1999).

***(ii)* Edible antibodies.** Transgenic plants are being looked upon as a source of antibodies also which can provide passive immunization by direct application. They provide as a tool for drug targeting. Gene technology has provided impetus to the utility of antibodies. The genes coding for both light and heavy chains have been expressed. Moreover, the modified genes capable of expressing Fab fragments (assembled light chain and shortened heavy chains) or scFV (single peptide chain where variable domains of heavy and light chains are covalently linked by a short flexible peptide) have also been expressed in bacteria and mammalian cells (Fig. 13.15; Table 13.14). Murine antibodies have been humanized by changing the constant and framework domains. Besides, the recent technology involving PCR and phage display allow cloning and screening of antibodies.

**> Table 13.14. Antibodies and antibody fragments produced in transgenic lants.**

| *Antibody* | *Antigen* | *Plant* |
|---|---|---|
| IgG (κ) | Transition stage analogy | Tobacco |
| IgM (λ) acetyl hapten | NP (4-hydroxy-3-nitrophenyl) | Tobacco |
| Single domain (dAb) | Substance P | Tobacco |
| Single chain Fv | Phytochrome | Tobacco |
| IgG simplex virus | Glycoprotein B of herpes | Soyabean |
| Fab : IgG (κ) | Fungal cutinase | Tobacco |
| IgG (κ) and SIgG/A hybrid | *Streptococcus mutans* | Tobacco |
| Single chain Fv | Abscisic acid | Tobacco |

*Source* : Many research papers.

In 1995, J.K.C. Ma and co-workers successfully produced multimeric secretary IgA (SIgA) molecules in plants which represent the predominant form of immunoglobulin in mucosal secretions. SIgA contained light and heavy chains, and domainized by a J-chain. The chains were protected by a fourth polypeptide. Thus, four transgenic tobacco plants were produced by genetic engineering which produced a murine monoclonal antibody like κ-chains, hybrid IgA-G antibody heavy chain, murine J-chain and rabbit secretary component. A series of sexual crosses were carried out to allow expression of all the four proteins simultaneously. The progenies produced a functional secretary immunoglobulin very efficiently. This demonstrates the potential of plants in assembly of antibodies, and flexibility of system (Fig. 13.15).

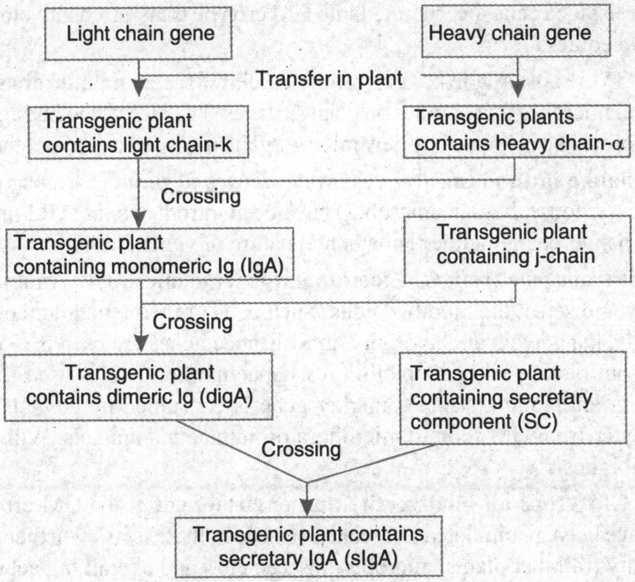


**Fig. 13.15.** Outline for production of secretory antibody in plants.

A hybrid monoclonal antibody (IgA/G) having constant regions of IgG and IgA fused, has been used successfully against human dental carriers caused by the bacterium, *Streptococcus mutans*. The secretary antibody generated SIgA/G in transgenic tobacco and the original mouse IgG was compared. It is interesting to know that both had similar binding affinity to surface adhesion protein of *S. mutans*. SIgA/G survived for 3 days in the oral cavity, whereas IgG survived only for one day. The plant antibody provided protection against the colonisation of *S. mutans* for at least four months.

***(iii)* Edible interferon.** Scientists at ICGEB have successfully produced transgenic tobacco and maize plants that secrete human interferon (IFN-γ). It was produced by transformation of nucleus and chloroplast through particle bombardment method. It has been found that interferon was 10-15 fold greater in chloroplast-transformed plants than nucleus-transformed plants.

## E. BIOETHICS IN PLANT GENETIC ENGINEERING

New variations arise in population due to mutation. However, the frequency of variation depends on the rate of mutation. But in nature, the frequency of mutation is very low (*i.e.* one in about $10^9$/ gene/generation).

Cultivation of genetically modified (GM) crops by the farmers is increasing fast throughout the world. Hopefully, the GM technology will support healthcare and industry and provide food, feed and fibre security at global basis. However, it should be used to increase the production of main staple food, increase the efficiency of production, reduce the environmental impact of agriculture and provide access to food for small scale farmers. The global community is facing the important challenges associated with public perception of transgenic crops. The major concerns about GM crops and GM foods are given below :

***(i)* The risk of transfer of allergies** : There is fear of transferring allergens (usually glycoproteins) from GM food to human and animals e.g. peanut and other nuts. GM food from peanut is now widely labelled, but what about GM crops (where there is no labelling) ?

***(ii)* Pollen transfer from GM plants** : There is a risk of *gene pollution* i.e. transfer of transgene of GM crop through pollen grains to related plant species and development of **super-weeds**. Will pest or herbicide-resistance gene incorporated into GM crop be transferred into closely related plants and increase the 'weediness' ?

***(iii)* Effect of GM crops on non-target and beneficial insects and microbes :** There are many non-target beneficial microbes that harbour on plant surfaces. The insects too harbour on flowers. Will the changed metabolities be colonised by new micro-organisms/insects or affect them to get altered ?

***(iv)* Risk of change in fundamental vegetable nature of plants :** Transgenes from animals (obtained from fishes, mouse, human, microbes) have been introduced into GM plant for **molecular farming**. There is risk of changing the fundamental nature of vegetables.

***(v)* Transfer of transgene from GM food to pathogenic microbes :** Antibiotic marker genes are used to identify and select the modified cells. Such cells grow on medium containing those antibiotics. Commonly, kanamycin and hydromyxin resistance genes are used in GM plants to confer resistance to these antibiotics, while ampicillin resistance marker gene is used for GM bacteria. If GM food containing antibiotic resistance marker gene is consumed by animals and humans, the transgene will transfer from GM food to microflora of human and animals. Will their gut microbe be resistant to antibiotics ?

***(vi)* Effect of GM crops on biodiversity and environment :** The GM crop is not naturally evolved but they have been manipulated artificially. However, there is risk whether they pose harmful effect on biodiversity (of other plants, microbes, insects, etc.) and overall impact on environment.

***(vii)* The GM crop may bring about changes in evolutionary pattern :** Evolution is going on naturally. Plants adapt the fluctuations occurring in the environment through changing their genes and developing better races to which one says the *evolved races*. Will transgene flow from GM crops to other non-GM plants and results in alteration in non-GM crops ? Will non-GM crop evolve through hybridisation with GM crop ?

## PROBLEMS

1. Discuss in detail the applications of micropropagation in horticulture and forestry.
2. Write an essay on the biotechnological applications of plant cell and tissue culture in agriculture.
3. In what way plant cell culture technique is helpful in industries?
4. Discuss in brief the benefits of plant cell culture with regard to plant pathology.
5. Write an essay on research work on plant cell and organ culture in India.
6. Write an essay on *in vitro* selection of cell lines for disease resistance.
7. Give a detailed account of production of secondary metabolites for cell culture.
8. What do you know about transgenic plants? Write in detail on selectable markers and their use in production of transgenic plants.
9. Write an essay on different methods used in production of transgenic plants.
10. Give a detailed account of molecular farming from transgenic plants with emphasis on immunotherapeutic drugs.
11. Write short notes on the following :

    *(i)* Micropropagation *(ii)* Encapsulated seeds, *(iii)* Cell culture and genetic engineering, *(iv)* Cell culture and plant pathology, *(v)* Industrial production of secondary metabolites through *in vitro* cultured cells, *(vi)* Production of virus-free plants, *(vii)* Biotransformation, *(viii)* Secondary metabolites from immobilized plant cells, *(ix)* Future of plant tissue culture in India, *(x)* Selectable markers, *(xi)* Transgenic plants, *(xii)* *gus* gene, *lac*Z gene, *npt*II gene, *(xiii)* Insect resistant transgenic plants, *(xiv)* Herbicide resistant transgenic plants, *(xv)* Edible vaccines, *(xvi)* Edible antibodies., *(xvii) Bioethics in Plant genetic engineering.*

CHAPTER

# 14

# Molecular Markers of Plant and Animal Genomes

In the beginning of 20th century scientists discovered that Mendelian factors of genetic traits lay on chromosomes and control inheritance. Later these were called genes which were advocated to be linearly arranged on the chromosome. Combinations of genes are inherited as groups and they get linked because they are closely arranged. In the absence of one gene phenotypes are altered. The function of a gene is marked by its presence (and expression of characters) or absence. Hence, the marker may be in polymorphic forms so that a chromosome bearing a mutant can be distinguished from normal chromosome.

In the marker, the polymorphism can be detected at three different levels: (*i*) morphological or phenotypic (the morphological marker corresponds to quantitative traits scored visually), (*ii*) biochemical due to differences in proteins (isozymes are successfully used as biochemical marker in some cases of plant breeding and genetics as neutral genetic markers), and (*iii*) molecular (*i.e.* DNA sequences among which the differences can be detected and monitored in the subsequent generations). In this chapter only molecular maps have been described in detail.

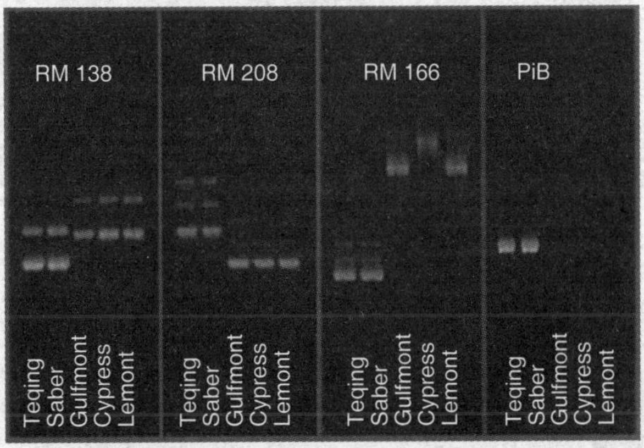


DNA microsatellite markers (RM 138, 208 and 166) revealing variations at DNA level.

## A. MOLECULAR MARKERS

In recent years recombinant DNA technology and PCR technology have helped in construction of genetic, cytogenetic and physical maps of genomes of plants and animals using molecular markers. The markers revealing variations at DNA level are referred to as the **molecular markers**. A large number of genetic polymorphism occurring at DNA sequence level can be used as molecular marker for evaluation of phenotypic variability. Molecular markers possess unique genetic properties on the basis of techniques used for detection. They are classified into the two major classes: *hybridization-based markers* and *PCR-based markers*. The molecular markers used for this effort are: (*i*) restriction fragment length polymorphisms (RFLPs), (*ii*) random amplified polymorphic DNAs (RAPDs), (*iii*) minisatellite or variable number of tandem repeats (VNTRs), and (*iv*) microsatellites or simple sequence repeats (SSRs). Besides, the contigs (contiguous sets of overlapping clones) used in cosmids and YAC (yeast artificial chromosome) vectors also helped in construction of physical maps of geneome. Use of RFLPs is the non-PCR-based approach and the others are PCR-based approaches.

A molecular marker must possess the following desirable properties:

(*i*) It must be polymorphic so that diversity must be measured.

(*ii*) It should be evenly distributed through out the genome.

(*iii*) It should be easily and fast detected.

(*iv*) It must distinguish the homozygotes and heterozygotes.

A comparison of the various marker systems in relation to their characteristics and applicability is provided in Table 14.1. For example for genetic linkage map development, any type of molecular markers may be used. However, codominant markers (*e.g.* RFLP, SSRs or SNPs), provide more genetic information in $F_2$ and backcross than markers detecting presence/absence or dominant polymorphism (*e.g.* RAPD or AFLPs). For comparative mapping within the and across crop species, the use of RFLP as anchor loci are the best choice as they detect evolutionary conserved loci in a more predictable manner than the loci detected by SSRs and AFLPs (Prasanna and Hoisington, 2003).

**> Table 14.1. A comparative properties of RFLP, RAPD, AFLP and simple sequence-repeats (SSRs).**

| *Characteristics* | *RFLP* | *RAPD* | *AFLP* | *SSRs* |
|---|---|---|---|---|
| Fingerprinting | ++ | –/+ | ++ | ++ |
| Genetic diversity | ++ | – | + | ++ |
| Mapping polygenic traits | ++ | –/+ | ++ | + |
| Comparative genome mapping | ++ | – | – | – |
| Principle | Restriction digestion | DNA amplification | DNA amplification | PCR of simple sequence repeats |
| Type of polymorphism | Single base changes; Insertion/deletion n | Single base changes; Insertion/deletion n | Single base changes; Insertion/deletion | Changes in length of repeats |
| Detection of allelic variants | Yes | No | No | Yes |
| No. of loci detected | 1-5 | 1-10 | 30-100 | 1 |
| Detection | Southern blotting | DNA staining | DNA staining | |
| DNA required | Relatively pure | Crude | Relatively pure | Pure |
| Quantity of DNA required | 2-15 μg | 10-50 μg | 2-15 μg | 2-15 μg |
| Use of radioisotopes | Yes | No | Yes/No | Yes/No |
| Part of genome surveyed | Low copy coding region | Whole genome | Whole | Whole |

| Dominant/codominant | Codominant | Dominant | Dominant/ codominant | Codominant |
|---|---|---|---|---|
| Level of Polymorphism | Medium | Medium | Medium | High |
| Need for sequence information | No | No | No | Yes |
| Probes/primers required | gDNA/cDNA | Random 9- or 10 mer oligonucleotides | Specific adaptors and primers | Specific 16-30 mer primers |
| Reliability | High | Intermediate | Medium/High | High |
| Recurring cost | High | Low | Medium | High |

Source: B.M. Prasanna and D. Hoisington (2003). Indian J. Biotech. 2.85.

## 1. Restriction Fragment Length Polymorphisms (RFLPs)

When the genomic DNAs of many individuals of one species are separately cleaved by restriction enzymes, passed through electrophoresis, blotted on nitrocellulose membrane and probed with a radiolabelled DNA, there arises polymorphism in hybridization patterns of digested DNAs. This show differences in sequences between the two individuals. These unique, common and repeated DNA sequences found in several individuals are termed as **restriction fragment length polymorphisms** (RFLPs). After digestion with one specific enzyme, variation obtained in one DNA fragment with that enzyme is referred to one RFLP. It may be explained by an imaginary example. There are three plants A, B and C, where A is the original plant, while B and C are the mutant ones (Fig. 14.1). These can be described on the basis of RFLP analysis. A new restriction site appeared in the 12 kb long fragment of mutant B resulting in 7 and 5 kb fragments. In mutant C, a 10 kb fragment appeared and 6 kb was lost. Restriction map can be constructed as shown in Fig. 14.1.

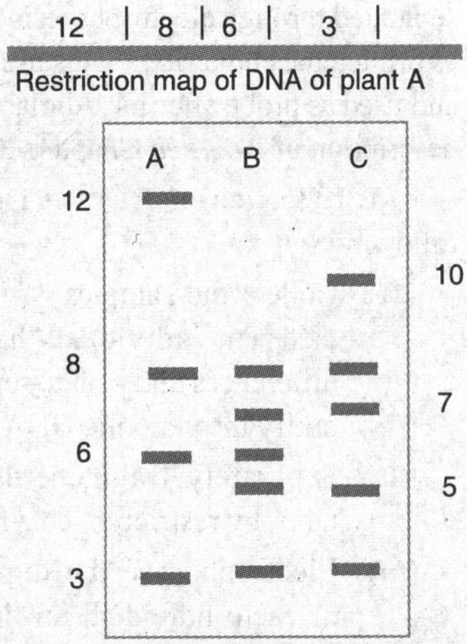


**Fig. 14.1.** Detection of RFLP in three plants (diagrammatic).

RFLPs are co-dominant markers that enable the heterozygotes to be differentiated from homozygotes at the species or population level (single locus probes) or individual levels (multi locus probes). However, it directly identifies a genotype or cultivar in any tissue at any developmental stage in a given environment. The RFLPs are detected by using the unique sequences of DNA called genomic probe, and RFLP map is constructed.

**(*a*) Preparation of Genomic DNA Probes:** There are several molecular DNA/RNA probes that recognise the complementary sequences in DNA/RNA molecules. Examples of molecular probes are antibody probe, cDNA probe, synthetic nucleotide probes, RNA probe (riboprobe) and genomic probe. Genomic probes are preferred over the other probes for RFLP analysis and construction of RFLP maps. The steps for construction of genomic DNA probe are briefly described as below:

(*i*) Extract DNA from plant or animal tissue to be studied.

(*ii*) Digest DNA with restriction enzyme such as *Eco*RI or *Hin*dIII which cleaves DNA at specific recognition site and produces fragments of specific sequences. Also *Pst*I is methyl-sensitive; therefore, the repetitive sequences are not cut with this enzyme.

(*iii*) Electrophorese the digested DNA fragments on agarose or PAGE so that fragments of different size may be separated.

(*iv*) Isolate the DNA fragment of specific size (0.5 – 2.0 kb) from a specific band (identify this fragment through Southern blotting technique as described in Chapter 6. If you want random probe, there is no need to hybridize the Southern blots).

(*v*) Join this DNA fragment to a vector *e.g.* pUC18 or pBluScript as described in Chapter 7. Transform the bacterial cells (*E. coli*) by the recombinant vector (billions of copies vector containing genomic DNA fragment are formed).

(*vi*) For preparation of cDNA probes repeat the steps (*ii*) to (*iv*).

The foreign DNA containing recombinant plasmid can be isolated from the transformed bacteria and used as DNA probe as desired: (*i*) the chimeric vector may directly be used as probe because the vector will not interfere in probe assay, (*ii*) the DNA fragment inserted into vector may be retrieved by using the same enzyme as used earlier for digestion of genomic DNA. The cleaved DNA is separated through electrophoresis and isolated from the gel, and (*iii*) by using the flanking sequences as primer, the chimeric DNA is used for PCR. The DNA products generated through PCR is separated and used as probe after its radiolabelling. Non-radioactive probes may also be prepared. For detailed description of PCR *see* Chapters 3 and 6.

**(*b*) Detection of RFLPs:** Detection of RFLP is outlined in Fig. 14.2 and described in the following steps:

(*i*) Collect the samples of tissues from different individuals having some differences and isolate separately and purify the genomic DNA.

(*ii*) Separately fragment the genomic DNA by restriction enzyme.

(*iii*) Electrophorese the digested DNA of each individual on the same gel slab (DNA of varying size form a continuous smear on the gel).

(*iv*) Transfer DNA fragments of gel to nitrocellulose membrane (*i.e.* carry out Southern blotting; *see* Chapter 6) (use unique DNA as probe for detecting the homologous DNA sequences).

(*v*) Observe the fragments of genomic DNA hybridized by probe by autoradiography. The hybridized fragments of genomic DNA is called RFLP.

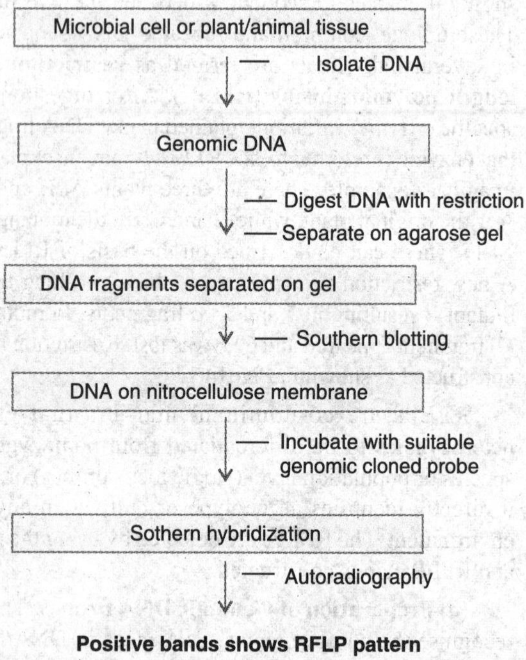


**Fig. 14.2.** Conventional RFLP analysis.

RFLPs are the first class of genomic markers which allow the construction of highly saturated linkage map. The comparative properties of RFLP with RAPD, AFLP and sequence-tagged microsatellite are given in Table 14.1.

Bosttein *et al.* (1980) for the first time suggested to use RFLPs for mapping of human genome. Later on its potential was recognized in plant breeding and experimentally tested in tomato and maize. Thereafter, RFLP linkage maps of a large number of plants have been prepared.

(*c*) **Uses of RFLPs:** The RFLPs are used in different ways in different cases. Some of the example of uses of RFLPs in are given below:

(*i*) **Indirect selection:** While selecting the conventional gene RFLP map supplements the conventional breeding protocols of plant breeding. The RFLP becomes closely related to a desired gene when the RFLP markers are intimately linked to that gene. This helps the selection of RFLP markers using the quantitative traits.

(*ii*) **Tagging the monogenic traits with RFLP markers:** If a single disease resistance gene is linked to nearly isogenic lines (NIL), it will help to detect the RFLP markers. When a donor parent bearing the resistance gene is crossed with a susceptible recipient parent (followed by repeated back cross), the NILs are obtained. Genes of high economic value are tagged in several crops following this technique.

(*iii*) **Indirect selection using quantitative trait loci (QTL):** Genetic traits of agronomic impotance are inherited quantitatively. Multiple factors act collectively on expression of this type of genetic variation of a trait. The loci of these multiple factors are designated as 'QTL' in which each gene affects the trait positively or negatively. A QTL may be defined as *a region of the genome that is associated with an effect on a quantitative trait. It may be a single gene or a cluster of tightly linked genes that affect the traits*. However, one or more genes that determine the traits must be present on a segment of specific chromosome marked by RFLP. One can score the RFLP markers easily.

In self-pollinated species as well as inbred lines of cross-pollinated species, one can study RFLP-QTL linkage relationship. Such parents are selected that differ for many RFLPs and quantitative traits. They are crossed to get $F_1$ progeny. After getting $F_2$ population you can study the presence of a chromosomal segment showing linkage between a RFLP marker and a QTL. Such linkage can also be studied in $F_2$ population under different environmental conditions.

Lander and Botstein (1989) advanced a method for mapping QTL by using RFLP maps. It exploits the complete linkage maps by 'interval mapping' of QTL and identifies the crosses for QTL mapping. The effect of each genome segment located between a pair of marker loci (but not QTL associated with a single RFLP) is assessed by interval mapping.

Consequently, the number of progenies that goes on decreasing is genotyped. The effect of each segment of the genome present between a pair of marker loci is assessed in interval mapping.

One can calculate the maximum likelihood equations (MLEs) after knowing the estimated phenotypic effect (*b*) of a single allele substitution at a putative QTL. The observed values of MLEs are compared with the assumpted values of MLEs *i.e.* no QTL is linked ($b$=0). The evidence for the presence of QTL is represented by LOD score (log of the score of odds for the presence of a linkage against the odds of not having this linkage. For example, if the score is 100:1 in favor of linkage, the **LOD score** is 2). The results are denoted as 'QTL likelihood maps'.

## 2. Random Amplified Polymorphic DNA (RAPD)

For the first time Williams *et al.* (1991) developed RAPD technique. It is a PCR-based technique that is widely used in molecular biology. PCR selectively amplifies the specific segments of DNA of an organism. If you have some information about a set of DNA fragments randomly distributed throughout the genome, you may synthesize a single short oligonucleotide primer (10 mers) and use at both ends of DNA segments before its amplification.

The DNA amplification product is generated from a region flanked by a part of 10 bp priming sites. Consequently random sample of DNA markers is obtained which is called **random amplified**

**polymorphic DNA** (RAPD). Genomic DNA of two individuals produces different RAPDs. A specific DNA segment (generated for one individual but not for other) shows DNA polymorphism. Hence, it can be used as genetic marker. Many random oligonucleotide primers are designed by using different combinations of nucleotides. These are available in market. Theoretically, many different gene loci can be analyzed because each random primer anneals to a different region of the DNA. An outline of RAPD analysis is given in Fig. 14.3.

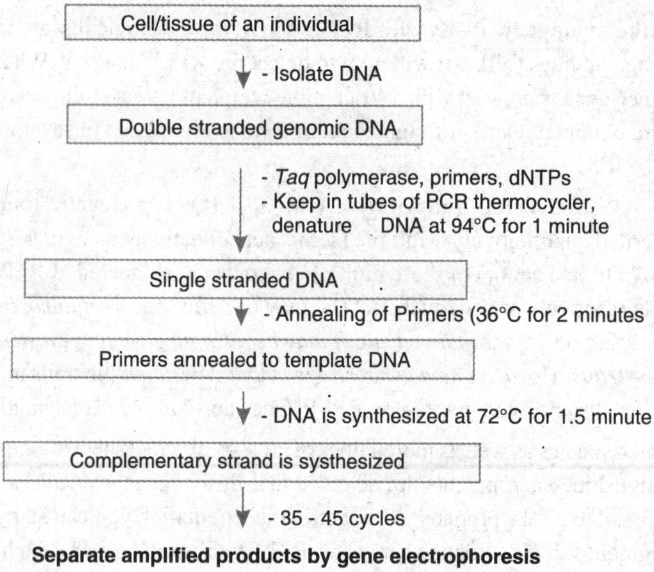


Fig. 14.3. Steps of RAPD analysis by using PCR.

Fig. 14.4 shows the amplification products generated from 14 individuals of the three population of *Syzygium tranvancorium* using RAPD primer, OPAB13.

**(*a*) Advantages from RAPD:** There are several advantages from RAPD over RFLPs: (*i*) no species specific probes are required for different species *i.e.* polymorphic marker-providing primers are same in *Arabidopsis*, maize, *Neurospora*, humans, etc., (*ii*) data can be collected quickly (about 5 times) by using RAPD, (*iii*) crude DNA preparation may be used for analysis of whole genome. However, it does not detect allelic variants, (*iv*) only small amount of DNA (15-25 μg) is required to work with populations, and (*v*) it does not require blotting or hybridization.

**(*b*) Application of RAPD:** RAPD is applied in several areas, a few of them are given below:

**(*i*) Preparation of genetic maps:** RAPD is used as genetic marker to construct genetic maps. Genetic maps of many organisms have been prepared by using RAPD such as *Arabidopsis, Helianthus,* pines, etc.

**(*ii*) Mapping the genetic traits:** Like RFLP, it is also used for indirect selection of segregating population through the tagging of genes of high economic values such as resistance traits against pathogens in many crop plants *e.g.* barley, maize, pea, rice, potato, tomato, wheat, etc.

**(*iii*) Fingerprinting:** Fingerprinting of individual organisms can also be done.

**(*iv*) Tagging of markers:** Certain genetic markers may be tagged at specific regions in the genome.

### 3. Minisatellite or Variable Number of Tandem Repeats (VNTRs) and Microsatellites or Simple Sequence Repeats (SSRs)

Among the segregating populations, the phylogenetic markers are very useful for mapping both the polygenic traits as well as Mendelial traits. The most widespread among the polygenic markers are the **variable number of tandem repeats (VNTRs)** or **microsatellites**. The locus-specific probe fails to cover it and shows highly polyallelic fragment length variation. For example, the tandem repeat loci associated with rRNA genes which are concentrated at nucleolar organizing regions (NORs) of certain chromosomes of an individual. Similarly, most VNTR loci get centred in proterminal regions of human (not mouse) chromosomes. Therefore, the desired density of markers is not provided. This difficulty is overcome in microsatellite or SSR loci.

Litt and Lutty (1989) coined the term **microsatellite**. Microsatellites are usually 1-6 bp long simple sequence repeats (SSRs). Different length of repeat motifs may be A, T, AT, GA, AAC, ATTT, GATA, etc. These occur randomly and most often every where in all eukaryotc DNA except yeasts. Due to their abundance, these are a DNA marker for genetic mapping. A high level of polymorphic (allelic variation) in the repeat number (10-50) of SSR could be demonstrated using PCR. The microsattelites are more common than the VNTR arrays but the latter is highly polymorphic. The VNTRs are less ameanable to PCR analysis because they have larger sequence motifs (about 1000 bp).

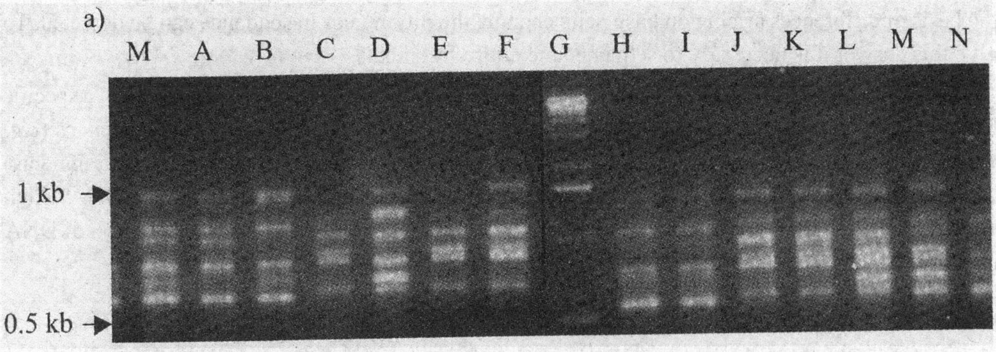


**Fig. 14.4.** Amplification products generated from 14 individuals of three different populations of *Syzygium tranvancorium* using RAPD primer, OPAB13; Lane M, molecular size DNA marker (1 kb ladder); Lanes M to N, individuals from different populations; Arrows indicate the expected molecular sizes on the kb ladder (Source: Anand *et al.*, 2004).

In contrast, the SSRs are more ameanable and can easily be cloned and characterized. Due to the variation in the number of repeat units, they show considerable stable polymorphism. Therefore, microsatellites are the ideal markers to consider high resolution molecular maps. PCR detects SSR length polymorphism at individual loci by using locus-specific flanking region primers. In this region the sequence is known. Moreover, high levels of polymorphism inherent to SSR has prompted an alternative approach, detection of SSR-derived polymorphism with sequencing.

### 4. Arbitrary Primed Polymerase Chain Reaction (AP-PCR)

It has earlier been described in Chapter 3.

### 5. Amplified Fragment Length Polymorphism (AFLP)

AFLP is a combination of both RFLP and RAPD which is very sensitive in detecting polymorphism through out the genome. Therefore, it is superior to RAPD and distinguishes heterozygotes to homozygotes. It is applied universally due to its reproducibility at high level. The basis of its working is PCR amplification of genomic restriction fragments that is produced by restriction enzymes (Vos *et al.*, 1995).

AFLP is a new technique of DNA fingerprinting that utilizes stringent reaction conditions for primer annealing. It may also be used for mapping of genome but it is an expensive technique. The principle of working of AFLP is as below:

(*i*) Cut DNA with two enzymes (*e.g. Eco*RI and *Mse*I) and ligate the double stranded oligonucleotide adaptors at both the ends of DNA fragments.

(*ii*) Carry out selective amplification of sets of restriction fragments with $^{32}PO_4$-labelled primers. The primers are designed according to adaptors ligated to 1-3 nucleotide(s) additionally. Only such fragments are amplified that contain restriction site sequence plus additional nucleotides.

(*iii*) The DNA products which have been amplified are separated on resolving gels. The isolated DNA fragments are observed through autoradiography. If radiolabelled nucleotides are not used in PCR, a fluorescent dye or silver staining method is used for visualization of amplified products.

### 6. Targeted PCR and Sequencing

It is another approach to design primers to target specific regions of the genome. Then the targeted product is compared to the similar products of other individuals by separating on agarose gel. On the basis of differences arising on base pairs change, the differences in sequence can be detected. To give this answer it is necessary to sequence the entire fragments.

*Sequence tagged site* (STS) is a unique marker sequence of 60-1000 bp which can detect a specific locus present on a chromosome (Olson *et al.*, 1989). By using specific probes or sequences (*e.g.* expressed sequence tag), specific PCR markers (matching the sequences of ends of DNA fragments) can be derived. They have been used in mapping projects. STSs are the physical mark of DNA which are detected by PCR. It represents the relative order how it is located within a given region of DNA molecule? A physical map of genome is prepared by using STSs.

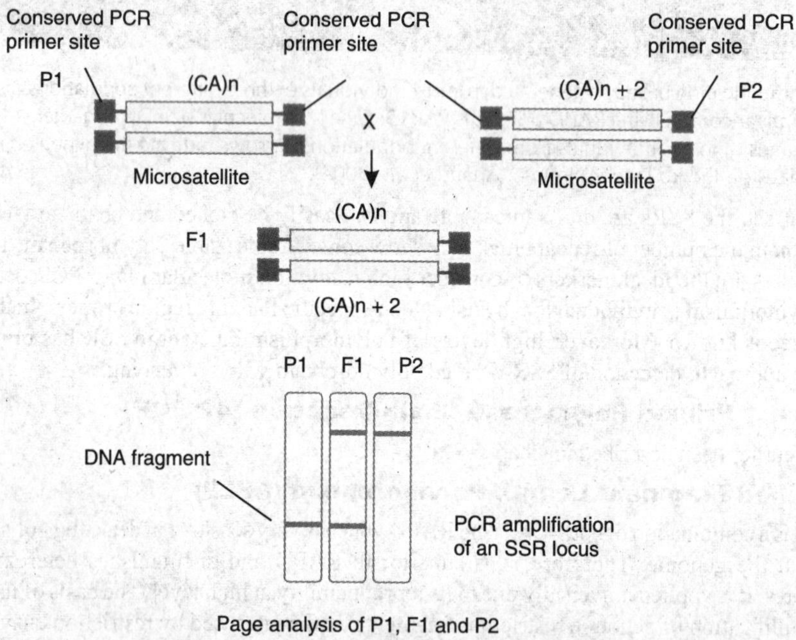


**Fig. 14.5.** Detection of SSR length polymorphism.

Examples of STSs are : (*i*) *sequence tagged microsatellites* (STMs) [tandem repeats of mono-, di-, tri- and tetra- nucleotides with different length of repeat motifs *e.g.* A, T, AR, GA, AGG, AAC, etc. The repeat number is denoted by *n* such as $(AT)_5$. Detection of SSR length polymorphism among parents and $F_1$ hybrids is given in Fig. 14.5], (*ii*) *sequence characterised amplified regions* (SCARs) (a fragment of genomic DNA present at a single locus that is identified by PCR amplification by using a pair of specific oligonucleotide primers), and (*iii*) *cleaved amplified polymorphic sequence* (CAPS) (revealing the restriction site polymorphism by digesting the PCR-amplified DNA fragments with restriction enzymes).

### 7. Computer Software for Construction of Linkage Maps

In recent years several types of computer software have been developed which can construct multipoint linkage maps of genomes of a variety of organisms including plants and animals. In 1995, several maps were prepared by using another computer software, LINKAGE. Moreover, another useful software were developed, in 1987. Several chromosomes of mouse, humans and many plants were mapped by using MAPMAKER and CRI-MAP. CRI-MAP infers genotypes and takes more genotypic data.

In addition, LOD scores are used for a large number of loci which finds out quickly the best genetic order. LOD scores have already been discussed.

## B. CONSTRUCTION OF MOLECULAR MAPS IN PLANTS

The molecular markers are used to construct genomic maps in microorganisms, plants and animals are discussed in this section. DNA-based markers are also being used to discover and exploit the evolutionary relationships between various genera within a family and various species within the genus. Genetic mapping of members of the agriculturally important grasses including rice, wheat, maize, sorghum and sugarcane with common DNA probes have revealed remarkable conservation of gene content and gene order. Among the flowering plants *Arabidopsis* is known to possess the smallest genome i.e. 145 Mb of DNA and 20,000 - 50,000 genes. Rice has 420 Mb genome, whereas wheat has about 16 billion nucleotide base pairs.

### 1. Construction of Genetic Maps Using RFLP Loci

You know that in genetic crosses two homozygous parents are crossed to produce $F_1$ progeny. The $F_1$ progeny is self-crossed to produce $F_2$ progeny and so on. The $F_1$ hybrids have a pair of homologous chromosomes, each of which has come from both the parents. This homologous chromosome differs from each other in respect to sequences of DNA.

The chromosomes of $F_1$ products undergo meiosis during gamete formation and crossover to form recombinants. The conventional genetic mapping is based on the recombination process. Moreover, the frequency of crossing over between the two chromosomal segments depends on their physical distances. In the closely linked chromosomal segments (as in case of linkage), the recombination frequency is very less as compared to the normal ones. The only way to observe recombination in offspring was observing the changes in flower colour, plant height, seed shape, size and colour, pest resistance, etc. These characters have been utilized to construct genetic map of plant which need much time.

For construction of RFLP map one has to look for RFLP markers similar to phenotypic characters observed in different generations. Because RFLP markers follow the same pattern of inheritance and segregates similar to genes following Mendel's laws of inheritance. Therefore, the RFLP markers can be analyzed following Mendelian rules and RFLP genetic maps of plants can be constructed accordingly (Fig. 14.6).

Recombination frequencies can be calculated from the results obtained from the backcross population and doubled haploids procured from pollens of $F_1$ population. Recombinant inbred (RI) lines have been produced in maize after carrying out several crosses through selfing in $F_2$ plants. These RI lines can be compared with those of mouse and doubled haploids. The data of recombination frequencies are analyzed by computer software such as MAPMAKER, LINKAGE, etc.

For some crops such as barley, maize, rice, tomato, etc. a well-defined outline is given for construction of RFLP map as below:

**(*a*) Selection of Parental Plants:** Such parental plants are selected that are genetically divergent and consist of enough RFLPs. A crop is selected where some desirable agronomic traits segregate. By using random selection of cloned probes the plants are surveyed. The types of plants to be considered as parental plant include both crop plants and related wild plants. After making selection of several plants their respective DNA is isolated from each plant, digested with restriction enzymes and screened for polymorphism following the methods as described earlier. Each accession is stored to analyse the alleles of RFLPs. Two such accessions are selected for crossing that consist of usable amount of polymorphism.

**(*b*) Production of a Mapping Population:** The selected parents are crossed to get $F_1$ plants.

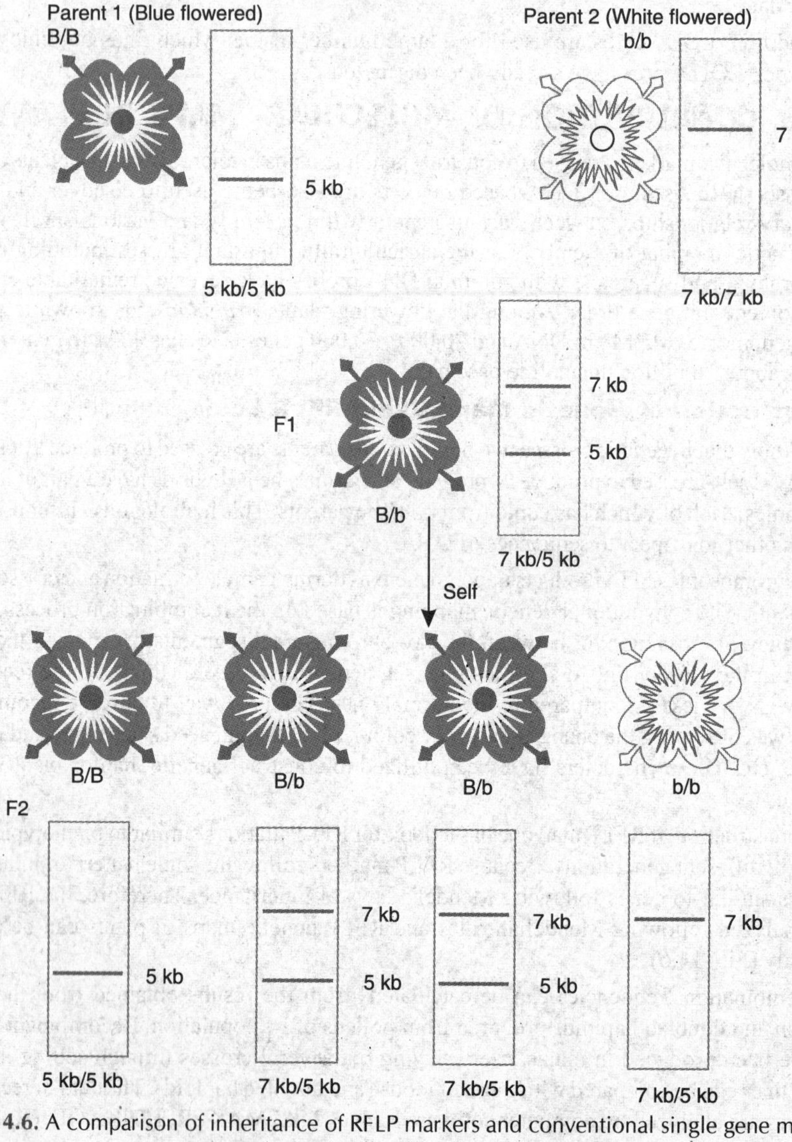


**Fig. 14.6.** A comparison of inheritance of RFLP markers and conventional single gene marker controlling the flower colour.

They are heterozygous and appear like the parental plants. Selfing of $F_1$ plants is done to produce $F_2$ population. Segregation of RFLPs is scored either in $F_2$ plants or $F_2$ plants derived after back-crossing (a cross between $F_1$ and one of the parents). But the selved $F_2$ population is preferred over the scored ones because it gives more information than that of backcross. For detailed map, a mapping population of about 50 $F_2$ plants is enough. Doubled haploids obtained from the pollen grains of $F_2$ plants can be used to estimate recombination frequencies (Fig. 14.7).

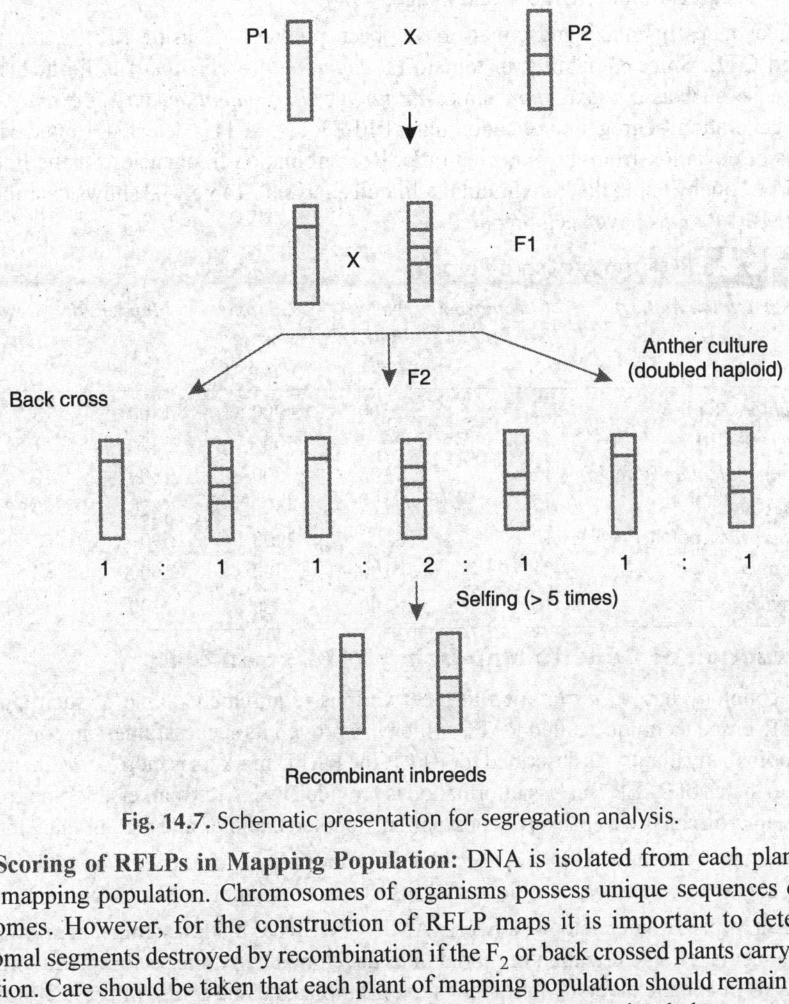


**Fig. 14.7.** Schematic presentation for segregation analysis.

***(c)* Scoring of RFLPs in Mapping Population:** DNA is isolated from each plant obtained from the mapping population. Chromosomes of organisms possess unique sequences of parental chromosomes. However, for the construction of RFLP maps it is important to determine the chromosomal segments destroyed by recombination if the $F_2$ or back crossed plants carry out sexual reproduction. Care should be taken that each plant of mapping population should remain alive. This will help to extract a large amount of repeated DNA. RFLPs are mapped as below:

***(i)* Screening for Polymorphisms:** Polymorphism is detected by using DNA probes and tested that which enzymes will recognize a polymorphism between the parents? The amount of variation present in plants govern the number of enzymes recognized. For example, only one or two enzyme(s) are required in inbred lines of maize. Eleven enzymes were required by rice, and so on.

***(ii)* Scoring:** The polymorphism probe/enzyme combination is scored in the mapping population after its detection. For getting scores, DNA spots are prepared on filters from agarose gel and hybridized with probes. Such DNA filters are termed as 'mapping filters'. Each mapping filter consists of DNA of each individual of $F_2$ plants which has been fragmented with *Eco*RI. Similarly, the other filter consists of the same of the other plant fragmented with *Pst*I. If a given probe/enzyme combination detects a polymorphism between the parental plants, the probe shall be scored on that set of mapping filter. Only three types of plants are found in F2 mapping population scored for a single RFLP: two homozygous and one heterozygous. Hence, each plant of mapping population is scored as a heterozygous or one of the homozygous after hybridizing with genomic probe.

***(iii)* Analysis of Linkage:** The data generated as above are used to prepare linkage map which depends on the degree of co-segregation of probes. Chi-square test is carried out to analyze the randomness of segregation of markers *i.e.* linkage.

Molecular maps of maize and tomato have been prepared by using RFLPs, morphological markers and QTL. Since variability in tomato (*Lycopersicon esculentum*) is limited, hence two different species such as *L. esculentum* and *L. hirsutum* or *L. esculentum* and *L. pennellii* have been used. In $F_2$ generation, segregation of codominant RFLP loci (1:2:1) is detected. Linkage is detected on the basis of deviation from the expected ratio. Recombination frequencies for the linked RFLP loci are worked out by using the 'maximum likelihood analysis'. Table 14.2 shows example of some crops where RFLP maps have been prepared.

**> Table 14.2. RFLP maps of some crops.**

| *Crop plants (common names)* | *Chromosome No. (n)* | *No. of RFLP loci* | *Genome size (bp × $10^6$)* | *Map length (cM)* | *Resolution (bp/cM × $10^6$)* |
|---|---|---|---|---|---|
| *Hordeum vulgare* (barley) | 7 | >500 | 4900 | 1400 | 3.5 |
| *Lens culinaris* (lentil) | 7 | 20 | 500 | 333 | 1.5 |
| *Lycopersicon esculentum* (tomato) | 12 | >1000 | 1000 | 1500 | 0.67 |
| *Oryza sativa* (rice) | 12 | 2275 | 430 | 1520 | 0.28 |
| *Solanum tuberosum* (potato) | 12 | 350 | 2000 | 690 | 2-9 |
| *Triticum aestivum* | 21 | >1000 | 6300 | 16000 | 2.54 |
| *Zea mays* (maize) | 10 | >1000 | 2500 | 2200 | 1.14 |

## 2. Construction of Genetic Map Using RAPDs and SSRs

PCR technology has been modified to meet various requirements. One of such attempts has been the PCR-based technique called RAPD which has proved its successfulness in constructing the genetic maps in many plants. As described for RFLP, the RAPD markers is more convenient and time-consuming than that of RFLP. But less information is provided by RAPD markers. For the construction of genetic maps you can use $F_2$ backcross population, doubled haploids and recombinant inbred lines.

Genetic maps by using RAPD and SSR loci of several plants have been prepared, for example maize, rice, sugarcane, soyabean, sunflower, citrus, pines, eukalyptus, peach, etc.

The Asian Maize Biotechnology Network (AMBIONET) has made a significant effort in relation to: (*i*) the molecular characterization of inbred lines developed by public sector institutions in India, (*ii*) the analysis of genetic diversity in Indian maize germplasm using SSRs (microsatellite markers), and (*iii*) the mapping of QTL conferring resistance in different downy mildews affecting maize in tropical Asia. Much of the SSR allelic variations in their inbred line analysis was contributed by the inbred lines developed at Punjab Agricultural University, Ludhiana. Cluster analysis of the genetic dissimilarity matrix for the genotype aided on determining genetic relationships.

***(a)* QTL Mapping in Maize:** Molecular markers have been used to identify and characterise the QTL associated with diverse traits in maize including grain yield, characters concerned with domestication, environmental adaptation, disease and insect pest resistance, and drought and heat stress tolerance.

***(i)* Mapping of QTL influencing resistance to downy mildews:** The major downy mildews that infect maize in the region are sorghum downy mildew (*Peronosclerospora sorghi*), Philippines downy mildews (*P. philippinensis*), sugarcane downy mildew (*P. sacchari*), Java downy mildew (*P. maydis*). Identification of molecular markers linked to downy mildew resistance gene should have a major impact on maize breeding across the tropical Asian region. As a network activity under the AMBIONET programme, four countries (India, Indonesia, Thailand and Philippines) undertook a QTL mapping project aimed at identifying downy mildew resistance gene. The mapping was based on evaluation of a set of recombinant inbred lines (RILs) derived from a cross of resistance susceptible variety of maize. During 2000-2001, 135 NIL families were evaluated for downy mildew reaction in South India against *P. sorghi*, in Western India against *P. heteropogoni*, in Thailand against *P. zea* and in Philippines against *P. philippinensis* (Prasanna and Hoisington, 2003).

In a genetic map the distance between markers is represented by a unit called **cM** (centi Morgan) and the physical distance is represented by another unit (megabase pairs = $10^6$ baise pairs).

The AMBIONET study led to the identification of QTL with significant effect on resistance to the five important downy mildew diseases affecting maize production in Asian region. Three QTL (two on chromosome 2 at 158 cM and 234 cM position, respectively, and one on chromosome 7) significantly influenced resistance only to a particular pathogen populations. Fig. 14.8 shows genotyping of a panel of backcross $F_1$ mapping population (for two polymorphic SSR loci, *bnlg490* (A) and *bnlg1655* (B). The mapping population was derived by using CM139 (elite susceptible ($P_1$) and NAI116 ($P_2$) used as susceptible and resistant parents, respectively. Analysis of genotypic and phenotypic data from this mapping population confirmed the effect of QTL detected on chromosome (Prasanna and Hoisington, 2003).

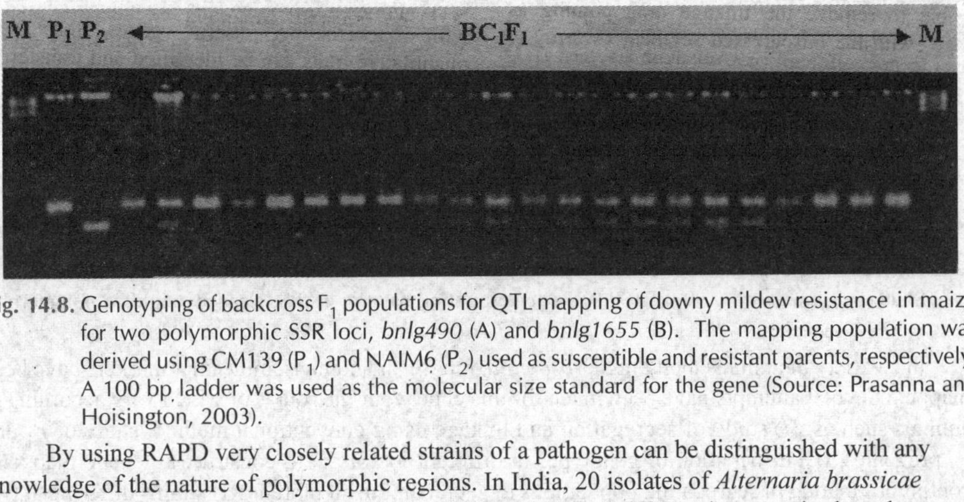


**Fig. 14.8.** Genotyping of backcross $F_1$ populations for QTL mapping of downy mildew resistance in maize for two polymorphic SSR loci, *bnlg490* (A) and *bnlg1655* (B). The mapping population was derived using CM139 ($P_1$) and NAIM6 ($P_2$) used as susceptible and resistant parents, respectively. A 100 bp ladder was used as the molecular size standard for the gene (Source: Prasanna and Hoisington, 2003).

By using RAPD very closely related strains of a pathogen can be distinguished with any knowledge of the nature of polymorphic regions. In India, 20 isolates of *Alternaria brassicae* collected from geographically distinct region of the world and different host species have been analyzed. RAPD analysis also differentiated genetic variation among seven isolates of *A. brassicicola* collected from France, Canada and England.

## 3. Physical Maps Using *In Situ* Hybridization (ISH)

*In situ* hybridization is a technique which detects and locates a given sequence of DNA directly within the cells or tissues. After preparation of smear slides are incubated in the hybridization buffer at high temperature so that the two strands of the cellular DNA could be separated. Thereafter, smear is put in a solution containing radiolabelled probe followed by autoradiography. ISH has been successfully used for mapping genes present on chromosomes by exploiting cDNA and gDNA (genomic DNA) probes. For instance, ISH carried out at metaphase of recombinant DNA in bread wheat cv. Chinese Spring. Showed the presence of 90% rRNA genes on chromosome 1B and 6B, whereas the rest of genes were present on chromosome 5D. Diploid genomes (*e.g.* A genome) which possess two chromosomal sites for rRNA genes, undergo changes when incorporated into 6X (hexaploid) bread wheat.

In wheat, barley and rye information on physical location of the 5S rRNA loci have been collected. After *in situ* hybridization, it could be found out that 5S rRNA loci are present at distant locus to secondary constriction on 1B chromosome of wheat, non-nucleolar chromosome in barley and 1R of rye.

In rye 5S rRNA is also present on 5 R chromosome. Further, in bred wheat physical map of 5S rRNA multigene family could be constructed by combining ISH technique with detection mapping. Moreover, the molecular markers could be located physically by ISH technique.

## C. CONSTRUCTION OF MOLECULAR MAPS IN ANIMALS

Molecular markers play a key role for livestock improvement through conventional breeding programme. The various possible applications of molecular markers are short range application or immediate and long range application (Table 14.3).

**> Table 14.3. Molecular markers useful in conventional live stock breeding.**

| *Application* | *Useful marker system* |
|---|---|
| *Short-range/immediate application* | |
| Parentage determination | DFP, microsatellite |
| Estimation of genetic distance | DFP, microsatellite, RAPD |
| Zygocity determination | RFLP, PCR RFLP, microsatellite |
| Sex determination | RFLP, PCR RFLP, microsatellite, DFP |
| Identification of disease carrier | RFLP, CAPS, microsatellite |
| *Long-term application* | |
| Gene mapping | RAPD, VNTR, minisatellite, microsatellite |
| Marker assisted selection | Any marker associated directly or indirectly |
| with genetic trait QTL | |

Source: Mitra *et. al.* (1999); DFP, DNA fingerprinting pattern; CAPS, cleaved amplified polymorphic sequence.

In a variety of animals including *Drosophila,* mouse and humans, molecular markers have been mapped. Several attempts have been made for this endeavour the routes of which vary according to animals such as: (*i*) study of segregation and linkage using conventional method, study of $F_1$ and $F_2$ progenies, (*ii*) distribution of genotypes at different loci in $F_2$ to construct a genetic map, (*iii*) continuous mating in segregating populations to get recombinant inbred (RI) strains differing at two or more loci that can provide information on recombination frequencies in recombinant inbreds, (*iv*) use of inter-specific crosses (*e.g. Mus musculus x M. spretus*) to overcome the problem for identification of allelic difference.

In humans controlled crosses as well as inter-specific crosses are not possible. Therefore, R1 lines cannot be obtained. But for linkage analysis LOD score can be calculated.

### 1. Molecular Genetic Maps in Humans (Homo sapiens)

During the 1980s, the concept of RFLP was proposed for disease diagnosis by using both gDNA and cDNA. RFLP markers are inherited in progenies in the form of the simple Mendelian codominant markers. Through pedigree analysis of several generations one can establish the linkage relationships among the RFGLP markers in humans. RFLP can be obtained as described earlier from lymphocyte DNA isolated from small amount of peripheral blood. Using standard LOD scores the data for restriction pattern (from every combination of probe-enzyme) of each individual can be generated and analysed for the presence of linkage.

In most of the human chromosomes, the markers are spaced about 15 cM apart in primary genetic maps. Hence, the high resolution maps containing markers spaced at 1-2 cM apart have been constructed by using the above primary genetic map. Moreover, the high resolution maps for some specific regions containing genes associated with certain diseases have been constructed. For construction of primary genetic map, 59 families were selected and cell lines developed. The laboratories that were collaborating to map the human genome were provided DNA from cell lines by the **Centre for Study of Human Polymorphism** (CEPH) of Paris. The same material was used in different laboratories also. Jean Dausset created the CEPH by spending the money which he obtained as Nobel prize. This facilitated to combine the information in a single linkage map. Resolution of 1 cM could be done by the end of 1993. Resolution map of 1 cM of human genome could contain 7000-8000 loci. Up to 1991, more than fifty percent of loci were mapped.

By studying co-segregation of disease with a marker in a family, genes for a specific disease present at a region of a chromosome can be found out with the help of primary linkage map. When the marker has been mapped and linkage of disease has been found out, the exact location of chromosome could be known through developing high resolution map of the region. The technique of 'chromosome walking', is employed to isolate the particular disease. For example, the gene for *cystic fibrosis* has been isolated following this method.

In addition, SSRs have also been used in preparation of genetic maps in humans. SSR sequences are developed by digesting the genomic DNA with restriction enzymes (*e.g. Sau*3AI or mixture of *Alu*I, *Rsa*I and *Hae*III), cloning in specific vectors, screening of genomic library for the presence of SSR through hybridization with repeating oligonucleotide probes such as (AC)n, (AAG)n, etc. In 1990, 'Human Gene Mapping 10.5' (HGM10.5) presented 368 SSRs. Each SSR consists of about 4 bp long motifs on about 20 bp long genome. It could be estimated that among 745 kb of genomic sequence of human one SSR occurs at every 6 kb distance.

### 2. Molecular Genetic Maps in Other Animals

Construction of molecular genetic maps in rodents like mouse will help the understanding of genetics and molecular biology of human diseases. Attempts are also being made to construct genetic maps of farm animals. This will help to improve the health of farm animals. Moreover, genetic maps of farm animals will support gene manipulation (through breeding programme) so that certain beneficial genetic traits may be increased. For detail see Chapter 11. Moreover, SSR maps of related species have been developed by using heterologous primers.

Beside, RFLP and RAPD are also used to prepare genetic maps of animals. By the end of 1989, a linkage map of mouse was prepared that consisted of 965 loci including phenotypic markers, biochemical markers, cloned genes and unknown DNA markers. In 1992, more than 300 SSR loci were mapped. *Mus musculus* Genome Organization (MUGO (U.S.A.) carried out most of the work on mouse.

### 3. Construction of Physical Maps using Molecular Markers

The cM units are proportionate to the recombination frequencies. The distance between centromere and a gene is denoted as a small fraction of the length of entire area of chromosome which is considered as unity. If a gene is located from the centromere at 1/10 locus of the whole area, the distance is expressed as 0.1

Sometimes, the distance in physical maps are not proportionate to the recombination frequencies. Hence the physical maps are more important than the genetic or linkage maps.

**(*a*) Physical Maps using Yeast Artificial Chromosome (YAC) and ISH:** The YAC vectors have been described in Chapter 5. Molecular maps in *Drosophila* has been constructed by using YAC.

*Drosophila* is selected as a model animal for mapping of a large number of eukaryotic genomes. Moreover, a significant number of specific regions of *Drosophila* chromosomes has been mapped by using chromosome walking. In YAC vectors a large sized DNA can be cloned as compared to cosmid vectors. The large clones of about 30 kb in length are prepared in cosmid vectors. To cover the whole genome of an organism (*e.g. Drosophila* or *Caenorhabditis elegans*) there is need of a few thousand clones. In this way to map the entire human genome, more than 10,00,000 clones are required in cosmids. It is very time-consuming process. DNA clones which are more than 10 times larger than these clones (3,00,000 to 1.5 million kb) can be cloned as YAC in yeast cells (Fig. 14.9). Thus the human genome can be represented by 10,000 YAC clones. For mapping of *Drosophila* genome a YAC library of 1,500 clones are enough. *Drosophila* YAC clones are often called DY clones.

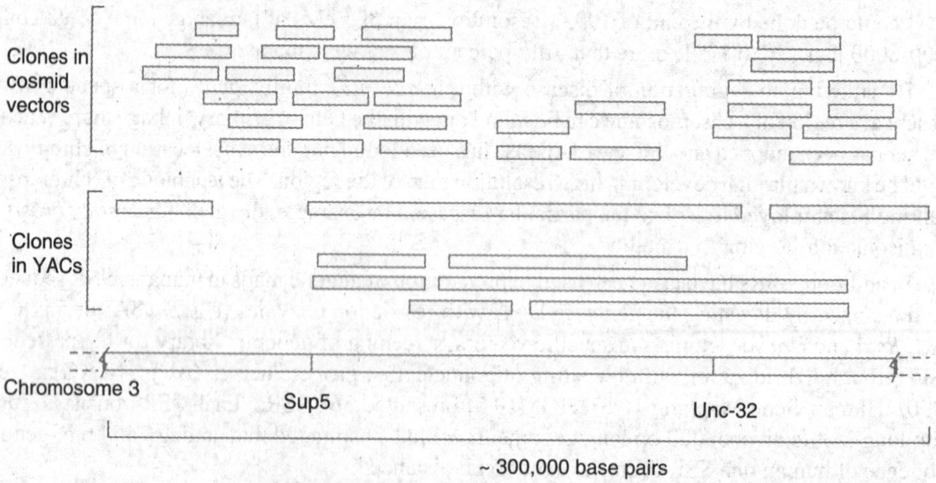


**Fig. 14.9.** Overlapping genomic DNA clones in the nematode worm *Caenorhabditis elegans* (only 0.3% of the total genome is shown) (diagrammatic).

The DY clones are prepared by isolating the large sized DNA from *D. melanogaster* DNA (using caesium chloride gradient centrifugation), separating DNA molecules (larger than 120 kb) in sucrose gradient (by size fractionation), ligating 120 kb long DNA in YAC vector, and transforming yeast cells by selected DY clones.

For the purpose of mapping DY, clones are separated from the yeast (where it has attached to yeast chromosome) using field inversion gel electrophoresis. Then *Drosophila* chromosome is separated from DY clones using restriction enzyme. Salivary gland chromosome of *Drosophila* is mapped following *in situ* hybridization (ISH) technique. Still there exists gap between them. Detailed restriction map (or sequence as per need) of probes is also done. In 1992 YAC contig of human was also published.

***(b)* Construction of Physical Maps by using Chromosome Walking:** A gene sequence of genomic library is identified by using a probe which hybridizes several clones. Each clone possesses some base pairs of a large fragment of a gene. When genome is partially digested, different genomes obtained from various types of cells results in several random fragments consisting of overlapping sequences. Entire sequence of no any fragment is represented in any probe. Therefore, the original genomic sequence may be constructed by using the overlapping sequences. Diagram of overlapping genomic DNA clones is shown in Fig. 14.9.

The large region of chromosome is reconstructed identifying the fragments containing an overlapping sequence following a technique called **chromosome walking** (Fig. 14.10).

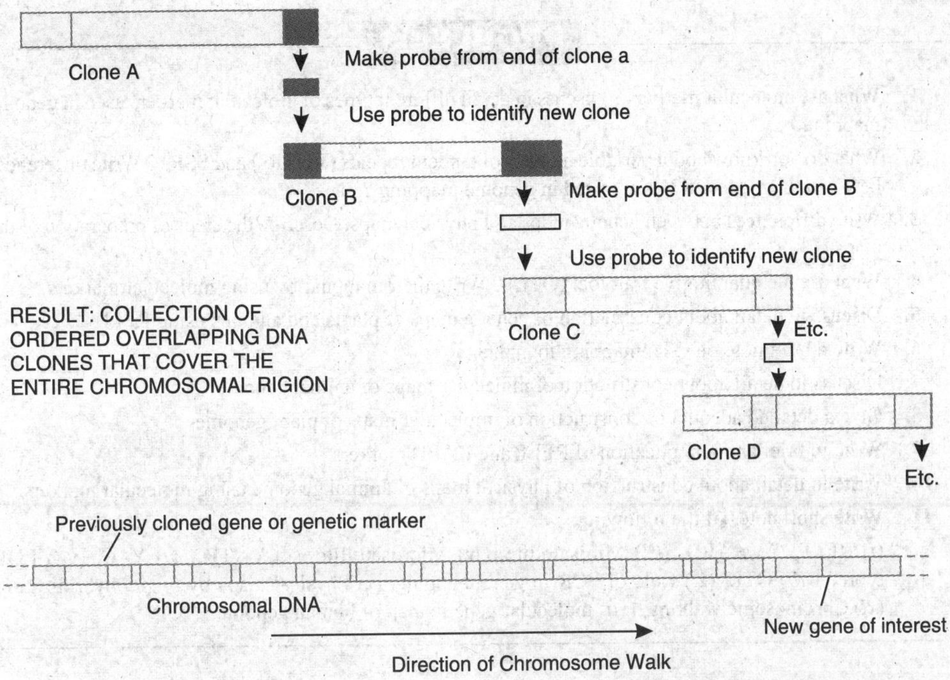


**Fig. 14.10.** Use of overlapping DNA clones to find a new gene by 'chromosome walking'.

Chromosome walking technique starts with: (*i*) selection of a desired DNA clone corresponding to a gene (or an RFLP marker), (*ii*) use of one end of this clone to prepare DNA probe, (*iii*) identification of overlapping sequences of clone through hybridization to find out an overlapping sequence of a clone in a genomic DNA clone library, (*iv*) purification of the DNA from this second clone and use of its far end to prepare a second clone (*i.e.* subclone) that contain overlapping sequence, (*v*) repetition of steps from (*ii*) to (*iv*), and preparation of the map of each selected clone and comparing to know the overlapping regions. In this way one can walk along one clone of a chromosome at a time (Fig. 14.10). Following this technique 1,000 kb long region of the chromosome has been mapped.

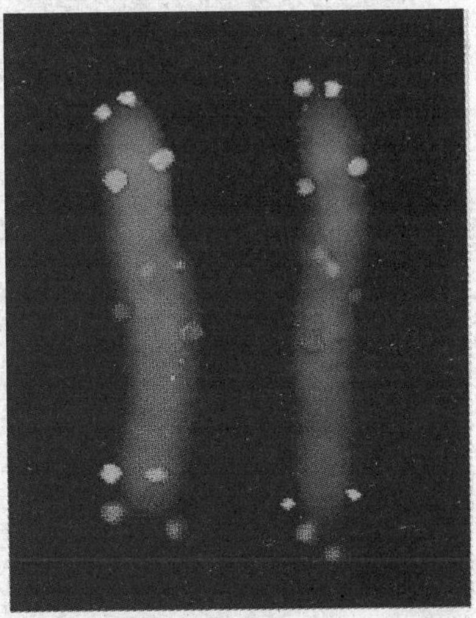

**Fig. 14.11.** *In situ* hybridization for locating the specific genes present on human chromosome 5. Six different probes were chemically labelled with fluorescent antibodies. At metaphase each probe represents two dots.

**(c) Construction of Physical Maps using *In Situ* Hybridisation (ISH):** Single copy DNA sequence of human genome has been successfully mapped by ISH technique. Methods of ISH are the same as desired for construction of physical maps of plants. Animal tissue to be worked out is gently fixed so that its RNA may remain in an unexposed form. When the tissue is incubated with a complementary DNA or RNA probe, it will be hybridized. Fig. 14.11 shows ISH of metaphase chromosome 5.

## PROBLEMS

1. What are molecular markers ? Discuss in detail different types of molecular markers used in genome mapping.
2. What do you know about variable number of tandem repeats (VNTRs) and SSRs ? Write differences between them. How SSRs are used in genome mapping ?
3. Write differences between genetic maps and physical maps. Describe the steps of preparation of the maps.
4. What are the quantitative trait loci (QTL) ? Write their mapping by using molecular markers.
5. Discuss in detail about construction of genetic maps of plants and animals using RFLP.
6. Write a brief note on QTL mapping in maize.
7. Discuss in detail about construction of molecular maps of animal genome.
8. Give a detailed account of construction of molecular maps of plant genome.
9. Write in brief about application of RFLP and RAPD markers.
10. Write in detail about construction of physical maps of animal genome using molecular markers.
11. Write short notes of the following:

    (*i*) RFLP, (*ii*) RAPD, (*iii*) Minisatellite, (*iv*) Microsatellite, (*v*) VNTR, (*vi*) YAC, (*vii*) LOD score, (*viii*) QTL, (*ix*) Role of PCR in gene mapping, (*x*) Physical maps by *in situ* hybridization, (*xi*) Chromosome walking, (*xii*) molecular genetic map of human genome.

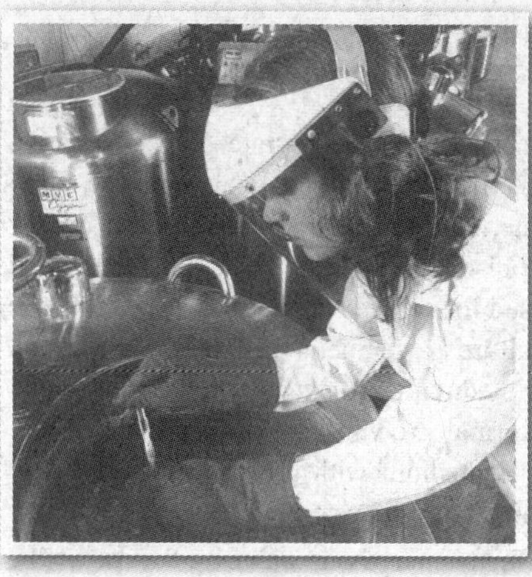

CHAPTER

# 15

# Cryopreservation

Cryopreservation (Gr. *Kryos* means frost) refers to 'preservation in the frozen state'. It means storage at very low temperature such as over solid carbon dioxide (– 79ºC), in deep freezers (– 80ºC), in vapour phase nitrogen (– 150ºC or in liquid nitrogen (– 196ºC).

The plant material is generally preserved and maintained in liquid nitrogen. There are several techniques for preservation of microbial cells including cryopreservation too. Moreover, cryopreservation of animal stock cells and cells lines is preferred to protect them from genetic drift (arising due to genetic instability), microbial contamination, cross contamination by other cell lines, incubator failure, senescence, etc. This Chapter deals with cryopreservation of plant and animal stock cells.

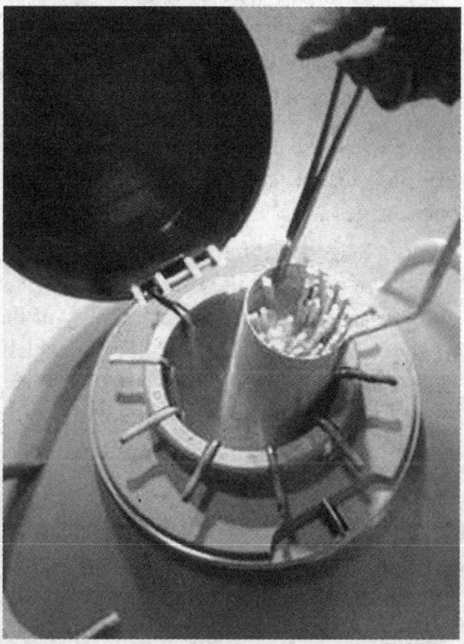

Sperm storage tank.

## A. CRYOPRESERVATION OF PLANT STOCK CELLS

Due to gradual disappearance of economic and rare plant species the necessity for storage of genetic resources of plant realm in general and agricultural plants in particular was realized by the biologists. The conventional methods of storage fail to prevent from losses caused by: (*i*) attack of pathogens and pests, (*ii*) climatic disorders, (*iii*) natural disorders, and (*iv*) political and economic causes. However, the conventional methods could not save the viability of short lived seeds of economic plants, for example, oil palm (*Elaeis guineensis*), rubber (*Hevea brasiliensis*), *Citrus* sp. and *Coffes* sp. (Dodds, and Roberts, 1985).

These materials are stored at low temperature, due to which growth-rate of cells retards; consequently biological activities are conserved for long time. **Cryobiology** deals with the study of metabolic activities and their responses in plant materials (and animal cells) stored at low temperature (–196°C) by using liquid nitrogen in the presence of cryoprotectants. Dodds and Roberts (1985) have discussed three principal methods used for growth suppression in plant tissue culture: (*i*) the alteration of physiological conditions of culture *i.e.* temperature or gas composition within the vessel; (*ii*) changing the composition of basal medium *e.g.* using sub or supra-optimal concentrations of nutrients (some factors essential for normal growth may be either omitted or employed at a reduced level) and (*iii*) supplementing the medium with growth retardants (*e.g.* abscisic acid) or osmoregulatory compounds (*e.g.* mannital, sorbital, etc.).

Storage at reduced temperature has been very effective for tissue culture of most of the plant species such as potato, cassava (*Manihot esculentum*), pea (*Pisum sativum*), chickpea (*Cicer arietinum*), rice (*Oryza sativa*), wheat (*Triticum vulgare*), coconut (*Cocos nucifera*), oil palm (*E. guineensis*) strawberry (*Fragaria vesca*) and sugarcane (*Saccharum officinarum*).

## 1. Difficulties in Cryopreservation

Several reviews available during the last two decades illustrate the significant progress made in this field and also the outline of the existing problem. The difficulties associated with cryopreservation are: (*i*) high specific feature of plant cells, such as their large size, strong vacuolization and abundance of water, (*ii*) cell damage during freezing and subsequent thawing caused by ice cyrstals formed inside the cells and by cell dehydration, and (*iii*) gradual formation of large crystals of more than 0.1 μm whose facets rupture many cell membranes. However, in the presence of cryoprotectants (the chemicals decreasing cryodestruction) and reduced temperature, free water has enough time to leave the cells. Therefore, it can freeze on the crystal surface in the solution. This results in marked dehydration and protoplast shrinkage. Excessive time and degree of plasmolysis are the reasons of cell destruction during slow freezing, since they cause irreversible contraction of the plasmalemma.

## 2. Methods of Cryopreservation

A standard protocol for cryopreservation of apical meristem is given in Fig. 15.1. The freezing-storage-thawing cycle is an external procedure which is accomplished in the following basic stages :

(*a*) **Selection of Materials.** A number of factors are taken into account for selecting the plant materials; the important ones are, nature and density of cells in the vials/ampules to be cryopreserved; because the cryoability of the cell cultures depends on these. Young meristematic, highly cytoplasmic and small cells which are non-vacuolated and thin walled and in small aggregates, are good materials to be selected for this purpose. Cell density in vials or amples should be high, as it shows prolonged survival at high cell density.

The ability of explant to survive at –196°C is influenced by the ability of their morphological and physiological conditions. Different types of explants which are used in cryopreservation are apical meristem and plant organs, ovules, anther/pollen, embryos, protoplasts, etc.

Log phase cell suspension of many plant species has successfully been cryopreserved. The cultured cells are not the ideal system that can be cryopreserved. In constrast the organised structures (e.g. shoot pieces, young plantlets, embryos) are preferred for cryopreservation.

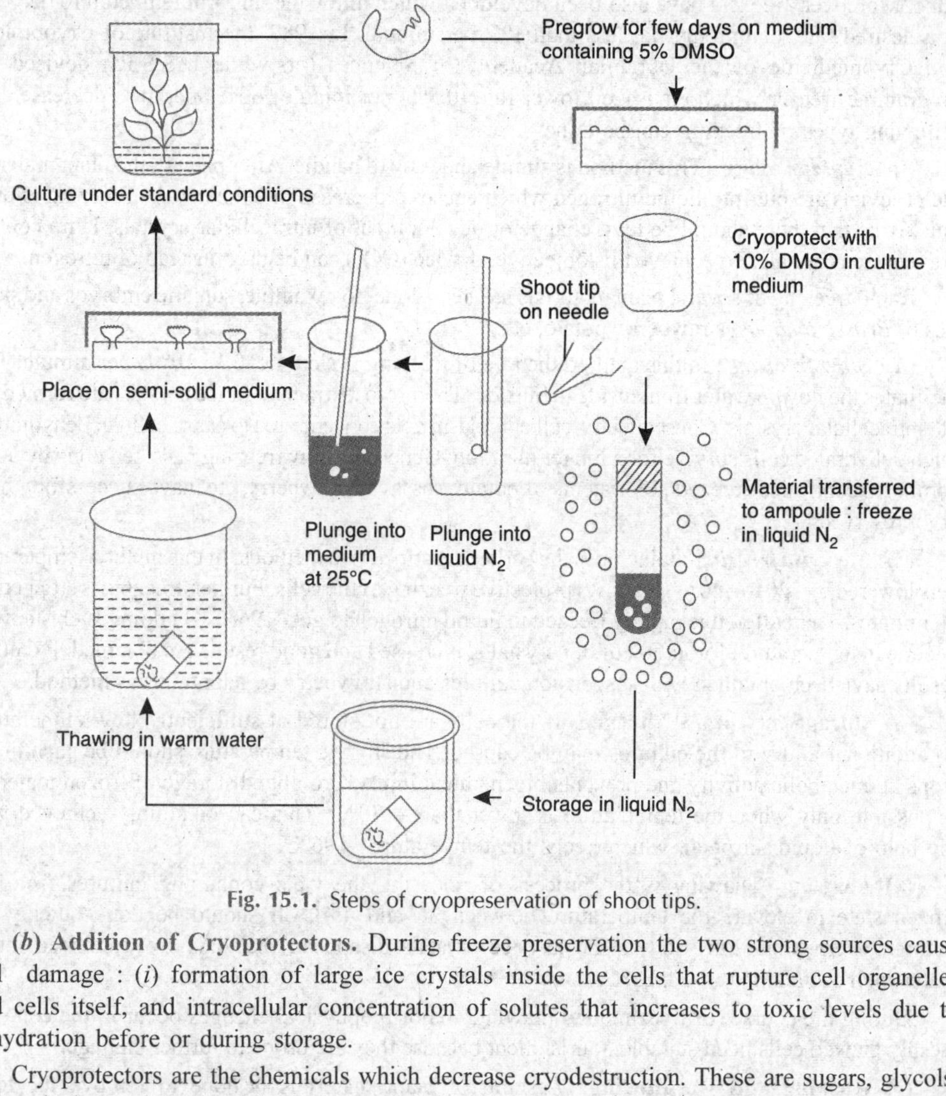


Fig. 15.1. Steps of cryopreservation of shoot tips.

**(*b*) Addition of Cryoprotectors.** During freeze preservation the two strong sources cause cell damage : (*i*) formation of large ice crystals inside the cells that rupture cell organelles and cells itself, and intracellular concentration of solutes that increases to toxic levels due to dehydration before or during storage.

Cryoprotectors are the chemicals which decrease cryodestruction. These are sugars, glycols, sugar alcohols, alcohols, polyvinylpyrrollidone, polyethylene glycol (PEG), polyethylene oxide (PEO), dextrans, hydroxystarch, glycerine, sucrose, and some amino acids (*e.g.* proline). Bajaj (1987) has advised to use a mixture of two or three cryoprotectants at low concentrations rather than a single cryoprotectant at a high concentration as it could be toxic. During treatment, the cultures should be maintained in ice to avoid deleterious effects.

Dimethyl sulfoxide (DMSO), sucrose, glycerol and proline are most frequently used cryoprotectants. DMSO has proved an excellent cryoprotectant because: (*i*) it has low molecular weight, (*ii*) it is easily miscible with solvent, (*iii*) it is toxic at low concentration, (*iv*) it is easily permeable into cells and easily washable.

The material to be preserved is put in culture medium and treated with a cryoprotectant. Such material is transferred to sterile cryovials or ampoules which are made up of polypropylene. Gradually 5-10% cryoprotectant is added into the ampoules. It is tightly closed with a screw cap.

**(*c*) Freezing.** Freezing should be done in such a way that it does not cause intracellular freezing and crystal formation. To avoid this problem, regulated rate of cooling or pre-freezing is done. Moreover, freezers have also been developed which allow the uniform temperature decrease at a desired rate, commonly not less than 1°C per minute. In 1987, the Institute of Cryobiology and Cryomedicine of the Ukrainian Academy of Sciences (erstwhile U.S.S.R.) devised the programme freezer which envisaged lower rate (0.5°C per minute) of temperature decrease. The following types of freezing can be done :

(*i*) ***Rapid freezing*** : This method is simple and easy to handle. After placing the plant material, the cryovials are put into liquid nitrogen which causes a decrease in temperature. Freezing is done quickly so that there should be least change or development of intracellular crystals. Ultra cooling prevents ice crystals. To achieve this objective, dry ice ($CO_2$) can be used instead of nitrogen.

Rapid freezing of several plant materials has been done, for example, somatic embryos and shoot tips of *Brassica napus*, strawberry, potato, etc.

(*ii*) ***Slow freezing*** : In this method the rate of freezing is slow i.e. 0.1 –10° C per minute. This facilitates the flow in water from inside to outside. Therefore, extracellular ice crystals are formed but not intracellular crystals. Consequently, cell cytoplasm gets concentrated to cause cellular dehydration. Such dehydrated cells survive for a longer duration. Generally slow freezing has been done by using computer-controlled freezers. Meristems of potato, cassava, strawberry, etc. have successfully been cryopreserved.

(*iii*) ***Stepwise freezing*** : It includes the earlier described two methods. In this method temperature gets lowered by –20 to –40°C. It allows protective freezing of the cells. Further, freezing is stopped for 30 minutes. Thereafter, it is rapidly freezed in liquid nitrogen to get –196°C. By doing such stepwise decline in temperature, formation of big crystal is increased and good results are obtained. Excellent results have been obtained with suspension cultures and strawberry by adopting this method.

(*d*) **Storage in Liquid Nitrogen.** If the cells are not stored at sufficiently low temperature, an additional injury to the cultures may be caused. The storage temperature should be such that it stops all metabolic activity and prevents biochemical injury. Prolonged storage of frozen materials is possible only when the temperature is lower than –130°C. This can be simply achieved with the help of liquid nitrogen, which keeps the temperature –196°C.

(*e*) **Thawing.** Thawing is the process of releasing the vials containing cultures from the frozen state to elevate the temperature between 35 and 45°C. It should be done quickly but without overheating. As soon as the last ice crystals disappear, the vials are tranferred into a water bath maintained at 20–25°C.

During the course of freezing and thawing, major biophysical changes occur in the cell. The freshly thawed cells need suitable nourishment because they are prone to further damage.

(*f*) **Washing and Reculturing.** Washing of plant materials is done to remove the toxic cryoprotectants. When low toxic or non-toxic cryoprotectants are used, the cultures should not be washed, but simply recultured. Washing becomes necessary only when cryoprotectants have toxic effects on cells. Washing follows the following procedure : dilution, resuspension, centrifugation and removal of cells. It is, however, possible that some cells die due to storage stress and the most stable ones survive. Therefore, determination of cell viability by culturing them on growth medium is essential.

Cell viability test can be performed done by using FDA staining and growth measurements. The parameters of growth measurements are counting cell number, dry and fresh weight, mitotic index, etc. Mitotic index (MI) is counted by using the following formula :

$$MI \frac{\text{Number of cells destined to cell division } (\textit{i.e.}\ P + M + A + T)}{\text{Total number of cells both dividing and undividing}}$$

where, P, cells in prophase; M, metaphase; A, anaphase and T, telophase (*g*) **Regeneration of Plantlets.** The viable cells are cultured on non-specific growth media to regenerate into plantlets. An extensive list of works on cryopreservation of cells, tissue, and organ culture of various plants *e.g.* potato, cassava, sugarcane, soyabean, groundnut, carrot, cotton, citrus, coconut, etc. has been given by many workers.

## 3. Plant Cell Bank/Germplasm Bank/Cell Cryobank

Cryopreservation of genetic stock *i.e.* germplasm (or vegetatively propagated crops, recalcitrant producing plants, rare plant species, medicinal, horticultural and forest plants, and VAM fungi) is a novel approach for their conservation in liquid nitrogen on a long term basis. To achieve this goal, a plant cell bank ( = germplasm bank and cell cryobank) has been suggested. Suggestions have also been made that germplasm bank should be attached to some of the International Research Institutes (*e.g.* IRRI) that would hold responsibility for the storage, maintenance, distribution (at national and international level), and exchange of these disease free germplasm of the important plants. Fig. 15.2. shows the potential and prospects of cryopreservation of plant cell, tissue and organ and establishment of germplasm bank.

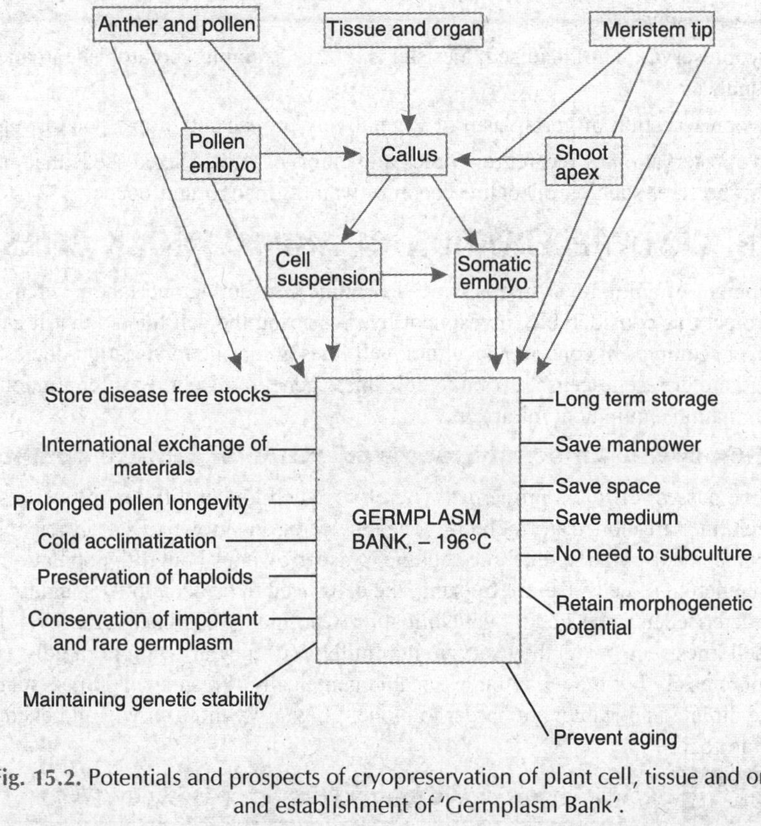


Fig. 15.2. Potentials and prospects of cryopreservation of plant cell, tissue and organ culture and establishment of 'Germplasm Bank'.

Facilities for storage of genetic stock of plants can be developed in large sized cylinders (30-50 litres capacity) where liquid nitrogen does not require refilling for 6-8 months. Thus, germplasm bank is such a device where facilities of cryopreservation of genetic resources of a variety of plants are available and on demand, the germplasm can be supplied nationally and internationally.

## 4. Pollen Bank

Besides germplasm bank, the storage of pollen grains in liquid nitrogen and establishment of pollen bank have also been suggested to retain their viability for various lengths of time. The freeze storage of pollen would enable: (*i*) hybridization between plants with flowers at different times, (*ii*) growth at different places, (*iii*) reducing the dissemination of diseases by pollination vectors, and (*iv*) maintenance of germplasm and enhancement of longevity.

### 5. Achievement Made Through Cryopreservation

Various forms of plant materials viz. cell suspensions clones, callii, tissues, somatic embryos, root/shoot tips, propagules (tubers) pollen grains, etc. have been preserved in liquid nitrogen for prolonged time and tested for their survival and regeneration potential.

No doubt, in most of the cases, the cells/tissues, organs regenerated into plants. A number of plant species that have been successfully cryopreserved have been described. Some of the observations made are as below :

(*i*) Cryopreservation of cell lines: For example, cell suspensions (soyabean, tobacco, dhatura, carrot) and somatic hybrid protoplasts (rice *x* pea, wheat *x* pea).

(*ii*) Cryopreservation of pollen and pollen embryos : For example, fruit crops, trees, mustard, carrot, peanut, etc.

(*iii*) Cryopreservation of excised meristems : For example, potato, sugarcane, chickpea, peanut, etc.

(*iv*) Cryopreservation of germplasm of vegetatively propagated crops: potato, sugarcane, etc.

(*v*) Cryopreservation of recalcitrant seeds and embryos: Large sized seeds that are shortlived and abortive, such as oil palm, coconut, walnut, mango and cocoa.

## B. CRYOPRESERVATION OF ANIMAL STOCK CELLS

Development of animal cell line is expensive, time consuming and labour intensive. It is essential to protect this considerable investment by preserving the cell line so that it can further be used whenever required. In continuous culture, cell lines are prone to variation due to selection in early passage culture, senescensce in finite cell lines, genetic instability in continuous cell lines, cross-contamination, equipment failure, etc.

### 1. Selection of Cell Line and Standaradisation of Culture Conditions

A cell line possess different properties. Therefore, a cell line of different properties is selected. Continous cell lines are cloned and suitable clone is selected and grown to get sufficient bulk of cells required for freezing. Similarly, the finite cell line is also grown to about fifth population doubling to get sufficient amount of cells. Before freezing, the cells need to be kept under standard culture conditions, characterised and checked for contamination. Continuous cell lines offer several advantages over finite cell lines such as: (*i*) they survive indifinitely, (*ii*) they grow more rapidly, (*iii*) they can be cloned more easily but they may be less stable genetically (Freshney, 2000). Usually the finite cell lines are diploid and stable, but harder to clone. They grow more slowly and eventually die or transfer (Table 15.1)

**> Table 15.1. Requirements before freezing (based on Freshney, 2002).**

| | | |
|---|---|---|
| Acquistion | Finite cell line | Freeze at early passage (<5 subcultures) |
| | Continuous cell line | Clone, select and characterise, amplify |
| Standardisation | Medium | Select optimal medium and adher to it |
| | Serum (if used) | Select a batch for use at all stages |
| | Substrate | Standardise on one type and supply |
| Ralidation | Provenance | Record details of life history and properties |
| Authentication | Check characteristics of cells line against provenance | |

| | |
|---|---|
| Transformation | Determine transformed status |
| Contamination | Test for microbial contamination i.e. sterility test of material |
| Cross-contamination | Adopt criteria to confirm identiity e.g. using DNA fingerpriniting, etc. |

The selection and their phenotypic expression of different cell types are influenced by the type of medium used. The best method to eliminate serum variation is to use serum-free medium. Unfortunately serum-free medium is not available for all cell types. Conversion is costly and time cosuming also. However, if serum is required, a batch is selected and used throughout every stage of cryopreservation.

## 2. Stages of Cryopreservation

When the cells are available from subculturing a primary culture or acquired cell lines, a few cryovials or ampoules are frozen. It is called *token freeze.* A *seed stock* is stored frozen after producing the propagated cell lines with the desired characteristics and free of contamination. The various stages in cell line preservation are given in Table 15.2.

**> Table 15.2.** **Stages in cell line preservation (based on Freshney, 2002).**

| *Stage* | *Source* | *No. of ampoules* | *Distribution* | *Validation* |
|---|---|---|---|---|
| Token freeze | Originator | 1-3 | None | Prevenance only |
| Seed stock | Originator stock token freeze | 12 | None | Viability, authenticity, transformation, contamination |
| Distribution stock | Test thaw from seed stock | 50-100 | Users, including other laboratories | Viability contamination, cross-contamination |
| User Stock | Distribution stock | - | None | Viability, contamination, cross-contamnation |

The seed stock of most cell lines should be protected. It should not be made available for general use. A *distribution stock* should be frozen and ampoules from this stock for a long period should freeze down their own *user stocks.* When the work is over, it should be discarded. The distribution stock sould be replenished from the seed stock after being over. When the seed stock falls below five ampoules, it should be replenished before issuing any other ampoules.

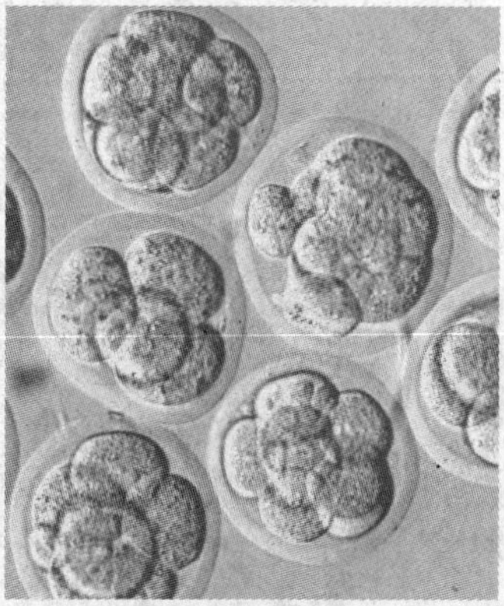
Cryo cells.

The cells suspension used for cryopreservation should have high cell density (106-107 cell/ml). By using glycerol (10-15%) or DMSO (5-10%) as preservative, cells are frozen slowly at about 1°C per minute. When temperature is below –70°C, the ampoules are rapidly transferred to liquid nitrogen. The ampoules are put in liquid nitrogen or in gas phase above the liquid. Then the ampoules cool rapidly to about –50°C but cooling rate falls after that. Therefore, the ampoules should be left overnight at –70°C before transferring to liquid nitrogen. However, when the cells are to be used, they are thawed rapidly and re-seeded at a high concentration to optimise their recovery.

### 3. Cell Banks

Several cell banks have been established for the secure storage and distribution of validated cell lines. Several cell lines may come under patent restriction (like hybridomas and other genetically modified cell lines); hence, they should be provided patent repositories with limited access. One must obtain the intial seed stock from a renowned cell bank where characterisation and quality control of cell lines are done. One must deposit the valuable cultures to a cell bank besides maintaining one's frozen stock. This will help to protect the loss of cell lines and its distribution to others too.

Data bank of cell lines has also been built up. This type of data bank provides a vast increase in the amount of available materials.

## PROBLEMS

1. What is cryobiology ? How can it help in storage of genetic resources ?
2. What do you know about germplasm bank and pollen bank ?
3. What are the difficulties associated with cryopreservation of genetic stock ?
4. Write short notes on the following :
   (*i*) Cryobiology, (*ii*) Pollen bank, (*iii*) Germplasm bank (*iv*) Stages of cryopreservation.
5. Write a Detailed note on cryopreservation of animal stock cells.

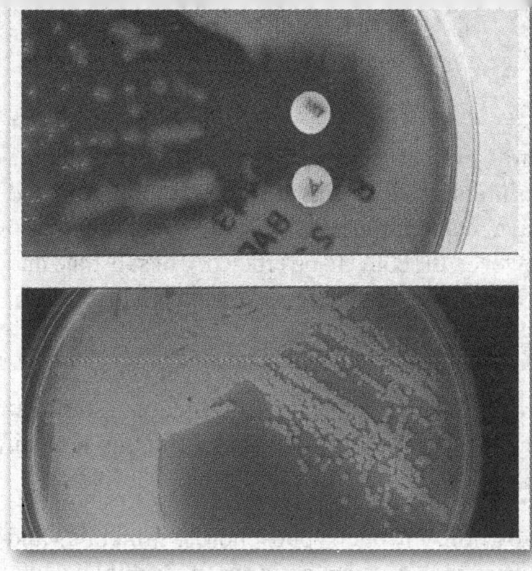

CHAPTER

# 16

# Features of Biotechnological Importance in Microorganisms

Until suitable methods for the isolation of specific microorganisms for specific product were known, they were grown merely by providing optimum growth conditions for a given product, for example, nutrients, temperature, etc. In recent years, a revolutionary success has been made for the production of alcohols, vitamins, organic acids, ketones, enzymes, antibiotics, etc. on large scale by: (*i*) characterisation of strains, (*ii*) improvement of potential mutant and (*iii*) their further use for mass production of chemicals. It is the inherent ability of micro-organisms that they utilize a variety of organic substrates and result in a wide range of products through metabolic process. Owing to presence of biochemical diversity, microorganisms have been commercially exploited by fermentation industry.

A microbial cell culture acts as a bioreactor or biofactory and converts the raw materials (using as substrate) into various products during specific time. A cell has its maximum capacity converting raw material into a certain product in a limited time. Therefore, the amount of product produced by bacterial cells within a given time can be calculated. Similarly, generation time (doubling time) of any microbial cell can also be calculated.

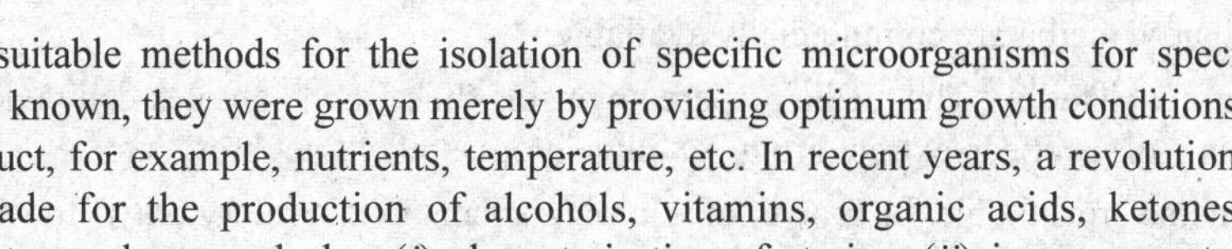


$$C_WH_XO_YN_Z + aO_2 \xrightarrow[\text{Optimum growth conditions}]{\text{Cells}} bCH_\alpha O_\beta N_\gamma + cCO_2 + dH_2O$$

**Fig. 16.1.** Cellular growth and production of product.

The optimum chemical reaction occurs at suitable pH temperature, solvent and atmospheric pressure. Such biochemical reactions also occur inside the cells of microorganisms. Therefore, microorganisms also need optimum conditions (such as pH, temperature, etc.) and growth media containing nutrients in particular substrate. The microbes utilise the nutrients at optimum conditions and convert into desired products (Fig. 16.1). After utilising a major amount of substrate the microbe also increase their biomass and growth and multiplication. However, the conversion efficiency of substrate into product falls in the range of 20-30% as compared to 80-90% conversion efficency of chemical reaction.

## A. TECHNIQUES OF MICROBIAL CULTURE

The technique of microbial culture is a multistep process and requires media formulation, sterilization, environmental control and operation of bioreactor, etc. These steps are discussed in this section.

### 1. Growth Media

Micro-organisms require several nutrients (e.g. carbon, nitrogen, phosphorus, minerals) and oxygen for growth and yield. The nutrient formulations which support optimum microbial growth and yield are called growth medial. On the basis of purity of chemical compounds used, media are grouped into the following three types :

(*i*) **Synthetic Media :** Microbes are cultured on a small scale in laboratory on artificially devised nutrient media by using pure chemicals. Such media are called synthetic media such as Czapek Dox agar medium for isolation of fungi.

(*ii*) **Semi-syntheic Media :** The media which contain pure form of chemicals as well as complex compounds are called semi-synthetic media; for example nutrient broth, brain heart infusion broth, etc. In these media the complex compounds are beef extract, yeast extract, peptone, potato or casein digest. Now these media are commercially available.

(*iii*) **Natural Media :** The media prepared by using the natural complex compounds are called natural media e.g. soyabean extract broth, V8 juice broth, soil extract broth, etc.

These media are suitable for growth of microorganisms in laboratory only. However, these are not used on a large scale because these are not economical. Such media are formulated for use on a large scale that are economical and of consistent quality, and available throughout the year. Besides, the raw materials are also pre-treated before use if desired. The following carbon and nitrogen sources used in media formulation are cheaper and economical and available throughout the year.

### 2. Sources of Nutrition

There are different sources of nutrients required by different types of microorganisms. These are given below :

(*i*) **Carbon Sources :** The carbon sources used for large scale microbial culture in fermentor are sugarcane molasses, beet molasses, vegetable oil, starch, cereal grains, whey, glucose, sucrose, lactose, malt, hydrocarbons, etc.

(*ii*) **Nitrogen Sources :** The nitrogen sources are corn steep liquor, slaughter-house wastes, urea, ammonium salts, nitrate, peanut granules, soyebean meal, soya meal, yeast extract, distilled soluble etc.

(*iii*) **Growth Factors :** There are certain micro-organisms which are not capable of synthesising vitamins or amino acids. Therefore, to achieve optimum growth, media are supplemented with the growth factors.

(*iv*) **Trace elements :** Micro-organisms also require certain trace elements (*e.g.* Zn, Mn, Mo, Fe, Cu, Co, etc.) in trace amount. Because these are associated with stimulation of metabolism or enzymes (metallo-enzymes) and proteins (leg-haemoglobin).

(*v*) **Inducers, repressors and precursors :** The catabolic enzymes are induced only in the presence of inducers. For example, yeast extract induces streptomycin. Production of catabolic enzymes and secondary metabolities is repressed by the presence of certain compounds in the culture medium. For example, rapidly utilised carbon sources repress the formation of amylases, griseofulvin, penicillin, etc. Besides, production of certain products is increased if precursor metabolite is supplemented in medium. For example, addition of phenyl acetic acid enhances penicillin fermentation.

(*vi*) **Antifoams** : Protein sources in culture media (*i.e.,* Products of medium are produced by micro-organisms) cause foaming. Mostly foaming creates problem in microbial process. Therefore, to check the foaming problem some antifoams i.e. fatty acids (such as sunflower oil or olive oil) are added in the culture medium.

(*vii*) **Water** : Water is the most important component of the living cells. Because all metabolic activites occur in cytosol. Water-soluble ionic forms of nutrients are absorbed by the cell. In laboratory single or double distilled water is used for preparation of culture media. But for large stage industrial production clean water of consistent composition is required. Dissolved chemicals and pH of water are measured. Water is also needed for ancillary activities for example, cleaning, washing, rinsing, cooling, heating, etc.

## 3. Procedures of Microbial Culture

Following are the procedures of microbial cultures :

(*a*) **Sterilisation** : Sterilisation is a process of complete eradication of micro–organisms from a given place or source. For small scale culture of micro–organisms in laboratory in culture tubes or flasks (in 100-1000 ml), the growth media are sterilised by autoclaving at 15 psi (*i.e.* pound per square inch) for 15-20 minutes. At this pressure, temperature reaches to about 120°C. According to requirement a pressure cooker or an autoclave is used. However, for large fermentation thousand to millions of litres of culture medium is used. Large sized fermentor and huge amount of medium are sterilised by using steam.

Besides, if medium is sterilised in a separate vessel, the fermentor must be sterilised by steam before passing the sterilised medium into it. Steam is sparged into the fermentor from its all entries. Steam pressure of 15 psi is maintained for 20-30 minutes. Then the steam is allowed to come out from air outlet. A sparger is attached at the bottom of fermentor. It is a tube containing several holes for coming out the steam. For detail description see preceding section and Fig. 16.2

(*b*) **Control of Environmental Conditions for Microbial Growth** : The success of fermentation to produce biomass and products depends upon the defined environmental conditions that exist inside the fermentor. Therefore, temperature, pH, agitation, $O_2$ concentration, etc. should be maintained during the process through careful monitoring of the fermentation. Microbial growth is significantly influenced by pH of the medium and temperature. Bacteria prefer neutral pH, while acidic pH favours the growth of yeast and fungi. Therefore, pH of the growth medium should be maintained before autoclaving as required by the microbe to be used in fermentor. During the process, suitable temperature of the medium should also be maintained, for example, psychrophiles at 5-15°C, mesophiles at 15-35°C and thermophiles at 45-65°C.

(*c*) **Aeration and Mixing** : Aeration and mixing are done in laboratory by keeping the flasks on shakers. Hence, it is called **shake culture.** On the other hand, in large sized fermentors the oxygen should be well mixed and dispersed in the medium, so that it may be available to microbial cells. Therefore, the fermentors are equipped with **stirrers** for oxygen mixing and **baffles** for increasing the turbulence. Thus adequate mixing is done by the stirrers and baffles. Sufficient mixing and oxygen requirement are also met out through forced aeration. The oxygen molecules come in the contact of cells and diffused inside the cell wall.

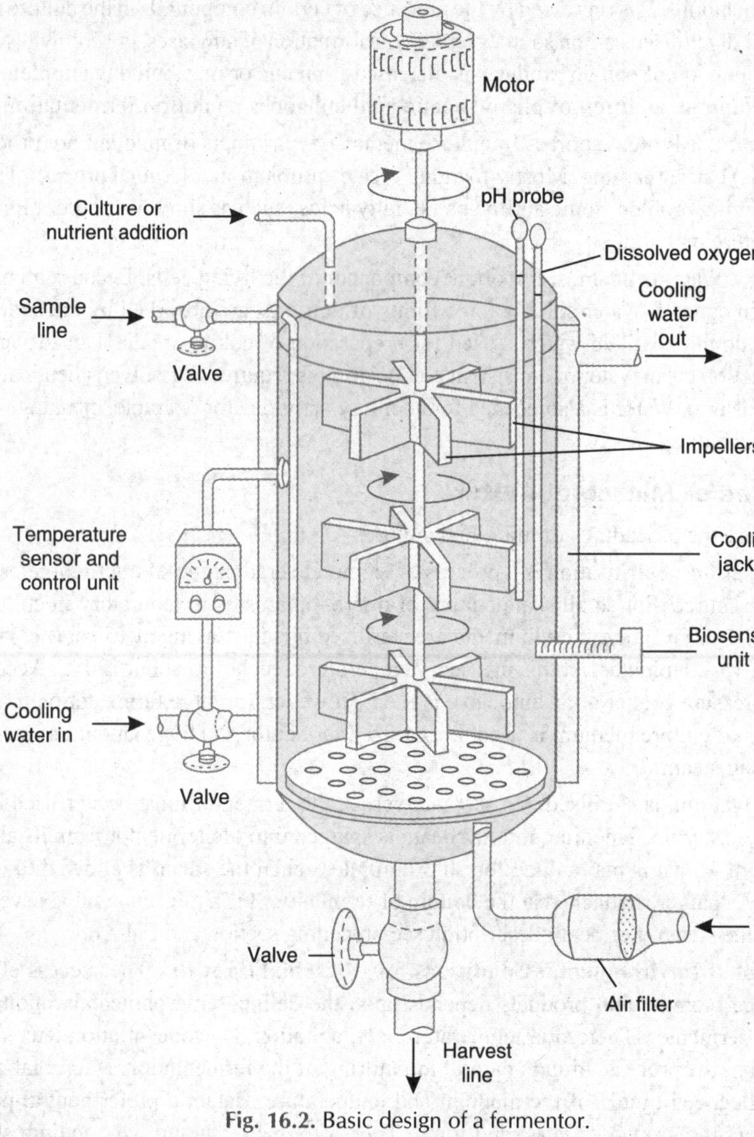


Fig. 16.2. Basic design of a fermentor.

**(*d*) Vessels for Microbial Cultures :** Microorganisms are grown on different types of vessels according to requirement. For example, in laboratory a single microbial culture is grown and maintained on **slant** of culture tubes. They are also grown in a simple Erlenmeyer flasks of different volumes (100-1000 ml). Growth in laboratory grown cultures can be improved simply by designing the flask using shakers at controlled temperature as given below :

**(*i*) Baffle Flasks :** When the sides of the flask are indented or V-shaped notch is produced, such flasks are called **baffle flasks.** The V-shaped notch or indentation increases turbulence of the agitated culture medium. This increases the efficiency of oxygen transfer and imporves growth of the micro-oganisms in the culture.

**(*ii*) Shakers :** There are different types of shakers used in laboratory. Shakers continuously agitate (100-120 throws/minute) the culture medium and efficiently transfer oxygen. This results in an improvement in microbial growth.

***(iii)* Fermentors (bioreactors):** Fermentors are the closed vessels which are used for production of products (cell mass and metabolites) on a large scale. Fermentors are of different capacities. Small scale fermentors (10-100 litres) are used by scientists in research laboratories for optimisation of different parameters of microbial growth and production of products. Besides, large scale fermentors (thousand to million litres capacity) are used in industries for production of commercial products.

Fig. 16.2 shows basic design of a fermentor. It is facilitated to meet several requirements such as pH and temperature control, aeration and agitation, drain or overflow and sampling facility. Interestingly, fermentor can be operated aseptically for many days under the controlled conditions.

Regular monitoring for physical, chemical and biological parameters is done through control systems of the fermentor, because these parameters influence the growth of microbial cells. Therefore, a maximum microbial yield can be achieved through monitoring these parameters.

As per requirement and use of types of microorganisms, there are different types of fermentors. But the most common among these is the **stirred tank fermentor** where the impellers are fitted to stir the medium. Stirring of medium is done forcibly. Otherwise the increasing concentration of microbial cells will deplete the dissolved $O_2$ concentration resulting in creation of anaerobic condition. Aerobic microbe will never grow under anaerobic condition.

Stirred Tank Fermentor.

Therefore, microbial growth will decrease with simultaneous decline in product production. Hence forced aeration favours rapid growth of microorganisms. Secondly, pH of the culture medium also declines sharply with rapidly growing microbial cells. Therefore, a pH probe fitted with the fermentor regularly monitors the pH and maintains at optimum by adding acid or alkali.

## B. FERMENTATION

The word fermentation is derived from a latin verb *'fervere'* which means to boil. However, events of boiling came into existence from the fact that during alcoholic fermentation, the bubbles of gas ($CO_2$) burst at the surface of a boiling liquid and give the warty appearance. The conventional definition of fermentation is the breakdown (metabolism) of larger molecules, for example, carbohydrates, into simple ones under the influence of microorgansims for their enzymes. This definition of fermentation had little meaning until the metabolic processes were known. In a microbiological way, fermentation is defined as "any process for the production of useful products through mass culture of micro-organisms" whereas, in a biochemical sense, this word means the numerous oxidation - reduction reactions in which organic compounds, used as source of carbon and energy, act as acceptors or donors of hydrogen ions. The organic compounds used as substrate give rise to various products of fermentation which accumulate in the growth medium (Riviere, 1977).

Almost in all organisms metabolic pathways generating energy are fundamentally similar. In autophototrophs, (*e.g.* some bacteria, cyanobacteria and higher plants) ATP is generated as a result of photosynthetic electron transport mechanisms, whereas in chemotrophs the source of ATP is oxidation of organic compounds in the growth substrates. The oxidation reaction may be accomplished in the presence of oxygen (in aerobes) or in absence of oxygen (in anaerobes). Thus, in aerobic microorganism the process of ATP generation is referred to as cellular respiration, whereas in anaerobes or aerobes functioning under anaeorobic condition, it is known as anaerobic respiration or fermentation.

Although, fermentation (*e.g.* brewing and wine production) was done for many hundred years, yet during the end of 15th century, brewing became partially industrialized in Britain. Antony van Leeuwenhoek (1632-1723) developed method to observe yeasts and other micro-organsim under the microscope but this study could not be further strengthened. By early 19th century Cagniard-Latour and Schwann reported that the fermentation of wine and beer is accomplished by yeast cells. It was L. Pasteur who observed micro-organisms associated with fermentation and causing many diseases in human beings. Detailed studies on fermentation product, culture improvement, recovery, and scale up of products were made after the world war I.

A fermentor is a vessel designed to carry out fermentation process, *i.e.* biological reactions under the controlled conditions. Hence it is also known as bio-reactor; while designing a fermentor, several criteria which help in maximizing the yield, are taken into account. These are: (*i*) long term operation in aseptic condition, (*ii*) adequate aeration and agitation, (*iii*) pH control system, (*iv*) sampling facility, (*v*) minimum labour in operation, harvesting, cleaning and maintenance, (*vi*) temperature control system, (*vii*) minimum evaporation losses from fermentor, (*viii*) suitable for a variety of processes. Fermentor is provided with limited amount of medium containing all the essential nutrients.

## C. TYPES OF MICROBIAL CULTURES

Differences lie in microorganisms with respect to their growth and production of products. Hence, the microorganisms are cultured in different types of vessels in various ways. Therefore, to get the desired product, microorganisms are grown as batch, fed-batch or continuous cultures.

### 1. Batch Culture

Batch culture is the simplest method. A desired microbe is grown in a closed culture system on a limited amount of medium of microbial culture. The laboratory grown microorganisms in ordinary flask is basically a batch culture.

In batch culture, growth phase of microorganisms passes through many stages. A microbe grows in the medium until the nutrients are exhausted or toxic metabolites secreted by it reach to inhibitory level. From the begining of inoculation to the end, microbial culture passes through several stages (Fig. 16.3). After inoculation, the microbe takes some time to adjust in the new environment according to size of fermentor, and hence does not grow in the medium. Thus, the time taken for adaptation before to come to its active growth is known as **'lag phase'**. The micro-organism grows luxuriently till nutrients are present.

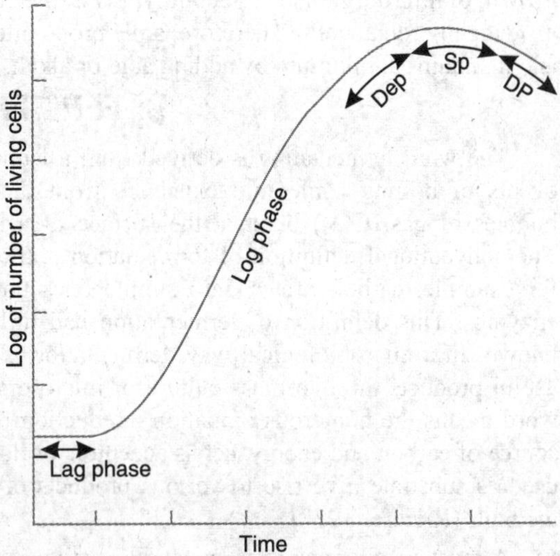


**Fig. 16.3.** Sigmoidal growth curve of a batch culture; DeP-deceleration phase; SP-stationary phase: DP-death phase.

Therefore , nutrient-dependent logarithmic or exponential active growth and, thereby, increase in biomass is known as **'log'** or 'exponential phase'. As soon as the levels of nutrients decrease, growth of culture is gradually slown down. The stage of retarding the growth to reach to stationary phase is known as **'deceleration phase'**. However, during stationary phase micro-organisms do not grow and thus, fail to increase their biomass. Ultimately the number of microbial cells declines due to accumulation of toxic metabolites. This stage is known as '**death phase**'. At the end, the amount of biomass depends upon the nutritional components and ability of micro-organisms to utilize the substrates and convert into biomass.

In 1965, J.D. Bu' Lock and coworkers have proposed different terminology for these growth stages. They used 'trophophase' for the log phase and 'idiophase' for the stationary phase of batch culture. This typical growth curve is known as 'sigmoidal growth curve'.

During the log phase of culture, growth rate of the micro-organisms reaches to its maximum (max). However, after depletion of a substrate, growth rate decreases and finally is ceased. The relationship of growth rate and concentration of the rate-limiting substrates can be explained by the following formula :

$$\mu = \frac{\mu \max [s]}{K_s + [s]}$$

where, $\mu$ = Growth rate constant,
$[s]$ = Concentration of limiting substrate,
$K_s$ = Saturation constant value of limiting substrate [s] at which the growth rate is half of the maximum growth rate ($\mu$ max).

***(a)* Microbial Growth Kinetics and Specific Growth Rate:** During log phase when cells utilise nutrients and grow to increase biomass, the growth behaves similar to autocatalytic reaction. At this phase growth rate is proportional to cell mass of that period. During this time the rate of 'cell mass increase' (*dx/dt*) is equal to the specific growth rate ($\mu$) and cell concentration.

$$\frac{dx}{dt} = \mu x \qquad ...(1)$$

where, $x$ = cell concentration (mg/cm$^{-3}$)
$t$ = time of incubation (hours)
$\mu$ = specific growth rate ($h^{-1}$)

Equation (1) may be written as

$$\mu = \frac{dx}{dt} . \frac{1}{x} \qquad ...(2)$$

The specific growth rate ($\mu$) acts as an index of rate of cell growth in that very environment. If you plot a graph between *dx/dt* and x, and determine the slope of straight line, you can calculate the specific growth rate. The value of specific growth rate can be converted into doubling time (*i.e.* time required by a cell to divide and double its number). It gives a better appreciation of the meaning of these values.

In the beginning of log phase ($t = 0$) the cell biomass is $X_0$. Assume that all the cells of batch culture require the same time to double their number (doubling time). Then on integration equation (2) you can calculate doubling time as below:

$$\mu.dt = \frac{1}{X} . \mu x$$

or $\log x = ut + K$ ($K$ = constant of integration) ...(3)

When $t = 0$ from equation (3), we get

$$\log x_0 = K$$

Putting the value of $K$ in equation (3), we get

$$\log X - \log X_0 = \mu . t$$

$$\log \frac{X}{X0} = \mu.t \qquad .....(4)$$

$$\therefore \quad td = \frac{1}{\mu} \ln \frac{X}{X0}$$

When $t = td$ (doubling time)

Then $X = 2X0$

$$\therefore \quad t_d = \frac{1}{\mu} \text{In } 2$$

or

$$t_d = \frac{\text{In } 2}{\mu}$$

If $t = td$ (doubling time), and $X$ = double of $X_0$, then

$$\text{In2} = \mu.td$$

The μ is inversely proportional to $t_d$. If the microbial culture have high $t_d$, it will have low μ and *vice versa.* Specific growth rate of $1\text{h}^{-1}$ is equivalent to $t_d$ of 0.693. Therefore,

$$t_d = \frac{0.693}{\mu} \qquad ...(6)$$

You can calculate the doublling time of the culture by calculating and substituting in equation (6)

There is another method of calculating the specific growth rate of a microbial culture. Number of microbial cells at different time is counted. If the medium is inoculated with $X_0$ cells at the time $t_0$, the number of cells after the time $t_t$ becomes $X$. The equation (1) may be written as

$$\ln \frac{X}{X_0} = \mu.t \text{ or } \ln X - \ln X_0 = \mu.t$$

After converting the natural logarithms (ln) to logarithm to the base 10, you get

$$\frac{\mu}{\dfrac{\log_{10} X - \log_{10} X_0}{2.303}} = -(t_t - t_0) \qquad ....(7)$$

or

$$2.303 (\log_{10} X - \log_{10} X_0) = \mu (t - t_0)$$

or

$$\mu = \frac{(\log_{10} X - \log_{10} X_0) \, 2.303}{t - t_0}$$

This equation is true for biomass concentration ($X$) as well as cell number ($N$). If at the time $t_0$ the microbial culture consists of $10^4$ cells/m1 and attains $10^{10}$ cells/m1 after 3 hours, the specific growth rate (μ) of the culture can be calculated using equation (3) as below:

$$\mu = \frac{(t - t_0) \, 2.303}{(10^{10} - 10^4)} \; (\log 2 = 0.301)$$

$$= \frac{(10 - 4) \, 2.303}{3} = \frac{6 \times 2.303}{3} = 2 \times 2.303 = 4.606 \; h^{-1}$$

The specific growth rate of microbial culture is 4.606 $\text{h}^{-1}$. Using the value of μ (*i.e.* 4.606 $\text{h}^{-1}$) in equaiton (6) you can derive $t_d$ of the culture :

$$t_d = \frac{0.693}{\mu} = \frac{0.693}{4.606} = 0.15 \text{ hour} = 4 \text{ minutes}$$

The $t_d$ of the microbial culture is 4 minutes.

It shows a reciprocal relationship between specific growth rate and doubling time $t_d$. If μ is high, the $t_d$ will be low and vice versa.

The μ represents the capacity of the microbial culture to grow in a specific environment. The μ of a microbial culture is measured during log phase of growth during which balanced cell growth occurs. In a batch culture the values of μ varies having maximum value at exponential (log) phase of growth. The environmental factors (*e.g.* pH, temperature, medium composition, aeration, etc.) that affect microbial growth also effect the specific growth rate of the culture. Representative values of $\mu_{max}$ of some microorganisms are given in Table 16.1

**> Table 16.1. Representative values of $\mu_{max}$ of some microorganisms.**

| *Mircro-organisms* | $\mu_{max}$ $(h^{-1})$ |
|---|---|
| *Benechea natriegens* | 4.24 |
| *Methylomonas methanolytica* | 0.53 |
| *Aspergillus nidulans* | 0.36 |
| *Penicillium chrysogenum* | 0.12 |

## 2. Continuous Culture

A continuous culture is that culture where a steady exponential phase for growth of culture retards due to depletion of nutrients, rather than by accumulation of toxic products; it is prevented by addition of fresh medium to the fermentor and removal of spent medium and microbial biomass from it as a result of which the exponential phase of culture is prolonged.

Continuous culture is an open process in which microbial cultures also grow continuously in log phase. One of the nutrients of culture medium is kept limited. Hence, at log phase the cell growth stops as the nutrients of limited quantity is exhausted. In continuous culture, fermented medium is continuosly removed from the fermentor. Therefore, to keep the culture always in log phase, fresh medium is added continuously to the fermentor (before diminishing the nutrients) at the time of removal of medium. Here the rate of supply of nutrients in the form of raw material and removal of products/cells should be volumetrically the same *i.e.* volume added is equal to volume removed. It means that volume of the medium always remains constant. This should be optimised with different microbial cultures and different growth media. If the working volume of the fermentor is $V m^3$, and the rate of flow in and out is $F\ m^3h^{-1}$, then the dilution rate ($D$) will be

$$D = \frac{F}{V}$$

$$\text{or} \quad F = DV \qquad \text{....(8)}$$

The unit of D is per hour ($h^{-1}$).

The output of biomass from a continuous culture system is given by the rate at which medium passes out of the outflow (*i.e.* the flow rate, *F*) multiplied by the concentration of biomass in that out flow (*i.e.X*).

$$\text{Thus,} \quad \text{output} = FX \qquad \text{....(9)}$$

Putting the value of F of equation (8) in equation (9), we get

$$\text{Output} = DVX \qquad \text{...(10)}$$

The productivity of this system (output per unit volume) is thus as below :

$$\text{Productivity} = \frac{DVX}{V} = DX \qquad \text{...(11)}$$

In continuous culture cells are grown at a particular growth rate. Then it is maintained for a long time. Most often the continuous culture is used for production of biomass of metabolies. Besides, liquid wastes are treated by using continuous culture. Microorganisms utilise organic materials of liquid wastes. Thus microbial biomass is produced in high amount. When such system is in equilibrium, cell number and nutrient status remain constant. At this stage the system is said to be in steady state.

### 3. Fed-batch Culture

Basically it is the batch culture which is fed continuously with fresh medium without removal of the original culture medium from the fermentor. It results in continuous increase in volume of medium in the fermentor.

In fed-batch culture the nutrients should be added at the same rate as they are consumed by the growing cells. Therefore, excess of nutrient addition should be avoided. In batch culture when high concentration of substrate inhibits microbial growth, the fed-batch culture is preferred over the former. Hence, in a fed-batch culture substrate is fed at such a concentration that remains below the toxic level. This activity accelerates the cell growth. A high cell density is achieved in fed-batch culture as compared to batch culture.

Fed-batch culture is an ideal process for production of intracellular metabolities in maximum amount. For example, alkaline protease used in biological detergents is produced by the species of *Bacillus.* Batch feeding of nitrogen sources (*e.g.* ammonia, ammonium ions and amino acids) keeps these substrates at low concentraton and induces protease synthesis.

## D. MEASUREMENT OF MICROBIAL GROWTH

To exploit the maximum capacity of biological process, full knowledge of microbial growth is important so that maximum amount of product could be produced. However, growth pattern of different groups of micro-organisms differ. For example, bacteria grow through binary fission, yeast grow by budding, fungi form hyphal branches and develop a complex network called mycelium. While viruses multiply intracellularly inside the living cells of suitable host. For example, bacteriophages multiply inside the bacterial cells, cyanophages inside cyanobacterial cells, mycophages inside fungal cells and actinophages inside cells of actinomycetes. Moreover, there lies host-specificity among the viruses.

### 1. Methods of Measuring Microbial Growth

There are different methods of counting microbial growth. These are based on different parameters of cells such as dry-weight and wet-weight measurement, absorbance, cell plate, density, turbidity, ATP measurement, viable count, ATPase activity and use of Coulter counter.

(*a*) **Wet Weight Measurement :** Measuring cell mass is an easy step of cell growth measurement. A known volume of culture sample from the fermentor is withdrawn and centrifuged. Wet weight of pellets is measured by using pre-weight filter paper. A pre-weighed filter paper of similar size is used to substract the weight of wet filter paper. Thus wet-weight of cells is calculated.

(*b*) **Dry Weight Measurement :** Dry weight measurement of cell material is similar to that of wet weight. Here dry weight of pre-weighed filter paper containing pellets of microbial cells is measured. Dry weight of filter is nullified by substracting the dry weight of only filter paper of similar size. Thus dry weight of microbial cells can be obtained. For example dry weight of about one million cells of *E. coli* is equal to 150 mg. Dry weight of bacterial cells is usually 10-20% of their wet weight.

(*c*) **Absorbance :** Absorbance is measured by using a spectrophotometer. Scattering of light increases with increases in cell number. When light is passed through bacterial cell suspension, light is scattered by the cells. Therefore, transmission of light declines. At a particular wavelength absorbance of light is proportional to the cell concentration of bacteria present in the suspension. Thus cell growth of an bacterial suspension at a particular wavelength at different intervals can be measured in terms of absorbance and a standard graph (between absorbance and cell concentration) can be prepared.

*(d)* **Total Cell Count :** Cell growth is also measured by counting total cell number of the microbes present in that sample. Total cells (both live and dead) of liquid sample are counted by using a special microscope glass slide called **Petroff-Hausser Counting Chamber.** In this chamber a grid is marked on the surface of the glass slide with squares of known area (Fig. 16.4). The whole grid has 25 large squares, a total area of 1 $mm^2$ and a total volume of 0.02 $mm^3$ (1/50 mm). All cells are counted in large square and total number per ml sample is measured. If 1 square contains 12 cells, the total number of cells per ml sample will be : 12 cells x 25 squares x $50 \times 10^3 = 1.5 \times 10^7$ cells.

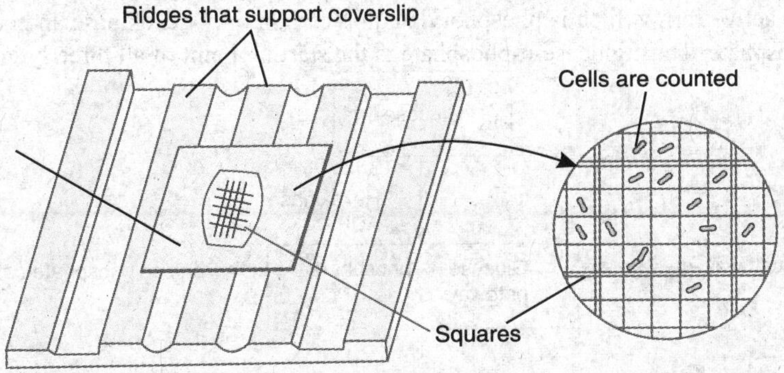


**Fig. 16.4.** Direct microscope counting by using a counting chamber.

If there is dilute culture, direct cell counting can be done. However, the cell culture of high density can be diluted. Otherwise clumps of cells would be formed which would create problem in exact counting of bacterial cells.

*(e)* **Viable Count :** A viable cell is defined as a cell which is able to divide and increase cell numbers. The normal way to perform a viable count is to determine the number of cells in the sample which is capable of forming colonies on a suitable medium. Here it is assumed that each viable cell will form one colony. Therefore, viable count is often called plate count or colony count. There are two ways of forming plate count. For detail see *Practical Microbiology* by R.C. Dubey and D.K. Maheshwari (2012)

## E. METABOLIC PATHWAYS IN MICROORGANISMS

Formation of vegetative cells in micro-organisms takes place only when there is continuous supply of energy. The cell components are synthesized by metabolism, which is "the ordered transformation of substances in the cell by a series of successive enzyme reactions through specific metabolic pathways". The metabolic pathways play dual role : *(i)* it provides precursors for the cell components and *(ii)* energy for energy requiring processes (Schlegel, 1986).

There is a variety of micro-organism which utilize organic compounds either in simple form ($C_1$) or complex form ($C_6$) or very complex form via a number of metabolic pathways. It is a unique feature of most of heterotrophic microorganisms that they secrete extracellular enzymes which act as chemical weapon for breaking down the substrates from complex forms to simple ones. In most of the microorganisms, the initial pathways, for example, glucose oxidation are common which are oxidized through glycolysis and Krebs cycle. However, in some microorganisms the metabolic pathways for utilization of substrates, for example, glucose differs. The metabolic pathways for the breakdown of hexoses are discussed in brief.

During photosynthesis, carbohydrates are synthesised in an adequate amount which serve a major portion of plant body and in turn, a source of nutrients for most of the microorganisms.

The macro-molecules are rendered from complex form to mono or dimeric forms by extracellular enzymes secreted by microorganisms. Once glucose enters into microbial cells, it is broken down into three carbon compound ($C_3$) through several routes such as glycolysis also known as fructose-1: 6-bisphosphate (FBP) pathway and Embden-Meyerhof-Parnas (EMP) pathway, pentose phosphate pathway or the Warburg-Dickens-Horecker pathway and 2-keto-3-deoxy-6-phospho-gluconate (KDPG) pathway or the Entner-Doudoroff pathway.

### 1. Glycolysis or EMP Pathway

It is most widely distributed catabolic pathway which proceeds through fructose-1 : 6-bisphosphate (FBP) hence, also known as FBP pathway (Fig. 16.5). Glucose comes in metabolically active form which is phosphorylated on carbon 6 by hexokinase and converted into glucose-6-phosphate. Thus, glucose-6-phosphate is the starting point of all three lytic mechanisms

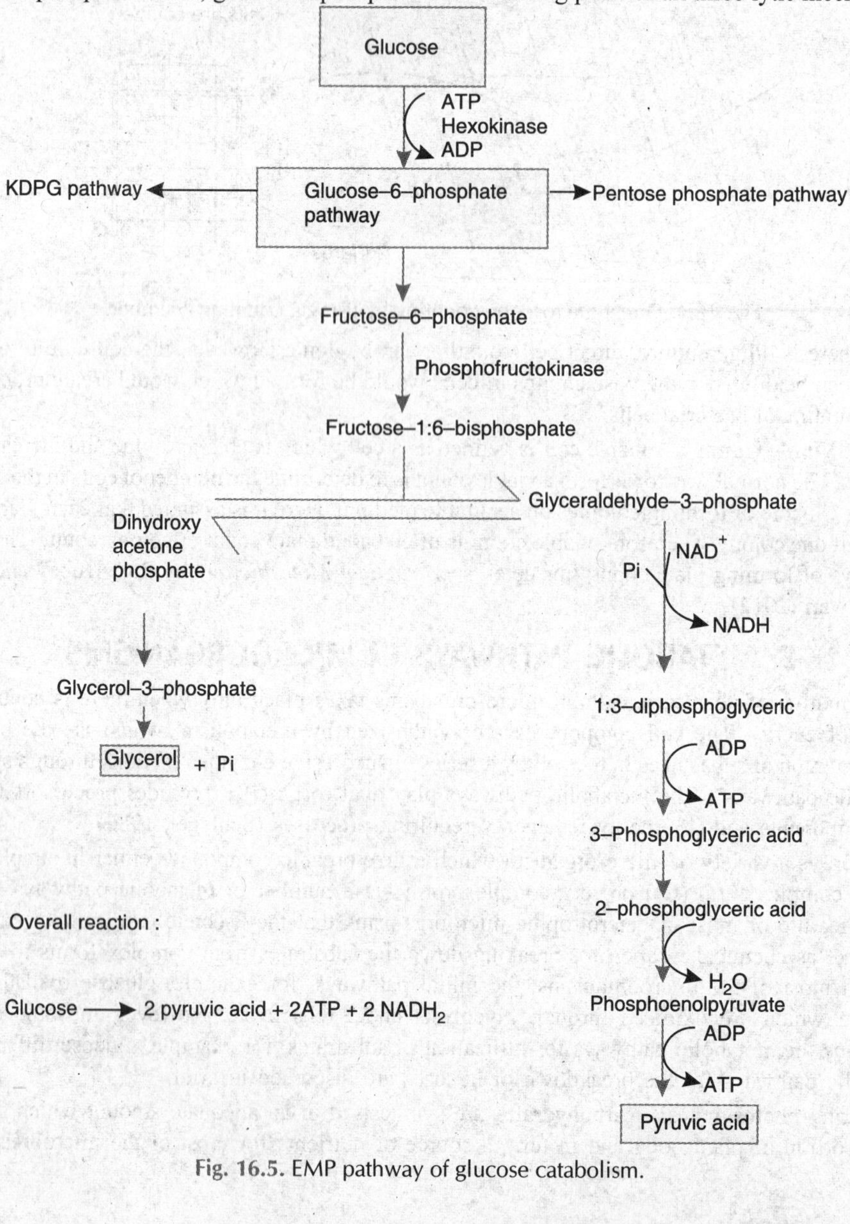


Fig. 16.5. EMP pathway of glucose catabolism.

(Schlegel, 1986). Glucose-6-phosphate is converted into fructose-1:6-bisphosphate which then is cleaved into triose phosphates. All the triose phosphates are converted into two molecules of pyruvic acid (pyruvate), and ATP and $NADH_2$.

## 2. The Entner-Doudoroff Pathway

Glucose is converted in its active form as glucose-6-phosphate. It is dehydrogenated to 6-phosphogluconate which removes water and yields 2-Keto-3-deoxy-6-phosphogluconate (KDPG) (Fig. 16.6). Due to formation of the intermediate product, the KDPG, this pathway is also known as KDPG pathway. The KDPG is then cleaved into pyruvic acid and glyceraldehyde-3-phosphate which is finally oxidized into pyruvic acid. In overall reaction one molecule of glucose yields two molecules of pyruvic acid and one mol of ATP, NAD(P) $H_2$ and $NADH_2$. This pathway is widely distributed in many bacteria of the genus *Pseudomonas.*

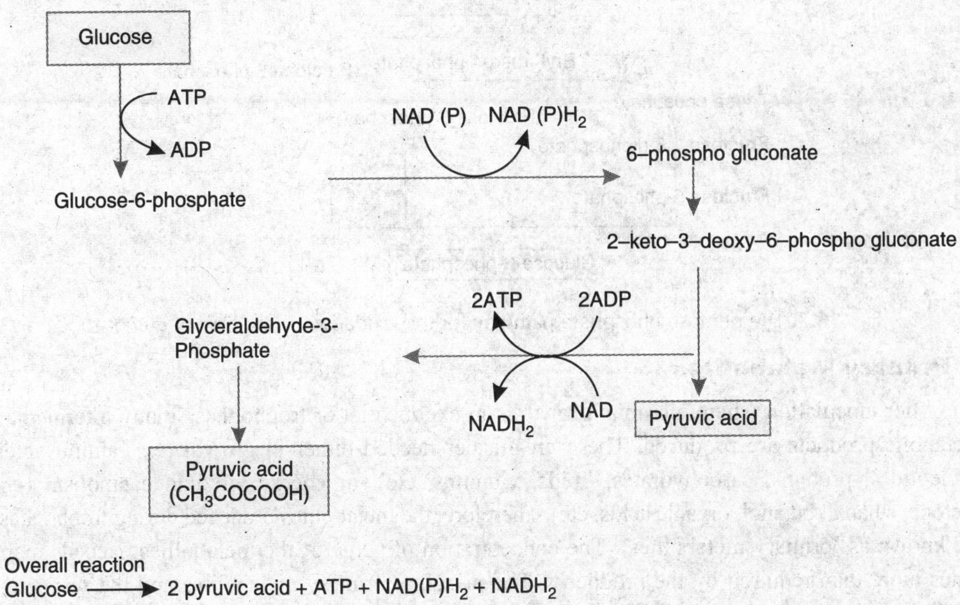


**Fig. 16.6.** The Entner-Doudoroff pathway for oxidative catabolism of glucose.

## 3. The Pentose Phosphate Pathway

This pathway forms a loop into the EMP pathway, for example, in heterofermenter lactobacilli. The bacteria do not synthesise aldolase which is needed to convert fructose biphosphate into two molecules of triose phosphate. Therefore, breakdown of glucose progresses through pentose phosphate pathway. Glucose-6-phosphate is converted to 6-phosphogluconate via dehydrogenation and hydrolysis. The 6-phosphogluconate yields ribulose 5-phosphate as the final oxidation product. Further conversion products are shown in Fig. 16.7.

# F. MICROBIAL PRODUCTS

Most of the natural products constituted by carbon, nitrogen, hydrogen, oxygen and phosphorus can be fermented under anaerobic conditions by microorganisms. There are many fermentation products used commercially (Fig. 16.8). A list of some microorganisms and their products are given in Table 16.2.

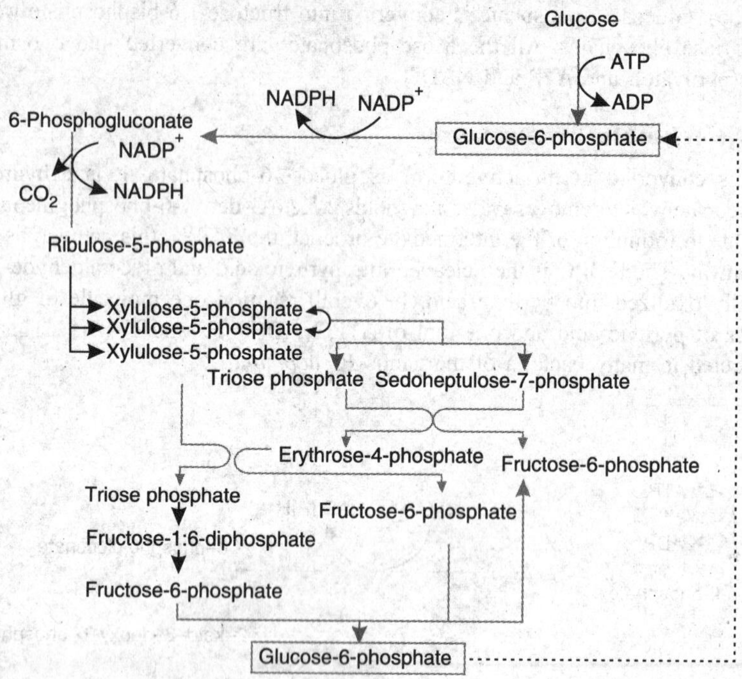


Fig. 16.7. The pentose phosphate pathway for the oxidative catabolism of glucose.

### 1. Primary Metabolites

After inoculation when microbial growth is in exponential or trophophase many intermediate metabolic products are produced. These are further needed either in growth (*e.g.* amino acids, nucleotides, proteins, carbohydrates, lipids, vitamins, etc), or energy yielding catabolism (*e.g.* acetone, ethanal, butanol, organic acids, etc). Therefore, the metabolites produced during trophophase are known as 'primary metabolites'. The concentration of some of the metabolites exceeds many times more than required by the producers. The principal primary metabolites and the respective micro-organisms are given in Table 16.2.

### 2. Secondary Metabolites

When the trophophase of growing culture is over, then starts the idiophase. Microbial products other than primary metabolites produced during idiophase by slow growing or non-growing cells of microorganisms are known as secondary metabolites or idiolites such as toxins, gibberellins, alkaloids, and antibiotics. The secondary metabolites play no role in growth of micro-organisms. It is produced by a limited number of micro-organisms (Table 16.2) when depletion of one or more nutrients is caused in culture medium.

### 3. Enzymes

Enzymes are naturally occurring biocatalysts which accelerate metabolic reactions. Various metabolic activities and production of primary and secondary metabolites are not possible without the involvement of enzymes. Enzymes produced during fermentation are mostly extracellular but a few are intracellular for example asparaginase, invertase and uric acid. Intracellular enzymes may be produced in industries, but with many difficulties. The important extracellular enzymes are amylases, cellulases, invertase, ß-galactosidase (lactase), esterase, lipases, proteases.

**> Table 16.2.** **Microorganisms and their products.**

| ***Microorganisms*** | ***Microbial products*** | | | |
|---|---|---|---|---|
| ***metabolites*** | ***Primary metabolites*** (1) | ***Secondary*** (2) | ***Enzymes*** (3) | ***Others*** (4) |
| **A. Algae :** | | | | |
| *Chlorella sorokiniana* | – | – | – | SCP |
| ***Spirulina maxima*** | – | – | – | SCP |
| *S. platensis* | – | – | – | SCP |
| **B. Bacteria :** | | | | |
| ***Acetobacter aceti*** | Acetic acid | – | – | – |
| *Acetobacterium woodii* | Acetic acid | – | – | – |
| *Bacillus brevis* | – | Gramicidin | – | – |
| *B. polymyxa* | – | Polymyxin B | Amylase | |
| *B. popilliae* | – | Endotoxin | – | – |
| *B. subtilis* | – | Bacitracin | – | – |
| *B. thuringiensis* | – | Endotoxin | – | – |
| *Clostridium aceticum* | Acetic acid | – | – | – |
| *Gluconobacter* | – | – | – | – |
| *suboxidans* | Vinegar | — | — | — |
| *Methylophilus methylotrophus* | Glutanic acid | — | — | — |
| *Pseudomonas denitrificans* | Vitamin B12 | — | — | Yoghurt |
| **C. Actinomycetes :** | | | | |
| *Micromonospora purpurea* | — | Gentamicin | — | — |
| *Nocardia mediterranei* | | Rifamycin | — | — |
| *Streptomyces aureofaciens* | — | Tetracycline | — | — |
| *S. tradiae* | — | Neomycin | — | — |
| *S. griseus* | — | Streptomycin | — | — |
| *S. noursei* | — | Nystatin | — | — |
| **D. Fungi :** | | | | |
| *Aspergillus niger* | — | Citric acid | — | — |
| *A. oryzae* | — | — | Amylase, Cellulase | SCP |
| | | | Amylase | Soya sauce |
| *Candida lipolytica* | — | — | Lipase | Soya sauce |
| *C. utilis* | — | — | — | SCP |
| *Cephalosporium acremonium* | — | Cephalosporin | — | — |

## 4. Microbial Biomass

Microbial cells which produce many commercial products, themselves serve as main source of biomass. Microbial biomass is exploited as microbial protein or single cell protein (SCP). SCP plays a significant role in supplying the protein in world food shortages. Production of SCP and its application have been described in detail in Chapter 18.

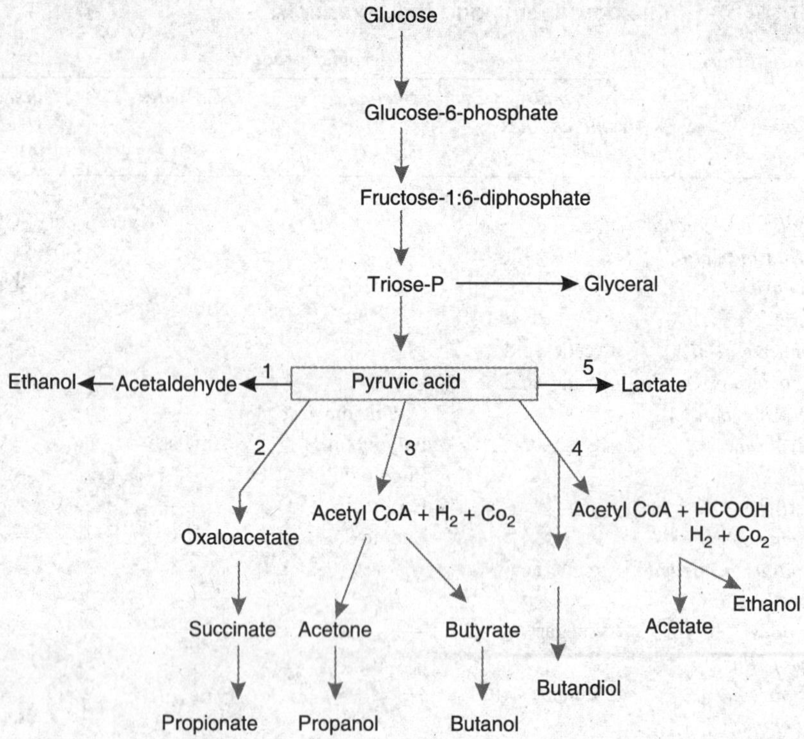


**Fig. 16.8.** Reduction of glucose into some important products during fermentation. 1-yeast; 2-propionibacteria; 3-clostridia; 4-coli-aerogenes groups; 5-lactic acid bacteria (modified after Schlegel, 1986).

## G. SCALE-UP MICROBIAL PROCESS

Microganisms are influenced by the continuously changing environment when moving inside the large fermentor. To avoid these problems the large scale should be taken as the point of reference. The possible effects should be studied by stimulation of the large scale variations in a small scale experimental set up. Limiting factors (*e.g.* mass transfer) are scaled down and can be studied and maintained in an economical way.

The desired microorganism can be used directly for production of products, because there are several risks associated with economics, production and quality of products. There must be more benefits on small investments but not *vice versa.* The laboratory processes need to be validated in laboratory at intermediate stage in a pilot plant. The pilot plant acts as a small model of commercial plant.

The results obtained from pilot plant are very useful in setting up a commercial scale plant. For setting up a commercial plant the capital investment is required in many areas such as design and construction of fermentor, utilities (*e.g.* steam water, electricity fitting, other infrastructure), manpower (technical and non-technical ) and marketing.

The results obtained from a pilot plant are theoretically extrapolated and applied for commercial plants too as far as capital investments are concerned. Finally, a techno-economic report is prepared based on the above data of capital investment. On the basis of these reports banks finance for establishment of commercial scale plant.

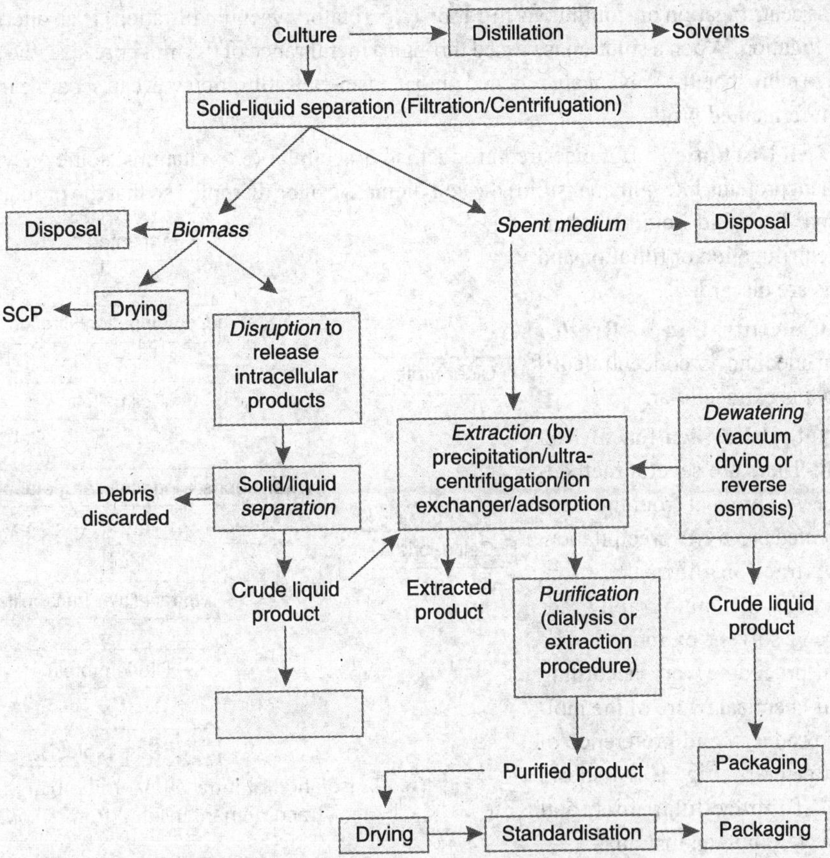


**Fig. 16.9.** Various stages of downstream processing.

## H. DOWNSTREAM PROCESSING

When fermentation is over, the desired microbial product is recovered from the growth medium. Then the product is purified, processed, and packed with equal efficiency and economy. Product recovery and purification is called **downstream processing**. The technology associated with down-stream processing is as important as technology-associated with the fermentation process itself. The operation of any fermentation production process integrates both the technologies.

Operation of downstream processing are summarised in Fig. 16.9. The volatile products can be separaed by distillation of the harvested culture without pre-treatment. Distillation is done at reduced pressure at continuous stills. At reduced pressure distillation of product directly from fermentor may be possible (as done for isolation of ethanol). The steps of downstream processing are as below :

***(a)* Separation of Biomass :** Usually the biomass (microbial cells) are separated from the culture medium (spent medium). If the product is biomass (single cell protein or vaccines), then it is recovered for processing and spent medium is discarded. Generally, cell mass is separated from the fermented broth by centrifugation or ultra-centrifugation. When there is no aeration and agitation, some of the microbial cells soon settle down in the fermentor.

Upon addition of floculating agents, settling may be more faster. For centrifugation process, settling of microbial cells is necessary. Otherwise, biomass separation may be affected. Ultrafiltration, continuous centrifugation or continuous filtration (*e.g.* rotatory vacuum filtration) is an alternative to the centrifugation. When a solution is passed through a membrance of 0.5 mm pore size, the particles having size more than the solvent are retained onto it. Using ultrafiltration you can separate microbial cells from fermented broth

(*b*) **Cell Disruption** : If the desired product is intracellular (*e.g.* vitamins, some enzymes and recombinant proteins like human insulin) the cell biomass can be disrupted so that the product should be released. The solid-liquid is seperated by centrifugation or filtration and cell debris are discarded.

(*c*) **Concentration of Broth** : The spent medium is concentrated if the product is extracellular.

(*d*) **Initial Purification of Metabolites** : There are several methods for recovery of product from the clarified fermented broth *e.g.* precipitation, solvent extraction, ultra-filtration, ion-exchange chromatography, adsorption and solvent extraction. The extraction procedure vary according to physico-chemical naure of the molecules of products, and preference of the manufacturers. Fig. 16.10 shows isolation of intracellular microbial product such as human insulin.

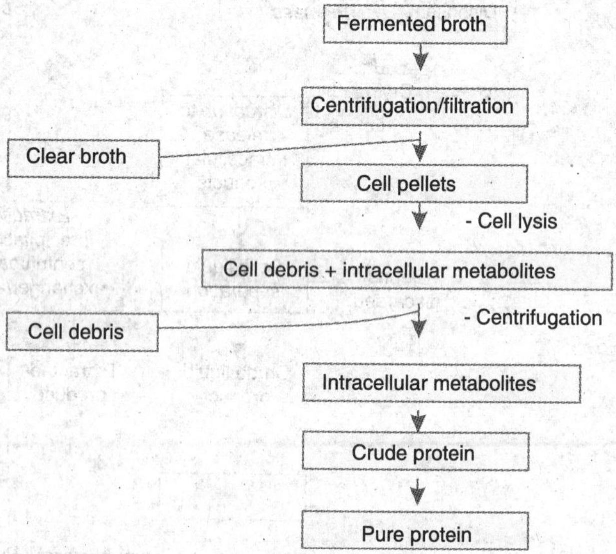


**Fig. 16.10.** Isolation of intracellular microbial product (recombinant insulin) from *E.coli*.

Steps of isolation of extracellular microbial metabolite (such as antibiotic which is secreted in broth) is given in Fig. 16.11

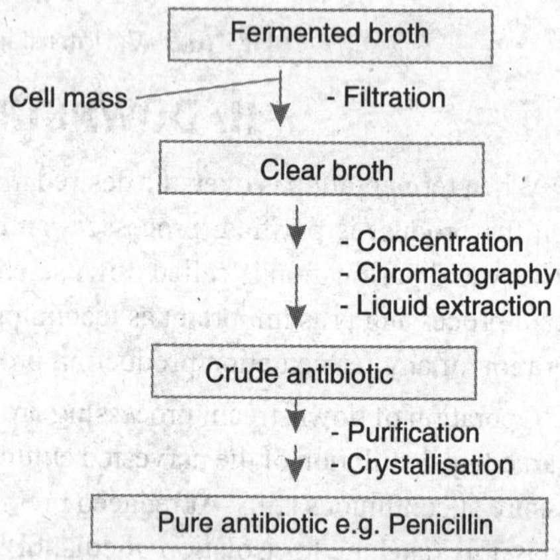


**Fig. 16.11.** Isolation of extracellular microbial product (e.g. penicillin from *Penicillum chrysogenum*).

(*e*) **Metabolite-specific Purification** : Specific purification methods are used when the desired metabolite is purified to a very high degree.

(*i*) **De-watering** : When a low amount of product is found in very large volume of spent medium, the volume is reduced by removing water to concentrate the product. It is done by vaccum drying or reverse osmosis. This process is called de-watering.

(*ii*) **Polishing of Metabolites** : It is the final step of making the product to 98-100% pure. The purified product is mixed with several cheaper inert ingredients called **excipients**. The formulated product is packed and sent to the market for the consumers.

# I. ISOLATION AND IMPROVEMENT IN MICROBIAL STRAINS

There is a large number of microorganisms found in a variety of habitats such as soil, water, air, volcano, arctic water and hot spring. Each microbe has some peculiar feature. All are not supposed to have novel and useful products. Hence, they are first isolated from their natural habitat. Then the strains are further improved using physico-chemical or molecular biological techniques, so that products could be produced at commercial scale. For example, *Thermus aquaticus* is a hyperthermophilic bacterium which grows on volcano. An enzyme, *Taq* polymerase, is isolated from this bacterium. It is unique enzyme which is used in PCR for polymerisation of DNA synthesis at high temperature.

## 1. Isolation of Strains

For isolation of micro-organisms, samples are collected from several sites of a habitat. Accordingly, nutrient media are prepared. Such specially designed media for culture of specific microorganisms are called *enrichment culture technique.* Equal amount of medium is poured into sterilised Petri plates. Known amount of serially diluted samples is poured onto the surface of agar medium. The inoculated plates are put into a BOD incubator at optimum temperature desired by the microbe (*e.g.* 25-35°C for mesophilic bacteria). The Petri plates are incubated for different period such as 24 hours for bacteria, 5-7 days for fungi and 14-21 days for actinomycetes, because growth rate of different micro-organisms differ. A flow diagram for isolation of micro-organisms is given in Fig. 6.10.

After incubation mixture of microbial colonies grows on surface of nutrient medium. Each colony is identified. The desired microbe is then isolated and purified through sub-culturing. A small amount of inoculum of desired microorganism is put onto the surface of fresh medium either in agar slants or Petri plates. These are further incubated as described above. Thus growth of the micro-organism is further improved.

The microorganisms isolated as above is further characterised to find out its specific properties such as production of desired proteins, vitamins, antibiotics, etc. For example, testing of antibiotic producing strains is done by dual culture method, where growth of one bacterium is inhibited by the metabolic product (*e.g.* antibiotic of thc other bacterium). Further testing of purified product is also carried out in laboratory by the microbiologists. Then the effect of antibiotics on certain metabilic activities *e.g.* respiration, protein synthesis, control systems, etc. are done. Using antibodies the microbial products are also assayed. Many probes have been developed through molecular biology which are used in diagnostics of micro-organisms in the group and detection of their products. One of such techniques such as microarray technique has been described in Chapter 8, Genomics and proteomics.

## 2. Strain Improvement of Microorganisms

A microbe isolated as above does not ensure that the product produced by it would be in sufficient quantity. Therefore, the strain of such organisms is improved by using classical (mutation and selection) and modern recombinant DNA technology to get the desired product in sufficient quantity. Penicillin production by *Penicillium chrysogenuim* is one of the good examples of antibiotics. During early days (1940s) penicillin was produced in low amount. It was administered only in army. Among those only officers at Brigadier levels were fortunate to get penicillin treatment but not the common soldiers. Today's strain of *P. chrysogenum* is capable of producing 1000 times more penicillin than the A. Fleming's strain. It has been possible through successive mutation and mutant selection of over-producers.

**(*a*) Mutation and Mutant Selection :** A cell divides and produces two identical cells of parental types. There is least probability of changes in inheritable characters. A strain that shows changed characteristics is termed as **mutant.** When the microbial culture is exposed to mutagenic agents such as ionising radiation, ultra violet light and various chemicals (*e.g.* NTG *i.e.* nitrosoguanidine, nitrous acid, etc.) the probability of muation is increased. When mutagenic dose is high, many of cells die.

The survivors may contain some mutants. A small portion of cells may be good metabolite producers also.

Geneticists make selection of superior producers from the inferior producers among the survivors.

(*i*) In 1957, Kinoshita isolated *Corynebacterium glutamicum* which was biotin auxotroph (*i.e.* biotin mutant) and excreted glutamic acid. But glutamic acid could be produced in high amount through biotin limitation and blocking the metabolic route that converts into TCA cycle intermediates. Besides the production of glutamic acid, mutants of *C. glutamicum* were also developed which produced high amount of other amino acids also.

(*ii*) In 1961, Nakayama and co-workers succeeded in isolating a homo-serine auxotroph of *C. glutamicum* which synthesised 44 g dm-3 lysine. Sano and Shiio (1970) developed a *C. glutamicum* strain for the production of lysine which was auxotroph of homo-serine and leucine.

Several antibiotics have been produced and commercialised through mutation and selection for example, penicillin, cephalosporin, chlortetracyclin, kanamycin, candidicin, kasugamycin, etc.

**(*b*) Recombination:** Recombination is defined as any process which helps to generate new combinations of genes that were originally present in different individuals. Recombination system may or may not be associated with sexual reproduction among the organisms. There are two approaches which have been made to produce recombinants (organisms having new combination of genes), protoplast fusion and recombinant DNA technology.

**(*i*) Protoplast Fusion :** Protoplasts are the cells devoid of cell walls. You can produce protoplasts using lysosome (cell wall degrading enzyme) in isotonic solution. Methods have been developed to fuse protoplast of two cells of different microorganisms. In 1982, Tosaka produced high lysine producing strain of *Brevibacterium flavum* by fusing protoplasts with another *B. flavum* strain (a non-lysine producer but high rate glucose consumer). Hamega and Ball (1979) fused protoplast of two different strains of *Cephalosporium acremonium* and produced a recombinant which had a high growth rate and synthesised a higher level of cephalosporin. In 1982, Chang and co-workers used protoplast fusion method in two strains of penicillin V-synthesising strains of *Penicillium chrysogenum.* They selected a recombinant of *P. chrysogenum* capable of synthesising only penicillin V with desired morphological type.

**(*ii*) Recombinant DNA Technology (= Genetic Engineering Technique) :** Recombinant DNA technology has earlier been described in detail in Chapters 3-6. The first commercial genetically engineered protein (human insulin called **Humulin**) was produced in 1982. Humulin is used by the persons suffering from diabetes mellitus. The efficiency of ammonia metabolism of *Methylophilus methylotrophus* (a bacterium used as single cell protein) has been improved by incorporation of a plasmid containing glutamate dehydrogenase gene from *E. coli.*

The improvement in production of commercially important enzymes has also been done by rDNA technology. In 1981, Colson and co-workers cloned a *Bacillus coagulans* gene coding for a thermostable α-amylase into *E.* coli where it was replicated and maintained at a high copy number resulting in high enzyme production. Similarly, penicillin acylase production in an *E.coli* strain has also been improved by introducing the relevant gene into a plasmid which was then incorporated into the original strain.

## PROBLEMS

1. What is fermentation ? Discuss the role of micro-organisms in fermentation process with suitable examples.
2. Discuss the microbial cultures and their significance.
3. Write an essay on metabolic pathways of micro-organisms.
4. Give a brief account of microbial products.
5. Write in detail the steps of microbial culture.
6. Give a detailed account of batch culture and microbial kinetics in batch culture.
7. Write in detail the methods of measurement of mcrobial growth.
8. What is downstream processing? Discuss giving suitable examples.
9. Write in detail isolation and improvement of microbial strains.
10. Write short notes on the following :

    (*a*) Fermentation, (*b*) EMP Pathway, (*c*) The Entner-Doudoroff pathway, (*d*) Pentose phosphate pathway, (*e*) Batch culture, (*f*) Primary metabolites, (*g*) Secondary metabolites, (*h*) Application of microbial culture technology, (*i*) Scale-up of microbial process.

CHAPTER

17

# Microbial Products (Primary and Secondary Metabolites)

In a bioreactor micro-organisms grow at optimum temperature and increase their biomass with utilisation of carbon and nitrogen sources along with other growth factors. When microbial cells enter in the log phase, they utilise nutrients very fast with subsequent production of primary metabolites. With gradual depletion of nutrients of growth medium, their growth retards and secondary metabolites are secreted. This chapter deals with primary and secondary metabolites secreted by the micro-organisms.

## A. PRIMARY METABOLITES

Primary metabolites are the microbial metabolites which are secreted during **trophophase** (active growth phase *i.e.* log phase). Hence, these are called growth-dependent metabolites.

It has earlier been mentioned that primary metabolites, secreted by micro-organism, in turn are required by micro–organisms. In this connection a detailed account of vitamins, organic acids, enzymes and alcohols are described in detail.

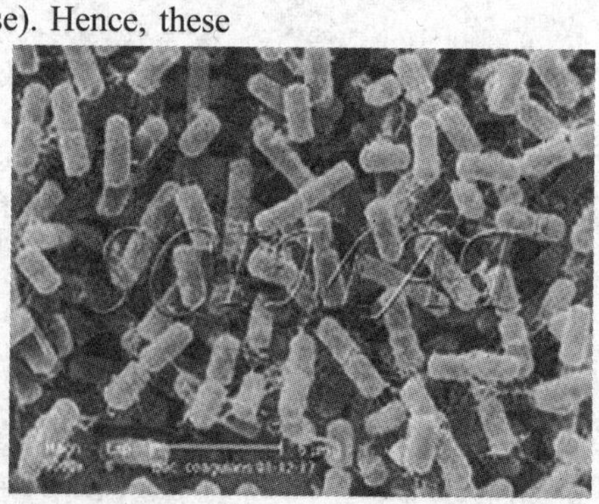

*Bacillus coagulans* a bacterium that produces vitamin $B_{12}$.

### I. VITAMINS

All phototrophic micro-organisms are capable of synthesizing vitamins and other growth stimulating compounds for their vegetative growth by using the chemical constituents from the culture medium. When synthesis of these compounds exceeds beyond their requirement, it accumulates in cultures; there from it is recovered. Now-a-days, commercial exploitation of such micro-organisms is being done which synthesize vitamins on large scale under different cultural conditions. Microbiologically produced some vitamins are: carotene, precursor of vitamin A (*Blakeslea trispora*), riboflavin (*Ashbya gossypii*), L-sarbose in vitamin C synthesis (*Gluconobacter oxidans*), and vitamin $B_{12}$ (*Bacillus coagulans, B. megaterium, Pseudomonas denitrificans and Streptomyces olivaceous*).

## 1. Vitamin $B_{12}$ (Cyanocobalamine)

In nature, vitamin $B_{12}$ is synthesized by micro-organisms. For industrial production of vitamin $B_{12}$ a number of bacteria and streptomycetes are used. The amount of vitamin $B_{12}$ produced by them has been estimated about 20 mg/litre. It is used in medicine and feed supplements, and is most essential for human growth. Daily requirement of vitamin $B_{12}$ is about 0.001 mg.

***(a)* Chemical Structure :** Vitamin $B_{12}$ (cyanocobalamin) contains a molecule of cobinamide linked to a nucleotide which has 5,6-dimethyl benziminazole as its base, instead of a purine or pyrimidine base (Fig. 17.1). The cobinamide molecule has a central atom of cobalt linked to a cyanide group and surrounded by four reduced pyrol rings joined to form a macroring. A number of carbon atoms carry methyl or other substituent group (Riviere, 1977).

***(b)* Commercial Production :** Commercially vitamin $B_{12}$ is produced in a continuous culture, where two fermenters are used in a series. In each fermenter culture is kept for about 60 hours. Precaution taken is that the first fermenter should be anaerobic, while the second aerobic one. In the second fermenter 5,6-dimethyl benziminazol is added continuously.

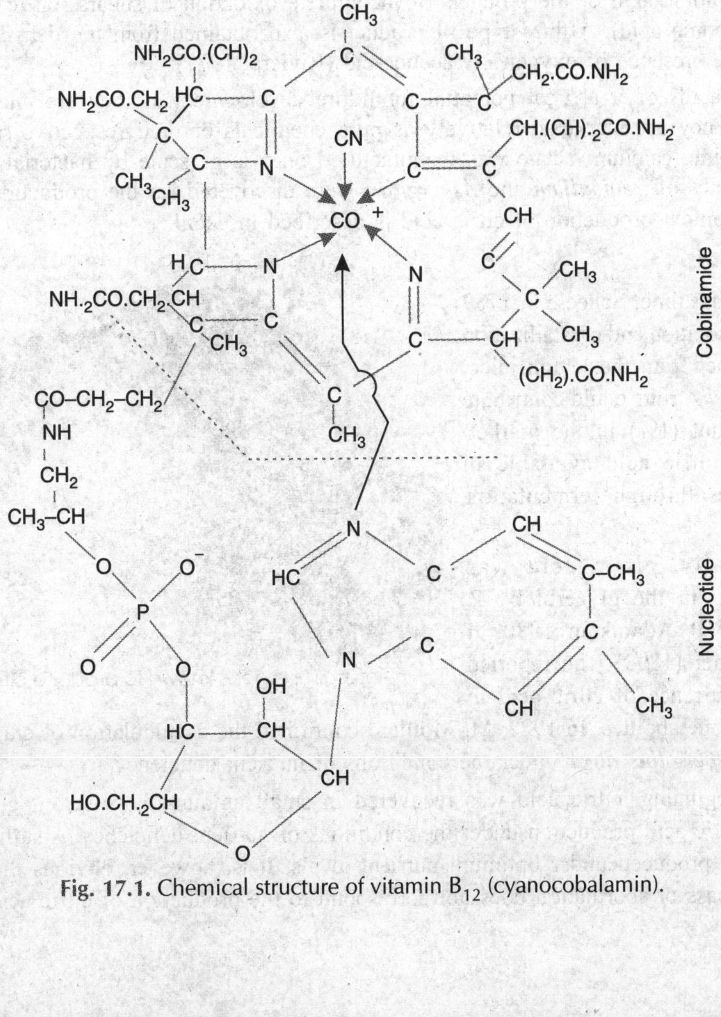


Fig. 17.1. Chemical structure of vitamin $B_{12}$ (cyanocobalamin).

Sterilized culture medium, containing glucose, corn steep, betain (5%), cobalt (5 ppm) and pH 7.5 in fermenter, is inoculated with *Propionibacterium freudenreichii* and allowed to anaerobic fermentation for about 70 hours. During this period cobinamide is produced which accumulates in the broth. Thereafter, 5,6-dimethyl benziminazole (0.1%) is added to it. The fermenter is then kept further for 50 hours for aerobic fermentation. During this period nucleotide is synthesized and linked with cobinamide molecule to yield about 20 ppm cobalamin. Culture is acidified to pH 2.0 to 3.0, gently heated to 100°C and filtered to remove cell debris. Finally potassium cyanide (5 ppm) is added to filtrate to give cyanocobalamin. Generally sodium sulphite is mixed with the solution so that cyanocobalamin could not be oxidized.

Using genetic engineering techniques, it has become possible to obtain mutants of micro-organisms, producing more vitamins than the natural ones. *P. denitrificans* is able to produce 50,000 times more vitamin $B_{12}$ than its parental strain (Sasson, 1984).

# II. ORGANIC ACIDS

A micro-organism grows in broth of the fermenter and during its tropophase of growth, produces organic acids. The organic acids are produced through the metabolisms of carbohydrates. They accumulate in broth of fermenter, wherefrom they are separated and purified.

The organic acids are either the terminal products of EMP pathway (glycolysis), *e.g.* lactic acid and propionic acid or the products of incomplete oxidation of sugars (citric acid, itaconic acid and gluconic acid). A third type of product is also obtained from the dehydrogenation of alcohol in the presence of oxygen *i.e.* acetic acid (Riviere, 1977).

Organic acids offer a great potential for future development as they are manufactured on large scale. They are marketed relatively as pure chemicals or their respective salts. In 1881, for the first time calcium lactate was manufactured on a large scale by bacterial fermentation. Later on species of *Penicillium* and *Aspergillus* were discovered for the production of acids. In the present context production of citric acid is described in detail.

## 1. Citric Acid

For the first time, Scheele in 1789 reported the isolation and crystallization of sour product from the lemon juice. However, *Citrus* fruits could contribute a small amount (1%) in the market. Now-a-days citric acid available in market comes through fermentation process.

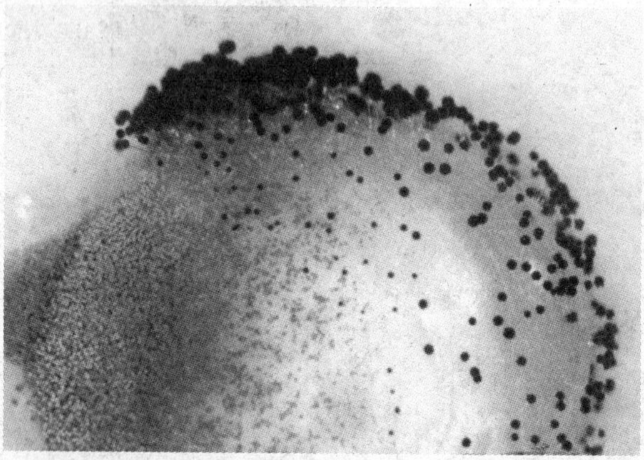

*Aspergillus niger* is known to produce citric acid.

Chemically, citric acid was synthesized from the glycerol by P. Grimaux and P. Adams in 1880. It was C. Wehmer (1893) who reported the wide occurrence of citric acid in the microbial metabolites. In 1922, M. Molliard confirmed the accumulation of citric acid in the cultures of *Aspergillus niger* under the conditions of nutrient deficiency.

In the beginning, citric acid was recovered in small amount; however, it gave the basic concept of citric acid production under the conditions of nutrient deficiency. A sufficient amount of biomass is produced under optimum nutrient levels. It is, however, obvious that production of mycelial mass or sporulation does not correspond to the production of citric acid in cultures.

**(a) Commercial production methods :** There are 3 methods for commercial production of citric acid:

(*i*) ***Koji fermentation Process*** : This method is used in Japan which accounts for about one fifth of citric acid produced per annum. Strains of *A. niger* are used in this method.

(*ii*) ***Liquid surface culture fermentation process*** : In this method, liquid cultures are inoculated with the spores of *A. niger* which germinate within 24 hours. Mycelia cover and float on whole surface of the solution.

(*iii*) ***Submerged culture fermentation process*** : In this case the mycelia of *A. japonicum* grow in solution of about 15 cm depth in tanks. Citric acid produced in this way is inferior to that produced by liquid surface culture fermentation. Moreover, it is produced in low amount and at high cost.

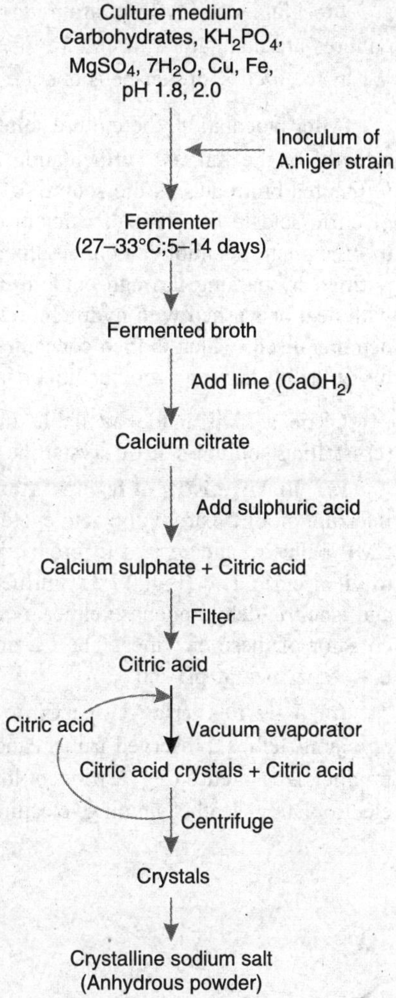


**Fig. 17.2.** Outline of citric acid production method.

Methods of citric acid production are given in Fig. 17.2. For commercial production strains of *A. niger* are selected from the hybrids or mutants developed by certain procedure. The strains should be such that could produce not less than 80 g citric acid per 100 g glucose. For large scale production, continuous culture is not suitable. Therefore, a multi tank system is required for continuous fermentation in any process in which cell growth and metabolic products occur at different.

Culture medium contains carbohydrates, $KH_2$ $PO_4$ (0.01-0.3%), Mg $SO_4$. $7H_2O$ (0.25%) and a few trace elements. Solutions of carbohydrates are high test cane syrup (concentrated cane juice), glucose or sucrose. High test syrup can also be beet molasses or cane molasses. If substrate is molasses, ferrocyanide (20-150 mg/litre) can be added to culture medium, before sterilization, to precipitate excess iron.

Instead of iron, clarified molasses can be passed through an ion exchange resin. Under laboratory condition highest yields are obtained by using sucrose passed through an ion exchange resin. At this stage, it is necessary to add copper (0.1 — 0.50 ppm) to the culture medium (Riviere, 1977). But the trace elements should be added only when fermenter is made up of steel or some impervious coating is applied (to ordinary fermenters). pH of the culture medium is maintained to about 3.5 by adding ammonium nitrate (0.25%). It avoids the formation of oxalic and gluconic acids. The culture medium is sterilized by passing through the pipes of a steam jacketed heat exchanger and cooled down to about 30ºC in another exchanger.

Inoculum of *A. niger* strain is prepared by culturing it on nutrient agar medium. Stock cultures are maintained in culture tubes kept in refrigerator or in the form of lyophilized spores. When required spore suspensions are prepared in the sterile distilled water.

After inoculation, the culture solution must be aerated by bubbling the air to allow maximum growth for the fungus. Fermentation process is completed in about 5-14 days at 27-33°C. The fermented broth acts as the source of citric acid. Lime ($Ca(OH)_2$) is added to allow precipitation of citric acid in the form of calcium citrate. Again the precipitate is treated with sulphuric acid to precipitate insoluble calcium sulphate. It is then filtered. The filtrate containing citric acid is purified by passing through column of carbon granules. The carbon granules should be treated with heat or washed with hydrochloric acid (HCl). Passing through ion exchange beds solution is demineralised, which is then concentrated under vacuum to form crystals. Crystals are recovered by centrifugation and mother liquor is returned to evaporator to get the remaining crystals.

Citric acid is made available in market in the form of anhydrous crystalline chemical (crystalling sodium salt of crystalline monohydrate) as powder.

***(b)* Biochemistry of fermentation :** Citric acid is produced during trophophase as a result of interruption of tricarboxylic acid cycle (Krebs cycle). In *A. niger* about 78% sugar passes through EMP pathway and results in production of acetyl CoA which condenses with oxaloacetic acid to yield citric acid (Fig. 17.3). Further citric acid cycle is interrupted by inhibition of aconitase and isocitric dehydrogenase either by copper or hydrogen peroxide. However, iron is an essential cofactor of these enzymes. The Cu and Fe in a ratio 0.3. : 2 (mg/litre) interrupt the activity of these enzymes at pH 2.0.

***(c)* Uses of Citric Acids :** Citric acid is used in food industry (*e.g.* fruit drinks, confectionery, jams, jellies, preserved fruits, candies, wines), pharmacy (*e.g.* blood transfusion, effervescent products), cosmetics (*e.g.* astringent lotions, shampoos and hair setting fluids), and industries (*e.g.* electroplating, leather tanning, cleaning of pipes, reactivation of old oil wells).

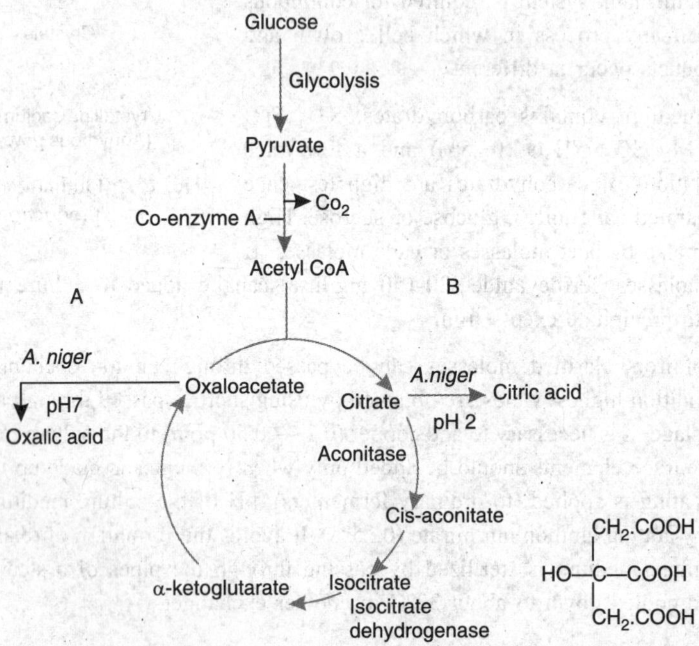


**Fig. 17.3.** Biochemical route of oxalic acid (A) and citric acid (B) formation.

# III. ALCOHOLS

For the first time, Louis Pasteur demonstrated the fermentation of sugar by micro-organisms and their regulation as well. He found that the yeast cells produce 20 times more cell materials from sugar under aerobic conditions than under anaerobic conditions. He explained that oxygen inhibits fermentation process. Yeasts are aerobic microorganism, but glucose fermentation takes place under anaerobic condition. In the presence of oxygen yeasts increase its biomass (cell materials) but under anaerobic condition they hardly grow, however, ferment glucose very efficiently. Thus, oxygen supresses fermentation process. This is known as "Pasteur effect" as the inhibition of fermentation by air was described by Pasteur about 100 years ago. This effect is applicable for all facultative anaerobic microorganisms.

In 1815, Gay-Lussac formulated the conversion of glucose to ethanol. During 1980s, lactic acid was produced by fermentation, as a substitute for tartaric acid used in baking powders; but it was not commercialized. Thereafter, in Germany glycerol was produced by fermentation, which was used in the manufacture of explosive materials. Similarly, explosive materials were produced in Britain by developing acetone-butanol fermentation process. This type of practice was also done in the U.S.A. Consequently ethanol was produced by fermentation and distillation process in many countries. After 1940, rapid development took place in the production and establishment of large industrial solvent fermentation and distillation plant. Recently, recombinant DNA technology has helped micro-organisms used in industrial process. After the U.S.A and Brazil, India is the worlds' third largest producer of fermentation ethanol. Now, we have over 100 distilleries with an installed capacity of about 700 million litres per annum. Production of ethanol per tonne of molasses is about 225 litres.

## 1. Microorganisms Used in Alcohol Production

There is a limited number of micro-organisms which ferment carbohydrates (pentose or hexose sugars) into alcohols and yield some by-products. Micro-organisms utilize various pathways. A summary of alcohol production through different routes of micro-organisms is given in Fig. 17.4.

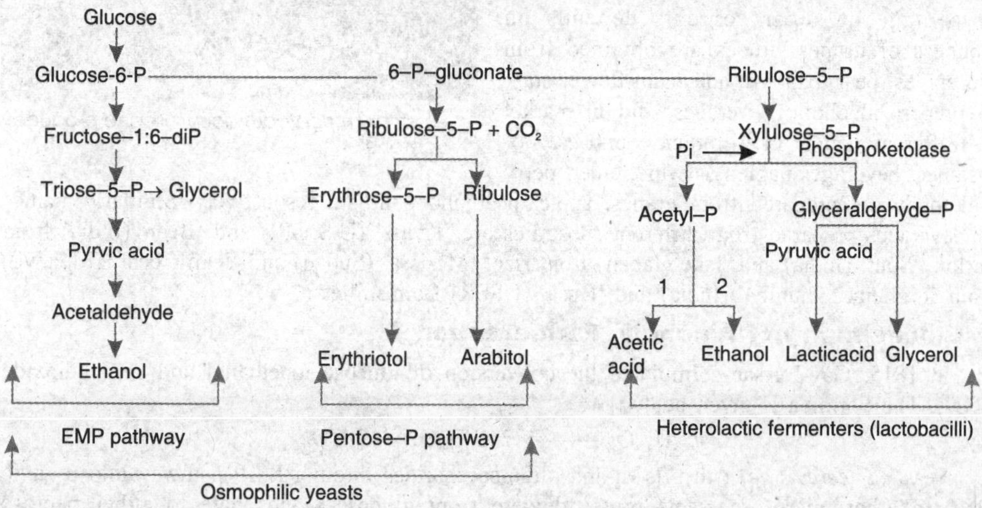


Fig. 17.4. Production of alcohols by micro-organisms 1-*Lactobacillus brevis*; 2-*Leuconostoc mesenteroides*.

Following are some of alcohol producing micro-organisms:

Bacteria : *Clostridium acetobutylicum, Klebsiella pneumoniae, Leuconostoc mesenteroides, Sarcina ventriculi, Zymomonas mobilis,* etc.

Fungi : *Aspergillus oryzae, Endomyces lactis, Kloeckera* sp., *Kluyreromyces fragilis, Mucor* sp., *Neurospora crassa, Rhizopus* sp., *Saccharomyces beticus, S. cerevisiae, S. ellipsoideus, S. oviformis, S. saki, Torula* sp., *Trichosporium cutaneum*, etc.

## 2. Fermentable Substrates

Ethyl alcohol is produced from such organic material that contains sugar or its precursor as fundamental units. The cost of substrates (raw materials) in fermentation is of major consideration, because it directly affects the cost of products. The fermentable substrates used for alcohol production are given under 'fermentable substrates' (*see* Chapter 20).

Before using in fermentation processes, the cellulosic, lignocellulosic and starchy materials are hydrolysed by enzymes or acids just to render the complex substances into a simple form (monosaccharides). Enzymes for hydrolysis are obtained from barley malt or moulds by heat treatment of acidified materials.

*Klebsiella pneumonia* has been demonstrated to be comparable of utilizing the wood hemicellulose hydrolysate which consists of pentose and hexose sugars, low molecular weight oligomers and uronic acids. The products of fermentation are butanol, ethanol acetone, etc. These products can be utilized as solvent.

*Clostridium acetobutylicum* anaerobically ferments the starchy substrates of grains and potatoes to produce acetone, ethanol and butanol. Recently, it has been demonstrated that *Schwaniomyces castellii* directly converts the soluble starch into ethanol.

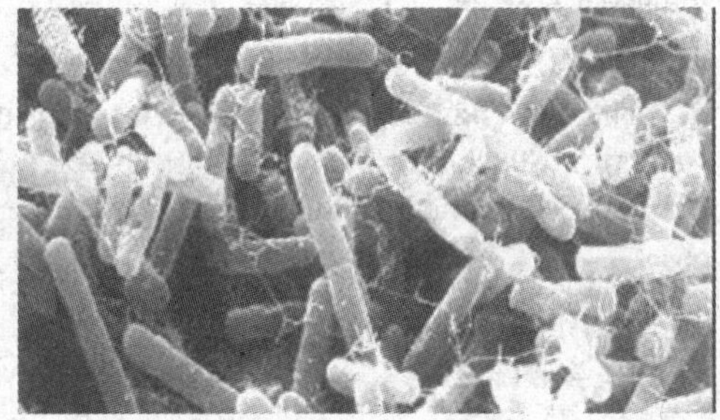

*Clostridium acetobutylicum* used to produce acetone

Sugarcane molassess, a by-product of sugarcane mill, contains high percentage of sucrose and fructose sugars. The fruit juice, for example, grapes (*Vitis vinifera*) contains high amount of sugars (8-13% glucose and 7-12% fructose) when grapes are ripen. Moreover, percentage of sugar contents depends on ripeness of grapes. Juices are obtained from the apples, pear, palmyra and palm-flower stalk to prepare alcoholic beverages sold in market by different names in different countries. For instance, beverages named as wine, cider, perry and toddy are prepared from grapes, apple, pear and palmyra, respectively. Similarly, names of beverages prepared from different substrates are 'Kvass' (U.S.S.R.) and 'Bear' (India) from barley, 'Sonti'(India) and Sake (Japan) from rice, 'Merissa' (Sudan) and 'Kaffir bear' (Malawi), from sorghum, 'Thumba' (India) and 'Busa' (USSR) from millets.

## 3. Biochemistry of Alcoholic Fermentation

In 1815, Gay-Lussac formulated the conversion of glucose to ethanol and carbondioxide ($CO_2$). The Formula is given below:

$$C_6H_{12}O_6 \longrightarrow 2\ C_2H_5OH + 2\ CO_2$$

Yeast (*S. cerevisiae*) converts di-and oligo-saccharides through EMP pathway into pyruvic acid. In other micro-organisms routes deviate from glucose-6-phosphate to either pentose phosphate pathway or the Entner-Doudoroff pathway. Even the fate of pyruvic acid differs in different alcohol producing microorganism. Therefore, differences lie in the alcoholic products. In the present context, ethanol formation is discussed with the examples of yeasts and bacteria.

**(*a*) Ethanol Formation by Yeasts :** Yeasts, especially strain of *S. cerevisiae* are the main producer of ethanol. They have been used as a major biological tool for the formation of ethanol since the discovery of fermentation process by the time of L. Pasteur. During 1890s fermentation of froth was discovered in sugar solution on addition of yeast extracts obtained by its grinding. This was the first evidence for a biochemical process of *in vitro* formation of ethanol in the absence of yeast cells. The extract supplied inorganic phosphate (Pi) which is incorporated in fructose-1:6-bisphosphate. Fructose-1:6-bisphosphate is accumulated due to lack of ATP utilization for energy requiring reactions in the cell free systems. Therefore, an excess of ATP is maintained. The reaction is given below:

Grapes are used for production of beverage by yeast.

$$2C_6H_{12}O_6 + Pi \longrightarrow 2C_2H_5OH + 2CO_2 + 2H_2O + \text{fructose-1:6-bisphosphate}$$

This equation is known as Harden - Young equation after the name of the discoverer.

Energetics of EMP pathway reveals that one molecule of glucose yields only 2 molecules of ATP from 2 molecules of ADP under anaerobic condition in contrast of 38 molecules of ATP through respiration:

$$C_6H_{12}O_6 + 2Pi + 2ADP \longrightarrow 2C_2H_5OH + 2CO_2 + 2H_2O + 2ATP + \text{energy.}$$

A total of 15.4 Kcal energy is evolved from one molecule of glucose as about 77 Kcal free energy is obtained from one molecule of ATP. Fermentation of carbohydrate is exothermic where certain amount of energy is lost to the environment as heat. However, temperature of fermenter gets increased. Increase in temperature (generally from 11-22°C) depends on size of the fermentor.

The Pasteur effect implies for inhibition of glycolysis by respiration which is explained by the respiratory chain and substrate level phosphorylation compete for ADP and phosphate. It is associated with several regulatory mechanism at a time which work together. Firstly, in phosphorylation, competition exists for ADP and Pi. In absence of these two, dehydrogenation of glyceraldehyde- 3-phosphate does not take place. Under aerobic condition, synthesis of ATP takes place as a result of competition of ADP and Pi by respiratory chain phosphorylation. Due to decrease in the intra-cellular concentration of ADP and Pi, consumption of glucose and production of ethanol are reduced accordingly. It has been demonstrated that on uncoupling of respiratory chain phosphorylation by 2, 4-dinitrophenol (DNP), ADP and Pi becomes available for the dehydrogenation of glyceraldehyde- 3 phosphate (Fig. 17.3). Consequently, consumption of glucose under aerobic condition increases to the level of anaerobic one.

Secondly, allosteric inhibition of phosphofructokinase by ATP is also responsible for the Pasteur effect. In brief, when aerobically cultured yeast or tissue cells are deprived off oxygen, electron transport phosphorylation is thereby blocked, the cellular concentration of ATP relative to adenosin monophosphate (AMP) declines. This results in an increase in phosphofructokinase activity due to which a faster flow of metabolites takes place through fructose -1:6-bisphosphate pathway. This type of regulatory mechanism occurs in the living cell. Upon aeration the concentration of glucose-6-phosphate and fructose-6-phosphate increases immediately whereas, concentrations of fructose-1:6-bisphosphate and triose phosphates are drastically decreased. Here phosphofructokinase acts as a valve controlled by adenylates and other metabolites (Schlegel 1986).

During fermentation of glucose, acetaldehyde, an intermediate product is formed which can bewtrapped with hydrogen sulphite which is non-toxic to yeast cells. On addition of hydrogen sulphite, during fermentation, acetaldehyde sulphite is precipitated. This results in production of glycerol, by diminishing the yield of ethanol and carbon dioxide. Therefore, fermentation of glycerol has been developed industrially by incorporation of hydrogen sulphite.

Glucose ⟶ Glycerol ($CH_2OHCHOHCH_2OH$)
↓
Pyruvic acid ($CH_2CO$ COOH) ⟶ Acetaldehyde ($CH_3$ CHO)
↓ ⟵ Hydrogen sulphite ($H_2SO_3$)
$CO_2$ + Acetaldehyde sulphite (ppt)

**(*b*) Ethanol Formation by Bacteria :** Among the bacteria discovered, only *Sarcina ventriculi* forms ethanol through fructose-1:6-bisphosphate pathway *i.e* EMP pathway and pyruvate decarboxylase as formed by yeasts. A rod shaped polarly flagellated and motile bacterium (*Zymomonas mobilis*) is known to metabolize glucose through the Entner-Doudoroff pathway (Fig. 17.4) and results in pyruvic acid. Pyruvic acid is then decarboxylysed by pyruvate decarboxylase to acetaldehyde and carbon dioxide. Acetaldehyde is reduced to ethanol. Thus, the fermentation products are ethanol, carbondioxide (and small amount of lactic acid). In some members of Enterobacteriaceae and Clostridia, ethanol is formed as a subsidiary product. Acetaldehyde is not directly produced from pyruvic acid by pyruvate decarboxylase, but originates through reduction of acetyl CoA.

Pyruvic acid —(Pyruvate decarboxylase)⟶ Acetaldehyde + $CO_2$
↓
Ethanol

The hetero-fermentative lactobacilli (*e.g. Leuconostoc mesenteroides*) use quite different pathway for alcohol production (Fig. 17.4). In the beginning of fermentation they utilize pentose cycle to result in xylulose-5-phosphate, which is then cleaved by phosphoketolase into acetyl phosphate and glyceraldehyde-3-phosphate. Acetaldehyde dehydrogenase and alcohol dehydrogenase reduce the acetylphosphate into ethanol. Similarly, glyceraldehyde-3-phosphate is converted via pyruvic acid to lactic acid (Schlegel, 1986).

## 4. Ethanol Fermentation Methods

Fermentation of ethanol is carried out in a large fermenter (size 1000 to 1.5 million $dm^2$). The inoculum of micro-organism is maintained in fermentor at the optimum growth conditions such as temperature, pH, oxygen and concentrations of carbohydrate, the substrates. Before starting the fermentation pure inoculum (starter inoculum) of species of *Saccharomyces* is prepared by inoculating the well defined and sterilized medium (Fig. 17.5). At the same time fermentation medium is formulated, sterilized and transferred to the sterile fermentor. Liquid medium in fermenter is inoculated with a small amount of inoculum of yeast. Growth conditions of liquid broth is maintained to provide optimum conditions such as temperature, pH, oxygen, etc. for the production of ethanol.

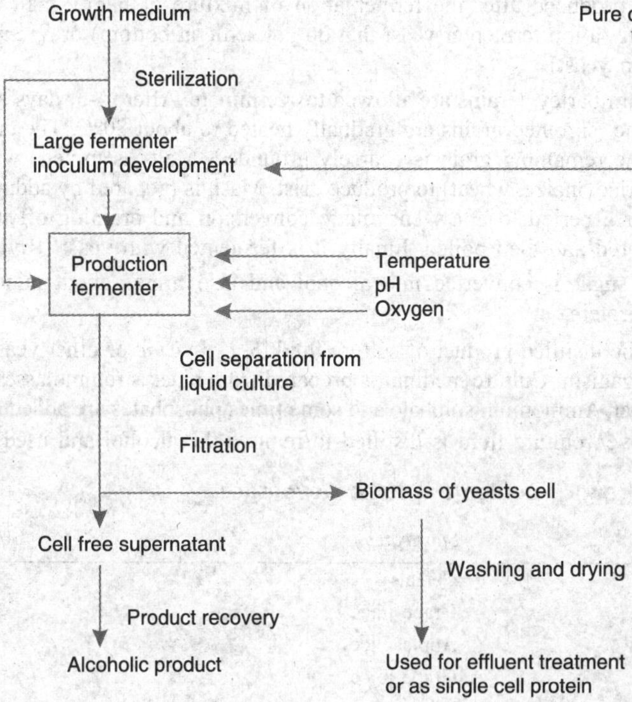


**Fig. 17.5.** Outline of alcohol production by yeast cells.

After inoculation at optimum growth conditions for different periods the culture fluid is filtered when growth of the micro-organism is over. Consequently, the yeast cells (biomass) are separated from the supernatant. From the supernatant, products are recovered and purified. The yeast mass is used for effluent treatment or as the source of single cell protein.

Commercial production of alcohol varied in scale. Production and product recovery depend on: (*i*) size of fermentor, (*ii*) optimum culture conditions, (*iii*) design for separation of solid and liquid, (*iv*) culture medium, (*v*) potential micro-organism, and (*vi*) joint efforts of microbiologists, geneticists, biochemists, engineers and chemists. Productive micro-organisms are improved through the improvement in growth medium, growth condition, mutation, recombination and process design.

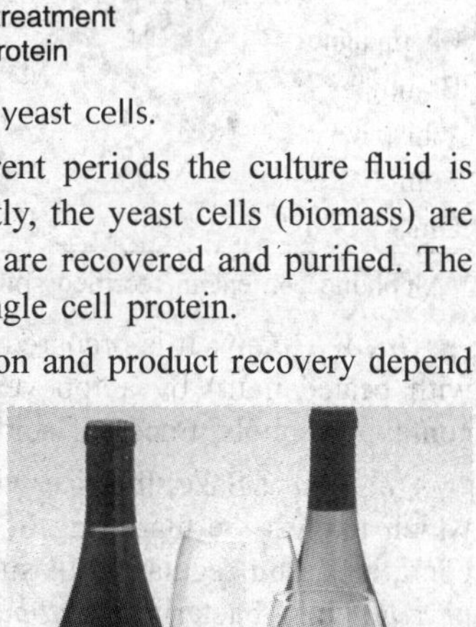

Wine is produced from fresh grape juice.

## 5. Alcoholic Beverages

Percentage of alcohol differs in different alcoholic beverages (Table 17.1). Some of them are briefly discussed:

**(*a*) *Wine*** : It is mainly an European drink produced from juice of fresh grapes (*Vitis vinifera, V. rotundifolia*). In ripen grapes, concentration of sugar (glucose and fructose) increases. Grape juice (27% sugar) is fermented by various strains of *S. ellipsoideus* into alcohol and also renders the chemical constituents which alters the flavour.

The fortified wines (brandy) are prepared on addition of extra ethanol to wines, when fermentation is over, for raising the concentration to about 20%.

***(b) Beer*** : Beer is produced after the fermentation of mixture of barley malt and starchy solutions by *S. cerevisiae* (a top fermenter yeast that do not settle at bottom) or *S. carlsbergensis* (the bottom fermentation yeast).

Malt is prepared from barley. Grains are allowed to germinate. After 4-6 days amylase and protease are formed. The sprouted grains are gradually heated to about 80°C. The dried rootlets are knocked off and the remaining grain is coarsely grounded. Malt is mixed with coarsely ground starchy cereals (rice, maize, wheat) to produce grist. Mash is prepared by adding hot water to the grist holding for a period to allow enzymatic conversion and draining off the resultant sweet wart. Wart is filtered and then boiled. Finally it is fermented with yeast (Riviere, 1977).

After fermentation sugar is converted into alcohol and also brings about minor chemical changes, for example, protein.

***(c) Rum*** : Rum is the distilled product of culture fluid. *S. cerevisiae* or other yeast is used as the fermenting micro-organism. Culture medium is prepared from black strap molasses containing 12-14% fermentable sugar. Ammonium sulphate and some times phosphates are added as nutrients. When fermentation is over, culture fluid is distilled to remove the alcohol and used as rum.

**> Table 17.1. Alcoholic contents in some beverages.**

| *Beverage* | *Substrates* | *Alcohol(%)** |
|---|---|---|
| Beer | Cereals | 4-8 |
| Wine | Grape juice | 10-22 |
| Cidar | Apple juice | 8-12 |
| Champagne | Grape juice | 12-13 |
| Brandy | Wine | 43-57 |
| Whiskey | Cereals | 51-59 |
| Rum | Molasses | 51-59 |
| Gin | Cereals | 51-69 |

* Alcoholic content in rectified spirit (prepared from cereals) is 95% (Source : Agrawal, S.P., 1985).

***(d) Whiskey*** : It is prepared through the fermentation of grain mash (cooked and saccharified with peated malt) by a top yeast (*S. cerevisiae*) when fermentation is over, the culture fluid contains alcohols, traces of acids and esters.

***(e) Sake*** : Sake, the rice wine, is manufactured from the starch. It is a complicated process which implies the mastering of different fermentation techniques in semi-solid and sub-merged conditions, and regulation of successive microbial populations. First of all a mould (*Aspergillus oryzae*) then bacteria (*Lactobacillus* and *Leuconostoc*) and finally yeast (*S. cerevisiae*) are mixed with the fermentable medium. The culture fluid contains about 20% alcohol; therefore, before marketing the concentration of alcohol is adjusted to 16% (Sasson, 1984).

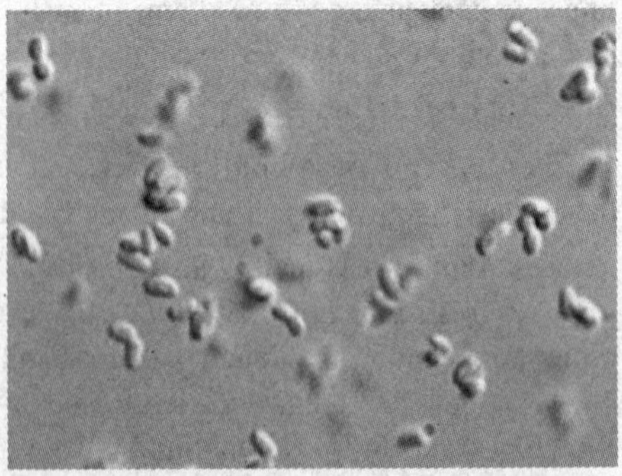

*Coynebacterium glutamicum.*

**(*f*) Uses of Alcohols :** Following are the uses of alcohols :

(*i*) Ethanol is used as solvent, extractant and antifreeze. It is also used as a substrate for the synthesis of many other solvents of dyes, pharmaceuticals, lubricants, detergents, pesticides, plasticizers, explosives and resins, and for the manufacture of synthetic fibres (Sasson, 1984). It has also been used as liquid fuel with the name 'gasohol' (*see* Chapter 20, Section 20.3.1).

(*ii*) N-butanol is used in the manufacture of plasticizers, brake fluids, urea-formaldehyde resins, extractants and pertrol additives.

(*iii*) Glycerol is used in medicals, biosynthesis of D-fructose via mannitol, and in food industry (because of its sweetness and high solubility). Mannitol is used in industry and research.

(*iv*) Butanol plus acetone acid 2, 3-butanediol are used as industrial solvent. Butanol plus acetone is used in the production of explosive materials (*e.g.* cordite) and 2, 3-butanediol in the synthesis of rubber.

(*v*) Ethanol is used as alcoholic beverages as described earlier.

## VI. AMINO ACIDS

For the first time in 1908, K. Ikeda working on flavouring components of kelp, discovered glutamic acid (L-glutamate) after acid hydrolysis and fractionation of kelp and neutralisation with caustic soda. These treatments enhanced the taste of kelp. This was the birth of the use of monosodium glutamate (MSG) as a flavour-enhancing compound. Soon the production of MSG was commercialised. Later on in 1957, S. Udaka and S. Kinoshita isolated a specific bacterium that excreted glutamic acid on a mineral salt medium. It could also be discovered that the isolated bacterium required biotin for secretion of glutamic acid. This bacterium was identified as *Corynebacterium glutamicum* which is a Gram-positive bacterium. Therefore, glutamic acid production by using *C. glutamicum* was boosted up. Table 17.4 shows some of the important amino acids commercially produced for a variety of purposes.

**> Table 17.2. Production and main uses of amino acids.**

| *Amino acid* | *Production method* | *Uses* |
|---|---|---|
| L-Glutamate | Fermentation | Flavour enchancer |
| L-Lysine | Fermentation | Feed additive |
| D/L -Methionine | Chemical synthesis | Feed additive |
| L -Asparate | Chemical synthesis | Feed additive |
| L -Phenyl alanine | Fermentation | Aspartama |
| L -Threonine | Fermentation | Feed additive |
| Glycine | Chemical synthesis | Feed additive, sweetner |
| L -Cysteine | Reduction of cysteine | Food additive |
| L -Arginine | Fermentation, extraction | Pharmaceutical |
| L -Leucine | Fermentation, extraction | Pharmaceutical |
| L -Valine | Fermentation, extraction | Pesticide, Pharmaceutical |
| L -Tryptophan | Whole cell process | Pharmaceutical |
| L -Isoleucine | Fermentation, extraction | Pharmaceutical |

### 1. Production of L-Glutamate

It has earlier been mentioned that L-glutamate was the first amino acid to be produced and commercialised. *E.coli* and *Bacillus subtilis* are also reported to secrete L-glutamate. However, *C.glutamicum* is still used for its successful production.

**(*a*) Metabolic Pathway of L-glutamate Production :** The generation of precursors or metabolites and reduced pyridine nucleotides, *C. glutamicum* utilises glycolysis, pentose phosphate pathway and citric acid cycle (CAC). But in the **anaplerotic reaction** of CAC, this bacterium shows a special feature. Glutamic acid is directly derived from α-ketoglutric acid (Fig. 17.6). Hence, a high capability for replenishing the CAC is a prerequisite for production of glutamic acid in high amount.

*C. glutamicum* secretes pyruvate dehydrogenase (PyrDH) which suffles acetyl-CoA into CAC. But the two other enzymes (*i.e.* pyruvate carboxylase and phosphoenolpyruvate carboxylase) supply oxaloacetate. Both carboxylases replace each other and facilitate the conversion of glucose into oxaloacetate. Glutamate dehydrogenase catalyses the reactive amination of α-ketoglutarte to yield glutamic acid (Fig. 17.6).

**(*b*) Production Strains :** For biotechnological production of glutamic acid, the intracellularly synthesised amino acid must be released from the cell; but it is not so. The charged glutamic acid is retained in the cytoplasmic membrane. If it is not retained in cell membrane, the cell will not remain viable. As it was mentioned earlier that glutamic acid is secreted outside when biotin is a limiting factor. This can be explained that : (*i*) active excretion of glutamic acid is mediated by a carrier, and (*ii*) its activity is triggered by the liquid environment of this carrier.

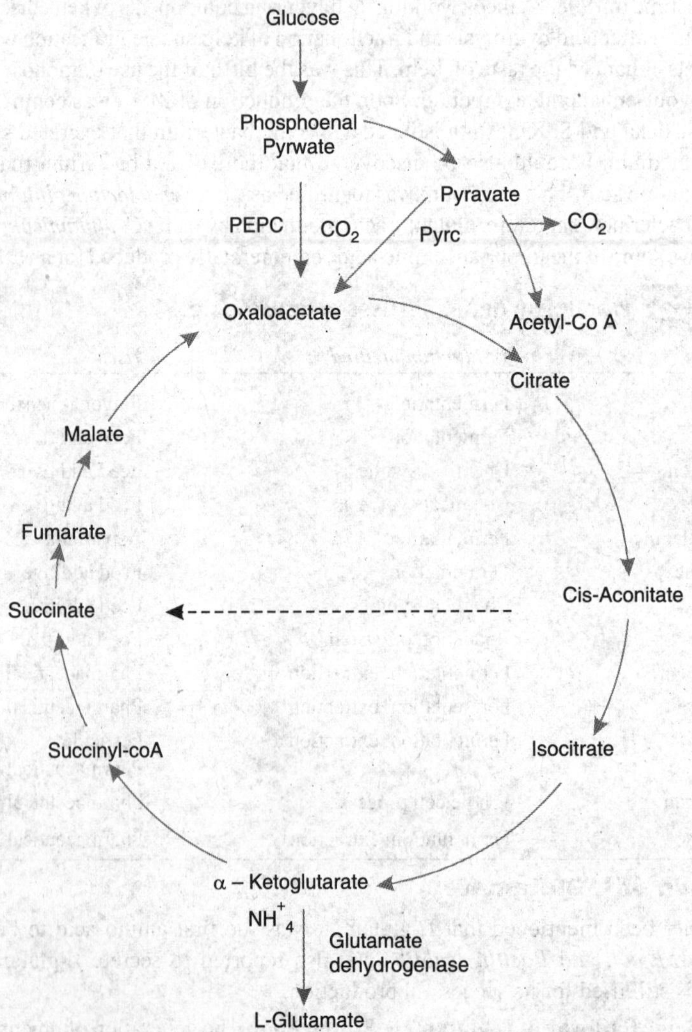


Fig. 17.6. Biochemical pathway for production of glutamic acid in *C. glutamicum;* PEPC, phosphoenol pyruvate carboxylase; PyrC, pyruvate carboxylase; PyrDH, pyruvate dehydrogenase. Dashed line shows that the reaction occurs to a least extent.

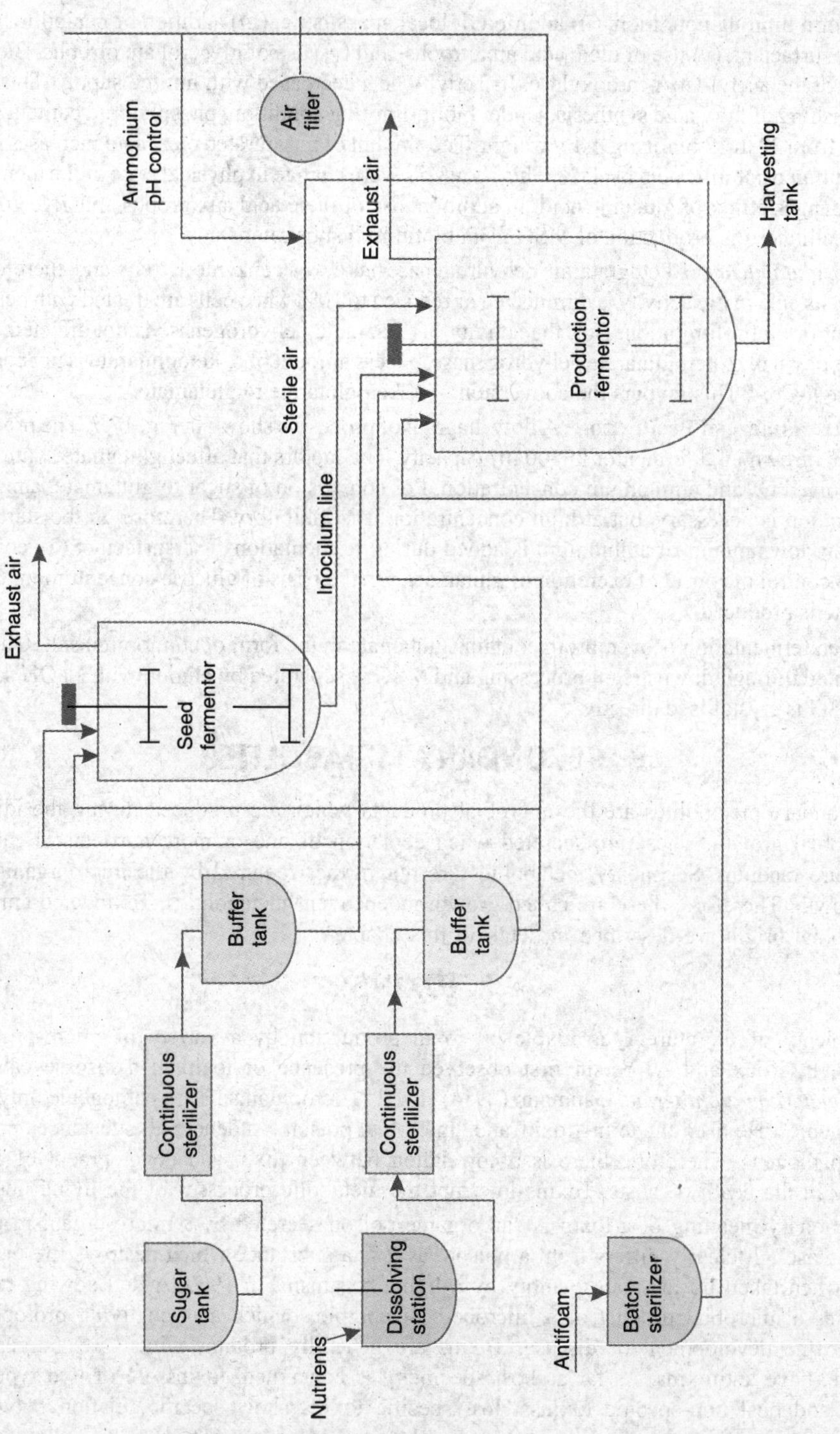


Fig. 17.7. A flow diagram of porduction of L-glutamate.

Triggering of active transport by the appropriate molecular environment of the cytoplasmic membrance is important. The means that trigger excretion of glutamate are : (*i*) growth under biotin limiting condition, (*ii*) addition of local anaesthetic, (*iii*) addition of penicillin, (*iv*) addition of surfactant, (*v*) use of oleic acid auxotrophs, and (*vi*) use of glycerol auxotrophs. Biotin is a cofactor of the acetyl-CoA carboxylase. Its activity gets decreased with limited supply. This results in diminishing of fatty acid synthesis. Under biotin-limiting condition, phospholipid content sharply declines from 32 to 17 nmol $mg^{-1}$ dry weight. The amount of unsaturated oleic acid increases relative to the saturated palmitic acid by 45%. This shows a severe change in physical state of the membrance which changes efflux of glutamic acid. In addition, use of oleic acid auxotrophs and glycerol auxotrophs facilitates the production of MSG from biotin-rich substrate.

In *C. glutamicum* α-ketoglutarate dehydrogenase has a weak enzyme activity and, therefore, it is unstable. Its enzymatic activity is diminished or reduced to 10% when cells are treated with penicillin, surfactant or biotin-limitation. But the activity of glutamate dehydrogenase is not affected. Due to lowering down of α-ketoglutarate dehydrogenase, excess amount of α-ketoglutarate is not converted to succinyl -CoA. This favours the conversion of α-ketoglutarate to glutamate.

(*c*) **Commercial Production :** A flow diagram of process is shown in Fig. 17.7. The production strains are grown in a fermentor in 500 $m^3$ capacity. The factors that affect glutamate synthesis are pH, dissolved $O_2$ and ammonium concentration. For conversion of sugar to glutamate, ammonium concentration is necessary, but at high concentration it is inhibitory. Therefore, in the start of fermentation, low amount of ammonium is added during fermentation. The surfactant (Tween 80) is added to control the onset of excretion of glutamate. On the basis of glucose concentration, 60-70% glutamate is produced.

When fermentation is over, broth contains glutamate in the form of ammonium salt. Glutamate is separated through downstream processing and MSG is separated by elution with NaOH solution. Then MSG is crystallised directly.

## B. SECONDARY METABOLITES

Secondary metabolites are the microbial products which are produced during the idiophase of microbial growth. These are secreted when depletion of one or more nutrients is caused in the culture medium. Secondary metabolites are not further required by the microorganisms for their growth. Therefore, these are called growth-independent metabolities. Toxins and antibiotics of microbial origin are described in detail in this chapter.

## I. TOXINS

A plenty of literature is available on toxin production by a variety of micro-organisms. In 1988, E. Roux and A. Yersin first observed the presence of toxin in a disease caused by *Corynebacterium diphtheriae.* Gaumann (1954) stated "microorganisms are pathogenic only if they are toxigenic". He used the term "toxin"and "microbial poison" to denote all substances produced by the pathogens. Therefore, there is a correlation between toxin and development of disease, resulting in the death of hosts. Toxins interrupt the metabolic processes of the living host cells.

Dictionary meaning of a toxin is "an organic poison secreted by a micro-organism causing some disease". It clearly differs from a poison, as it is a substance which destroys life or injures health, when taken up in small quantity by a living organism. In 1960, A.R. Ludwing regarded a toxin as a microbial product or a microbe-host complex which acts on living protoplasts to influence the development of disease. Toxins are generally proteins which are antigenic and therefore, have antitoxins. As far as host specificity is concerned, toxins are of two types, host specific and host non-specific toxins. Host specific toxins show specific relation between its production and pathogenicity, whereas host non specific toxins do not.

Toxins secreted by micro-organism lead to the development of infection in plants/animals, as the affected cells lose their resistance. Toxin production and disease development in plants are not the aim of this chapter.

## 1. Bacterial Toxins

Toxins secreted by bacteria also play a key role in disease development in animals. In recent years, use of toxin secreting micro-organisms in control of disease causing flies, has become a field of major interest in formulation of biological pesticides. However, screening programme for discovery of new secondary metabolite from *Streptomyces* having insecticidal activity has been carried out in Japan.

**(*a*) Chemical Nature of the Toxin :** It is mentioned that generally toxins are proteinaceous in nature; they interrupt the metabolism of host cells. It differs from enzyme in action as the latter destroys structural integrity of the cell.

It is interesting to note that crystals of toxin secreted by *B. thuringiensis* are water soluble, and heat stable. This toxin has been termed as ß-exotoxin which is of molecular weight of 900. Fig. 17.5 shows that ß-exotoxin is an adenine nucleotide with an unusual sugar derivative called allomucic acid.

**(*b*) Production of ß-exotoxin :** Production of ß-exotoxin on a large scale is easy. Pure culture of *B. thuringiensis* is grown on a nutrient medium (in submerged culture) supplemented with starch because this bacterium secretes amylolytic enzymes. When bacterial growth is over the culture fluid is filtered. The final product is obtained from the filtrate by vacuum drying.

**(*c*) Microbial Insecticides :** ß-exotoxin is active against flies and mosquitos (diptera), beetles (coleoptera), termites (orthoptera), butterflies and moths (lepidoptera). Mammals including man are not affected because of secretion of acid in stomach and most probably degradation by pepsin to non-toxic form. Similarly, an anaerobic spore forming rod shaped bacterium, *B. popilliae* feeds upon the larvae of the Japanese beetle, *Popillia japonica.* A detailed account of microbial insecticides is given in Chapter 21.

## 2. Fungal Toxins (Mycotoxins)— the Biotechnological Hazards of Fungi

Undoubtedly, fungi play a significant role in production of many products such as enzymes, organic acids, antibiotics, SCP, cheese, etc. However, the biotechnological hazards caused by them have also been realised in recent years. Mycotoxin refers to secondary metabolites of certain fungi which are toxic to human animals and plant health as well. Chemically, mycotoxin are non-antigenic compounds of low molecular weight. These toxins are generally detected in milk, cheese, corn, peanuts, cotton seeds, copra, almond, figs, spices and other foods and feeds. Depending on concentrations, these may be carcinogenic, mutagenic, teratogenic or oestrogenic with the result of reduced immune response and acute disease syndrome.

Mycotoxicosis is the poisoning by ingestion of mycotoxins through food contaminated by toxigenic fungi. Historically, mycotoxicosis has been known for hundreds of years ( 9th and 18th century) due to death of thousands of people in Europe caused by food-borne and a toxins (ergot) producing fungus (*Claviceps pupurea*) in rye grains. In 1978, Ethiopian epidemic occurred due to consumption of grains contaminated with *C. purpurea* infected wild oats where about 50% infected people died and the remaining suffered from many side effects.

*Aspergillus flavus* is a toxigenic fungus.

In India, a similar episode is reported following human ingestion of bajra (Pearlmillet) infected with *C. fusiformis.* Moreover, the cases of liver carcinoma in North Bihar have been found more frequent in post monsoon period (Oct-Jan.) and these have been suspected to be due to consumption of mouldy maize. In the coastal line of Karnataka high incidence of liver enlargement was found associated with heavy fungal infestation of foodgrains.

Though there is a large number of mycotoxins yet some of them which contaminate different food materials are aflatoxins, ochratoxins, zearalenone, citrinin, sterigmatocystin, trichothecenes, patulin, penicillic acid, etc. (Table 17.3; Fig. 17.8). In 1988, a new group of water soluble toxin, fumonisins, produced by certain stains of *Fusarium moniliforme* was discovered in 1988. These were associated with contaminated corn and commercial corn based food stuff. If contaminated corns are used in fermentation for ethanol production, a little degradation of toxin occurs. However, most of the fumonisin $B_1$ could be recorded through distillation in spent grains, and stillage. When spent grains are used as animal feed without detoxification, it causes serious problems. Similarly, patulin was found in detectable levels in wines, beers and fruit juices.

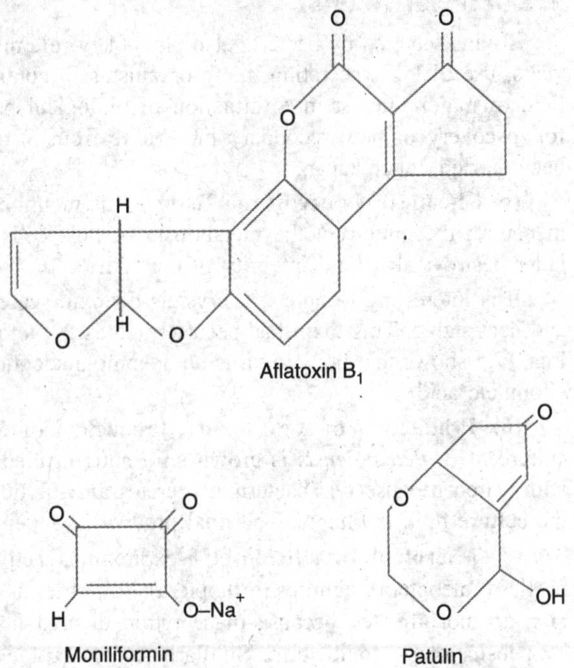


**Fig. 17.8.** Chemical structure of some mycotoxins.

**(*a*) The Toxigenic Fungi :** Mostly all the fungi are equipped with toxin producing ability depending on environmental conditions. However, a few of them associated with food and feed materials pose a great hazard, for example, species of *Alternaria, Aspergillus, Claviceps, Fusarium, Penicillium, Boletus, Agaricus, Amantia, Myrothecium, Pithomyces,* etc.

The biochemical pathways through which mycotoxins are produced are : (*i*) the polyketide pathway (aflatoxins), (*ii*) the terpene pathway (trichothecenes), (*iii*) the tricarboxylic acid pathway (rubratoxin), and (*iv*) the amino acid pathway (gliotoxin). Discussion of these metabolic pathways in detail is not the aim of the author, therefore, these are not dealt with.

**(*b*) Mode of Action of Mycotoxins :** Mode of action involves the biochemical reactions of mycotoxins with molecular receptors in animal cells. The molecular receptors are DNA, RNA, functional proteins, enzyme co-factor and membrane constituents. The reactions between mycotoxins and their receptors may be either co-valent irreversible or non-covalent reversible. In the first reaction, the reactive forms of mycotoxins conjugate with receptors to form adducts, whereas in the second reaction mycotoxin - receptor complexes get dissociated as the metabolic processes remove the toxin from the receptor sites.

After ingestion mycotoxins enter in human body and encounter various molecules. Toxins interact with gastrointestinal microflora, epithelial cells of intestine, liver, bile, blood, kidney, reproductive and nervous systems, skin and lungs. Some of the effects of mycotoxins are briefly described below:

**(*i*) Effect on energy production :** There are several mycotoxins which inhibit certain enzymes involved in Krebs' cycle. Moniliformin, a toxin of *Fusarium*, inhibits the oxidation of pyruvate and a-ketoglutarate. It also causes disturbances in intracellular osmoregulation.

***(ii)* Inhibition in synthesis of DNA, RNA, proteins and immune systems :** Inhibition in the synthesis of these macromolecules lead to cell death. Aflatoxin $B_1$ inhibits DNA synthesis in liver cells. This is caused due to covalent binding of aflatoxin $B_1$ to DNA and proteins. Aflatoxin $B_1$ is also known as to inhibit the synthesis of nuclear RNA in liver cells of rat. Since the mycotoxins inhibit DNA synthesis, the other products expressed by genes are also inhibited. Aflatoxin $B_1$ causes delay in interferon production in turkeys. However, at high doses, it reduces IgG and IgA in chicks with the consequences of decreased aquired immunity. Aflatoxin $B_1$ reduces the cell mediated immune response in animals. A reduced responses to phytohaemagglutmin stimulation in human lymphocytes has been observed *in vitro*.

**Table 17.3. Natural occurrence of mycotoxins on different commodities, and mycotoxicoses.**

| *Mycotoxins/name* | *Mycotoxicoses* | *Commodities* |
|---|---|---|
| **A. Aspergillus Toxins** | | |
| Aflatoxin $B_1$, $B_2$, $G_1$, $G_2$ | Liver cancer | Corn, peanuts, milk, treenuts |
| Sterigmatocystin | Carcinogenesis*, | Green coffee, grains, fruits |
| Ochratoxin | Renal tumour | Corn, coffee, wheatflour, bread |
| **B. Fusarium Toxins** | | |
| Moniliformain | Onylai diseases | Rice |
| Fumonisins | Leukoencephalo-malacia *, promote cancer * | Corn, wheat flour, corn flakes |
| Trichothecenes (T-2 toxin, Deoxyninalenol) | Dermatites, oesophagal cancer, digestive disorder | Corn, wheat, baby cereal, wine, commercial cattle feed |
| Zearelenone | Cervical cancer, abortion* | Corn meal, corn flakes, walnut |
| **C. Penicillium Toxins** | | |
| Citreoviridin | Cardiac beriberi | Mouldy pecan fragments |
| Citrinin | Kidney damage* | Dry fruits, rice, corn |
| Cyclopiazonic acid | Kodua poisoning | Cheese crust, corn |
| Patulin | Capillary damage in vital organs | Cider, apple juice, wine, jam, grape juice, scented supari |
| Penicillic acid | Edema* carcinoge-nesis* | Corn, bean, apple |
| Penitrem A | Bloody diarrhoea* death * | Mouldy cream cheese |
| Rubratoxin B | Liver disease * | Mouldy grains |
| **D. Other Mycotoxins** | | |
| *Agaricus toxins* | | |
| (*i*) Agaritin, amotoxin | Kwashiorkor rooms | *Agaricus bisporus*, frozen, mush |
| (*ii*) Amanitins | Mushroom poisoning | Mushrooms |
| *Ergot alkaloid* | | |
| (*i*) Ergosine, ergometrin, ergocristine | Ergotism | Flour of rye, wheat, triticale, wheat cereals, baby cereals. |

* Symptoms observed in animals; Source : Many research and review papers.

Ochratoxin A inhibits the activity of phenylalanyl - tRNA synthetase which is required in the first step of protein synthesis. It also reduces the renal mRNA coding for certain enzymes such as phosphoenolpyruvate carboxylase. Due to inhibition in protein synthesis several consequences occur. One of these is the changes in composition of serum protein which results in the suppression of non-specific humoral substances.

***(iii)* Effects on Nervous Systems.** Mycotoxins are grouped into three on the basis of mode of action :

- Mycotoxins causing paralysis and inhibition in respiratory system *e.g.* citrioviridin. They kill nerve cells disrupting energy supply as they inhibit ATPase activity.
- Mycotoxins inducing trembling in animals e.g. fumitremorgin A, penitrem A. They alter functional states of neurotransmitters and disrupt nervous system.
- Mycotoxin causing vomiting in animals *e.g.* vomitoxins such as trichothecenes. They act on chemoreceptor trigger zone in medula oblongata and change the biogenic amines.

***(iv)* Effects on Hormones' Activities.** In target cells, steroid hormones regulate the functions. Steroid forms the complexes with receptor which are then activated and transported to the nucleus. They bind to the activator sites of chromatin and induce protein synthesis selectively. Aflatoxin $B_1$ binds covalently with acceptor sites of chromatin and thus, reduces the nuclear acceptor sites of hormone-receptor complexes. Consequently, hormonal activities are reduced.

***(v)* Carcinogenic Effects.** Aflatoxins, $B_1$, $G_1$, and $M_1$ and sterigmatocystin, versicolorin A, luteoskyrin are known as carcinogenic mycotoxins. These are genotoxic also. The chemicals which cause gentle damage and initiate the carcinogenic process are known as genotoxic or initiators, whereas those which promote transformation of genetically modified cells to cancerous cells are known as promoters. Most of the chemicals are both genotoxic and promoters.

Aflatoxin $B_1$ causes liver cancer. It binds with DNA at the guanine base in liver cells, corrupting the genetic code that regulates cell growth.

***(vi)* Effects on Reproductive Systems.** Urogenital systems of swine cattle and poultry animals are known to be affected by zearalenone which at 1 ppm produces hyperastrogenism in pigs. In young male swine, it produces testicular atrophy, and mammary gland enlargement. It has also been observed that at high concentrations zearalenone is associated with infertility, teat enlargement and under secretion in cattle. Zearalenone causes embryonic death and inhibition of development in swines.

Ergot ingestion may cause abortion in animals. Moreover, ergot is also associated with reduced weight gain and milk production in animals.

***(c)* Control and Management of Mycotoxins :** Although total elimination of mycotoxin from human food and animal feed is far difficult, yet the hazards from high concentrations can be minimised in some cases by adopting certain precautionary steps.

(*i*) Residues of susceptible crops serves as a substrate for inoculum production of certain toxigenic fungi such as *Claviceps, Aspergillus flavus,* etc. Therefore, management of crop residues and use of resistant varieties should be done.

(*ii*) Crop rotation should be adopted to lower the primary inoculum of toxigenic fungi.

(*iii*) Since grain and oilseed crops are susceptible to fungal deterioration, these must be harvested when the crops have reached to their optimum maturity. At this time, presence of low moisture content in grains will minimise the risk of field contamination. As most of the contaminants get associated from the field.

(*iv*) Prior to storage grains and oilseeds must be sundried to lower the mositure content. Fungicides and insecticides of low toxicity must be used.

(v) Electronic or hand sorting of contaminated peanuts and other seeds, wherever possible, should be done.

(*vi*) Biological method of detoxification should be adopted by using certain strains of yeasts, moulds or bacteria *e.g. Flavobacterium aurantiacum* (NRRL B-184) from the liquid medium. These catalyse the hydration of aflatoxins (Marth and Doyle, 1979).

(*vii*) Chemical detoxification is also one of the methods for inactivating the aflatoxins. Ammoniation (use of ammonia gas) resulted in significant reduction in levels of aflatoxins in contaminated peanuts, cotton seed meals, corn, etc. (Council for Agricultural Science and Technology, Report No. 116, 1989).

## II. ANTIBIOTICS

In the beginning of 20th century, the idea of growth inhibition of one micro-organism present in the vicinity of other one came into existence. Later, it was demonstrated that growth inhibition of the former micro-organism was mediated by secretion of toxic metabolites by the latter. This toxic metabolite was termed as 'antibiotic' and the phenomenon of act of growth inhibition by antibiotics as 'antibiosis'. The antibiotics are defined as "the complex chemical substances, the secondary metabolites which are produced by microorganisms and act against other micro-organisms".

In nature, there is universal distribution of antibiosis among the microorganisms owing to which they are involved in antagonism. Those microorganisms which have capacity to produce more antibiotics can survive for longer time than the others producing antibiotics in less amount.

However, antibiotics produced by micro-organisms have been very useful for the cure of certain human diseases caused by bacteria, fungi and protozoa. Due to continuous endeavour made in this field, the antibiotics discovered at present are about 5,500. Total world production of antibiotics is more than one million tonne per annum. This success has been possible only due to continuous researches made during the last four decades. A list of micro-organisms producing antibiotics and their applications are given in Table 17.4.

**> Table 17.4.** Antibiotics produced by microorganisms.

| *Micro-organism* | *Antibiotics** | *Applications* |
|---|---|---|
| **Bacteria :** | | |
| *Bacillus brevis* | Tyrothricin ($G^+$, $G^-$) | Mouth and throat infection |
| *B. polymyxa* | Polymyxin B (AT) | UTI, gastroentritis |
| *B. subtilis* | Bacitracin ($G^+$) | Dermatitis, superficial pyogenic infection, dysentery |
| *Streptococcus cremoris* | Nisin ($G^+$) | In cheese, food preservation (non-medical use) |
| **Actinomycetes:** | | |
| *Micromonospora purpurea* | Gentamicin ($G^+$, $G^-$) | UTI, abscess |
| *Nocardia, mediterranei* | Rifamycin (My) | Meningitis |
| *Streptomyces griseus* | Streptomycin ($G^+$, $G^-$, My) | Tuberculosis |
| *S. aureofaciens* | Tetracyclines *($G^+$, $G^-$) | Cholera, tetanus, UTI |
| *S. erythreus* | Erythromycin ($G^+$) | Cholera, tetanus, arthritis |
| *S. noursei* | Nystatin (AF) | Skin lesions |
| *S. spheroides* | Novobiocin ($G^+$) | Abscess |
| **Fungi:** | | |
| *Cephalosporium acremonium* | Cephalosporins ($G^+$, $G^-$) | UTI, pneumonia, meningitis |
| *Penicillium chrysogenum* | Penicillin ($G^+$) | Pneumonia, pharyngitis |
| *P. griseofulvum* | Griseofulvin (AF) | Skin and hair lesions |
| *P. notatum* | Penicillin ($G^+$) | Fever, pneumonia, genital infection |

UTI, Urinary tract infection; * Antibiotic spectrum; $G^+$, Gram positive bacteria; $G^-$, Gram negative bacteria; My, Mycobacteria, AF, Antifungal; AT, antitumour.

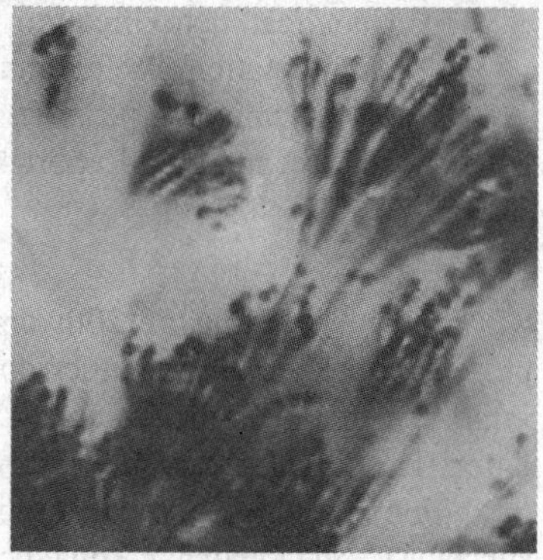

*Penicillium chrysogenum.*

Among the antibiotics discovered so far, there are four major groups which are most extensively used throughout the world : the penicillins (Fig. 17.9), cephalosporins, tetracyclines and erythromycins (Fig. 17.10).

It has been already mentioned that antibiotics are produced in culture medium during idiophase due to depletion caused by one or more nutrient (s) in the medium. In 1979, D. Perlman has described that the biosynthesis of antibiotics may be regarded as a result of a series of in-born errors of metabolisms. These errors may be exaggerated by subjecting the original micro-organims to mutagenic substances. However, high yield of commercially important antibiotics owe much to the selection of such mutant stains as improvment of strain of *Penicillium chrysogenum* to yield benzylpenicillin about 20 mg/ml over the normal rate 720 μg/ml.

Moreover, researches done on this aspect have shown (in U.K.) that the synthesis of some of the antibiotics in *Streptomyces* was mediated not by a plasmid. Therefore, there is possibility to produce new antibiotics by transfer of plasmids into a single cell of *Streptomyces.*

At present, there are thousands of antibiotics produced by micro-organisms of which only a few hundreds have been marketed so far. Among these the penicillins, cephalosporins and tetracyclines have been commercialized. In this context penicillin production is described in detail.

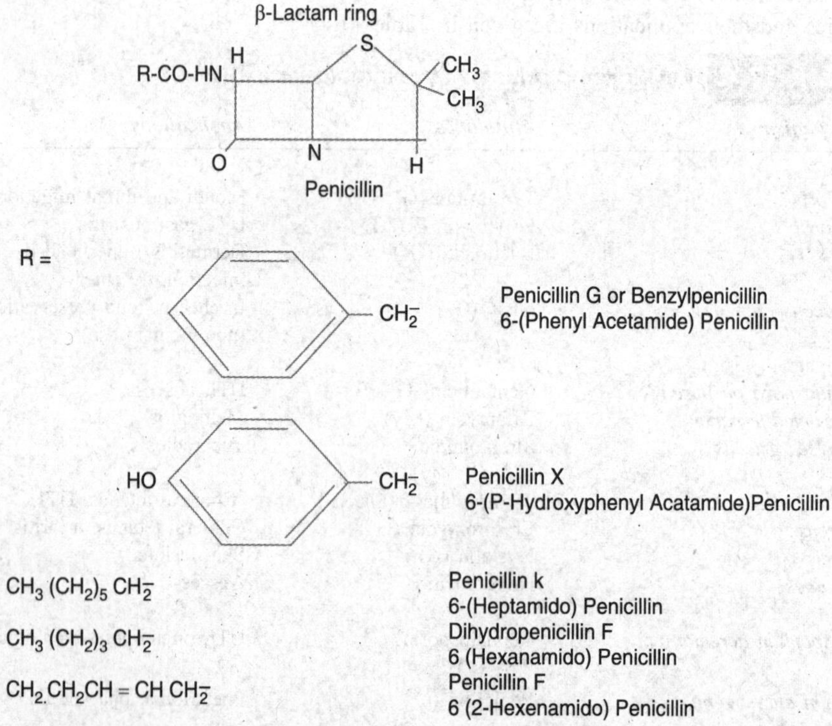


Fig. 17.9. Chemical structure of penicillins.

## 1. Penicillins

It was Alexander Fleming (1929) who first discovered the bacteriostatic principle from a fungus and named it penicillin. He observed that a fungal contaminant prevented the growth of staphylococci,

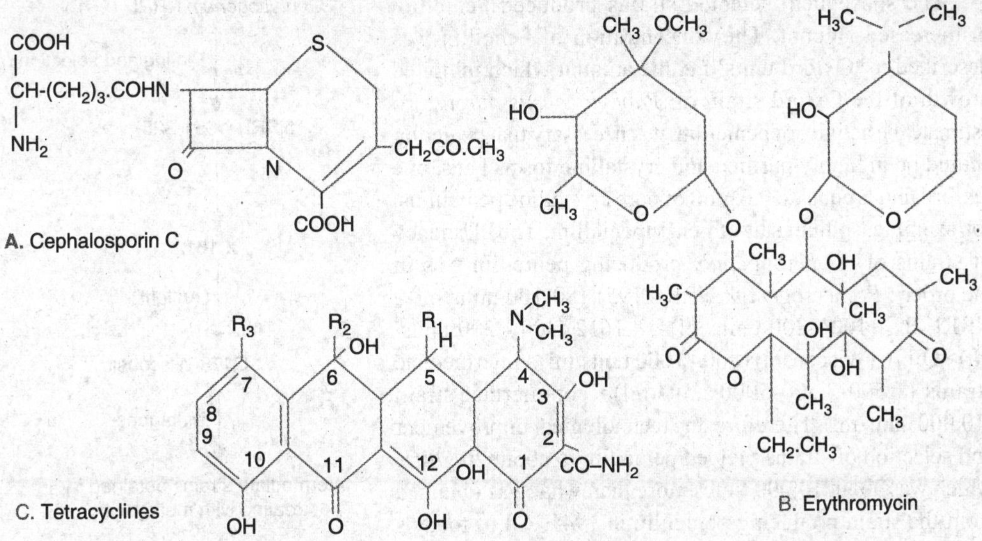


| $R_1$ | $R_2$ | $R_3$ | |
|---|---|---|---|
| H | $CH_3$ | H | — Tetracycline |
| H | $CH_3$ | Cl | — Aureomycin |
| OH | $CH_3$ | H | — Terramycin |
| H | H | Cl | — 6-Dimethyl-7--chlortetracycline |

**Fig. 17.10.** Chemical structure of cephalosporin C (A), erythromycin (B) and tetracyclines (C)

which was later on identified as *Penicillium notatum.* In 1932, P.W.J. Clutterbuck and coworkers studied the chemical nature of penicillin. They found that penicillin was an organic acid which was dissolved into organic solvents from aqueous solutions at low pH. It was vulnerable to hydrogen ion ($H^+$) and heat. After evaporation of solution to dryness, the biological activity was lost. Further studies done on *P. notatum* confirmed that this mould could produce about 2 ppm active substance.

The biological activity and recovery of penicillin were investigated by Chain *et al.* (1940). They cultured Fleming's fungus in surface culture in a small pilot plant scale and recovered penicillin in 1,000 fold amount only by keeping low temperature during the extraction. They also produced dry powder in the form of salt of penicillin. This was the first attempt to extract penicillin salt in the form of dry powder which could have the curative properties.

It was the time of World War II when significance of penicillin was realized. England had no money to expend on penicillin production. Thus, the Oxford group was in crisis. Dr. Florey and Dr. Heatley came to U.S.A. America took up the problem and gave high priority on antibiotic production. This is why, by the time of attack of France in 1944, an adequate amount of penicillin was available to save the life of wounded people.

**(*a*) Strain Improvement** : The fungus *P. notatum* originally used by Fleming for penicillin production, gave poor result. Moreover, many strains of this fungus were developed which produced many fold more penicillin than the original one. Besides *P. notatum, P. chrysogenum* was also tested which gave good results in submerged culture condition. One of these strains was *P. chrysogenum* NRRL 1951. Mutations in *P. chrysogenum* are generally induced by ultraviolet (UV) radiation or other mutagenic chemicals (*e.g.* N-methyl-N-nitro-N-nitrosoguanidine (NTG). The subsequent strains isolated from NRRL 1951 by treating with mutagens as shown in Fig. 17.11.

The subsequent selected strains produced penicillin in increased amount. The concentration of pencillin was described in 'Oxford units' *i.e.* the amount which inhibited growth of the Oxford strain of *Staphylococcus aureus.* To estimate the activity of penicillin, it is necessary that penicillin should be in highly purified and crystalline form. Thus, one Oxford unit is equal to 0.6 g/ml of pure crystalline penicillins, for example, sodium salt of benzylpenicillin. The efficiency of strains of *P. chrysogenum* producing penicillin was in the order : *P. chrysogenum* NRRL 1951 (80-100 units/ml)> NRRL $B_{25}$ (100 - 200 units/ml)> × 1612 (300 - 500 units/ml)> Q 176 (Wisconsin) (800 - 1000 units/ml)> intermediate strains (1,500-2,500-5000 units/ml)> commercial strain (10,000 units/ml). Therefore, the technological improvement and selection of strains yielded penicillin to about 20 g/litre which was about 10,000 times more than what was obtained from the strain producing penicillin in 1941. All of today's industrial strains were derived from the Wisconsin Q -176 and Wisconsin 51-20 improved strains.

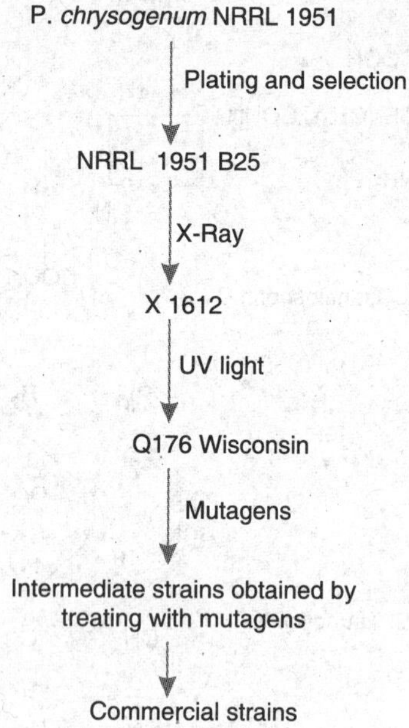


**Fig.17.11.** Development of commercial strain from *Penicillium chrysogenum*

(*b*) **The Chemical Nature of Penicillins :** Molecular structure of penicillins reveals that they include the 4 membered ß-lactam ring. The amide bond of ß-lactam ring is readily broken in both acidic and alkaline medium and can be hydrolysed by penicillinase (ß-lactamase) synthesized by many bacteria. The naturally occurring penicillins differ from each other in R groups. When different R group attaches with penicillin it results in different types of penicillin (Fig. 17.9).

The amber coloured droplets seen on the surface of this culture of the mold is antibiotic penicillin.

(*c*) **Fermentation Medium :** For the first time, surface culture of *P. notatum* was carried out in Czapek-Dox broth. Rate of penicillin production increased on supplementing it with the organic materials such as yeast extract, casein or beaf extract. In addition, increased yield was also obtained with amending the cotton seed, peanut meal and soybean meal. Several natural penicillins (*e.g.* penicillin F, penicillin K and dihydropenicillin F) are produced when a suitable medium is fermented by *P. chrysogenum* in the absence of supplemented precursor. Production of these natural penicillins is also stimulated by the presence of fatty acids in the medium. Medium supplemented with phenylacetic acid, phenylacetamide or ß-phenylethylamine stimulate the production of penicillin G. In 1949, C.T. Calam and D.J.D. Hockenhull developed a chemically defined medium which supported production of antibiotic by many strains of *P. chrysogenum.* Chemical composition of media is given in Table 17.3. Production of antibiotic was also increased by: (*i*) keeping the pH of fermentation medium between 6.8 and 7.4, (*ii*) adding buffering agents *e.g.* $CaCO_3$ and phosphate to the medium and sterile $H_2SO_4$ and NaOH when required, (*iii*) keeping the temperature at 25 ± 0.5°C during incubation, and (*iv*) agitating the culture for aeration in a large fermentor.

(*d*) **Fermentation Process :** Penicillins are produced on large scale in a commercially devised fermenter which provides optimum growth conditions to *P. chrysogenum* for maximum yield. Following are the steps for fermentation of benzyl-penicillin *i.e.* Penicillin G.

(*i*) Inoculate 100 ml medium in 500 ml Erlenmeyer flask with spores of *P. chrysogenum* strains and incubate at 25°C by keeping them on a rotatory shaker.

(*ii*) After 4 days, transfer the content of flask to another flask (4 litre capacity) containing 2 litres medium and incubate for 2 days as earlier.

(*iii*) Transfer the content to a stainless steel tank (800 litre capacity) containing 500 litre medium. This tank is equipped in such a way that it could provide the optimum conditions for fungal growth.

(*iv*) After 3 days, use the contents for inoculation of about 1, 80,000 litre medium kept in a fermentor (2,50,000 litre capacity). The latter is equipped with automatic devices to optimum growth conditions (*e.g.* temperature, pH, $O_2$, etc.)

(*v*) Filter the content of fermentor after 6 days of incubation. Filtrate contains penicillin. Extract the penicillin into amyl-or butyl-acetate. From it transfer the penicillin into aqueous solvent by extracting with phosphate buffer.

**> Table 17.5. Chemical composition of fermentation medium for the production of penicillins.**

| *Constituents* | *C. H. medium (g/l)** | *Calam's medium (g/l)*** |
|---|---|---|
| Lactose | 30 | 35 |
| Glucose | 10 | 10 |
| Starch | 15 | - |
| Corn steep solids | - | 35 |
| Ammonium sulphate | 5 | - |
| Ethyl amine | 3 | - |
| Vegetable oil | - | 2.5 |
| Citric acid | 10 | - |
| Acetic acid | 2.5 | - |
| Phenyl acetate | 0.5 | - |
| $KH_2PO_4$ | - 4 | |
| $CaCO_3$ | - | 10 |

* C.T. Calam and D.J.D. Hockenhull (1949) ; ** C.T. Calam (1976).

From a butanol-water mixture crystallize the potassium penicillin G. Again purify potassium penicillin G before its use.

(*e*) **Antibiotic Producing Companies :** There are thousands of antibiotic–producing companies in Indian and abroad. Name of a few companies are given below :

**Indian Companies**

1. Astra-IDL Limited, 32/1-2 Crescent lower, Crescent Road, Bangalore - 560001.
2. Byer (India) Ltd., Express Towers, Nariman Point, Bombay-400021.
3. Biochem Pharmaceutical Industries, A - Aidum Building, I-Dhobi Talao, Bombay-400002.
4. Cadila Chemicals Pvt. Ltd., Maninagar, Ahmedabad-380008.
5. Glaxo India Ltd., Annie Besant Road, Bombay-400025.

6. Hindustan Antibiotic Ltd., Pimpri, Pune-411018.
7. Hoechst India Ltd., Hoechst House, Nariman Point, 193 Backbay Reclamation, Bombay-400021.
8. John Wyeth (India) Ltd., Steelcrete House, Dinshaw Wacha Road, Bombay 400020.
9. Maharastra Antibiotics & Pharmaceuticals Ltd., Nagpur.
10. Pfizer Ltd., Express Towers, Nariman Point, Bombay-400021.
11. Sandoz (India) Ltd., Sandoz House, Annie Besant Road, Worli, Bombay-400021.
12. Sarabhai Chemicals, Wadi, Vadodara.

**Foreign Companies**

13. American Cyanamid, U.S.A.
14. China National Chemical Import & Export Corporation, China.
15. Glaxo Laboratories Ltd., England.
16. Hoechst AG, W.Germany.
17. Imperial Chemical Industries Ltd., England.
18. Pfizer, Inc. U.S.A.
19. Wyeth Laboratories, U.S.A.

## PROBLEMS

1. What are vitamins ? Give a detailed account of vitamin $B_{12}$.
2. Give a detailed account of commercial production of citric acid.
3. Discuss in brief biochemical pathways of organic acid production with special reference to citric acid.
4. What is fermentation ? Give an illustrated account of design and functioning of a fermentor.
5. Write an eassy on microbial secondary metabolites.
6. What are the toxins? How do they differ from antibiotics?
7. What is microbial insecticide? Upto what extent it is useful in biocontrol of insect pests?
8. Discuss in brief the culture selection method of *Penicillium.*
9. Give a detailed account of formulation of growth medium and fermentation process of Penicillin.
10. Write short notes on the following :
(*a*) Microbial secondary metabolites, (*b*) Toxins (*c*) Microbial insecticide, (*d*) *Bacillus thuringiensis,* (*e*) Antibiotics, (*f*) Alexander Fleming (*g*) Penicillins (*h*) *Penicillium chrysogenum* (*i*) Antibiotic producing companies in India and abroad, (*j*) Mycotoxins and health hazards, (*k*) Fermentor, (*l*) Techniques of microbial culture, (*m*) Scale-up, (*n*) Downstream processing, (*o*) Isolation and improvement of microbial strains. (*p*) Glutamic acid production

CHAPTER

# 18

# Single Cell Protein (SCP) and Mycoprotein

The dried cells of micro-organisms (algae, bacteria, actinomycetes and fungi) used as food or feed are collectively known as 'microbial protein'. Since the ancient times a number of micro-organisms have been used as a part of diet. Fermented yeast (*Saccharomyces* sp.) was recovered as a leavening agent for bread as early as 2500 B.C. (Frey, 1930). Fermented milk and cheese produced by lactic acid bacteria (*Lactobacillus* and *Streptococcus*) was used by Egyptians and Greeks during 50-100 B.C. *e.g. Lactobacillus* and *Streptococcus.* Cultured dairy products contain $10^7$ to $10^{10}$ lactic acid bacteria per gram of product. During the first century B.C. the palatability of edible mushrooms was also realized in Rome. In 16th Century blue-green algae (*e.g. Spirulina*) was consumed as a major source of protein. *Spirulina* provides protein, iron, vitamins, minerals and most of the essential micro-nutrients.

Spirulina tablets are commercially available in market.

The term 'microbial protein' was replaced by a new term "single cell protein" (SCP) during the First International Conference on microbial protein held in 1967 at the Massachusetts Institute of Technology (MIT), Cambridge, Massachusetts, U.S.A. Criteria for coining this term was the single celled habit of micro-organisms used as food and feed.

In 1973, when Second International Conference was convened at MIT, some actinomycetes and filamentous fungi were reported to produce protein from various substrates. Since then many filamentous fungi have been reported to produce protein. Therefore, the term SCP is not logical, if an organism produces filaments.

Since the 1920s, filamentous fungi have been used for the production of protein. For such fungi, the term 'fungal protein' has been used by many workers. Recently, the term 'mycoprotein' has been introduced by Ranks Hovis McDougall (RHM) in the United Kingdom for protein produced on glucose or starch substrates.

Importance of mass production of micro-organisms as a direct source of microbial protein was realized during World War I in Germany and consequently, baker's yeast (*S. cerevisiae*) was produced in an aerated molasses medium supplemented with ammonium salts. During World War II (1939-1945) the aerobic yeasts (*e.g. Candida utilis*) were produced for food and feed in Germany. Since World War II, considerable effort has been made to develop technologies for mass cultivation of SCP by formulating different types of growth media and improved culture of micro-organisms. In the late 1950s, British Petroleum started producing the SCP from hydrocarbons since the crude oil contains 10-25% (paraffins) and established a first large scale plant in Sardinia at the end of 1975. It had a capacity of 1,00,000 tonnes SCP per annum. Large scale production has been envisaged in England and Rumania with the annual production of 60,000 tonnes bacterial mass in England. The erstwhile U.S.S.R. was the World's largest producer of SCP in 1980. The production was estimated to 1.1 million tonnes of SCP per annum.

In India, little attention has been paid on the production of SCP, though mushroom cultivation started in the early 1950s. However, work on mushroom culture at Solan (Himachal Pradesh) from 1970 onward has brought satisfactory results. Recently, National Botanical Research Institute (NBRI), Lucknow and Central Food Technological Research Institute (CFTRI), Mysore, have established Centres for mass production of SCP from cyanobacteria. At the NBRI, SCP is produced on sewage which is further utilized as animal feed.

Therefore, in the light of protein shortage, micro-organisms offer many possibilities for protein production. They can be used to replace totally or partially the valuable amount of conventional vegetable and animal protein feed. For this, development of technologies to utilize the waste products would play a major role for the production of SCP.

## A. ADVANTAGES OF PRODUCING MICROBIAL PROTEIN

Roth (1982) has described a number of advantages in the production of microbial protein, compared to protein problems of conventional crops used as food and feed. These include :

(*i*) Rapid succession of generations (algae, 2–6) h; yeast, 1–3h; bacteria, 0.5–2h);

(*ii*) Easily modifiable genetically (*e.g.* for composition of amino acids);

(*iii*) High protein content of 43-85 per cent in the dry mass ;

(*iv*) Broad spectrum of original raw material used for the production, which also includes waste products;

(*v*) Production in continuous cultures, consistent quality not dependent on climate in determinable amount, low land requirements, ecologically beneficial. Other advantages are : (*a*) high solar energy conversion efficiency per unit area (net production in cultivated land-290 $gC/m^2/day$, lakes and streams - 225 $gC/m^2/day$ ; estuaries - 810 $gC/m^2/day$), (*b*) easy regulation of environmental factors *e.g.* physical, nutritional, etc. which maximize solar energy conversion efficiency and yield (*c*) cellular, molecular and genetic alterations and (*d*) algal culture in space, which is normally unused instead of competing for land.

## B. MICROORGANISMS

Many groups of micro-organisms are used as sources of proteins. Some of the micro-organisms with their carbon and energy sources are given in Table 18.1.

**>Table 18.1. Single cell protein (SCP) and mycoprotein produced on the selected substrates.**

| *Microgial groups* | *Microorganisms* | *Protein (% /100g., on dry weight basis)* | *Substrates* |
|---|---|---|---|
| **Algae** | *Chlorella pyrenoidosa* (36)[a] | - | $CO_2$ (10%), light |
| | *Scenedesmus acutus* (20)[b] | - | $CO_2$ sunlight |
| | *Spirulina maxima* (15)[b] | 53 | $CO_2$ (5%), combustion gases, bicarbonate, sunlight (in pond) |
| **Bacteria** | *Achromobacter delvacvate* | - | Diesel oil in fermenter |
| | *Bacillus megaterium* | - | Collagen meat packing waste in fermenter |
| | *Cellumonas* sp. (0.45)[c] | 87 | Bagasse |
| | *Methylomonas clara* (0.5)[c] | 13 | Methanol |
| | *Pseudomonas* sp. (1.0)[c] | - | *n*-alkanes fuel oil |
| **Actinomycetes** | | | |
| | *Nocardia* sp. (0.98)[c] | - | *n*-alkanes |
| | *Thermomonospora fusca* (0.4)[c] | 5.6 | Cellulose pulp |
| **Fungi** | | | |
| **Yeasts** | *Candida lipolytica* (0.88)[c] | 65-69 | *n*-alkanes |
| | *C. utilis* (0.39)[c] | - | Potato starch waste |
| | *B. utilis* | 54 | Sulphite liquor |
| | *Saccharomyces cerevisiae* (0.5)[c] | 53 | Molasses |
| | " " (0.5)[c] | 45 | Beer |
| | *S. fragilis* | 54 | Milk whey |
| | *Rhodotorula glutinis* | - | Domestic sewage |
| | *Torulopsis* sp. | - | Methanol |
| **Moulds** | | | |
| | *Aspergillus niger* | 50 | Molasses |
| | *Trichoderma viride* | 64 | Straw, starch |
| | *Paecilomyces varioti* | 55 | Sulphite waste liquor |
| **Mushrooms** | | | |
| | *Agaricus campestris* | 36-45 | Glucose |
| | *Morchella crassipes* | 31 | Glucose, cheese whey, sulphite, liquor. |

(*a*) yield g/day (on dry weight basis); (*b*) yield g/$m^2$/day (on dry weight basis); (*c*) yield dry weight basis (g/g substrate used). Values in parentheses denote yield of biomass.

## C. SUBSTRATES USED FOR PRODUCTION OF SCP

A variety of substrates are used for SCP production. However, availability of necessary substrates is of considerable biological and economic importance for the production of SCP. Algae which contain chlorophylls, do not require organic wastes. They use free energy from sunlight and carbon-dioxide from air,while bacteria (except photoautotrophs) and fungi require organic wastes, as they do not contain chlorophylls.

The major components of substrates are the raw materials which contain sugars (sugarcane, sugarbeet and their processed products), starch (grains, tapioca, potato, and their by-products), lignocelluloses from woody plants and herbs having residues with nitrogen and phosphorus contents and other raw materials (whey and refuses from processed food). Organic wastes are also generated by certain industries and are rich in aromatic compounds or hydrocarbons. Recent price-increase in petroleum and refined petroleum products has made hydrocarbons and chemicals derived from them (such as methanol and ethanol) less attractive as raw materials for SCP production than renewable sources such as agricultural wastes or by-products. A detailed account of these wastes is given in Chapter 23.

## D. NUTRITIONAL VALUE OF SCP

Nowadays, considerable information is available on the composition of microbial cells *e.g.* protein, amino acid, vitamin, and minerals (Litchfield, 1979). Commercial value of SCP depends on their nutritional performance and nevertheless, it has to be evaluated to the prevalent feed protein. SCPs either from alkanes or methanols, are characterized by good content and balance in essential amino acids.

Composition of growth medium governs the protein and lipid contents of microorganisms. Yeasts, moulds and higher fungi have higher cellular lipid content and lower nitrogen and protein contents, when grown in media having high amount of available carbon as energy source and low nitrogen.

Ignoring a few extreme values, the mean crude protein in dry matter of algae and yeasts, on conventional substrates, lies between 50 and 60 per cent, for alkane yeasts between 55 and 65 per cent, and for bacteria about 80 per cent. A high content of nucleic acid free protein is extremely important for the economic efficiency of the procedure in SCP production. Because of high protein and fat contents, the contribution of carbohydrates to the nutritional value of SCP is not of prime importance.

The crude ash content is determined in particular by the nutrient salts of the fermentation medium. Estimation of crude protein is based on total nitrogen which is multiplied by the factor 6.25. The protein content of micro-organisms computed in this manner does not give the exact figure of protein content, as in the estimation of total nitrogen, the value of nucleic acid is also included which is somewhat erroneous.

The most important measure of nutritional value is the actual performance of SCP products as determined in feeding studies. The determinants of the utility of SCP product for application as food for human beings and feed for animals differ. Table 18.2 shows nutrition value of food proteins from different sources. For human beings, protein digestibility and protein efficiency ratio (PER), biological value or net protein utilization (NPU), determined in rats, are the parameters for food application, whereas for animals, metabolizable energy, protein digestibility and feed conversion ratio (weight of ration consumer/weight gain) are the measures or performance in broiler, chickens, swine and calves (and egg laying in hens).

Digestibility (D) is the percentage of total nitrogen consumed, which is absorbed through the alimentary tract. It is calculated as below :

$$D = \frac{Ni - Fn}{Ni}$$

where,

Ni = nitrogen ingested from SCP.

Fn = nitrogen content in faeces after feeding SCP

Biological value (BV) is the percentage of total nitrogen assimilated which is retained by the body, taking into account the simultaneous loss of endogenous nitrogen through excretion in

**> Table 18.2. Nutritional value of food protein.**

| Food protein | Analytical composition (%)[A] | | Essential amino acids (g/100g crude protein)[B] | | Biological coefficient (%)[A] | | Digestibility of crude protein (%)[B,*] | Metabolisable energy for (Kcal/kg)A |
|---|---|---|---|---|---|---|---|---|
| | Total nitrogen | Crude protein (N X 6.25) | Lysine | Threonine | NPU | NPV | | |
| Algae | 8.0 | 45 -71 | 5.7 | 5.2 | — | — | 82 | — |
| Dried skimmed milk | 5.7 | 35.9 | 8.0 | — | 87 | 31.2 | — | 2510 |
| Soybean meal | 7.0 | 44 | 6.4 | 4.0 | 64 | 65 | — | 2240 |
| Alkane yeast (toprina - LBP) | 11.2 | 7.0 | 7.4 | 4.2 | 91 | 96 | 92 | 2540 |
| Bacteria from methanol | 11.5 | 72 | 6.2 | 4.6 | 84 | 88.4 | 91 | 3468 |

*Source* : A, Senez (1986); B. Roth (1982); * Digestibility determined by pig feeding; NPU = Net protein utilization; NPV = Net protein value

urine. This is expressed by the following formula :

$$BV = \frac{Ni - (Fn + Un)}{(Ni - Fn)} \times 100$$

where, Un = nitrogen content in urine after feeding SCP.

Protein efficiency is the proportion of nitrogen retained when protein under test is fed compared with that retained when a reference protein (*e.g.* egg albumin) is fed (Riviere, 1977).

Nutritional values of SCP product are given in Table 18.3 which indicate that protein digestibility range from good to very good and is true for bacteria and yeasts growth on un-conventional substrates. However, there are certain problems which warrant the use of SCP products as human foods such as: (*i*) high content of nucleic acid leading to development of kidney stone and gout if consumed in high quantity, (*ii*) possibility for the presence of toxic secondary metabolites and (*iii*) poor digestibility and stimulation of gastrointestinal and skin reactions.

## E. GENETIC IMPROVEMENTS IN MICROBIAL CELLS

At present, production of SCP by mass culture of micro-organisms is in its infancy. It needs much boost to solve the problem of starvation in the coming decades. One of the ways to enhance productivity and quality of SCP product is the genetic improvement of micro-organisms. At Sosa Texcoco, Mexico, researches are in progress on production of genetically engineered cells, which can grow in alkaline environment even upto pH 8.0-10.0 and under the artificial conditions as well. In 1981, O. Ciferri developed mutants of *S. platensis* , which had about 40 times more longer pools of certain amino acids than found in wild alga.

Moreover, transfer and expression of beneficial genes in the micro-organisms have opened a new era for the production of algal proteins and other compounds to be used in food and feed. In 1981, D. Rochaix and J. Van Dillerviger have successfully introduced genes of *S. cerevisiae* into *Chlamydomonas reinhardii* cells and got expression of fungal genes in algal cell.

## F. PRODUCTION OF ALGAL BIOMASS

Algae (cyanobacteria and unicellular eukaryotes) grow autotrophically and synthesize their food by taking energy from sunlight or artificial light, carbon source from carbon dioxide, and nutrients from carbohydrates present in growth medium. In a few countries, cultivation of algae is carried out in large trenches *i.e.* particularly in sewage oxidation ponds by using sunlight or in an artificial illumination conditions for use in life supportive systems for extended space exploration.

*Chlorella* strains are being used for a variety of applications in biotechnology. Due to their very high protein contents, they serve to improve protein deficiency and can be used as feed for production of animal protein. In many countries strains of *Chlorella* are utilized for sewage oxidation and waste water treatment.

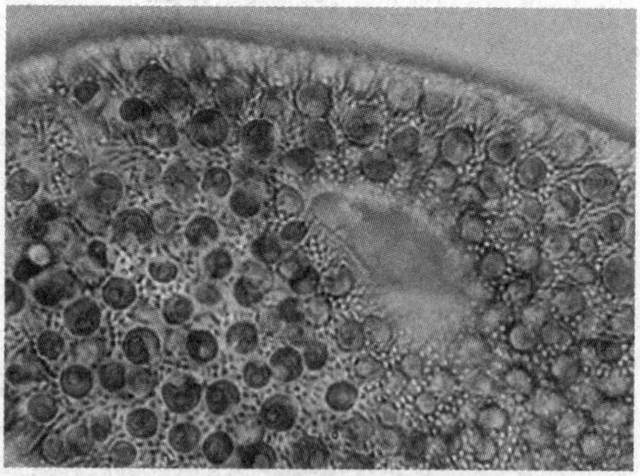

*Chlorella* has multipurpose use in biotechnological applications.

For cultivation of algae on sewage wastes, oxidation ponds are prepared, where sewage is allowed to accumulate. It is awaited till mixed cultures of algae grow (or inoculated with a singly prepared algal culture). For example, in Japan mixed culture of *Chlorella ellipsoides* and *Scenedesmus obliguus* was developed in open pond systems.

### 1. Factors Affecting Biomass Production

The factors affecting the yield of biomass : (*i*) illumination time, (*ii*) light intensity, (*iii*) supply of $CO_2$. Concentrations of $CO_2$ differ in different conditions, for example, an alkaline lake. Lake Texcoco in Mexico, has high concentration of sodium carbonate. On the other hand, algal growth is limited as a result of liberation of $CO_2$ and ammonia by bacterial activity, (*iv*) nitrogen sources (ammonium salts or nitrates are the suitable nitrogen sources which increase biomass yield), and (*v*) agitation of growing cells to maintain cells in suspension.

Biomass yield ranges between 12 and 15 $g/m^2/day$ (on dry weight basis) as obtained with *Spirulina maxima* and *Scenedesmus quadricauda* grown in out door pond conditions.

Mass cultivation of algae has been started in many countries, such as Japan, West Germany (now Germany), Mexico, *Chechoslovakia*, India, etc. In India, National Environmental Engineering Research Institute (NEERI), Nagpur has developed a technique of cultivating algae in sewage oxidation pond systems. This practice is also in use at NBRI (Lucknow), Hyderabad and other centres. Interestingly, experiments conducted at the CFTRI, Mysore have shown that the microalgae *e.g. Scenedesmus acutus* and *Spirulina platensis* could be cultivated on a large scale and used as food and feed as they are rich in protein and their nutrient value is comparable to conventional foods. A flow diagram of use of different groups of algae at various stages in waste water ponds and possible application of algal biomass is shown in Fig. 18.1.

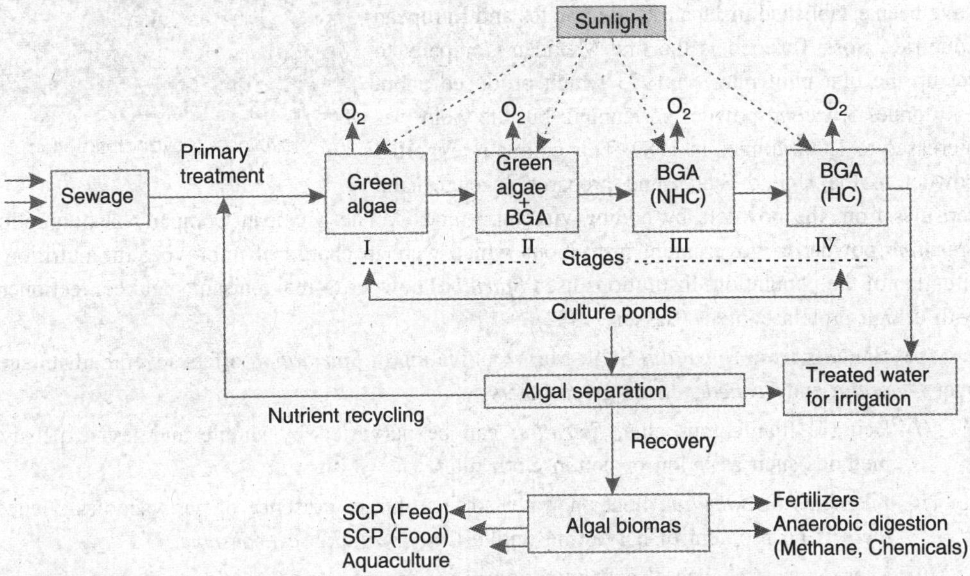


Fig. 18.1. Flow diagram of cultivation of algae in sewage oxidation ponds and possible application of algal biomass (modified after M. Vijayan, 1988).

### 2. Harvesting the Algal Biomass

Harvesting of algal culture becomes problematic because of settling down of cells at bottom and mixing of the algal cultures. The cells are recovered by concentration, dewatering and drying. Sometimes flocculants *e.g.* aluminium sulfate and calcium hydroxide and cationic polymers are added to the medium but they cannot be separated from the harvested cells. Therefore, this method warrants the application of SCP products in food and feed. Methods of separation and concentration also follow centrifugation, flocculation and centrifugation plus flocculation, but it is not economically feasible so far.

Harvesting the cyanobacteria, for example, *Spirulina* sp. is less troublesome as their spiral filaments float on the surface of water because of gas filled vacuoles in their cells which result in floating algal mats. Cells are able to fix atmospheric nitrogen. Algal mats are filtered and suspension of *Spirulina* is dried with hot air to get fine powder. Algal yield from stabilization pond is around 114 tonnes/ha/year. In California, 70 tonnes/ha/year of *Scenedesmus* was obtained from sewage. From Bangkok much high yield of about 170 tonnes/ha/year has been reported.

### 3. SCP Product of Spirulina

In 1821, Bernal Diaz del Castillo, for described the biscuits with the name 'tecuitlatl'. It contained dried mats of *S. maxima* collected from the Lake Texcoco. The biscuit was sold in the markets of Mexico. In 1964, J. Leonard (a Belgian botanist) took part in an expedition and noticed dried biscuits of blue green colour in several village markets. The biscuit *dihe* consumed by local population consisted of *Spirulina platensis*.

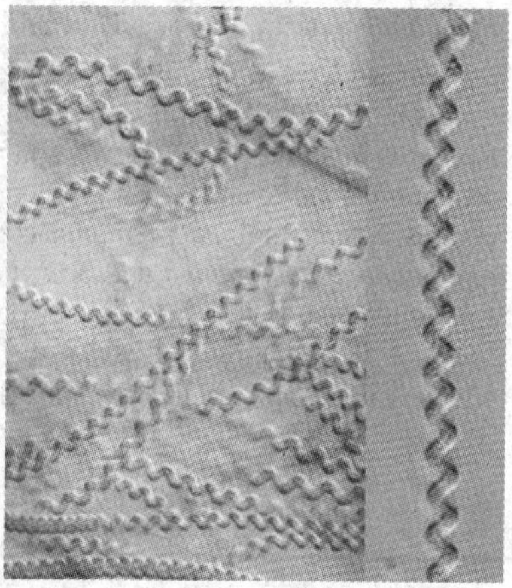

*Spirulina* - a cyanobacterium.

Many pilot plants for the production of *Spirulina* powder have been established in Japan, United States and European countries. Sosa Texcoco is the first Mexican Company to set up the first pilot plant in 1973 which produced about 150 tonnes *Spirulina* powder per annum; but the yield was increased to 1,000 tonnes in 1982. This company exported powder to the United States and prepared lozenges and capsules from the powders by adding vitamin A and C. The Mexican Company also supplied *Spirulina* powder to government institutions which were in charge of improving the nutritional situation of the population. Institutions used *Spirulina* powder to make biscuits and confectionery with a high protein content (Sasson, 1984).

**(*a*) Benefits from *Spirulina* SCP:** Mass Cultivation of *Spirulatina* offers several advantages over *Chlorella* and *Scenedemus* as given below:

(*i*) Being a filamentous alga, *Spirulina* can be harvested by simple and less expensive methods such as nylon or cotton cloth filter.

(*ii*) Filaments of *Spirulina* float on water surface due to presence of gas vacuoles. Hence, there is no problem of harvesting unlike *Chlorella* and *Scenedesmus*.

(*iii*) There is least chance of contamination in growth tanks of *Spirulina* as it grows at high alkaline pH 8-11.

(*iv*) Heat drying is sufficient for *Spirulina* as it has thin cell wall, whereas spray drying is required for *Chlorella* and *Scenedesmus* which is expensive.

(*v*) Researches done by UNIDO programme in Mexico (1980), Mexican National Institute of Nutrition, Hyderabad (1988, 1990) under Indian Council of Medical Research on several aspects of possible adverse changes in multigeneration feeding tests on laboratory animals and humans have shown no adverse effects.

(*vi*) *Spirulina* is highly digestive (85-95%) due to thin wall and low nucleic acid contents (4%). It contains high percentage of digestible proteins (62-72%), vitamins, amino acids and other nutrients (Table 18.3). The aminogram of *Spirulina* is comparable to the FAD, milk and egg protein pattern.

**Table 18.3. Composition of multin (*i.e.* dried powder of *Spirulina fusiformis*) (constituents are in per 100 of powder)*.**

| A. Major constituents (%) | | | C. Minerals (mg) | | |
|---|---|---|---|---|---|
| Total protein | .. | 64.6 | Calcium | .. | 6.58 |
| Fat | | 6.7 | Phosphorus | | 977 |
| Crude fibre | .. | 9.3 | Iron | .. | 44.7 |
| Carbohydrates | | 16.1 | Sodium | | 796 |
| Calories | .. | 346 | Potassium | .. | 1.28 |
| **B. Vitamins** | | | **D. Essential amino acids (%)** | | |
| Beta-carotene | .. | 320,000IU | Lysin | .. | 2.99 |
| Biotin | | 0.22 mg | Cystine | | 0.474 |
| Cyanocobalamin ($B_{12}$) | | 65.7 mg | Methionine | .. | 1.38 |
| Folic acid | | 17.6 mg | Phenylalanine | | 2.87 |
| Riboflavin | .. | 1.78 mg | Threonine | | 3.04 |
| Thiamin | | 0.118 mg | | | |
| Tocopherol | .. | 0.773 IU | | | |

* Analysed by Michelson Laboratories Inc., California, U.S.A. (1988)

(*b*) **Mass Cultivation of *Spirulina* SCP:** At present two types of farms for mass production of *Spirulina* SCP are under operation. A third type (*i.e.* enclosed system using transparent tube, biocoil or photobioreacter) is under development (Henrikson, 1990).

(*i*) **Semi-natural lake system.** Sosa Texcoco Lake (Mexico) and Lake Chand (Africa) offer an ideal environment for the natural growth of *Spirulina*. The product is expensive but of low quality due to contamination and pollution by uncontrolled natural factors. SCP of these lakes are good for fish and animal feed. Researches are in progress to refine the powder and make the products of good quality.

(*ii*) **Artificially built cultivation system.** The climatic conditions of most of the developing countries are such that favour mass outdoor production of *Spirulina*. Therefore, on the basis of water quality and nutrient status, this system can be grouped into the following two :

- **Clean water system.** This system is more expensive because of construction of artificial cultivation farms. These have shallow raceway ponds circulated with paddle wheels and high quality of nutrients. For the fast growth of *Spirulina* in clear water, addition of $NaNO_3$ and $NaHCO_3$ is necessary. pH of the water must be initially maintained to 8.5. It is a self pH adjusting alga which elevates the pH between 10-10.5. At this pH levels there is the least chance of contamination.

The Earthrise Spirulina Farm of California is the world's largest food grade *Spirulina* farm having the size of about 10 hectares with a capacity of production to about 120 tonnes/annum. The other big farms are operating in Japan, Thailand, Mexico, Taiwan, Israel, Vietnam, China and India. In India, food grade *Spirulina* is cultivated at two main centres, one at Shri A.M.M. Murugappa Chettiar Research Central (MCRC), Madras, and the other at Central Food Technology and Research Institute (CFTRI), Mysore. Madras centre is the biggest food grade *Spirulina* farm in India. Its annual production capacity is of about 75 tonnes. The products are marketed in India and abroad as health food, baby food and multivitamin tablets.

- **Waste water system.** This system is applicable in highly populated countries like India where wastes are generated in high quantities and pose environmental problem. In this system, human and animal wastes and sewage are used for growth of *Spirulina*. The wastes are added into the digester to settle down the solid particles. The liquid effluent is used as a source of the nutrients and added in artificially constructed ponds. As desired $NaNO_3$ and $NaHCO_3$ are also mixed. *S. platensis* is found to grow better in sewage amended with $NaCO_3$ and nutrients in different proportion and also in diluted sewage. When full growth of *Spirulina* is over, it is screened from the pond and added to aquaculture to feed fish or dried in a small solar drier for human food.

This system is most suitable for third world countries where wastes are the major sources of pollution. R.D. Fox of France has developed the 'Integrated Health and Energy System Project' for poor countries to grow *Spirulina* and fight against the problems of food, malnutrition and environmental pollution. Today, a large number of these projects are in operation in the villages of developing countries like India, West Africa and South America. In India, the first integrated village system was established by Indo-France Govt. at Centre of Science for villages (CVS), Wardha (Maharastra).

The CVS distributes *Spirulina* cookies and nodules to malnourished children of local villages. This has shown very encouraging results.

**(c) Requirements for Growth of *Spirulina:*** Following are the requirements for growth of *spirulina*:

***(i)* Algal tanks.** Generally, circular or rectangular cemented tanks are constructed. The circular tanks are more preferred over the rectangular one because of ease in handling. Size may be according to convenience and yield needed. Depth should be about 25cm. Open tanks are suitable for tropical and sub-tropical regions.

***(ii)* Light.** Low light intensity is required at the beginning to avoid photolysis. *Spirulina* exposed to high light intensity is photolysed.

***(iii)* Temperature.** Temperature for optimum growth should be between 35-40°C.

***(iv)* pH.** *Spirulina* grows at high pH ranging from 8.5 to 10.5. Initially, culture should be maintained at pH 8.5 which automatically is elevated to 10.5.

***(v)* Agitation.** Agitation of culture is very necessary to get good quality and better yield. The culture is agitated by brush, paddle power, pipe pumps, wind power, rotators, etc.

***(vi)* Harvesting.** The filaments of *Spirulina* float on surface of water forming thick mat. Therefore, it can be harvested by fine mesh steel screens, nylon or cotton cloths, etc.

***(vii)* Drying.** As it has a thin wall, sun drying is the most suitable and economical. Various trials done at CFTRI, Mysore and MCRC, Madras with sun drying have given good results.

***(viii)* Yield.** An average yield of 8-12 g *Spirulina* powder/$m^2$/day has been obtained in India and other countries. This is equivalent to 20 tonnes /ha/annum. In warmer climate, the yield can increased to about 20 g/$m^2$/day.

***(ix)* Avoiding contamination.** Although there is the least chance for contamination, yet regular monitoring of algal culture is necessary. Because the microbial load is likely to affect the quality and safety of the product. At MCRC and CFTRI the cultures of *Spirulina* were found either within or very close to safety limits of Indian Standard Institute (ISI) for baby food, to about $5x10^5$ propagules per gram. Dried *Spirulina* powder is packed in aluminium bags or sealed in bottles and sent to market.

(*d*) **Uses of *Spirulina* Single Cell Protein:** Single cell protein of *Spirulina* is used as below:

(*i*) **As protein supplemented food.** Since *Spirulina* is a rich source of protein (60-72%), vitamins, amino acids, minerals, crude fibres, etc., it is used as supplemented food in diets of under-nourished poor children in developing countries. The United Nations, Mexican National Institute of Nutrition, French Petroleum Institute and National Institute of Nutrition, Hyderabad have formulated four algal recipes as a weaning substitute for infants. In India, the Village Health and Energy System Projects are operated at CVS (Wardha). At MCRC, the products are distributed to the local under-nourished children. It has been found that 1 g of *Spirulina* tablets contains as much nutrition as one kg assorted vegetable.

(*ii*) **As health food.** *Spirulina* is very popular as health food. Most of Sosa Texcoco products are exported to U.S.A., Europe and France where it is sold in health and food stores. It is the part of the diet of the U.S. Olympic team. Jaggers take *Spirulina* tablets for instant energy. Since it provides all the essential nutrients without excess calories and fats, it is taken by those who want to control obesity. The MCRC has for the first time launched the project as health and baby food, and multivitamin powder and tablet under trade name 'Multin' and 'Multinal'.

(*iii*) **In the therapeutic and natural medicine.** *Spirulina* possesses many medicinal properties. Therefore, it is used as social and preventive medicine also. It has been recommended by medicinal experts for reducing body weight, cholesterol and pre-menstrual stress and for better health. It lowers sugar level in blood of diabetics due to the presence of gamma-linolenic acid and prevents the accumulation of cholesterol in human body. It is a good source of β-carotene (a precursor of vitamin A) and, therefore, helps in monitoring healthy eyes and skin. β-carotene is known as the best anti-cancer substance. In 1989, UN National Cancer Research Institute announced that substances from blue-green algae are active against AIDS and cancer virus. In Vietnam, its tablets are used to increase lactation in nourishing mothers.

(*iv*) **In cosmetics.** *Spirulina* contains high quality of proteins and vitamin A and B. These play a key role in maintaining healthy hair. Many herbal cosmeticians are making efforts to develop a variety of beauty products. Phycocyanin pigment has helped in formulating biolipstics and herbal face cream in Japan. These products can replace the present coaltar-dye based cosmetics which are known as carcinogenic.

## G. PRODUCTION OF BACTERIAL AND ACTINOMYCETOUS BIOMASS

Bacteria are widely used as a source of single cell protein because of their short life cycle (20-30 minutes) and capacity to utilize a wide range of organic substrates as a source of energy. Actinomycetes also utilize these renewable sources as they have more or less same generation rate as bacteria.

Since the establishment of British Petroleum in 1960, a significant progress is made in the production of microbial products by using gaseous and liquid hydrocarbons and chemicals derived from them, for example, methanol, ethanol, etc. The Shell Research Limited, U.K. conducted research on pilot plant scale process for the production of bacterial SCP from methane by using *Methylococcus capsulatus* or mixed culture of *Pseudomonas* sp., *Hyphomicrobium* sp., *Acinetobacter* sp. and *Flavobacterium.* Several processes for the production of bacterial SCP from gaseous hydrocarbons have been developed at Kyowa Hakko Kogyo Company, Limited in Japan. *Brevibacterium ketoglutamicum* ATCC No. 15587 was able to utilize methane,ethane, propane, *n*-butane, iso butane, propylene, butylene or mixture of these hydrocarbons. A pilot scale process for the production of *Achromobacter delvacvate* from diesel oil is developed at the Chinese Petroleum Corporation, Taiwan.

*Streptomyces.* sp. is capable of growing on methanol. *Theromonospora fusca,* a thermophilic species, degrades 60-65 per cent paper mill fines resulting in 30 per cent protein product. Nowadays, cellulose degrading thermophilic actinomycetes offer a great opportunity to yield SCP from cellulosic wastes.

### 1. Procedure for Production

In 1982, F.X. Roth has described the following steps for the production of bacterial biomass : (*i*) supply of a nutrient substrate (*ii*) formulation of a suitable medium, (*iii*) multiplication of micro-organisms through fermentation, (*iv*) separation of cellular substances from the left over medium, and (*v*) further treatment to kill and dry the bacterial biomass.

Imperial Chemical Industry (ICI) is a world's leader in biotechnology, as far as the production of a bacterial biomass, pruteen, from *Methylophilus methylotrophus* is concerned. Pruteen was produced on methanol but it served as high grade protein for animal feed. It contains 72 per cent protein, 86 per cent total lipid and an amino acid profile high in lysin and methionine. The conversion of methanol to SCP by *M. methylotrophus* is represented by the following equation :

$$1.72\ CH_3OH + 0.23\ NH_3 + 1.51\ O_2 \longrightarrow 1.0\ CH_{1.68}O_{0.36}N_{0.22} + 0.72\ CO_2 + 2.94\ H_2O.$$

Production of SCP by other bacteria and actinomycetes is shown in Table 18.1.

### 2. Factors Affecting Biomass Production

Growth of bacteria and actinomycetes are affected by many factors, as a result production of biomass on a given substrate significantly changes. These are : (*i*) suitable strain of bacterial culture, (*ii*) genetic stability of bacterial strain, (*iii*) absence of bacteriophage, (*iv*) suitable pH 5-7 of growth medium, (*v*) temperature (15-35°C according to strain), (*vi*) oxygen/agitation to create aerobic condition; (*vii*) organic substrate and nitrogen concentrations. Optimum C:N ratio which favours high protein contents in cells and inhibits accumulation of lipid is 10:1, and (*viii*) maintenance of sterile conditions throughout the growth.

### 3. Product Recovery

Like algal biomass, there are many problems related to the recovery of bacterial cells as they are very small and have cell density in the order of 10-20 g/litre. Centrifugation cost is also high. Processes for cell recovery have been devised in such a way that could cut down the cost. Many pilot plants use flocculants and many have set up decanter type centrifuge. For example, Hoechst (Germany) has developed a device for separation of *Methylomonas clara* from methanol containing culture medium which is based on electrochemical, coagulation and centrifugation. The cells are washed and spent medium is again treated by conventional treatment process as it contains inorganic salts and small amount of cells. Cell biomass is then spray-dried.

## H. PRODUCTION OF YEAST BIOMASS

In the previous chapter, much has been discussed about the use of yeasts in fermentation since centuries. Consumption of baker's yeast (*S. cerevisiae*) as food in Germany during World War I increased its importance. Since then, rapid development took place in biotechnological applications of *S. cerevisiae,* as far as culture development, process optimization and scale up of products are concerned. World production of yeast biomass is of the order of 0.4 million metric tonnes per annum including 0.2 million tonnes baker's yeast alone.

Yeasts synthesize amino acids from inorganic acids and sulphur supplemented in the form of salts. They get carbon and energy sources from the organic wastes, *e.g.* molasses, starchy materials, milk whey, fruit pulp, wood pulp and sulphite liquor.

Yeast cells.

A comparative composition of nutrients is given in Table 18.3. It is obvious that biomass of *S. cerevisiae* produced on sugarcane molasses differs from that of beer. Yield of yeast biomass is greatly affected by many factors similar to bacteria. Yield corresponds to growth nutrients, organic wastes, temperature, culture, oxygen, etc. Yield of yeasts is given in parentheses in Table 18.2. J.C. Bennett and coworkers in 1969 have given the typical equations for the growth of yeasts on carbohydrates or hydrocarbons:

***Carbohydrates*** :

$$8n\ CHO + 0.8\ nO_2 + 0.19n\ NH_4 + \text{trace elements} \longrightarrow n(CH_{1.7}O_{9.5}N_{0.12}\ Ash) + 0.8n\ CO_2 + 1.3\ nH_2O + 80{,}000n\ KCal.$$

***Hydrocarbons*** :

$$2nCH_2 + 2nO_2 + 0.19n\ NH_4 + \text{trace elements} \longrightarrow n(CH_{1.7}O_{0.5}N_{0.19}\ Ash) + nCO_2 + 1.5\ H_2O + 200{,}000n\ KCal.$$

### 1. Factors Affecting Yield of Yeast Biomass

Like bacteria, growth and yeild of yeasts are also affected by the following factors: (*i*) organic substrate and nitrogen ratio (optimum C : N ratio favouring high protein content should be between 7:1 and 10:1), (*ii*) pH of nutrient medium (pH should be in the range of 3.5 to 4.5 to minimize growth of bacterial contaminants), (*iii*) temperature (it differs from organism to organism). Most yeasts have specific growth rate in the range of 30°C to 34°C. Some strains also grow in the range of 40-45°C, (*iv*) oxygen (for growth on carbohydrates), $O_2$ required should be 1 g/g of dried cells, and for growth on n-alkanes it should be about 2 g/g dried cells), (*v*) maintenance of sterile condition through out the process, and (*vi*) suitable strain of yeast.

### 2. Recovery of Yeast Biomass

Yeast cells are small in size (5-8 m), the density of which reaches to 1.1 g/ml. Post-fermentation treatment of food yeast is shown in Fig. 18.2. Yeast cells are recovered by decantation-centrifugation (including washing) drying treatment methods. After washing undesirable traces of medium are removed which are again recycled for economic reasons. As a result of final harvesting by rotary vacuum filter a cake containing 20-40 per cent dry matter is obtained which is then dried to get a product of 6-10 per cent water content (Riviere, 1977).

## I. PRODUCTION OF FUNGAL BIOMASS, THE MYCOPROTEIN (OTHER THAN MUSHROOMS)

During the World War II, attempts were made to use the cultures of *Fusarium* and *Rhizopus* grown in fermentation as protein food. The inoculum of *Aspergillus oryzae* or *Rhizopus arrhizus* is chosen because of their non-toxic nature (Riviere, 1977). Saprophytic fungi grow on complex organic compounds and render them into simple forms. As a result of growth, high amount of fungal biomass is produced. Mycelial yield vary widely depending upon organisms and substrates. Strains of some species of moulds, for example, *Aspergillus niger, A. fumigatus, Fusarium graminearum* are very hazardous to human, therefore, use of such fungi should be avoided or toxicological evaluations should be done before recommending to use as SCP. Protein contents of moulds are given in Table 18.1.

In 1982, D.S. Chahal has described the increasing popularity of myco-protein because of the following reasons: (*i*) some of the filamentous fungi grow as fast as most of the single celled organism, (*ii*) the finished product of filamentous fungi is fibrous in nature and can be easily converted into various textured foods. In comparison, protein is extracted from single celled organisms and spun into fibrous form, (*iii*) filamentous fungi have a greater retention time in the digestive system than single celled organisms, (*iv*) protein content can be as high as 35-50 per cent with comparatively less nucleic acid than single celled organism, (*v*) digestibility and net protein utilization without any pretreatment is higher than single celled organisms, (*vi*) the overall cost of protein production from filamentous fungi is more economical as compared to that of single celled organism, (*vii*) filamentous fungi have greater penetrating power into insoluble substrates and are therefore, more suitable for solid state fermentation of lignocellulosic materials, (*viii*) most of filamentous fungi have a faint mushroom like odour and taste which may be more readily acceptable as a new source of food than the yeast odour and green colour associated with yeasts and algae respectively, (*ix*) the biomass produced by filamentous fungi can be used as such without any further processing because it provides carbohydrates, lipids, minerals, vitamins and proteins. In addition, nucleic acid contents of fungal protein is lower than that of yeast and bacteria.

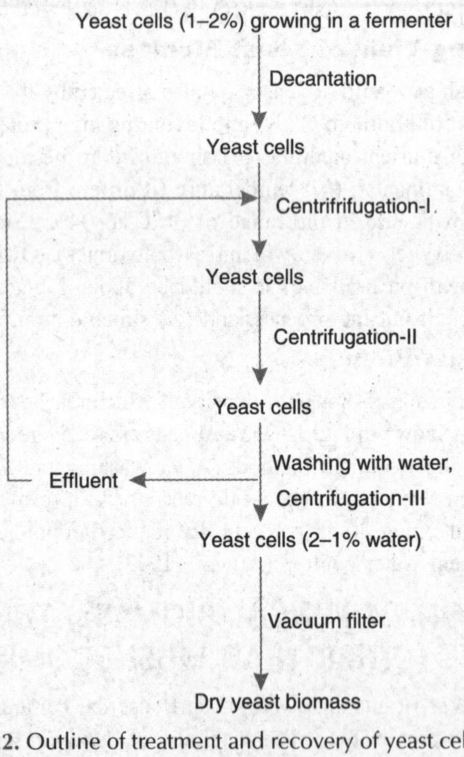


Fig. 18.2. Outline of treatment and recovery of yeast cells.

## 1. Growth Conditions

The following factors (like bacteria) affect the growth of moulds : (*i*) carbon, nitrogen (C : N) ratio is required to be in the range of 5:1 to 15:1, (*ii*) ammonium salts are used as a source of nitrogen in continuous culture and phospheric acid for phosphorus. Moreover, most of fungi, for their growth require minerals, such as potassium, sulphur, magnesium, calcium, iron, manganese, zinc, copper and cobalt. Their concentration differs with respect to species, (*iii*) pH of growth medium ranges from 3.0 to 7.0 but pH 5-6 or below is desirable for most of fungi because bacterial contaminants do not grow, (*iv*) temperature ranges from 25°C to 30°C with certain exceptions and (*v*) oxygen is required for good growth of fungi. During agitation, mycelial mat forms pellets.

## 2. Organic Wastes

There are many sources of organic wastes on which fungi grow. Large quantity of ceulosic and lignocellulosic materials are present in agriculture, forestry and industry wastes. *Trichoderma reesei, T. viride* and *T. harzianum* have cellulase secreting ability which catalyse the conversion of cellulose to cellubiose, and cellubiose, in turn to glucose. Before microbial invasion, cellulosic materials are treated with alkali so that, glucosidic bonds could be broken. Thus alkali treatment and hydrolysis of cellulosic material increases the bio-degradability of cellulose by many fungi, for example, *A. fumigatus, Chaetomium* sp., *Geotrichium candidum, Penicillium* sp. and *Trichoderma* sp.

The waste pods or carob beans contains sugars which are used as substrate for the growth of fungi *e.g. A. niger, Fusarium moniliforme* etc. Tate and Lyle Limited, U.K. has developed a process for utilization of carbohydrates in carob bean extracted by *A. niger*. A yield of 0.45 g/g substrate of *A. niger* grown on carob bean extract was obtained.

The pellets of fungal mat are recovered from the growth medium by decantation and filtration process. For example, mycelial biomass of *A. niger,* which does not form pellets was separated by rotary vacuum filters. The mycelial biomass was dried in tray or belt driers.

## 3. Traditional Fungal Foods

Nowadays, interest in traditional fermentation technology for food processing has greatly increased because of emphasis placed upon plant materials as human foods. Most of the fermentation processes were developed in ancient time before the recognition for existence of microorganisms. How surprising! the early man developed the methods and maintained the cultures through the centuries without having the concept of microorganisms. Nowadays, these foods are an important component of daily diets of many of the world's population especially in Japan, Indonesia, India, Pakistan, Thailand, Phillippines, Taiwan, Korea, China and encompassing areas and part of Africa.

These Foods were produced long before the written history. Though production of shoyu, miso and sake was being practised for the ancient time, yet it gained additional impetus with the spread of Buddhism (635 BC-550 AD) which favoured the vegetarian diet and taught the sanctity of all life. That is why this concept has fostered the more widespread consumption of sauces and pastes such as shoyu and miso that provide meaty flavor. The first scientific report on application of oriental fermentation to modern industrial process are about 100 years old. Between 1878 - 1914 a number of fermented foods and drinks, and isolation of micro-organisms were done. Studies of these could not be strengthened due to World War I and II; the same could be resumed in 1950s. However, these techniques could be industrialized in recent years and many more fermented foods are produced at village level also. Some of the oriental fermented foods, substrates and cultures of microorganisms are given in Table 18.4.

Some times it becomes problematic if there is presence of toxins in oriental fermented fungal food. Biotechnological hazards of mycotoxins have been discussed in Chapter 16. However, more than 73 strains of *Aspergillus* including *A. orizae, A. niger, A. usami,* etc. have been isolated from oriental fermented fungal foods. In some samples, as such no aflatoxins were found in foods, but 30 per cent of these strains produced compounds that were very similar to those of aflatoxins in fluorescent spectra of $R_f$ value in thin layer of chromatograms. These fluorescent compounds were found most toxic. Some strains have been found to produce aspergillic acid. In Koji, fermentation β-nitropropionic acid was produced by one of the strains. *A. flavus* or *A. parasiticus* contaminated culture of *A. oryzae* used for preparation of soy sauce has been found to produce aflatoxin. Therefore, precautions should be taken while inoculation done. The preparation of food by fermentation has certain advantages. Fermentation produces certain enzymes, destroys or makes undesirable flavour and odours, adds flavour, preserves food, synthesises desirable constituents such as vitamins and antibiotics, increases digestibility, changes physical states, produces colour and reduces cooking time. The properly fermented foods are non-hazardous to health.

**> Table. 18.4. Traditional fermented foods, substrates and microorganisms used for production.**

| *Food/country* | *Substrates* | *Micro-organism used* |
|---|---|---|
| Dosa/India | Rice, dehulled black gram | Yeast, *Leuconostoc mesenteroides* |
| Idli/India | do | do |
| Lao-chao/China, Indonesia | Waxy variety of rice | *Amylomyces rouxii* NRRL3160, *Rhizopus chinensis, R. oryzae* |
| Miso/Japan, China, | Whole soyabean, rice or barley | *Asperqillus oryzae* NRRL5593, *Saccharomyces rouxiii* |
| Mahewu/S. Africa | Maize | *Lactobacillus delbrueckii* |
| Natto/Japan | Whole soyabeans | *Bacillus natto* (= *B. subtilis*) |
| Ontjam/Indonesia | Peanut press cake | *Neurospora intermedia* NRRL6025 |
| Shoyu (soy sauce)/ Japan, China | Wheat, soyabeans | *A. oryzae, A. sojae* |
| Sufu(fu-ru)/China | Soyabean curd (tofu) | *Actinomucor elegans, Mucor dispersus* |
| Tempe/ Indonesia | Whole soyabeans | *Rhizopus oligosporus* NRRL2710 |

*Source* : G.L. Wang and C.L. Hesseltine (1981, 1983)

(*a*) **Shoyu (soy sauce)** : Shoyu is a Japanese term which denotes its preparation by fermentation not by chemical treatments of soybeans. It was applied thousands of years ago. Historically, 'sho' or 'misho' refers to animal sauces which were first made from salted marine animals as well as meat from wild game. These terms were developed probably later to denote the 'meaty flavoured' vegetarian sauces. The term koji refers to any substrates overgrown with filamentous fungus and used as a source of enzymes.

During the making of soy sauce fernentation of wheat and soy is carried out in large steel tanks.

The Japanese Agricultural Standard defines five types of shoyu of which koikuchi types represents 85 per cent of all the shoyu consumed in Japan.

Koji is prepared by roasting wheats at 170 -180°C for a few minutes and finally coarsely crushing into 4-6 pieces. Soyabeans (seeds or defatted soyabean grits ) are soaked in water for 15 h and autoclaved for 1 h (Fig. 18.3). Crushed wheat and autoclaved soyabeans are mixed properly so that the surface of soyabeans could be coated with ground wheat. Finally, it is inoculated with seed culture (*Aspergillus oryzae* or *A. sojae*). Wheat provides nutrients to the fungus to grow well on the surface of soyabeans. In Japanes shoyu, nearly equal amount of soyabean and wheat are used, whereas in Chinese shoyu the amount of wheat remains low.

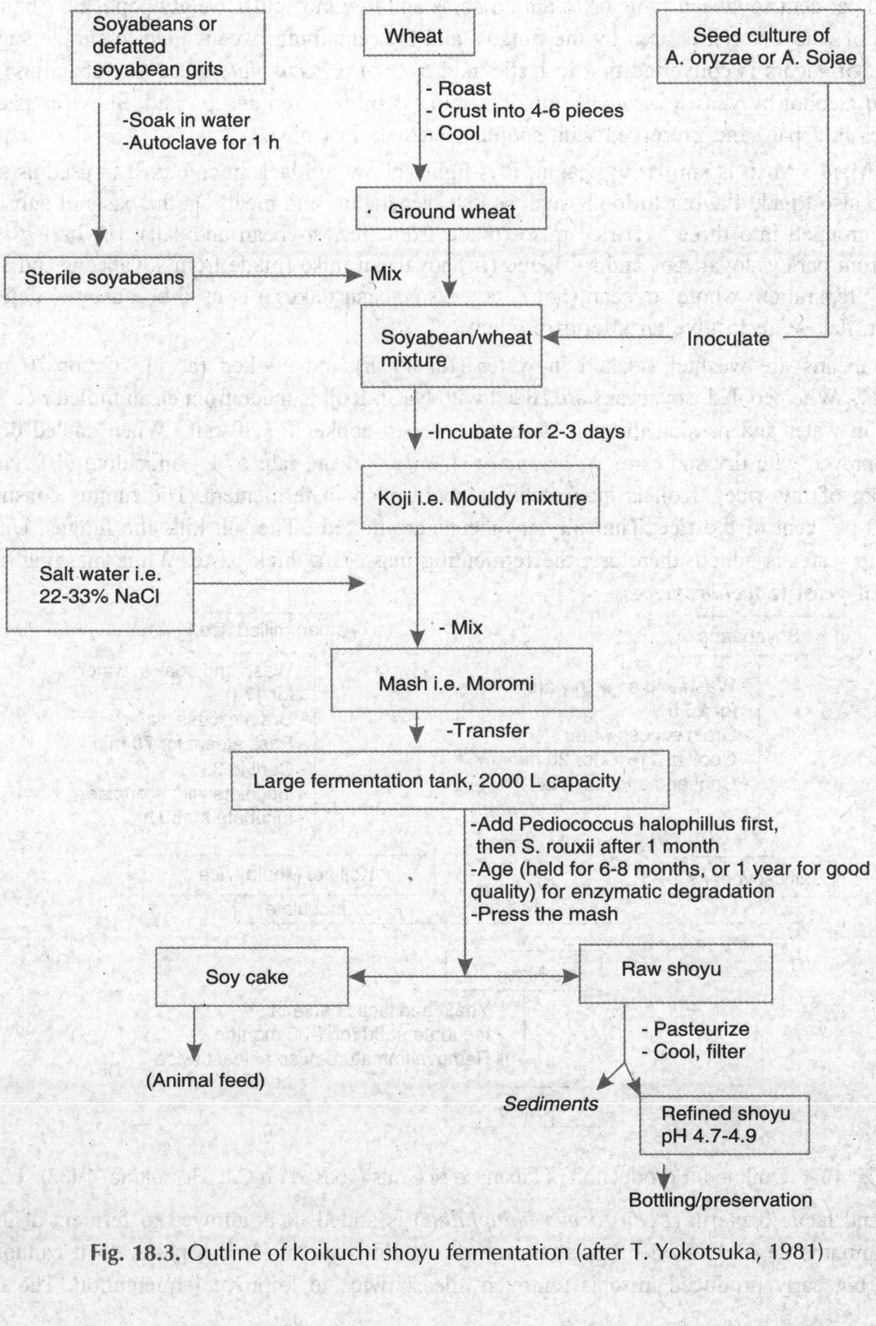


**Fig. 18.3.** Outline of koikuchi shoyu fermentation (after T. Yokotsuka, 1981)

Koji fermentation is carried out on a large perforated stainless steel plate (5 × 12 m) for 2-3 days. Proper temperature (30°C) and moisture (40-43 per cent) are maintained during this period. The koji is overgrown by the fungus. This facilitates to produce desirable enzymes. Now, koji is mixed with salt water (22-23 per cent salt, w/v) to make mash, the 'moromi'. The mash is transferred to deep fermentation tank (50-300 kilolitre capacity) and held for 4-8 months, depending on temperature, with occasional air compressing. It is agitated at certain intervals. These promote microbial growth. At this time the fungus is killed but its enzymes hydrolase about 10 per cent soyabean proteins to amino acids and low molecular weight peptides. About 20 per cent of wheat starch is used by the fungus, and the remaining is converted to simple sugars. One half of sugars is converted first to lactic acid by *Pediococcus halophilus* (lactobacillus) and then into alcohol by yeast (*S. rouxii*). Finally, liquid is pasteurized and packed. Shoyu is packed in bottles in Japan, and preserved with sodium benzoate in China.

***(b)* Miso :** Miso is similar to peanut; it is light yellow to black in colour. It is used as soup base and also to add flavour to foods such as fish, vegetables and meat. On the basis of substrate miso is grouped into three : (*i*) rice miso (made from rice, soybean and salt), (*ii*) barley miso (made from barley, soyabeans and salt), and (*iii*) soyabean miso (made from soyabeans and salt). In miso preparation, whole soybean (not defatted soyabean flakes ) is used because the defatted soyabean flakes are to give an inferior product.

Soyabeans are washed, soaked in water (for 17 h) and cooked (at 115°C for 20 min) (Fig. 18.4). When cooled, soyabeans are added with NaCl. Koji is made from clean milled rice after soaking in water and passing through steam is an open cooker for 40 min. When cooled down, rice in sprayed with dry spores of A. *oryzae* or *A. sojae* (at the rate of 1 g inoculum ($10^9$ viable spores)/kg of raw rice). Koji is incubated for about 50 h in fermenters. The fungus consumes about 10 per cent of the rice. The salty soyabeans are mixed . The salt kills the fungus. Unlike shoyu, no water is added, therefore, the fermenting mash is a thick paste. While mashing, pure culture of yeast (*Saccharomyces*

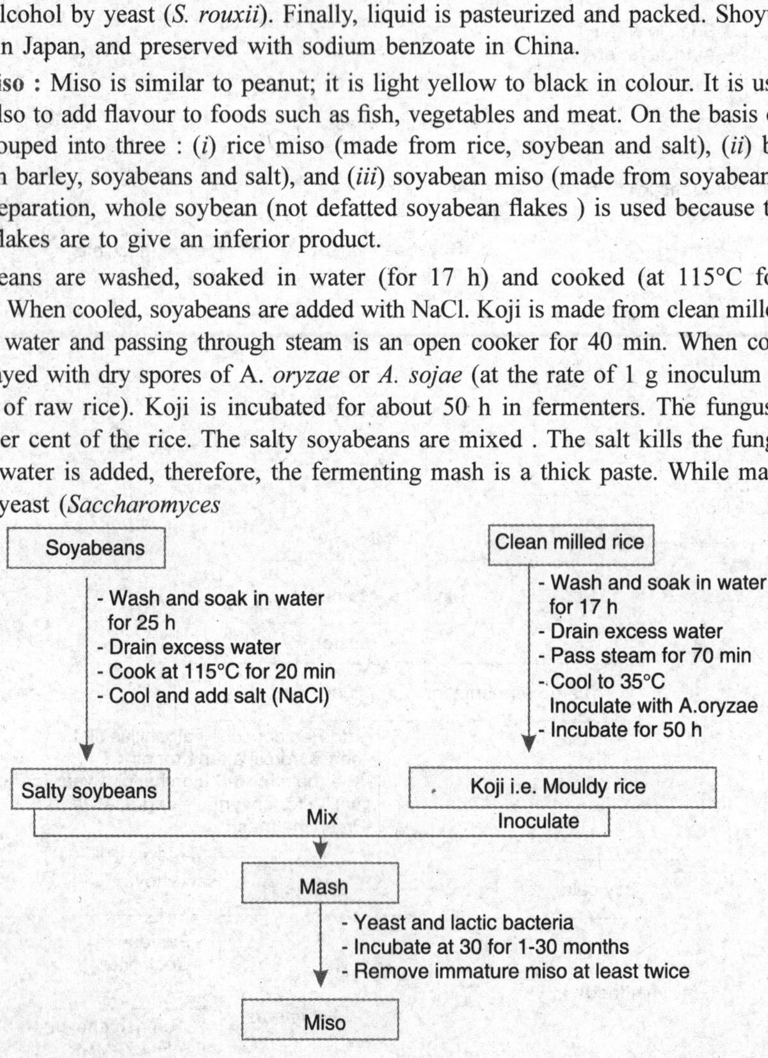


**Fig. 18.4.** Outline for production of Japanese rice miso (based on C.L. Hesseltine, 1983).

*rouxii*) and lactic bacteria (*Pediococcus halophillus*) is added. It is allowed to ferment at 30°C for 1-3 months depending on miso types desired. In this case mash is a paste, so it cannot be agitated but early produced miso is removed atleast twice to improve fermentation. The final product, miso, is obtained. In the final product 60 per cent of protein including amino acids and peptides is water soluble, and 75 per cent carbohydrate is reducing sugar. Miso is becoming an important food among the vegetarians in the United States because of its varying variety of flavours *i.e.* sweet, salty and meaty.

Sake: a Japanese traditional drink.

(*c*) **Sake :** By fifth century A.D., sake brewing was well established in Japan and became the traditional drink. Since then several refinements have been done time to time. Rice is milled and polished which, after washing and steeping, is steamed for 20-30 minutes (Fig. 18.5). The steamed rice is cooled to about 35°C and mixed to get even temperature and moisture. The steamed rice is inoculated with spores of *A. oryzae* and mixed well. Koji is distributed in boxes having wire mesh bottom to control the temperature and moisture. The boxes are incubated for 40 h. When temperature reaches to 40-42° it favours for optimum amylolytic activity of the fungus. Fermentation of

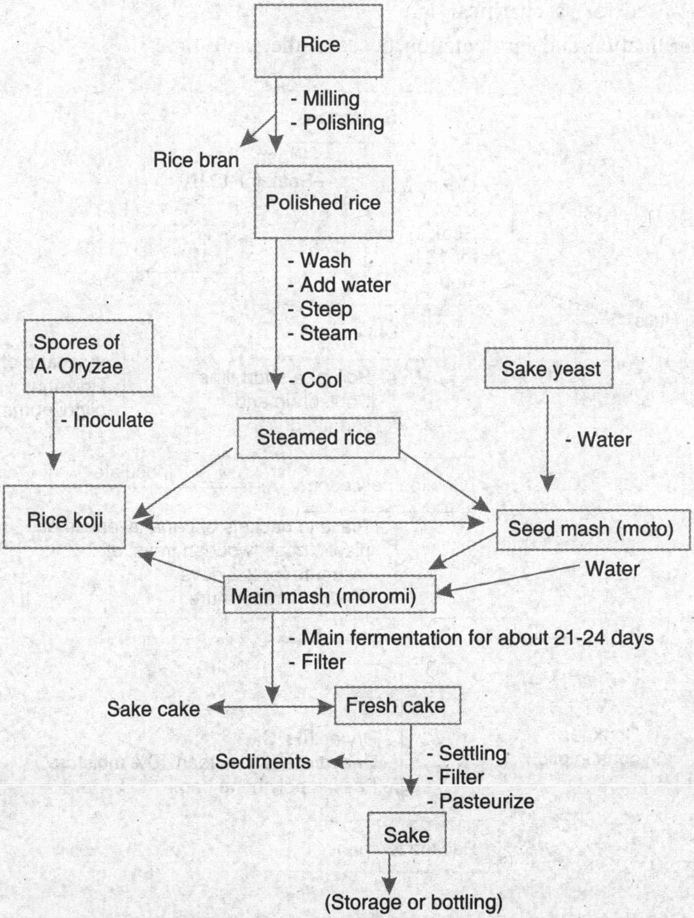


**Fig. 18.5.** Outline of brewing process of Japanese sake (after K. Kodama and K. Yoku, 1977).

starch occurs. Koji is spread over clean cloths to lower the temperature until used for mashing. The Koji mould brings about saccharification of rice starch resulting in production of sugar. The koji is mixed with water, steamed rice and sake yeast (*S. sake;* syn. *S. cerevisiae* ) culture. It is called as yeast seed mash or 'moto'. It is a pure yeast culture to be used as starter inoculum for main fermentation. The yeast seed mash, koji and water are mixed to get fresh sake, the main mash or 'moromi'. Fermentation of main mash is done for about three weeks. Sugars formed by koji fungus is fermented to alcohol by the sake yeast. In 1972, Therefore, it is filtered to get fresh sake. It is again filtered, pasteurized and bottled or stored as desired. In 1972, Imayasu has described the following four characteristics of sake :

Tempeh is the fermented soy cake which is highly rich in protein.

(*i*) In Japan, sake preparation is based on fermentation of nonglutinous rice, whereas, elsewhere in Asia, it is made from glutinous rice.

(*ii*) Koji is used for saccharification.

(*iii*) Saccharification and fermentation occur at the same time.

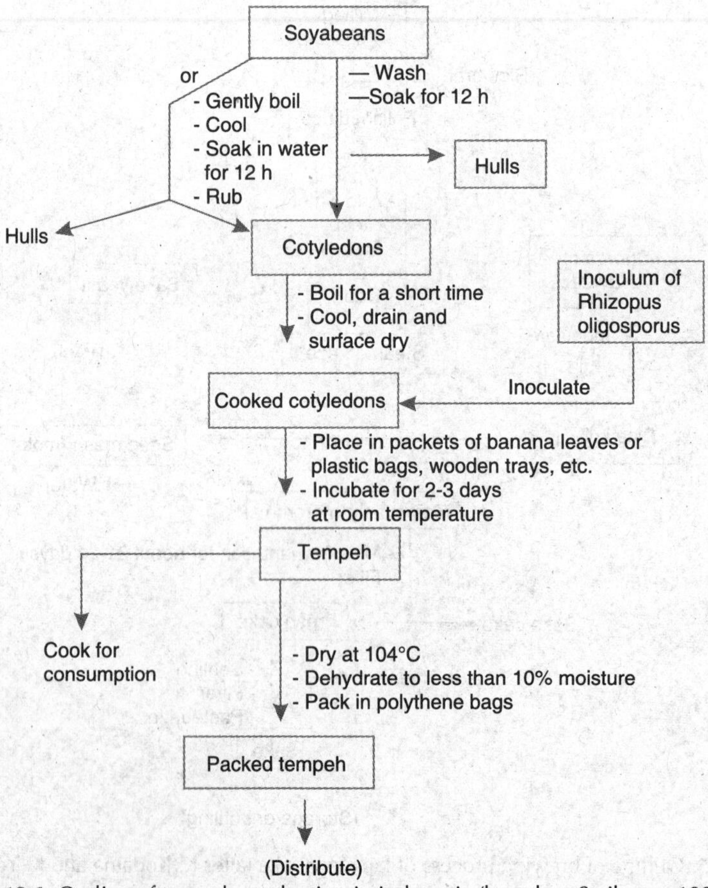


**Fig. 18.6.** Outline of tempeh production in Indonesia (based on Steikraus, 1983).

(*iv*) Sake has the highest amino acid content of all alcoholic drinks. Amino acids come from the enzymatic digestion of rice proteins by the mould enzyme.

**(*d*) Tempeh :** Tempeh is a traditional meat analogue of Indonesia. It is a white and mouldy cake produced by fungal fermentation of soaked, dehulled, hydrated and partially cooked soybean cotyledons. It can be sliced and fried in fat or cut into pieces, and used as a protein rich meat analogue.

Soyabeans are washed and soaked for 12 h. During this period it undergoes bacterial acid fermentation. Consequently, pH is lowered to 5 or below. The other method is to bring the soyabeans to boil and then soak in water for about 12 h. This helps to dehull the soyabeans. Soyabeans are rubbed gently. The hulls float on water surface from where they are removed. The cotyledons are boiled for a very short time to get them artificially cooked. They are drained, cooled, surface dried and inoculated with pure culture of *Rhizopus oligosporus.* The cotyledons are inoculated before or after placing them in fermentation containers. Traditionally, the inoculated cotyledons are wrapped in small pockets made of wilted banana leaves or other large leaves (Fig. 18.6). These packets are incubated in a warm place for 2-3 days.

But for a few years, the commercial tempeh industry in Indonesia has adopted wooden trays (35 × 80 × 1.3 cm). The inoculated cotyledones are also placed in perforated plastic bags, plastic tubes, etc. The wooden trays are lined with perforated plastic sheets. The perforations allow the exchange of gases. These bags are incubated at 30-37°C with high moisture contents. Fungal mycelium grows well and completely covers the cotyledons. It is ready for cooking. Also, it is dried at 104°C and dehydrated to less than 10 per cent moisture and packed in polythene bags. During fermentation, proteins are partially hydrolysed, lipids are hydrolysed, riboflavin doubles and niacin increases 7 times.

In addition to meaty flavour and source of protein, tempeh is also known to prevent bacterial infection of intestine. *R.oligosporus* is an antibacterial agent active against *Streptococcus, Leuconostoc, Staphylococcus aureus, Bacillus, Clostridium.*

## J. MUSHROOM CULTURE

Mushrooms are the members of higher fungi, belonging to the class Ascomycetes (*e.g. Morchella, Tuber,* etc.) and Basidiomycetes (*e.g. Agaricus, Auricularia, Tremella,* etc.). They are characterized by having heterotrophic mode of nutrition. The edible mushroom refers to both epigeous and hypogeous fruiting bodies of microscopic fungi that are already commercially cultivated or grown in half culture process or implemented under controlled conditions. They are rich in protein and constitute a valuable source of supplementary food. Some of them are deadly poisonous, for example *Amanita verna, A. virosa,* etc.

Use of mushrooms can contribute possitively in facing the challenge of world-wide food shortage, originating with rapidly expanding human population at the rate of more than 2 lakh per day. Mushrooms are rich in protein and other accessory compounds. Their value as food accessory is beyond the computation of the chemists and physiologists. They are among the most appetizing of the table delicacies and aid greatly to palatability of many food when cooked with them, for example *Morchella esculenta, Agaricus bisporus* (syn. *A. compestris*), *Tuber* sp. etc.

An edible button mushroom.

Mushroom cultivation was thought to be very simple, but infact it is a complicated business. Yield is affected by compost, spawn, temperature, moisture, etc. There has been extensive concern in recent years, as far as production of food protein from domestic, agricultural and industrial wastes is concerned. On the other hand, high and sophisticated technology in the yeast and algal cultures for SCP demands complicated input requirements and large capital. The immediate products may even need to be processed before being accepted as human food. The great value in promoting the cultivation of mushrooms lies in their ability to grow on cheap carbohydrate materials and to transform various waste materials.

### 1. Historical Background

For the first time, cultivation of white button mushroom (*A. bisporus*) started in France around 1630. In the beginning, it was grown in open conditions. Around 1810, a French gardener (Chambry) cultivated them in underground querries in Paris. The possibility of continuous production was demonstrated when he cultivated *A. bisporus* in a cropping house in England. He was able to produce about 1.5 lb/sq. ft. By 1925, mushroom was grown in caves in Holland. The U.S.A. took up this work in the late 19th century. After the Second World War mushroom cultivation spread in about 80 countries. Nowadays, edible mushrooms are eaten in Africa, Australia, Switzerland, Italy, France, Germany, Japan, Europe, India, Bhutan, Pakistan, Afganistan, Tibet and China.

**> Table 18.5. Distribution of edible mushrooms in India.**

| *Mushroom* | *Distribution* |
|---|---|
| *Agaricus bisporus* | Solan (Himachal Pradesh), Punjab |
| *A. compestris* (= *Psalliota compestris*) | Punjab, W. Bengal, Bihar, Jammu |
| *Amanita vaginata* | Uttar Pradesh, Deoban |
| *Cantharellus cibarius* | W. Bengal, Kashmir, Solan |
| *Heterobasidium annostum* | U.P. Punjab, H.P. Assam |
| *Laccaria laccata* | Sikkim, Mussorie, Assam |
| *Lycoperdon perlatum* | North-west Himalayas, Punjab, Darjeeling, H.P., Sikkim |
| *Morchella esculenta* | Punjab, Kashmir, H.P., Kumaon hills (U.P) |
| *M. conica* | H.P., Dehradun, Siwalik hills, |
| *M. deliciosa* | Kashmir, Kumaon hills, H.P. |
| *Pleurotus sojar-caju* | W. Bengal, foot hills of Himalaya |

Distribution of edible mushrooms in India is given in Table 18.5. The common mushroom *A. bisporus* is abundant in cattle fields in Punjab. It is used by many people. The morel (*M. esculenta*) is found in Kashmir and hills of Kumaon region in U.P. Bhoteans consume *Hypoxylon vernicosum*. Kashmiri guchhi (*Morchella* spp.) is very popular which is sold even at the rate of Rs. 1000/kg dry mushroom.

In India, mushroom cultivation started long before a century, as the *Volvariella valvacea* was cultivated on paddy straw. Therefore, this mushroom is also known as the paddy straw mushroom. In 1950s, an attempt was made to cultivate mushroom in Coimbatore. In 1962, *Pleurotus flabellatus* (*Dhingri coroyester*) was successfully cultivated in Mysore. Besides many attempts, its cultivation could not be popularized upto the late 1960s. For the first time an attempt was made for artificial cultivation of *A. bisporus* at Solan (Himachal Pradesh) where synthetic compost preparation technology was developed, by using horse dung and wheat straw. Rapid development took place at this centre. Modern Spawn Laboratory and Air Conditioned Cropping rooms were constituted under the guidance of an expert from Food and Agricultural Organizations (FAO).

From 1974, a co-ordinated scheme was launched at Solan, Banglore, Ludhiana and New Delhi. FAO deputed its expert for improving the cultivation technology. Dr. W.A. Hayes came to

India, who recommended for incorporation of molasses and brewer's grain in the preparation of synthetic compost. This increased the mushroom yield. In 1977, State Department of Horticulture (H.P.) launched a project of Rs. 1.27 crore, under which a Central Mother Unit (CMU) for bulk pasteurization of compost and casing soil was established. C.M.U. supplies about 80 tonnes of pasteurized compost per month to growers in Solan, Shimla and Sirmur districts. During 1966-70 mushroom cultivation was introduced in Kashmir valley, where by the end of 1975, the number of growers increased to 90. This took up its cultivation as cottage industry in Srinagar and Jammu region.

In 1974, Uttar Pradesh Department of Agriculture (UPDA) started mushroom cultivation on exploratory trial at Vivekanand Parvatiya Krishi Anushandhan (VPKA), Almora. U.P. Govt. also sanctioned a project for mushroom cultivation to the Department of Botany, Kumaun University, Nainital. At Almora Centre, two crops in a year are raised (*i.e.* in February-April and September-November) in natural conditions. The compost is prepared from agro-wastes, *i.e.* straw of wheat, barley and oat and dehulled corn cobs, grasses, fresh leaves, etc.

## 2. Present Status of Mushroom Cultivation in India

Since 1983, a large number of growers started mushroom cultivation during winter around Delhi, Chandigarh and some districts of Haryana (*e.g.* Sonepat, Rohtak, Karnal) and Punjab (*e.g.* Ferozpur, Patiala, Ludhiana and Jalandhar). In Bhiwani district of Haryana, mushroom cultivation is gaining much popularity. Under the guidance of specialists of Krishi Gyan Kendra and Scientists of Haryana Agriculture University (Hissar) the farmers in villages Tagrana and Bamla have undertaken cultivation of mushrooms. These villages have been declared as mushroom villages.

Since mushrooms have a very short life, it should reach to consumers within a short time or immediately canned. This will lead to proper marketing of mushrooms. In India, over-production and improper care for marketing have resulted in the increase in price. The cost of production in India at present is comparatively higher than other countries (Rs. 10/kg in North Indian plains and Rs. 15/kg in hills) and hence cannot compete in international market. The yield obtained so far is low due to: (*i*) improper infrastructure of preparation of pasteurized compost, (*ii*) use of ordinary buildings lacking proper temperature control as cropping rooms, (*iii*) use of low yielding strains, (*iv*) inadequate supply of quality spawn, (*v*) lack of trained manpower, and (*vi*) inadequate research support. The technology for cultivation and processing of mushroom has been developed at CFTRI (Hyderabad), RRL (Jammu) and NBRI (Lucknow). RRL and NBRI are distributing mushroom spawns in rural areas for mass cultivation. CFTRI has developed technique for processing and drying mushrooms.

## 3. Nutritional Value

Generally, mushroom contains 85-90 per cent water of its dry matter. However, amount of water is greatly influenced by relative humidity and temperature during growth and storage. Protein is the most critical component which contributes to a lot of nutritional value of food. Not all, only 34-89 per cent mushroom proteins are digestible. Amount of protein varies from 34 per cent to 44 per cent of total dry weight in *Agaricus* sp. The crude fat content ranges from 1-20 per cent of total dry weight. Besides protein, a large variety of free and combined fatty acids also occur in *A. bisporus* with high concentration of palmic acid, stearic acid and oleic acid. Fresh mushroom contains relatively large amount of carbohydrates, *i.e.* 3-28 per cent, particularly pentoses, hexoses, disaccharides, and trehlose (a mushroom sugar). They appear as a good source of several vitamins (thiamin, riboflavin, niacin, biotin, ascorbic acid, vitamin A, B, C, D, and minerals (sodium, potassium, calcium, iron, etc.), essential amino acids (methionin, citralline, ornithin) and several undesirable elements (cadmium, selenium, chromium, etc.).

## 4. Cultivation Methods

Different countries have adopted different cultivation methods in accordance with facilities available and cost affecting gross production. The methods are : garden and field cultivation in Europe, and cave and house cultivation in America.

**(*a*) Garden and Field Cultivation:** In this method; there is no need of constructing houses. Small ridges are made in garden or fields. Soil inoculated with spawn is covered with leaf litter just to check from drying the mycelia of spawn.

**(*b*) Cave Cultivation :** In this method, small tunnels are prepared in rocky areas and mushroom farms are established. Moreover, mines after their use are taken to develop into mushroom farms. Inside the tunnels and mines small flat beds of 16 × 16 feet size are prepared. On these beds suitable crops of mushrooms are raised.

**(*c*) House Cultivation Method :** Houses of different sizes (50-150 x 18-24 feet) are constructed which may be above ground, or partly above ground and temperature and moisture control systems. Inside the house small beds are prepared in tiers on either sides and in middle portion. Compost mixed with soil or compost alone is spread over beds. The major steps of mushroom cultivation are: (*i*) obtaining pure culture of a suitable mushroom by tissue or spore culture method on the specific culture media, (*ii*) preparation of spawns, for example, grain or straw spawn, (*iii*) preparation of substrate *i.e.* compositing, and (*iv*) spawning, spawn running and cropping. The first two steps come under laboratory methods and the later (*iii*) and (*iv*) under mushroom house method. The major steps of laboratory and mushroom house methods are shown in Fig. 18.7 and each step is described in detail as below:

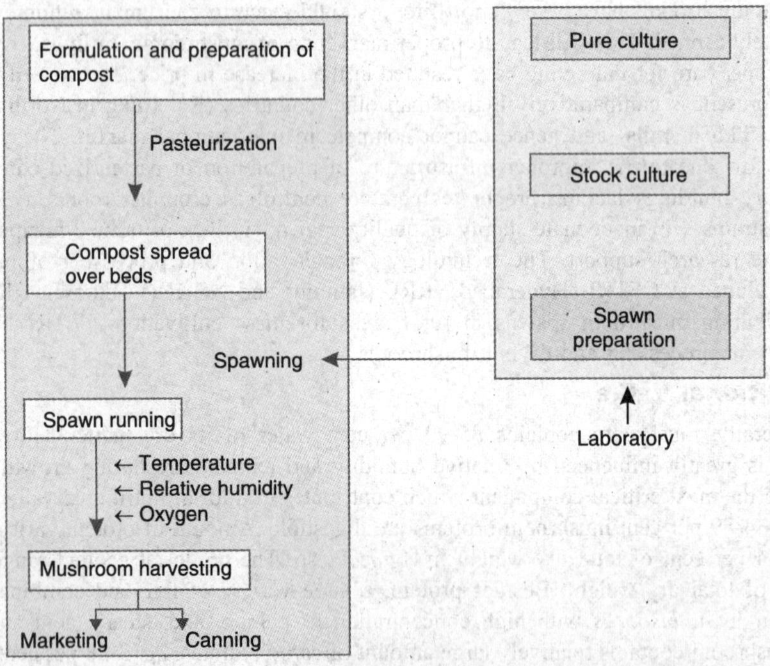


**Fig. 18.7.** Procedure for mushroom cultivation.

**(*a*) Obtaining Pure Culture :** To get pure culture, mushrooms are either isolated from nature, purified and characterized in laboratory before their use or procured from national or international mushroom culture centres. When they are isolated from the nature the method of isolation is totally microbiological one for which aseptic conditions are essential.

**> Table 18.6.** **Nutrient media used for isolation of mushrooms.**

| | | |
|---|---|---|
| 1. Potato dextrose agar (PDA) medium: | Potato | 200g |
| | Dextrose | 20 g |
| | Agar | 15 g |
| | Distilled water | 1 litre |
| Peel potato, wash, boil (in distilled water for 30 min), filter and finally raise the volume to 1 litre; add dextrose and agar to it and autoclave the medium. | | |
| 2. Malt extract agar medium: | Malt extract | 20 g |
| | Agar | 15 g |
| | Distilled water | 1 litre |

Sterilized and cooled potato dextrose agar (PDA) or malt extract medium (see Table 18.6) is poured into sterile Petri dishes and when solidified they are inoculated by a piece of tissue or spore(s) of mushroom. In tissue culture method fresh mushroom is removed from the bed (or stipe is collected after cropping), washed in running water to remove adhering soil particles, dried with blotting paper, gently washed with 70 per cent ethyl alcohol and finally cut from centre into two halves. A small portion of pseudoparenchymtous tissue from the centre of stipe is transferred onto Petri dishes. Petri dishes are incubated at suitable temperature for the growth of hyphae. The spore culture (single or multispore) is described elsewhere in detail.

**(*b*) Preparation of Spawns :** Spawn is a fungal growth medium impregnated with mycelial fragments of mushroom which serves as inoculum for mushroom cultivation. There is a great problem in preparing pure spawn of a particular strain of a mushroom because of fungal, bacterial or viral contamination.

Many substrates are used for spawn making either alone or in combinations, for example, rice straw cuttings, cotton waste, hulls of cotton seed and rice, and grains of sorghum and rye. For the selection of substrates to be used in making spawn, care is taken for cost and availability of raw materials and mycelial growth on it as well. The steps of grain spawn (*e.g.* rye/sorghum/wheat grains) or straw spawn (paddy/wheat straw) preparations follow, (*a*) cooking the grains in water until they swell/cutting of straw into 5 cm long pieces and soaking in water for 5-10 minutes, (*b*) decantation of water, mixing of 2 per cent lime (calcium carbonate), (*c*) transferring into glass tubes/flasks, (*d*) plugging with cotton, (*e*) autoclaving at 121°C for 30 minutes and cooling down to 30-40°C, (*f*) inoculating the substrate with pure culture of mushroom as described earlier, and (*g*) incubation at suitable temperature for proper infestation of mycelium for their use as spawn.

**(*c*) Formulation and Preparation of Composts :** Methods of preparation of composts for mushroom cultivation is known as composting. Initially, composting was restricted to industrial levels by using horse manure, but now can easily be applied to other substrates as the methods of formulation and preparation of composts are the same. The purpose of compost preparation is to provide medium for the rapid growth of mycelium. Therefore, physical and chemical compositions are developed in such a way that can alter the gross microbial community and promote maximum growth and yield of mushroom. No such chemicals or conditions should be present in compost that inhibit mycelial growth. H.S. Sohi (1980), former Director of National Centre for Mushroom Research and Training, Solan (H.P.), has described many compost formulations used in India and abroad (Table 18.7).

**> Table 18.7. Compost formulations used in India and abroad.**

| *Countries* | *Centres* | *Composition (kg)* | |
|---|---|---|---|
| India | 1. Central Mushroom Unit (Solan, H.P.) | Wheat straw | 1,000 |
| | | Chicken manure | 400 |
| | | Brewer's grain | 72 |
| | | Urea | 14.5 |
| | | Gypsum | 30 |
| | 2. Mushroom Farm (Chail, Haryana) | Horse manure | 1,000 |
| | | Wheat straw | 500 |
| | | Chicken manure | 300 |
| | | Brewer's grain | 60 |
| | | Urea | 7 |
| | | Gypsum | 30 |
| | 3. South India | Paddy straw | 1,000 |
| | | Urea | 70 |
| | | Cotton seed | 12 |
| | | Rice bran | 100 |
| | | Gypsum | 24 |
| Netherland | | Horse dung | 1,000 |
| | | Chicken manure | 100 |
| | | Urea | 15 |
| | | Gypsum | 25 |
| Taiwan | | Rice straw | 1,000 |
| | | Ammonium sulphate | 19.8 |
| | | Super phosphate | 19.8 |
| | | Urea | 4.95 |
| | | Chalk | 29.7 |

*Source* : Sohi, H.S. (1988)

While formulating the substrate care is taken for ease in microbial degradation. The main constituents of straw or plant waste are cellulose, hemicellulose and lignin. The first two are carbohydrates which upon decomposition result in glucose units.

Substrates are filled in small trays or wooden boxes to make beds. Use of trays of plastic bags makes air spaces. In India, tray culture is being replaced with plastic bags and shelf beds due to high cost of wood and carry over of pests and diseases. The use of plastic bags enables the growers to undertake mushroom growing with a lesser initial capital investment. Moreover, in Sonepat (Haryana) areas most of growers have adapted shelf bed system on a bamboo platform in Kuchcha mud houses covered with saccharam stalks or dried sorghum. Trays or plastic bags are transferred in the room for its partial pasteurization at low temperatures. Maximum care should be taken at this stage. Room temperature is maintained around 50-55°C for about 6 hours by live steam generated from a boiler. Later on, it is cooled down to a preferable temperature for casing the compost. Casing is the covering of compost when spread over the beds with a thin layer of soil or soil like materials. It gives support to mushrooms, maintains temperature and prevents drying the compost. Various casing materials including farm yard manure + soil (1: 1 w/w), moss + garden soil (2 : 1 w/w) have been tried.

**(*a*) Spawning, Spawn Running and Cropping :** Inoculation by spawn of compost in beds is known as spawning. Bed material is inoculated by a small amount of spawn by removing it from container and spreading over bed material. Room temperature and humidity is controlled for maximum mycelial growth and spawn running. Environmental conditions greatly influence when not suitable for spawn running.

Mushroom crop becomes mature at different intervals, producing flushes. After five to six flushes, the culture is renewed. Mature mushrooms are picked up without disturbing the neighbouring ones. During this period also, environmental conditions are left undisturbed. Harvested mushrooms are sent to market or canned for their use as food.

## 5. Control of Pathogens and Pests

A number of harmful fungi and bacteria are encountered in composts which deplete the nutrients present for growth of mycelia, and attack fruit bodies at different growth stages of crops, resulting in serious crop losses. Myco-viruses are also known to attack mushrooms. Biological control of nematodes through plant extracts and nematophagous fungi are being tried at Solan. Leaf extract of castor, madar (*Calotropis*) were lethal to nematode (*e.g. Aphelenchoides composticola*) followed by neem, chrysanthemum, bhang (*Canabis sativa*), safeda and marigold, respectively. Incorporation of dried plant materials of above plants to compost increased yield to 7-12 per cent through favourable shift in the compost mycoflora.

## 6. Cultivation of Paddy Straw Mushroom (*Volvariella*)

The paddy straw mushroom, *Volvariella* sp. prefer to grow on paddy straw hence it is known as paddy straw mushroom. It's cultivation was started in China during 18th century. In India, it was cultivated for the first time in Coimbatore.

It grows at temperature from 30-40°C on paddy straw. Other substrates which have been tasted, are sugarcane bagasse, cotton wastes, water hyacinth, etc.

**Fig. 18.8.** Diagram of preparation of paddy straw beds for the cultivation of *Volvariella* spp.

Species of *Volvariella,* which are grown in India are three viz., *V. esculenta, V.displasia* and *V. valvacea.* In 1943, K.M. Thomas and coworkers for the first time, gave the method of cultivation of *V. esculenta* on three twisted bundles of 10 kg paddy straw soaked with water and placed on wooden plateform (75 × 75 × 30 cm) in a haphazard manner. The paddy bed (prepared separately) was spawned and covered with the third twist. The whole bed was covered with the polyethylene sheet to maintain moisture (Fig. 18.8).

In 1982, Nita Bahl gave a new technique (polybag method) for cultivation of *V. volvacea.* The polybag method follows the following steps : (*i*) chop the paddy straw,and soak in water for 24 h, (*ii*) cut waste paper into small pieces and soak for the same period as above, (*iii*) decant water after 24 h, (*iv*) mix thoroughly the chopped straw, paper pulp and spawn, (*v*) fill the mixture in polyethylene bags, (*vi*) puncture the bags with needle to facilitate the exchange of air and drainage of water, (*vii*) tie the mouth of bag and keep them at 35-40°C, (*viii*) cut and remove polyethylene bag carefully to leave its contents undisturbed, (*i*) maintain the humidity of the contents to about 85-90 per cent by spraying water. After 10-15 days of incubation, small pins appear to grow.

### 7. Cultivation of White Button Mushroom

The following steps are adopted for cultivation of the white button mushroom :

(*i*) Formulation of compost (*see* Taiwan, Table 18.7) and methods of composting;

(*ii*) Spawn preparation,

(*iii*) Spawning of compost,

(*iv*) Casing of compost, and

(*v*) Harvesting the crops.

It requires a temperature of 15-18°C during cropping. Therefore, its cultivation is gaining much popularity in hilly regions in our country. A photograph of white button mushroon is given in Fig. 18.9.

Fig. 18.9. White button mushroom (*Agaricus bisporus*) growing on compost kept in polyethylene bag.

### 8. Cultivation of Dhingri (Pleurotus sajor-caju)

*Pleurotus* is also one of the important edible mushrooms gaining popularity in recent years. It is found growing naturally on dead organic materials rich in cellulose. Its several species are edible such as *P. sajor-caju, P. sapidus, P. flabellatus, P. ostriatus, P. corticatus, P. florida,* etc. These species can be cultured successfully on various agricultural, domestic, industrial and forestry waste materials. It is very versatile in nature as far as substrate preference and growth are concerned. However, it can be grown on paddy straw, gunny bags, rice husk, copped *Parthenium* stem, etc. The steps for cultivation of *dhingri* start with preparation of substrate for growth.

*Pleurotus ostreatus* is an important edible mushroom.

Paddy is cut into 2.5 cm long pieces and soaked in hot water at 60°C for about 30 minutes. Excess water is drained off from straw. About 4 kg of wet straw is transferred into the large sized polythene bags. About 5 grams of bengal-gram powder with half bottle of spawn of fungus are mixed with straw. This mixture is filled in large-sized polythene bags. Mouth of the bags is tied and kept on a raised platform (Fig. 18.10) in well ventilated cropping room or in open when properly protected for about 15 days. At this time when mycelia are visible inside the polythene bags over the surface of paddy straw, the polythene

Fig. 18.10. *Pleurotus sajor-caju* growing on paddy straw.

bags are cut and gently removed. Now paddy straw forms a compact mass and does not lose its make up. This composite mixture is watered daily just to maintain moisture. The temperature where the compost has been kept should be between 20°C and 25°C with relative humidity of 75 per cent. After 15 days first flash of dhingri becomes apparent. These are harvested when become young. A photograph of *P. sajor-caju* growing on paddy straw is shown in Fig. 18.10.

### 9. Recipes of Edible Mushrooms

Mushrooms are taken out from the mushroom house (or procured from the market), washed with water to remove the adhering soil particles and outer portion is gently removed if stipe is hard. It is air-dried, if required dilute, vinegar is mixed to check the blackening due to cutting with knife. Delicious recipes of mushrooms are prepared in different ways as required. Some of the recipes are described as below :

Mushroom soup ready to serve.

**(*a*) Mushroom Puree :** Chop mushroom (250 g), put into molten vegetable oil in pan, add butter (50 g) or water (50 ml), cover pan, cook for 30 minutes on mild heat, boil 300 ml milk in another pan, rub together one table spoonful butter and flour, add milk to it, stir, cook until thickened, add mushroom and press the whole thing with a sieve, season the mixture of milk, flour and mushroom with spices, add salt and use as puree.

**(*b*) Mushroom Paneer :** Cut mushroom (400 g), paneer (200 g) and tomato (50 g) into small pieces, chop onion, transfer chopped onion to molten vegetable oil kept in a pan, fry until golden brown, add cut mushroom and paneer, cook till water dries, add salt and spices, serve hot.

**(*c*) Mushroom Rice (Pulao) :** Fry chopped onion in vegetable oil in a pan till it turns brown, add chopped mushroom (250 g) and properly washed rice, cover with hot water as required, add spices and salt to taste, and cook for 15 minutes, serve it when ready.

**(*d*) Mushroom Omelette :** Wash three mushroom chop and put in pan, add 25g butter and little amount of salt, simmer it, fry the omelette mixture in a separate pan, add simmered mushroom, and vegetable oil, cook and serve it.

**(*e*) Mushroom Soup :** Cut a small onion and fry until golden brown, add a little amount (25 g) of flour, mix and wait for a minute, add small amount of water, stir, add required quantity of hot water or meat cube, heat for 5 minutes, add sliced mushroom (500 g), season with spices, add salt, cook for 5 minutes and serve hot.

About 100-200 g mushrooms (on dry weight basis ) are required for a normal human of weighing about 70 kg to maintain the nutritional balance of body.

## K. PROBIOTICS

Probiotics (*pro* = for, and *bios* = life) *are live microorganisms that may confer a health benefit on the host when administered in adequate amounts*' (FAO/WHO). In 1965, D.M. Lilly and H. Stilwell for the first time used the term 'probiotic' for those microorganisms that affect the other microorganism. The substances secreted by one microorganism stimulate the other microorganism. In 1971, G. Sperti used this term to describe tissue extracts which stimulated microbial growth. Later in 1974, R. Parker advanced the meaning of the term by adding the word 'organisms', and described probiotics as '*organisms and substances that have a beneficial effect on the host animal by contributing to its intestinal microbial balance*'. In 1989, R. Fuller greatly improved the definition and described probiotics as 'live microbial feed supplements which beneficially affect the host animal by improving its intestinal microbial balance'. The most commonly used microbes as probiotics are the lactic acid bacteria (LAB) and bifidobacteria, but certain yeasts like *Saccharomyces boulardii* and bacilli may also be used.

Probiotics are commonly consumed as fermented foods with specially inoculated live cultures, such as in yogurt, soy yogurt, dahi, yakult or as dietary supplements. Probiotics are also delivered in faecal transplants, in which stool from a healthy donor is delivered like a suppository to an infected patient.

## 1. Hisory of Probiotics

The ability of beneficial bacteria to transform milk into a longer lasting food like curd was recorded before 7,000 BC during pre-vedic period. Long before the microorganisms were known and the fermented products were used to treat colds, fevers constipation and diarrhea.

Élie Metchnikoff (1845-1916), a Nobel laureate'

In the beginning of the 20th century, a Russian Nobel laureate Élie Metchnikoff suggested that the gut flora can be modified and harmful microbes can be replaced with useful microbes. Metchnikoff proposed the hypothesis that the ageing process results from the activity of proteolytic microbes producing toxic substances (phenols, ammonia). These compounds were responsible for 'intestinal auto-intoxication' which caused the physical changes associated with old age. Milk fermented with lactic-acid bacteria inhibits the growth of proteolytic bacteria because of the low pH produced by the fermentation of lactose. Metchnikoff himself introduced in his diet sour milk fermented with the bacteria called 'Bulgarian Bacillus' (later called *Lactobacillus delbrueckii* subsp. *bulgaricus*) and found his good health.

In 1917, during an outbreak of shigellosis, Alfred Nissle from Germany isolated a strain of *Escherichia coli* from the faeces of a soldier who was not affected by the disease. Nissle used the *E. coli* Nissle 1917 strain in acute gastrointestinal infectious salmonellosis and shigellosis. In early 1930's, Minoru Shirota a researcher at the Faculty of Medicine at Kyoto University (Japan) developed a fermented milk product called Yakult. Thereafter, intestinal lactic acid bacterial species having beneficial properties have been introduced as probiotics, such as *Lactobacillus rhamnosus, Lactobacillus casei and Lactobacillus johnsonii, etc.*

## 2. Probiotics, Prebiotics and Synbiotics

Probiotics are the live microorganisms that favorably influence the host's health by improving the indigenous microflora'. Prebiotics are non-digestible food components that increase the growth of these probiotic microorganisms in the gastrointestinal tract. In 1995, Marcel Roberfroid for the first time identified the prebiotics. The prebiotic substances derived from whey or lactose are : lactitol, lactulose, lactosucrose and galactooligosaccharides. These compounds are not digested in the small intestine; instead they pass to the colon where they serve as substrates for fermentation by probiotic bacteria. They stimulate the microflora in the large intestine

Recently, many new products have been introduced in the United States that combine probiotic bacteria and prebiotic ingredients. This combination is called 'synbiotics', *i.e.* nutritional supplements combining probiotics and prebiotics in a form of synergism. They target similar benefits to humans by improving digestive health through nourishing a healthy colon. Both prebiotics and probiotics must be ingested in sufficient amount so that it may have an impact; both should not have excessive sugar, calories, carbs, etc. Some of the similarities and differences between prebiotics and probiotics are given in Table 18.8.

**> Table 18.8. Similarities and differences between prebiotics and probiotics.**

| | *Prebiotics* | *Probiotics* |
|---|---|---|
| 1. | Prebiotics are a very special form of dietary fiber | Probiotics are living bacteria that benefit colon health |
| 2. | Prebiotic fiber is not affected by heat, cold, acid or time | Probiotics must be kept alive to create health benefits. They can be killed by heat, acid or storage |
| 3. | Prebiotics nourish thousands of good bacterial species already living in the colon | Probiotics contain from one to a few species of bacteria which are added to the colon upon ingestion |
| 4. | Prebiotic fiber is a naturally-occurring substance found in thousands of plant species in very small amounts | Probiotics occur naturally in fermented foods like yogurt or sauerkraut. Some companies have also engineered proprietary bacteria and patented them |
| 5. | Prebiotics foster an environment in the colon which is hostile to bad bacteria | Probiotics may have impact on bad bacteria by crowding them out |

### 3. Mechanism of Action

Probiotics improves the health through a variety of mechanisms, such as : (*i*) regulation of intestinal microbial balance, (*ii*) posing antimicrobial activity, (*iii*) immuno-modulation, and (*iv*) pathogen exclusion. Each of the mechanisms is further subdivided on the basis of different function. For epithelial barrier function, probiotics act through general antimicrobial effects, effects on intestinal permeability in the presence of invasive bacteria, effects on epithelial cell inflammatory responses, and effects on epithelial cell survival. Health benefits stemming from probiotic consumption is given in Fig. 18.11.

**(a) Regulation of Intestinal Microbial Balance:** There is an optimum balance among gut flora in a healthy gut; the predominant beneficial bacteria that live in gut are *Lactobacilli* and *Bifidobacteria*. Altered equilibrium of microbial community may change luminal immune and inflammatory responses, and metabolism of epithelial cells. Hence, probiotics can regulate intestinal homeostasis by using different mechanisms, such as : (*i*) reduction in adherence and invasion of pathogens, (*ii*) reinforcing tight junctions and intestinal barrier function by activating various intracellular signalling pathways and rearrangement of tight junction proteins, and (*iii*) modulation of luminal metabolism by production of short-chain fatty acids, which later modulate gene expression of epithelial cells by influencing epigenetic modulations. These changes stimulate epithelial proliferation and barrier function.

**(b) Antimicrobial Activity:** The antimicrobial properties of probiotics are due to production of some active metabolites that inhibit or kill the potential pathogens. Probiotics ferment sugars and produce organic acids such as lactic acid and acetic acid. The mixtures of organic acids have a powerful antimicrobial activity at low pH. Secondary metabolites, such as ethanol, acetaldehyde, acetoin, carbon dioxide, reuterin, reutericyclin and other germicidal compounds are produced by probiotics. These low molecular weight components works as growth inhibitory factors against pathogens. Some probiotics strains produce proteinaceous antibiotic substances called bacteriocins that exert antimicrobial action in cell wall of the target organisms. Hydrogen peroxide is produced by lactic acid bacteria that oxidise sulphydryl group of surface proteins and enzymes of target organisms at low amount.

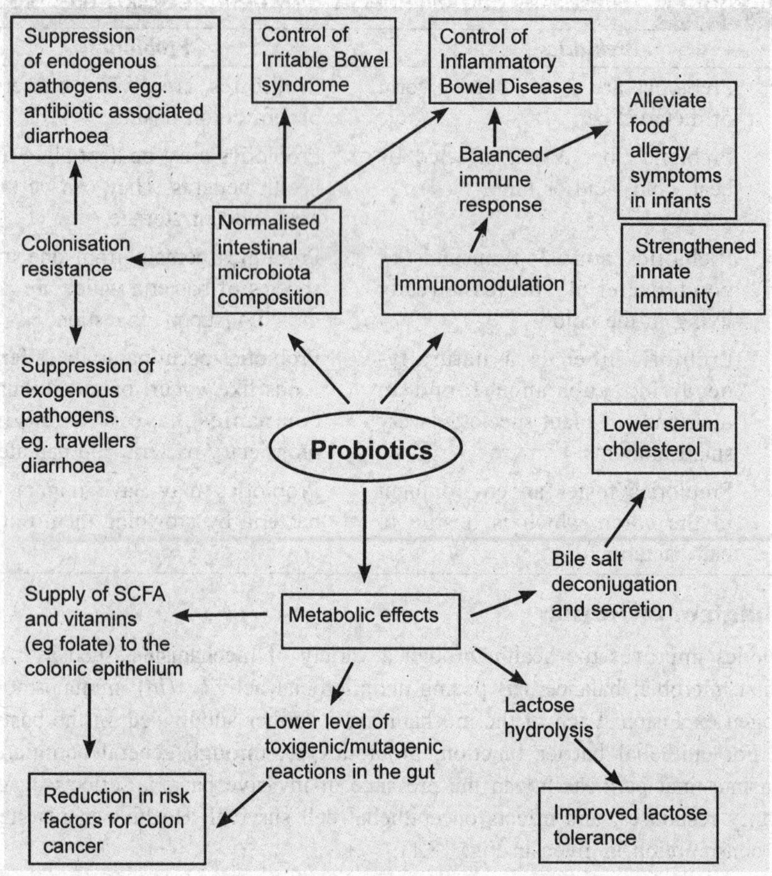


**Fig. 18.11.** Health benefits obtained from probiotic consumption.

**(c) Immuno-modulation:** Probiotics exert immunomodulatory effect after interacting with epithelial and dendritic cells (Fig. 18.11). The effects are strain-dependent, induced by cytokines secreted by lymphocytes or dendritic cells. Interaction of probiotics with these cell surface receptors activates some intracellular signalling pathways and induce immune systems. For example, probiotics stimulates production of pro-inflammatory cytokines (TNF-α, IL-8, IL-12, and IFN-γ) from dendritic cells. But they regulate the activation of innate immunity and inhibit excessive inflammation.

**(d) Pathogen Exclusion:** There is some process by which an organism fails to colonise an environment due to the presence of another organisms that get established suitably. Probiotics also share the same receptor sites on host cells with pathogens. Due to this property they exclude the pathogens from host intestine, urogenital tract and other host sites. Probiotics excludes intestinal pathogens by several mechanisms, such as : (*i*) steric hindrance and competitive depletion of essential nutrients, (*ii*) competitive exclusion by adhered probiotics by competition for receptor sites and by displacement of adhered pathogens, and (iii) non-competitive exclusion by induction of secretion of antimicrobial components from host cell and by regulation of epithelial barrier function.

## 4. Types of Probiotics

On the basis of purpose and function, probiotics are of several types. In spite of their natural occurrence, amount and probiotic types in our body can also be changed by eating foods and taking food supplements. Mainly probiotics found in many foods and supplements are of following types :

- *Bifidobacterium* : It includes *bifidum* and *longum* strains that can be used for control of mineral absorption and regulation of other bacteria.
- *Enterococcus faecium* : This bacterium may affect cholesterol levels and may relieves symptoms associated with antibiotic diarrhea.
- *Lactobacillus* : It includes *acidophilus*, *bulgaricus* and *rhamnosus* strains; these are the most common type of probiotic.
- *Streptococcus thermophilus* : It helps in lactose digestion and aid people who have lactose-intolerance.

Moreover, probiotics are recommended by many physicians to offset some antibiotic-associated side effects. But one must consult with physicians for most suitable types of probiotics. In 2007, M.E. Sanders has given a list of microbial strains used as probiotics affecting human health (Table 18.9).

**Table 18.9. Microbial strains and their probiotic research on human health** (source: Functional foods & Nutraceuticals Magazine: 36–41).

| *Microbial strains* | *Potential effect in humans* |
|---|---|
| *Bacillus coagulans* GBI-30, 6086 | Improves abdominal pain, and increases immune response to a viral infection |
| *Bifidobacterium animalis* subsp. *lactis* BB-12 | Affects on the gastrointestinal system |
| *Bifidobacterium longum* subsp. *infantis* 35624 | Provides relief from abdominal pain/discomfort, bloating and constipation |
| *Lactobacillus acidophilus* NCFM | Reduces the side effects of antibiotics |
| *Lactobacillus johnsonii* La1 | Reduces incidence of *Helicobacteri pylori* that causes gastritis, and reduces inflammation |
| *Lactobacillus reuteri* | Affects on gingivitis, periodontitis, reduces risk factors for caries |
| Mixture of*Lactobacillus rhamnosus* GR-1 and *Lactobacillus reuteri* RC-14 | Oral ingestion resultes in vaginal colonisation and reduces vaginitis |
| Mixture of *Lactobacillus acidophilus* NCFM & *Bifidobacterium bifidum* BB-12 | Reduces *Clostridium difficile* –associated disease |
| Mixture of *Lactobacillus acidophilus* CL1285 & *Lactobacillus casei* LBC80R | Affects digestive health, in vitro inhibits *Listeria monocytogenes* and *L. innocua*, *Escherichia coli*, *Staphylococcus aureus*, *Enterococcus faecalis* and *Enterococcus faecium*, and reduces lactose intolerance and immune stimulation |

## 5. Probioics Products in India

The Probiotic Association of India (PAI) has been constituted that provides new finding of probiotic research, product development, health claims associated with probiotic foods. It also provides the forthcoming scientific activities happening across the world including India.

The arena of probiotics is the rapidly expanding. India can play a key role as being the largest producer of milk and having world's highest cattle population. Indian probiotic industry is in its infancy stage and presently accounts for only a small fraction, *i.e.* less than 1% of the total world market in the probiotic industry. India is emerging as a major probiotic market of the future with annual growth rate of 22.6% until 2015. Some of the Indian probiotic industry are : Amul, Mother Dairy, Yakult Danone and Nestle along with some other small industries in different regions operating on their own cost. The Indian probiotic market turnover is expected to reach $8 million by the year 2015. In India Probiotics generally come in two forms, milk and fermented milk products. The milk products occupy 62% of the market share and the fermented milk products have 38% market share.

In 2007, Amul was the first to make probiotic ice creams prolife in February at National level; it also introduced probiotic lassi. Mother Dairy has the largest milk (liquid/unprocessed) plants in Asia selling more than 25 lakh liters of milk per day. The company's probiotic products are : Beta-Activ Probiotic Dahi, Beta-Activ Probiotic Lassi, Beta- Activ Curd and Nutrifit (strawberry and mango). Probiotic products are contributing to 15% of the turnover of their fresh dairy products.

Nestle NESVITA was India's first dahi with probiotics for healthy digestion. Yakult Danone India Pvt Ltd (YDIPL) is a 50:50 joint venture between Japan's Yakult Honsha and the Danone Group of French, and is offering Yakult (a probiotic drink made from fermented milk, *Lactobacillus* and some sugar).

In India, the major players in the probiotics drug market include : Ranbaxy (Binifit), Dr. Reddy's Laboratories, which has four probiotic brands, Zydus Cadila, Unichem, JB Chem, and Glaxo SmithKline. While probiotics in the form of drugs are widely accepted; probiotic foods are still viewed with scepticism.

## 6. Global Scenario of Probiotics

The growth of probiotic products in the developed world has been quite amazing. The global probiotics market is expected to be worth US $ 31.1 billion by 2015 with the Europe and Asia accounting for nearly 42 and 30% of the total revenues, respectively. In 2000, pro/ prebiotic yogurt accounted for one-quarter of global yogurt sales according to Euromonitor International's packaged food data; but in 2010, it accounted for one-third. Ingestion of specific strains of the healthy microflora forms the probiotic therapy for normalisation of unbalanced indigenous intestinal microflora.

Obviously probiotics hold great promise and can serve as candidate biotherapeutics in the management of inflammatory metabolic disorders, such as cardiovascular diseases like atherosclerosis, hypertension stroke, etc. Several food companies are expanding their market profile to the most promising growth markets for probiotics. Some of the probiotic products available worldwide are presented in the Table 18.10.

**Table 18.10.** **Sellers, exporters and manufacturers of probiotics.**

| *Country* | *Manufacturers* |
|---|---|
| Australia - *Queensland* | Coral Cay Health |
| Bangladesh – *Dhaka* | ROM International |
| Canada – *Ontario* | NicaVita corporation |
| China – Shandong | Zibo Excellent Fodder & Feed Additive Co., Ltd. |
| India : | 1. De Generic Bio-Tech Pvt LTd |
| *Andhra Pradesh* | 2. Sks Bioproducts |
| | 3. Unique Biotech Ltd. |
| *Tamil Nadu* | 4. Agriya Impex |
| | 5. Natural Nutrition Marketing Pvt. Ltd. |
| *Maharastra* | 6. Pavan Turnkey Projects Co., Ltd. |
| | 7. Zytex India Pvt. Ltd. |
| South Africa - *North-West* | Affmech cc |
| Thailand - *Nakhon Pathom* | Nuevotec Co., Ltd. |
| UK - *South East of England* | Life On Healthcare Ltd |

## 7. Cautions about Probiotics

Though ingestion of probiotics is safe, but it may cause few side effects. People have been eating yogurt, cheeses, and other foods containing live cultures for around the world centuries. Still, probiotics may be dangerous for people with weakened immune systems or serious illnesses. It has been reported that patients with severe pancreatitis had a higher risk of death when were given probiotics.

## PROBLEMS

1. What is Single Cell Protein (SCP)? How many types of micro-organisms are used for the production of single cell protein products?
2. Write a brief note on advantages of using single cell protein, and organic waste utilized for production of single cell protein (SCP).
3. Write an eassy on nutritional value of single cell protein (SCP) or probiotics.
4. How are the algae produced on waste water and used as single cell protein?
5. Give a brief account of bacterial and actinomycetous biomass production.
6. Give a detailed account of yeast biomass.
7. What is edible mushroom? What role could they play in facing the challenge of world food shortage? Discuss in light of their nutritional status.
8. Discuss in brief the current situation of mushroom cultivation in India.
9. How many types of methods are used in mushroom culture? Discuss in brief any one studied by you.
10. What is spawn? How is it prepared?
11. Write an eassy on formulation and preparation of composts in India studied by you.
12. How are the pathogens and pests associated with compost controlled?
13. Describe in brief the cultivation of paddy straw mushroom.
14. What are the methods for mass cultivation of *Spiralina* SCP ? Discuss the benefits from *Spirulina* SCP.
15. Write short notes on the following :

    (*a*) Single cell protein (SCP), ((*b*) *Spirulina* as algal biomass (*c*) Bacterial biomass, (*d*) Yeast biomass, (*e*) Nutritional value of single-cell protein (SCP), (*f*) Genetic engineering and single-cell protein, (*g*) Mushroom Culture in India (*h*), Spawning, (*i*) Composting, (*j*) Fungul protein, (*k*) Recipes of Mushrooms, (*l*) Traditional fungal foods. (*m*) Sake, (*n*) shoyu, (*o*) Miso, (*p*) Tempeh, (*q*) Mass cultivation of *Spirulina* SCP, (*r*) Uses of *Spirulina* SCP, (s) Dhingri cultivation.

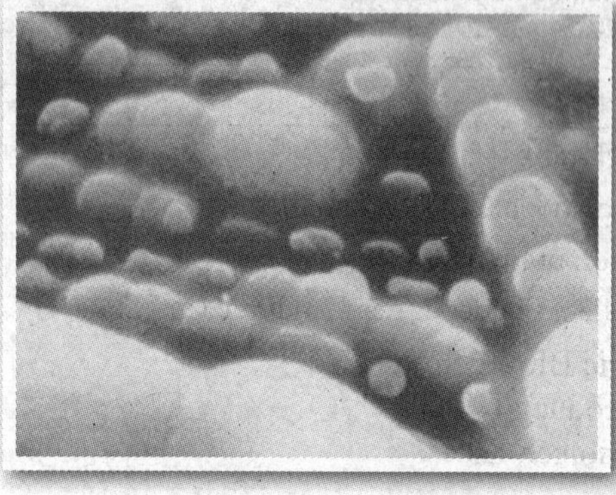

CHAPTER

19

# Biological Nitrogen Fixation

Nitrogen is available in atmosphere in high amount (78%) in the form of gas. It is converted into combined form of organic compounds by some prokaryotic microorganisms through biological reactions. The phenomenon of fixation of atmospheric nitrogen by biological means is known as 'diazotrophy' or 'biological nitrogen fixation' and these prokaryotes as 'diazotrophs' or 'nitrogen fixers'. For the first time Beijerinck (1888) isolated *Rhizobium* from root nodules of leguminous plants. Thereafter, in 1893, S. Winogradsky discovered a free-living nitrogen fixing bacterium, *Clostridium pasteurianum.* Then a large number of nitrogen fixers were discovered from different sources and associations. For example, *Frankia* from nodules of non-legumes (*e.g.* alder, *Casuarina,* etc.), *Nostoc* from lichens, *Anabaena* from *Azolla* leaves, and coralloid roots of *Cycas.* The diazotrophs may be in free living or in symbiotic forms.

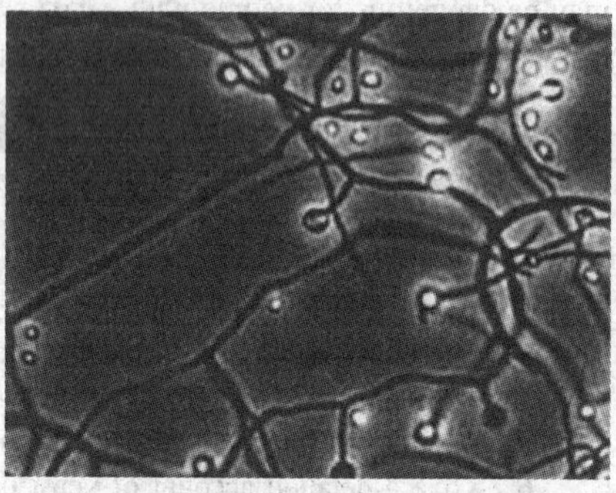

Culture of *Frankia* showing vegetative branched septate hyphae as well as thick-walled cells specialized in nitrogen-fixation.

It is estimated that the free living microorganisms contribute about 1–3 kg N/ha/annum. During 1974, the total global nitrogen fixation was about 175 × $10^6$ tonnes, of which 90 × $10^6$ tonnes N was fixed in cultivated soil (Schlegel, 1986) and about 45 × $10^6$ tonnes by rhizobial symbiosis.

In recent years, to meet the demand of nitrogen fertilizers, the number of chemical industries has been increased. These industries depend on fossil fuels; hence the cost of fertilizers is governed by the cost of fuels. In order to reduce dependance on chemical fertilizers, an alternative method is to be developed. Artifical inoculation of rice and other crop fields with cyanobacteria has attracted much attention to promote rice production in developing countries, where fertilizers are expensive and in short supply. In this regard biological nitrogen fixation has been dealt with free living and symbiotic microorganisms.

# A. NON-SYMBIOTIC NITROGEN FIXATION

## 1. Diazotrophs

Microorganisms which pass independent life and fix atmospheric nitrogen are known as free living diazotrophs. There are two groups of such micro-organisms : bacteria and cyanobacteria (blue-green algae). Based on the mode of nutrition (carbon, nitrogen and oxygen and requirement of reducing groups) bacteria are divided into (*i*) aerobic bacteria (*Azonomas, Azotobacter, Beijerinckia, Mycobacterium, Methylomonas*), (*ii*) facultative anaerobic bacteria (*Bacillus, Enterobacter, Klebsiella*, etc.), (*iii*) anaerobic bacteria (*Clostridium, Desulfovibrio*, etc.) and (*iv*) photosynthetic bacteria (*Rhodomicrobium, Rhodopseudomonas, Rhodospirillum, Chromatium, Chlorobium*, etc.).

Among cyanobacteria, both heterocystous and nonheterocystous forms fix atmospheric nitrogen, for example, *Anabaena, Anabaenopsis, Aulosira, Calothrix, Cylindrospermum, Gloeocapsa, Lyngbya, Nostoc, Oscillatoria, Plectonema, Scytonema, Stegonema, Tolypothrix, Trichodesmium*, etc.

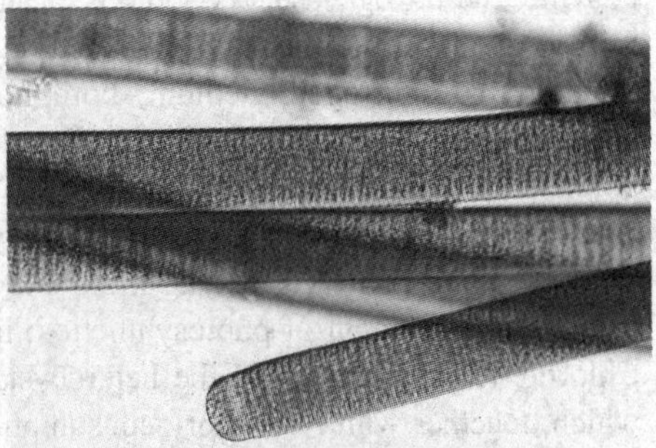

*Oscillatoria* is a cyano bacterium without heterocyst.

## 2. Ecology of Diazotrophs

Free living bacteria and cyanobacteria prefer a variety of habitats with varying degree of nutrients, pH, oxygen, etc. Photosynthetic nitrogen fixing bacteria are divided into : (*i*) non-sulfur purple, (*ii*) purple sulfur and (*iii*) green bacteria. Since water is a major component of cytoplasm, adequate amount of water must be required for their vegetative growth. The amount of water governs the concentration of oxygen; therefore, oversupply of water limits gas exchange, lowers the available supply, and finally creates anaerobic condition. Due to water-logging conditions, number of anaerobic bacteria increases and that of aerobic ones decreases in soil.

Environmental factors which influence number, community size, vegetative growth and activity of microorganisms are temperature, organic matter, pH, inorganic fertilizers, light, oxygen, season, soil and depth. In water-logging fields (anaerobic condition) such as flooded soils, lakes, ponds, ricefields, etc. non-sulfur purple bacteria grow luxuriantly.

The azotobacters are the most intensively investigated heterotrophic group. They are the aerobic bacteria possessing the highest respiratory rates. Members of these genera are mesophilic, which require optimum temperature of about 30°C for their growth. Density of azotobacters ranges from $10^3$ to $10^6$ per gram soil. Other dominant $N_2$ fixing aerobic bacteria present in soil are *Beijerinckia* and *Derxia*. *Beijerinckia* grows luxuriantly in acid soil in tropical region. However, *Derxia* can tolerate a pH range of 5.0 to 9.0. The studies on facultative anaerobes have not been given due consideration. The presence of 20 to 18 $x$ $10^3$ *Klebsiella* cells and less than $10^3$ *Enterobacter* and *Bacillus* cells per gram soil would be significant to utilize an adequate amount of $N_2$.

In water-logging conditions the number of clostridia (e.g. *Clostridium acetobutylicum, C. butyricum, C. pasteurianum*) increases in soil in a range from $10^2$ to $10^6$ cells per gram. Some of the $N_2$ fixing bacteria are given in Table 19.1.

Cyanobacteria are found commonly in well drained paddy and other crop fields. Some of them possess heterocysts (e.g. *Anabaena, Aulosira, Cylindrospermum, Nostoc, Tolypothrix*, etc.) and some do not (e.g. *Lyngbya, Oscillatoria, Plectonema*, etc.).

### 3. Special Features of Diazotrophs

**(*a*) Sites of $N_2$ Fixation** : Sites of $N_2$ fixation vary in different microorganisms as given below:

(*i*) **Cyanobacteria***:* Many cyanobacteria capable of fixing nitrogen are filamentous and contain pale and thick walled cell (intercalary, lateral or terminal, single or in chains) called **heterocyst.** (Fig. 19.1). These are the sites of nitrogen fixation. Heterocysts are formed in the absence of utilizable combined nitrogen, such as ammonia, because it inhibits heteroyst differentiation and $N_2$ fixing enzyme, the nitrogenase. As soon as cyanobacteria growing on ammonia-supplemented media are transferred to nitrogen medium, both heterocysts and nitrogenase develop parallelly. Besides, some strains, high light intensity inhibits $N_2$ fixation. Heterocyst lacks oxygen evolving photosystem II, ribulose biphosphate carboxylase and may lack or have reduced amount of photosynthetic biliproteins. Chlorophyll-*a* is present in the heterocysts. Wall of heterocyst contains $O_2$ binding glycolipids which, together with respiratory consumption, maintain the anaerobic conditions (highly reduced atmosphere) necessary for $N_2$ fixation. In contrast, in vegetative cells adjacent to heterocysts, both photosystem I and II are present; therefore, oxygen evolution takes place by these cells.

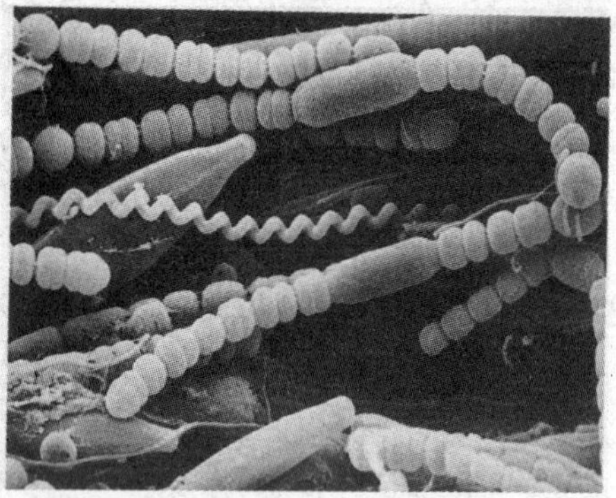

**Fig. 19.1.** A trichome of *Anabaena* showing heterocyst.

In other group of cyanobacteria such as *Lyngbya, Oscillatoria, Plectonoma,* and bacteria where heterocyst does not develop, $N_2$-fixation process takes place in internally organized cells having reduced levels of oxygen.

(*ii*) **Bacteria:** Root nodules developed by *Rhizobium* (in leguminous plants) and actinorhiza formed by *Frankia* (in non-leguminous plants) are the site of nitrogen fixation. However, lichens (a symbiotic structure formed by cyanobacteria and fungi) are the site of $N_2$ fixation in lower group of micro-organisms.

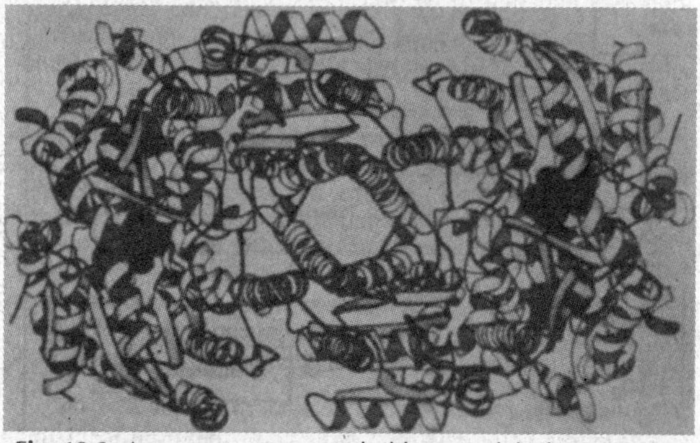

**Fig. 19.2.** A computer generated ribbon model of component I of nitrogenase (Mo-Fe-protein).

**(*b*) Presence of Nitrogenase and Reductants** : All diazotrophs possess an enzyme nitrogenase which helps in conversion of $N_2$ to $NH_3$. Structure of nitrogenase is given in Fig. 19.2. It consists of two brown metalloproteins whose joint action is essential for reduction of $N_2$ to $NH_3$.

- **Component I:** Component I, *i.e.* molybdo-ferro-protein (Mo-Fe-protein) which is also known as **nitrogenase** has a molecular weight of about $2.2 \times 10^5$ Dalton. Nitrogenase contains molybdenum (2 atoms/mol), iron (32 atoms/mol) and a sulphur (30 atoms/mol). It is a tetramer and made up of two sub-units i.e. $\alpha2$ and $\beta2$. Two metal centres are present on this sub-units i.e. *P centre* (P cluster pair) and *M centre* $Fe_4S_4$ centres (i.e.,

8Fe:8S centres) are found from A to P cluster. One Mo-Fe-Cofactor subunit contains 1Mo, 7Fe, 9S and one homocitrate. Thus, it is organised as 2Fe;4S and Mo:3Fe:3S cluster (Fig. 19.2). It is a larger unit than component II and non-sensitive to cold but losses activity at 0°C.

- ***Component II*** **(Fe-protein)** : It is also called 'dinitrogenase reductase.' It has molecular weight of about $5 \times 10^4$ Dalton. It consists of iron (4 atoms per molecule) and sulphur (4 atoms/mol) ($Fe_4S_4$) and has one iron-sulphur clusture. It is less the stable than the component I. The Fe-S clusture acts as redox site. The nucleotide-binding sites are situated at the margins of two subunits. Both the subunits are linked together by two pairs of cysteine and a site for nucleotide-binding and cleaving (Fig. 19.3).

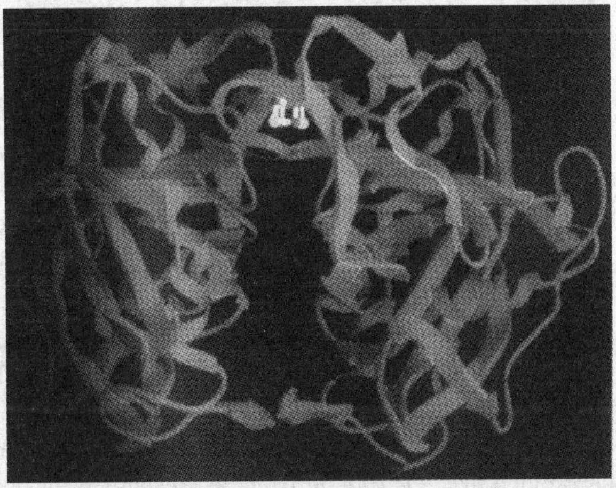

**Fig. 19.3.** A computer generated ribbon model of component II of nitrogenase (Fe-protein).

Thus, nitrogenase is an equilibrium mixture of Mo-fe-protein and Fe-protein in the ratio 1: 2 (Fig. 19.4) as below :

Mo-Fe-Protein+2 (Fe-protein)
Nitrogenase

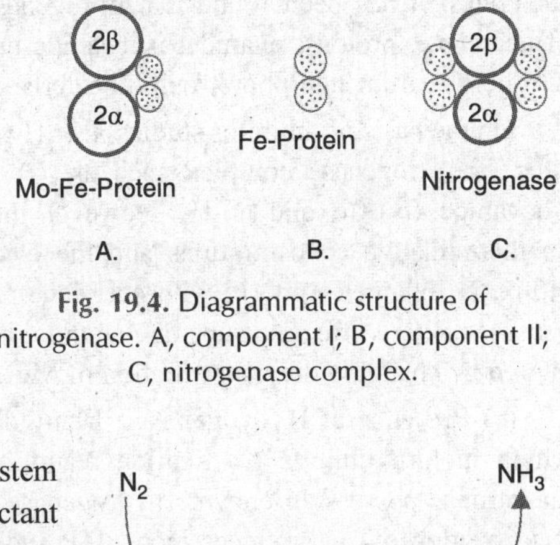


**Fig. 19.4.** Diagrammatic structure of nitrogenase. A, component I; B, component II; C, nitrogenase complex.

The specific need of these components for organisms is probably due to the presence of the elements in two protein components of nitrogenase. All diazotrophs contain similar proteins and suggesting thereby a similar, not identical, reaction sequence. This is why the molecular weight of these two components in different microbial cells differ.

In addition to nitrogenase, the $N_2$ reducing system requires Mg ATP as a source of energy and a reductant (electron donor) to catalyse the reduction of substrates. Here, ATP functions as carrier of energy (Fig. 19.5). Energy released in metabolic oxidation of carbohydrates is utilized in phosphorylation of ADP in the presence of inorganic phosphate (Pi) to reduce the energy rich ATP species. Whenever energy is needed, ATP undergoes enzymatic hydrolysis to form ADP + Pi. $Mg^{++}$ functions as catalyst. Overall reaction calaysed by nitrogenase may be written as below :

$$N_2 + 12\text{ Mg ATP} + 6(H^+ + e^-) \longrightarrow 2NH_3 + 12\text{ (Mg ADP + Pi)}$$

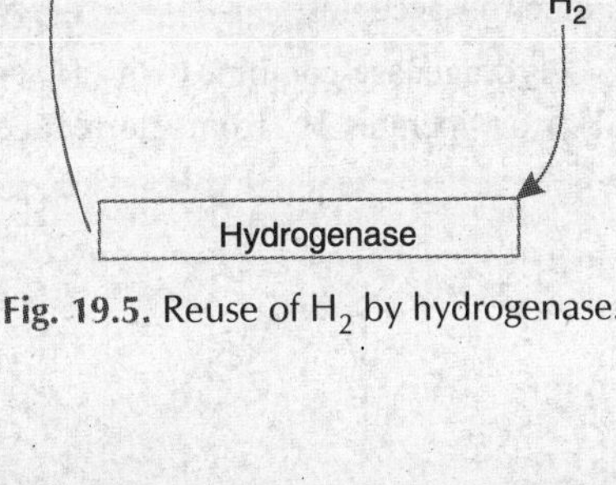


**Fig. 19.5.** Reuse of $H_2$ by hydrogenase.

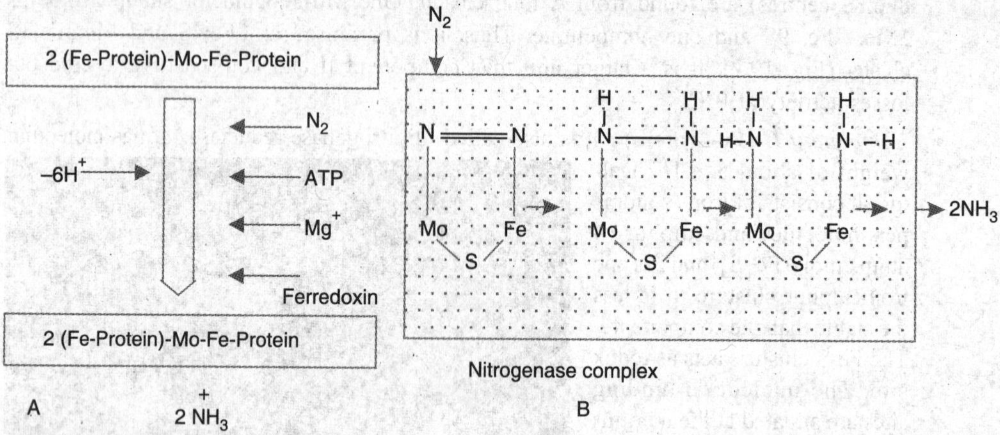


**Fig. 19.6.** Nitrogen fixation by nitrogenase complex (A) and mechanism of $N_2$ conversion by component I (B).

In *Rhodospirillum rubrum* the physiological donors (e.g. ferredoxin I and II) of nitrogenase has been demonstrated to transfer reducing equivalent from the illuminated chloroplasts to crude preparation of *R. rubrum* nitrogenase. Recently, it has been shown that pyruvate- 2-oxoglutartate and to a lesser extent, oxaloacetate supported nitrogenase activity in crude extract of *R. rubrum.*

In addition to reduction of nitrogenase to ammonia, nitrogenase catalyses a number of other substrates. It has been found that nitrogenase isolated from *R. rubrum* catalysed the reduction of acetylene, protons, cyanide, isocyanide and azide; all of which behaved as the substrates of nitrogenase from non-phototrophic bacteria.

On the basis of inhibitor studies, I.C. Hwang and coworkers in 1973 have proposed substrates sites for nitrogenase complex such as: (*i*) $N_2$, (*ii*) acetylene, (*iii*) azide, cyanide and methyl isocyanide, (*iv*) Co, and (*v*) $H^+$. However, the allocation of electrons to these substrates depends on their relative concentrations and the electron efflux through the complex. At low electron efflux, $N_2$ competes poorly with acetylene and $H^+$ as a substrate but at high electron efflux $N_2$ becomes a very effective substrate as the dominant electron acceptor. The electron efflux can be altered by changing the concentration of Mg -ATP or the ratio of the Fe-protein to Mo-Fe-protein.

**(*c*) Presence of Hydrogenase :** Many diazotrophs evolve hydrogen ($H_2$) during $N_2$ fixation which in turn inhibits $N_2$ fixation reaction. To get protection from inhibition by $H_2$ many diazotrophs possess an enzyme hydrogenase to recycle $H_2$ produced by nitrogenase (Fig. 19.5). Reutilization of $H_2$ produces more ATP and improves the efficiency of $N_2$ fixation.

Present knowledge indicates that the hydrogenase of purple bacteria is membrane-bound or at least membrane-associated. It is cold labile and very sensitive to $O_2$. The purified enzyme has a molecular weight of about 6,500 and most probably consists of a single polypeptide chain with 4 iron atoms and 4 acid labile sulphur atoms per molecule. It does not use $NAD^+$ or $NADP^+$ as electron acceptors.

Hydrogenase combines with $H_2$ and reduces into a number of substrates (S) to form products ($SH_2$) or liberates $H_2$ from the reduced compounds as below:

$$H_2 + S \xrightleftharpoons{\text{Hydrogenase}} SH_2$$

Hydrogen production and $N_2$ fixation have shown the close relationship in some of studies, where both the processes are catalysed by the same enzyme or enzyme complex (e.g. *Azotobacter vinelandii*). $H_2$ evolved by nitrogenase of *A. chroococcum* is diffused very quickly to be able to inhibit $N_2$ redution. Therefore, $H_2$ production is an intrinsic activity of nitrogenase. Possession of this activity has been suggested to function as a safty valve through which the cell can dispose excess reducing power.

When substrate differs from $N_2$, nitrogenase of non-phototrophic bacteria catalyses the reduction of these *viz.*, acetylene, protons, cyanide, isocyanide and azide, and produces $H_2$ which is of physiological importance. Some electrons are diverted to produce $H_2$ *in vitro* even in the presence of nitrogen. This phenomenon is attributed to Mo-Fe-protein being less reduced when present in excess amount or low concentration of ATP.

Ammonium salts are required for vegetative growth but in excess amount, it inhibits nitrogenase and declines $N_2$ fixation process. It is obvious that ammonia is generated in microbial cells from $N_2$ and at the same time is inhibitory to nitrogenase activity. Therefore, to continue $N_2$ fixation process, the necessary step is to convert ammonia rapidly to organic nitrogen compounds or be removed from intracellular site of $N_2$ metabolism.

**(*d*) Presence of Self Regulatory Systems :** In diazotrophs nitrogenase activity is regulated by a complex mechanism developed in different microorganisms in different ways. Oxygen is required by aerobic bacteria and evolved by cyanobacteria. However, $O_2$ inhibits nitrogenase activity. To carry out $N_2$ fixation process efficiently, diazotrophs possess certain physiological adaptations as the defence mechanisms so that they can protect nitrogenase and carry out the $N_2$ fixation process.

Azotobacters are among the most $O_2$ tolerant free living diazotrophs. The molecular features that allow this tolerance to $O_2$, include an $O_2$ excluding and protective protein that complexes with nitrogenase *in vitro* during the period of $O_2$ stress. In Azotobacters protection is given to cell by their high respiratory rate. Due to which $O_2$ is prevented from reaching the site of nitrogenase reaction.

Another important feature is the presence of an alternative system (in *A. vinelandii*) for reducing $N_2$ to $NH_3$ when organisms are grown under conditions of Mo-deprivation. This system does not require the conventional *nif* HDK genes which encode nitrogenase Fe-protein and Mo-Fe-protein.

In cyanobacteria, creation of $O_2$ tension is remarkable as the photoautotrophs evolve $O_2$ which inhibits nitrogenase activity. Hydrolysis of ATP in $N_2$ fixing system provides an anhydrous atmosphere around the active site of nitrogenase and keeps enzyme in its most active form and activates electrons.

In heterocystous cyanobacteria, the compartmentation of photosynthesis and $N_2$ fixation takes place. $N_2$ is reduced only in heterocysts where photosynthetic systems (thylakoids) are in reduced amout or absent, therefore, nitrogenase is not inhibited by $O_2$ evolved during photosynthesis.

The heterocysts lack $O_2$ evolving photosystem II apparatus, ribulose biphosphate carboxylase, and may lack or have reduced amount of photosynthetic biliproteins. The walls of heterocyst contain $O_2$ binding glycolipids which, together with respiratory oxygen consumption, maintain the conditions necessary for $N_2$ fixation. In contrast, the vegetative cells of filaments/trichomes contain both photosystem I and II and, therefore, evolve oxygen.

$N_2$ fixation has been reported in non-heterocystous species, for example in *Oscillatoria* and *Lyngbya* where such cells are located in the centre of the colony. These cells do not produce $O_2$ photosynthetically and thickened specialized cell walls as in heterocysts. However, $N_2$ fixation takes place in the same way as in heterocysts. Presence of leghaemoglobin in root nodules denotes regulation of $O_2$ tension and carry out $N_2$ fixation process.

### 4. Mechanism of $N_2$ Fixation and Ammonia Assimilation

Two metalloproteins *i.e.* larger Mo-Fe-protein and smaller Fe-protein components are involved in $N_2$ fixation. Fe-protein interacts with ATP and $Mg^{++}$, and receive electron from ferredoxin or flavodoxin when it is oxidized. Mo-Fe-protein of nitrogenase complex combines with the reducible substrates *i.e.* $N_2$ and yields two molecules of $NH_3$. It appears that $N_2$ is reduced step-wise without breaking N-N bond until the final reduction and production of ammonia are accomplished. Finally two molecules of $NH_3$ are released from the enzyme (Fig. 19.6).

Mechanism of electron transport is shown in Fig. 19.7. Fe-protein (oxidized form) gets electrons from ferredoxin (when it combines with $2H^+$ and yields $H_2$) and energy from ATP. $Mg^{++}$ activates this reaction. Finally, electron is transferred to oxidize Mo-Fe-protein which becomes reduced and Fe-protein is oxidized. It is the reduced form of Mo-Fe-protein which combines with $N_2$ and other substrates to result in $NH_3$ and other various products with respect to substrate. $H_2$ produced during this reaction is further utilized by some microorganisms which possess hydrogenase. Reutilization of $H_2$ increases nitrogenase activity by protecting it from inhibition of $H_2$.

Ammonia is further synthesized into a number of metabolic products in microbial cells. However, ammonia is not accumulated in the cell, although a few species may create it; rather it is incorporated into organic forms by combining with an organic acid (α -keto-glutaric acid) to give rise to amino acid *e.g.* glutamic acid. The ammonia may also combine with organic molecules to yield alanin or glutamine.

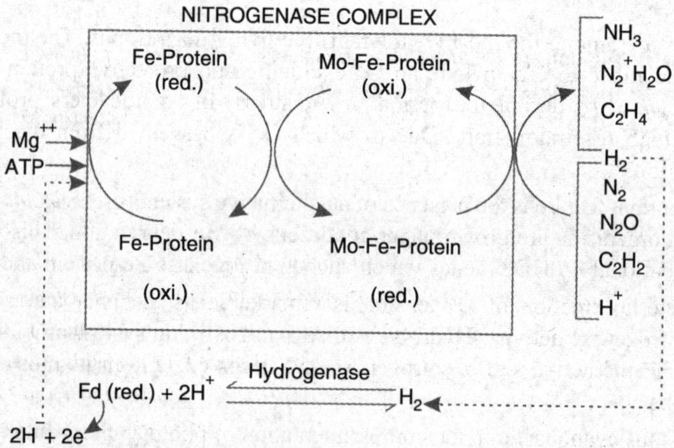


Fig. 19.7. Mechanism of electron flow during $N_2$ (and other substrate) reduction.

ATP is the source of reducing power produced through respiratory process (NADH → ATP) or photosynthetic process (NADPH → ATP). ATP reduces ferredoxin. The source of ATP and reductants differ in different nitrogen fixers. In *Clostridium pasteurianum* pyruvate acts as electron donor to ferredoxin. Sequential reduction of $N_2$ to $NH_4^+$ is shown in Fig. 19.8. ATP required for $N_2$ reduction comes from oxidative metabolism of pyruvate. *Clostridium* is an obligate bacterium. It ferments glucose, and produces butyrate and acetate which releases source of reducing power (ATP).

**$N_2$ fixation in cyanobacteria :** In heterocystous cyanobacteria, heterocycsts are the site for nitrogen fixation. In 1980, R. Haselkorn and coworkers have found that in isolated heterocysts of *Anabaena variabilis,* nitrogenase activity in light depends on $H_2$ supply, whereas in dark condition nitrogenase activity depends on $O_2$ and $H_2$ supply. Glutamine is the main product of $N_2$ fixation

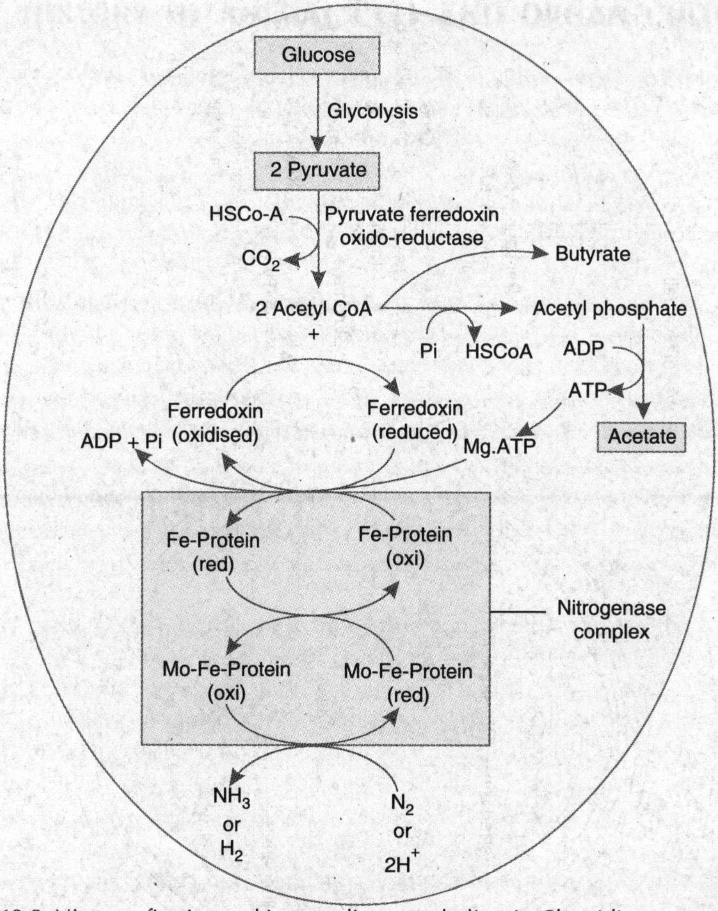


**Fig. 19.8.** Nitrogen fixation and intermediary metabolism in *Clostridium pasteurianum.*

which is translocated from heterocyst to vegetative cells (Fig. 19.9.). However, transfer of glutamate from vegetative cell to heterocyst governs the formation of glutamine.

Maltose is a product of photosynthesis which moves to heterocyst from vegetative cells. Here it is metabolised to glucose 6-phosphate and oxydised by oxidative pentose pathway. In this pathway the reduced NADPH combines with $O_2$ and creates an environment for reduction of ferredoxin which in turn is also reduced by PSI. Electrons ($e^-$) are donated to nitrogenase by the reduced ferredoxin. Nitrogenase reduces $N_2$ to $NH_4^+$ and releases $H_2$. The uptake hydrogenase is present only in heterocyst which reduced $H_2$. A high amount of glutamine synthatase (GS) and low amount of GOGAT (glutamine oxo-glutarate amino transferase) is present in heterocyst. Glutamate react with $NH_4^+$ and forms glutamine. Glutamine enters into vegetative cell and reacts with $\alpha$-ketoglutarate to form two molecules of glutamate; one molecule returns into heterocyst and second molecule takes part in production of other metabolites.

In *Clostridium pasteurianum,* pyruvic acid is produced from glucose-6-phosphate. However, in *R. rubrum* pyruvic acid supports nitrogenase activity releasing ATP. Thus, nitrogen fixed by microorganisms is released into their surrounding.

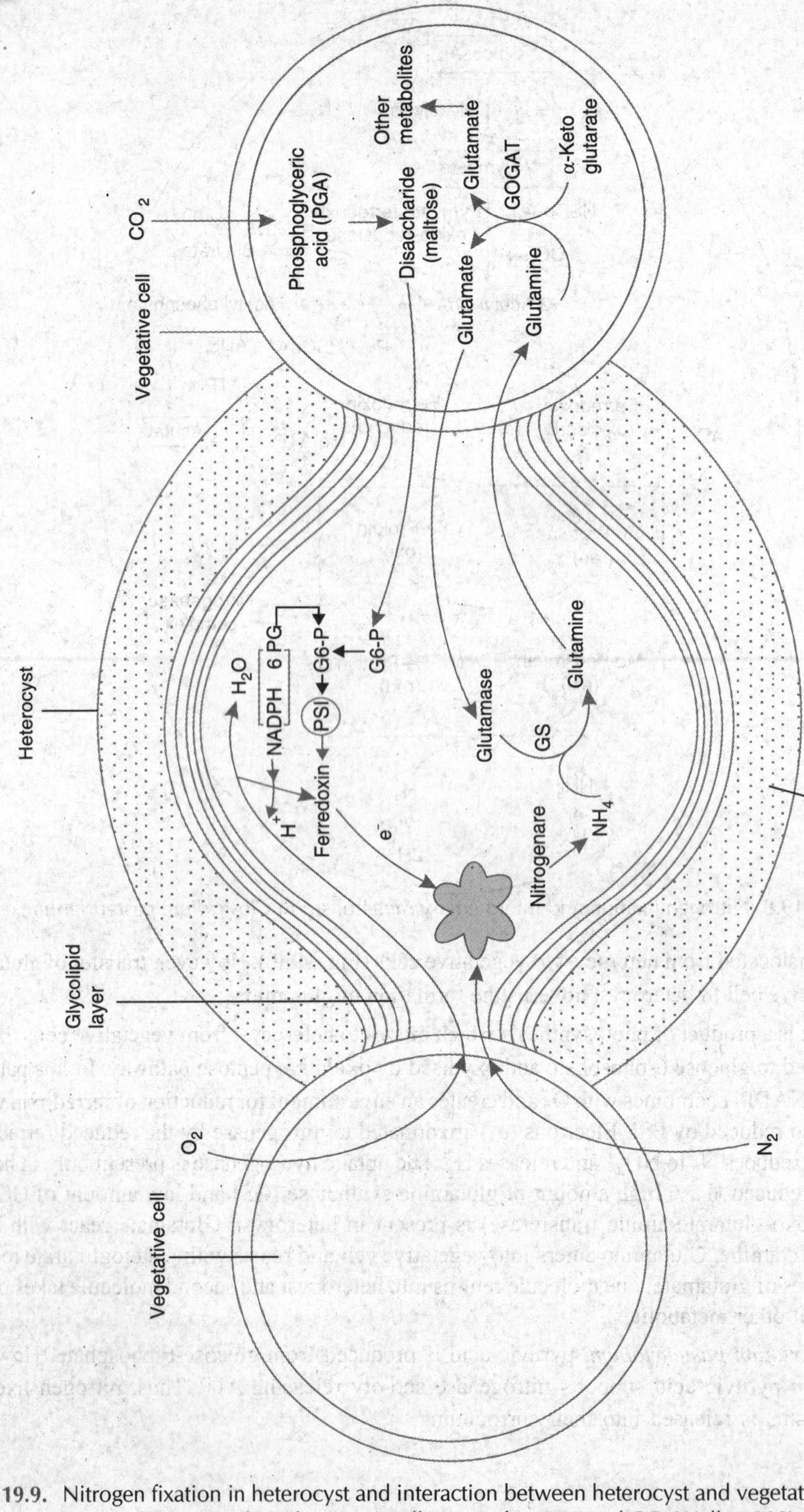


**Fig. 19.9.** Nitrogen fixation in heterocyst and interaction between heterocyst and vegetative cells in a cyanobacterium (based on R. Haselkorn *et al.*, 1980 and C.P. Wolk, 1980).

## B. SYMBIOTIC $N_2$ FIXATION

*Rhizobium* bacteria is found in symbiotic association in root nodules of legumes.

There are some microorganisms which establish symbiotic relationships with different parts of plants and may develop (or may not) special structures as the site of nitrogen fixation. Microsymbionts are given in Table 19.1. Non-nodule forming diazotrophs, for example, *Azotobacter, Beijerinckia, Derxia* are known to be intimately associated with the roots of certain plants. *Azotobacter paspali* is restricted to the roots of the tropical grasses, *Paspalum notatum* and rare to other species of *Paspalum* or other genera. *Beijerinckia* shows host specificity with sugarcane root. On the other hand, *Azospirillum* is known to be associated with the roots of corn, wheat, sorghum, *Digitaria decumbens, Panicum maximum* and *Melinis multiflora,* and a large number of mono-and dicot plants.

*Azospirillum* is Gram-negative aerobes ; it is curved and rod shaped, and has polar flagellum. Bacteria are associated with the grass roots in such a way that a gentle washing do not dislodge the nitrogen metabolizing activity. Based on acetylene reduction it has been calculated that *A. paspali* contributes 15-93 kg N/ha/annum on *P. notatum* roots, and *Beijerinckia* assimilates about 50 kg N/ha/annum on sugarcane root. Other bacteria on corn roots may fix about 2.4 kg. N/h/day. *Azospirillum* increases yield of cereals amounting to a saving of fertilizer nitrogen equivalent to 20-40 Kg/ha.

In addition to these bacteria, *Frankia, Rhizobium* sp. and cyanobacteria undergo symbiosis by getting established inside the plant tissues and may or may not develop special symbiotic structures.

The classical examples of symbiotic association developed by *Rhizobium* sp. are found in about 13,000 or more leguminous plants, both cultivated and non-cultivated herbs, shrub and trees. Among leguminosae, the largest number of plants is in papilionoidae. Moreover, species of *Anabaena, Nostoc, Tolypothrix,* etc. develop symbiotic association with fungi (symbiotic structure is lichen), bryophytes, pteridophytes, gymnosperms and angiosperms and fulfil the requirement of nitrogen deficiency (Table 19.1). However, actinorrhizic nodules are developed by *Frankia,* a member of Actinomycetes, on roots of about 170 species of woody dicot non-leguminous plants like *Alnus, Myrica, Casuarina,* etc. Nodules are of two types: (*i*) *Alnus* type where nodules show dichotomous branching to form a corraloid root, and (*ii*) *Myrica/Casurina* type in which case the apex of each nodule produces a normal but negative geotropic root. The function of nodules is to facilitate gas diffusion to the nitrogen fixing endophyte in the nodule under low $O_2$ tension.

> **Table 19.1.** Nitrogen fixing bacteria (diazotrophs).

| *Association with host* | | *Major characteristics* |
|---|---|---|
| **Free-living diazotrophs** | | |
| *Azotobacter chroococcum* | | Obligate aerobe, heterotroph |
| *A. vilandii* | | Obligate aerobe, heterotroph |
| *Azospirillum lipoferum* | | Obligate aerobe, heterotroph |
| *Clastridium pasteurianum* | | Obligate anaerobe, heterotroph |
| *Desulfovibrio desulfuricans* | | Obligate aerobe, heterotroph |
| *Derxia gummosa* | | Obligate aerobe, heterotroph |
| *Klebsiella pneumoniae* | | Facultative anaerobe, heterotroph |
| *Enterobacter* sp. | | Facultivate anaerobe, heterotroph |
| *Methanobacterium formicicum* | | Obligate anaerobe, heterotroph |
| *Nostoc, Anabaena, Gloeocaps* | | Obligate aerobe, photoautotroph |
| *Rahodopseudomonas* | | Photosynthetic anaerobe |
| **Symbiotic nitrogen fixers** | | |
| *Rhizobium* | | Legume symbiosis, forms root nodules which are the site of $N_2$ fixation, heterotroph |
| *R. leguminosarum* | | Symbiosis with pea |
| *Bradyrhizobium japonicum* | | Symbiosis with soybean |
| *Azorhizobium coulinodans* | | Stem nodule in *Sesbania* |
| *Mesorhizobium ciceri* | | Symbiosis with black-gram (*Cicer aeritinum*) |
| Actinomycetes, *Frankia* sp. | | Root nodules Non-leguminous angiosperms *e.g. Alnus, Casuarina, Myrica, Discaria.* |
| **Cyanobacteria** | | |
| | | *Fungi:* |
| *Anabaena, Nostoc* and *Tolypothrix* | Lichens (*Collema, Peltigera, Usnia*) | Ascomycetous and basidiomycetous fungi |
| *Nostoc* sp. | — | Bryophytes: *Anthoceros, Blasia* |
| *Anabaena azollae* | — | Pteridophytes : *Azolla* |
| *A. cycadae* | Coralloid roots | Gymnosperms : *Cycas* |
| *Nostoc* | — | Angiosperm : *Gunnera* |

*Rhizobium* is a free-living, Gram negative nonsporulating, aerobic and motile rod shaped bacterium (0.5-0.9 μm *x* 1.2 - 3 μm) which resides in soil and is capable of infection, nodulation, establishing symbiosis and $N_2$ fixation. It grows on organic nutrients. *Rhizobia* are more prominant in the rhizosphere of leguminous plants. Based on cross-inoculation studies, six species of *Rhizobium* are defined according to the legume host(s) which they nodulate (Table 19.2). A cross-inoculation group refers to a collection of leguminous species that develop nodules on any member of that particular plant group. Therefore, a single cross inoculation group ideally includes almost all host species which are infected by an individual bacterial strain. Six inoculation groups have been specified so far.

Rhizobia are nitrogen-fixing bacteria that form root nodules on legume plants. Most of the species are in the family Rhizobiacae in the alpha-proteobacteria and are in either the *Rhizobium*,

*Mesorhizobium, Ensifer*, or *Bradyrhizobium* genera. Classification of *Rhizobium* is given as below:

Domain: Eubacteria
Kingdom: Bacteria
Phylum: Proteobacteria
Class: Aplpha Proteobacteria
Order: Rhizobiales:
Family: Rhizobicacea
Genus: *Rhizobium*

In 2012, B.S. Weir published 'the current taxonomy of rhizobia' (http://www.rhizobia.co.nz/taxonomy/rhizobia). According to the current list the $N_2$-fixing species of rhizobia put into the order Rhizobiales are divided into several genera such as : *Rhizobium, Mesorhizobium, Ensifer (formerly Sinorhizobium), Bradyrhizobium, Phyllobacterium, Microvirga, Azorhizobium, Ochrobactrum, Methylobacterium, Cupriavidus*, etc.

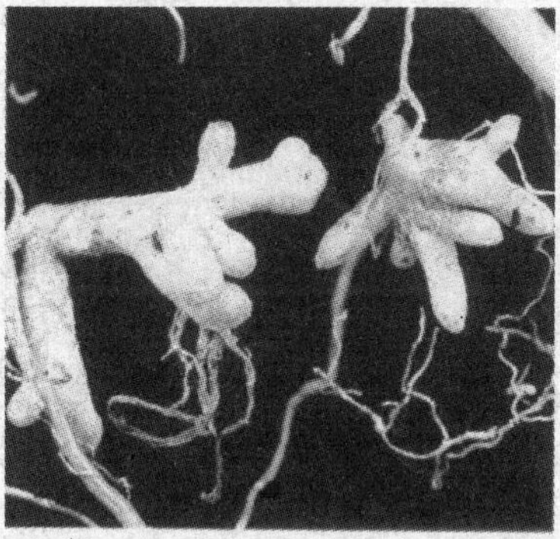

Nodules on the roots of a bean plant resulting from invasion by nitrogen fixing bacteria.

## 1. Establishment of Symbiosis

Establishment of *Rhizobium* inside the host root and development of nodules are a complex process which follow many events such as recognition and infection of host root, differentiation of nodules, proliferation of bacteria and conversion into bacteroids in nodules. These steps are briefly described as below :

**(*a*) Host Specificity and Curling of Root Hairs:** A variety of microorganisms reside in close vicinity, the rhizophere, of roots. Depending on environmental conditions and host susceptibility, the phenomenon of host recognition by the specialized microbe is achieved. The level of specilization for symbiosis differs in different microbial groups and even in the same group as well. As a result of host recognition, *Azospirillum* sp. in some cases are intimately associated with their host and have been isolated from the rhizoplane also. The apparently differntiated structures are not formed, but pictures of root hair deformation are known. Moreover,

**Table 19.2. Species of *Rhizobium* and cross inoculation groups of host plants.**

| *Rhizobium sp.* | *Host genera* | *Cross inoculation groups* |
|---|---|---|
| *R. japonicum* | *Glycine* | Soybean group |
| *R. leguminosarum* | *Pisum, Lathyrus, Vicia,* Lens | Pea group |
| *R. lupini* | *Lupinus, Ornithopus* | Lupin group |
| *R. meliloti* | *Melilotus, Medicago, Trigonella* | Alfalfa group |
| *R. phaseolus* | *Phaseolus* | Bean group |
| *R. trifolii* | *Trifolium* | Clover group |
| Other species of *Rhizobium* | *Arachis, Crotalaria, Vigna, Pueraria* | Cowpea group |

*Azospirillum* can also invades cortical and vascular tissue of host, and enhances the number of lateral root hairs. This results in an increase in mineral uptake, which may be attributed to phytochrome production rather than $N_2$ fixation. Host specificity in *Azospirillum* has also been noted but it differs from that of *Rhizobium.*

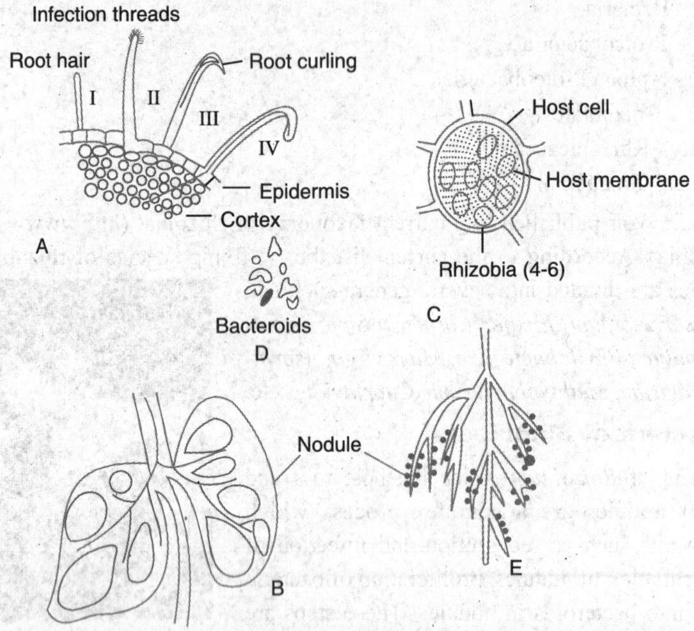


Fig. 19.10. Root nodule formation in a leguminous plant. A- transverse section of a root showing root hairs and its invasion by infection threads of *Rhizobium* sp.; B- vertical section through nodules; C- a cell of the infected host filled with rhizobia; D- various shape of bacteroids; E- root system of a leguminous plant showing numerous nodules.

Host plant secretes exudates in rhizosphere ; subsequently compatible strains of rhizobia are stimulated over the other microbes in soil. Root exudates contain growth stimulating substances like biotin, thiamine, amino acids, etc. Bacteria grow near the root surrounded by mucigel. Mucigel denotes microbial cells and their products together with associated microbial cells and mucilages, organic and inorganic matter in the rhizosphere of root region. The initial host response is observbed as root curling which is known as Shepherd's Cook (Fig. 19.10.A). Root curling takes place due to secretion by *Rhizobium* of a curling factor which includes cytokinin, polymixin B, etc.

**Lectin-mediated root hair binding:** According to current theories the host-microsymbiont specificity is governed by a specific plant protein called lectins (phytoagglutinins) involved in recognition of compatible symbiont. It is supplemented by secretion of specific polysaccharides by the symbiont termed as callose which helps in attachment of rhizobia on host root surface.

Moreover, considerable amount of work has been done on *R.trifolii.* The rhizobial cells that infect clover posses the cross-reactive antigen (CRA). Clover roots absorb CRA and infective *R. trifolii* cells. These sites on roots are called 'receptor sites' that accumulate at the tip of a root hair. Receptor sites decrease towards the base of root hairs. *R. trifolii* cells are agglutinated by a lectin called 'trifolin' but inhibited by 2-deoxyglucose. Hence, the major subject of research is the nature of carbohydrate receptor for lectin present on *Rhizobium* cells.

Some workers are of the opinion that lectin theory of *Rhizobium*-plant recognition does not explain the rhizobial infection into legume roots and establishe symbiosis. But current theory holds that plant roots produce various types of flavonoids that stimulates the release of Nod factor by *Rhizobium.* There are many different Nod factors that control infection specifcity. *Rhizobium* secretes proteins called 'rhicadhesins' which help it to get attached on the surface of root hair. Rhizobial infection depends on secretion of specific flavonids by roots. For example, white clover exudes dihydroxyflavone and alfalfa secretes luteolin along with lectins. Certain *nod* genes of *Rhizobium* is induced by these exudates. This results in secretion of host determinant compounds on the cell wall of *Rhizobium*.

*Rhizobium* polysaccharide is synthesised by many genes. Mutant rhizobia defective in polysaccharide synthesis poorly infect the roots and form nodules. After addition of exopolysaccharide (EPS) from parents to plants in the presence of $EPS^-$ mutant rhizobia, induction of $N_2$ fixing ability by rhiozobia could be retrieved. This shows that polysaccharides are directly involved in the beginning of symbiosis (Fig.19.11)

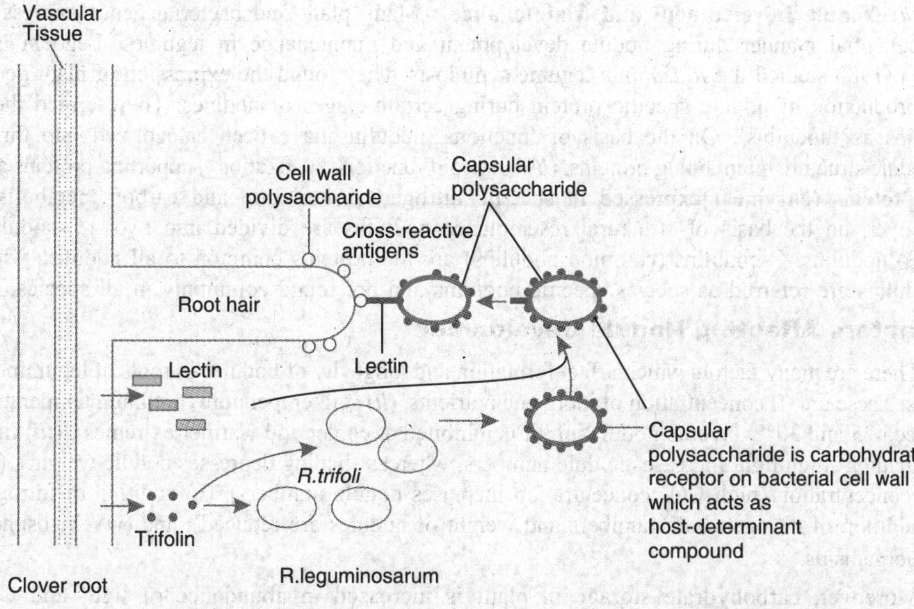


**Fig. 19.11.** Lectin-mediated binding of *Rhizobium trifolii* cells to the wall of root hair of clover plant.

***(b)* Infection of Root Hairs :** Rhizobial aggregates have been observed at distinct sites on curled root hairs (Fig. 19.10 A II). The infection thread is formed by a process of invagination of the hair cell walls in the region of curling. This process of invagination is being repeated at each cell penetrated by infection thread. Root hair wall invaginates until it develops a tube like structure. Deformed hairs are penetrated in the first stage of infection (Fig. 19.10A III). Not all, only a few proportion (about 5%) of young root hairs develop nodules. Following the penetration, a hypha like infection thread is formed. The infection threads resemble to invading fungal hyphae. It is a unique structure which contains a cellulose sheath deposited by the host cell enclosing a strand of hemicellulosic substance in which the bacteria are embeded. If the population of rhizobia is even in the infected tubes, it can be observed under light microscope also. The infection thread grows towards root cortex and ramify throughout the central part of cortex (Fig. 19.10 A IV). It is noteworthy that plasmoptysed hairs have not been observed to form infection threads, although they can be invaded by bacteria. Plasmoptysis may be an important mechanism of root exudation.

***(c)* Nodule Formation :** As the infection thread continues to grow through the root tissue, inner cortical cells are stimulated by bacteria through growth hormone to divide and form an organized mass of infected plant tissue which protrudes from the root surface as a visible nodule (Fig. 19.10B). Rhizobia are released from the infected threads, within some of the cortical cells of nodule and multiply thereby rapid cell division, and ultimately occupy the major central portion of root nodule. The peculiar feature of cells of central nodule is that it contains tetraploid chromosome number. The chromosome doubling of nodule cells may be attributed to secretion of stimulating chemicals by rhizobia.

In the nodules the rhizobia are found to be enclosed in a membrane derived from hypertrophied cells of the host (Fig. 19.10C), where they multiply into 4-6 in number, and finally become enlarged and pleomorphic to form bacteroids (Fig. 19.10D). In morphology, bacteroids appear as swollen, irregular, star shaped, club-shaped, branched or Y-shaped structures. Size and shape of the bacterioids vary in different species of *Rhizobium.* On accomplishment of nodule formation process, the root hairs of plant contain a clusture of nodules (Fig. 19.10E).

**(*d*) Nodule Development and Maintenance :** Many plant and bacterial genes express in an sequential manner during nodule development and maintenance in legumes. Legochi and Verma (1980 studied the *Rhizobium*-legume symbiosis. They found the expression of plant genes and production of nodule specific protein during certain stages of nodules. They termed these proteins as 'nodulins'. On the basis of functions, nodulins have been categorized into three: (*i*) nodule structure maintaining proteins, (*ii*) bacteroid function ($N_2$-fixation) supporting proteins and (*iii*) proteins (enzymes) expressed in specific nitrogen assimilation and carbon metabolism. Moreover, on the basis of structural resemblance nodulins are divided into two : C-nodulins and S- nodulins. C-nodulins (common nodulins) are the proteins common to all nodules, while S-nodulins are referred as species specific nodulins and not found commonly in all species.

## 2. Factors Affecting Nodule Development

There are many factors which affect formation and longevity of nodules in roots of leguminous plants. These are (*i*) concentration of inorganic nutrients, (*ii*) soil temperature (optimum temperature between 25 and 30°C favours nodulation; it is inhibited at cooler and warmer extremes), (*iii*) light and shading (high light increases nodule numbers, whereas shading depresses nodule weight), (*iv*) $CO_2$ concentration (high $CO_2$ concentration increases nodule numbers), (*v*) addition of nitrogen (on addition of nitrogen both numbers and weight of nodules are reduced), and (*vi*) rhizosphere microorganisms.

Moreover, carbohydrate storage in plant is increased in abundance of light and $CO_2$ concentration as a result of which the numbers and weight of nodules are increased. Nitrogen retards carbohydrate concentration; therefore, carbohydrate : nitrogen ratio (CNR) theory may be applicable to support nodulation.

## 3. Mechanism of Nitrogen Fixation in Root Nodules

A characteristic feature of the healthy root nodules of leguminoids plants is the presence of a special pink or red pigment like haemoglobin often known as leghaemoglobin (LHb).

**(*a*) Importance of Leghaemoglobin:** LHb has characteristics similar to myoglobin or a variety of haemoglobin found in animal. It is red in colour due to presence of iron. For the first time LHb was isolated and crystallized from soybean root nodules.

LHb is found only in healthy nodules. The unhealthy plants or white nodules donot develop LHb; therefore, $N_2$ fixation does not takes place in such nodules. LHb is present outside the bacteroid membrane (peribacteroid space) but in their close contact. Recent studies suggest that the peribacteroid membrane may separate the bacteroids from the oxygen buffering system. LHb regulates $O_2$ concentration as bacteroids are aerobic and consume $O_2$. Indirectly LHb promotes $O_2$ utilization in bacteroids and favours nitrogen fixation. The conditions where host plants suffer from $O_2$ deficiency, LHb serves to facilitate the movement of $O_2$ to $O_2$ poor tissues in a proper way. It has been found that at ambient atmospheric concentration (0.2 atm.), $O_2$ becomes a limiting factor for $N_2$ fixation and the reduced $O_2$ diffusion results in prompt inhibition of nitrogenase activity in nodules.

Oxygen levels above 0.5 atm. inhibit $N_2$ fixation due to inactivation of nitrogenase by excessive oxygen. LHb combines with $O_2$ to form oxyleghaemoglobin (OLHb) and makes it available at the surface of bacterial membrane where $O_2$ is diffused into it under low concentration of $O_2$ in root nodule. The haeme protein is highly labile to oxygen. It combines with $O_2$ and converted into *oxyleghaemoglobin.*

$$O_2 + \text{leg-haemoglobin} \rightleftarrows \text{Oxyleghaemoglobin}$$

The oxyleghaemoglobin maintains very low concentration of free oxygen and thus creates oxygen-tense environment. This condition allows nitrogenase to function properly. At the same moment oxygen is delivered to cytochrome oxidase. The nodules respire aerobically. Hence, oxygen is used as terminal electron acceptor.

Thus LHb acts as a carrier of $O_2$ and helps in accomplishment of bacterial respiration and consequent provision of ATP for $N_2$ fixation. Without LHb, diffusion of $O_2$ through thick nodule tissue would be completely inadequate to meet the necessary ATP requirement. Therefore, the efficiency of the nodules to fix $N_2$ can be examined by estimating the concentration of LHb.

LHb is found only in root nodules of legumes. It is not found in actinorrhizic nodules *i.e.* nodules formed by *Frankia* in roots of non-leguminous plants. Therefore, presence of $O_2$ buffering system has not been reported so far in actinorrhizic nodules.

LHb facilitates $O_2$ uptake by terminal oxidases and increases ATP production for nitrogenase activity. It also creates an $O_2$ free environment around the active site of nitrogenase which fails to function in the presence of $O_2$ (Fig. 19.12).

***(b)* How Does Nitrogenase Work ?:** Both the metalloproteins, nitrogenase (Mo-Fe-protein) and nitrogenase reductase (Fe-protein) are essential for nitrogenase activity. Fe-protein interacts with ATP and $Mg^{++}$, and Mo-Fe-protein catalyses the reduction of $N_2$ to $NH_3$, $H^+$ to $H_2$ and acetylene to ethylene. The reduced ferredoxin or flavodoxin serves as a source of reductant for electron transfer during $N_2$ fixation (Fig. 19.12).

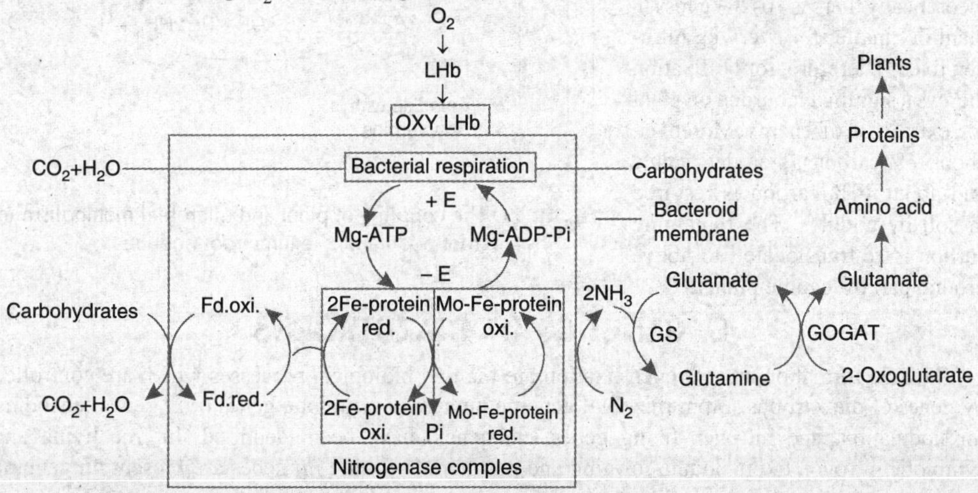


**Fig. 19.12.** Biological nitrogen fixation in root nodule GS, glutamine synthetase; GOGAT, glutamine oxoglutarate aminotransferase.

From reduced form of ferredoxin (Fd res.) electrons ($e^-$) flow to Fe-protein which reduce to Mo-Fe-protein (in nitrogenase enzyme complex) with subsequent release of inorganic phosphate, Pi. This enzyme complex gets energy from Mg- ATP which in turn is produced after bacterial respiration. Finally, Mo-Fe-protein passes on the electron to reducible substrate *i.e.* $N_2$ (or other substrates). The equation of $N_2$ fixation in nodules of legumes may be written as :

$$N_2 + 16\ ATP + 8e^- + 10H^+ \xrightarrow{Mg^{++}} 2NH^+_4 + H^+_2 + 16ADP + 16Pi$$

It is obvious that ammonia is the first stable product of $N_2$ fixation. But it is not clear whether neutral ammonia or cationic ammonium ($NH_4^+$) is formed. Soon after formation, it is transferred through three layered bacteroid membrane (two of bacteria and third of host origin) to host cells, where it is enzymatically converted into many products.

### 4. Energy and Oxygen Relation in Symbiotic Association

Energy is required for the reaction of $N_2$ fixation and host plant supplies the energy. The rhizobia use photoassimilated carbon via the Krebs cycle and generate energy as ATP. Therefore, rhizobia are a carbon drain on their host (Fig. 19.13). In return plant gets fixed nitrogen. For example, one third of the net photosynthate (total assimilated carbon respired photoassimilate) can be derived in pea (*Pisum sativum*) through phloem to the root nodules.

Such energy supply as photosynthate acts as regulator of $N_2$ fixation, It has been estimated that theoretically there is 10-20 kg loss in plant dry matter for every kg nitrogen fixed. Therefore, for $N_2$ fixation there is a significant burden on plant for extra cost of energy Moreover, about 15% carbon is stored in nodule and about 40% carbon is respired in soil by nodules. The remaining carbon is re-translocated to above ground part of legume plants.

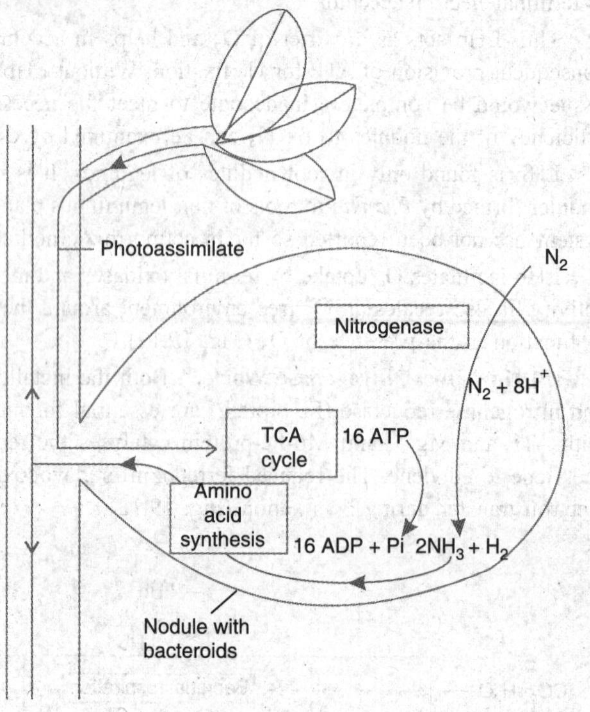


**Fig. 19.13.** The coupling of plant and microbial metabolism in root nodule of a legume root nodules.

## C. GENETICS OF DIAZOTROPHS

Nodule formation and nitrogen fixation are the two biological processes which are controlled by genes of diazotrophs. From rhizobial genome numerous symbiotic genes (*nod* genes) encoding for nodulation, and nitrogen fixing genes (*nif* genes) have been identified. In free living and symbiotic nitrogen fixers nodule-forming and non-nodule formig *nif* genes are present on genome or megaplasmid in their cells. Organization, structure and function of *nod* genes, *nif* genes and *Hup* genes (hydrogen uptake genes) are briefly described below :

### 1. Nod Genes

Nodule forming species of *Rhizobium* consists of an extremely large plasmid known as 'megaplasmid' in cell which possesses numerous genes coding for nodulation (*nod* genes) and nitrogen fixation (*nif* genes). Both *nod* genes and *nif* genes are closely located. However, a physiological map of 135 Kb segment of megaplasmid was established which contained *nod-nif* regions.

The identified *nod* gene region is of 11.5 Kb in length. Nucleotide sequence and proteins encoded by 8.5 Kb fragment in *E. coli* has been determined. The *nod* genes consist of 4 genes designated as *nod* A, B, C and D; the 4 genes code for proteins of 196, 217, 426 and 311 amino acid residues, respectively. Based on comparisons of nucleotide and amino acid sequences of *nod* A, B and C genes and amino acid sequence of *nod* A, B and C genes between different species of *Rhizobium,* about 69-72% homologous region has been characterized which is known as *common nod genes*.

The structural and functional conservation of *common nod genes* has become a tool to identify common *nod* genes of other species of *Rhizobium.* Recently, about 25 Kb fragment of megaplasmid containing all essential *nod* genes has been identified and a recombinant plasmid (pPP346) has been constructed. *A. tumifaciens* cells became able to develop nodules on alfalfa roots when this plasmid was incorporated.

## 2. *Nif* Genes

A region lies on genetic material of free living and symbiotic $N_2$ fixers which consists of nitrogen fixing *nif* genes. In *Rhizobium leguminosarum, R, meliloti* and other species *nif* genes are located on a megaplasmid adjacent to *nod* genes,whereas in cyanobacteria *e.g. Anabaena* 7120 and most of free living bacteria *nif* genes are localized on the chromosome. Early studies on *nif* genes were carried out in *Klebsiella pneumoniae* and the function was confirmed by transferring into *E. coli* cells. Thereafter, *nif* genes in *Rhizobium*, cyanobacteria and other $N_2$ fixers were discovered. The *nif* gene products and their function in *K. pneumoniae* is given in Table 19.3.

**Table 19.3. Nif gene products and their funçtion in Klebsiella pneumoniae**

| *Genes* | *Function* |
|---|---|
| Q | Not known |
| B | Synthesis or processing of FeMoCo |
| A | Regulatory-*nif* A product activates the other operons |
| L | Regulatory |
| F | A flavoprotein involved in electron transfer in nitrogenase |
| M | Processing of Fe-protein |
| V | Influences specificity of MoFe-protein |
| S | Not known |
| U | Not known |
| X | Not defined genetically; presence deduced from physical map and cloning of *nif* DNA |
| N | Like B |
| E | Like B |
| Y | Discovered in the same way as X |
| K | Codes for B-subunit of MoFe-protein |
| D | Codes for subunit of MoFe-protein |
| H | Codes for subunit of Fe-protein |
| J | May be involved in electron transfer to nitrogenase |

Researches were done on physical mapping to know the product of *nif* gene clustuer. It was found that there are several parts of *nif* genes forming a gene clusture of 24 Kb nucleotides which are located between the genes encoding for histidine (his) and shikimic acid (shi A). (Fig. 19.14). This clustuer is organized in 7 operons *i.e.* transcription units (*e.g.* QB AL FM VSUX NE YKDH J). On the basis of mutational studies performed in all *nif* genes, the nature of various products of *nif* genes was determined. Now it is confirmed that *nif* HDKY operon encodes nitrogenase, whereas *nif* LA has regulatory function. In *nif* HDKY genes H, D and K encode for subunit of Fe-protein, Mo-Fe-Protein and subunit of Mo-Fe-protein, respectively.

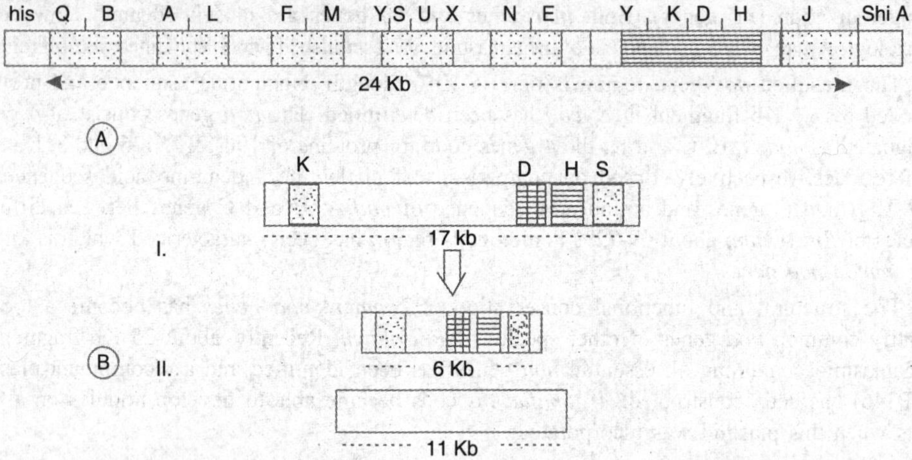


**Fig. 19.14.** A diagram of *nif* gene cluster of *Klebsiella pneumoniae* (A) and *Anabaena* sp. (B). I-*nif* genes in vegetative cells; II-*nif* genes in heterocyst after rearrangement.

Nitrogenase acts only in the absence of ammonia and other nitrogen compounds as they inhibit its expression. Glutamine synthatase (GS) also losses its function. Glutamine synthatase gene encodes GS enzymes which is quite apart from *nif* genes.

Some filamentous cyanobacteria are composed of entirely vegetative cells. In absence of combined nitrogen source some photosysthetic vegetative cells differentiate into heterocysts at regular intervals along the filaments, terminal or lateral singly or in chains. In heterocysts, during differentiation many morphological, biochemical and genetical changes occur. At this time induction of nitrogenase takes place.

In *Anabaena* 7120 *nif* H, D and K have been identified with DNA probe of *K. pneumoniae* through hybridization. However, during heterocyst differentiation two DNA rearrangements occur. In *Anabaena,* organization of *nif* H, D and K genes differs in vegetative cells from that of *K. pneumoniae* where three genes (S, H and D) are contiguous and form an operon. *Nif* H is in close proximity to *nif* D, while *nif* K is 11 Kb apart from *nif* D (Fig. 19.14 B).

Organization of *nif* genes of vegetative cells and heterocysts was compared. It was found that in heterocysts the *nif* K and *nif* H were closely attached; size of *nif* gene clustuer reduced from 17 Kb (in vegetative cell) to 6 Kb (in heterocyst) (Fig. 19.14 B). This was proved by hybridizing the *nif* gene fragment with some of the *nif* gene probes by using restriction enzymes.

In recent years, *Azospirillium* has attracted attention of workers for being as a possible source of biofertilizer due to presence of *nif* HDK clusture like *K. pneumoniae.* Hybridization experiment between *nif* probe of *K. pneumoniae* and total DNA of many strains of *Azospirillum* has confirmed the presence of *nif* HDK and *nif A* genes. Like *Rhizobium* sp., *Azospirillum* sp. also contain a megaplasmid and the sequence homologous to *nod* genes originated from a common ancestors

**(*a*) Cloning of *nif* Genes :** In many countries researches on *nif* gene transfer into higher plants, especially in monocots, and gene expression in them are in progress. However, success has been made in *nif* gene cloning into *E. coli.* Since *nif* genes are prokaryotic in origin, the best strategy of their transfer into non-leguminous crops would be to transfer *nif* genes into chloroplast. The transcriptional and translational machinery of chloroplast bears several prokaryotic features. These attempts would be successful because the chloroplasts are geared to the production of ATP and reducing power, as both of which are required for nitrogen fixation. But the major problems for doing so are (*i*) lack of chloroplast transferring techniques, and (*ii*) protection of nitrogenase from $O_2$ evolved during photosynthesis. Some more aspects of *nif* gene cloning have been discussed elsewhere (*see* Chapters 7 and 13).

### 3. Hydrogen Uptake or Hup Genes

In some species of *Rhizobium,* hydrogen uptake (*Hup*) genes have been reported which displayed the ability to recycle $H_2$ (produced as a result of conversion of $N_2$ to $NH_3$) back to nitrogenase complex. This mechanism helps the plant to harvest the energy as being lost by the plants (Fig. 19.5).

Most of legumes loss 30-50 per cent of their energy as $H_2$ gas which inturn reduces the efficiency of $N_2$ fixation. Recently, Indian Scientists at IARI (New Delhi) have produced a genetically engineered $N_2$ strain by transferring the *Hup* genes of *R. leguminosarum* into *Rhizobium* strain (which did not contain *Hup* genes). This strain infected the roots of chick-pea and developed root nodules. *Hup* system recycled $H_2$ and reduced energy losses by 8-13 per cent. This is the world's first case of interspecific transfer of *Hup* genes. The successful transfer and expression of *Hup* genes has increased the possibility of improving symbiotic energy efficiency of chick-pea- *Rhizobium* system.

### 4. Nodulin Genes

Some specific polypeptides called nodulins are secreted by the chromosomal DNA in the root nodules of the plant. About 20 nodulins are secreted inside the nodule. But the function of only a few of them is known in detail. Basically all the nodulins are divided into three groups: I, II and III.

**Group I nodulins** (e.g. nodulins 23, 24, 26 and 36) includes structural polypeptides of nodules.

**Group II nodulines** (e.g. uricase, glutamine synthetase) are the enzymes associated with uride metabillism that convert $NH_3$ into urides. In turn, urides are connected into amino acids and nucleotides. For commencement of $N_2$ fixation continuously, they reduce the level of $NH_3$ inside the root nodule.

**Group III nodulins** include proteins that regulate the reduction of nitrogen *e.g.* leg-haemoblobin.

## PROBLEMS

1. What do you know about non-symbiotic nitrogen fixation? Write in brief the special features of diazotrophs.
2. Write an essay on "Biofertilizers as an alternative of chemical fertilizers".
3. Oxygen inhibits $N_2$ fixation. How does the process of $N_2$ fixation go on in nonheterocystous cyanobacteria?
4. What is nitrogenase? Discuss in detail the structure and function of nitrogenase.
5. Write an essay on 'development of self protection mechanism in diazotrophs'.
6. What do you know about leghaemoglobin? What role do they play in $N_2$ fixation?
7. Give an illustrated account of process of nodulation. How does a microsymbiont recognise its host?
8. Give an illustrated account of mechanism of $N_2$ fixation in free living microbes.
9. Write an eassy on 'symbiotic $N_2$ fixation'.
10. What do you know about genetics of diazotrophs? Give a breif account of structure of *nif* genes in bacteria and cyanobacteria studied by you.
11. Write short notes on the following :
    (*a*) Cyanobacteria, (*b*) Heterocysts, (*c*) Nitrogenase, (*d*) Leghaemoglobin, (*e*) *Rhizobium*, (*f*) Megaplasmid, (*g*) *Frankia*, (*h*) *Nif* gene cloning, (*i*) *Hup* gene, (*j*) Nodulin genes, (*k*) Energy and oxygen relations in symbiotic association, (*l*) Lectin-mediated root hair binding by *Rhizobium*, (*m*) Nodulin genes.

CHAPTER 20

# Biofertilizers (Microbial Inoculants)

During the early of World war I, Fritz Haber (a German Chemist) chemically synthesised nitrogen and hydrogen into ammonia. The sole method of nitrogenous fertiliser production in the world is still the 'Haber-Bosch process'. This process requires the maintenance of temperature up to 800°F, a catalyst and high pressure (as compared to atmospheric pressure).

Nitrogenous fertilizers produced in industry by Haber-Bosch process consume high energy (about 13,500 K Cal/kg N fixed). In such industries, fossil fuel is the source of energy. In recent years, due to Gulf crisis, the cost of crude oil increased about three fold within a year. Therefore, fossil fuel (oil and coal) based method of farming has becoming more expensive accordingly. To combat with this problem, however, it is necessary to develop an alternative method of supplying nutrients to plants.

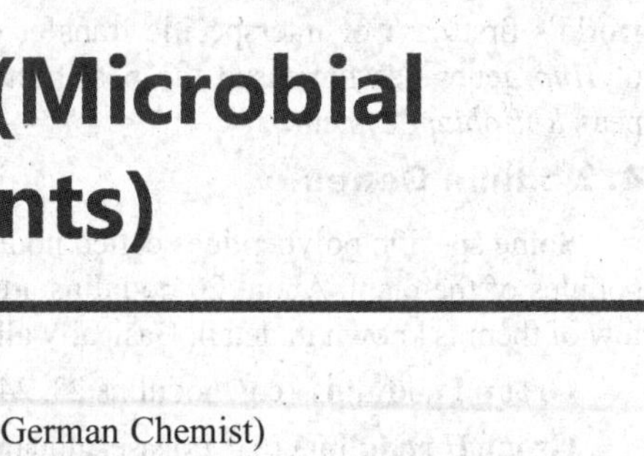

Biofertilizers induce plant growth and increased crop yield.

In recent years, use of microbial inoculants as a source of biofertilizers has become a hope for most of countries, as far as economical and environmental view points are concerned. Biologically fixed nitrogen is such a source which can supply an adequate amount of nitrogen to plants and other nutrients to some extent. It is a non-hazardous way of fertilization of field. Moreover, biologically fixed nitrogen consumes about 25 per cent to 30 per cent less energy than normaly done by chemical process.

Therefore, in developing countries like India, it can solve the problem of high cost of fertilizers and help in saving the economy of the country. The term 'biofertilizers' denotes all the "nutrient inputs of biological origin for plant growth" (Subba Rao, 1982). Here biological origin should be referred to as microbiological process synthesizing complex compounds and their further release into outer medium, to the close vicinity of plant roots which are again taken up by plants. Therefore, the appropriate term for biofertilizers should be **microbial inoculants** as suggested by N.S. Subba Rao in 1982. As bacteria and cyanobacteria (also *Frankia*) are known to fix atmospheric nitrogen, both bacteria and cyanobacteria are widely used as biofertilizer.

During 1989-1999 (10 years), a National Project on Development and Use of Biofertilizers was implanted by the Ministry of Agriculture, Govt. of India through the National Biofertilizer Development Centre (Ghaziabad, U.P.). Biofertilizer-wise crop response showed that the contribution of nitrogenous biofertilizers was in the order of *Azolla* > *Rhizobium* > cyanobacteria > *Azospirillum* > *Azotobacter.* Use of phosphatic biofertilizers (PSM) also showed increased crop response. But nitrogenous biofertilizers + PSM resulted in higher crop response

Beside benefiting N and P nutrition, the microbes benefited plant growth and increased crop yield due to secretion of phytohormones like auxins. Giberellins, cytokinins (*Azospirillum, Azotobacter, Rhizobium,* cyanobacteria, *Bacillus* and *Pseudomonas*). Some microbes also acts as plant growth-promoting rhizobacteria which control plant pathogens also. Biofertilizers and their beneficiary crops are given in Table 20.1.

**> Table 20.1. Biofertilizers and their beneficiary crops.**

| *Name of biofertilizers* | *Beneficiary crop* |
|---|---|
| **A. Nitrogen Biofertilizers** | |
| *Rhizobium* | Crop specific biofertilizers for legumes like groundnut, soybean, red-gram, green-gram, black-gram, lentil, cowpea, Bengal-gram and fodder legumes |
| *Azotobacter* | Cotton, vegetables, mulberry, plantation crops, rice, wheat, barley, ragi, jwar, mustard, safflower, sunflower, tobacco, spices, condiments, ornamental flowers |
| *Azospirillum* | Sugarcane, vegetables, maize, perlmillet, rice, wheat, fodder, oilseed, fruits, flower |
| Cyanobacteria | Rice |
| *Azolla* | Rice |
| **B. Phosphate Biofertilizers** | |
| Phosphate solubilising microorganisms | All crops |

## A. BACTERIA INOCULANTS

Many free-living and symbiotic bacteria are discussed in Chapter-19 which fix atmospheric $N_2$. Therefore, certain measures are adopted to increases number of such bacteria in soil which may increase the gross yield of nitrogen. The two methods, bacterization and green-manuring are the most widely used techniques.

**Bacterization** is a technique of seed-dressing with bacteria (as water suspension) for example, *Azotobacter, Bacillus, Rhizobium* etc. It has been proved that bacteria can successfully be established in root region of plants which in turn improve the growth of hosts. Bacterial fertilizers named 'azotobakterin' (containing cells of *Azotobacter chroococcum*) and 'phosphobacterin' (containing cells of *Bacillus megaterium* var. *phosphaticum*) have been used in U.S.S.R. and East European countries, respectively. These increased the crop yield about 10-20 per cent. Subsequently bacterization of seeds in Russia, Czechoslovakia, Rumania, Poland, Bulgaria, Hungary, England and India has clearly demonstrated the increase in crop yield such as wheat, barley, maize, sugarbeet, carrot, cabbage and potato. In rhizosphere, bacteria secrete growth substances and antibiotic secondary metabolites which contribute to seed germination and plant growth.

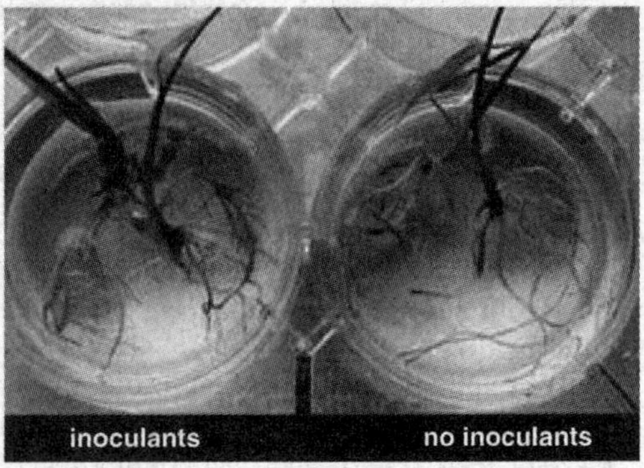


*Azospirillum* enhances plants growth and crop yield by fixing nitrogen and phytohormone synthesis.

Moreover, informations gathered on associate symbiosis (*i.e.* symbiosis between roots of grasses and *Azospirillum*) has increased the interest on this bacterium to be used as seed inoculant for cereals. In recent years, free-living bacteria (*Azotobacter*), associate (*Azospirillum*) and symbiotic (*Rhizobium*) bacteria, and phosphate solubilizing ones (*e.g. Bacillus megaterium, B. polymyxa,* and *Pseudomonas striata*) are gaining much popularity. Such practices are being encouraged to save the chemical fertilizers, national economy and the environment.

## 1. Rhizobial Inoculants

For the first time Nobbe and Hiltner in 1895 introduced a laboratory culture of rhizobia with the name 'Nitragin'. It was the first rhizobial industry in the world. Later on much time was devoted for isolation and characterisation of strains and optimisation of growth conditions. Rapidly expanding knowledge of *Rhizobium*-legume symbiosis expanded rapidly and contributed towards better usage of this symbiosis for sustainable crop yield.

**(*a*) Isolation of *Rhizobium*** : Different species of rhizobia reside in soil as well as in root nodules of legumes. Therefore, they can be isolated either from soil or root nodules. But root nodules are preferred for getting contamination-free rhizobia. Healthy root nodules are taken out from the mature plant and washed with sterile distilled water to remove soil particles. Nodules are externally sterilised either by mercurric chloride or sodiul hypochlorite solution or 90% ethanol. For detail see *Practical Microbiology* by Dubey and Maheshwari (2012).

Clean nodules are crushed in sterile water and suspension is streaked on Petri plates containing YEMA (yeast extract mannitol agar; the constituents are : yeast extract, 1.0 g; mannitol, 10.0 g; $K_2HPO_4$, 0.5 g; $MgSO_4.7H_2O$, 0.2 g; NaCl, 0.1 g; agar, 20 g; distilled water, 1 litre; *p*H, 0.5-7). YEMA plates are incubated for 3-4 days at 28-30°C. White, translucent, glistening and elevated (with entire margin) colonies appear on the surface of medium.

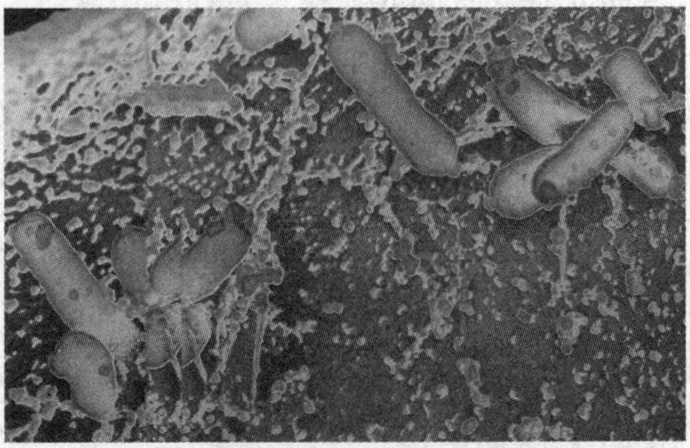

*Rhizobium* cells used as biofertilizer.

**(*b*) Identification of *Rhizobium*** : Both *Rhizobium* and *Agrobacterium* grow on YEMA medium. One can distinguish these two genera using the following tests:

**(*i*) CRYEMA test** : CRYEMA medium is prepared by mixing 2.5 ml congo red dye with one litre of YEMA medium. Bacterial colonies growing on YEMA are streaked over CRYEMA medium and plates are incubated at 28-30°C for about a week. *Rhizobium* utilises congo red dye very slowly and form white, circular and raised colonies. In contrast *Agrobacterium* displays colony characteristics like *Rhizobium* but the colony colour is similar to congo red. Hence, white coloured colonies are isolated and rhizobial inoculants are produced.

**(*ii*) Microscopic observation :** Bacterial colonies growing on CRYEMA are stained with carbol fuschin and observed microscopically, The β-polyhydroxybutarate (PHB) of *Rhizobium* is stained. Thus the colonies PHB are picked up to prepare *Rhizobium* inoculant.

**(*iii*) Glucose-peptone agar (GPA) test :** It is the confirmative test of *Rhizobium.* A master plate is prepared using the bacterial colonies on YEMA. Replica plating from master plate is done on Petri plate containing GPA medium. *Agrobacterium* grows well on GPA medium but *Rhizobium* fails to grow on this medium. Hence, comparing with master plate, *Rhizobium* colonies are picked up for preparation of inoculum in mass.

**(*iv*) Salt tolerance test :** *Agrobacterium* is able to grow on YEMA medium containing 2% NaCl, whereas *Rhizobium* cannot grow on such medium.

**(*v*) Lactose test :** *Agrobacterium* utilises lactose by secreting the bacterial enzyme ketolactose, whereas rhizobia cannot use lactose. Colonies containing lactose agar are streaked and incubated for 4-10 days. Thereafter, plates are flooded with Bendicts's reagent. Formation of yellow colour shows the presence of *Agrobacterium.*

**(*c*) Starter Culture of *Rhizobium*:** YEM broth medium is prepared and autoclaved by transferring in a flask. Thereafter, pure *Rhizobium* colony is transferred into sterilised YEM broth. Inoculated YEM broth is incubated on a rotary shaker at 28-30°C. After 4 days sufficient number of cells are present in YEM broth. It is called **mother culture** or **starter culture.** Later on the starter culture is transferred to a seed tank fermentor and incubated for 4-9 days. It can be maintained for a long time after subsequent sub-culturing. It has to be preserved for long time. It may be done following microbiological technique. For detail see *A Textbook of Microbiology* by Dubey and Maheshwari (2012).

**(*d*) Mass Culture of *Rhizobium*:** For mass cultivation of rhizobial inoculant, broth medium is prepared in large quantity and transferred in a large production fermentor. The *p*H of the medium is adjusted to 6.5 to 7.0 by using KOH or $H_2SO_4$ solution. Following are the steps of mass cultivation of *Rhizobium:* (*i*) sterilize the growth medium and inoculate with broth of mother culture prepared in advance, (*ii*) incubate for 3-4 days at 30 - 32°C, (*iii*) test the cultures for its purity and transfer to a large fermenter,wait for 4-9 days for bacterial growth (for good bacterial growth make the device for its aeration), (*iv*) allow to grow the bacteria either in a large fermenter containing broth or in small flasks as per demand, and (*v*) check the quality of broth.

**(*e*) Measuring Cell Counts in Broth:** Cell counting of the fermentor broth must be done by using the serial dilution plate method. Rhizobial cell counts of different strains vary. *R. japonicum* counts ($5 \times 10^9$ cells/m1) may be attained in 96 hours with a lag phase of 48 hours. However, the lag phase can be reduced by increasing the inoculum level. For example, by using the initial inoculum of *R.: meliloti* in the fermentor by 5% of volume of medium a lag phase of 4 hours can be achieved to get cell counts of $5 \times 10^9$.

When cell count has reached to $10^8 - 10^9$ cells/m1, the broth is taken out from the fermentor and used as inoculant. If broth is left in fermentor, the cell death will start due to lack of nutrient in broth. Objective should be such that in minimum time viable cell counts to about $2 - 4 \times 10^9$ cells/ml could be attained by regulating certain factors such as aeration, volume of medium, initial level of inoculum bacterial strain, temperature and incubation time.

**(*f*) Preparation of Carrier-Based Inoculum and Curing:** A carrier is an inert material used for mixing with broth so that inoculants can easily be handled, packed, stored/transported and used.

For this purpose a variety of carriers are blended with broth. Characteristics of carriers used for rhizobial inoculants in India are given in Table 20.2.

**> Table 20.2. Characteristics of carriers used for rhizobial inoculants in India.**

| *Carrier* | *Availability* | *Location* | *Water holding capacity (%)* | *Organic Matter (%)* |
|---|---|---|---|---|
| Charcoal | Plenty | Every where | 90 | 77 |
| Farm yard manure | Plenty | Every where | 40-50 | 30 |
| Lignite | 3 million tonnes annually | Neyveli | 92 | 28 |
| Peat | 3-5 million tonnes (total) | Ootacamund (Nilgiri Hills) | 52-120 | 21-56 |

Source: K.V.B.R. Tilak (1991)

A variety of carriers are used,for example, peat, lignite, farmyard manure, charcoal powder, etc. In India powdered farmyard manure and charcoal powder are good carrier and an alternative to peat and lignite. Good quality of carrier culture is that which contains sufficient amount of rhizobial cells *i.e.* $1000 \times 10^6$ to $4000 \times 10^6$ rhizobia/g carrier.

The carrier is powdered and dried in sun to get 5% moisture level. Then it is screened through 100-200 mesh sieve and neutralised by mixing with calcium carbonate powder. If the carrier is neutral, there is no need of mixing $CaCO_3$ powder. The carrier is sterilised by autoclaving at 15 PSI (pound per square inch) for 3 hours and dried. The harvested broth obtained as above is mixed with carrier (kept in trays or tubes) with hand or mechanically. The moisture content is maintained to about 35-40% on dry weight basis.

After proper mixing carrier containing inoculant is left for 2-10 days by covering the trays with polythene at 22-24°C. During this period *Rhizobium* multiplies in the remaining broth. This process is called **curing.** Care should be taken that rhizobial count should be $3 \times 10^8$ cells/g of carrier, Thereafter, *Rhizobium* inoculant can be used directly or packed and stored.

**(*g*) Packing and Storage :** The cured carriers are packed in polythene bags and kept at a constant room temperature for about a week to facilitate the rhizobial cells to get established. Care should be taken to store the packets at about 4–15°C in a cold room so that rhizobial cells may remain viable for more than 6 months. The packets should not directly be exposed to sunlight or heat otherwise rhizobial cells will be killed. The rate of cell death is influenced by type of inoculum and age of the cell culture.

**(*h*) Quality Control of Rhizobial Inoculants :** Inoculum quality means the number of effective rhizobial cells present in carrier. Quality control includes: (*i*) strain testing, selection skill and control of inoculum density in broth before mixing with carrier, (*ii*) control of carrier-based culture after manufacture, and (*iii*) control of culture during storage period (Tilak, 1991).

There are different types of inoculants available in market produced by different companies. They have their own quality control. The Indian Standard Institution (ISI) (now named as Bureau of Indian Standard, BIS) have prepared certain quality standards which are being followed by the manufacturers in private sector. An ISI mark is issued to manufacturers for the certified inoculants. For rhizobial cultures the relevant clauses from ISI specifications are given below :

(*i*) The inoculants shall be carrier-based

(*ii*) The inoculants shall contain at least $10^8$ viable cells of *Rhizobium* per gram of carrier within 15 days of manufacture, and $10^7$ cells within 15 days before the expiry period date after storing at 25-30°C (*iii*) The expiry period of inoculant (from the date of manufacture) must be 6 months

(*iv*) The inoculum must be contamination-free

(*v*) The inoculant must display effective nodulation on the host plants given on the packet

(*vi*) The manufactures must control the quality of broth and maintain records of the test done specifying records fo isolation and identification of *Rhizobium,* record of nodulation ability, and effectiveness of pure cultures

(*vii*) Each packet must furnish the following information:

- Product's name
- Name of host plants for which to be used
- Name and address of manufactures
- Type of carrier used
- Batch number
- Date of manufacturer and date of expiry
- Net quantity meant for 0.4 hectare
- Instructions for storage

(*viii*) The packets must be marked with ISI mark

(*ix*) The manufacturers must store the inoculants at about 15°C

**(*i*) Methods of seed inoculation with rhizobial culture.** Seed inoculation with aqueous suspension of carrier culture during sowing has revealed the luxuriant nodulation and good yield of crops. The steps of seed inoculation with rhizobial culture are given in Fig. 20.1. Dissolve 10 per cent sugar or *gur* (jaggery) in water by boiling it for some time. Leave the content to cool down. Gum arabic solution (10% ) may also be added to the solution. This serves as sticker for *Rhizobium* cells to seeds. Mix this carrier based culture of *Rhizobium* to form the inoculum slurry. For one hectare, 400 g charcoal based culture would be sufficient for mixing the seeds. Transfer the inoculum slurry on seeds and mix properly. The number of rhizobial cells/ seed should be between $10^5$ to $10^6$. Spread the seeds in shade for drying on cement floor or plastic sheets.

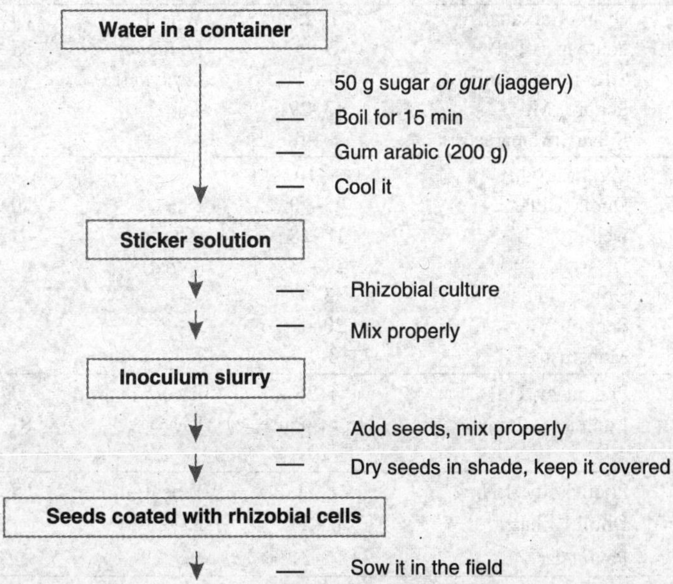


**Fig. 20.1.** Procedure for seed inoculation with rhizobial culture.

While using rhizobial cultures, certain precautions are taken into account. For example, use of culture before expiry date, use of small amount of pesticides when required, immediate sowing of seeds after mixing, etc. Seeds must be stored at 4°C when not used immediately to protect the rhizobial cells.

(*j*) **Pelleting.** When soil has the adverse conditions such as dryness, acidity, excess fertilizers and pesticides, etc., the rhizobial cells are protected by adopting special method of inoculation. One of these methods is pelleting *i.e.* preparation of pelleted seeds. This method involves the procedure as described earlier. High amount of gum arabic (40%) or carboxymethylcellulose (20%) is added to the inoculum slurry before mixing with seeds. Finally, pelleting agent is mixed when inoclated seeds are moist (before seed drying ) to get the seeds evenly coated. The commonly used pelleting agents are calcium carbonate, rock phoshate, charcoal powder, gypsum and betonite.

(*k*) **Effects of Rhizobial Inoculants on Crop Yield**. Effect of rhizobial culture on yield of different pulses, and subsequent crops grown after harvesting the pulse crops are given in Table 20.3 based on the study made under All India Co-ordinated Pulse Improvement Research Programme of I.C.A.R., New Delhi. Yield of pulse crops can be substantially increased by rhizobial inoculation. Legume crops get benefit from rhizobial symbiosis. In addition, certain amount of N is left over in soil which is taken by the other plants also. The effect of residual nitrogen of many legumes on the yield of subsequent crops of wheat or rice has been found. More yield of subsequent crops in *Rhizobium* inoculated fields has been noticed than in uninoculated control.

**Table 20.3. Response of different pulse crops to rhizobial cultures under different agroclimatic condition and residual effect on yield of subsequent crops.**

| Crops | Location | Crop response (range in % increase in grain yield (q/ha) over UI control)* | Yield of subsequent crops (% increase in yield over control, soil pH 7.3)** | | |
|---|---|---|---|---|---|
| | | | Crops | Yield (q/ha) | % increase over control |
| Arhar | Hisar, Haryana | 5 -25 | Wheat | UI–20.75 | 16.4 |
| (*Cajanus* | Pantnagar, U.P. | 2–25 | (*Triticum* | RI–24.15 | |
| *cajan*) | S.K. Nagar, Gujarat | 9–21 | *aestivum*) | | |
| | Sehore, M.P. | 13–29 | | | |
| | Rehari (Maharastra) | 3–40 | | | |
| Chickpea | Varanasi, U.P | 4–19 | | | |
| (*Cicer* | Dholi, Bihar | 25–40 | Rice | UI-25.15 | 7.9 |
| *aritinum*) | Delhi | 18–28 | (*Oryza* | RI–27.15 | |
| | Hisar | 24–43 | *sativa*) | | |
| | Dohad, Gujarat | 33–67 | | | |
| | Sehore | 20–41 | | | |
| | Maharastra | 8–12 | | | |
| Lentil | Pantnagar, U.P | 4–26 | Rice | UI–22.57 | 13.2 |
| (*Lens* | Ludhiana, Punjab | No response | Rice | RI–25.55 | |
| culinaris | | | | | |
| Urd bean | Pudukkotti, T.N. | 4–21 | Wheat | UI–20.75 | 2.4 |
| (*Vigna munga*) | Dholi, Bihar | 11–29 | RI–21.25 | | |
| | Pantnagar | 17–21 | | | |

* Rewari (1984, 1985); ** Subba Rao and Tilak (1977); UI, Uninoculated control; RI, Inoculated with *Rhizobium* culture.

## 2. *Azotobacter* Inoculants

For the frst time Beijerinck isolated and described *Azotobacter chroococcum* and *A. agilis.* Later on *A. agilis* was renamed as *Azomonas. Azotobacter* is a free-living nitrogen fixing bacterium which resides in soil and rhizosphere and fixes nitrogen in association with the other hosts. It is a heterotrophic bacterium that derives energy by degrading plant residues. Besides, strict anaerobic nitrogen fixing species are the genera: *Chlorobium, Clostridium, Chromatium* and *Desulfovibrio.* Species of *Azotobacter* are found in slightly acidic (*A. beijerincki*) to neutral and alaline (*A. chroococcum*) soils. *Azotobacter* fixes 20–40 kg N/ha/annum. It also produces IAA, gibberellic acids, vitamins, etc. Hence, it is recommended as biofertilizer for rice, wheat, millets, cotton, etc. (Table 20.1).

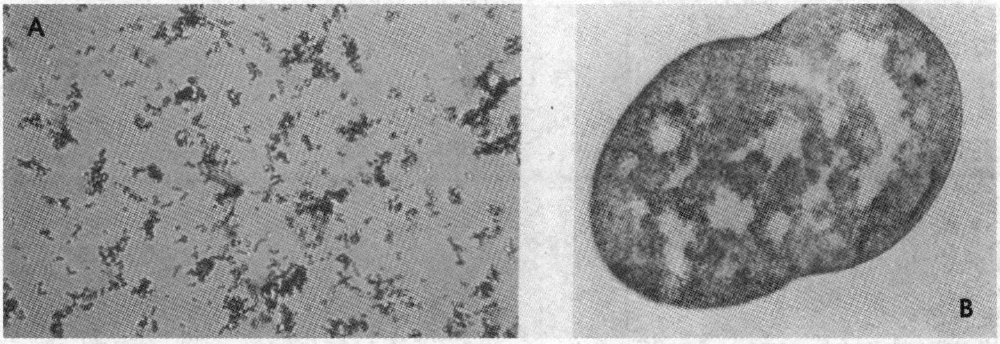


(A) *Azotobacter chroococcum* cells (B) Electron micrograph of *A. chroococcum* cell.

*A. chroococcum* is a soil and rhizosphere inhabitant, while *A. agilis* is water-borne species. Thereafter, several other species vere described such as *A. beijerincki, A. insignis, A. paspali, A. macrocytogenes, A vinelandii,* etc. The species of *Azotobacter* are placed in the family Azotobacteriaceae.

**(*a*) Characteristics of *Azotobacter.*** *Azotobacter* is a Gram-negative, rod-shaped, aerobic bacterium (2-7 × 1-2.75 μm). Size and shape of each of cells vary with the type of species. Cells show varying morphology. Each cell consists of peritrichous flagella but motility of different species varies. One of the unique features of *Azotobacter* is that it forms an insolube black-brown pigment containing melanin due to its oxidation by tyrosinase, The optimum environmental conditions for *Azotobacter* are: temperature (25-30°C), high humidity, aeration, *p*H (7.2-7.6), high salt concentration, etc.

Each cell may produce a resting structure called **cyst.** Cysts vary from the vegetative cells in having speherical cells with contracted cytoplasm and double coloured thick wall. Cysts germinate to liberate cells through breaking cyst coat. Encysted cells can tolerate the adverse environmental conditions such as desiccation, and exposure of UV light.

**(*b*) Isolation of *Azotobacter:*** The species of *Azotobacter* are isolated by soil dilution plate technique on the nitrogen-free medium (Table 20.4). For detail steps see *Practical Microbiology* by Dubey and Maheshwari (2010). After 3 days incubation, flat soft, milky and mucoid colonies of *Azotobacter* grow on agar surface of the plates. Besides, it can also be isolated directly from soil by spreading a lump of soil on nitrogen-free nutrient medium. After incubation for 3 days colonies of *Azotobacter* grow on the surface of agar medium.

**(*c*) Mass Production of *Azotobacter* Inoculant:** As described earlier *Azotobacter* is transferred to a flask containing sterile Jensen's medium (Table 20.4). The flask is incubated on a rotary shaker or batch fermentor for a few days at 30°C. The pure cells of *Azotobacter* developed in broth acts as starter culture which may be used for mass production of inoculant. All precautions of sterilisation steps are followed as described for *Rhizobium.*

**> Table 20.4. Composition of media for mass cultivation of bacteria (g/l).**

| *Chemical composition* (Jensen's medium) | *Azotobacter chroococcum* | *Azospirillum* *Pseudomonas striata* | *Bacillus megaterium,* |
|---|---|---|---|
| $KH_2\ PO_4$ | 1.0 | 0.50 | - |
| $MgSO_4.\ 7H_2O$ | 0.50 | 0.10 | 0.10 |
| NaCl | 0.50 | 0.02 | - |
| KCL | - | - | 0.20 |
| $MnSO_4\ .H_2O$ | - | 0.01 | - |
| $CaCO_2$ | 2.00 | - | - |
| KOH | - | 4.00 | - |
| $(NH_4)_2\ .\ SO_4$ | - | - | 0.50 |
| $MnSO_4$ | - | - | 0.002 |
| $FeSO_4$ | 0.10 | - | 0.002 |
| $FeSO_47.\ H_2O$ | - | 0.05 | 5.00 |
| $Ca_3(Po_4)_2$ | - | - | - |
| $Na_2MoO_4$ | - | 0.002 | 5.00 |
| $CaCl_2$ | - | 0.01 | - |
| Sucrose | 20 | - | - |
| Glucose | - | - | 10 |
| Malic acid | - | 5.00 | - |
| Yeast extract | - | - | 0.5 |
| pH | - | 6.0 -7.0 | - |

For 1 litre of starter culture, 100 litres of medium is transferred to sterilised Jensen's medium in a bioreactor with proper maintenance of temperature of 30°C and continuous agitation for aeration. *Azotobacter* multiplies in medium. When inoculum density has reached to $10^8$ - $10^9$ cell/ml broth, it should be harvested so that carrier-based inoculant should be prepared.

Suitable carrier is dried and powdered passing through a sieve. Calcium carbonate powder is added to neutralise the carrier followed by sterilisation by autoclaving. The harvested broth is poured over the carrier in such a way that 40% moisture is maintained. The inoculant is mixed and curing is done for a week. Then carrier-based inoculant is packed in polythene bags so that it can be stored and sent in market for sale.

**(*d*) Application of *Azotobacter* Inoculant in Field:** *Azotobacter* inoculant can be applied in different ways for different types of plants in varying conditions as beriefly discussed below:

**(*i*) Foliar application:** Foliar application of *Azotobacter* ($10^8$ - $10^9$ cells/ml broth diluted to 1:10 ratio with water) biofertilizer individually or in combination with half of normal dose of chemical nitrogen fertiliser, increased leaf production and leaf quantity in mulberry. However, combined effect of *Azotobacter* and *Beijerinckia* potentiated the improvement of mulberry leaf production beyond the amount obtained through N P K application. In this case foliar application was better than soil application.

**(*ii*) Seed treatment:** The method of seed treatment of *Azotobacter* inoculant is the same as described for *Rhizobium* inoculant.

**(*iii*) Seedling treatment:** This method is applicable for transplantation crop. Roots of the seedlings to be transplanted are kept in slurry of carrier-based inoculum for 10-30 minutes. Then the seedlings are trasnplanted immediately. A substantial number of *Azotobacter* inoculant gets adsorbed on the root surface of seedlings.

***(iv)* Pouring of slurry:** Carrier-based inoculant is diluted in water. Small amount of slurry is poured near the root zones. Slurry may also be mixed with farmyard manure (FYM) and administered near root zone.

***(v)* Top dressing:** Carrier-based inoculant is diluted in water @ 2 kg/ha and mixed with FYM @ 20-25 kg/ha. After transplantation of rice, it is important to broadcast by dressing.

***(e)* Crop Response after Field Application:** At the Indian Agricultural research Institute (New Delhi), field trials were conducted on different types of crop plants such as maize, sorghum, cotton, vegetable crops, wheat, rice, etc. at different places. Always better responses of crop plants to *Azotobacter* inoculant was recorded. An increase in yield of some of the crops with *A. chroococcum* inoculant by 6.7 - 71.7% over the in inoculated control has been reported (Table 20.5). The beneficial effects of *Azotobacter* are due to $N_2$ fixation, synthesis of growth promoting substances and antifungal antibiotics. A significant increases in dry matter of sorghum and maize with combined treatment of *A. chroococcum* + *Azospirillum brasilense* has been recorded (Tilak, 1991).

**> Table 20.5.** Effect of *Azotobacter* inoculant on crop yield.

| *Crops* | *% increase in yield over uninoculated control* | |
|---|---|---|
| | *Azotobakterin* * | *Azotobacter chroococcum* ** |
| Cotton | – | 6.7 – 26.6 |
| Barley | 9 . 0 | – |
| Maize | 8 . 0 | 36.5 – 71.7 |
| Oat | 12 . 0 | – |
| Potato | 8 . .0 | – |
| Sorghum | – | 9.3 – 38.1 |
| Sugar beat | 7 . 6 | – |

* Mishustin and Shilnikova (1969); ** Shende and Apte (1982).

## 3. Azospirillum Inoculants

In 1925, Beijerinck for the first time described a nitrogen fixing bacterium found in the root of digit grass in Brazil and named *Spirillum lipoferum,* In 1956, the Russian investigators also reported some of the nitrogen fixing spirilla. During 1963, its nitrogen fixing ability could be proven by several workers. In 1978, Tarrand and co-workers renamed *Spirillum* as *Azospirillum* (nitrogen fixing *Spirillum*). During 1970s, this bacterium could also be isolated from Indian soils and rhizosphere regions of many plants. Soil *p*H governs the distribution of *Azospirillum.* Soil pH between 5.6 and 7.2 promotes nitrogenase activity, wherease pH below 5-6 does not encourage nitrogenase activity in soil. By using semi-solid sodium malate enrichment medium *Azospirillum* could be isolated from *Digitaria decumbens.* The essential requirements for its isolation are the surface sterilisation of roots by 70% ethanol and creation of micro-aerophilic conditions in the medium.

It is an associative symbiont because it effectively colonises the roots and infects cortex also. It is present inside and outside the roots without developing any apparent structure on roots. It has been found to occur in xylem vessels of black-gram and sugarcane. Moreover, it is capable of fixing 20-40 kg nitrogen under micro-aerophilic conditions. So far there are five species of *Azospirillum: A. lipoferum, A. brasilense, A. amazonens, A. halopraeferns* and *A. irakense.* The first two species are most commonly used as biofertilizers.

(*a*) **Isolation of *Azospirillum* :** *Azospirillum* can be isolated from plant roots as well as soil samples. For isolation from plants roots, a suitable host plant is selected and taken out from soil. Its root system is washed with running tap water and roots are cut into 0.5 cm long pieces. Roots are sterilised with 0.1% $HgCl_2$ solution for one minute. Therefore, these are serially washed with sterile distilled water. One or two pieces of roots are placed on sterile and cool down semi-solid agar medium containing sodium malate (see Table 20.5) in screw-capped tubes. Inoculated tubes are incubated at 28-30°C for 2 days. Just 1-2 cm below from the surface of medium, white pellicles of *Azospirillum* can be obserbed. Thereafter, the nitrogen-free malate medium becomes blue in colour which represents the presence of *Azospirillum.*

(*b*) **Characteristics of *Azospirillum* Strains :** *Azospirillum* is a Gram-negative, motile (with a long polar flagellum and occasionally with peritrichous flagella), vibroid bacterium. It consists of poly-β-hydroxybutyrate (PBH) granules. It behaves as highly aerobic when grown in ammonium-containing medium, and found as micro-aerophilic when grown in nitrogen-free medium. The carbon sources that provide energy are malate, succinate, lactate and pyruvate. Hence, it grows well on media containing these carbon sources. It grows moderately on galactose or acetate-containing medium and poorly on glucose or citrate-containing medium.

(*c*) **Mass Cultivation of *Azospirillum* Inoculants:** The starter culture of *Azospirillum* is prepared by transferring its loopful colony to ammonium chloride-containing Okon's medium in culture flask. On this medium it proliferates profusely under aerobic condition when the flasks are incubated on a rotary shaker. After 3 days of incubation at 35°C on rotary shaker the cells can be harvested for inoculation of Okon's medium filled in a fermentor (@ 1% to total volume of medium. Optimum temperature (35°C) and sufficient aerobic conditions are maintained in fermentor for a few days. At certain intervals, broth is tested for its purity and cell number. The broth inoculum is harvested when cell number reaches to $10^9$/ml.

(*d*) **Preparation of Carrier-based Inoculants:** Several low cost and locally available carriers have been evaluated for *Azospirillum;* for example farmyard manure (FYM), FYM + soil, FYM + charcoal, peat, etc. Properly sterilised soil and FYM in the ratio of 1 : 3 proved as the best carrier for *Azospirillum.* The bacterial cells can survive in FYM + soil for more than 6 months and restore about $10^6$ cells/g of carrier material (Tilak, 1991).

The harvested broth inoculum is mixed with FYM + soil carrier till 40% moisture is obtained. The carrier-based inoculant is packed in polythene bags. The bags are stored at 4°C for about 7 days. Method of carrier-based inoculant production is the same as described for *Rhizobium* and *Azotobacter.*

(*e*) **Application of *Azospirillum* Inoculants in Field:** *Azospirillum* inoculant is applied in field for various crops as given in Table 20.1. However, it is applied in field as given below:

(*i*) **Seed treatment:** Slurry of *Azospirillum* inoculant is prepared by mixing with water in a container. Seeds to be sown in field are soaked in slurry @ 2 kg inoculant per hectare overnight. Then seeds are sown in field.

(*ii*) **Seedling treatment:** Slurry of *Azospirillum* inoculant is prepared by mixing 1 kg of inoculant with 40 litres of water. The roots of transplanted seedlings are dipped in slurry for 15-20 minutes. Then seedlings are transplanted in field. The remaining slurry is spread in field.

(*iii*) **Top dressing:** Carrier-based inoculant is mixed with FYM and soil in the ratio of 3:25:25 (w/w). This mixture is top dressed through out field especially young seedlings of rice. Sudhakar *et al.* (2001) found a better yield in mulberry through foliar application of *Azospirillum* + *Azotobacter* incoculant than soil application. Positive effects of foliar application such as plant growth, plant biomass and yield were also recorded.

**(f) Crop Response:** All azospirilla are nitrogen fixing bacteria with nitrogenase activity like the other nitrogen fixing bacteria. Field experiments have shown that *Azospirillum* consists of 1-10% of the total bacterial population in the rhizosphere soil. The *Azospirillum*-inoculated plants take up minerals (NPK) from soil at the faster rates than uninoculated plants. Hence, inoculated plants accumulate more N, P and K in their stems and leaves. Then minerals are transferred to fruits also. Table 20.6 shows the effect of seed -inoculation with *A. brasilense* on grain yield of plants. *Azospirillum* benefits the plant growth and increases the crop yield by improving root development, mineral uptake and plant-water relationship. Beside nitrogen fixation, it also produces the growth-promoting substances like IAA and giberellins that enhance the plant growth. In sugarcane, *Azosprirllum* can save about 25% recommended dose of nitrogen fertiliser. Table 20. 7 shows the demand of *Azospirillum* in Tamil Nadu for some selected crops.

Increased yield in pearlmillet obtained by inoculation of *Azospirillum* was due to the production of indole acetic acid, gibberellins and cytokinin like substances by the bacterium and their subsequent effect on the plant (Table 20.6).

In 1989, A.C. Gaur and A.R. Algawadi studied the interaction of $N_2$ fixing bacteria (*A. chroococcum*) and phosphate solubilizing bacteria (*B. megaterium* and *P. striata*) and their effect on sorghum and rice crops in a green house experiment by dual inoculation of sorghum with *A. brassilens* and *P. striata*. They found a significant increase in root nitrogenase activity, dry matter and seed yields as compared to single inoculation of both the organisms and control. Similarly, nutrient uptake and yield of rice increased when inoculated together with *A. chroococcum* and *P. striata* as compared to uninoculated control or single inoculated experiments.

**> Table 20.6. Effect of seed inoculation with *Azospirillum brasilense* on grain yield (q/ha).**

| *Plant species* * | *Treatments* | | | | *References* |
|---|---|---|---|---|---|
| | *Control (N-free)* | *A. brasilense* | *Urea*$^a$ | *Urea + A. brasilense* | |
| *Eleusine coracana* (fingermillet) | 16.54 | 19.47 | 23.07 | 28.09 | Subba Rao *et al.* (1985) |
| *Pennisetum americanum* (pearlmillet) | 12.4 | 14.3 | 20.4 | – | Tilak and Subba Rao (1987) |
| *Sorghum biocolor* ** (sorghum var. CSH 5) | 19.23 | 22.78 | 30.48 | 31.05 | Tilak (1991) |

* Values of 1, 2 and 3 are all India mean of 4.4 and 9 centres respectively,

** Average of 4 years' results; a, 40 Kg N/ha.

Field experiments conducted at IARI, New Delhi, have shown that upon seed inoculation of sorghum and cotton by *A. chroococcum* the yield was increased by 38 per cent and 27 per cent respectively, whereas increase in yield of wheat was 10 per cent at the same conditions. These experiments suggest that small and marginal farmers can follow the seed bacterization and increased crop yield. Effects of these bacteria on yield of crops are given in Tables 20.6. and 20.7.

**Table 20.7.** *Azospirillum* **demand in Tamil Nadu for some selected crops.**

| *Crop* | *Area* | *Recommended dose (kg/ha)* | *Required quantity (tonnes/annum)* |
|---|---|---|---|
| Rice | 2261 | 4.0 | 9044 |
| Millets | 790 | 2.5 | 1975 |
| Cotton | 228 | 2.5 | 570 |
| Sugarcane | 282 | 9.0 | 2538 |
| Sunflower | 18 | 2.5 | 45 |
| Termeric | 6 | 5.0 | 30 |
| Tobacco | 8 | 2.5 | 20 |
| Coffee | 32 | 10.0 | 32.0 |
| Tea | 65 | 10.0 | 650 |
| Total | | | 15,192 |

Source : G. Prasad and K. Govindarajan (2001).

## 4. PHOSPHATE SOLUBILISING MICROORGANISMS (PHOSPHATE BIOFERTILIZERS)

Phosphorus is the second vital nutrient next to nitrogen required for growth of microorganisms and plants. But most of phosphorus is not available to paints. Only 1-2% phosphorus is supplied to aboveground parts of the plants. Therefore, to meet out the phosphorus demand of plant, exogenous source of phosphorus is applied to plant as chemical fertilizers. One of the most common forms of phosphate fertilizers is the superphosphate (single or triple). The basic raw material for phosphate fertilizer is the rock phosphate. But rock phosphate is not recommended to apply directly due to agronomic problems being as raw material.

There are several phosphate solubilising microorganisms (PSM) present in soil, for example the species of *Pseudomonas, Bacillus, Micrococcus, Flavobacterium, Aspergillus, Penicillium, Fusarium, Sclerotium,* etc. They can utilise tri-calcium phosphate [$Ca_3(PO_4)_2$], apatite, rock phosphate, $FePO_4$ and $AlPO_4$ as sole phosphate source present in medium. The indication of utilisation is that they produce clearing zones around each colony. They secrete organic acids such as acetic acid, lactic acid, succinic acid, propionic acid, formic acid, etc. Consequently, bound form of phosphates are solubilised and charged molecules of phosphorus ($PO_4^{-3}$) are absorbed by the plants. Therefore, the PSM save 30-50 kg/ha of super-phosphate and increase crop yield up to 200-500 kg/ha.

**(*a*) Isolation of PSM:** For isolation of PSM, Pikovskaya medium is prepared, mixed with 0.5% gum arabic, autoclaved and dispensed in Petri plates. A small amount of soil (1 g) is collected from field and serially diluted in known volume of water. Each plate is inoculated with 1 ml of soil-water suspension. Plates are incubated at 28°C for about 4-5 days. Only PSM grow and form colony which can be identified due to formation of clear zone around each colony. Because PSM utilise $Ca_3(PO_4)_2$ and form clear zone. Such colonies are picked up, purified and preserved for further use. Starter culture is prepared by inoculating the fresh Pikovskaya broth and incubated on a rotary shaker at 28°C.

**(*b*) Mass Production of PSM:** The PSM are produced on a large scale in a large fermentor containing Pikovskaya medium. Starter culture is transferred in a bioreactor @ 1 litre/100 of medium and grown at 28°C for 10-15 days. Culture broth is harvested when cells have attained $10^8$-$10^9$ cells/ml.

**(*c*) Production of Carrier-based Inoculants:** Different carriers (*e.g.* wood charcoal, peat mixture or mixture of wood charcoal and soil) are used for inoculum production. The carrier is powdered, neutralised, sterilised and mixed with broth inoculant. Carrier and inoculant are properly mixed till 40% moisture is attained. This mixture is left for curing by leaving it in a sterile chamber. Then it is filled in polythene bags @ 200 g/packet) and stored at 15-20°C.

***(d)* Crop Response Against PSM:** Slurry is prepared by diluting the PSM in water and treating with gum arabic and $CaCO_3$. Seeds to be sown in field are mixed in slurry for bacterisation and dried in shade. The PSM adher on seed surface. Then the bacterised seeds are sown in field.

Scientists at the Indian Agricultural Research Institute (New Delhi) have investigated the new efficient bacteria (e.g. *Pseudomonas striata* and *Bacillus polymyxa*) and fungi (e.g. *Aspergillus awamori*) that can be used as PSM. PSM carrier-based inoculant has been prepared. This preparation is known as 'IARI Microphos Culture' (Tilak, 1991).

The PSM solubilise 20-30% phosphate which is then absorbed by the plants. Consequently, plant growth is increased. The PSM can be used for all types of plants because they are heterotrophs and show host-specificity. Beneficial effect of seed inoculation with phosphobacterin (*Bacillus megaterium* var. *phosphaticum*) has been recorded on barseem and wheat. On the application of PSM, crop yield of vegetables was increased. But this increase got more potentiated on co-inoculation of PSM with *Azospirillum.* Percent yield increase in vegetables due to biofertilizers varied from 9 to 23%.

## B. GREEN MANURING

In 1989, S.K. Ghai and G.V. Thomas have defined green manuring as "a farming practice where a leguminous plant which has derived enough benefits from its association with appropriate species of *Rhizobium* is ploughed into the soil and then a non legume is grown and allowed to take the benefits of the already fixed nitrogen". The practice of green manuring was started from several century B.C. in India and China. During the course of time availability of chemical fertilizers decreased the significance of green manuring. In recent years, due to hike in price of chemical fertilizers, the practice of green manuring is reemphasized.

Some of the leguminous plants to be used as green manure are: cultivated annual legumes (*e.g. Crotalaria juncea, C. striata, Cassia mimosoides, Cyamopsis pamas, Glycine wightii, Indigofera linifolia, Sesbania rostrata, Vigna radiata*), perennial legumes (*e.g. Acacia nilotica, Cassia hirsuta, Sesbania aegyptica, Leucaena leucocephala*), and wild annual legumes (*e.g. Cassia cobanensis, Lathyrus sativus, Mimosa invisa, Mucuna bracteata*). Contribution of N by some of the nodulated legumes used as green manure is given in Table 20.8. In a report of International Rice Research Institute (Philippines) it has been suggested that fast growing tropical legumes can accumulate more than 80 Kg N/ha when grown as green manures.

**>Table 20.8. Contribution of nitrogen by some of the nodulated legumes when used as green manure/cover crop.**

| *$N_2$ fixing system* | *Amount of N contributed (q/ha)* |
|---|---|
| 1. *Green manure legumes :* | |
| *Sesbania aculeata - Rhizobium* | 70–120 |
| *Leucaena leucocephala – Rhizobium* | 500–600 |
| Beans (broad beans / lupines /soybean / lentil, etc.) – *Rhizobium* | 60–210 |
| Fodders (*Trifolium / Medicago/ Melilotus*, etc. ) – *Rhizobium* | 100–300 |
| 2. *Cover crop legumes :* | |
| *Lablab purpureus – Rhizobium* | 240 |
| *Glycine jawanica - Rhizobium* | 210 |

| | |
|---|---|
| 3. *Non-legumes :* | |
| *Casuarina equisitifolia – Frankia* | 100 |
| *Alnus – Frankia* | 30–300 |
| 4. *Others :* | |
| *Azolla – Anabaena* | 25–190 |
| Grasses – *Azospirillium* | 15–100 |

In India, for a small and marginal farmers, green manuring may be important because of high cost of chemical fertilizers. Moreover, reclamation of "Usar lands" can also be done by green manuring. In addition to nitrogen, green manures also provide organic matter, N, P, K and minimize the number of pathogenic microorganisms in soil.

## C. CYANOBACTERIAL INOCULANTS

Role of the blue-green algae (*e.g. Aulosira, Anabaena, Cylindrospermum, Nostoc, Plectonema, Tolypothrix*) in the paddy fields was realised much earlier. In water-logging condition, the cyanobacteria multiply, fix atmospheric $N_2$ and release it into the surroundings in the form of amino acids, proteins and other growth promoting substances. Recent works done at Central Rice Research Institute, (Cuttack), Indian Council of Agricultural Research (New Delhi) and other centres, *e.g.* Centre of Advanced Study in Botany, Banaras Hindu University, Varanasi are of much importance.

### 1. Algalization

It were Japanese workers (Watanabe and coworkers) who developed techniques for mass cultivation of blue-green algae to be used as biofertilizer in paddy fields. G.S. Venkataraman in 1961 coined the term *'algalization'* to denote the process of application of blue-green algal culture in field as biofertilizer. He initiated algalization technology in India and demonstrated the way how this technology could be transferred to farmer level who hold small lands.

At present, algalization is being followed in Tamil Nadu and Uttar Pradesh, and tried in Jammu and Kashmir, Andhra Pradesh, Karnataka, Maharastra and Haryana. Algalization is also being practiced in China, Egypt, Philippines and erstwhile U.S.S.R. In 1990, Department of Biotechnology, Government of India, New Delhi establishment four major centres in different paddy growing areas of the country for acceleration and extension of works on algal biofertilizers. The programme was launched under 'Technology Development and Demonstration Projects on Cyanobacterial Biofertilizers' in U.P. (Lucknow), Tamil Nadu (Madurai), West Bengal (Calcutta) and New Delhi. The main objectives of the programme were (*i*) to develop low cost indigenous technology for mass production of cyano bacteria, (*ii*) to isolate regional specific fast growing and better $N_2$ - fixing strains, (*iii*) to develop starter inoculum, (*iv*) to demonstrate the farmers in field, and (*v*) to study the benefits on both economy and ecology. As a result of these studies cyanobacterial biofertilizer was found very useful, especially for small and marginal farmers of the country with the view point of both economy and ecology.

D.B.T. Centre of U.P. (Lucknow) has reported the increase in yield of paddy (about 12.5 q/ha) to be due to cyanobacterial biofertilizer (Fig. 20.4). Moreover, its use continued with chemical fertilizer gave the best results. Because the cyanobacterial biofertilizer not only increase the productivity and quality of paddy, but also minimise the harmful effects of the chemical fertilizers.

**(*a*) Isolation of Cyanobacteria (Blue-Green Algae, BGA):** The cyanobacteria are isolated on Fogg's medium. About 5 kg soil from top layer of paddy field is collected, powdered and transferred into a flask containing Fogg's medium (0.2 g $KH_2PO_4$, 0.2 g $MgSO_4 \cdot 7H_2O$, 0.1 g $CaCl_2$, 0.1 mg $Na_2MoO_4$, 0.1 mg $MgCl_2$, 0.1 mg $H_3Bo_3$, 0.1 mg $CuSO_4$, 0.1 mg $ZnSO_4$, 1.0 ml Fe-EDTA, 1 litre distilled water, *p*H 7). After proper agitation, flask is incubated at room temperature under the influence of 12 hours of light (1,500 lux) and dark regime. A loopful culture is transferred to 10 ml sterile distilled water, which is then serially diluted. A drop of water from each dilution is inoculated into Fogg' medium in Petri plates. Plate are incubated for further growth of algae. Culture of cyanobacteria are observed microscopically and each species are purified for further use.

(*b*) **Preparation of Starter Culture :** Each pure culture of cyanobacteria is separately grown in Fogg's medium in flasks. The flasks are incubated in light for algal growth. The cultures of cyanobacteria can be used as starter culture for mass cultivation of BGA inocula.

(*b*) **Mass Cultivation of Cyanobacterial Biofertilizers :** For outdoor mass cultivation of cyanobacterial biofertilizers, the regional specific strains should be used. However, many germplasm collection laboratories have been established by the D.B.T. in different parts of the country for the development of starter inoculum. Mixture of 5 or 6 regional acclimatized strains of cyanobacteria, *e.g.* species of *Anabaena, Aulosira, Cylindrospermum, Gloeotrichia, Nostoc, Plectonema, Tolypothrix* are generally used for starter inoculum. The following four methods are used for mass cultivation : (*i*) cemented tank method., (*ii*) shallow metal troughs method, (*iii*) polythene lined pit method, and (*iv*) field method. The polythene lined pit method is most suitable for small and marginal farmers to prepared algal biofertilizer. In this method, small pits are prepared in field and lined with thick polythene sheets (Fig. 20.2). Mass cultivation of cyanobacteria is done by using any of the four methods under the following steps:

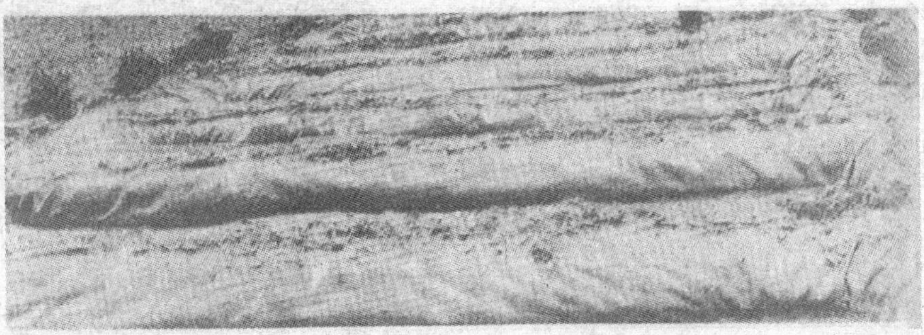

**Fig. 20.2.** Polythene lined pits for mass production of cyanobacterial biofertilizers (Courtesy : CST, UP (DBT Centre), Lucknow).

(*i*) Prepare the cemented tanks, shallow trays of iron sheets or polythene lined pits in an open area. Width of tanks or pits should not be more than 1.5 m. This will facilitate the proper handling of culture.

(*ii*) Transfer 2 -3 Kg soil (collected from open place for 1m $^2$ area of the tank) and add 100 g of superphosphate. Water the pit to about 10 cm height. Mix lime to adjust the pH 7. Add 2 ml of insecticide *e.g.* malathion to protect the culture from mosquitoes. Mix well and allow to settle down soil particles.

(*iii*) When water becomes clear, sprinkle 100 g of starter inoculum on the surface of water.

(*iv*) When temperature remains between 35-40° during summer, optimum growth of cyanobacteria is achieved. Always maintain the water level to about 10 cm during this period.

(*v*) After drying, the algal mat will get separated from the soil and forms flakes. During summer about 1 kg pure algal mat per m$^2$ area is produced. These are collected, powdered, kept in sealed polythene bags and supplied to the farmers.

(*vi*) The algal flakes can be used as starter inoculum if the same process is repeated.

Moreover, the cyanobacterial inoculants can be stored for more than 3 years without any loss in viability.

(*d*) **Field Application of BGA Inoculants:** For one hectare of paddy field, 10 kg of BGA inoculant is applied. However, after 10 days of transplantation of rice seedlings, powder of BGA flakes is dispersed which grows luxuriantly in water.

(*e*) **Crop Response :** It has been found that BGA inoculant increases crop yield by 34% and saves nitrogen fertiliser by 30%.

## D. AZOLLA AS BIOFERTILIZERS

*Azolla* is an aquatic heterosporous fern which contains an endophytic cyanobacterium, *Anabaena azollae,* in its leaf cavity. The significance of *Azolla* as biofertilizer in rice field was realized in Vietnam. Recently, it has become very popular in China, Indonesia, Philippines, India and Bangladesh.

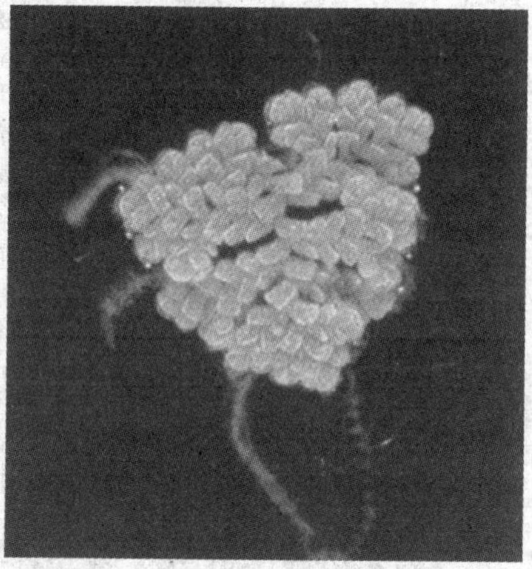

*Azolla* is widely used as biofertilizer.

A total of six species of *Azolla* are known so far *viz., A. caroliniana, A. filiculoides, A. mexicana, A. microphylla, A. nilotica, A. pinnata* and *A. rubra.* Out of these *A. pinnata* is commonly found in India. The global collections of several species of *Azolla* are maintained at CRRI (Cuttack). Within the leaf cavity filaments of *Anabaena azollae* are present. Dr. P.K. Singh, at CRRI has done an outstanding work on mass cultivation of *Azolla* and its use as biofertilizer in rice and other crop fields.

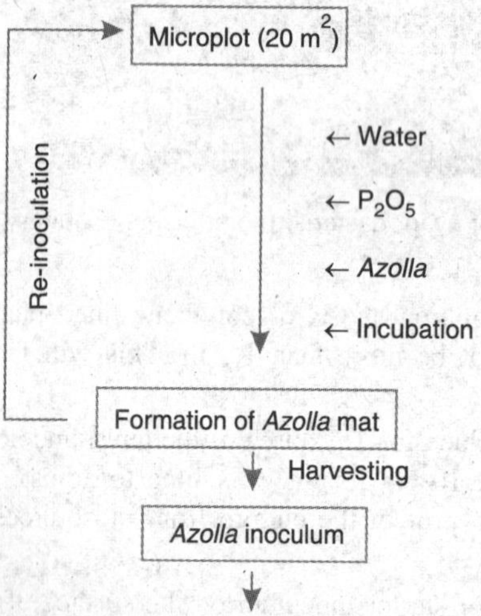


**Fig. 20.3.** Mass cultivation of *Azolla* in India.

### 1. Mass Cultivation of Azolla

Methods of mass cultivation of *Azolla* are shown in Fig. 20.3. Microplots (20 m$^2$) are prepared in nurseries in which sufficient water (5-10 cm) is added. For good growth of *Azolla*, 4-20 Kg $P_2O_5$/ha is also amended. Optimum pH (8.0) and temperature (14-30°C) should be maintained. Finally, microplots are inoculated with fresh *Azolla* (0.5 to 0.4 Kg/m$^2$). An insecticide (furadon) is used to check the attack of insects. After three week of growth mat formed by *Azolla* is harvested and the same microplot is inoculated with fresh *Azolla* to repeat the cultivation.

*Azolla* mat is harvested and dried to use as green manure. There are two methods for its application in field: (*a*) incorporation of *Azolla* in soil prior to rice plantation, and (*b*) transplantation of rice followed by water draining and incorporation of *Azolla.* However, reports from the IRRI (Philippines) reveal that growing of *Azolla* in rice field before rice transplantation increased the yield equivalent to that obtained from 30Kg N/ha as urea or ammonium phosphate.

Moreover, *Azolla* shows tolerance against heavy metals viz. As, Hg, Pb, Cu, Cd, Cr, etc. It tolerates low concentration but at high levels a setback in biochemical pathways is caused. *A. pinnata* absorbs heavy metals into cell walls and vacuoles through evolution of specific metal resistant enzymes. Therefore, heavy metal resistant species such as *A. pinnata* can also be incorporated as green manure in rice field near the polluted areas where heavy metal concentration is between 0.01 and 1.5 mg/litre. Due to development of chemical industries and discharge of effluents into water bodies, heavy metal concentration is gradually increasing day by day. Industries where work of electroplanting,fertilizers, tanning, etc. are done, they act as a chief source for soil and water pollution. For example, disturbed vegetation in aquatic system around Damodar river valley in India has received a great attention.

## E. FRANKIA-INDUCED NODULATION

A member of actinomycetes, *Frankia* is known to form a symbiotic structure with the roots of non-leguminous plants which is called **actinorhiza** ('actino' = actinomycte *Frankia* named after the name of the discoverer Frank in 1880s; 'rhiza' = plants bearing nodules). Moreover, about 24 genera from 8 families of angiosperms are known to bear actinorhizal root nodules. Some of the important genera are: *Alnus, Hippophae, Allocasuarina, Casuarina, Gymnostoma, Coriaria,* etc.

**(*a*) Isolation of *Frankia* :** In 1978, for the first time Callaham and co-workers in Torrey's laboratory (U.S.A) isolated *Frankia* from *Comptonia peregrina.* Thereafter, it has been isolated from several host plants. The actinorhizal nodules are large sized; therefore, they get contaminated and inhibit the growth of pure culture of *Frankia.* Hence, un-suberised young nodules are preferred to isolate *Frankia.*

Actinorhizae. *Frankia*- induced actinorhizal nodules in *Ceanothus* (Buckbrush).

The isolated nodules are surface sterilised with osmium tetraoxide under mild vacuum in a fume hood so that nodules may not be injured by the chemical. The nodules are washed several times to remove traces of chemical. Then the nodules dissected so that *Frankia* cells may be released. Due to maceration process nodules release toxic phenolic compounds. Therefore, the homogenate should be passed through activated charcoal so that phenolic compound may get adsorbed to it.

The homogenates are passed through nylon membrane or subjected ot sucrose density centrifugation. This separates vesicle clusters from the contaminating microorganisms. The separated fractions are plated on BAP medium (0.591 g $K_2HPO_4$, 0.952 g $KH_2PO_4$, 0.267 g $NH_4Cl$, 0.095 g Mg $SO_4$. $7H_2O$, 0.01 g $CaCl_2$. $2H_2O$, 0.01 g FeNa-EDTA, 0.48 g Na-propionate, 1ml trace element solution, 1 ml vitamin solution, 1 litre distilled water, pH 6.7). The Petri disshes are sealed by paraffin and incubated at 28-30°C. Colonies of *Frania* appear at the edge of nodule pieces. The colonies appear easily when medium contain activated charcoal because it eliminates toxic phenolic compounds. On agar plate colonies are diffuse with a loose hyphal network around the centre or compact hyphal network that bear many sporangia.

**(*b*) Culture Characteristics:** *Frankia* grows very slowly and has a very long log phase (14 days). Also it shows a slow log phase Therefore, most of the hypae are autolysed. This makes difficult to grow *Frankia* in a mass scale. It is microaerophile and requires and pH rage of 6.7-7.0. Hyphae are thin slender poorly branched, hyaline or pigmented depending of the medium. Spores are produced inside the sporangia in clusters intrahyphally or terminally. In nitrogen-free media, round or spherical vesicles are formed as a result of swelling of the tips. These are considered as the sites for nitrogen fixation like heterocysts in cyanobacteria in culture media and inside the nodules. Oxygen-protecting ability resides inside these vesicles. Since it is a slow grower, it takes about 2 months to come in a visible shape in culture. Colonies shows polymorphisms ranging from starfish, diffuse to compact structures (*c*) **Infection of Host Roots:** Cells of *Frankia* are embedded in mucigel of root region. The spores also get attached on the surface of roots haris. This results in root hail curling. Consequently spore germinate and infect the deformed curled root hairs. Thin and higly branched hyphae can be observed inside the root hairs. Therefore, hyphae penetrate the cortex. Due to secretion of growth substances like auxins, cytokinis, givverellins, etc. mitotic activity is enhanced which results in formation of nodules. Root nodules bears *Frankia* cells present in clusters and appear like vesicles (Fig. 20.4)

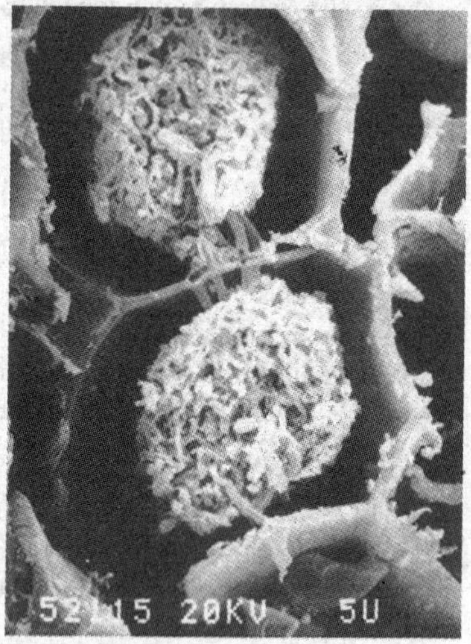


**Fig. 20.4.** Scanning electron micrograph showing vesicular structure of *Frankia* present inside the nodule cells of the host.

(*d*) **Benefits of *Frankia* Inoculation:** It has been found that the *Alnus* tree increases the nitrogen content of soil by about 61.5 to 157 kg N/ha/annum and *Casuarina* by about 60 kg N/ha/annum. Thus the non-leguminous plants also increase the nitrogen economy of the soil. Besides, the nodulating *Alnus glutinosa* can accumulate about 300 mg nitrogen per plant are compared to nodule-free plants. Field experiments have been conducted to estimate the potentialities of *Frankia* in pure culture forms. Due to slow growth of *Frankia*, continuous culture for field experiment with *Casuarina* can be carried out. *Frankia* cells can be mobilised in alginate or montmorillonite clay or kaolinite clay and entrapped beads was used for field inoculation. Consequently, increased nitrogen contents in *Frankia* inoculated *Casuarina* trees was recorded.

## F. MYCORRHIZAL FUNGI AS BIOFERTILIZERS

Mycorrhiza (fungus roots) is a distinct morphological structure which develops as a result of mutualistic symbiosis between some specific root - inhabiting fungi and plant roots. Plants which suffer from nutrient scarcity, especially P and N, develop mycorrhiza *i.e.* the plants belong to all groups *e.g.* herbs, shrubs, trees, aquatic, xerophytes, epiphytes, hydrophytes or terrestrial ones. In most of the cases plant seedling fails to grow if the soil does not contain inoculum of mycorrhizal fungi.

In recent years, use of artificially produced inoculum of mycorrhizal fungi has increased its significance due to its multifarous role in plant growth and yield, and resistance against climatic and edaphic stresses, pathogens and pests.

### 1. Mechanism of Symbiosis

The mechanism of symbiosis is not fully understood. In 1949, E. Bjorkman postulated the carbohydrate theory and explained the development of mycorrhizas in soils deficient in available P and N, and high light intensity. At high light intensity, surplus carbohydrates are formed which are exuded from roots. This in turn induces the mycorrhizal fungi of soil to infect the roots. At low light intensity, carbohydrates are not produced in surplus, therefore, plant roots fail to develop mycorrhizas.

### 2. Types of Mycorrhizae

By earlier mycologists the mycorrhizas were divided into the following three groups :

(*a*) **Ectomycorrhiza.** It is found among gymnosperms and angiosperms. In short roots of higher plants generally root hairs are absent. Therefore, the roots are infected by mycorrhizal fungi which, in turn, replace the root hairs (if present) and form a mantle. The hyphae grow intercellularly and develop Hartig net in cortex. Thus, a bridge is established between the soil and root through the mycelia.

Ectomycorrhiza of oak.

**(*b*) Endomycorrhiza.** The morphology of endomycorrhizal roots, after infection and establishment, remain unchanged. Root hairs develop in a normal way. The fungi are present on root surface individually. They also penetrate the cortical cells and get established intracellulary by secreting extracellular enzymes. Endomycorrhizas are found in all groups of plant kingdom.

**(*c*) Ectendomycorrhiza.** In the roots of some of the gymnosperms and angiosperms, ectotrophic fungal infection occur. Hyphae are established intracellularly in cortical cells. Thus, symbiotic relation develops similar to ecto- and endo-mycorrhizas.

In 1991, G.C. Marks classified the mycorrhizas into seven types on the basis of types of relationships with the hosts: (*i*) vesicular-arbuscular (VA) mycorrhizas (coiled, intracellular hyphae, vesicle and arbuscules present), (*ii*) ectomycorrhizas (sheath and inter-cellular hyphae present), (*iii*) ectendomycorrhizas (sheath optional, inter and intra-cellular hyphae present), (*iv*) arbutoid mycorrhizas (seath, inter-and coiled intracellular hyphae present), (*v*) monotropoid mycorrhizas (sheath, inter- and intra- cellular hyphae and peg like haustoria present), (*vi*) ericoid mycorrhizas (only coiled intracellular hyphae, long coiled hyphae present), and (*viii*) orchidaceous mycorrhizas (only coiled intracellular hyphae present). Type (*i*) is present in all groups of plant kingdom; Types (*ii*) and (*iii*) are found in gymnosperms and angiosperms. Types (*iv*), (*v*) and (*vi*) are restricted to Ericales, Monotropaceae and Ericales respectively. Types (*vii*) is restricted to Orchidaceous only. Types (*iv*) and (*v*) were previously grouped under ericoid mycorrhizaes.

## 3. Methods of Inoculum Production and Inoculation

Methods of inoculum production of VAM fungi differ; however, some of these two are briefly described here.

**(*a*) Ectomycorrhizal fungi :** The basidiospores, chopped sporocarp, sclerotia, pure mycelial culture, fragmented mycorrhizal roots or soil from mycorrhizosphere region can be used as inoculum. The inoculum is mixed with nursery soils and seeds are sown.

**(*i*) MycoRhiz®:** Institute for Mycorrhizal Research and Development, U.S.A., Athens and Abbott Laboratories (U.S.A) have developed a mycelial inoculum of *Pisolithus tinctorius* (Pt) in a vermiculite-peat moss substrate with a trade name 'MycoRhiz'which is now commercially available on large quantities. In 1982, about 1.5 million pine seedlings were produced with MycoRhiz in the U.S.A.

For the production of biofertilizer, Pt is grown at 25°C for three weeks in Petri plates containing Modified Melin-Norkrans (MMN) agar medium. Then the mycelial discs (8 mm diameter) are cut and transferred into a deep tank aerated submerged culture container containing MMN medium for mass culture.

In 1977, Abbott inoculum was produced in a vertical deep tank (solid-substrate fermentor). The substrate contained vermiculite and MMN liquid medium with slightly more than recommended amount of carbohydrates and organic and inorganic nitrogen. The substrate was transferred in a fermentor, steam-pasteurised (72°-84°C), cooled and inoculated with the starter mycelia of Pt. The culture was incubated in the fermentor, removed and dried. Inoculum was not leached.

Results of bare-root seedlings tests (the fast assay technique) and the container test showed that the most effective inoculum had: (*i*) abundant hyphae of Pt inside the vermiculite particles, (*ii*) pH between 4.5 and 6.0, (*iii*) minimum microbial contamination, and (*iv*) low amount of residual glucose as a consequence of leaching the inoculum before drying. Leaching removes many other nutrients as well. An inoculum broadcast rate of 1.08 litres per m2 soil surface gave the best result.

(*ii*) **Mycobeads:** Mycobeads are the hydrogel bead inocula of ectomycorrhizal fungi first developed by I.C. Tommercup and co-workers in 1987. The mycelia of ectomycorrhizal fungi immobilised in hydrogel beads are being used in Australia for tailoring the seedlings. The beads are 2 mm in diameter and are kept under axenic condition in sterile de-ionised water at 4°C unitl used. The beads of *Laccaria laccata, Discolea maculata, Hebeloma* sp. and the other ectomycorrhizal fungi are marketed by Interbec Australia Ltd. The beads remian viable for about 6 months. The fungal mycelia initiate mycorrhiza whitin 5-10 days after inoculation. The efficacy of mycobeads has been found very effective for *Eucalyptus* spp. in steam pasteurised potting mixture consisting for sand (50%), compost saw dust (40%) and peat (10%).

Mycobeads ensures high uniformity as propagules. It is cost-effective also. These can be stored for a long time and, therefore, can be used where there is a long gap between production and use.

(*b*) **VA mycorrhizal fungi :** There are six genera of VA-mycorrhizal fungi of which two are sporocarpic (*Glomus* and *Scleocystis*) and the rest of four are non-sporocarpic (*Gigaspora, Scutellospora, Acaulospora* and *Entrophospora*). Currently two terms are used for VA-mycorrhizal fungi: AM fungi and VAM fungi. The term AM is used for fungi forming arbuscules only, whereas VAM for those forming both vesicles and arbuscules. According to the published literature, *Gigaspora* forms arbuscules, while *Scleocystis, Aculospora, Scutelospora* and *Entrophospora* form both vesicle and arbuscules. Spores of some VA-mycorrhizal fungi are given in Fig. 20.5.

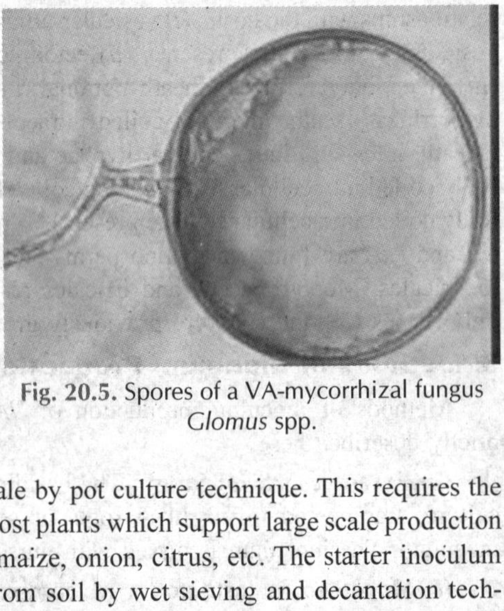

**Fig. 20.5.** Spores of a VA-mycorrhizal fungus *Glomus* spp.

VA mycorrhizas can be produced on a large scale by pot culture technique. This requires the host plants, mycorrhizal fungi and natural soil. The host plants which support large scale production of inoculum are sudan grass, strawberry, sorghum, maize, onion, citrus, etc. The starter inoculum (spores) of VA mycorrhizal fungi can be isolated from soil by wet sieving and decantation technique. VA mycorrhizal spores are surface sterilized and brought to the pot culture. Commonly used pot substrates are sand: soil (1 : 1, w/w) with a little amount of moisture. An out line for inoculum production is given in Fig. 20.6.

There are two methods of using the inoculum : (*i*) using a dried spore-root- soil to plants by placing the inoculum several centimeters below the seeds or seedlings, (*ii*) using a mixture of soil-roots, and spores in soil pellets and spores adhered to seeds with adhesives.

Commercially available pot culture of VA mycorrhizal hosts grown under aseptic conditions can provide effective inoculum. Various types of VA mycorrhizal inocula are currently produced by Native Plants, Inc (NPI), Salt Lake City.

In India, Forest Research Institute, Dehra Dun has established mycorrhizal bank in different states of the country. Inocula of these can be procured as needed and used in horticulture and

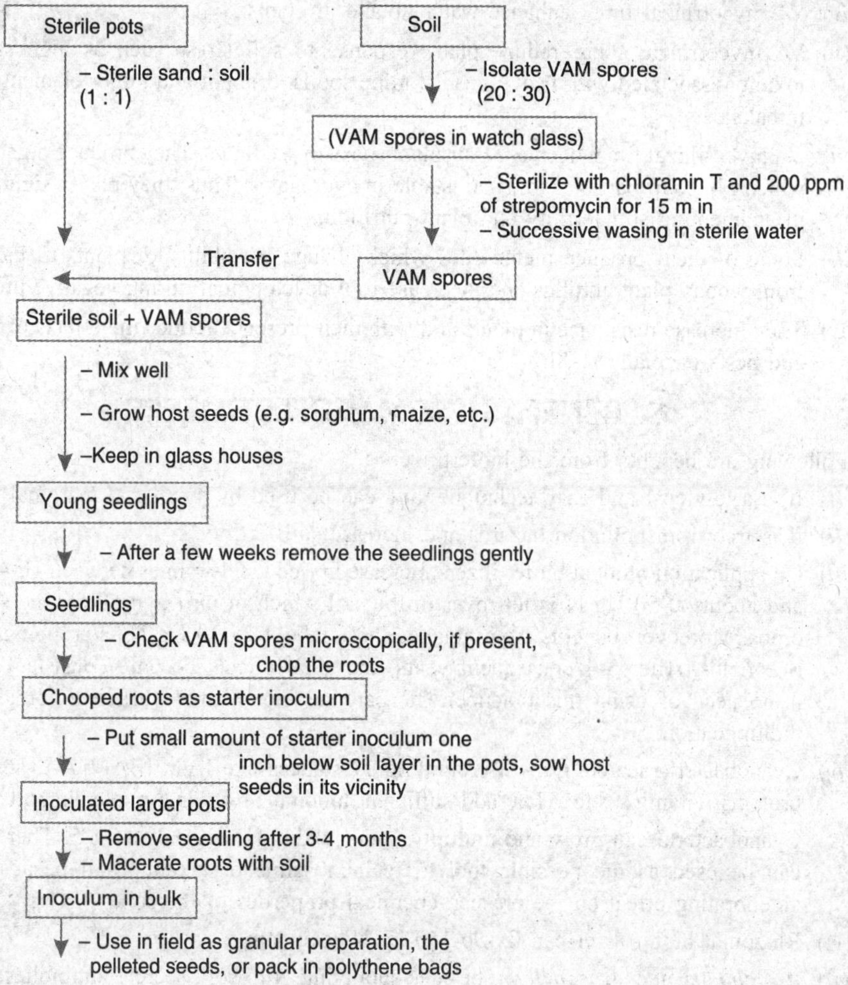


**Fig. 20.6.** Method of production of VAM (VA mycorrhizal) inoculum for application in fields.

forestry programmes. Similarly, IARI, New Delhi is also producing and suppling VA-mycorrizal inoculants.

## 4. Benefits from Mycorrhizas to Plants

Development of mycorrhizal symbiosis is a unique relation between plant roots and mycorrhizal fungi. Both host plants and mycorrhizal fungi are benefited from each other. Some of the important benefits are disscussed below:

(*i*) They increase the longevity of feeder roots, surface area of roots by forming mantle and spreading mycelia into soil and, in turn, the rate of absorption of major and minor nutrients from soil resulting in enhanced plant growth.

(*ii*) They play a key role for selective absorption of immobile (P, Zn and Cu) and mobile (S, Ca, K, Fe, Mn, Cl, Br, and N) elements to plants. These are available to plants in less amount.

(*iii*) Some of the trees like pines cannot grow in new areas unless soil has mycorrhizal inocula because of limited or coarse root hairs.

(*iv*) VA mycorrhizal fungi enhance water uptake in plants.

(*v*) VA mycorrhizal fungi reduce plant response to soil stress such as high salt levels, toxicity associated with heavy metals, mine spoils, drought and minor element (*e.g.* Mn) imbalance.

(*vii*) VA mycorrhizal fungi decrease transplant socks to seedlings. They produce organic 'glues' which bind soil particles into semistable in aggregates. Thus, they play a significant role in augmenting soil fertility and plant nutrition.

(*viii*) Some of them produce metabolites which change the ability of plants to induce roots from woody plant cuttings and incerease root development during vegative propagation.

(*ix*) They increase resistance in plants and with their presence reduce the effects of pathogens and pests on plant health.

## G. BENEFITS FROM BIOFERTILIZERS

Following are benefits from the biofertilizers :

(*i*) It is a low cost and easy technique, and can be used by small and marginal farmers.

(*ii*) It is free from pollution hazards and increase soil fertility.

(*iii*) On application of algal biofertilizers increase in rice yields ranges between 10-45 per cent and about 40-50 Kg N is left over in the soil which in turn is used for the subsequent crops. Moreover, benefits from algalization is about 25-30 Kg N/ha/cropping season in rice fields. After 3-4 consequent years, the algal effects become consistent and there is no need of using this practice. The parental inoculum is sufficient for growth and multiplication.

(*iv*) Cyanobacteria secrete growth promoting substances like IAA, IBA, NAA, amino acids, proteins, vitamins, etc. They add sufficient amount of organic matter in soil.

(*v*) Cyanobacteria can grow and multiply under wide pH range of 6.5-8.5. Therefore, they can be used as the possible tool to reclaim saline or alkaline soil because of their ameliorating effect on the physico-chemical properties of the soil.

(*vi*) Rhizobial biofertilizer can fix 50-150 kg N/ha/annum.

(*vii*) *Azotobacter* and *Azospirillum,* besides supplying N to soil, secrete antibiotics which act as pesticides.

(*viii*) *Azolla* supplies N, increases organic matter and fertility in soil and shows tolerance against heavy metals.

(*ix*) The bioferilizers increase physico-chemical properties of soils such as soil structure, texture, water holding capacity, cation exchange capacity and pH by providing several nutrients and sufficient organic matter.

(*x*) The mycorrhizal biofertilizers make the host plants available with certain elements, increase longevity and surface area of roots, reduce plant response to soil stresses, and increase resistance in plants. In general, plant growth, survival and yield are increased.

## H. COMMERCIAL PRODUCERS OF BIOFERTILIZERS

Following are some of the companies producing microbial inoculants (the addresses are subject to change):

**A. Indian Companies: Trade name**

1. Bacfil, 25 N.K. Road, Lucknow 'Rhizoteeka'

| | | |
|---|---|---|
| 2. | Microbes India, 87 Lenin Sarani, Calcutta-13 | — |
| 3. | Rallis India Ltd, 87 Richmond Road, Bangalore-25 | — |
| 4. | Indian Organic Chemicals Ltd, 15 Mathew Road, Bombay - 4 | 'Nodin', 'Natrin' |

**B. Foreign Companies:**

| | | |
|---|---|---|
| 1. | Union Chemique S.A., Belgium | 'Nodosit' |
| 2. | Phyluxia Allami, Hungary | 'Rhizonit' |
| 3. | Laboratorie de Microbiologie, France | 'N-germ' |
| 4. | Root Nodue Pvt. Ltd., Australia | 'Nitrogerm' |
| 5. | Agricultural Laboratories, Austria. | 'Nodulud' |
| 6. | Radicin Institute, Germany. | 'Radicin Impfsfoff' |
| 7. | Abbott Laboratories , U.S.A. and Institute for Mycorrhizal Research and Development, U.S.D.A., Athens. | 'MycoRhiz'. |
| 8. | Interbec Australia Ltd. | 'Mycobedds' |

## I. WORKS DONE ON BIOFERTILIZERS IN INDIA

During the 7th five year plan, Ministry of Agriculture, Department of Agriculture and Co-operation, sanctioned a scheme **'National Project on Development and use of Biofertilizers'** to promote the use of biofertilizers in Indian agriculture for augmenting crop nutrients through soil microbes. The National Biofertilizer Development Centre (NBDC), Ghaziabad and six Regional Biofertilizer Development Centres (RBDCs) at Bangalore, Nagpur, Jabalpur, Bhubaaneshwar, Hisar and Imphal have been conducting biofertlizer field trials on various crops. This project started doing its services after 1989. In the project 1245 field demonstrations were carried out, and briefly described as below:

**(*a*) Availability of Quality Inoculants:** Before the inception of this project in 1983, availability of biofertilizers in the country was in a very low quantity. The project was aimed at to provide *Rhizobium* for legumes and oil seed crops. and *Azotobacter* and *Azospirillum* for cereals, millets, sugarcane and vegetable, and phosphate solubilising microorganisms (PSM) for all major crops and soils. During this period a large number of Private Companies and Central Govt-funded biofertilizer production units were set up to produce biofertilizers.

**(*b*) Organisation of Training Programmes for Extension Officials and Orientation Course on Quality Control of Biofertilizers:** For Indian agriculture biofertilizer technology is new. It is basically a microbial technology. The officials working with biofertilizers at state levels are trained in the project about the current development on biofertilizer technology so that new information should be transmitted to farmers. Quality control protocols of biofertilizers are highly technical and most of people are not aware of microbial protocols used to test the quality of biofertilizers. Therefore, time to time training is organised at Regional and National Centre(s).

**(*c*) Quality Control:** In India biofertilizers are being produced mostly by public sector and small biofertilizer producers. All the centres are engaged in testing the quality of biofertilizers produced by different production units following microbiological protocols.

**(*d*) Organisation of Field Demonstration and Farmers Fair :** For proper use of biofertilizer technology, the biofertilizers are tested in the farmer's field so that the farmers must be convinced about the potentialities of biofertilizers to increase the yield, savings of chemicals and economic returns in the form of input.

(*e*) **Mass Publicity of Biofertilizers:** For the creation of awareness on the use of different biofertilizers, different centres are involed in popularising the benefits of biofertilizers to the farmers. It can be done by distributing literature and through All India Radio, Doordarshan and mass media programme.

(*f*) **Distribution of Mother Culture:** Different centres are distributing authenticated microbial strains to biofertilizers poroduction units at nominal rates. The microbial repository is being maintained at each of the Regional and National Centre(s) where the agriculturally important microbial cultures are being maintained.

(*g*) **Research and Development Activities:** Biofertilizer technology involves the use of specific microorganisms which finally reach to rhizosphere region of different crops in different agricultural fields. Some times, inappropriate microbial strain may fail. Therfore, to ensure the right functions of the strains developed, different technologies are being developed to improve the quality of biofertilizers.

The NBDC (Ghaziabad) acts as nodal agency of the Government to provide technical advice and support on various aspects of biofertillzers. The ulitmate objective is to popularise the use of biofertilizers so that maximum number of farmers may be benefited.

## PROBLEMS

1. What is a biofertilizer? Write a brief note on its significance.
2. What is bacterization? What types of microorganisms are used as microbial inoculants?
3. Give a detailed account of mass cultivation of *Rhizobiu*m and its use as biofertilizer.
4. Why is *Azotobacte*r termed as associative symbiont?
5. What is algalization? Upto what extent the use of blue-green algae as biofertilizer is successful in India?
6. *Azoll*a is a member of water fern; why is it called as biofertilizer but not other water ferns?
7. Discuss in detail about rhizobial inoulauts, its identification, mass production and field application
8. Give a detailed account of *Azotobacter* inoculant preparation and its field response.
9. Give a detailed note on *Azospirillum* inoculant and its field application.
10. Write in detail about *Frankia* - induced nodulation.
11. Write an essay on works done on biofertilizers in India.
12. Write short notes on the followings :
    (*a*) Bacterization, (*b*) Azotobakterin (*c*) Phosphobacterin (*d*) Biofertilizers, (*e*) Green manuring (*f*) *Rhizobium* (*g*) *Azospirillum* (*h*) *Pseudomonas*, (*i*) *Azoll*a (*j*) Mycorrhizal biofertilizers, (*k*) Benefits from biofertilizers, (*l*) Quality control of rhizobial inoculants, (*m*) Nodulation test, (*n*) *Azotobacter* inoculant, (*o*) MycoRhiz®, (*p*) Mycobeads, (*q*) Mother culture, (*r*) PSM, (*s*) Frankia

CHAPTER

# 21

# Biopesticides (Biological Control of Plant Pathogens, Pests and Weeds)

Since the time immemorial, when man started cultivation, he had a little knowledge of plant disease. Many control measure were adopted time-to-time, which included cultural practices and use of organic and inorganic chemicals. In recent years, the increasing informations, on hazardous effects of synthetic pesticides on plant and animal health, have alarmed the scientists to seek an alternative methods, which should not cause pollution and should be non-phytotoxic. Biological control method is such a technique which involves disease control by some biological agent(s) (living micro- or macroorganisms, other than disease causing organisms (the pests) and damaged plants (the hosts).

The biocontrol agents provide several advantages such as : (*i*) the conservation of natural enemies of insect pests, (*ii*) they are environment friendly, (*iii*) they maintain ecological balance and check environment pollution, (*iv*) their targets are mainly the invertebrate insect pests, (*v*) they pose no toxic effects on plants, animals or human beings as well as no health hazards to mankind, (*vi*) their use is sustainable and cost effective.

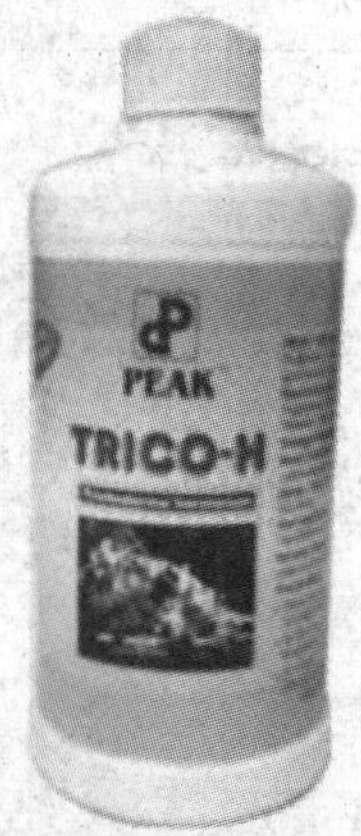


Trico-H is a liquid based bio-fungicide containing the antagonistic *Trichoderma harzianum that* attacks fungal pathogens before they reach to the roots.

## A. BIOLOGICAL CONTROL OF PLANT PATHOGENS

Biological control has been defined by S.D. Garrett in 1965 as "any condition under which or practice whereby, survival or activity of a pathogen is reduced through the agency of any other living organisms except man himself with the result that there is reduction in incidence of disease caused by the pathogen". Pathogens cause many diseases on different parts of the plants. A pathogen is any disease-causing agent. But according to the Federation of British Plant Pathologists (1973), a pathogen means "an organism or a virus, capable of causing disease in a particular host or range of hosts". A unified concept of biological control is given by K.F. Baker and R.J. Cook in 1974 as "the reduction of inoculum density or disease-producing activity of a pathogen or parasite in its active or dormant state, by one or more organisms, accomplished naturally or through manipulation of environment, host or antagonist, or by mass introduction of one or more antagonists."

An antagonist is an organism which has inhibitory relationships with other organism(s). Some antagonists are given in Table 21.1. The aim of biological control is the reduction of disease by: (*i*) reduction of inoculum of the pathogen through disease survival between crops, decreased production or release of viable propagules or decreased spread by mycelial growth (*ii*) reduction of infection of the host by the pathogen and (*iii*) reduction of severity of attack by the pathogen.

**The Inoculum and Inoculum Potential.** Inoculum is a portion of infectious material of an individual pathogen that can cause disease when present in contact with host. In 1956, S.D. Garrett has defined the inoculum potential of a fungus as "the energy for growth of a fungus available for colonization of a substrate to be colonized."

## 1. Historical Background

History of biological control of plant pathogens dates back to 1900s when M.C. Potter in 1908 demonstrated the inhibition in activity of plant pathogens by an accumulation of its own metabolic compounds. In 1926, G.B. Sanford suggested the control of potato scab by green manuring. He proposed two concepts for disease control: (*i*) saprophytic micro-organisms can control the activity of plant pathogens, and (*ii*) the microbial balance of soil can be changed by altering the soil conditions.

**> Table 21.1.** Types of antagonists to be used as biopesticides.

| *Bacteria* | *Actinomycetes* | *Fungi* | *Soil amoeba* |
|---|---|---|---|
| | | **1. Mycoparasites :** | |
| *Aerobacter cloacae* | *Streptomyces* sp. | *Aspergillus* | *Arachnulla* |
| *Agrobacterium radiobacter* | *Micromonospora globosa* | *Chaetomium* | *Arcella* |
| *Bacillus megaterium* | | *Gliocladium* | *Gephyramoeba* |
| *B. subtilis* | | *Penicillium* | *Geococcus* |
| *Bacterium globiformae* | | *Pythium* | *Sccamoeba* |
| | | *Spicaria,* | |
| | | *Talaromyces,* | *Vampyrella* |
| | | *Trichoderma* | |
| | | **2. Nematophagous fungi :** | |
| | | *Arthrobotrys, Dactylaria,* | |
| | | *Dactyleela, Phialospora* | |

Addition of fresh organic material will promote the activity and multiplication of saprophytes which by competition for nutrients and oxygen, and by their excretion will depress the activity and multiplication of the pathogens. In 1927, W.A. Millard and C.B. Taylor also reported the control of scab of potatoes grown in sterilized soil inoculated with *Streptomyces scabies* through simultaneous inoculation of the soil with *S. praecox,* a vigorous saprophyte. G.B. Sanford and W.C. Broadfoot in 1931 then provided an experimental evidence for Sanford's original hypothesis and demonstrated that infection of wheat seedlings by *Ophiobolus graminis* in sterilized soil could be completely suppressed by antagonistic action of various individually co-inoculated species of fungi and bacteria.

R. Weindling in 1932 reported the parasitation by *Trichoderma viride* of many soil fungi. Control of rhizoctonia damping off of citrus seedlings has also been reported by R. Weindling and O.H. Emerson in 1936.

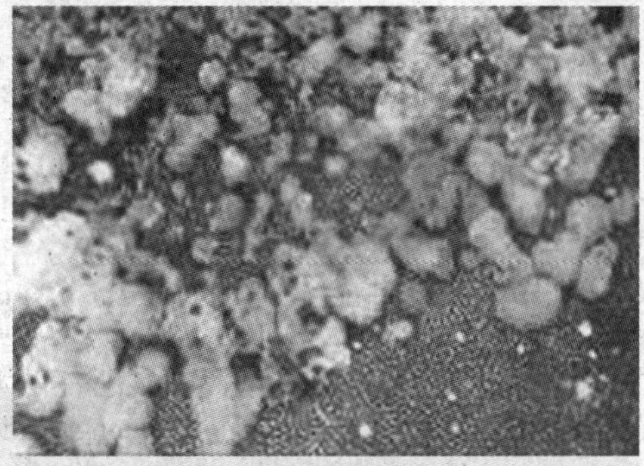

*Trichoderma viride.*

Significance of antibiotic production by soil micro-organisms and their possible role in biological control by antibiosis and fungistasis have been discussed by several workers.

In recent years, significant researches done on mycoparasitism, mycophagy, nematophagy and antibiosis have led to rapid development of training of antagonists to be applied in biological control of plant pathogens/disease in many ways.

## 2. Phyllosphere and Phylloplane, and Rhizosphere and Rhizoplane Regions

Phyllosphere is the immediate vicinity of leaf surface, as rhizosphere is the zone of immediate vicinity of roots, where microbial communities are constantly in dynamic state due to exo-and endogenous sources of nutrients. As rhizoplane denotes root surface, similarly phylloplane refers to leaf surface of plants. Nutrients stimulatory for microorganism on phyllosphere and phylloplane regions consist primarily of plant materials (*e.g.* pollen grains, old petals, etc.) or insect excreta. Stimulation of necrotrophic (destructive) fungi by pollen grains and aphids honey dews is well documented. The antagonistic activities of some phylloplane fungi (of mustard and barley) have been investigated against *Alternaria brassicae and Drechslera graminea.* The antagonists are *Aureobasidium pullulans, Epicoccum purpurascens, Cladosporium cladosporioides* and *Alternaria alternata.* The most significant effects were observed when the spores of leaf surface fungi or their metabolites were sprayed on leaves prior to inoculation of the pathogens. For detail see *A Textbook of Microbiology* by Dubey and Maheshwari (2012).

In 1993, R.R. Pandey and coworkers studied the antagonistic activities of some phylloplane fungi of guava against its fungal pathogens *i.e. Colletotrichum gloeosporiodes* and *Pestalotia psidii.* They found a pronounced inhibition in lesion development after application of spore suspension (3 x $10^5$ propagules/ml) of *A. pullulans, C.cladosporioides*, *E. purpurascens, F. oxysporum* and *Trichoderma harzianum* against *C. gloeosporioides*, and *A. niger, A. terreus, C. roseo-griseum* and *T. harzianum* against *P. psidii.*

Similarly, a variety of soil microorganism are present in the rhizosphere and rhizoplane regions of a plant. Their number remains many times more than the non-rhizosphere one. The increase in microbial number and their activity have been referred to as 'rhizosphere effect'. It is caused by secretion of growth promoting substances (root exudates) and casting of sloughed off root tissues by plants in soil during their growth phases.

In 1988, R.C. Dubey and R.S. Dwivedi studied the population dynamics of microfungi in root region of soybean during different growth stages of the plant. They recorded a gradual increase in numbers of soil fungi, both qualitatively and quantitatively, from seedling to flowering stage, thereafter their frequency declined at the senescence stage. Nowadays, manipulation of soil environment has become a tool for biological control of soil borne plant pathogens.

Several methods have been developed which bring about artificial manipulation of rhizosphere, phyllosphere and soil environment. Consequently number of antagonistic microorganism is increased. This can be done by: (*i*) artificial introduction of antagonists in soil or spraying these antagonists on the aerial parts of plants, (*ii*) modification of soil environment by organic amendments, (*iii*) green manuring, changing soil pH, C : N ratios, temperature, and (*iv*) adding the selective chemicals or heat treatment of plant tissues.

## 3. Antagonism—The Mechanism of Biocontrol

Biological control is principly achieved through antagonism (the inhibitory relationships between microorganisms including plants) which involves : (*i*) amensalism *i.e.* antibiosis and lysis, (*ii*) competition, and (*iii*) parasitism and predation.

**(*a*) Amensalism (Antibiosis and Lysis):** Amensalism is a phenomenon where one population adversely affects the growth of another population whilst itself being unaffected by the other population. Generally amensalism is accomplished by secretion of inhibitory substances. Antibiosis is a situation where the metabolites secreted by organism A inhibit organism B, but organism A is not affected. It may be lethal also. Metabolites penetrate the cell wall and inhibit its activity by chemical toxicity. Generally antimicrobial metabolites are produced by underground parts of plants, soil microorganisms, plant residues, etc. Fig. 21.1 shows *in vitro* interaction of colony of *Celletotrichum gloeosporioides* (a fungal pathogen associated with fruit rot of guava) and *Fusarium oxysporum* (a saprophyte) and formation of inhibition zone between the colonies.

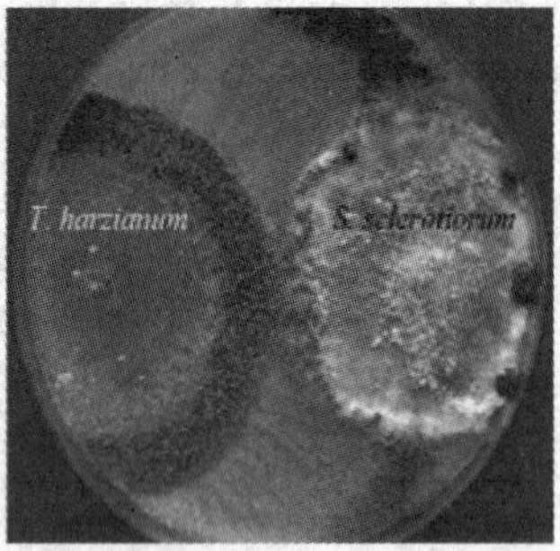


**Fig. 21.1** Mycoparasitism of *Trichoderma harzianum* on Sclerotinia sclerotiorum

Substances noxious to certain soil-borne plant pathogens are secreted by roots of maize, clover, lentil (glycine, phenylalanin) and other legumes, flax (hydrocyanic acid), pine (volatile mono-and sesquiterpenes) and by other plant roots. Other plant residues are the source of phenolic and non-volatile compounds. Similarly, antimicrobial substances (antibiotics) produced by microorganisms (soil bacteria, actinomycetes, fungi) are aldehydes, alcohols,acetone, organic acid, nonvolatile and volalite compounds which are toxic to microbes. Changes in microbial structures (cell wall, hyphae, conidia, etc.), may occur when microorganisms lack resistance against the attack by deleterious agents or unfavourable nutritional conditions. A chemical substance (*i.e.* melanin) is present in their cell walls to resist the lysis. Moreover, cell wall constituents, for example, xylan or xylose containing hetero polysaccharides, may also protect fungal cells from lysis.

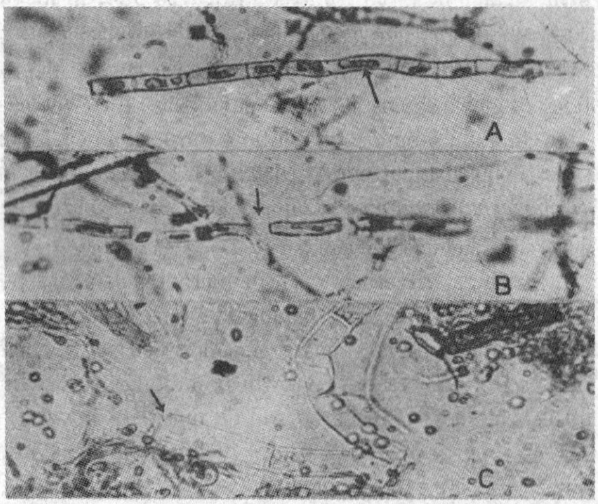


**Fig. 21.2.** Antagonistic effect of fungi on *Macrophomina phaseolina*. A-B, coagulation of cytoplasm of *M. phaseolina* due to toxic metabolites of *A. niger;* C, hyphal bursting of *M. phaseolina* due to toxic metabolites of *T. viride* (R.C. Dubey and R.S. Dwivedi, 1988).

In 1988, R.C. Dubey and R.S. Dwivedi studied the antagonistic potentiality of some rhizospheric microfungi of soybean against a charcoal rot pathogen, *Macrophomina phaseolina.* They observed colony growth inhibition, coagulation of cytoplasm, hyphal discolouration and dissolution of hyphal cell wall of *M. phaseolina* due to secretion of toxic metabolites by *Aspergillus niger* and *Trichoderma viride* (Fig.21.2).

**(*i*) Cell wall degrading enzymes:** The potent antagonists *e.g. Trichoderma harzianum* and *T. viride* are known to secrete cell wall lysing enzymes, ß-1, 3-glucanase, chitinase, and glucanase. However, production of chitinase, and ß-1, 3-glucanase by *T. harzianum* inside the attacked sclerotia of *Sclerotium rolfsii* has also been reported.

**(*ii*) Siderophores.** Siderophores are the other extracellular metabolites which are secreted by bacteria (*e.g. Aerobacter aerogenes, Arthrobacter pascens, Bacillus polymyxa, Pseudomonas cepacia, P. aeruginosa, P. fluorescens, Serratia,* etc.), actinomycetes (*e.g. Streptomyces* spp.) yeasts (*e.g. Rhodotorula* spp.), fungi (*Penicillum* spp.). and dinoflagellates (*Prorocentrum minimum*). Siderophores are commonly known as microbial iron chelating compounds because they have a very high chelating affinity for $Fe^{3+}$ ions and very low affinity with $Fe^{2+}$ ions. Siderophores are low molecular weight compounds. After chelating $Fe^{3+}$ they transport it into the cells. In 1980, J.W. Kloepper and coworkers were the first to demonstrate the importance of siderophore production by PGPR in enhancement of plant growth. Siderophores after chelating $Fe^{3+}$ make the soil $Fe^{3+}$ deficient for other microorganisms. Consequently growth of other microorganisms is inhibited. When the siderophore producing PGPR is present in rhizosphere, it supplies iron to plants. Therefore, plant growth is stimulated.

In recent years, role of siderophores in biocontrol of soil-borne plant pathogens is of much interest. Microbiologists have developed the methods for introduction of siderophore producing bacteria in soil through seed, soil or roots.

In 2012, Poonam Dubey, G.P. Gupta and R.C. Dubey from Gurukul Kangri University, Haridwar have reported that the culture filtrates of *Bradyrhizobium* strains VR2 caused the maximum growth inhibition of *Macrophomina phaseolina* in dual culture causing several deformities in hyphae and sclerotia such as fragmentation, shrinkage and lysis of hyphae, cytoplasm vacuolation, and loss of mycelial pigment. Such pathogens cannot survive in nature in the presence of aggressive antagonists.

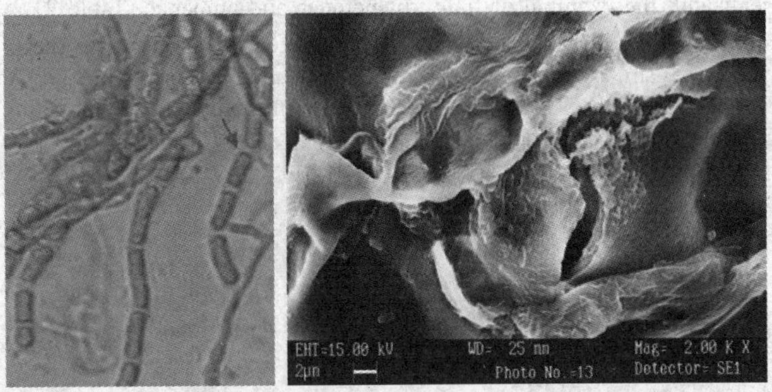


**Fig. 21.3.** Post-interaction events between *Bradyrhizobium* strain VR2 and *M. phaseolina* [courtesy: Poonam Dubey *et al.* New York Sci. J. 2012; 5(8)].

**(*iii*) Toxic metabolites :** Working at Gurukul Kangri University Haridwar, Gupta *et al.* (2001) have reported that a fluorescent *Pseudomonas* $GRC_2$ isolated from potato rhizosphere possessed a strong necrotrophic effect against two fungal pathogens, *Macrophomina phaseolina* and *Sclerotinia sclerotiorum.* Scanning electron photomicrographs from the zone of interaction showed the loss of sclerotial integrity, hyphal shrivelling, mycelial and sclerotial deformity and hyphal lysis in *M. phaseolina*. Hyphal perforations, lysis and fragmentation were observed in *S. Sclerotiorum*. These effects were caused due to secretion of toxic secondary metabolites by *Pseodomonas* $GRC_2$.

**(*b*) Competition :** Among micro-organisms competition exists for nutrients, including oxygen and space but not for water potential, temperature, or pH. Amensalism involves the combined action of certain chemicals such as toxins, antibiotics and lytic enzymes. Success in competition for substrate by any particular fungal species is determined by competitive saprophytic ability and inoculum potential of that species. Competitive saprophytic ability is "the summation of physiological characteristics that make for success in competitive colonization of dead organic substrates".

In 1950, S.D. Garrett has suggested four characteristics which are likely to contribute to the competitive saprophytic ability: (*i*) rapid germination of fungal propagules and fast growth of young hyphae towards a source of soluble nutrients, (*ii*) appropriate enzyme equipment for degradation of carbon constituents of plant tissues (*iii*) excretion of fungistatic and bacterio-static growth products including antibiotics, and (*iv*) tolerance of fungistatic substances produced by competitive microorganisms. Possession of any of the four characteristics and inoculum potential is sufficient for a microbe to get success in microbial competition.

Biological control of *Fomes annosus* by inoculating the freshly cut stumps of pine with *Peniophora gigantea* is a result of competition, as is the control of *Pseudomonas tolaasii* on mushroom by other bacteria. The fate of plant pathogens in competition for food and space depends on other factors also such as, cellulolysis rate that mediates the speed of saprophytic tissue penetration. Species of *Trichoderma and Gliocladium* are the two potent antagonists which produce antibiotics to destroy mycelia of other fungi. Some examples of successful competitors competing for nutrients with other microorganism are : bacteria vs *S. scabies* (for oxygen), soil amoebae vs *Gaeumannomyces graminis* var. *tritici* (wheat roots), *T. viride* vs *Fusarium roseum* (wheat straw), *Chaetomium* sp. vs *Cochliobolus sativus* (wheat straw,), *Arthrobacter globiformis* vs *Fusarium oxysporum* f. *lini* (glucose and nitrate).

**(*c*) Predation and Parasitism :** Predation is an apparent mode of antagonism where a living microorganism is mechanically attacked by the other with the consequences of death of the farmer. It is often violent and destructive relationship. Parasitism is a phenomenon where one organism consumes another organism, often in a subtle, non-debilitating relationship. These aspects are dealt with the example of fungi, nematodes and amoebae (Table 21.2).

**(*i*) Mycoparasitism :** When one fungus is parasitized by another one the phenomenon is called as mycoparasitism. The parasitising fungus is called hyperparasite and the parasitized fungus as hypoparasite (Fig. 21.4). Mycoparasitism commonly occurs in nature. As a result of inter-fungus interaction *i.e.* fungus-fungus interaction, several events take place which lead to predation viz., coiling, penetration, branching, sporulation, resting body production, barrier formation and lysis (Fig. 21.4)

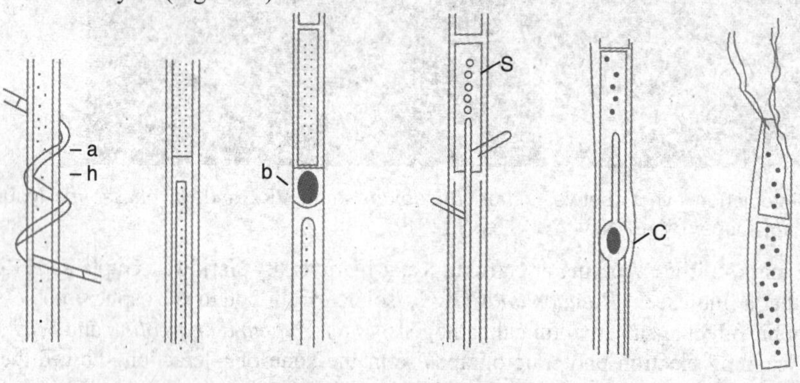


Fig. 21.4. Post-interaction events during mycoparasitism. A, coiling (a, antagonist; h, host hypha); B, penetration; C, barrier formation (b) by host; D, branch formation and sporulation (s) by antagonist; E, chlamydospore (c) formation; F, lysis of host hypha (diagrammatic, after R.C. Dubey and R.S. Dwivedi, 1986).

In coiling (A) an antagonist, the hyperparasite (*a*) recognises its host hyphae *i.e.* the hypoparasite (*h*) among the microbes and comes in contact and coils around the host hyphae. The phenomenon of recognition of a suitable host by the antagonists has been discovered in recent years. In 1985, MS. Manocha has given a molecular basis of host specificity and host-recognition by mycoparasites. Cell wall surface of host and non-host contains D-galactose and N-acetyl D-galactosamine residues as lectin binding sites. With the help of lectins present on the cell wall, an antagonist recognises the suitable sites (residues of lectins) and binds the host hypha. As a result of coiling, the host hypha loses the strength. If the antagonist has capability to secret cell wall, degrading enzymes, it can penetrate the cell wall of host hyphae and enter in lumen of the cells. The event of entering in lumen of host cell is known as penetration (Fig. 21.4. B). Several cell wall degrading enzymes such as cellulase ß-1, 3-glucanase, chitinase,etc. have been reported.

Sometimes host develops a resistant barrier ( Fig. 21.4B) to prevent the penetration inside the cell. Cytoplasm accumulates to form a spherical, irregular or elongated structure, so that the hypha of antagonist could not pass towards the adjacent cells of the hypha (Fig. 21.4C). Depending upon nutrition, the antagonist forms branches and sporulates (s) inside the host hypha (Fig. 21.4D). Until the host's nutrients deplete,the antagonist produces resting bodies, the survival structures, for example, chlamydospores (*c*) inside the host hypha (Fig. 21.4E). Finally post-infection events lead to lysis of the host hypha (Fig. 21.4F) due to loss of nutrients and vigour for survial. Example of parasitism and post-infection events are given in Table 21.2 and Fig. 21.4.

**> Table 21.2. Examples of predation and parasitism**

| *Mode of antagonism* | *Plant pathogens* | *Antagonists (hosts)* | *Post-infection events* |
|---|---|---|---|
| Mycoparasitism : | *Botrytis alli* | *Gliocladium roseum* | Penetration of hyphae' |
| | *Cocchliobolus sativus* | *Myrothecium verrucaria and Epicoccum purpurascens* | Antibiosis and penetration |
| | *Rhizoctonia solani* and *Fomes annosus* | *Trichoderma viride* | Coiling, cytoplasm coagulation |
| | *Sclerotium rolfsii* | *T. harzianum* | Coiling, penetration and lysis |
| Nematophagy : | *Heterodera rostochiensis* | *Phialospora heteroderae* | Penetration of cysts and egg killing |
| Mycophagy : | *Cocchliobolus sativus* | Soil amoebae | Perforation in conidia |
| | *Gaeumannomyces graminis* var. *tritici* | Soil amoebae | Penetration and lysis of hyphae |

*Source* : Various mycological/microbiological research and review papers.

***(ii)* Nematophagy.** This is the phenomenon of eating upon nematodes by fungi. However, several nematode eating *i.e.* nematophagous fungi (NF) are known which develop different kinds of trap (T), arrest the pathogenic nematodes (N) and finally kill them (Fig. 21.5). Morphological and biochemical aspects of trap formation is known. Examples of nematode trapping fungi are *Arthrobotrys, Dactylaria, Dactyleela,* etc. *Phialospora heteroderae* penetrates the cysts and kills the eggs of *Heterodera rostochiensis.*

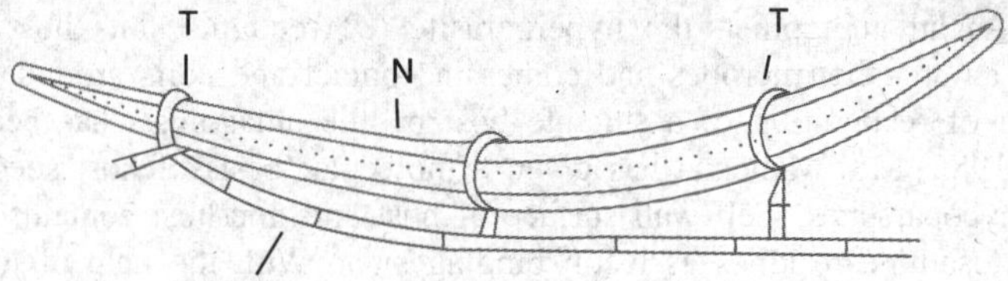


**Fig. 21.5.** *Nematophagy* : N, a nematode; NTF, a nematode trapping fungus; T, a ring like trap formed by NTF (diagrammatic).

*Paecilomyces lilacinus* and *Verticillium chlamydosporium* have shown a greater efficiency in biocontrol of endoparasitic nematodes. *P. lilacinus* has demonstrated tremendous potential as biocontrol agent of some important parasitic nematodes such as *Meloidogyne incognita, Heterodera pallida, Globodera pallida.* It frequently infects the eggs and occasionally the females of *M. incognita.* When *P. lilacilus* spores are applied in different dosages. Population of *M. arenaria* got reduced. This fungus is known to control the nematode and reduce root-knot formation in tomato. The fungal hyphae first network around the egg and finally penetrate the eggs. It has been found that maximum fungal growth occurs at 36°C.

Besides the fungi eating on nematodes, a spore forming bacterium, *Bacillus penetrans* kills the nematode and, therefore, is used for the control of *Meloidogyne* sp. *B. penetrans* is resistant to nematicides. Being an obligate parasite, this bacterium can not be grown in axenic culture. The bacterium shows host specificity and its spores survive for a long time. These spores adhere to the surface of infectious second-stage female larvae and eat on it. Adherance is followed by infection but it is not apparent until the adult stage comes.

**(*iii*) Mycophagy** : Mycophagy is the phenomenon of feeding on fungi by amoebae. In recent years, mycophagy has become a new field of research as far as biocontrol of soil-borne plant pathogens is concerned. Many soil amoebae are known to feed on pathogenic fungi. For example, take-all disease of wheat caused by *Gaeumannomyces graminis* var. *tritici* was very severe in Australia. Finally it was found that some natural soil exhibited suppressiveness against this disease. Microbiological analysis of soils showed the presence of several amoebae in the soil. These amoebae played a significant role in take-all decline of wheat. The antagonistic soil amoebae *e.g. Arachnula, Archelle, Gephyramoeba, Geococcus, Saccamoeba, Vampyrella,* etc. make perforation on the hyphal wall of *Cochliobolus sativus, G. graminis* var. *tritici, Fusarium oxysporum* and *Phytophthora cinnamomi* and on the conidial wall of *C. sativus* and *F. oxysporum* and develop round cysts on the lysed hyphae.

S. Chakraborty and coworkers in 1983 described the following three major steps of feeding on fungal propagules by soil amoebae ( Fig. 21.6 A-C ).

- **Attachment** : Attachment of trophozoites of amoebae (*a*) to fungal propagules *i.e.* conidia (*c*) or hyphae (*h*) appears to be a matter of chance. It takes place by chemotaxis or thigmotaxis (Fig. 21.6A).
- **Engulfment:** Fungal propagules (*e.g.* spores, conidia, fragments of hyphae) are fully engulfed by amoebae (*B*). The small trophozoites attached the hyphal wall or spore and make perforations on it.
- **Digestion:** The completely or partially engulfed propagules/cyloplasm of the host fungi are digested in a large central vacuole formed inside the cyst (Fig. 21.6 C).

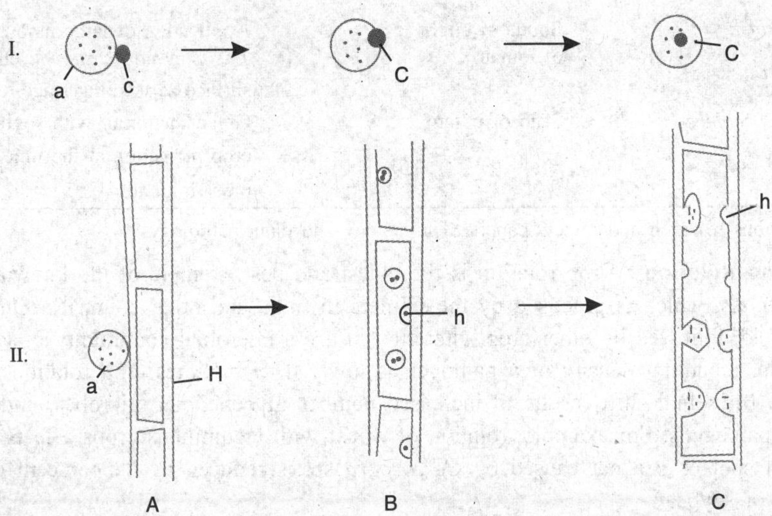


Fig. 21.6. Mycophagy (A–C). Feeding of amoebae (a) upon fungal conidia (Ic) and fungal hypha (II h). A, attachment of amoeba; B, engulfment (I) and hole formation (II); C, digestion of conidium (I) (diagrammatic).

## 4. Applications of Biological Control in Field

Practical application is the translation of *in vitro* experiment into *in vivo* conditions. Owing to variations in environmental factors, sometimes, antagonistic potential of certain biocontrol agents fails in *in vivo* conditions. Now-a-days researches on induction of antagonistic potential and production of mutants adjustable to stress conditions are being carried out in most of the countries.

It is obvious that a suitable antagonist, present in rhizosphere, fails to antagonise the pathogen and control the disease, possibly due to its low inoculum potential. Some of the methods are now available which: (*i*) can decrease virulence of the pathogens, (*ii*) increase hosts's resistance, and (*iii*) stimulate the antagonistic potentialities and intensify their activity. These may be accomplished by one or more methods (Table 21.3) as discussed below :

**> Table 21.3. Examples of biological control of soil-borne plant pathogens.**

| *Plant pathogens* | *Diseases* | *Methods applied for control* |
|---|---|---|
| *Cephalosporium gregantum* | Brown stem of soybean | Five years cropping of maize before soybean crop. |
| *Fusarium oxysporum* f. *pisi* | Fusarium wilt of pea | Crop rotation with cereals or rape |
| *Meloidogyne incognita* | Root knot of tomato | Incorporation of chopped alfalfa hay, oat straw at high dosages into soil a month before planting |
| *Pythium ultimum* | seedling blight of lettuce | Incorporation of cereals or cruciferous cover crop into soil two weeks prior to planting |
| *Phytophthora palmivora* | Seedling root knot of papaya | Planting seeds in small quantity in pathogen free virigin soil placed in infected field |
| *Rhizoctonia solani* | Black scurf of potato | Treatment of infected tubers with *Gliocladium virens* |

| | | |
|---|---|---|
| *Sclerotium rofsii* | Southern blight of tomato | Application of large amount of *Trichoderma harzianum* onto *soil* surface at seedling stage of tomato |
| *Streptomyces scabies* | Scab of potato | Green manuring with soybean cover crop, irrigation at the time of tuber initiation |

*Source* : Various research and review papers of mycology and plant pathology.

**(*a*) Crop Rotation :** Crop rotation is the oldest and best example of the known biological control methods, which is practised by the farmers in India and other countries. On rotating a crop in a field, followed by other crop, alteration in gross microbial community in soil is done. By doing so, inoculum density of a pathogen is lowered. Sometimes crop rotation has a long term effect. Microbial shift results in increased number of beneficial microbes, and decreased number of pathogens. For example, rotation of wheat with leguminous crops affects take-all of wheat. Leaf spot of pea nut caused by *Cercospora* sp. is reduced by 88 per cent on rotation with maize or soybean.

**(*b*) Irrigation :** Irrigation practice is a versatile mean of manipulation of microbial community in soil. Many microorganisms, at high moisture level do not grow due to oxygen stress in soil. Consequently, anaerobic bacteria rapidly grow and participate in the antagonism. For example, practice of raised soil beds for controlling verticillium wilt of cotton, and root knot of cotton (caused by nematodes) are done in some countries.

**(*c*) Alteration of Soil pH :** Many antagonistic fungi luxuriantly grow at low soil pH 4.0 and many more do not. By addition of certain fertilizers, such as ammonium sulphate, sodium nitrate, calcium nitrate, ammonium chloride and lime, soil pH is considerably changed and there develops a chance for change in community size. For example, soil acidity favours *Fusarium solani f.s. phaseoli* and *Rhizoctonia solani* but suppresses *S. scabies* and *Verticillium albo-atrum.* Take-all of wheat is more severe at alkaline pH. Fertilizers applied on forage directly affect root exudation and in turn on rhizosphere microflora.

**(*d*) Organic Amendments :** Organic materials viz., crop residue, chitin, chopped leguminous hay, green manures, oil cake powders, rice husk, wheat and paddy straw and compost manures, when applied in field soil, bring about changes in the spectrum of micropial community and intensify their activities. The control of *F. solani f. phaseoli* causing disease on bean root with a high C: N ratio has been reported. Take-all of wheat was controlled by amending soil with gelatin. The amendments and crop residues can and should be one of our best weapons for elevating soil-borne plant disease without seriously polluting the soil environment with pesticides.

**(*e*) Treatment of Soil with Selected Chemicals :** The idea of recolonization and survival of treated soil with some chemicals by microorganism was given. In soil treated with carbon disulphide for the control of *Armillaria mellea* from infected citrus root, population of *T. viride* rapidly increased. Consequently, *A. mellea* was killed directly by fungicidal action of carbon disulphide and necrotrophic activity of *T. viride.*

In 1981, P.P. Pathak and R.S. Dwivedi have reported that population of antagonistic fungi, such as *Aspergillus terreus, Penicillium citrinum Cephalosporium roseogriseum, T. viride* etc. increased in soil treated with fungicides, insecticides and herbicides; consequently the wilt of tomato caused by *Fusarium oxysporum f. lycopersici* decreased. Increase in numbers of *Aspergillus, Penicillium* and *Trichoderma* in soil treated with benomyl have also been reported. However, in biocontrol systems, pesticide-stimulated antagonists in soil may help in the integrated system of disease control.

*(f)* **Introduction of Antagonists :** In this method aim lies in the introduction of such antagonists that can intensify microbial interaction resulting in control of disease and or disease causing organism. A potential antagonist is isolated from a specialized niche (a niche is the habitat of organisms where they live and function with respect to other organisms and the environment), artificially multiplied on nutrient media and introduced in the same habitat for microbial interactions and control of a particular disease. It is unlikely that an antagonist can be applied for a number of disease/pathogen in a varied habitats. The means of introducing antagonists are seed inoculation, vegetative part inoculation and soil inoculation.

In recent years, several commercial products from microorganisms such as Suppresivit, Tris 002, Ecofit, SoilGard, Protus, etc. (Table 21.4) have been prepared by several companies to use them for the control of plant pathogens and diseases.

**> Table 21.4. Some of the biocontrol products available commercially**

| *Trade name* | *Active ingredients* | *Formulation** | *Target pathogen/effects* | *Manufacturer/ Distributor* |
|---|---|---|---|---|
| SUPRESIVIT | *Trichoderma harzianum* PV5736-89 | WP | Damping off of ornamentals, fungal diseases of peas | FYTovita, Czech Republic |
| TRI 002 | *T. harzianum* KRL AG-2 | G | Plant growth stinulation, strengthening of plants against pathogens | Bioworks, Inc., Geneva, USA, PLANT SUPPORT, B.V., The Netherlands |
| ECOFIT | *T. viride* | WP | Species of *Fusarium, Pythium* and *Rhizoctonia* | Hoechst Schering AgrEvo, Mumbai, India |
| SOILGARD 12G | *Gliocladium virens* | G | Damping off and root rot pathogens | Thermo Trilogi, USA |
| PROTUS | *Talaromyces flavus* | G | Stimulates plant growth, strengthens plant against the pathogens | Prophyta Gmbh, Germany |

* Wettable powder; G, granules.

*(i)* **Seed inoculation.** Nowadays seed inoculation technology is developed in India and other countries as well. Different strains of *Rhizobium* spp. are now available which have been developed in a view of increased nodulation, nitrogen fixation and yield. It helps in disease control also. However, seeds of barley, wheat and oat soaked in aqueous suspension of *Bacillus subtilis* ($10^6 - 10^7$ cell/ml) and grown in fields infested with *R. solani, Pythium* sp. and *Fusarium* sp. gave a good result; wheat and oat showed increased tillering and dry weight, and barley and oat gave up to 90 per cent increase in yield. Similarly water suspension of *B. subtilis* ($10^6 - 3 \times 10^7$ cell) and *Streptomyces* sp. ($5 \times 10^4 - 3 \times 10^5$ cells) applied on seeds of oat resulted in an increased yield by 40 per cent and 45 per cent respectively.

Proper inculation of seed is essential for nitrogen production.

**Use of Biofertilizers for Biocontrol of Plant Diseases :** Diatotrophic bacteria have been reported to control may plant disease. Treatment with *Azotobacter* of rice, soybean and cotton seeds resulted in reduced seedling mortality by inhibiting the pathogen. The biocontrol mechanisms have been explained to be due to two facts: (*i*) antagonism, and (*ii*) development of host resistance against pathogens. Antagonism takes place by secretion of antibiotic substances and toxic metabolites, release of acids that make unfavourable *p*H for growth, competition for nutrients or stimulation of host defence mechanism or direct parasitism.

In 2001, P. Sudhakar and coworkers found that the culture filtrates of *Azotobacter chroococcum, Azospirillum brasilense* and *Beijerinckia indica* inhibited the spore germination of three fungal pathogens of mulberry (e.g. *Phylloctinia corylea, Pseudocercospora mori* and *Cerotelium fici*) *in vitro. A. chroococcum* caused maximum inhibition followed by *A. barasilense* and *B. indica.* A similar result was also obtained with foliar application in field condition.

V. K. Deswal, R.C. Dubey and D.K. Maheshwari in 2003, working on a symbiotic nitrogen fixer, siderophore- and IAA-producer, phosphate solubiliser and antagonistic strain of *Bradyrhizobium* (*Arachis*) sp., found an enhanced seed germination, seedling growth and biomass yield. This bacterium inhibited the growth of a fungal pathogen (*Macrophomina phaseolina*) that causes charcoal rot in peanut seedlings.

Moreover, activity of antagonists can be stimulated by adopting both seed dressing and chemical treatment methods where individually seed treatment (seed dressing with microorganism) or chemical treatment fails to control the disease.

**(*ii*) Vegetative part inoculation.** When the vegetative parts of plant are wounded or plant is felled, the fresh wound releases an adequate amount of nutrients which serve as substrate for the microorganisms coming to its contact. Thus, the wounds support for rapid growth and sporulation of pathogens. Also the antagonists colonize the wound in the same manner. Therefore, inoculation of wounds with antagonistic saprophytes for disease control are now been practiced in many countries. Some of the examples are discussed in this connection.

Research work of J. Risbeth (1963) gives a most successful example for control of invasion of vegetative parts of pine (freshly cut pine stumps) by *Fomes annosus,* by sprinkling pellets of spores of *Peniophora gigantea.* This method is widely practiced in England.

The freshly cut surfaces of carnation cuttings when transplanted into soil were infested by *Fusarium roseum, f. cerealis*, a major uncontrolled disease. The protection of carnation was provided to the growing plants by inoculating the freshly cut surfaces with suspension of *B. subtilis.*

*Trichoderma viride* and *B. subtilis* are used to protect pruning wounds on apple tree shoots from the invasion by *Nectria galligena* causing cankers on it. Inoculation of drill wounds in red mapple with conidia of *T. harzianum* in glycerol gave the complete protection against invasion by hymenomycetes within 21 months after treatment. Control of *Agrobacterium tumefaciens* causing crown gall on many plants was done by its another species, *A. radiobacter* on artificially made wounds on roots of tree seedlings. *A. radiobacter* produced an unusual antibiotic substance which selectively inhibited the stain of *A. tumefaciens.* This antibiotic is known as nucleotide bacteriocins. A bacteriocin can be defined as "an antibiotic produced by certain strains of bacteria which is active against other strains of the same or closely related species".

**(*iii*) Soil inoculation :** Soil is a unique habitat which harbours a vast majority of microorganisms in a continuous dynamic state by actions and interactions. Now-a-days researches on introduction of antagonists in soil for disease control are in progress. Damping off of seedlings caused by *Pythium ultimum* and *R. solani* was reduced by introduction of *Bacillus* or *Streptomyces* species into the steamed soil.

The satisfactory requirements for microbiological preparations at commercial levels are: (*i*) a standardized and dosable content of antagonist, (*ii*) the antagonists must remain viable for several months, (*iii*) excessive contamination by foreign microorganisms should be avoided because of the possible influence on the activity of the antagonists and (*iv*) the antagonists must be absolutely non-pathogenic on plants.

**(*g*) Use of Mycorrhizal Fungi :** Use of vesicular-arbuscular mycorrhizal (VAM) fungi in biocontrol of soil-borne plant pathogens has increased its significance in recent years. A VAM fungus, *Glomus mosseae,* increased the resistance of tobacco roots to infection by *Thielaviopsis basicola* by increasing the arginine contents in roots. This amino acid inturn inhibited chlamydospore formation by the pathogen. *G.mosseae* reduces the effcts of three pathogens (*e.g. Macrophomina phaseoli, R. solani* and *F. solani*) individually on soybean in autoclaved soil.

VAM fungi reduced the severity of disease caused by *Olpidium brassicae* on lettuce and tobacco and *Rhizoctonia solani* on *Brassica naprus. Glomus fasciculatus* when added in soil reduced severity of *Sclerotium rolfsii* on peanut. VAM fungi are known to reduce the inoculum multiplication and infection of nematodes, viruses and bacteria as well. In addition to certain exceptions, VAM fungi increase the growth and yield of certain crop plants.

Ectomycorrhizal fungi also warrant the penetration of soil-borne pathogens in host roots. Several ectomycorrhizal fungi such as *Lacaria laccata, Leucopaxillus cereails, Suillus luetus*, etc. are known to escape infection by *Phytophthora cinnamomi* on roots of *Pinus taeda.*

**(*i*) Mechanism of Interactions and Protection.** VA mycorrhizal fungi retard the development of pathogens in root systems and increase disease severity in non-mycorrhizal roots. This systemic influence can be attributed to better nutrition, enhanced plant growth and physiological stimulation in mycorrhizal plants. Roots colonized by a VAM fungus exhibit high chitinolytic activities. These enzymes can be effective against the other fungal pathogens. Under direct influence of mycorrhizal fungi , root tissues become more resistant to pathogenic attack. This induced resistance is strictly limited to the site of host-endophyte interaction and will only affect soil-borne pathogen. Thus, application of selected VA mycorrhizal fungi offers the possibility of increasing resistance against soil-borne pathogens.

Ectomycorrhizal fungi develop a mantle around the roots which is composed of tightly interwoven hyphae in several layers. It completely covers the root meristem and cortical tissues. Thus, the mantle provides physical protection against the pathogens. Thickness of mantle varies with fungus types.

## 5. Genetic Engineering of Biocontrol Agents

Occurrence of a number of interactions among the microorganisms provides a broad opportunity for genetic engineering of microbial biocontrol agents directed at plant pathogens or other microorganisms against weeds and pests. Several biocontrol agents have been successfully employed for such purposes in experimental or commercial agriculture.

The effectiveness of biocontrol agents can be intensified by gene splicing, gene cloning and transformations. It is, however, obvious that bacterial pathogenicity on plant is determined by several genes. Certain genes encode enzymes involved in the biosynthesis of phytotoxins, growth hormones or enzyme capable of degrading plant cell wall or other constituents. Since these genes are positively needed for pathogenesis, their inactivation (*e.g.* through deletion-replacement technique) would destroy the organisms, their pathogenic potential or reduce virulence.

In India, at National Institute of Immunology (NII), New Delhi some genes of Baculoviruses have been engineered to make them 'suicide' (self destroying) after they have infected and killed the insect pests.

Genetic material of Baculoviruses is covered by a protein coat called polyhedrin. It is synthesised by a gene of the viruses. Scientists at NII have made a suicide squads of Baculoviruses by removing the polyhedrin gene. After application, the genetically engineered Baculoviruses infect and kill the larvae, and millions of viruses, after completing their life cycle, are released into the environment. Soon after release, viruses die within a few hour as they lack polyhedrin, therefore, fail to survive. By making the viruses crippled before application, possible risk associated with virus release can be completely eliminated. Field trials of these have also been conducted at Tamil Nadu Agricultural University, Coimbatore.

## B. BIOCONTROL OF INSECT PESTS

A variety of synthetic insecticides have been evaluated which generally affect non-target organisms (where there is no target to kill the non-pathogens). As a result of their use in agriculture, many beneficial organism are killed. In turn, they cast hazardous effects on man also. Who can forget *en masse* killing of more than 8,000 people within a night, many animals and plants, when methyl isocyanate (MIC) gas leaked out in the night of 2/3 December, 1984 from the underground reservoir of Union Carbide Factory of Bhopal. Many gas-affected people are suffering even today. Therefore, use of insecticides should be discouraged as far as possible.

There is a vast majority of microorganisms such as viruses, bacteria, fungi, protozoa, and mycoplasma, known to kill the insect pests. The suitable preparations of such microorganisms for control of insects is called as "microbial insecticides". The microbial insecticides are non-hazardous, non-phytotoxic and selective in their action. Pathogenic microorganisms which kill insects are viruses (DNA containing viruses *e.g.* baculovirus iridovirus, entomopoxvirus), bacteria (*Bacillus thuringiensis, B. popilliae, B. sphaericus, B. moritai*) and fungi (*e.g. Aspergillus ,Coelomomyces, Entomophthora, Fusarium, Hirsutella, Paecilomyces*).

### 1. Microbial Pesticides (Microbial Insecticides)

(*a*) **Bacterial Pesticides :** *B. thuringiensis* is a widely distributed bacterium. It can be isolated from soils, litters, and dead insects. It is a spore forming bacterium and produces several toxins viz., α-, β-and γ-exotoxins and δ-endotoxin which can be obtained in crystalline forms. β-exotoxin contains adenin, ribose, glucose, etc., whereas δ-endotoxin is composed of a glycoprotein subunit. These toxins have insecticidal properties. Structural formula of ß- exotoxin is given in Fig. 21.7. *Bacillus thuringiensis* has been found a strong antagonist to be used as a biocontrol agent. It is an aerobic and spore forming bacterium, pathogenic to larvae of lepidoptera. After ingetion of spores, larvae are damaged, as the rod-shaped bacterial cell secretes at the opposite end, a single large crystal in the cell which is called parasporal body. This crystal (toxin) is proteinaceous in nature. In nature this crystal is short-lived because it is damaged by day light It gets dissolved in alkaline solution and in alkaline juice of caterpillar's digestive cavity (Riviere, 1977).

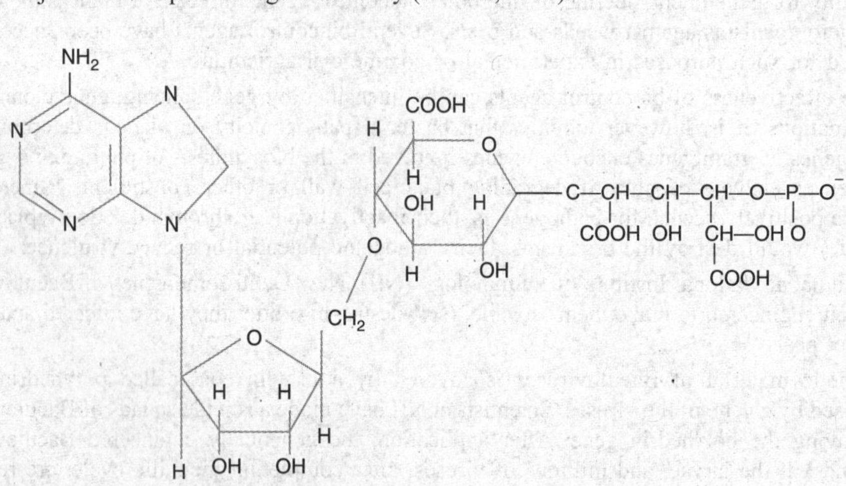


Fig. 21.7. Chemical structure of β-exotoxin.

This bacterium consists of several different strains (subspecies) each of which produces a different toxin that can kill certain specific insects. This classification is based on flagellar agglutination which demosntrates the existence of flagellar antigen (H-antigen). For example, *B. thuringiensis* subsp. *kurstaki* is toxic to lepidopteran larvae including moth, butterfly, skipper, cabbage worm, and spruce budwrom. *B. thuringiensis* subsp. *israelensis* kills diptera such as mosquitoes and black flies. *B. thuringiensis* subsp. *tenebrionis* is effective against coleoptera (beetles) such as potato beetle and boll weevil. Besides, there are more different strains each is toxic to different insects.

In 1979, about 2 million hectares of forest was sprayed with *B.thuringiensis* subsp. *kurstaki.* In 1986, about 74% forest was sprayed by the bacterium. For biocontrol of insect pests, about $1.3\text{-}2.6 \times 10^8$ spores of *B. thuringiensis* subsp. *kurstaki* per square foot of target are are sprayed. Spores should be sprayed in such a way that this time may coincide with peak of larval population of target insect so that a large number of insects may be killed.

Moreover, presence of plasmids in bacterial cells and synthesis of toxins in the form of crystals have been reported. The expression of crystal formation by the plasmid of *B. thuringiensis* in the transformed cells of *B. subtilis* and *E. coli* through gene cloning has also been confirmed.

Microbial pesticides are produced by many companies by using viruses, bacteria and fungi. *B. thuringiensis* preparations have been produced in the U.S.A., France, U.S.S.R., etc. in the form of wettable powder and water suspensions. But in U.S.A. the commercial formulation of ß-exotoxin from *B. thuringiensis* has been banned.

**(*i*) Use of transgenic plants : a new area of biocontrol.** Instead of spraying cell suspension of *B. thuringiensis*, scientists have now started producing transgenic plants where killer genes of bacterium are introduced. Recently, the scientists at the US Multinational Monsanto Co., Australia, Cotton Seed Distributors and the Council of Scientific and Industrial Research and Organisation (CSIRO) of Australia have produced a genetically engineered species of cotton known as 'killer cotton'. It kills the predators especially the bullworm (*Heliothis* sp.).

The leaves of transgenic cotton plants, produced through cell culture, secrete the lethal toxins. When the bullworm eats upon cotton leaves, toxins is taken up by them. Toxin is activated in their guts resulting in death of bullworm. Toxin does not harm the spiders, humans and the other mammals.

In 1980s, one groups of scientists isolated delta endotoxin gene of *B. thuringiensis* var Kurstaki HD-1 and inserted into *Pseudomonas fluorescens* and *Agrobacterium radiobacter*. These bacteria live in rhizosphere region of plants. The delta-endotoxin gene of the bacteria was introduced into root cells of tomato through bacteria. The transgenic tomato killed three pests of tomato. Similarly, transgenic tobacco plants were produced through cell culture which killed hornworm (*Manduca sexta*).

*B. thuringinensis* genes have also been cloned in *E. coli*. Such transgenic *E. coil* produced larvicidal proteins in large amount which accumulate in cells to form large crystals. These are visible under light microscope.

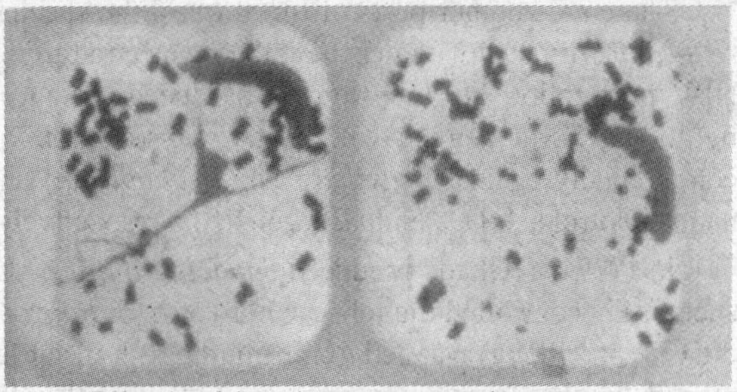

Effect of beet armyworm larve on tobacco leaves from normal plants.

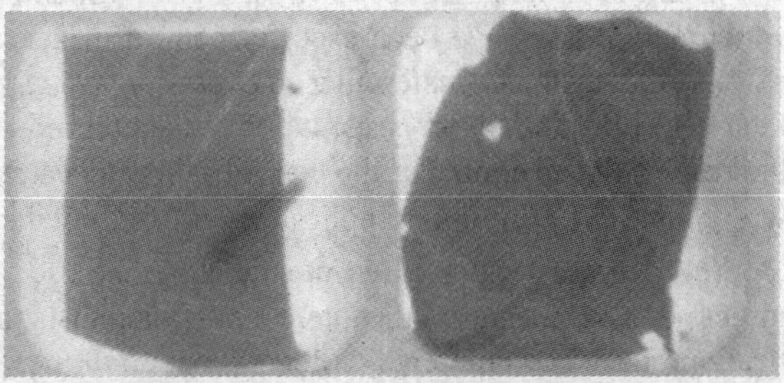

Tobacco leaves taken from transgenic plants Express Bt-toxin in their chloroplasts.

**(b) Viral Pesticides** : A number of viruses has been discovered which belong to groups Baculoviruses and cytoplasmic polyhedrosis viruses (CPV). Preparations of viruses or their products have been developed as effective biopesticide and being successfully used for the control of insect pests in agriculture, forestry and horticulture. This method of disease control is free from pollution, toxicity or any hazards related to plant or animal health. However, these viruses are specific and have no harmful effects on useful insect pollinators, insects yielding useful products, warm blooded animals and even man. After application viruses get entered into the mouth and digestive tract of insect pest and kill them.

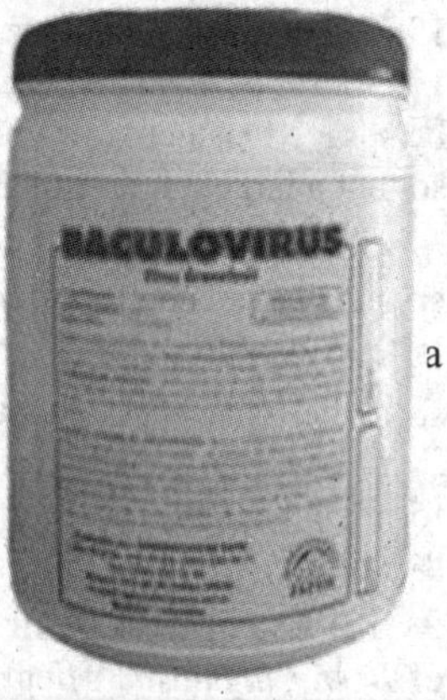


Baculovirus-based viral biopesticides for insect control

Nuclear polyhedrosis viruses (NPVs) which belong to subgroup of Baculoviruses have been used for the preparation of potential pesticides. *Heliothis* sp. is a cosmopolitan insect pest attacking at least 30 food and fibre yielding crop plants. They have been controlled by application of NPVs of *Baculovirus heliothis*. In 1984, a nucleopolyhedrovirus was used as a pest control agent for the Douglas fir tussock moth, which was a notable success. This resulted in an increased effort to understand the molecular biology of the baculovirus which in turn led to industrial interest in the development of baculovirus pesticides in the 1990's.

In 1975, Environmental Protection Agency, U.S.A. registered the *B. heliothis* preparations. At present it is marked as wettable powder under the name 'Elcartm' (Sandoz Inc.), 'Biotrol' VHZ (Nutrilite Products Inc.) and 'Viron/H' (International Minerals Chemical Corp.).

*Lymantria dispar*, commonly known as *gypsy moth*, is a serious pest of forest trees. It has been successfully controlled by *gypsy moth Baculovirus i.e. NPV preparations*. NPVs product has been registered under the trade name '*Gypchek*' (the Environmenta Protection Agency, U.S.A) to control gypsy moth population on deciduous hardwood trees.

There are another pests of forest trees, the species of sawfly (*Neodiprion sertifer, N. lecontei, N. pratti pratti,* etc), which have been successfully controlled by their NPVs. In the U.S.A., a NPV preparation of *N. sertifer* is produced commercially under the trade name 'Polyvirocide' by Indian Farm Bureau Cooperative Association. Similarly, NPVs have also been produced commercially against *Brassica pest* (*Trichoplusia*) and cotton pest (*Spodoptera litura*) under the name 'biotrol - VTN' and `Biotrol-VSE' respectively (Nutrilite Products Inc., U.S.A.).

In 1977, M. Okada produced *Spodoptera litura* nuclear polyhedrosis virus on a mass scale and sucessfully controlled the tobacco cutworm (*S. litura*) The artificial diet containing nuclear polyhedra ($5 \times 10^7$ PIB/g) was fed to the middle fifth instar larve that died in sixth instar larva. The larval mortity by ultra low volume spraying ($2 \times 10^8$ PIB/ml, 0.51/10 day) caused maximum mortality.

There is another group of virus, CPVs, which are present in more than 200 insects, out of which only a few have been used for control of insect pests. For the first time in France, a CPV of pine processionary caterpillar (*Thaumetopoea pityocamps*) was applied as pesticide used for the control of pine forest pests. In 1974, Japan produced a commercial preparation of CPV under the name 'Matsukemin' for the control of a pine caterpillar (*Dendrolimus spectabilis*). This insect was controlled fully upon application of $10^{11}$ polyhedral inclusion bodies per hectare.

**(c) Mycopesticides** : Recent studies on the use of entomopathogenic fungi for the control of insect pests has increased the interest of mycologists throughout the world. Fungal preparations are being produced commercially. Mode of action of these fungi is different from viruses and bacteria. The infective propagules *e.g.* conidia, spores, etc. of the antagonistic fungi reach the haemocoel of the insect either through integument or mouth. They get attached to epicuticle, germinate and penetrate the cuticle either by germ tubes or infection peg. They multiply in haemocoel followed by secretion of mycotoxins which result in death of insect hosts. Thereafter, mycelia spread saprophytically and grow outside the integument which later on produce conidiophores and conidia.

In England, *Verticillium lecanii* has shown as a potential antagonist against aphid pests, for example *Macrosiphonrella sanbornii, Branchycaudus helichrysi* and *Myzus persicae* affecting *Chrysanthemum* in green houses. In USSR, spraying of aphids and spider mites with *Entomophthora thaxteriana* and *E. sphaerosperma* resulted in 95 per cent mortality within 24 h by secreting the mycotoxins. *E. thaxteriana* suspension when applied on aphids of apple trees resulted in about 74 per cent mortality without harmful effects on natural predators. Also, use of 1.5 kg/ha of 'Boverin' (*i.e.* 30 × $10^9$ conidia of *Beauveria bassiana* /g) with reduced dosages of chemical insecticide *e.g.* chlorophos, has been recommended in Russia for the control of Colorado beetle. A commercial preparation of *B.bassiana* has also been produced under the product name 'Biocontrol FBB' (Nutrilite Products Inc., USA).

## C. BIOLOGICAL CONTROL OF WEEDS

Weeds are the unwanted plant which grow in agricultural fields, ponds,lakes, etc. and have bad effects on the flora and fauna present/growing in their vicinity. However, in agricultural fields, nutrients supplied to specific crop are absorbed by weeds and results in poor supply of nutrients to crop plants. Similarly, in paddy fields, they do so. Larvae of the harmful insects *e.g.* malarial mosquito survives around the aquatic weeds growing in ponds or lakes. Although synthetic herbicides have been formulated which have phytotoxic and mutagenic effects on many agricultural plants and in turn enter in animal and human system. Therefore, use of synthetic herbicides is being discouraged in many countries. An alternative method of control of weeds has been developed which is the use of "microbial herbicides" or "bioherbicides".

Some of the noxious weeds are : *Aeschynomene virginica, Canabis sativa, Chondrilla juncea* (skeleton weed), *Convolvulus arvensis, Cyperous rotundus, Eichhornia crassipes* (water hyacinth), *Hydrilla verticillata, Lantana camera, Nymphaea odorata, Panicum dicotomiflorum, Pistia statiotes, Rubus* sp., *Rumexa crispus, Solanum dulcambra, Xanthium strumarium.* Pathogens and insect pests are applied in control of most of the weeds in both agricultural and aquatic systems. Pathogens of some of the aquatic and terrestrial weeds are given in Table 21.5.

### 1. Mycoherbicides

Biological potential of plant pathogens has been realised in recent years. They can be trained to behave as biocontrol agents. Some of them are host specific and even beyond the species to the variety or biotype level (Freeman, 1982). Plant pathologists are making efforts to characterize plant pathogens and their introduction in those areas where weed is a serious problem. If the pathogens could minimize the disease they can be recommended as mycoherbicides.

Currently mycohrbicides are produced by submerged liquid fermentation. However, several potential fungal pathogens do not sporulate in submerged liquid fermentation. Many other methods such as solid state, airlift fermentation and plant substrate remains have been widely accepted. But lack of commercial scale technology for large scale prodcution, long term storage and stability are the serious limitations in the development and use of potential mycoherbicides.

Formulation of mycoherbicides are done by blending the fungal spores with inactive ingredients such as wheat straw, kaolin clay, sodium alginate, cornmeal, infected oat, etc.

In India, biocontrol of water hyacinth has been much emphasized. In the U.S.A. plant pathologists and weed scientists made the joint efforts for biocontrol of weeds by using fungal pathogens. Control of *A. viroinica* growing in rice field by *Collectotrichum gloeosporioides* has been reported. University of Arkansas and the Upjohn Company have made a cooperative effort to prepare the microbial herbicide by using C. *gloeosporioides.*

In 1979 G.E. Templeton and coworkers sprayed the spores of *C. gloeosporioides* ($1.5 \times 10^6$/ml) by aeroplane, in fields at the rate of 96 litre water per hectare and estimated about 95 – 100 per cent killing of weeds in 17 rice fields. In 1975, C.C. Orr and coworkers used a host specific pathotype of *Phytophthora citrophthora* as a biocontrol agent for *Morrenia odorata* (milk weed vine) in *Citrus* spp. Templeton reviewed the biocontrol of weeds where fungi have been used as myco-herbicides.

A B International Institute of Biological Control, U.K. has been developing a mycoherbicide for the control of *Rottboellia cochinchinensis* which is a serious weed. An isolate of *Colletotrichum* from Thailand was selected against *R. cochinchinensis*. Field trails have demonstrated that the mycoherbicide has potential for controlling this weed, particularly in combination with low dosages of chemical herbicides.

**> Table 21.5. Pathogens of aquatic and terrestrial weeds.**

| *Plant species* | *Pathogens* |
|---|---|
| **A. Aquatic weeds** | |
| Algae (Cyanobacteria) | Cyanophages (viruses) and bacteria |
| Alligator weed (*Atternathea philoxeroides*) | Stunt virus |
| Hydrilla (*Hydrilla verticillata*) | *Aspergillus* sp., *Penicillium* sp., *Phytophthora parasitica, Trichoderma* sp., |
| *Salvinia* (*Salvinia* sp.) | *Alternaria* sp. |
| Water hyacinth (*Eichhornia crassipes*) | *Acremonium natum, Alternaria eichhorniae, Bipolaris stenospila, Cercospora piaropi, C. rodmanii, Fusarium roseum, Rhizoctonia solani* |
| Water lettuce (*Pistina stratiotes*) | Virus |
| **B. Terrestrial Weeds** | |
| *Sida spinosa* | *Colletotrichum malvarum* |
| *Sorghum halepense* | *C. graminicola* |
| *Crotalaria spectabilis* | *C. dematium* |
| *Diospyros virigiana* | *Acremonium diospyri* |
| *Parthenium hysterophorus* | *Selerotium rolfsii, Fusarium oxysporum.* |

In 1977, T.E. Freeman enumerated five characteristics of plant pathogens that make them desirable candidates as biocontrol agents for aquatic weeds : (*i*) they are numerous and diverse, (*ii*) they are often host specific, and their use of control againt noxious species would not endanger desirable plant species, (*iii*) they are easily disseminated and self maintaining; therefore, there is no need of reapplication, (*iv*) they have ability to exert a limiting influence on plant populations without eliminating the species, and (*v*) they are non-pathogenic to animals.

## 2. Insects as Biocontrol Agents

The technique of using insects for disease control originated in China. Chinese used predator ants to control certain insect pests of citrus. This practice continued through the ages, and even in modern time, it is still in use. Citrus growers purchased colonies of predatory ants (*e.g. Oecophylla smaragdina*) to introduce in citrus leaf feeding insects.

Moreover, insects play a valuable role in the eradication of valuable weeds but complete eradication has not been achieved by this method. Insects attack the host and multiply rapidly until much plant growth is destroyed. Due to food shortage the number of insects decreases, consequently weeds reappear. Again weed is invaded by insects causing a resultant decreases in weed number. Therefore, the ratio of weed number and insect population fluctuates. A few cases are described as below :

(*i*) *Senecio jacobaea*, a serious weed in wetter areas has been controlled by introducing *Tyrea jacobaea,* cinnaber moth, in California (U.S.A.) and by *Pegohylemyia seneciella* (seedfly) in Australia.

(*ii*) *Hypericum perforatum* (goat weed), a serious weed found in California and pacific North-West is controlled by *Chyroline hyperici* (goat weed bettle).

(*iii*) *Eichhornia crassipes* (water hyacinth), is controlled by an insect, *Neochetina olchhorniae.* This insect was introduced from Latin America to India. The female lays eggs on petiole of water hyacinth. Larvae feed upon petiole and the adults feed on leaves resulting in destruction of the whole plant.

(*iv*) *Lantana*, a pasture pest and poisonous plant is found throughout the world. In Hawaii, caterpillars of *Plusia verticillata* was introduced for its control. Larvae of seedfly, *Agromyza lantanae*, eat many berries and cause others to dry so that birds could not carry them. The lace bug, *Teleonemia scrupulosa* was most effective insect to control *Lantana* but this insect could not yield much success.

## PROBLEMS

1. Define the biological control. How far is it effective in disease control ?
2. What is an antagonist? Explain in brief with suitable examples.
3. What do you know about inoculum and inoculum potential ? How does inoculum density play a role in the infection ?
4. What is the mechanism of biological control? Discuss in detail with suitable examples.
5. How can biological control be applied in field? Explain with examples.
6. What is the mode of introduction of antagonists in field for disease control ?
7. How do mycorrhizal fungi help in biocontrol of plant pathogens ? Discuss with suitable examples.
8. What is the microbial insecticide? Discuss in brief its field application and future prospects.
9. What is microbial herbicide? How will you control aquatic weeds by using microorganisms?
10. Write an essay on biocontrol of weeds.
11. Write short notes on the following :

    (*a*) Pathogens, (*b*) Antagonists and antagonism, (*c*) Antibiotics and antibiosis (*d*) Nematophagy (*e*) Mycophagy, (*f*) Mycoparasitism, (*g*) Microbial competition, (*h*) Seed inoculation, (*i*) Crop rotation (*j*) Lectins and Mycoparasitism (*k*) Biocontrol-1 (*l*) Microbial insecticides, (*m*) myco-herbicides, (*n*) *Bacillus thuringiensis,* (*e*) *Trichoderma viride,* (*f*) Viral pesticides, (*g*) use of biofertilizers for biocontrol of plant diseases.

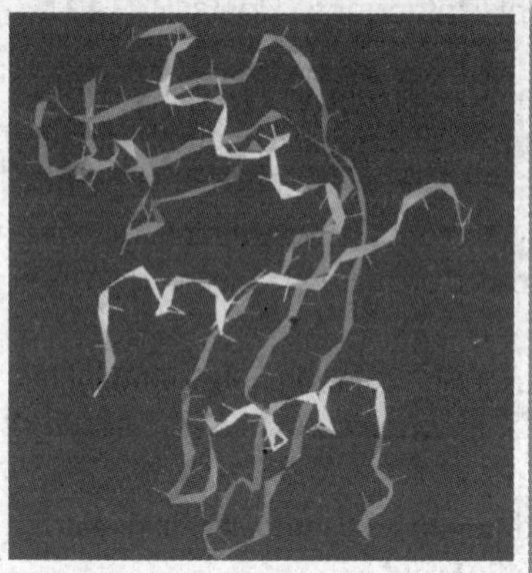

CHAPTER

# 22

# Enzyme Biotechnology

Enzyme is a biocatalyst which accelerates biological reactions. However, the concept of biocatalysts is very wide. It includes the pure enzyme, crude cell extract, viable plant cells, viable animal cells, viable microbial cells and intact non-viable microbial cells. Source of enzymes used in commerce is plant and animal cells. The sources of enzymes are microorganisms, higher plants and animals. Animal enzymes used currently are lipases, tripsin, rennets, etc. Most prevalent plant enzymes are papain, proteases, amylases and soybean lipoxygenase. These enzymes are used in food industries, for example, papain, extracted from papaya fruit is used as a meat tenderizer and pancreatic protease in leather softening and manufacture of detergents (Sasson, 1984).

Enzymes are usually produced by microorganisms and are widely used in industry for different commercial products.

Microbial enzymes have gained much popularity. Production of primary and secondary metabolites by microorganisms is possible only due to involvement of various enzymes. They are of two types: the extracellular and the intracellular enzymes. The former is secreted out the cell and the later remain within the cell. There is a wide range of extracellular enzymes produced by pathogenic and saprophytic microorganisms such as cellulase, polymethylgalacturonase, polyglacturonase, pectinmethylesterase, etc. These enzymes help in establishment in host tissues or decomposition of organic substrates. The intracellular enzymes such as invertase, uric oxidase, asparaginase are of high economic value and difficult to extract as they are produced inside the cell (Riviere, 1977). They can be obtained by breaking the cells by means of a homogenizer or a bead mill and extracting them through the biochemical processes. The process of enzyme purification is difficult as the cell debris and nucleic acid are not easily removed.

Microbial enzymes have two advantages over the animal and plant enzymes. Firstly, they are economical and can be produced on large scale within the limited space and time. The amount produced depends on size of fermenter, type of microbial strain and growth conditions. It can be easily extracted and purified. Secondly, there is technical advantages in producing enzymes via using microorganism as: (*i*) they are capable of producing a wide variety of enzymes, (*ii*) they can grow in a wide range of environmental conditions, (*iii*) they show genetic flexibility that is why they can be genetically manipulated to increase the yield of enzymes, and (*iv*) they have short generation times (Trevan., *1987*).

Enzyme containing detergents have been known since 1913 but their use was limited because of its unstability in detergent formulations. In 1965, a new stable enzyme *e.g.* protease was introduced for application in detergent production. In 1970s, the first commercial process was used for production of fructose from glucose through the isomerisation of glucose. Even in brewing industry, malt is used as the source of enzymes.

At present, more than 2,000 enzymes have been isolated and characterized, out of which about 1,000 enzymes are recommended for various applications. Among them about 50 microbial enzymes have industrial applications. Some of the enzymes are given in Table 22.1. In recent years, application of enzymes in industries has much significance. In 1981, the total world production of enzymes was estimated about 65,000 tonnes which valued about $4 \times 10^8$ US dollars.

## A. MICROORGANISMS PRODUCING ENZYME

There is a large number of microorganisms which produce a variety of enzymes. Enzymes differ with respect to substrates. Some of the microorganisms producing enzymes are listed in Table 22.1.

**> Table 22.1. Microorganisms producing enzymes.**

| *Microorganisms* | *Enzymes* |
|---|---|
| **Bacteria:** | |
| *Bacillus cereus* | Penicillinase |
| *B. coagulans* | α-amylase |
| *B. licheniformis* | α-amylase, protease |
| *B. megaterium* | Penicillin acylase |
| *Citrobacter spp.* | L-asparaginase |
| *Escherichia coli* | Penicillin acylase, ß-galactosidase |
| *Klebsiella pneumoniae* | Pullulanase |
| **Actinomycetes:** | |
| *Actinoplanes sp.* | Glucose isomerase |
| **Fungi:** | |
| *Aspergilus flavus* | Urate oxidase |
| *A. niger* | Amylases, protease, pectinase, glucose oxidase |
| *A. oryzae* | Amylases, lipases, protease |
| *Aureobasidium pullulans* | Esterase, invertase |
| *Candila lipolytica* | Lipase |
| *Mucor micheli and M. pusillus* | Rennet |
| *Neurospora crassa* | Trysinase |

| | |
|---|---|
| *Penicillium funiculosum* | Dextranase |
| *P. notatum* | Glucose oxidase |
| *Rhizopus* sp. | Lipase |
| *Saccharomyces cerevisiae* | Invertase |
| *S. fragilis* | Invertase |
| *Trichoderma reesei* | Cellulase |
| *T. viride* | Cellulase |

## B. PROPERTIES OF ENZYMES

The enzymes bear the following properfies:

### 1. Presence of Species Specificity

Macromolecules including proteins differ in different species *i.e.* they are species specific. It is attributed that the phylogenetic development which has given rise to microbiological variation is caused by variation in these molecules. Enzyme types (protease, α-amylase, lactase) which are found in many species will have properties which vary as much as the other properties of the organisms, for example, protease of two closely related species differs in several ways inspite of some similarities.

### 2. Variation in Activity and Ability

Most of microbial enzymes applied in various ways are extracellular in their origin; they are influenced externally by temperature, pH, etc. However, their optimum stability and activity are very much close to optimum conditions for microbial growth. For example, optimum pH and temperature for amylase activity of a thermophilic microbial species *e.g. Bacillus coagulans* differ from that of mesophilic species of some microbe (*B. licheniformis*). Unlike extracellular enzymes, the intracellular enzymes are little influenced by external environmental factors.

Activity and stability of enzymes also differ. Xylose isomerase is stable at pH range from 4.0 to 8.5 but shows optimum activity at pH between 5.5. to 7.0. Similarly, temperature also influences enzyme activity.

On increasing the temperature enzyme activity gradually increases, but at certain stages temperature inactivates the rate of reaction and finally enzyme is denatured (at high temperature) as it is proteinaceous in nature. Thermal stability in the target enzyme may be a useful attribute during production of enzyme itself as heat may be used to destroy contaminant enzyme activity (Trevan, 1987). In addition to pH and temperature, the stability of enzyme is also increased by many factors such as: (*i*) high concentration of respective enzymes (as protein aggregates and protects them), (*ii*) presence of their substrate and/or product (*e.g.* amylase shows more stability in the presence of starch than in its absence), (*iii*) presence of ions (*e.g.* α-amylase is denatured within 4 h in the absence of $Ca^{++}$ ), and (*iv*) reduced amount of water content in reaction mixture (for example, at natural conditions ß-glactosidase results in production of glucose and galactose from hydrolysis of lactose in whey. But the same enzyme produces some glucose and galactose and mixture of trisaccharides from the same concentrated whey.

### 3. Substrate Specificity

Organic matter contains the various constituents such as cellulose, hemicellulose, lignin, etc. in a complex matrix. In nature these are decomposed and mineralized by a variety of microorganisms. However, it is not possible for a single microbe to decompose all the constituents. For example, a cellulose decomposer will fail to decompose the lignin because of the presence of only cellulose. Therefore, on the decomposing materials community dynamics of microorganisms *i.e.* changing community of microbes with time exists till the disappearance of complex organic matter.

It is also possible that a particular microbe develops potentiality to secrete an enzyme in higher amount and utilize the substrate more rapidly than others. This inherent capacity makes the microbes capable to compete in the microbial competition for substrate utilization. Due to possession of this activity *i.e.* high enzyme producing ability, exploitation of microorganisms is done. For example, *Trichoderma reesei* secretes cellulase in high amount; therefore, this fungus is used for commercial production of cellulase.

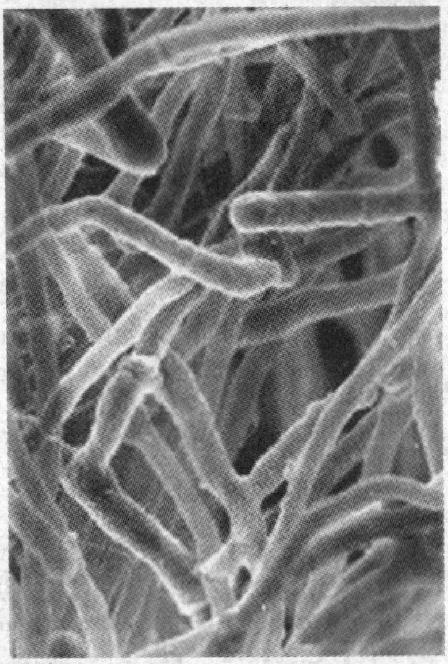

Fungus *Trichoderma reesei* is used for commercial production of cellulase.

### 4. Activation and Inhibition

Some enzymes obtained from different sources show difference in responses to a given activator of inhibitor. For example, ß-galactosidase isolated from fungi does not require cobalt, whereas the same of bacterial origin requires cobalt as a confactor. Thus cobalt activates ß-galactosidase isolated from bacteria and inhibits it when obtained from fungi. Examples of some activators of enzymes used commercially are: proteins (for proteases), starch (for $\alpha$-amylase), cellulose (for cellulase) and pectin (for pectinase).

## C. METHODS OF ENZYME PRODUCTION

Production of enzymes takes place in the following steps:

### 1. Isolation of Microorganisms, Strain Development and Preparation of Inoculum

Microorganisms are isolated on culture media following the microbiological techniques. Aim for isolating a suitable microorganism lies in (*a*) production of enzyme in high amount and other metabolites in low amount, (*b*) completion of fermentation process in short time, and (*c*) utilization by the microorganisms of low cost culture medium. Once a suitable microorganism is obtained its enzyme producing ability is optimized by improving strains and formulating culture medium (pH and temperature). Strains of microorganisms are developed by using mutagens *i.e.* mutagenic chemicals and ultraviolet light. Procedure for development of antibiotic producing strains is given in Chapter 17.

### 2. Medium Formulation and Preparation

Culture medium is formulated in such a way that should provide all nutrients supporting for enzyme production in high amount but not for good microbial growth. For this purpose, an ideal medium must have a cheap source of carbon, nitrogen, amino acids, growth promotors, trace elements and little amount of salts. Care must be taken to maintain pH during fermentation. For a specific microbe pH, temperature and formulation of culture medium is optimized prior to inoculation. Production of enzymes increases with the concentration of culture medium. The following are typical constituents of media for enzyme fermentation:

**Carbohydrates** : Molasses, barley, corn, wheat and starch hydrolysate.

**Proteins** : Meals of soybean, cotton seed, peanut and whey, corn steep liquor and yeast hydrolysate.

Central Food Technological Research Institute (CFTRI), Mysore, has developed technology for conversion of tapioca starch to glucose by using fungal enzymes. Enzymes have also been isolated from bacterial cultures which convert glucose to fructose in starch hydrolysate. At this institute, about 15 tonnes of high fructose syrup (dry weight basis) could be obtained from tapioca starch from one hectare land as against only 5 tonnes of the same from sugarcane.

### 3. Sterilization and Inoculation of Medium, Maintenance of Culture and Fluid Filteration

Medium is sterilized batch-wise in a large size fermenter. For this purpose, continuous sterilization method is now becoming popular. After medium is sterilized, inoculation with sufficient amount of inoculum is done to start fermentation process. Fermentation process is the same as described for antibiotic production. Traditional method of enzyme production has been the surface culture technique where inoculum remains on upper surface of broth. Now-a-days submerged culture method is most widely practised because of less chances for infection and possibility for more yield of enzymes. The former technique is still in use for production of some of the fungal enzymes, for example, amylase (from *Aspergillus* sp.), protease (from *Mucor* sp. and *Aspergillus* sp.) and pectinase (from *Penicillium* sp. and *Aspergillus* sp.). Use of continuous culture technique for cellulase production by *Trichoderma* has been suggested.

The growth conditions are maintained in fermenter at optimum level. These factors differ microbe to microbe and even in the same species of a microbe. A little amount of oil is added to fermenter to control foaming as it happens during fermentation. After 30–150 h incubation, extracellular enzymes are produced by the inoculated microbe in culture medium. Most of enzymes are produced when exponential phase of growth completes but in a few cases, they are produced during exponential phase. Besides extracellular enzymes, other metabolites (10–15 per cent) are also produced in the fermented broth. These metabolites are removed after enzyme purification.

When fermentation is over broth is kept at 5°C to avoid contamination. Recovery of enzymes from the fermented broth (fluid) of bacteria is more difficult than from that of filamentous fungi. Fungal broth is directly filtered or centrifuged after pH adjustment. Therefore, the bacterial broth is treated with calcium salts to precipitate calcium phosphate which help in separation of bacterial cells and colloids. Then the liquid is filtered and centrifuged to remove cell debris.

### 4. Purification of Enzymes

Enzyme purification is a complex process. For detailed description readers are advised to study elsewhere. The main steps of purification are: (*i*) preparation of concentrated solution by vacuum evaporation at low temperature or by ultrafiltration, (*ii*) clarification of concentrated enzyme by a polishing filtration to remove other microbe, (*iii*) addition of preservatives or stabilizers, for example, calcium salts, proteins, starch, sugar, alcohols, sodium chloride (18–20 per cent), sodium benzoate, etc., (*iv*) precipitation of enzymes with acetone, alcohols or organic salts, *e.g.* ammonium sulfate or sodium sulfate, (*v*) drying the precipitate by free drying, vacuum drying or spray drying, and (*vi*) packaging for commercial supply.

## D. IMMOBILISATION OF ENZYMES

Many enzymes secreted by microorganisms are available on a large scale and there is no effect on their cost if they are used only once in a process. In addition, many more enzymes are such that they affect the cost and could not be economical if not reused. Therefore, reuse of enzymes led to the development of immobilisation techniques. It involves the conversion of water soluble enzyme protein into a solid form of catalyst by several methods (see 17.4.2). It is only possible to immobilise microbial cells by similar techniques.

Thus, immobilisation is "the imprisonment of an enzyme in a distinct phase that allows exchange with, but is separated from the bulk phase in which the substrate, effector or inhibitor molecules are dispersed and monitored". Imprisonment refers to arresting the enzyme by certain means where polymer matrix is formed. The first commercial application of immobilised enzyme technology was realised in 1969 in Japan with the use of *Aspergillus oryzae* amino acylase for the industrial production of L-amino acids. Consequently, pilot plant processes were introduced for 6-amino penicillanic acid (6 APA) production from penicillin G and for glucose to fructose conversion by immobilised glucose isomerase.

## 1. Advantages of Using Immobilised Enzymes

The advantages of using immobilised enzymes are: (*i*) reuse (*ii*) continuous use (*iii*) less labour intensive (*iv*) saving in capital cost (*v*) minimum reaction time (*vi*) less chance of contamination in products, (*vii*) more stability (*viii*) improved process control and (*ix*) high enzyme: substrate ratio. The first immobilised enzymes to be scaled up to pilot plant level and industrial manufacture were immobilised amino acid acylase, panicillin G-acylase and glucose isomerase. Some other industrially important enzymes are aspartase, esterase and nitrilase.

## 2. Methods of Enzyme Immobilisation

There are five different techniques of immobilising enzymes: (*i*) adsorption, (*ii*) covalent bonding, (*iii*) entrapment, (*iv*) copolymerisation or cross-linking, and (*v*) encapsulation (Fig. 22.1). For the purpose of immobilisation of enzymes carriers *i.e.* the support materials such as matrix system, a membrane or a solid surface are used.

**(*a*) Adsorption:** An enzyme may be immobilised by bonding to either the external or internal surface of a carrier or support such as mineral support (aluminium oxide, clay), organic support (starch), modified sapharose and ion exchange resins. Bonds of low energy are involved *e.g.* ionic interactions, hydrogen bonds, van der Waals forces, etc. If the enzyme is immobilised externally, the carrier particle size must be very small in order to achieve an appreciable surface of bonding. These particles may have diameter ranging from 500 Å to about 1 mm. Due to immobilisation of enzymes on external surface, no pore diffusion limitations are encountered. In addition, the enzyme immobilised on an internal surface is protected from abrasion, inhibitory bulk solutions and microbial attack, and a more stable and active enzyme system may be achieved. Moreover, in internal pore immobilisation the pore diameters of carriers may be optimised for internal surface immobilisation (Fig. 22.1A).

There are four procedures for immobilisation by adsorption: (*i*) *static process* (enzyme is immobilised on the carrier simply by allowing the solution containing the enzyme to contact the carrier without stirring), (*ii*) *the dynamic batch process* (carrier is placed into the enzyme solution and mixed by stirring or agitated continuously in a shaker), (*iii*) *the reactor loading process* (carrier is placed into the reactor that will be subsequently employed for processing, then the enzyme solution is transferred to the reactor and carrier is loaded in a dynamic environment by agitating the carrier and enzyme solution), and (*iv*) *the electrodeposition process* (carrier is placed proximal to one of the electrodes in an enzyme bath, the current put on, the enzyme migrates to the carrier and deposited on the surface).

**(*b*) Covalent Bonding:** Covalent bond is formed between the chemical groups of enzyme and chemical groups on surface of carrier. Covalent bonding is thus utilised under a broad range of pH, ionic strength and other variable conditions. Immobilisation steps are attachment of coupling agent followed by an activation process, or attachment of a functional group and finally attachment of the enzyme (Fig. 22.1B).

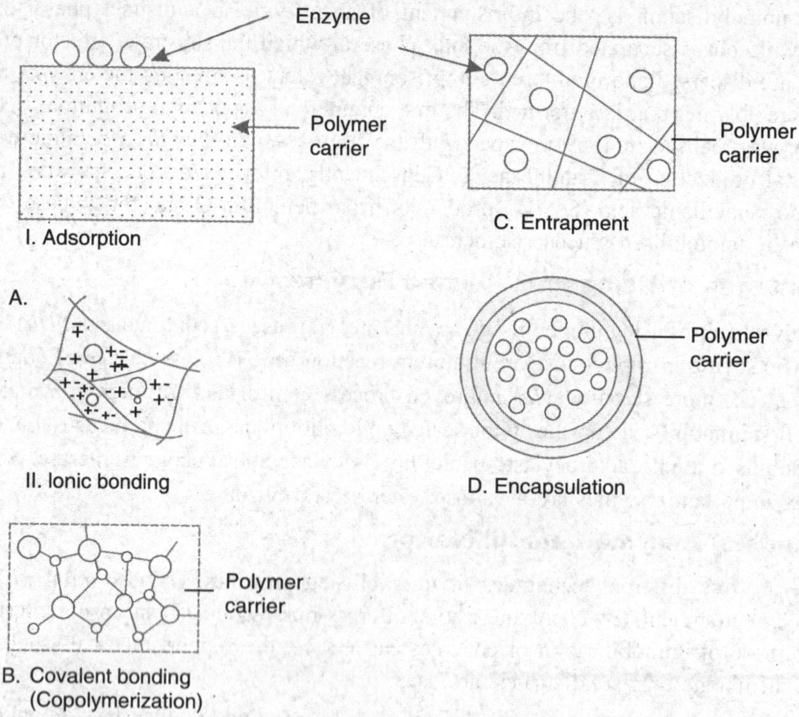


**Fig. 22.1.** Means of enzyme immobilization.

The different types of carriers are used in immobilisation such as carbohydrates proteins and amine-bearing carriers, inorganic carriers, etc. Covalent attachment may be directed to a specific group (*e.g.* amine, hydroxyl, tyrosyl, etc.) on the surface of the enzyme. Hydroxyl and amino groups are the main groups of the enzymes with which it forms bonds, whereas sulphydryl group least involved. Different methods of covalent bonding are: (*i*) *diazoation* (bonding between the amino group of the support *e.g.* aminobenzyle cellulose, aminosilanised porous glass, aminoderivatives and a tyrosyl or histidyl group of the enzyme), (*ii*) *formation of peptide bond* (bond formation between the amino or carboxyl group of the support and amino or carboxy group of the enzyme), (*iii*) *group activation* (use of cyanogen bromide to a support containing glycol group *i.e.* cellulose, syphadex, sepharose, etc.), and (*iv*) *polyfunctional reagents* (use of a bifunctional or multifunctional reagent *e.g.* glutaraldehyde which forms bonding between the amino group of the support and amino group of the enzyme).

The major problem with covalent bonding is that the enzyme may be inactivated by bringing about changes in conformation when undergoes reactions at active sites. However, this problem can be overcome through immobilisation in the presence of enzyme's substrate or a competitive inhibitors or protease. The most common activated polymers are celluloses or polyacrylamides onto which diazo, carbodimide or azide groups have been incorporated.

**(*c*) Entrapment:** Enzymes can be physically entrapped inside a matrix (support) of a water soluble polymer such as polyacrylamide type gels and naturally derived gels *e.g.* cellulose triacetate, agar, gelatin, carrageenan, alginate, etc. (Fig. 22.1C). The form and nature of matrix vary. Pore size of matrix should be adjusted to prevent the loss of enzyme from the matrix due to excessive diffusion. There is possibility of leakage of low molecular weight enzymes from the gel. Agar and carrageenan have large pore sizes (< 10 μ).

There are several methods for enzyme entrapment: (*i*) *inclusion in gels* (enzyme entrapped in gels), (*ii*) *inclusion in fibres* (enzyme entrapped in fibre format), and (*iii*) *inclusion in microcapsules* (enzymes entrapped in microcapsules formed monomer mixtures such as polyamine and polybasic chloride, polyphenol and polyisocyanate). The entrapment of enzymes has been widely used for sensing application, but not much success has been achieved with industrial process.

**(*d*) Cross-linking or Copolymerisation:** Cross-linking is characterised by covalent bonding between the various molecules of an enzyme via a polyfunctional reagent such as glutaraldehyde, diazonium salt, hexamethylene disocyanate, and N-N′ ethylene bismaleimide. The demerit of using polyfunctional reagents is that they can denature the enzyme. This technique is cheap and simple but not often used with pure proteins because it produces very little of immobilised enzyme that has very high intrinsic activity. It is widely used in commercial preparation.

The basic approaches made to immobilise enzymes by cross-linking are: (*i*) formation of an insoluble aggregates by cross-linking of enzymes (*e.g.* papain) with glutaraldehyde, (*ii*) adsorption of enzyme followed by cross-linking *e.g.* cross-linking of trypsin adsorbed to the surface of colloidal silica particles, and (*iii*) impregnation of porous material with enzyme followed by cross-linking of enzyme in pores *e.g.* papain in collodion membrane.

**(*e*) Encapsulation:** Encapsulation is the enclosing of a droplet of solution of enzyme in a semipermeable membrane capsule. The capsule is made up of cellulose nitrate and nylon. The method of encapsulation is cheap and simple but its effectiveness largely depends on the stability of enzyme although the catalyst is very effectively retained within the capsule. This technique is restricted to medical sciences only (Fig. 22.1D).

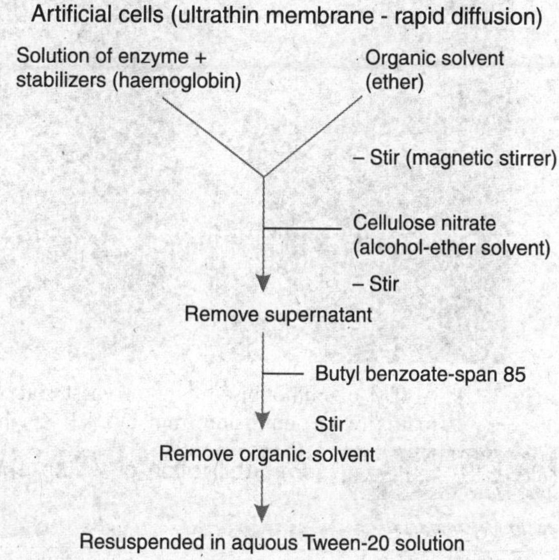


Fig. 22.2. Enzyme encapsulation.

In this method a large quantity of enzyme is immobilised but the biggest disadvantage is that only small substrate molecule is utilised with the intact membrane. The method of enzyme encapsulation is given in Fig. 22.2.

### 3. Immobilisation of Cells

In the field of enzyme technology, immobilisation of whole cells is now a well developed method. Successful performance of several industrial plants has been demonstrated. In cell immobilisation technology the main important feature is that enzymes are active and stable for a long period of time. It keeps within the cellular domain together with all cell constituents whether the cells are dead or viable but in resting state.

The methods of whole cell immobilisation are same as described for enzyme immobilisation, *i.e.* adsorption (Fig. 22.3), covalent bonding, cell to cell cross-linking, encapsulation, and entrapment in polymeric network. Since a long time adsorption of cells to preformed carrier has been done, for example use of woodchips as carrier for *Acetobacter* has been in practice for vinegar production since 1823. Preformed carrier of ones choice is used (Table 22.2). Cell attachment to the surface of preformed carrier is done by covalent bonding.

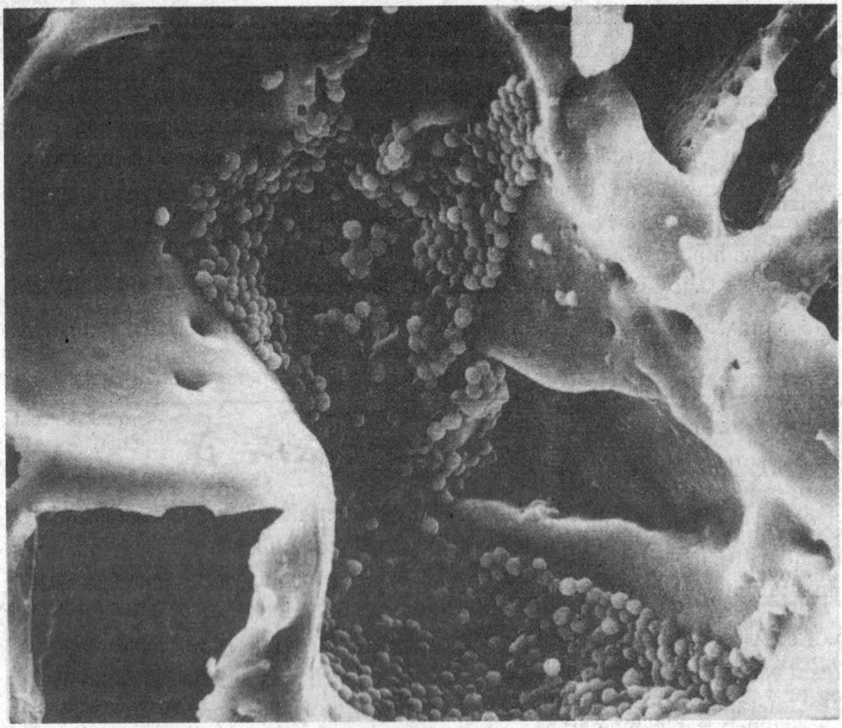

**Fig. 22.3.** A SEM of immobilised cells of yeast (*Saccharomyces cerevisiae*) with in a solid phase matrix (with permission from Dr. A.J. Knights, Univ. Wales, Swansea).

**> Table 22.2. Immobilisation of whole cells by different method.**

| *Support material* | *Cells* | *Reaction* |
|---|---|---|
| **Adsorption** | | |
| Gelatin | *Lactobacilli* | Lactose/lactic acid |
| Porous glass | *Saccharomyces carlsbergensis* | Glucose/ethanol |
| Cotton fibres | *Zymomonas mobilis* | Glucose/ethanol |
| Vermiculite | *Z. mobilis* | Glucose/ethanol |
| DEAE-cellulose | *Nocardia erythropolis* | Steroid conversion |
| **Covalent Bonding** | | |
| Cellulose + cyanuric chloride | *Scerevisiae* | Glucose/ethanol |
| Ti (IV) oxide, etc. | *Acetobacter* sp. | Wort/vinegar |
| Carboxymethylcellulose + carbodiimide | *Bacillus subtilis* | L-histidine/uronic acid |
| **Cross-linking of cell-to-cell** | | |
| Diazotized diamines | *Streptomyces* | Glucose/fructose |
| Glutaraldehyde | *E. coli* aspartic acid | Fumaric acid/L- |

| | | |
|---|---|---|
| Flocculation by chitosan | *Lactobacillus brevis* | Glucose/fructose |
| **Entrapment** | | |
| Al alginate | *Candida tropicalis* | Phenol degradation |
| Ca alginate | *S. cerevisiae* | Glucose/ethanol |
| Mg pectinate | Fungi | Glucose/fructose |
| K-carrageenan | *E. coli* | Fumaric acid/L-spartic acid |
| Chitosan alginate | *S. cerevisiae* | Glucose/ethanol |
| **Encapsulation** | | |
| Cellulose acetate | *Comamonas* sp. | 7-ACA production |
| Ethylcellulose | *Streptomyces* sp. | Glucose/fructose |
| Polyester | *Streptomyces* sp. | Glucose/fructose |
| Alginate-polylysine | Pancreas cells | - |
| Alginate-polylysine | Hybridoma cells | Monoclonal antibodies |

*Source:* M. Moo-Young (1985).

# E. ENZYME ENGINEERING

Since genes encode enzymes, the changes in gene certainly bring about alteration in enzyme structure. In addition to methods available for gene manipulation alteration of genes by site-directed mutagenesis (*see* Chapter 6) for enzyme engineering has become much popular. Thus, site-directed mutagenesis produces single amino acid substitutions in the primary structure of enzymes.

In recent years, the term "enzyme engineering" is used to denote the modification of enzyme structure by alteration of genes which code enzymes. An enzyme produced by a modified gene is structurally new which has great promise to create a new enzyme. Enzyme engineering embraces the (*i*) production of new enzymes, (*ii*) study of structural feature related to stability (*iii*) increase in stability by changing amino acid composition, and (*iv*) production of stable enzymes by genetically engineered microbial cells.

# F. APPLICATIONS OF ENZYMES

The biocatalysts (enzymes and cells) are used in multifarious ways in different field. Trevan (1987) has grouped the applications into four broad categories: (*i*) therapeutic uses, (*ii*) analytical uses, (*iii*) manipulative uses, and (*iv*) industrial uses.

## 1. Therapeutic Uses

Enzymes are used for this purpose where some inborn errors of metabolism occur due to missing of enzyme. Where specific genes are introduced to encode specific missing enzymes. However, in most of cases certain diseases are treated by administering the appropriate enzyme. For example, virilization of a disease developed due to loss of an hydroxylase enzyme from adrenal cortex and introduction of hydroxyl group (–OH) on 21-carbon of ring structure of steroid hormone. Steroids are compounds having a common skeleton in the form of perhydro-1, 2-cyclo-pentano-phenanthrene (Fig. 22.4). The missing enzyme synthesizes aldosterone (male hormone) in excess leading to masculinization of female baby and precocious sexual activity in males in about 5–7 years.

Similarly, treatment of leukaemia (a disease in which leukaemic cells require exogenous asparagine for their growth) could be done by administering asparaginase of bacterial origin.

Human and domestic animals suffer from worms. Deworming medicines taken to kill them are sometimes digested by their secretions. In recent years deworming medicines have been prepared which are not inhibited by the worm secretions but digest them; for example plant protease **papain** from papaya and **ficin** from fig.

In patients bleeding occurs during the operation or removal of tooth. Therefore, the enzyme thrombin is administered topically to the patients so that bleeding should not occur. Thrombin is obtained from beef plasma which converts the fibrinogen to fibrin and small peptide. These are insoluble with and for blot clots.

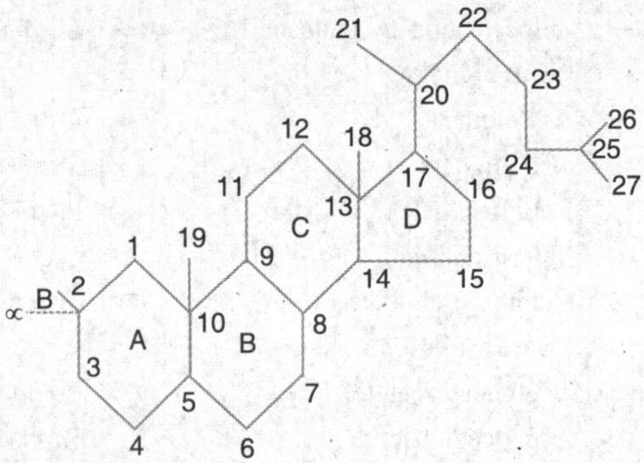


Fig. 22.4. A common skeleton of the steroids.

## 2. Analytical Uses

Use of enzymes for analytical purposes is also important. Generally end point and kinetic analysis are possible. End point analysis refers to total conversion of substrates into products in the presence of enzymes in a few minutes while kinetic analysis involves the rate of reaction and substrate/product concentration. Moreover, analysis of antibodies, immunoglobins, necessary for human use poses a great promise. The usable enzymes are alkaline phosphatase, B-galactosidase, β-lactamase, etc.

Another use of enzyme is in biosensor, a device of biologically active material displaying characteristic specificity with chemical or electronic sensor to convert a biological compound into an electronic signal. It is constructed to measure almost anything from blood glucose. A simple carbon electrode, an ion sensitive electrode, oxygen electrode or a photocell, may be a sensor (Trevan, 1987).

## 3. Manipulative Uses

A variety of enzymes isolated from different sources are now-a-days applied in genetic engineering as one of the biological tools. Some of them are available in market (Chapter 4). Examples and brief discussion of these enzymes are given these.

## 4. Industrial Uses

Enzymes are used in industries in different ways.

**(*a*) In Dairy Industry:** For a long time calf rennet has been used in dairy industry. In recent years, calf rennets are replaced by microbial rennets (*e.g. Mucor michei*). They are acid aspartate proteases. They slightly differ from calf rennets as they depend for reaction with casein on $Ca^{++}$, temperature, pH, etc.

Lactase (produced by *Bacillus stearothermophilus*) is used for hydrolysis of lactose in whey or milk, and lipase for flavour development in special cheeses.

**(*b*) In Detergent Industry:** During normal washing proteinaceous dirt often precipitates on solid cloths and proteins facilitate to adhere the dirt on textile fibres and make stains on cloths. These stains are difficult to remove from clothes. Nevertheless, it can be easily removed by adding proteolytic enzymes to the detergent. It attacks on peptide bonds and therefore, dissolves protein. The alkaline serine protease obtained from *B. licheniformis* is most widely preferred to use in detergent. In addition, the serine protease of *B. amyloliquefaciens* is also used for this purpose. It contains α-amylase, hence to some extent it may be advantageous.

**(*c*) In Starch Industry:** It has been mentioned earlier that hydrolysis of starch began in early 1960s to prepare dextrose and glucose syrups. Furthermore, for complete acid hydrolysis of starch to dextrose glucoamylase was coupled with bacterial a-amylase. Currently, various enzymatic processes are applied for various products (Fig. 22.5).

Glucose isomerase is an important enzyme used commercially in conversion of glucose to fructose via isomerization. Fructose is used in the preparation of fructose syrup.

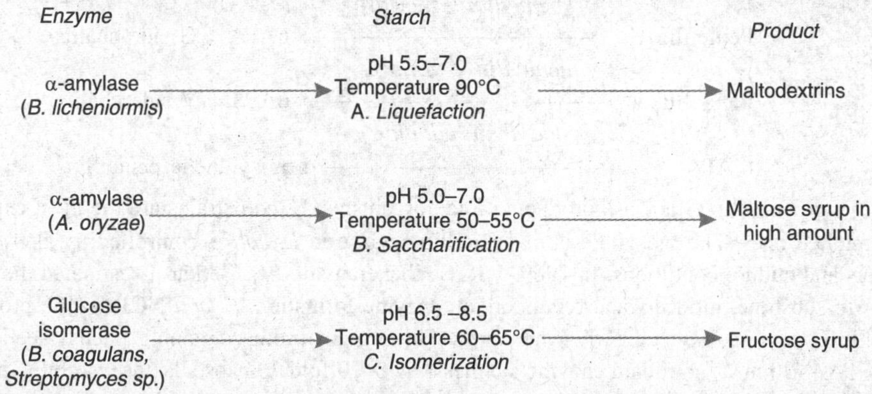


**Fig. 22.5.** Enzymatic process applied in starch indusry.

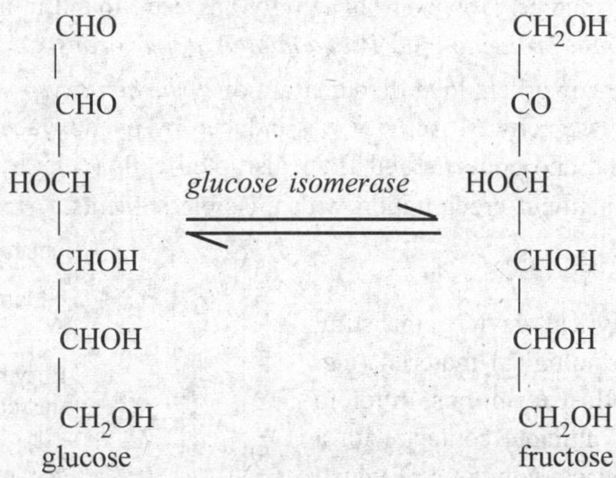


The reaction mixture at the end contains 42% fructose, 52% glucose and 6% dextrins. The mixture is sweeter than glucose and as sweet as sucrose. Now, techniques have been developed to obtain 55% fructose concentration in syrup.

**(*d*) In Brewing Industry:** Enzymes used in brewing industry are α-amylase, ß-glucanase and protease which are required for malt in substitution of barley. Source of these enzymes is *B. amyloliquefaciens*. α-amylase is not required for liquefaction or brewing adjuncts and ß-glucanase alleviates filtration problems due to poor malt quality and neutral protease helps in the inhibition of alkaline protease by an inhibitor.

**(*e*) In Wine Industry:** Pectic enzymes are used in wine industry for high yield of products of improved quality. The pectic enzymes are pectin transeliminase (PTE), polymethyl galacturonase (PMG), polygalacturonase (PG), pectine esterase (PE), etc. However, pectic enzymes give a good result when combined with other enzymes *e.g.* protease glucoamylase, etc.

**(*f*) In Pharmaceutical Industry:** Penicillin G/V acylase, glucose isomerase, etc. are widely used in pharmaceuticals for the production of semisynthetic penicillins and fructose syrup, respectively. All penicillins consist of an active beta lactam ring *i.e.* 6-amine penillanic acid (6 APA) group combined with different side chains (R group). Penicillin G/V acylase removes G/V group from penicillin G/V resulting in separation of 6 APA and R groups. Finally, new synthetic side chains are coupled with 6 APA to synthesize new semisynthetic penicillins. Enzyme reaction are as below:

$$\text{Penicillin G} \xrightarrow{\textit{penicillin G acylase}} \text{6APA + G side chain}$$

$$\text{Penicillin V} \xrightarrow{\textit{penicillin V acylase}} \text{6APA + V side chain}$$

$$\text{6 APA} \xrightarrow{\textit{side chain addition}} \text{semisynthetic penicillins}$$

Among penicillin G acylase producing microorganisms, *E. coli* strains are the most explored and exploited ones. The biosynthesis of penicillin acylase in *E. coli* is controlled by altering the nutrients and culture conditions. In 1989, V.K. Sudhakaran and P.S. Berkar investigated the effect of growth substrate, inducers and regulators on enzyme formation. *E. coli* NCJM-2400 produced penicillin G acylase intracellularly when grown in nutrient broth containing phenyl acetic acid (PAA). PAA (20 mM) stimulatd enzyme synthesis by 8–10 fold. Glucose, lactose, sorbitol, acetate and lactate (all 0.1%) catabolically repressed the enzyme formation. Phosphate and yeast extract were found essential for both the growth and enzyme biosynthesis. Penicillin V acylase occurs in fungal and actinomycetes sources. However, its activity has been found in many bacteria such as *Bacillus sphaericus, Erwinia aroideae,* and *Pseudomonas acidovorans.*

In 1986, D.A. Lowe and coworkers identified a strain of *Fusarium oxysporum* which showed intracellular penicillin V acylase activity. Activity was induced by phenoxyacetic acid in culture. Enzyme was partially purified and concentrated from disrupted cells (cells hydrolysed with 5% penicillin V solutions) by fractional precipitation with miscible solvents.

## G. BIOSENSOR

A biosensor is an analytical device consisting of an immobilised layer of biological material (*e.g.* enzyme, antibody, organelle, hormones, nucleic acids or whole cells) in the intimate contact with a transducer, *i.e.* sensor (a physical component) which analyses the biological signals and converts into an electrical signal. A sensor can be anything, a single carbon electrode, an ion-sensitive electrode, oxygen-electrode, a photocell or a thermistor.

The principle of the biosensor is quite simple (Fig. 22.6). The biological material is immobilized as described earlier on the immobilization support, the permeable membrane, in the direct vicinity of a sensor. The substances to be measured pass through the membrane and interact with the immobilized material and yield the product. A product (*i.e.* the monitored substrate) may be heat, gas (oxygen), electrons, hydrogen ions or the product of ammonium ions. The product passes through another membrane to the transducer. The transducer converts product into an electric signal which is amplified. The signal processing equipment converts the amplified signals into a display most commonly the electric signal

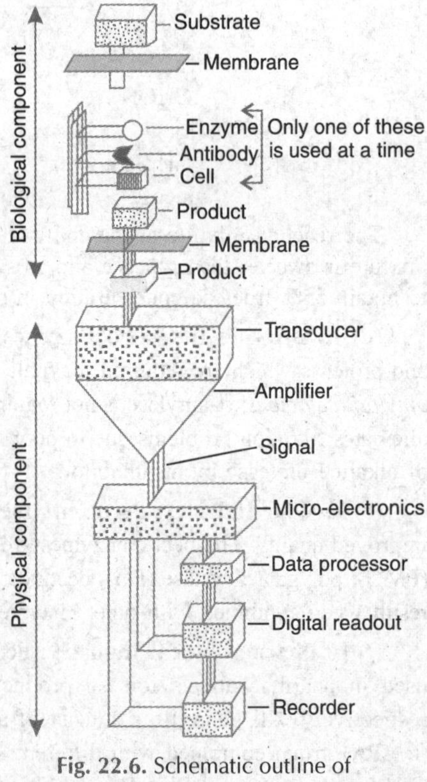


Fig. 22.6. Schematic outline of biosensor.

which can be read out and recorded.

A glucose-electrode is shown in Fig. 22.7. It can be built up by immobilizing glucose oxidase in polyacrylamide gel around a platinum oxygen-electrode separated by a teflon membrane. KCl solution is placed around the platinum-oxygen electrode. From upper surface glucose oxidase is intimately covered by a cellulose acetate membrane.

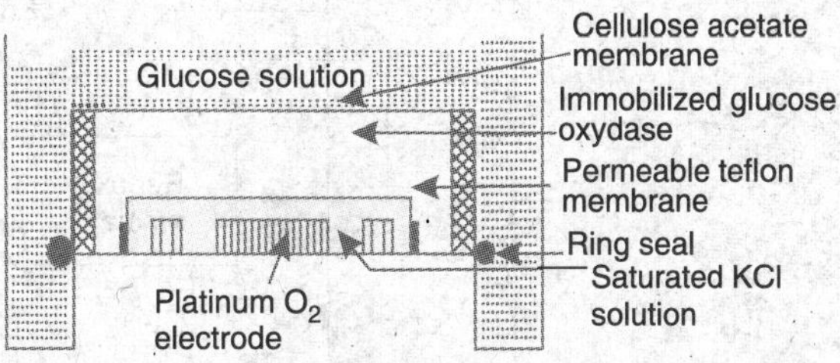


**Fig. 22.7.** A glucose electrode.

When glucose solution is brought into the contact of membrane, glucose and oxygen pass through membrane into the enzyme layer and, as a result of oxidation-reduction reactions, converted into gluconic acid and hydrogen peroxide in the presence of water, oxygen and glucose oxidase. Consequently, oxygen concentration in the gel around the electrode is lowered down. Hydrogen peroxide brings about a change in current *i.e.* measurable signal. Electrode records the rate of reactions. The rate of diminition of oxygen concentration is proportional to glucose concentration of the sample. It responds linearly to glucose concentrations over a range of $10^{-1}$ – $10^{-5}$ mol $dm^{-3}$ with a response time of 1 minute.

In 1987, for the first time Yellow Springs Instruments Co., USA developed a biosensor for diagnostic purposes for measuring glucose in blood plasma. It is a hand–held machine which measures six components of blood plasma for example, glucose, urea, nitrogen, sodium, potassium and chloride.

An indigenous glucose sensor has been developed by the scientists at Central Electro-chemical Research Institute (CECRI), Karaikudi. It gives electrical signal for a glucose concentration as low as 0–15 millimoles.

## 1. Types of Biosensor

Biosensor are of different types based on the use of different biological material and sensor devices; a few of them are discussed below:

***(a)* Electro-chemical Biosensor.** This type of biosensor has been developed by using electronic devices such as field effect transmitors or light emitting diode; the former measures charge accumulation on their surface and the later photoresponse generated in a silica based chip as an alternating current. Hence, the field effect transmitor measures a biochemical reaction at the surface and induce into current. Moreover, the field effect transmitors can be modified to ion sensitive, enzyme sensitive or antibody sensitive ones by using selective ions, enzymes or antibodies respectively.

***(b)* Amperometric Biosensor.** Amperometric biosensors are those which measure the reaction of anylate with enzyme and generate electrons directly or through a mediator. The amperometric biosensors contain either enzyme-electrode or without a mediator, or chemically modified electrodes. The oxygen and peroxide based biosensor and others (Table 12.4) discussed earlier are enzyme electrode biosensor. Some advancement has been brought into this type of biosensor by using a mediator. In addition, more advanced types are the direct electron transfer systems. Principle of a mediated biosensor is shown in Fig. 22.8

In this biosensor, a redox reaction catalysed by an enzymes is directly coupled to an electrode where enzyme is presented with the oxidizable substrate. The electrons are transferred from the substrate to the electrode via enzyme and redox mediator. In this biosensor the oxidase replaces the oxygen requirement of the enzymes.

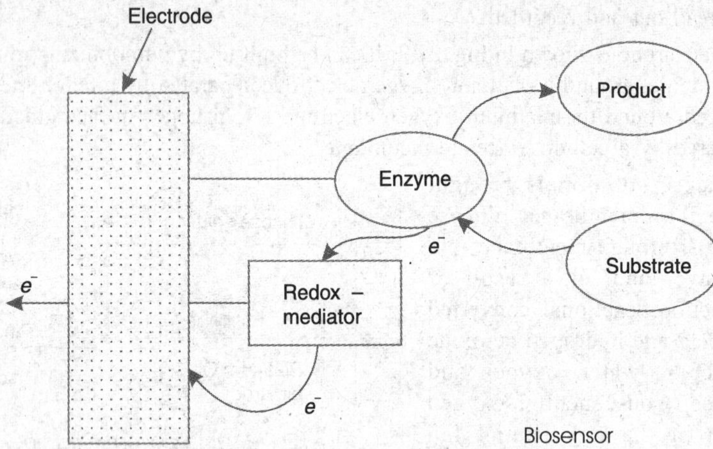


Fig. 22.8. Mediated biosensor.

(*i*) **Enzyme Electrodes.** Enzyme electrodes are a new type of biosensors which have been designed for the amperometric assay of potentiometric assay of substrates such as urea, amino acid, glucose, alcohol, and lactic acid. The electrode consists of a given electrochemical sensor in close contact with a thin permeable enzyme membrane capable of reacting with the given substrates. The enzyme is embedded in the membrane and produce $O_2$, $H^+$, $NH_4^+$, $CO_2$ or other small molecules depending on enzymatic reactions. This is detected by the specific sensor. The magnitude of the response determines the concentration of substrates.

(*c*) **Thermistor Containing Biosensor:** Thermistor is used to record even a small temperature changes (between 0.1–0.001°C) during biochemical reactions. By immobilizing enzymes like cholesterol oxidase, glucose oxidase, invertase, tyrosinase, etc. thermistors have been developed (Gronow *et al.* 1988). Moreover, thermistors are also employed for the study of antigen-antibody with very high sensitivity ($10^{-13}$ mol dm$^{-3}$) in case of thermometric Enzyme Linked Immunoabsorbant Assay (ELISA).

(*d*) **Bioaffinity Sensor:** Bioaffinity sensors are developed recently. It measures the concentration of the determinants, *i.e.* substrates based on equilibrium binding. This shows a high degree of selectivity. These are of diverse nature because of the use of radiolabelled, enzyme labelled or fluorescence-labelled substance. Principle of bioaffinity sensor is given in Fig. 22.9.

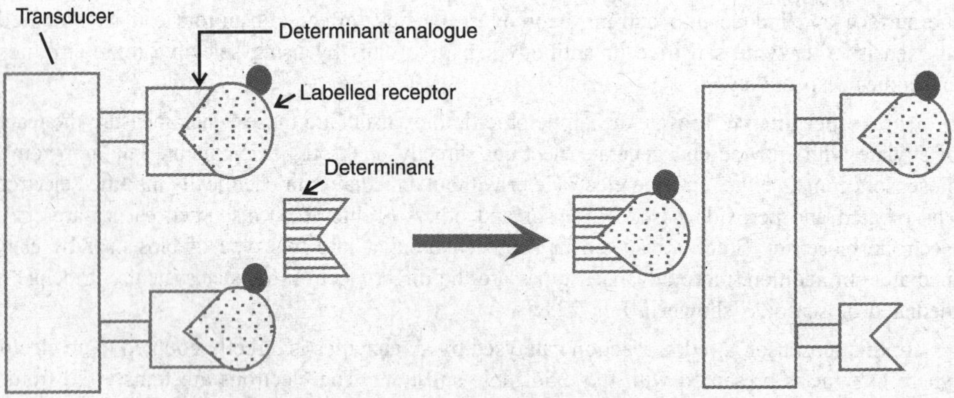


Fig. 22.9. Principle of bioaffinity sensor.

In this biosensor, a receptor is radiolabelled and allowed to bind with determinant analogue immobilized onto the surface of a transducer. When concentrations of a determinant are increased, the labelled receptor forms an intimately bound complex with determinant.

**> Table 22.3. Typical enzymes based biosensors**

| *Substance* | *Enzymes time* | *Response* | *Range* |
|---|---|---|---|
| Amines (for meat freshness) | Monoamine oxidase | 4 min | 50 – 200 μ mol $dm^{-3}$ |
| Cholesterol | Cholesterol oxidase | 2 min | $10^{-2} - 3 \times 10^{-5}$ mol $dm^{-3}$ |
| Carbon monoxide | CO : acceptor | 15 sec | 0 – 65 μ mol $dm^{-3}$ |
| Glucose | Glucose oxidase | 20 sec | $2 \times 10^{-3} - 3 \times 10^{-6}$ mol $dm^{-3}$ |
| Penicillin | Penicillinases | 25 sec | 1 – 10 m mol $dm^{-3}$ |
| Sucrose | Invertase | 6 min | $10^{-2} - 2 \times 10^{-3}$ mol $dm^{-3}$ |
| Uric acid | Uricase | 30 min | $5 \times 10^{-3} - 5 \times 10^{-5}$ mol $dm^{-3}$ |

*Source*: Gronow *et al.* (1988)

Finally, radiolabelled receptor-determinant complex is removed from the immobilized determinant analogue resulting in the increased concentrations of labelled receptor. This is measured by a reduction in signal of the labelled receptor. Gronow *et al.* (1988) have discussed that the possibilities of this type of biosensors are the use of lectin receptors for saccharide estimation, hormone receptors for hormone, drug receptors for drug, antibodies receptors for antigens and nucleic acid (as gene probe) for inherited diseases and fingerprinting.

(*e*) **Whole Cell Biosensors — (Microbial Biosensors).** In this device, either immobilized whole cell of microorganisms or their organelles are used. These react with a large number of substrates and show generally slow response. Immobilized *Azotobacter vinelandii* coupled with ammonia electrode shows sensitivity range between $10^{-5}$ and $8 \times 10$ mol $dm^{-3}$. It measures the concentration of nitrate within 5–10 $min^{-2}$. Examples of microbial biosensors are given in Table 22.4.

**> Table 22.4. Microbial biosensors containing oxygen electrode.**

| *Microorganisms* | *Sensoring for time (min)* | *Response* | *Range* |
|---|---|---|---|
| *Brevibacterium lactofermentum* | Assimilase sugars | 1–10 | Linear above 1 m mole $dm^{-3}$ |
| *Bacillus subtilis* | Mutagen screening | 90–100 | 1–6 μ g $cm^{-3}$ |
| *Methylomonas flagellata* | Methane | 1 | upto 6.6 m mol $dm^{-3}$ |
| *Trichosporon brassicae* | Acetate | 6–10 | Linear upto 22.5 mg $dm^{-3}$ |
| *T. brassicae* | Ethanol | 10 | below 22.5 mg $dm^{-3}$ |

(*f*) **Opto-electronic Biosensor.** In these biosensors either enzymes or antibodies are immobilized on the surface of a membrane. For measuring colour, biosensor with enzyme and dye is immobilized to a membrane. When a substrate is catalysed to yield product, changes in pH of the medium occur. This results in changes in dye-membrane complex. These changes in colour are measured by using a light emitting diode and a photodiode.

### 2. Applications of Biosensor

In the beginning biosensor was applied in the field of medicine and industry. But in recent years, biosensors are becoming popular in many areas due to the small size, rapid and easy handling, low cost, and greater sensitivity and selectivity. Application of biosensor in some of the areas is described as below:

**(*a*) Uses in Medicine and Health.** Biosensors have tremendous potential for its application in the field of medical science. In 1979, the first glucose analyser using biomolecule for the detection of blood glucose was commercialized by Yellow Springs Instruments Co., USA. A device, a minipump filled with insulin, has been constructed to deliver insulin to diabetics based on glucose levels of blood. When biosensor provides informations, the device delivers accurate amount of insulin required by the diabetics. Mitomycin, an aflatoxin, causes cancer in inborne infants. Therefore, mutagenicity of such chemicals can be detected by using the biosensor. Similarly, any other abnormal toxic substance produced in body due to infectious disease can also be detected.

**(*b*) Uses in Pollution Control.** Biosensors are very helpful in environmental minitoring and pollution control, since they can be miniaturized and automated. As far as quality control of drinking water is concerned, the minitoring biosensors are successful in monitoring of pesticides in water. In Japan, a biosensor coupled with oxygen electrode and immobilized *Trichosporon cutaneum* is used for measuring biological oxygen demand (BOD) (Gronow *et al.,* 1988).

The whole cell biosensor developed by immobilising *Salmonella typhimurium* and *Bacillus subtilis* in conjugation with oxygen electrode can be used to measure mutagenicity and carcinogenecity of several chemical compounds.

**(*c*) Uses in Industry.** Generally, spectrophotometer and autoanalyzer are used to estimate the substrates utilized and the products formed in the fermented broth. In addition, there are a lot of problems associated with these. So the biosensors can be designed to measure the fermentation products to improve the feedback control, to carry out rapid sampling and rejection of below standard raw materials to improve the efficiency of workers. Isaokarube and coworkers of Tokyo University Research Centre of Advance Science & Technology have recently developed an ion sensitive field effect transistor (ISFET). This device is highly sensitive to change the ion concentration. Using this biosensor, it is possible to measure the odour, freshness and taste of foods. In determining fish freshness either ATPase, aminoxidase or putrescine oxidase is used. ATPase detects the presence of ATP in fish muscle. As ATP is not present in staled food, therefore, signals do not occur. Recently, a biosensor has been developed at Cransfield Institute of Technology, UK which measures cholesterol levels in butter. The enzyme cholesterol oxidase, when immobilized on the electrodes, reacts with cholesterol of food.

**(*d*) Biosensor for Military.** The darker side of biosensor application is to provide support to military with such a biosensor that can detect toxic gases including chemical warfare agents. Such biosensors have advantages over the traditional methods of sensing of chemicals.

## H. BIOCHIPS–THE BIOLOGICAL COMPUTER

Biochip is the result of marriage of microchips business with biotechnology. In future, there is the possibility of developing biological computers.

Until the development of silicon microchips, setting up of computers was very costly and space occupying. But recently, one can have a computer to be fit on desk top. These affordable prices are maily due to the development of silicon microchips which brought into a rapid revolution in technology. Further reduction in size of computers and improvement in computing powers will not be possible because the silicon microchip technology has certain limitations as below:

(*i*) There is inherent limit beyond which circuits cannot be squeezed onto a silicon chip. For example the width of the circuit cannot be shorter than the wavelength of light. Light is used to etchout circuits during the manufacturing of silicon chips.

(*ii*) Close placing beyond a limit of many electrical circuits on the same microchip results in 'electron tunnelling' which creates short circuits ruining the whole system.

(*iii*) After cramming together of a large number of circuits, heat is generated by the electric current. This may cause total failure of the system.

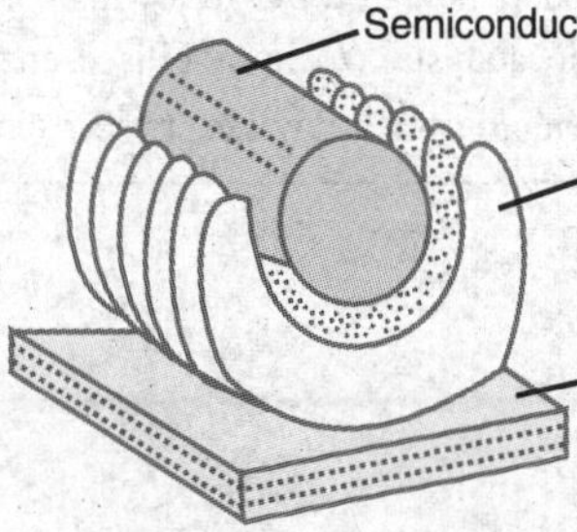


Fig. 22.10. Diagrammatic representation of biochip.

## 1. Principles of Biochips

One of the important features of macromolecules (*e.g.* proteins) is their self shaping into predetermined three dimensional structure. This property of proteins helps in biochip designing because the circuits can be crammed around three dimensional protein structure. While designing the biochips, a semiconducting organic molecule is inserted into a protein framework; the whole unit is fixed onto a protein support (Fig. 22.10). In biochips the electrical signals can pass through the semiconducting organic molecule in the same way as in silicon microchip. It has many advantages over silicon microchip as below:

(*i*) In biochip the width of electrical circuits should not be more than that of one protein molecule which is smaller than the smallest silicon microchip.

(*ii*) The problem of electron tunnelling would be to certain extent less acute in biochip than silicon microchip.

(*iii*) The protein molecule possess less electrical resistance, therefore, less heat will be generated during the course of production of electrical signals. Consequently, a large number of circuits can be placed together as it is not possible in silicon microchips.

## 2. Application of Biochips

There are many areas where biochip holds a great promise as given below:

(*i*) Biochips can respond to natural nerve impulses making looks more natural when implanted into the artificial limbs.

(*ii*) It is also possible that they can also be used as a heart-heat regulator. This will solve the problems of users of costly pace makers.

(*iii*) It can also help blind or deaf. It can be designed in such a way that can sense light and sound, and convert them to electrical signals. These signals after reaching brain stimulate sight and sound.

(*iv*) It can be designed in accordance with the need of military and protect the silicon-based computers. It can keep immune to the disastrous effects of electromagnetic waves which are generated due to nuclear explosion and protect the silicon-based computers.

## PROBLEMS

1. What is a biocatalyst? What role does it play in biotechnology?
2. Write an eassy on microorganisms producing useful enzymes.
3. What are the properties of enzymes?
4. What is immobilization of enzymes? What are the advantages in using them?
5. Give an illustrated account of methods and effects of enzyme immobilization.
6. Write an eassy on application of enzymes.
7. Write short notes on the following.
   (*a*) Biocatalysts; (*b*) Immobilised enzymes; (*c*) Encapsulation; (*d*) Enzyme engineering
   (*e*) Penicillin G/V acylase; (*f*) Biosensor; (*g*) Biochip, (*h*) Immobilisation of cells.

CHAPTER

23

# Biomass : A Renewable Source of Energy

Undoubtedly, energy manifests the whole cosmos. Much has been described about energy, its manifestation and conversion into various forms in *Pre-vedic* literature. Really, '*Ekoham bahusyamah*' explains the conversion of energy from one form to other forms. As the germ of civilization started sprouting in man, at the same time, he thought over the sources for fulfilling his needs. The existence of fire was realized when it sparked from stone friction. Since then it became an integral part of our livelihood. It is necessary to go to the science of physics to define energy. The ideas have developed from centuries of observations by humans. Energy is defined as the ability to work.

Fossil-fuel based industry.

## A ENERGY SOURCES : A GENERAL ACCOUNT

Energy *per se* is an integral component of any socio-economic development for raising the standard of living as also improving the quality of life of the people in general. Moreover, energy has played much role in the dawn of human civilization. It is obtained in different forms such as nuclear energy, fossil fuels energy (coal, oil and gas), and non-fossil and non-nuclear energy.

## 1. Nuclear Energy

In recent years, we have much hopes for getting nuclear energy. It is made available through the two processes: (*a*) nuclear fission, where a nucleus of an element is broken into two nuelei or more and releases sufficient amount of energy, and (*b*) nuclear fusion, in which case energy is released as a result of joining of two very small nuclei.

For getting energy of the first kind, nuclear reactors are set up in the developed and some of developing countries like India. Energy is generated from uranium for the peaceful work. Inspite of development of nuclear waste management technology, still there is fear for the disposal of radioactive nuclear waste. Radioactive chemicals are long lived, and if entered into human systems it can cause death also. The second method of nuclear energy generation is still in infantary stage. It may take more than 50 years to be developed.

## 2. Fossil Fuel Energy

The living plants buried during the carboniferous period (about 330–350 million years ago) have been a source for fossil fuels (coal, oil and gases). Coal is the major reserve, followed by oil and natural gass. It is widely distributed and occurs in high quantity. It may reach its peak of production in another 150 years. Oil stands second to coal; its price is increasing day by day due to high cost of extraction and purification. However, during World War II, oil was cheaper than coal. But its price increased gradually with oil-based economics in most of the developed countries. Since 1973, when oil producing countries in the middle-east decided to reduce the oil production and raised the oil price, there has been very significant effect on the economies of non-oil producing nations.

With the onset of oil crisis, most of the countries became aware of their total dependence on only one form of energy. In fact, it was soon realized that mankind had been living in a petroleum society and that this crisis threatened the life style, as also the national security because most of the defence systems use petroleum as energy. It also become clear that such an abrupt crisis would not have affected any nation if a broad energy policy, involving many sources, had been followed.

## 3. Non-fossil and Non-nuclear Energy

In addition to energy sources as described above, the star (sun), planet (earth), satellite (moon) and water and wind are the other sources of energy owing to which our existence is possible. Global power potential of some renewable resources is given in Table 23.1.

**> Table 23.1. Global power potential of renewable sources (D.M. Gates, 1985)**

| *Resources* | *Total watts (W)* | | *Useful power watts* | | *Potential (quads* per year)* |
|---|---|---|---|---|---|
| Wind | 1.3 | $10^{15}$ | 1.3 | $10^{14}$ | 3900 |
| Hydro | 9.0 | $10^{12}$ | 2.9 | $10^{12}$ | 86 |
| Waves | 7.0 | $10^{13}$ | 2.7 | $10^{12}$ | 80 |
| Ocean thermal | 5.0 | $10^{13}$ | 2.0 | $10^{12}$ | 59 |
| Geothermal | 2.7 | $10^{13}$ | 1.3 | $10^{11}$ | 4 |
| Tidal | 3.0 | $10^{12}$ | 6.0 | $10^{10}$ | 1.9 |
| Current | 1.0 | $10^{11}$ | 5.0 | $10^{10}$ | 1.6 |

*1 quad = 1 quadrillion BTU (British Thermal Unit); 1 quadrilloin = $10^{15}$

Energy from tides (due to moon) and geothermal one (hot interior of the earth) have least contribution. Recently, in Gujarat, the Central Electricity Authority at the Kutch Tidal Power

Project, Navlakhi, is investigating for the possibility of electricity generation from tidal wave energy, along specific areas of the Indian coastline. It is hoped that 900 MW electricity can be generated. Investigations for assessment of tidal power potential in the gulf of Kutch were conducted by the National Hydroelectric Power Corporation (NHPC) in association with National Institute of Oceanography, Geological Survey of India (G.S.I.) and Central Water and Power Research Station.

Energy from wind can also be utilized by technological breakthrough. National Aeronautical Limited, Bangalore is engaged in research and development of power generation through wind mills. In many states, windmills have been set up for irrigation purposes. However, by using wind energy, power generation plants can also be set up. Much work is to be done on utilization of wind energy. Sun has the most unlimited source of energy *i.e.* the solar energy. It is due to constant nuclear reactions going on naturally in it. A total of 35 per cent energy is reflected by air molecules, dust, cloud, etc. present in the atmosphere, and through surface reflection as well.

Total energy present on earth surface ($3 \times 10^{24}$ J/year) comes from sun at the rate of $173 \times 10^{12}$ KW, of which about 0.1 per cent energy per annum is utilized by plant life, which results in net annual production of $2 \times 10^{11}$ tonnes of organic matter having energy content of $3 \times 10^{12}$ J. Total annual energy use, however, is the order of $3 \times 10^{20}$ J.

There are three routes of utilization of solar energy: (*i*) solar thermal route, (*ii*) solar photovoltaic route, and (*iii*) biological route. The first two routes are out of scope of this text, whereas the biological route or photosynthesis is very important as far as renewable source of energy is concerned.

## B. BIOMASS AS A SOURCE OF ENERGY

The Forest contributes a huge biomass

The peculiar feature of plants is that they possess various photosynthetic pigments in thylakoids present in cells either in free state or in chloroplasts. The photosynthetic pigments are chlorophyll a, chlorophyll b, chlorophyll c, xanthophylls and carotenoids. The presence of these pigments varies with group of the plants. In prokaryotic photoautotrophs where chloroplasts lack, photosynthetic pigments are present in thylakoids. The different pigments absorb light of different wavelength. During photosynthesis in plant cells, in the presence of chlorophyll $CO_2$ is converted into complex carbohydrates with the evolution of oxygen. The reaction is shown below:

$$CO_2 + H_2O \xrightarrow[\text{chlorophyll}]{\text{light}} (CH_2O)_n + O_2$$

During photosynthesis, solar energy is trapped into light harvesting molecules in the chloroplasts by reduction of $CO_2$ into carbohydrates, fats and proteins. Radiant energy stored in plant is known as primary production which later on creates plant biomass or biomaterial. The rate of storage of photosynthetic products is known as 'primary productivity'. The energy remaining as organic matter (after respiration) is called as net primary productivity (NPP). NPP is expressed as KCal or g/m$^2$/yr. NPP accumulates over time as plant biomass. It is expressed as dry weight of organic matter per unit area (g/m$^2$). Biomass differs from production, which is the rate of organic matter production by photosynthesis. NPP and biomass of some world ecosystems are given in Table 23.2. The biomass is all forms of matter derived from biological activities and present on the surface of soil or at different depth of vast body of water, lakes, rivers, sea and ocean. Biomass includes wood, crops, herbaceous plants, residues from agricultural and forest products, manure, fresh water and marine plants and microorganisms as well. Besides plant material, biomass also includes all animal waste, manure, etc. because in essence, the latter are basically plant based (Khashoo, 1988). Different types of biomass of various sources are given in Table 23.3.

**> Table 23.2. Net primary productivity and plant biomass of world ecosystems.**

| *Ecosystems (in order of productivity)* | *Mean net primary production per unit area (g/m$^2$/yr)* | *Mean biomass per unit area (kg/m$^2$)* |
|---|---|---|
| **Tropical forests** | | |
| Rain | 2,000.0 | 44.00 |
| Seasonal | 1,500.0 | 36.00 |
| **Temperate forests** | | |
| Evergreen | 1,300.0 | 36.00 |
| Deciduous | 1,200.0 | 20.00 |
| Boreal forest | 800.0 | 20.00 |
| Savanna | 700.0 | 4.00 |
| Wood land and shrub land | 600.0 | 4.00 |
| Tundra and alpine meadows | 144.0 | 0.67 |
| Desert shrub | 71.0 | 0.67 |

The route of photosynthesis differs in certain group of plants. In $C_3$ plants, where the first photosynthetic product is 3 carbon compound, $CO_2$ combines with a 5-carbon compound (ribulose biphosphate, RuBP) to form phosphoglyceric acid (a 3 carbon compound). However, in $C_4$ Plants, where the first photosynthetic product is a 4 carbon compound, $CO_2$ combines with phosphoenolpyruvate (a 3 carbon compound), instead of RuBP and produces oxaloacetic acid, a 4 carbon compound.

Moreover, photosynthetic efficiency affects the accumulation of biomass as it depends on plant efficiency and light intensity. We are fortunate enough to have abundant sun shine in our country. Total energy received in India is about $60 \times 10^{13}$ MWH, with 250–300 days of useful sun shine per year in most part of the country.

Net primary productivity and biomass accumulation are high due to the high rate of $CO_2$ assimilation and low rate of photo-respiration between 25–30°C. Photo-respiration accounts for 50 per cent reduction in efficiency of the process, as it has a rapid rate of photo-respiration and low rate of $CO_2$ fixation.

Photo-respiration differs from the normal respiration. The later takes place in dark and is associated with oxidation of compounds produced during photosynthesis, which occurs in peroxisomes. Photo-respiratory $CO_2$ arises from glycolate pathway, not from glycolysis. In this pathway, 4 molecules of glycolates are converted into one molecule of glucose and 2 molecules of $CO_2$; $CO_2$ is released when glycine is converted into serine. Glycolate is oxidised to glycolate by enzyme glycolate oxidase.

**> Table 23.3.** **Biomass as the source of energy.**

| *Sources of biomass* | *Forms of biomass* | *Conversion process* | *Forms of energy* |
|---|---|---|---|
| A. **Plantations:** | | | |
| Silviculture | Fire wood | Combustion | Heat (fire) |
| (Energy plantations) | Fuel wood | Destructive distillation | Charcoal |
| Agriculture | Carbohydrate | Fermentation | Ethanol |
| (Energy crops) | Hydrocarbon | Fermentation | Fuel oil |
| Aquatic biomass | Aquaculture | Fermentation | Methanol |
| Weeds | Whole plant body | Fermentation | Methane |
| B. **Residues/wastes/weeds:** | | | |
| Rural/urban/industrial wastes | Wastes | Combustion | Fire/fuel |
| | | Pyrolysis | Fuel oil |
| | | Fermentation | Methan and ethanol |
| Forestry wastes | Wastes | Combustion | Fire/fuel |
| | | Pyrolysis | Oil gas |
| | | Gasification | Gas |
| | | Fermentation | Methane, ethanol |
| Agricultural wastes | Wastes | Fermentation | Methane |
| Weeds and aquatic biomass | | Fermentation | Methane |
| Cattle | dung | Combustion | Fire/fuel |
| | | Fermentation | Methane (Biogas) |

In recent years, attempts have been made to convert $C_3$ plants into $C_4$ ones by introduction of characteristics of $C_4$ plants in $C_3$ ones. Regulation of photo-respiration by chemical or genetical manipulation offers a possibility of increasing crop-productivity. Some metabolic inhibitors are reported which catalyse photo-respiration. In 1974, Zelitsch at the Connecticut Agricultural Research Station, New Haven (USA) demonstrated that an anologue of glycolate and glyoxalate, inhibited glycolate synthesis and photo-respiration by 50 per cent, and increased 14 $CO_2$ assimilation by 50 per cent. In 1973, Peter Carlson at the Brookhaven National Laboratory obtained the mutants of tobacco cells by growing them on nutrient medium supplemented with sulphoxymine. The mutant cells displayed a low photorespiration rate.

## 1. Composition of Biomass

Plant cell wall is constituted by mainly 6 components: (*i*) cellulose, (*ii*) hemicellulose, (*iii*) lignin, (*iv*) water soluble sugars, amino acids and aliphatic acids, (*v*) ether and alcohol-soluble constituents (*e.g.* fats, oils, waxes, resin and many pigments), and (*vi*) proteins. These components build up plant biomass. Proportion of these constituents vary in different groups of plants and even in the same group. If the concentration of sugar is high, the biomass will be sugary *e.g.* sugarcane, and sugar beet. Similarly, high amount of starch present in biomass yields the starchy biomass *e.g.* potato and tapioca. Variation in chemical constituents (cellulose, hemicellulose and lignin) in some plants is shown in Table 23.4.

**> Table 23.4.** **Chemical composition of lignocellulosic material, and cotton (on % dry weight basis)**

| *Sources of biomass* | *Cellulose* | *Hemicellulose* | *Lignin* | *Protein (N × 6.25)* |
|---|---|---|---|---|
| Birch angiosperm* | 44.9 | 32.7 | 19.3 | 0.5 |
| Spruce gymnosperm* | 46.1 | 24.6 | 26.3 | 0.2 |
| Crop residues** | 30–45 | 16–27 | 3–13 | 3.6–7.2 |
| Wood residues*** | 45–56 | 10–25 | 18–30 | – |
| Cotton | 89.0 | 5.0 | 0.0 | 1.3 |

(*a*) **Cellulose:** Cellulose constitutes the major portion of plant cell wall, the fundamental unit of which is glucose. Formation of cellulose is a complex process. From each glucose unit, one molecule of water is removed to yield an anhydrous glucose:

$$C_6H_{12}O_6 - H_2O = C_6H_{10}O_5 + H_2O$$

The anhydrous glucose units are linked end to end with ß-1,4-linkage to form the long chain polymer of cellulose $(C_6H_{10}O_5)n$. Here n represents the degree of polymerization, the number of which varies from 5000 to 10,000 (Fig. 23.1). Enzymatic hydrolysis of cellulose and production of glucose are in the preceding C.

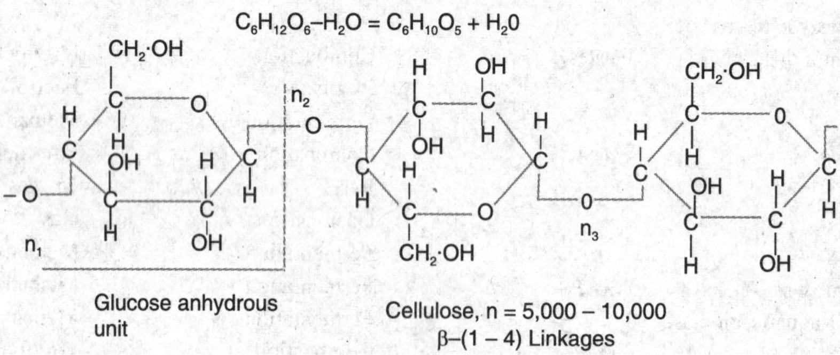


Fig. 23.1. Structure of cellulose chain.

(*b*) **Hemicellulose:** Hemicellulose is also constituted by sugars (xylans) which comprises of 20-25 per cent plant biomass on dry weight basis. In addition, it also contains glucose and several other hexoses (galactose and mannose) and pentoses (xylose and arabinose) (Fig. 23.2). The proportion of these constituents varies plant to plant. Degree of polymerization to yield hemicellulose does not exceed beyond 50. The polymers has branched chains. It occurs as amorphous mass around the cellulose strands. Hemicellulose are insoluble in water but easily solubilised in alkali.

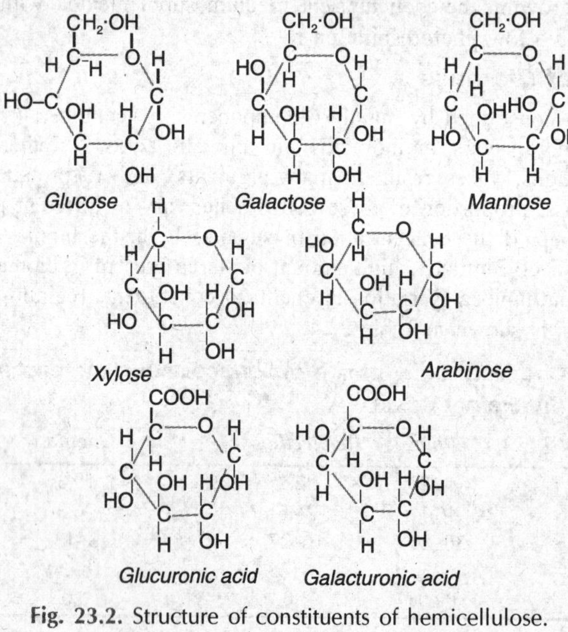


Fig. 23.2. Structure of constituents of hemicellulose.

(*c*) **Lignin:** The third constituent, lignin, is a complex and high molecular weight polymer (Fig. 23.4). It is formed by de-hydrogenation of p-hydroxy-cinnamyl alcohols (Fig. 23.3) such as p-coumaryl (I), coniferyl (II), and sinapyl (III) alcohols. Presence of these alcohols differs in different plant groups. For example, gymnosperm lignin is formed from coniferyl alcohols, angiosperm lignin is formed from the mixture of coniferyl and cinapyl alcohols, and grass lignin from mixtures of coniferyl, sinapyl and coumaryl alcohols. Lignin is phenolic in nature; it is very stable and difficult to isolate. It occurs between the cells and cell walls. It is deposited during lignification of the plant tissue and gets intimately associated within the cell walls with cellulose and hemicellulose and imparts the plant an excellent strength and rigidity.

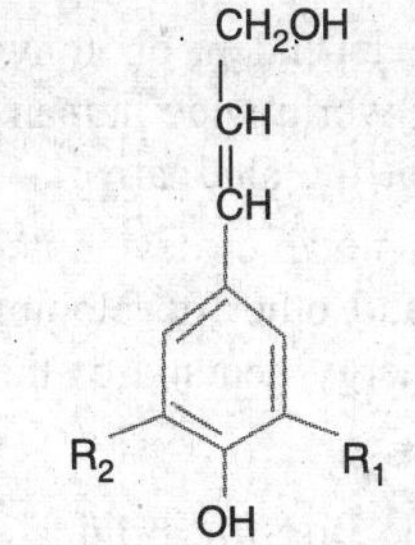


I. $R_1 = R_2 = H$

II. $R_1 = OCH_2$ ; $R_2 = H$

III. $R_1 = R_2 = OCH_3$

**Fig. 23.3.** Structure of p-hydroxy cinnomyl alcohol unit forming lignin.

As a result of photosynthesis, an enormous amount of plant biomass is accumulated in terrestrial and aquatic systems, which are then utilized into different ways as the source of energy.

## 2. Terrestrial Biomass

Since long, terrestrial biomass has been used to fulfil the need of food, feed, vegetables, fibre, furniture and cooking purpose as well. Traditionally the need of fire/fuel was fulfilled by trees, remains of agricultural crops, and fossil fuels (coal, and petroleum). During the course of

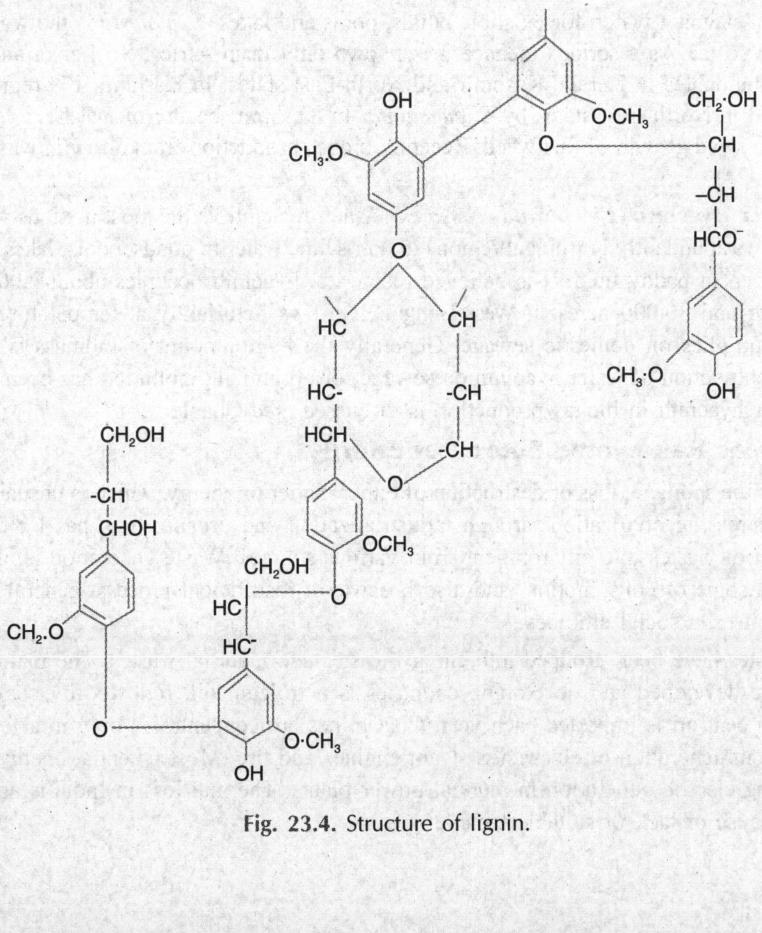

**Fig. 23.4.** Structure of lignin.

time, we totally became dependent on conventional energy sources of fossil fuel and electricity. But gradually increasing world wide human population and diminishing stock of fossil fuel have challenged us to seek out the alternative sources of energy.

The "photosynthetic model of development" has been emphasised in recent years. This model is applicable for India and other developing countries. However, the extent and nature of this model may vary with energy demand of that country.

## 3. Aquatic Biomass

It is obvious that the first life originated in water. Therefore, water bodies support a vast community of plant and animal. Many aquatic plants become troublesome for aquatic animals and human as well such as the aquatic weeds like water hyacinth, *Salvinia, Hydrilla, Lemna, Pistia, Wolffia,* etc.

*Salvinia*: a source of aquatic biomass.

In addition to higher plants, the lower plants (especially blue-green algae and green algae) have much future prospects, as far as production of biomass conversion of aquatic biomass into biogas/ hydrocarbon and abatement of pollution (in sewage oxidation) are concerned.

***(a) Salvinia*:** *Salvinia,* a member of Pteridophyta, is commonly known as water fern. It grows luxuriantly in stagnant water, for example ponds, pools and lakes. *S. molesta* is the world's worst weed known so far. As a serious menace it is known only from Africa, Sri Lanka and India. In India, it predominates in Kerala, Kashmir and North-East states. In Kashmir it is represented by *S. natans* and in North-East states by *S. cucculata.* In Kashmir, beauty of Dal Lake is gradually fading due to rapid growth of this weed. Recently biogas production from *Salvinia* was suggested (*see* Chapter 24).

**(*b*) Water Hyacinth (*Eicchornia crassipes*):** Water hyacinth is the most noxious weed of the world. It grows abundantly in tropical regions in non-saline water in ponds, pools, lakes reserviors, rivers and even in paddy field. It is believed that water hyacinth occupies about 2,00,000 acres land in Bihar and 30,000 acres in West Bengal. It grows luxuriently at temperature 28–30°C. It rapidly multiplies on domestic sewage. Generally the huge amount of biomass is of no use. Nowadays, cultivation of water hyacinth on sewage for minimizing pollution has been suggested. Use of water hyacinth in biogas production is discussed (*see* Chapter 24).

## 4. Wastes as Renewable Source of Energy

Waste is the spoilage, loss or destruction of either matter or energy, which is unsuable to man. Gradually increasing civilization through industrialization and urbanization, has led to increase in generation of wastes into environment from various sources. Waste generation is, therefore, a necessary outcome of consumption, and also because of insufficient process, general ignorance, wasteful habits and social attitudes.

The wastes have been grouped into energy wastes and material wastes. The main source of energy in the developed and developing countries is petroleum oil, followed by coal. In India, about 50 per cent oil is imported each year. Coal mines are concentrated only in a few regions. Coal is used in generation of electricity, steam engines and fire. Most potential energy of coal is wasted during electric generation in thermal power plants. Thermal loss in India is about 20–30 per cent because of lack of suitable technologies.

Chemically, the material wastes are of various types: (*i*) inorganic wastes (those generated by metallurgical and chemical industries, coal mines, etc.), (*ii*) organic wastes (agricultural products, dairy and milk porducts, slaughter houses, sewage, forestry, etc.), and (*iii*) mixed wastes (those discharged from industries dealing with textiles, dyes, cake and gas, plastic, wool, leather, petroleum, etc.). The inorganic wastes may be recovered by chemical/ mechanical treatment, whereas organic and mixed wastes require biological as well chemical treatments.

Moreover, the wastes occur in three states, the solid, liquid and gaseous ones. The solid wastes can be burnt, thermally decomposed, anaerobically digested to get methane and other combustible gases or biologically converted to a variety of products. Liquid wastes are most troublesome, because of the presence of non-retractable chemicals, and their further return to environment through surface waters. Gaseous wastes include the toxic gases such as $NO_2$, $NO_2$, $NH_3$, $CO_2$, $CO_2$, $SO_2$, etc. When concentration of these gases increases in the atmosphere they cause gaseous pollution, which has its bad impact on plant and animal lives.

The organic wastes and residues become a source of renewable energy in multifarious ways. Some of the renewable sources of the organic materials are given in Table 23.5.

**(*a*) Composition of Wastes:** Waste is a general term which embraces all types of wastes irrespective of constituents and phases. Therefore, composition of waste differs with differing nature, phases and sources. It may be inorganic, organic or mixed types. Organic wastes play a major role in being renewed and becoming a source of energy. Composition of organic materials is given under 'composition of biomass' (*see* earlier section).

**> Table 23.5.** **Some rnewable sources of biomaterials**

| *Plants* | *Residues/wastes* | *Products* |
|---|---|---|
| Cashewnut | Shell, testa | Gum, tannin |
| Coconut | Coir, shells and pith | Coir board, coir fibre, xylose |
| Cotton | Fibre | Fuel |
| Maize | Cobs | Single cell protein (SCP), Fuel |
| Paddy | Husk, bran, straw | Bran oil, vitamin E, SCP, |
| Sugar beet | Pulp | Ethanol fuel |
| Sugarcane | Bagasse, molasses | SCP, fuel/fire, alcohol |
| Sunflower | Husks | SCP, Fuel (alcohol) |
| Tapioca (cassava) | Tubers | Ethanol |
| Tea | Tea waste | Caffeine |
| Trees | Seeds and leaves | Many products |
| Wheat | Straw | SCP, fuel |

**(*b*) Sources of Wastes:** Industries generate various types of wastes/ by-products which contain sufficient amount of energy. Some of the industrial wastes are described below:

**(*i*) Paper Mill:** The wastes are bisulphite liquor and lignocellulosic pulp.

**(*ii*) Chemical Industries:** The chemical wastes are maleic anhydride and phthalic anhydride.

**(*iii*) Oil Refineries:** They produces wastes as gas, oil, paraffins (*n*-alkanes), olefins (*n*-alkenes) and other hydrocarbons.

**(*iv*) Cotton Mills:** Cotton mills produce the cotton seeds and fibres as wastes.

**(*v*) Food Industry:** Waste materials of food industries are the collagen meat packaging waste and lactoserum (a by-product of cheese making food industry).

**(*vi*) Dairy:** Dairy industry is one of the important industries which requires special attention, as far as treatment and disposal of wastes are concerned. Dairy wastes contain milk whey, butter milk, unused skim milk, plant washings and traces of detergents. The waste is a dilute solution or suspension containing lactose, protein, fat and minerals. Therefore, dairy wastes serve a food substrate for production of single cell protein, lactic acid, vitamins, ethyl alcohol and alcoholic beverages.

**(*xii*) Sugar Mills and Distilleries.** Molasses and bagasse are the wastes generated from sugar mills.

**(*c*) Sugarcane (*Saccharum officinarum*):** Sugarcane belongs to family Gramineae. It is a tall, perennial grass, the stems of which are the source of cane syrup.

India is the largest producer of sugarcane in the world. Sugarcane is grown in many states of the country. There are about 400 sugar mills and 127 distilleries in working order; besides, about 88 new sugar mills are coming up soon. Enormous quantity of wastes in the form of bagasse, molasses, and press mud are produced by sugar mills.

Sugarcane: a plant of Indian origin

Bagasse is the cellulosic material of sugarcane produced after extraction of sugary juice. It is used for the production of fuel, alcohols, single cell protein, and in minipaper mills also. Molasses is an important by-product of sugar mills which contains between 50–55 per cent fermentable sugars, mainly sucrose, glucose and fructose. Every tonne of sugar produced gives about 190 litres of molasses, and about 280 litres of ethanol can be produced per tonne of molasses. Molasses is used for the production of animal feed, liquid fuel and alcoholic beverages.

About 732 million litres of alcohol per annum is produced from distilleries. For every litre of ethanol (rectified spirit) about 12–15 litres of effluent, reddish brown to dark in colour, is generated which has a high biological oxygen demand (BOD) of about 40,000 mg/l. It is because of the presence of organic materials in high amount. Production of single cell protein, ethanol and biogas from the distillery effluent is shown in Fig. 23.5. For the utilization of distillery wastes works are being done at National Sugarcane Institute (Kanpur), National Environmental Engineering Research Institute (Nagpur), and Department of Microbiology, Hissar University (Hissar).

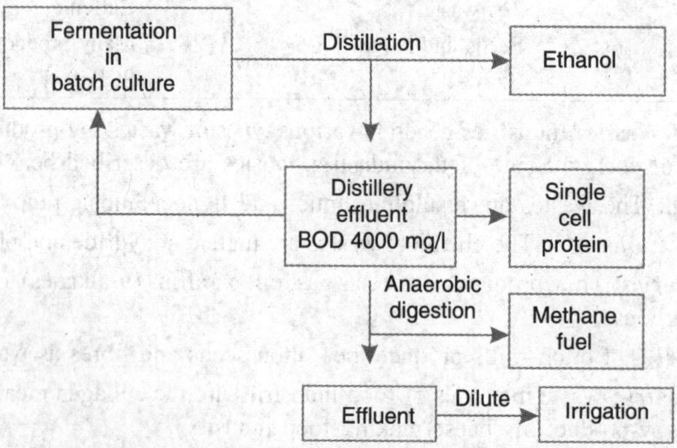


Fig. 23.5. Production of single cell protein, ethanol and biogas from the distillery wastes.

Importance of using bagasse as an alternative source of energy for sugar manufacturers has been realized in recent years. The Indian Renewable Energy Development Agency (IREDA) sanctioned around 97 projects in this field. Uttar Pradesh Government launched a programme for producing alternative energy from sugar cane waste in sugar mills. The first project of its kind in the country has been installed with the Sheh Road Sugar Mill in Bijnor. It will produce 6 MW electricity from the sugarcane waste. In this way fuel consumption would be reduced from 40 to 58 per cent.

**(*d*) Sugar Beet (*Beta vulgaris*).** Sugar beet is a member of Chenopodiaceae. It is a biannial herb with fleshy leaves and swollen roots. For sugar production good crops of sugar beet can be had in Rajasthan, Punjab, Haryana and Uttar Pradesh.

In India, a large area is occupied by *usar* (sodic) land. It is a characteristic feature of sugar beet that it can grow in sodic soil. It is a short duration plant. The underground parts, *i.e.* tubers contain high contents of sugar. Like sugarcane, it is also a good source of ethanol.

**(*e*) Paddy (*Oryza sativa*):** In agriculture, a huge amount of residues/wastes are produced which, however, are thrown into field because of non-availability of technologies for utilization at village level. Some of the agricultural wastes produced from various plants are given in Table 23.3. However, residue/wastes of some crops are utilized in multifarious ways, and their future prospects are discussed. Paddy plant produces paddy and straw; the products of paddy are rice, husk and bran. Husk is the outer most hard coat of paddy. Bran is the thin papery layer present between husk and rice.

Out of total (about 80 million tonnes) production of paddy, about 16–18 million tonnes of husk (*i.e.* 20–25 per cent of paddy) are produced per annum. Due to lack of utilization technology bran is thrown as waste, and probably in some part it is used as cattle feed. Recently, bran has become a source of rice bran oil, whether edible or non-edible. Paddy husk is used as fuel in rice mills and villages as well.

Recent analysis of paddy husk of different species has shown the presence of high caloric value *i.e.* 3200–3500 kcal/kg, which can replace about 10 million tonnes of coal per annum. It has been found that 1 tonne of husk can replace 450 litres of furnace oil (Gupta, 1984).

To meet the demand of energy and minimize the pressure on fossil fuel, many countries like Japan, China, Philippines, Indonesia, India have given special attention to develop technologies for utilization of paddy husk. Recently, the Central Fuel Research Institute (CFRI), Dhanbad has developed a process to produce oxalic acid from cellulosic matter and silica from mineral matter.

India has started a thermal power plant from paddy straw, which is first of its kind in the world which can generate 62 million unit of energy per annum. A rice straw-fired Thermal Power Plant was set up at a village, Jalkheri (Punjab) in March 11, 1989. It has been done as a joint venture of the Department of Non-conventional Energy Source (DNES), Punjab State Electricity Board and Bharat Heavy Electricity, Limited. Moreover, Punjab Agricultural University (Ludhiana) and IIT (Kanpur) have also developed a husk-fired plant. Similarly, Paddy Processing Research Centre (Tiruvarur) has developed methods for parboiling and milling operation in rice mill to extract about 92 per cent potential energy of husk.

**(*f*) Tapioca (*Manihot esculenta*):** Tapioca, commonly known as cassava, is a member of Euphorbiaceae. It is a small shrub producing tubers inside the soil. Tubers are rich in starch. Therefore, this plant is exploited commercially for starch, sago and flour. It is an important source of alcohol also.

For these purposes, tapioca is cultivated in Kerala, Tamil Nadu, and Karnataka. Of the total production, Kerala alone contributes 80 per cent tapioca in the country. It grows in alkaline soil in marginal and infertile lands.

Tapioca can yield alcohol about 180 litres/tonne in contrast with sugarcane which produces about 70 litres/tonne. Central Tuber Crop Research Institute (CTCRI), Trivandrum has carried out work on alcohol production from tapioca. Starch is extracted from the tubers which is then converted by alpha amylase and gluco-amylase into sugar. Sugar is then fermented into alcohol.

**(*g*) Forestry:** Forests contribute a considerable amount of biomass which could be variously used as a source of energy. Lignocellulosic contents also varies from 60–75 per cent of dry weight. Soft wood has higher lignin to cellulose ratio than hard wood. Residues/wastes generated from forestry are wood chips, saw dust, dried tree branches, tree twigs, tree-bark, leaf litter, etc.

**(*h*) Municipal Sources:** High amount of municipal wastes are generated from cities, as a result of anthropogenic activity which are thrown near cities in open lands or in rivers. These wastes again become a serious hygenic problem as it contain high amount of organic matter and pathogenic microorganisms. Municipal and domestic wastes include sewage and sludge, garbage, horse dung, cattle dung and wastes from animal slaughter houses.

**(*i*) Sewage and Sludge:** Sewage is a product of water which are thrown away after its use. Treatment of sewage results in generation of another waste, which is called as sludge. Sludge is a solid matter in the settling tanks of sewers, and other treatment operations in a sewage treatment plant.

Sludge digestion of Municipal Sewage

There are 142 class I cities in India which produce around 9,000 million litres of sewage per day. However, there is no sewage treatment facilities in about 70 class I cities such as Srinagar, Ranchi, Dhanbad, Bhopal, Jabalpur. Although methods for single cell protein production from sewage oxidation pond and irrigation of agriculture crops by treated water have been developed (see Fig. 22.1), yet this facility is available only at limited places. Moreover, if 900 million litres of sewage is converted into sullage gas per day from the major cities, 20 per cent of their energy demand could be met (*see* Chapter 24).

**(ii) Urban (City) garbage:** Garbage potential of Indian cities is quite high. It is estimated that production of city garbage in India is about 41,000 tonnes per day, the annual production is about 15 million tones. Nonetheless, city garbage produced in Mumbai, Chennai and Kolkata is comparable to that of developed countries. In India, garbage is thrown near the city. In some cities like Chennai, Kolkata, Delhi, Baroda, Jodhpur, etc. Municipal solid waste composting plants are in operation.

Moreover, the Central Mechanical Engineering and Research Institute, Durgapur has established a first pilot plant to produce electricity by using city garbage. The plant has a capacity to use about 500 kg garbage/ha, as a result of which about 5 Kwh electricity can be generated. Process of electricity production is the conversion of garbage, anaerobically, into biogas and in turn into electricity.

# C. BIOMASS CONVERSION

Biomass can be converted into energy by the following ways:

## 1. The Non-Biological Process (Thermo-chemical Process)

There are different non-biological routes for biomass conversion into energy viz., direct combustion, gasification, pyrolysis and liquefaction.

**(*a*) Direct Combustion:** Biomass from plants (wood, agricultural wastes) or animal (cow dung) origin are directly burnt for cooking and other purposes. In recent years "hog fuel" production technology has been developed which is being utilized for generation of electricity.

Now-a-days municipal, agricultural and light industrial wastes are used for conversion into energy by direct burning in refuse fired energy systems.

In India, short term rotation plantation and social forestry are being boosted up to meet energy demand for cooking in rural sector as the industrial wastes and urban garbage are used for generation of electricity and engine in industry.

**(*i*) Hog Fuel:** The mixture of wood and bark waste burnt directly is collectively termed as 'hog fuel'. Hog fuel combustion technology has been developed recently in the USA This fuel is produced in large sized boilers made up of steel. Boilers are designed time to time to develop a good control system of combustion. In the USA a boiler has been modified by increasing the size from 15,000 lb/h to 500,000 lb/h with certain other improvements. In the USA a cogeneration technology has been developed to generate electricity from hog fuel, and to use the exhaust heat in the form of process steam for manufacturing operations. This is done by burning hog fuel to make high pressure in the hog fuel boiler (600–1,200 lb/inch$^2$), passing this steam through back pressure or extraction turbine driving a generator, and then using the steam exhausted from the turbine at low pressure (50–300 lb/inch$^2$) for process heat. When wood containing 55 per cent moisture is used, the power generation has about 60 per cent efficiency compared to that 39 per cent from fossil fuels and 33 per cent from nuclear fuel. In order to economically deploy cogeneration systems, steam usage should be more than 70,000–120,000 lb/h which is equivalent to 3–5 MV of back pressure.

**(*b*) Pyrolysis:** Pyrolysis is defined as the destructive distillation or decomposition of organic matter, for example, solid residues, wastes (saw dust, wood chips, wood pieces) in an oxygen-deficient atmosphere or in absence of oxygen at high temperature (200–500°C or rarely 900°C). Products of pyrolysis are gases, organic liquids and chars, depending on the pyrolysis process and temperature of reaction. The condensable liquids separate into aqueous (pyroligneous acid), oil and tar fraction (if the substrate is wood). The composition of gas is carbon monoxide (28–33 per cent), methane (3–5–18 per cent), higher hydrocarbons (1–3 per cent) and hydrogen (1–3 per cent). During pyrolysis, hydrogen content of gas increases with increasing the temperature.

After pyrolysis, the amount of different products varies with the nature of wood, type of equipment and systems employed. For example, low temperature favours liquids and char; low heating rates favour gas and char and short gas residence time favours liquids. In contrast high heating rate, favours liquids and the long gas-residence time favours gas. Thus, the liquid is obtained before the solid is completely burnt to yield gases. The liquid is very useful for high energy fuel.

Pyrolysis has been employed to produce charcoal for the last few decades. Charcoal is a smokeless and low sulphur fuel used mostly for cooking purpose. Besides wood, other wastes/ residues used in pyrolysis are cotton, bagasse, ground nut shell, etc.

**(*c*) Gasification:** Gasification is a process of thermal degradation of carbonaceous material under controlled amount of air or pure oxygen, and high temperature upto around 1,000°C. As a result of gasification, high amount of gases is produced. Gasification of biomass is done in a gasifier designed in various ways. Success for gasification process is based on its designing. Therefore, the design of a gasifier is an important factor in controlling gas quality. Gas is used in a controlled manner for irrigation, pumping and electricity generation.

The advantages of gasification of wood over coal are: (*i*) much low oxygen requirements, (*ii*) practically no steam requirements, (*iii*) low cost for changing $H_2/CO_2$ ratios which are high in wood gas, and (*iv*) no or very little desulfurization cost.

When gasification of farm wastes (manure) takes place, the phenomenon is known as hydro-gasification, because gasification of organic wastes occurs in the presence of hydrogen at 500–600°C.

At present there are 400 gasifiers in the country, based on using waste wood. A few pilot gasifier plants, using non-wood biomass, have also been designed. For gasification purpose, wood wastes from agriculture, pharmaceuticals, coconut coir, saw dust and tree-trimings are used. Efficiency of thermal conversion of wood to gases is 60–80 per cent.

The multiple use of five HP gasifier programme was also emphasized in the seminar. The gasifier used for pumping water for irrigation can also be used to light house-lamps and street lamps (with a single gasifier).

Three to five kilograms of biomass per hour is needed for generating power for an hour. In a year it has to run for at least 15,000 hours for which 6 tonnes of biomass is required. Therefore, it is very essential to make sure that the biomass is continuously available. Recently, United Nations Department Programme (UNDP) has recognised the India's five HP gasifiers as the best ones.

**(*d*) Liquefaction:** Liquefaction involves the production of oils for energy from wood or agriculture and carbon residues by reacting them with carbon monoxide and water/steam at high pressure (4,000 $lb/in^2$) and temperature (350–400°C) in the presence of catalysts. By this method about 40–50 per cent oil can be obtained from wood. This oil serves a good source of fuel.

## 2. The Biological Process (Bioconversion)

Bioconversion involves the conversion of organic materials into energy, fertilzier, food and chemicals through biological agency. The term biological agents means the microorganisms *i.e.* bacteria, actinomycetes, fungi and algae. In broad sense bioconversion involves 2 steps: photosynthetic production of biomass, and its subsequent conversion into more useful energy forms (gaseous, liquid or solid fuel; heat and electricity).

The average production of waste materials in India is about $1,540 \times 10^6$ tonnes/year. Process of bioconversion of biomass into various utilizable forms are briefly discussed.

**(*a*) Enzymatic Digestion:** This process involves the conversion of cellulosic and lignocellulosic materials into alcohols, acids and animal feeds by using microbial enzyme *e.g.* cellulose, hemicellulase, amylase, pectinase, etc.

**(*i*) Degradation of cellulose:** It is clear that cellulose is a polymer of ß-1, 4 linked anhydrous glucose units, comprising of 40–60 per cent of cell wall materials of plants. Micro organisms, which produce cellulases and other enzymes in high amount are given in Chapter 22 (See Table 22.1). In recent years, *Cellulomonas, Trichoderma reesei, T. viride* and other microorganisms are used for the production of cellulases in high amount.

There are 3 enzyme components of cellulase: ß-1, 4-endoglucanase, ß-1, 4-exoglucanase and ß-1, 4-glucosidase. ß-1, 4-endoglucanase randomly attacks along cellulose chain; ß-1, 4-exoglucanase splits from non-reducing end of cellulose and ß-1, 4-glucosidase *i.e.* cellobiase cleaves 2 molecules of glucose from cellobiose (Fig. 23.6).

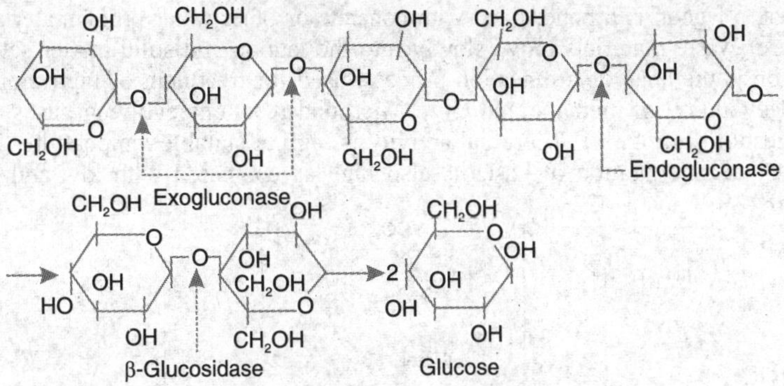


**Fig. 23.6.** Enzymatic breakdown of cellulose by exoglucanase, endoglucanase and ß-glucosidase.

Following is the sequence of cellulose degradation.

Cellulose chain → Smaller polymers and soluble oligomers →

ß-1, 4-endogluconase, ß-1, 4-exogluconase → Cellobiose and Glucose → ß-1, 4-glucosidase (cellobiase) → glucose (2 molecules)

(*ii*) **Degradation of hemicellulose:** Sugars constituting hemicellulose are discussed earlier. An analogous system of enzymes is involved in the degradation of hemicellulose. This enzyme-system consists of 3 enzymes: exoxylanase endoxylanase and ß-xylosidase (which split xylose and other short chain xylobioses). Steps of hemicellulose break down are given in Fig. 23.7.

(*b*) **Anaerobic Digestion:** An aerobic digestion is a partial conversion by microorganisms of organic substrates into gases in the absence of air. The gases produced are collectively known as 'biogas'. Anaerobic digestion is accomplished in 3 stages: solubilization (of complex substrates by enzymes into simple forms *i.e.* fatty acid, sugars, amino acids), fermentation (of hydrolysed organic substrates into simplest forms *e.g.* organic acids) and methanogenesis (production of methane from simple substrates by methanogenic bacteria under anaerobic conditions). Anaerobic digestion is carried out in a digester, which is a brick-lined or concrete-lined chamber covered completely to prevent the entry of air. A detailed account of biogas production is given elsewhere (*see* Chapter 24).

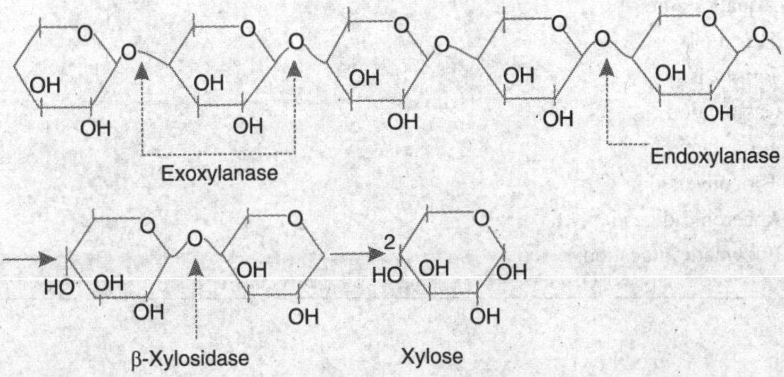


**Fig. 23.7.** Enzymatic breakdown of hemicellulose by exoxylanase, endoxylanase and ß-xylosidase.

In contrast, a lagoon is a pond, lined with concrete or other water proof material and open to the atmosphere. Waste materials move slowly into the lagoon and solid matter settles at bottom. Lagoonification is the anaerobic digestion process, used for treatment of high moisture material, as a result of evolution of methane and $CO_2$. Microbial reactions evolving gases are same as in anaerobic digestion (Fig. 23.8). Microbial activity is high at suitable temperature (29–35°C) and, therefore, the rate of evolution of gases is also high as compared with low and high extremes of temperature.

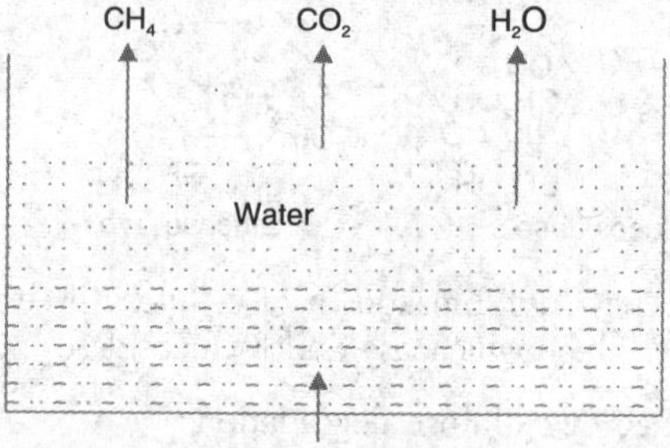


Fig. 23.8. Diagram of lagoonification showing evolution of gases after anaerobic decomposition of organic matter by benthic bacteria.

**(c) Aerobic Digestion:** Aerobic digestion involves the conversion of organic substrates by microorganisms into utilizable forms in the presence of air, for example composting (biological decomposition of organic wastes/residues under controlled conditions to result in release of C, N, P, K, etc.) and oxidation systems (of sewage in oxidation ponds by bacteria and algae) to produce gases, single cell protein, fertilizers, etc. (*see* Chapter 18).

## PROBLEMS

1. What are the sources of energy? Discuss the role of biomass as the source of energy.
2. What are the wastes? How they can be used as renewable source of energy?
3. Write an eassy on sources of wastes and their future scope in India.
4. What are the routes of conversion of biomass? Discuss in brief any one studied by you.
5. What is bioconversion?
6. Write short notes on the following :
   (*a*) Biomass,
   (*b*) Waste and energy,
   (*c*) Aquatic biomass,
   (*d*) Hog fuel,
   (*e*) Pyrolysis,
   (*f*) Gasification,
   (*g*) Liquefaction,
   (*h*) Bioconversion,
   (*i*) Anaerobic digestion,
   (*j*) Enzymatic digestion.

CHAPTER

# 24

# Biomass Energy (Bio-Energy)

Biomass based energy (bio-energy), derived from sun via biological routes, has been fulfiling the human need for centuries as fossil fuel. With the expanding world population, increasing energy demand and diminishing stock of fossil fuel, most of the countries have paid much attention on the bioenergy programmes.

However, there is basic need to develop technologies and standardize the techniques, to reach to village level. Among the various non-conventional sources of energy, forest biomass plays a significant role in solving the fuel wood crisis. It is obvious that one-seventh of world's fuel supplies are biomass. This figure is equivalent to 20 million barrels of oil per day.

The area of bio-energy involves sophisticated research and development work in botany (plant physiology and genetics), agriculture, forestry,phytochemistry, biochemistry, microbiology, fuel research, heat physics and many aspects of engineering, so that a complete package from agriculture, processing, conversion, engine systems and delivery to users is developed.

Plant biomass – an energy source.

Many countries viz., Bangladesh, China, France, Germany, India, Indonesia, Japan, Malaysia, Pakistan, Sweden, U.K., U.S.A., are promoting research and development activities in bioenergy. In addition, several international agencies, such as Food and Agricultural Organisation of the United Nations (FAO), International Development Research Centre (IDRC), United Nations Educational, Scientific and Cultural Organisations (UNESCO), United Nations Environmental Organisations (UNEP), United Nations Industrial Development Organisation (UNIDO) and World Health Organisation (WHO) are also engaged in bioenergy research and development programmes.

## A. ENERGY PLANTATIONS

Energy plantation is the practice of planting trees, purely for their use as fuel. Terrestrial biomass *i.e.*, the wood plants has been used since long time to generate fire for cooking and other purposes. In recent years, to meet the demand of energy, plantation of energy plants has been re-emphasized. It is well-known fact that trees have been intensively cut in Gangetic plains and coastal belts, leaving the area totally denude. The same practice has also been done in Shiwalik region and foot-hills of Himalayas.

According to a report, if fuel/fire wood plants were not raised rapidly, by 2,000 AD more than 250 millions people would not be able to manage fuels for cooking purpose and, therefore, they would be forced to burn animal dung which, however, depends on availability of animals and agricultural crop residues.

India is the biggest fuel wood producing country in the world, but the per capita fuel wood production is very low, *i.e.* 290 $m^2$/head. The need of fire wood for cooking purpose in India is 0.8 kg per capita per day. But this value does not hold good for cold areas where firewood is also required for keeping houses warm.

In India, conditions of hills are quite different from that of plains. The villagers hardly get fire-wood plants, as they have to go in interior of forest and collect wood-falls. Even they have very limited right to fell trees and therefore, they depend on wood falls and loppings of minor tree branches. If one talks about to meet their needs by making available technologies developed for plains, it is not feasible. For example, they cannot be motivated to use solar cooker, because of being solely traditional to which religious factors have been associated. Even gobar gas plant cannot be useful in hills, because of prevailing low temperature in mountain belt. Therefore, the search for renewable source of energy is highly desirable for survival of population in hills and for reducing the pressure on forests.

Recently, energy plantation has got much boost in our country. Government has started many plans, for example, social forestry, silviculture, agro-horticulture practices, and afforestation in waste lands. A large area of land is available in our country which is of no use. A total of 7 million hectares are *usar.* The alkali soils found in U.P. are commonly known as *usar* soils. They contain hard and compact surface layer with clay loam texture. Percentage of exchangeable sodium in these soils is more than 15%. The pH ranges between 8.5 to 10.5. Due to high amount of sodium, plant growth is badly inhibited.

Land is available in east, west and northern part of India of which 1.3 million hectare land exists in U.P. alone. In addition, marginal lands in the form of long strips along railway lines, highways and other roads are also available.

About 200 billion tonnes of carbon per year is fixed photosynthetically into the terrestrial and aquatic biomass. This amount of biomass contains 3,000 billion giga ($10^9$) joules energy every year. It has been estimated that at present, only one seventh of the World's total energy comes from biomass and a large amount of it remains untapped.

Afforestation and forest management systems will have to be developed for getting maximum biomass. These must include social forestry, silviculture (short-rotation forestry) tree-use systems, coppicing system, drought, salt-, pollutant - resistant plantations and high density energy plantations (HDEP). HDEP is the practice of planting trees at close spacing. This leads to rapid growth of trees due to struggle for survival. It provides quick and high returns, and opportunities for permanent income and employment.

Annual plants should be grown to meet the demand of energy. Keeping in view the climatic and edaphic factors, plantation of deciduous trees should be encouraged, as their growth is faster than the coniferous ones. The species to be planted should have the following characters: (*i*) fast growth, (*ii*) stress resistance, (*iii*) less palatable to cattle and other animals, (*iv*) early propagable, (*v*) high caloric value, (*vi*) absence of deleterious volatiles when smokes come out, (*vii*) high yield of biomass, and (*viii*) disease/pest resistant.

### 1. Social Forestry

Plantation through social forestry has been much emphasized by the Government of India to meet the demand of fuel and fodder in the rural areas. It will certainly decrease the gradually increasing pressure on the forests. This includes planting trees along road sides, canals, railway lines and waste lands in villages. Some of important plants are: *Acacia nilotica, Albizia lebbek, A. procera, Anthocephalus chinensis, Azadirachta indica, Bauhinia variegata, Butea monosperma, Cassia fistula, Dalbergia sissoo, Eucalyptus globulus, E. citriodora, Ficus glomerata, Lagerstroemia speciosa, Madhuca indica, Morus alba, Populus ciliata, P. nigra, Terminalia arjuna, Toona ciliata, Salix alba and S. tetrasperma.*

### 2. Silviculture Energy Farms (Short Rotation Forestry)

Silviculture energy farms employ techniques more similar to agriculture than forestry. The chief objective of energy plantation is to produce biomass from the selected trees and shrub species in the shortest possible time (generally 5-10 yrs) and at the minimal cost, so as to satisfy local energy needs in the decentralised manner. This would certainly relieve the pressure on the consumption of fossil fuel like kerosene and prevent the destruction of plant cover which is one of the primary components of the life support system.

**(*a*) Advantages of Short Rotation Management:** Following are the advantages from short rotation management from production of fuel biomass :

(*i*) High yield per unit of land area,
(*ii*) Smaller land requirements for given biomass output,
(*iii*) Shorter time span from initial stand establishment to harvestable crop,
(*iv*) Increased labour efficiency through mechanization and other methods similar to those used in agriculture,
(*v*) Ability of most short rotation species to regenerate by coppicing; and
(*vi*) Ability to take advantages of cultural and genetic advances quickly.

Therefore, production of plant-based renewable energy is a part of the dynamic agro-botany-forestry system and the need is to integrate both modren biology and culture in agriculture.

## B. PETROLEUM PLANTS

There are some species of certain families which accumulate the photosynthetic products (hydrocarbons) of high molecular weight (10,000). They are commonly known as petroplants or petroleum plants. In 1979, Dr. M. Calvin of the University of California reported for the first time the collection and use of photosynthetically produced hydrocarbons from the plants (Calvin, 1979). Furthermore, he suggested it as a substitute for conventional petroleum sources. Calvin and coworkers screened most of the plants of Euphorbiaceae, especially *Euphorbia* (containing 2,000 species) which reduce $CO_2$ beyond the carbohydrates.

Rubber plant yields hydrocarbon.

The petroplants have lactiferous canals in their stem and secrete a milky latex. The latex can be either continuously tapped like *Hevea* latex and stored or extracted from the biomass by using the organic solvents. The product rich in hydrocrackable hydrocarbon is called as `biocrude'. Biocrude yields about 70.6% energy; out of which 22% as kerosene and 44.6% as gasoline.

About 400 plant species, belonging to different families are known which grow in different part of the country. It is hoped that petroplants can yield petroleum more than 40-45 barrel/acre.

## 1. Hydrocarbon from Higher Plants

Euphorbiaceae has been extensively screened, which has shown the fruitful results. In addition, the most useful plant families to be investigated are Asclepiadaceae, *Apocyanaceae, Leguminosae, Sapotaceae, Moraceae, Dipterocarpaceae, Compositae*, etc.

However, the members of Euphorbiaceae possess high amount of hydrocarbons. Plants producing rubber and other hydrocarbons are given in Table 24.1.

**> Table 24.1. Plants producing hydrocarbons.**

| *Plant group/families* | *Common names* | *Botanical names* |
|---|---|---|
| Algae (Chlorophyta) | | *Botryococcus* sp. |
| | | *Chlorella pyrenoidosa* |
| Euphorbiaceae | Hevea rubber | *Hevea brasiliensis* |
| | Rubber plant | *Euphorbia abyssinica* |
| | " " | *E. resinifera, E. lathyris,* |
| | Sehund | *E. tirucalli* |
| Compositae | Guayule | *Parthenium argentatum* |
| | Russian dandelion | *Taraxacum koksaghyz* |
| Asclepiadaceae | Aak | *Calotropis procera* |
| Leguminoseae | – | *Copaifera langsdorfii, C. mutijuga* |
| | Samprani | *Hardwickia pinnata* |
| Dipterocarpaceae | Gurjun | *Dipterocarpus turbinatus* |
| Myristicaceae | – | *Dialynthera otoba* |
| Pittosporaceae | – | *Pittosporum resiniferum* |

**(*a*) Hevea Rubber:** Rubber plant, (*Hevea brasiliensis*) commonly known as *Hevea* rubber is the principal source of rubber which is restricted in distribution in South-East Asia. This plant meets one-third of total world demand of rubber. The synthetic rubber elastomers from petroleum have not replaced the demand of natural rubber, due to its low cost build up, resilience, elasticity and good performance in automobiles and aeroplanes. Rubber is tapped from stem of trees by making incision and collecting the latex from it. The latex is further processed to get rubber.

**(*b*) *Euphorbia*:** In Italy, Euphorbia Gasoline Refinery was set up to tap vegetative gasoline. *Euphorbia lathyris* is an annual herb and *E. tirucalli* is a perennial one. *E. lathyris* can produce 20 t dry matter/ha/yr. Chemical analysis of this plant in organic solvents revealed that heptan extract and ether soluble fraction constituted about 8% terpenoid extract. By using zeolite catalyst, it could be converted into high grade transportation fuel. Of the 85% converted materials, about 10% is in the form of natural gas and 75% in gasoline-like fractions. Calvin and co-workers estimated that 10 tonnes of biomass could yield 5.3 barrels of crude extract convertable to gasoline.

Till now, much emphasis has been given on crop plants. At present, there is need to encourage cultivation of petroplants, in waste lands. There is need for joint efforts of botanists, phytochemists, engineers, economists, and geneticists to make successful research on these lines.

**(*c*) Guayule and Russian Dandelion:** Guayule (*Partheniuum argentatum*) and *Taraxacum koksaghyz* of family compositae are sources of rubber. Guayule a shrub is indigenous to North Central Mexico and South-West U.S.A. Guayule generally grows in arid, semi-arid and desert areas. The U.S. Government encouraged the cultivation of this plant after World War II to reform the economy of the country. It can tolerate temperature ranging from 32–38°C, and can grow in Indian conditions.

Like *Hevea,* guayule contains cis-polyisoprene and identical physical properties. There is need to develop technologies for the production of hydrocarbon to be used as alternative fossil fuel.

**(*d*) Aak (*Calotropis procera*):** Aak, a shrub of 1–2.5 meters in height, grows in hot and dry parts of India on waste dry places, river beds, roadsides and forest clearings. It secretes latex which causes irritation to skin. Latex contains high amount of extractable hydrocarbons. The ratio of C, H, O in the hexane extract has been found as 78.03%, 11.22% and 10.71% respectively. The ratio of C and H is similar to crude oil, fuel oil and gasoline. Hydrocarbon yield and energy value of *C. procera* are comparable to those of *E. lathyris*. Therefore, this plant can be used as a substitute of petroleum. Researches on it are being done at the Central Arid Zone Research Institute, Jodhpur.

In India, cultivation of petroleum plants needs to be encouraged and suitable technologies should be developed for extraction of crude oil to be used as fuel. NBRI (Lucknow), and Indian Institute of Petroleum (Dehra Dun) have started preliminary screening programme of such plants. Over 400 plant species are to be tested for growth conditions, habitat performance, biomass yield and hydrocarbon content.

## 2. Algal Hydrocarbons

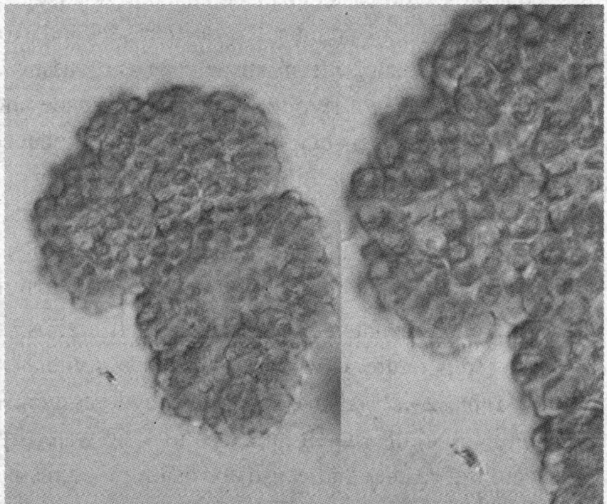

*Botryococcus braunii* that yields hydrocarbon.

Dead algal scum of *Botryococcus braunii*, an unicellular alga of Chlorococcales of green algae, contains about 70% hydrocarbons. Percentage of hydrocarbon may vary. The algal hydrocarbons closely resemble the crude oil, and therefore, can be used as a good source of direct production of hydrocarbons. *B. braunii* grows in fresh or brackish water as well as in tropical and temperate zones. When in full growth, it becomes apparent in water as the small dots. The alga appears in two forms, as far as pigmentation and structure of synthesized hydrocarbons are concerned. The first form is of green colour and contains linear hydrocarbons with an odd number of carbon atom (25-31) low in double bonds. The second form of alga is red in colour which contains hydrocarbons with 34-38 carbon atoms and several double bonds, the 'botryococcenes'. Significance of these two forms are not known (Sasson, 1984).

This alga is composed of proteins, carbohydrates, and lipids, the percentage of which varies. However, it has proved to be a source of hydrocarbons. As a result of metabolic activity, the hydrocarbons are synthesized during growth phase of the alga.

Hydrocarbon is accumulated as globules on outer walls and cytoplasm of the cells. On cell wall, a major portion of hydrocarbon (95%) is located, whereas a small amount (0.7%) of it is present within the cells. Hydrocarbons are recovered from the cells by centrifugation. The cells are again added in the fresh culture medium as inoculant. For the production of hydrocarbons in high amount, it is necessary to increase the algal biomass. However, it could be achieved by characterizing the culture medium and light and shade conditions for its growth and biomass production.

*Chlorella pyrenoidosa,* a fresh water alga, is known to be converted into hydrocarbons. Hydrogenation is done in a steel reactor at high temperature (> 400°C) and pressure (12,000 p.s.i., pound per square inch) in the presence of a catalyst (cobalt molybdate). The alga is suspended in a mineral oil in the reactor. Hydrogenation is carried out for about one hour. Consequently, 50% of algal biomass is converted into oil with a little amount (12–14%) of a byproduct, ammonium carbonate. Oil is a clear golden liquid which is separated from the reactor, blended with light gas oil in refineries and processed before its use.

## C. ALCOHOLS : THE LIQUID FUEL

### 1. General Account

Historical background and methods of alcohol production from various substrates are discussed in Chapter 17. In this section, use of alcohols as liquid fuel is described.

During World War II, mixing of methanol and ethanol in different amount with petrol was carried out in Europe. Thereafter, blending of alcohol with petrol to be used as transport fuel started after the oil crisis in 1973. Use of 5% alcohol (ethanol) in petrol was made compulsory in Brazil by 1931. This country has been the second World sugar producer, therefore, large quantity of alcohol is produced from sugar cane and cassava. In 1985, Brazil launched a programme of blending 20% ethanol (produced from sugar cane and cassava) with petrol and thus saved about 40% of its petrol consumption. In 1990, it produced about 20.5 billion litres of alcohol from molasses and saved 11 billion dollars of foreign exchange. This country marketed about one million cars running on alcohol alone (an additional heat exchanger is required to start cars as the ignition point of alcohol is higher than that of gasoline). In 1986, Brazil gave employment to 3.5 million people from 14 billion litres of tipsy fuel.

The other advantage of ethanol is that there is safety for cars if used as fuel. The flashpoint (*i.e.* the temperature at which a substance ignites) of ethanol is three times higher than that of petrol (Table 24.2). Also it emits less hydrocarbon. But the disadvantage is that in cold engine starting will be difficult. It may also pick up water from air. Therefore, mixture of enthanol-water will not burn readily and cause corrosion problems in tanks and engines. Ethanol also reacts with metals such as alloy of maganesium and aluminium. An alternative to using pure ethanol as a fuel is to mix it with petrol. In mixed condition, it becomes a part of energy. Such a mixture containing 20% ethanol is currently marketed in the USA as 'gasohol'. The drawback with this approach is that the ethanol must be distilled to 100% purity before mixing with petrol, otherwise it would be separated from the petrol.

> **Table 24.2.** Some of the physical and chemical properties of petrol and ethanol.

| *Physico-chemical properties* | *Petrol* | *Ethanol* |
|---|---|---|
| Freezing point (°C) | < – 130 | – 117 |
| Boiling point (°C) | 35–200 | 78 |
| Energy value (MJ $kg^{-1}$) | 44.0 | 27.2 |
| Density (kg $l^{-1}$) | 0.74 | 0.79 |
| Flashpoint (°C) | 13 | 45 |
| Latent heat of vaporisation (MJ $kg^{-1}$) | 293 | 855 |
| Octane number | 80–100 | 99 |

In 1980, U.S.A. commercialized the 'gasohol'. This work boosted up alcohol production in the country even from cereals. Moreover, addition of methanol (a wood alcohol) in petrol has already caught on in the U.S., with President Bush endorsing the production of 10,00,000 alternative fuel vehicles.

India is fortunate enough for having many sources of biomaterials to be used in ethanol production. The Government is facing a crisis in case of molasses. Every year, potatoes have rotten for lack of buyers. Cassava is grown on large scale in Kerala and some parts of Tamil Nadu. A large number of distilleries are in operation and many more to be set up. Utilization of sugary and starchy materials for the production of ethanol would be a good step to cut down the oil price and meet the fuel demand in country.

Conversion of coal to oil also represents an important facet of the general energy problem. The Central Fuel Research Institute suggested that replacement of oil by coal will help in saving 50% cost of fuels. In our country, coal is a major source of energy. It is gasified and converted to methanol with a high degree of efficiency. Methanol can be directly converted to gasoline. Using this method, New Zealand has set up commercial plants and supplying one-third of national requirements of gasoline.

## 2. Ethanol Production

Ethanol is produced by chemical as well as biological routes. Through the chemical route, synthetic alcohol is produced by catalytic hydration of ethylene ($C_2H_2$) with water, using phosphoric acid at 70 atmosphere pressure and 300°C. This process involves the use of petroleum as fuel to generate high pressure and temperature for producing alcohol. Ethylene is derived from both natural and coke oven gases, and the waste gases released in refining petroleum to produce gasoline.

Biological route is an alternative way to produce alcohol. The biomass to be converted into liquid fuel can be derived from agriculture, municipal or forestry wastes (*see* Chapter 13). The agricultural crops contain sugar (*e.g.* sugar cane, sugar beet), starch (maize, tapioca) or cellulose (cotton, species of populus) in high amount. Chemical composition of cellulosic and lignocellulosic material is shown in Chapter 23.

## 3. Fermentable Substrates

Following are the types of substrates used for alcohol production :

**(*a*) Sugary Materials:** Examples of sugary materials are sugarcane and its by products/ wastes (molasses, bagasse) and sugar beet, tapioca, sweet potatoes, fruit juice, sweet sorghum, etc. Sugar cane molasses is largely being used in many country for alcohol production.

**(*b*) Starchy Materials:** Starchy materials used in ethanol production are tapioca, maize, wheat, barley, oat, sorghum, rice and potatoes. But tapioca and corns are the two major substrates of the interest (*see* Chapter 23). It has been estimated that 11.7 kg of corn starch can be converted into about 7 litres of ethanol.

(*c*) **Lignocellulosic Materials:** The sources of cellulosic and lignocellulosic materials are the agricultural wastes and wood. However, yield of ethanol from lignocellulose is low because of lack of suitable technology and failure of conversion of pentoses into ethanol. On the basis of technology available today about 409 litres of ethanol can be produced from one tonne of lignocelluloses. Structure of cellulose, hemicellulose and lignin is given in Figs. 23.1-4 and anaerobic hydrolysis of these is discussed in the preceding section. Production of ethanol from lignocelluloses follows the following steps: (*i*) hydrolysis, (*ii*) fermentation, and (*iii*) recovery.

## 4. Hydrolysis of Lignocellulosic Materials

Hydrolysis of lignocelluloses is done to release the monomer sugars. These materials are hydrolysed by acids and/or enzymes. Acid hydrolysis is performed by using dilute sulphuric acid (0.1–0.2%) through layers of saw dust or wood chips under pressure and a high temperature (180–121°C). Sugars produced during hydrolysis are destroyed by acid. Therefore, it must be quickly removed. Acid portion should be neutralized with lime before the solution is fermented. The other chemicals used in hydrolysis are hydrochloric acid, sodium hydroxide, ammonia, etc. Substrate hydrolysis coupled with high pressure and temperature for varying times, depending on nature of substrate, has shown a good result.

Enzymatic hydrolysis of cellulosic and lignocellulosic materials is performed by using the respective enzymes *i.e.* cellulases, hemicellulases, pectinases, lignases, etc. These enzymes are produced on a large scale from microorganisms (*see* chapter 22). Enzymatic hydrolysis of cellulose, hemi-cellulose and lignin is discussed elsewhere (*see* chapter 23).

Upon hydrolysis of carbohydrates of plant materials, the following sugars are liberated in the solution which in turn are fermented into ethanol :

(*i*) Cellulose ⟶ hexoses ⟶ glucose ⟶ Ethanol

(*ii*) Hemicelluloses ⟶ hexoses ⟶ glucose, galactose, fructose, manose

⟶ pentoses ⟶ xylose, arabinose ⟶ Ethanol

⟶ uronic acids

(*iii*) Starch ⟶ hexoses ⟶ glucose ⟶ Ethanol.

(*a*) **Effect of Substrate Composition on Hydrolysis:** Composition of substrates directly governs the hydrolysis, such as insolubility,high crystallinity and lignin coating (over the cellulose microfibrils in cell wall).

Crystallinity of cellulose directly affects its hydrolysis by cellulases. Cellulases will readily degrade the amorphous portion of cellulose, but not the crystalline portion, until the cellulase complex is rich in $C_1$ enzyme. Lignin hinders substrate hydrolysis. As the amount of lignin is high in substrate, the yield of ethanol would be low. Treatment of substrate with alkali (2% NaOH at 70°C for 90 minutes, washed and sterilized) would be more effective than without treatment. Moreover, steam explosion (*i.e.* steam treatment at 190°C) of lignocellulosic materials is another pre-treatment which has proved to be the best for hydrolysis.

## 5. Fermentation

The soluble hexoses and pentoses obtained in solution after hydrolysis of sugary-starchy-cellulose are subjected to anaerobic fermentation by using bacteria, yeast and filamentatous fungi (*see* Chapter 17). Theoretically, maximum conversion of glucose to ethanol is 51% by weight. Yeast ferments glucose to 10–20% ethanol. An outline of ethanol production is given in Fig. 24.1. The micro-organisms associated with fermentation utilize various metabolic process, depending on substrates *i.e.* glucose, xylose, etc.

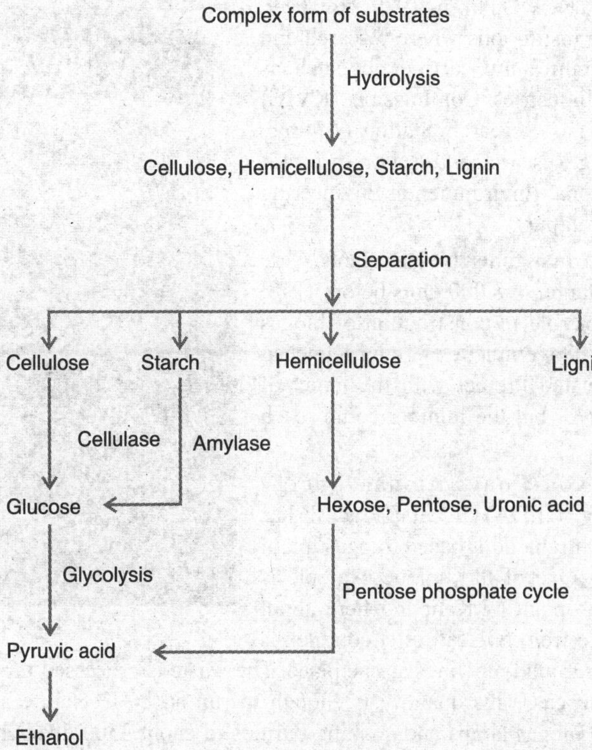


Fig. 24.1. Ethanol formation from the fermentable complex substrates.

### 6. Recovery of Ethanol

When fermentation is over, the microbial mass (cells, mycelia and conidia) is separated from the fluid. The fluid contains the mixture of ethanol, water and small amount of alcohols and ether. Ethanol is recovered by distillation process, *i.e.* vaporization of ethanol/water mixture.

## D. GASEOUS FUELS – BIOGAS AND HYDROGEN

### 1. What is Biogas?

In 1776, for the first time, the Italian Physicist, Volta, demonstrated methane in the marsh gas, generated from organic matter in bottom sediments of ponds and streams. Under anaerobic conditions, the organic materials are converted through microbiological reactions into gases (fuel) and organic fertilizer (sludge). The mixture of gases is composed of 63% methane, 30% $CO_2$, 4% nitrogen and 1% hydrogen sulphide and traces of hydrogen, oxygen and carbon monoxide. Methane is the main constituent of biogas. It is also referred to as biofuel, sewerage gas, Klar gas, sludge gas, will-o-the wisp of marsh lands, fool's fire, *gobar* gas (cow dung gas), bioenergy and fuel of the future. About 90% of energy of substrate is retained in methane. Biogas is used for cooking and lighting purposes in rural sector. It is devoid of smell and burns with a blue flame without smoke.

## 2. Biogas Technology in India

Many developing countries are encouraging for installation of biogas plants to meet the demand of fuel. India is one of the pioneer country in biogas technology where biogas research and plant construction has been carried out over the past 30 years. The experiments were initiated at the IARI., New Delhi, in 1939. However, there are many other institutions where research and development programmes are carried out such as Khadi and Village Industries Commission (KVIC) Bombay, the Gobar gas Research Station (Ajitmal, Etawah), the Planning Research and Action Institutes (Lucknow), and National Environmental Engineering Research Institute (Nagpur).

A Biogas Plant.

In 1951, for the first time, biogas plants were constructed with the target of 8,000 units before 1973. KVIC played a major role in construction of biogas plants during 1960s. Government of India launched an All India Coordinated Project with the target of 1,00,000 units by 1978, but the number could reach to 50,000 units only.

Upto 1993, Non-Conventional Energy Development Agency (NEDA) of Uttar Pradesh has installed about 100 night soil based biogas plant throughout the state. One of the unique example is the Rajapurwa biogas plant of Kanpur where about 1400 kg human waste from 50-seat toilet complex is pooled through underground pipelines at one place. The waste is processed through a digester into about 55 $m^3$ methane gas. This quantity is enough to run an 8 HP engine and pumpset and to provide fuel to some biogas lamps and cooking burners to about 3500 dwellers. For maintenance of the plant, each family pay some nominal charges.

The sewage plant at Okhla (New Delhi) has 15 digesters of 5665 $m^3$ capacity each, and produces 17,000 $m^3$ gas per day. The gas generated is equivalent to about 10,000 litres of kerosene per day, where 1 $m^3$ biogas has energy potential equal to about 2/3 litres of kerosene.

The Dadar Sewage Treatment Plant (Bombay) produces about 2,800 $m^2$ biogas per day from sewage. From this plant pipe lines are connected to about 500 houses, located about 5 km away from the station.

In Sonepat District of Haryana, over 300 biogas units of 2 to 10 $m^3$ capacity have been set up for cooking and lighting purposes. The Haryana Government provided 25% subsidy per plant for setting up the units.

National Sugar Institute (Kanpur) has developed methods for production of biogas from bagasse and other agricultural residues. This plant was built up in 1960s. In this plant , 12 steel digesters,each having capacity of 50 $m^3$, are set up to which about 14 tonnes of mixture of agricultural wastes and cattle manure and 28 tonnes of water are fed to begin the biogas production.

Government policy is to provide direct subsidy of 25% of the cost of plants and encourage banks to pay loans for remaining costs to aspirants. High construction costs, and cost of cement and steel digesters and rates of interest on loans are some of the factors that discourage villagers to setting biogas plants. It has been pointed out that in villages, support of minimum number of cattle (at least 5) is necessary to run a small home digester plant. Five cattle generate dung to produce about 2 $m^3$ biogas plant to meet the demand of cooking and lighting for a family of

4–5 people. In addition, attempts are also being made to develop a small engine powered with methane gas.

## 3. Benefits from Biogas Plants

In Asia, biogas is used mainly for cooking and lighting purposes. In addition, there are many other advantages in installing the biogas plants. It is used in internal combustion engines to power water pumps and electric generators. It is also used as fuel in fuel type refrigerators. Sludge is used as fertilizer. The most economical benefits are minimizing enviornmental pollution and meeting the demand of energy for various purposes.

## 4. Feedstock Materials

There are two sources of biomass *i.e.* plant and animal for biogas production. Biomass from plant origin is aquatic and terrestrial ones, derived from various sources (*see* Chapter 23). Biomass from animals are cattle dung, manure from poultry, goats and sheep, and slaughter house and fishery wastes. In most of biogas plants cattle dung is used for gas production. There are many cases where gobar gas plants are in operation, and sullage gas production is started. Beside cattle dung (gobar), agricultural wastes containing cellulosic and lignocellulosic materials are also being used.

## 5. Biogas Production : Anaerobic Digestion

Anaerobic digestion is described in Chapter 23. Anaerobic digestion is carried out in an air tight cylindrical tank which is known as digester. A digester is made up of concrete bricks and cement or steel. It has a side opening (charge pit) into which organic materials for digestion are incorporated. There lies a cylindrical container above the digester to collect the gas.

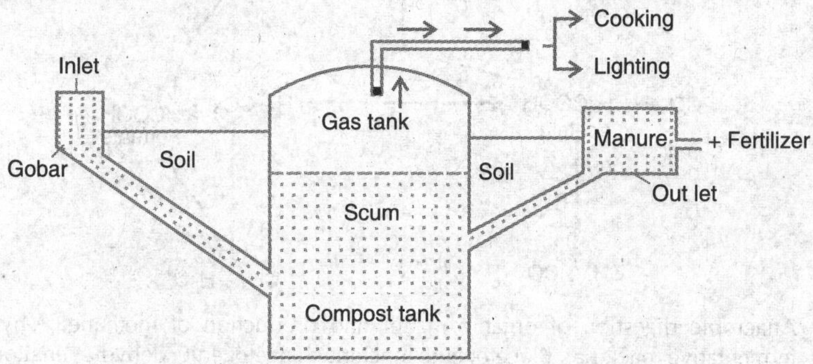


Fig. 24.2. Diagram of a Janata Vayu Gas (gobar gas) Plant.

In India, single stage digester is set up in gobar gas plant. However, in other countries single-stage, two stage and multistage digester(s) are set up to accomplish digestion at high rate. A diagram of a single stage digester for biogas/gobar gas production is shown in Fig. 24.2. In biogas plant, a concrete tank is built up which has the concrete inlet and outlet basins. Fresh cattle dung is deposited into a charge pit, which leads into the digestion tank. Dung remains in tank. After 50 days, sufficient amount of gas is accumulated in gas tank, which is used for house-hold purposes. Digested sludge is removed from the basin and is used as fertilizer. Usually digesters are burried in soil in order to benefit from insulation provided by soil. In cold climate, digester can be heated by the installations provided from composting for the agricultural wastes.

Anaerobic digestion is accomplished in 3 stages, solubilization, acidogenesis and methanogenesis. These stages are characterised by 3 groups of bacteria.

**(*a*) Solubilization:** It is the initial stage, when feedstock is solubilized by water and enzymes. The feedstock (cattle dung and other organic polymers) is dissolved in water to make slurry.

The complex polymers are hydrolysed into organic acids and alcohols by hydrolytic fermentative bacteria which are mostly anaerobes (Fig. 24.3A).

**(*b*) Acidogenesis:** During this stage, the second group of bacteria *i.e.* facultative anaerobic and $H_2$ producing acidogenic bacteria convert the simple organic material via oxidation/ reduction reactions into acetate, $H_2$ and $CO_2$. These substances serve as food for the final stage. Fatty acid is converted into acetate, $H_2$ and $CO_2$ via acetogenic dehydrogenation by obligate $H_2$ producing acetogenic bacteria. There is other group of acetogenic bacteria which produce acetate and other acids from $H_2$ and $CO_2$ via acetogenic hydrogenation (Fig. 24.3B I and II).

**(*c*) Methanogenesis:** This is the final stage of anaerobic digestion where acetate and $H_2$ plus $CO_2$ are converted by methane producing bacteria (methanogens) into methane, carbon dioxide, water and other products (Fig. 24.3C III and IV).

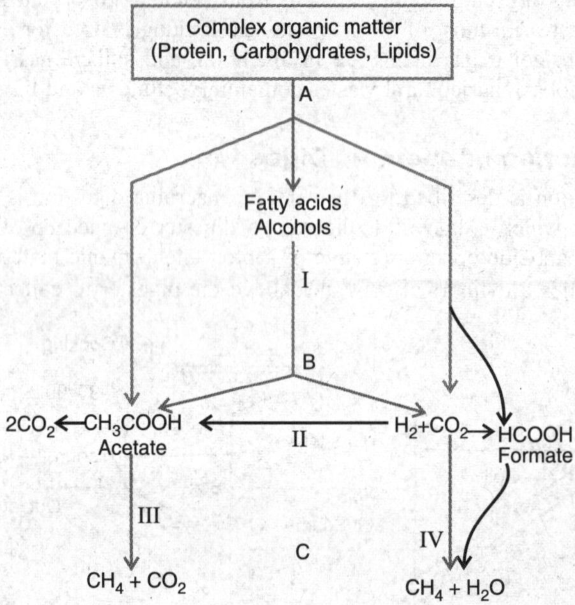


**Fig. 24.3.** Anaerobic digestion of organic matter and production of methane. A-hydrolytic and fermentative bacteria; B-acetogenic bacteria (I-acetogenic dehydrogenation by proton reducing acetogenic bacteria; II-acetogenic hydrogenation by acetogenic bacteria); C-methanogenesis by acetoclastic methanogens (*i.e.* acetate respiratory bacteria) (III), and hydrogen oxidizing methanogens (IV).

Different species of methanogens are involved in breakdown of complex organic matter into acetate or other organic acids. Acetate is one of the substrates of methanogenic bacteria. Hydrogen with $CO_2$ is a general substrate for methanogenesis. Numbers of these bacteria differ with type of substrates. For example, counts of $10 - 10^6$ per ml and $10^5 - 10^8$ per ml of hydrogen utilizing bacteria were determined from the pig waste and sewage sludge digesters, respectively.

**(*i*) Methanogens.** Methanogens are a unique group of bacteria. They are obligate anaerobes and have slow gowth rate. They play a major role in breakdown of substrate into gas form. They are the only organisms which can anaerobically catabolize acetate and $H_2$ to gaseous products in the absence of exogenous electron acceptors other than $CO_2$ or light energy. In their absence, effective degradation would cease because of accumulation of non-gaseous, reduced fatty acid and alcohol products of the fermentative and other $H_2$ using bacteria that have almost the same energy content as the original organic matter. In morphology, they are of different types such as cocci, bacilli, spirilli and sarcina.

All the bacteria require $H_2$ and formate (except *Methanobacterium bryantii, M. thermoautotrophicum and M. arboriphilus*) for growth and methane production,whereas *M. barkeri* requires (besides $H_2$) methanol ($CH_3OH$), methyl amine ($CH_3NH_2$) and acetate for their growth. Thus, the methanogens are either autotrophs or utilize simple organic compounds as formate, acetate, and methyl amine and occupy the terminal position in anaerobic food chain.

***(ii)* Mechanism of methane formation.** Metabolically the methanogens are very peculiar. Carbon dioxide fixation, Calvin cycle, serine or hexulose pathway are absent in them. The mechanism of methane formation is not well understood. Several new coenzymes are involved which are not present in any other group of bacteria. These coenzymes are methyl coenzymes M, hydroxy methyl coenzyme M, coenzyme F420, coenzyme F430, component B, corrinoids, methanofuran or carbon dioxide reducing factor and methanopterin and formaldehyde activating factor.

A detailed description of biochemistry and pathways of methane formation is out of scope of this text. Therefore, only reactions are given in this connection. Primary reaction in which carbon monoxide takes part is as below :

$$CO + H_2O \longrightarrow CO_2 + H_2$$

The secondary reaction takes place in the presence of sufficient hydrogen:

$$CO_2 + 4H_2 \longrightarrow CH_4 + 2H_2O$$

Other reactions showing methane formation from various substrates are given below:

$$\underset{\text{(Methanol)}}{4CH_3OH} \longrightarrow 3CH_4 + CO_2 + 2H_2O$$

$$\underset{\text{(Formate)}}{4HCOOH} \longrightarrow CH_4 + 3CO_2 + 2H_2O$$

$$\underset{\text{(Acetate)}}{CH_3COOH} \longrightarrow 12\ CH_4 + 12\ CO_2$$

## 6. Biogas Production from Different Feedstocks

Although there are various types of biomaterials differing in chemical composition, yet gobar (cow dung) is most popular in India. Besides gobar, other materials which would be successful in the coming years are agricultural wastes, municipal wastes, industrial wastes and some of the aquatic biomass.

***(a)* Salvinia :** Significance of *Salvinia* is described earlier (*see* Chapter 23). This fern can be a feedstock material for biogas production. Fermentation of *Salvinia* starts within 7-9 days on putting under water in suitable container. Biogas yield is about 0.1 litres /kg fresh weight for 4 weeks. Air dried weed produces about 1 litre/kg for 90 days and fermentation is continued for 3 months. Thereafter, gas yield gradually declines.

Daily requirement of gas is 0.4 litre per capita. Two tonnes of air dried *Salvinia,* therefore, can meet the fuel requirement of a small family for 3–5 months. Special advantage of using *Salvinia* is that unpleasant odours do not come out. This will certainly provide cheap and clean burning fuel and reduce the increasing pressure on forests.

***(b)* Water Hyacinth :** This weed can be used for the production of laboratory and domestic fuel. Its decomposition rate is higher than that of cow dung for gas production. Water hyacinth is totally decomposed within 3 days in summer while cow dung takes 8 days.

The ratios between total gas evolved by water hyacinth and cow dung under the identical conditions in summer and winter are 5:3 and 5:1, respectively.

**(*c*) Municipal Sewage:** The potential of municipal sewage in India and its future prospects are described elsewhere (*see* Chapter 23). New techniques have been developed to make available sullage gas for cooking purpose or industrial activities. The Okhla Sewage Disposal Works (New Delhi) receives 25 $m^3$ sewage per second and generates 18,000 $m^3$ of gas. In energy terms, it is equivalent to about 4.161 million litres of kerosene/year. From the plant, gas pipe lines are given to about 700 families of the surrounding villages. It is hoped, if whole sewage generated from the city is digested anaerobically to yield biogas, it could meet about 20% fuel demand of the city.

## E. HYDROGEN–A NEW FUEL

Hydrogen is the simplest molecule present in the universe. Production and use of hydrogen represent a potential alternative source of fuel. It can be easily collected, stored (as gas, liquid or hydrides of metals) and transported (by truks, ships or trains). Hydrogen can be piped or transmitted over wires for a distance of over 10 km. After the use, hydrogen does not pollute the environment.

For the production of hydrogen, water serves as a source of raw material. The bond between hydrogen and oxygen in water can be broken by providing necessary energy by heat, electricity or light photons as below :

$$H_2O \xrightarrow[Catalyst]{hv\,(photons)} \tfrac{1}{2}\, O_2 + H_2$$

Based on the types of energy used, the follwoing categories of splitting water have been made :

(*i*) Electrolysis: Electrical splitting

(*ii*) Thermolysis : Splitting of water by heat,

(*iii*) Thermochemical lysis : Splitting of water by both heat and chemical catalysts.

(*iv*) Photolysis : Splitting of water by light.

The first three approaches have been already in practice, whereas photolysis of water has much future prospect as far as seeking of alternative source of energy is concerned.

Solar energy available on earth surface constitutes an abundant and free energy source. It can be converted directly into heat, mechanical energy, electricity or fuel. Nowadays several processes have been envisaged for the conversion of optical energy to chemical energy. These processes are given below :

**(*i*) Photo-chemical process:** Hydrogen is produced by using photocatalysts such as compound salts and photosynthetic dyes.

**(*ii*) Photo-electro-chemical (PEC) process:** In this process, semiconductor photocatalysts (*e.g.* Sr. Ti $O_3$) are used for the production of hydrogen. Based on semi-conductor system PEC cells have been designed. PEC cells offer the unique advantages for converting optical energy to chemical energy. It is the most efficient chemical system designed so far.

**(*iii*) Photobiological process :** This process involves the splitting of water by using natural or synthetic chlorophyll, algae and bacteria.

### 1. Photobiological Process of $H_2$ Production

Biophotolysis of water refers to breakdown of water and production of oxygen and hydrogen by biological process. Hydrogen production is brought about by the following means :

**(*a*) Hydrogenase and $H_2$ Production:** In the early 1960s, production of hydrogen was demonstrated by using chloroplasts isolated from spinach (*Spinacia oleracea*) in the presence of artificial electron donors and bacterial extracts containing hydrogenase. Electron donors (organic compounds) transfer electrons to photosystem I of the chloroplast from where electrons are received by electron carriers (*e.g.* ferredoxin). Hydrogenase accepted electrons from the electron carrier as shown below. The organic compounds which acted as electron donor also served as a source of hydrogen (Sasson, 1984).

Electron donor → Photo system I → Electron carrier $H^+$ → Hydrogenase → $H_2$

Hall *et al.* (1980) demonstrated the production of 50 micromolecule of $H_2$/mg chloroplast/h for about 6 hours from a mixture of chloroplast and bacterial (*Clostridium*) hydrogenase at pH 7 and temperature 25°C.

In the visible light, hydrogenase separates high energy electrons from ferredoxin and facilitates their transfer to $H^+$; ultimately $H_2$ is evolved. Those plants which produce carbohydrates lack hydrogenase. This enzyme is restricted in a variety of bacteria, cyanobacteria and other green algae. Hydrogenase of these plants produces $H_2$ in the absence of $CO_2$.

Sensitivity of hydrogenase to $O_2$ varies with species. It is, however, unlikely to develop an efficient system for hydrogen production if considerable amount of $O_2$ is produced. Therefore, for the production of hydrogen from cyanobacteria addition of a suitable inhibitor which may stop photosystem II and produce free hydrogen has been suggested. Research works are being done on the production of hydrogen from organic waste by using many anoxygenic phototrophic bacteria, water-using cyanobacteria or green algae grown in light condition. The algae which possess hydrogenase are given in Table 24.3.

**> Table 24.3. Algal species containing hydrogenase**

| *Groups* | | *Species* |
|---|---|---|
| Blue-green algae (Cyanobacteria) | : | *Anabeana azollae, A. cylinderica, Anacystis elongata, Nostoc muscorum, Spirulina platensis, Synechococcus elongatus* |
| Green algae | : | *Chlamydomonas moewusii, Chlorella fusca, C. homosphaera, C. kessleri, C. sorokiniana, Ulva lactuca* |
| Brown algae | : | *Ascophyllum nodosum* |
| Red algae | : | *Ceramium rubrum, Chondrus crispus, Corallina officinalis Porphyra sp, Porphyridium cruentum* |

There are many bacteria which contain hydrogenase as they possess nitrogenase for nitrogen fixation. Efficiency of nitrogen fixation in heterocystous cyanobacteria and root nodule bacteria (*e.g. Rhizobium japonicum*) is governed by the efficiency of hydrogenase (*see* Chapter 18).

Moreover, $N_2$ fixation in root nodule bacteria does not operate efficiency and about half the electrons are used for $H_2$ production. In USA, it has been estimated that the soybean plantations leak about 30 billion $m^3$ $H_2$/annum which has an energy equal to about 300 billion squar feet of natural gas. From this system production of an adequate amount of $H_2$ can be done by allowing the ATP to produce $H_2$ under conditions of low partial pressure of $N_2$, low pH and high temperature.

In the 7th International Symposium on Biotechnology, held in New Delhi (from February 19-25,1984), Dr. A. Mitsui of the University of Miami (U.S.A.) emphasized the production of $H_2$ by using blue-green algae (*e.g. Synechococcus*). This alga fixes atmospheric nitrogen under aerobic conditions and also produces $H_2$ and $O_2$ by splitting water in light conditions. Mitsui's group is currently working on sulphur-tolerant photosynthetic bacteria for growth and $H_2$ production.

(*b*) **Halobacteria** : Recently, a group of microorganism, the purple bacteria, has been discovered which is another potential source of hydrogen. Two species of *Halobacterium, e.g. H. halobium* and *H. curtirubrum,* are known. Halobacteria are rod-shaped and physiologically a unique bacteria, as they are highly halophilic (salt loving). They differ from other bacteria in respect of cell-wall and energy producing mechanisms. They require the high concentrations of salt (*e.g.* 3-4 M sodium chloride) and low amount of oxygen. Therefore, they can grow in saline water *i.e.* saline lakes and sea. In salt saturated (> 3M NaCl) and oxygen poor environment they utilize sunlight to produce ATP and preserve their structure. When salt concentration reaches below 3 M the rod-shaped cells become irregular or spherical. Below 1 M NaCl concentration both cell membrane and cell disintegrate. It is not known how NaCl helps in growth and integrity of the cells. No salt, like sodium chloride has been found to support growth of the cells.

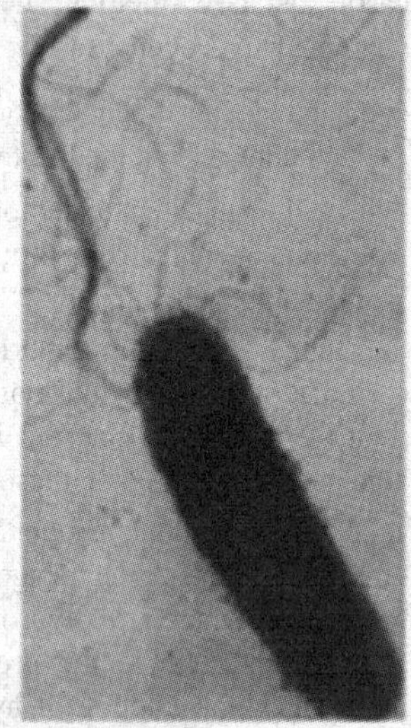
Halobacteria

(*i*) **Bacteriorhodopsin.** In saline condition, the bacteria survive only by incorporating a purple pigment to cell membrane in sunlight. This pigment is commonly known as bacteriorhodopsin (bR). It is linked with retinal in 1:1 ratio. Retinal is the aldehyde form of vitamin A which after reduction yields vitamin A, the retinol (alcohol form). The pigment is constituted by protein (75%) and lipids (25%) and occupies about 50% area of the total cell surface. The pigment occurs on cell membrane in separate patches of 3 molecules. Each patch contains about 1,20,000 molecules of the pigment. A total of 40,000 patches are found on each purple membrane. The three molecules work synergistically. The bR molecules function normally even after destroying the bacterium only if the pigment is not disorganised.

Presence of light plays a major role in the synthesis of the pigment. *Halobacterium* synthesizes a purple membrane in the presence of light and red membrane in the absence of light.

W. Stoeckenius and coworkers of the University of California (U.S.A.) have shown that in the presence of light, *Halobacterium* survives even in anaerobic condition, but in the same condition, in dark it dies rapidly. Bacteriorhodopsin exists in two forms according to light and shade conditions. In dark it exists as bR 560 and in light as bR 570. The two forms are inter-conversible, which undergo fastly a series of photoreaction cycle when they are illuminated.

Bacteriorhodopsin acts as a light driven proton pump and the changes occur in about 10 milli second. During photo-reaction cycle, bR molecules take up the protons on the inner surface of membrane and release them on outer surface. Proton concentration gradient develops on cell membrane and electric potential is generated in the same way on inner and outer cell membrane. Consequently light energy is converted into electrochemical gradient.

Each bR molecule pumps about 200 $H^+$ per second under light condition. Thus, the number of protons pumped per second per sheat is $24 \times 10^6$ ($1{,}20{,}000 \times 200 = 24 \times 10^6$). It has also been estimated that each molecule of the unit patch of the bR pigment is capable of generating about 200 mV photopotential if the rate of proton pump per second is 200.

(*ii*) **Biocells.** In 1981, S. Prentis has found that membrane fragments of some of strains of *H. halobium* contain 1,00,000 bR molecules and a 10 litre bacterial culture would yield 0.5 gram of purple membrane. Now, it has become possible to sandwich the purple membrane and lipid between two platinum electrode. In this device, the photovoltages are generated as a result of photochemical conversion of purple pigment after illumination. This voltage can be measured. This device is termed as biocell. The device of preparing biocells can be improved by replacing the electrodes (*e.g.* Ag-AgCl electrode or others). Therefore, due to increasing energy demand utilization of solar energy by designing biocells offers a great potential. However, much work is needed to make it cheap and popular.

## PROBLEMS

1. How are energy plantations helpful in meeting the demand of fuel in village sectors in hills and plains in India ?
2. What are the aim of social forestry ?
3. What is short rotation energy farms ? What are the benefits from them ?
4. How do petroplants differ from other plants ? What significant role could they play for energy requirement ?
5. Write an eassy on liquid fules.
6. What is biogas ? Give a detailed account of biogas technology in India?
7. What are the feed stock materials used in biogas production? Discuss in brief any one of them used in biogas production.
8. What are the sources of hydrogen used as gaseous fuel ? Discuss its future importance.
9. Write short notes on the following :

   (*a*) Silviculture energy farms, (*b*) Algal hydrocarbons, (*c*) Gasohol, (*d*) Biogas programme in India, (*e*) Methanogens and methanogenesis (*f*) Aquatic plants and biogas production, (*g*) Halobacteria, (*h*) Role of hydrogenase.

CHAPTER

# 25

# Environmental Biotechnology

The area of polluted soil and water is expanding day-by-day due to rapid increase in world population. About 100 million people are added to the world population each year. Therefore, more and more industries, food, health care, medicines, vehicles, etc. are required to meet the demand of people. This is possible by involving biotechnology. However, it is probable that the modern biotechnology based on the sophisticated art of genetic engineering could solve all these problems.

Biooxidation of gold ores.

The environmental biotechnology employs the application of genetic engineering to improve the efficiency and cost which are central to the future widespread exploitation of microorganisms to reduce the environmental burden of toxic substances. It is hoped that in future the application of microorganisms coupled with genetic engineering techniques will make a major contribution to improve the quality of our environment. However, is there risk associated with release of genetically engineered microorganisms into the environment! Uses of microorganisms in environmental clean up with special reference to bioremediation types, processes and methods with several examples (*e.g.* hydrocarbons, heavy metals, dyes, xenobiotics, etc), utilization of sewage and agro-wastes, and the benefits and hazards associated with release of GMMs for our clean environment have been discussed in this chapter. The benefits expected from the release of GMMs into the environment are summarised in Table 25.1.

**> Table 25.1. Benefits from release of genetically modified microorganisms into the environment. (Source: V. Velkov 1996).**

| |
|---|
| **Protection of environment** |
| - Bioremediation of polluted environment |
| **Control of global environmental process** |
| - Reversal of land desertification |
| - Reversal of green house effect |
| **Agriculture** |
| - Increasing efficiency of plant nutrition |
| - Pest control (safe biopesticides) |
| - Protection of plants from climatic stress |
| - Protection of plants from tumour formation and disease |
| **Food Industry** |
| - Microorganisms producing enzymes for food industry |
| - Microorganisms with improved efficiency of fermentation |
| - Improved microorganisms for milk industry |
| **Health care** |
| -Microoganisms as live attenuated vaccines |

## A. BIOREMEDIATION

Bioremediation is the use of living microorganisms to degrade environmental pollutants or prevent pollution. It is a technology for removing pollutants from the environment, restoring contaminated sites and preventing future pollution. However, it has global, regional, and local application. The basis of bioremediation is the enormous natural capacity of microorganisms to degrade organic compounds. This capacity could be improved by applying the GMMs.

In Japan, academic, industrial and governmental research is tightly coordinated for global application of environmental biotechnology. Researchers are exploiting large scale application of bioremediation that can affect desert formation, global climate change and the life cycle of materials. Attempts are being made to develop microorganisms that can help reverse desert formation. This work is based on developing biopolymers that retain water and reverse desert formation. *Alcaligens luteus* is being used to produce 'superbioabsorbent', a polysaccharide which is composed of glucose and glucuronic acid. These can absorb and hold more than thousand times of its own weight of water.

Using the informations from fundamental research biore-mediation technology has been used to remove environmentally hazardous chemicals, accumulated in their cells or detoxify them into non-toxic forms. Several members of algae, fungi and bacteria are known to solubilize, transport and deposit the metals, and detoxify dyes and complex chemicals.

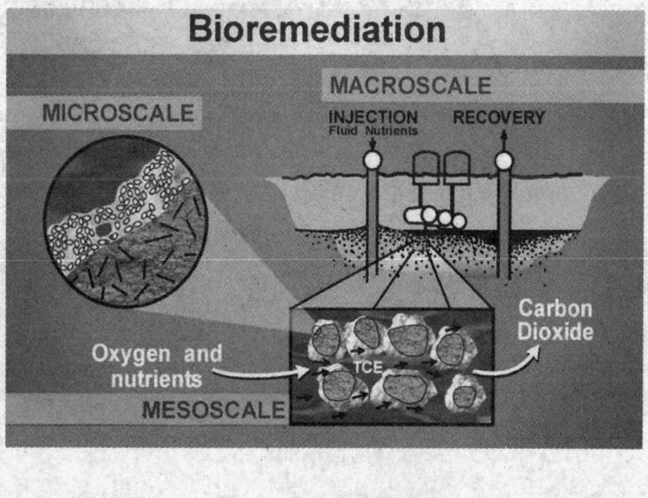


The toxic waste materials remain in vapour, liquid or solid phases, therefore, bioremediation technology varies accordingly whether the waste material involved is in its natural setting or is removed and transported into a fermenter (bioreactor). On the basis of removal and transportation of wastes for treatment, basically there are two methods: *in situ* bioremediation and *ex situ* bioremediation.

## 1. In situ Bioremediation

*In situ* bioremediation is the clean up approach which directly involves the contact between microorganisms and the dissolved and sorbed contaminants for biotransformation. Biotransformation in the surface environment is a very complex process. Potential advantages of *in situ* bioremediation methods include minimal site disruption, simultaneous treatment of contaminated soil and ground water, minimal exposure of public and site personnel, and low costs. But the disadvantages are (*i*) time consuming method as compared to other remedial methods, (*ii*) seasonal vatiation of microbial activity resulting from direct exposure to prevailing environmental factors, and lack of control of these factors, and (*iii*) problematic application of treatment additives (nutrients, surfactants and oxygen). The microorganisms act well only when the waste materials help them to generate energy and nutrients to build up more cells. When the native microorganisms lack biodegradation capacity, genetically engineered microorganisms (GEMs) may be added to the surface during *in situ* bioremediation. But stimulation of indigenous microorganisms is preferred over addition of GEMs. There are two types of *in situ* bioremediation: intrinsic and engineered *in situ* bioremediation.

**(*a*) Intrinsic Bioremediation:** Conversion of environmental pollutants into the harmless forms through the innate capabilities of naturally occurring microbial population is called intrinsic bioremediation. However, there is increasing interest on intrinsic bioremediation for control of all or some of the contamination at waste sites. The intrinsic *i.e* inherent capacity of microorganisms to metabolize the contaminants should be tested at laboratory and field levels before use for intrinsic bioremediation. Through site monitoring programmes progress of intrinsic bioremediation should be recorded time to time. The conditions of site that favour intrinsic bioremediation are ground water flow throughout the year, carbonate minerals to buffer acidity produced during biodegradation, supply of electron acceptors and nutrients for microbial growth and absence of toxic compounds. The other environmental factors such as pH, concentation, temperature and nutrient availability determine whether or not biotransformation takes place. Bioremediation of

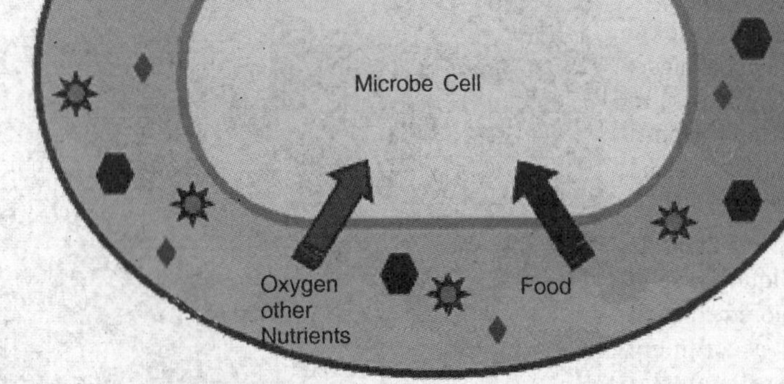


Secretion of enzymes by microbial cell in bioremediation.

waste mixtures containing metals such as Hg, Pb, As and cyanide at toxic concentration can create problem.

The ability of surface bacteria to degrade a given mixture of pollutants in ground water is dependent on the type and concentration of compounds, electron acceptor and duration of bacteria exposed to contamination. Therefore, ability of indigenous bacteria degrading contaminants can be determined in laboratory by plate count and microcosm studies.

***(b)* Engineered *in situ* Bioremediation:** Intrinsic bioremediation is satisfactory at some places, but it is slow process due to poorly adapted microorganisms, limited ability of electron acceptor and nutrients, cold temperature and high concentration of contaminants. When site conditions are not suitable, bioremediation requires construction of engineered systems to supply materials that stimulate microorganisms. Engineered *in situ* bioremediation accelerates the desired biodegradation reactions by encouraging growth of more microorganisms via optimising physico-chemical conditions. Oxygen and electron acceptors (*e.g.* $NO_3^{1-}$ and $SO_4^{2-}$)and nutrients (*e.g.* nitrogen and phosphorus) promote microbial growth in surface.

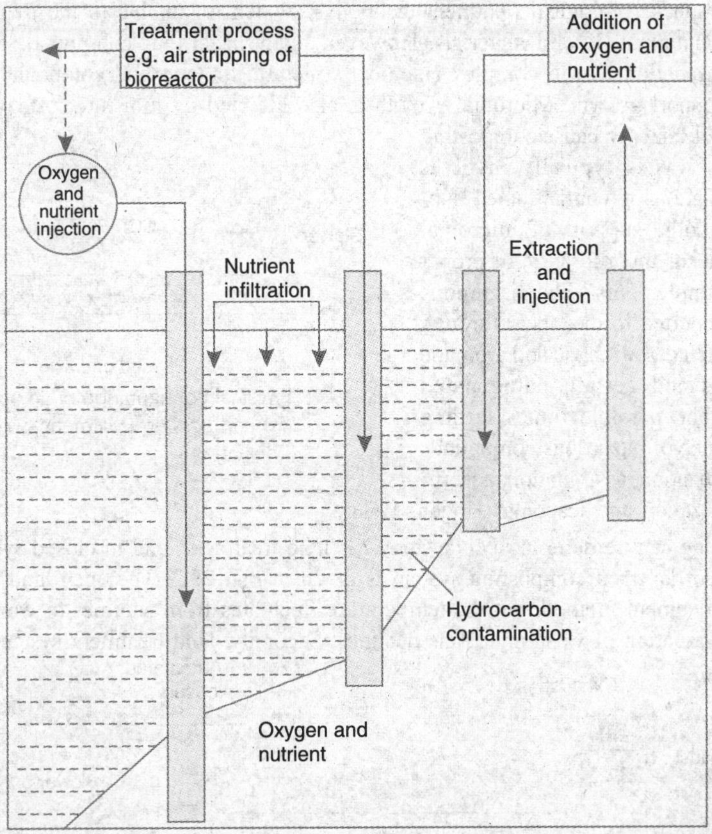


**Fig. 25.1.** Surface treatment using above-ground reactor, injection of oxygen, acid nutrient and extraction walls.

When contamination is deeper, amended water is injected through wells. But in some *in situ* bioremediation systems both extraction and injection wells are used in combination to control the flow of contaminated ground water combined with above-ground bioreactor treatment and subsequent reinjection of nutrients spiked effluent are done (Fig. 25.1).

## 2. Ex situ Bioremediation

*Ex situ* bioremediation involves removal of waste materials and their collection at a place to facilitate microbial degradation. *Ex situ* bioremediation technology includes most of disadvantages and limitations. It also suffers from costs associated with solid handling process *e.g.* excavation, screening and fractionation, mixing, homogenising and final disposal. On the basis of phases of contaminated materials under treatment *ex situ* bioremediation is classified into two: (*i*) solid-phase system (including land treatment and soil piles), *i.e.* composting, and (*ii*) slurry-phase systems (involving treatment of solid-liquid suspensions in bioreactors).

**(*a*) Solid-phase Treatment:** Solid-phase system includes organic wastes (*e.g.* leaves, animal manures and agricultural wastes), and problematic wastes (*e.g.* domestic and industrial wastes, sewage sludge and municipal solid wastes). The traditional clean-up practice involves the informal processing of the organic materials and production of composts which may be used as soil amendment.

**(*i*) Composting.** Composting is a self-heating, substrate-dense, managed microbial system, and one solid-phase biological treatment technology which is suitable to the treatment of large amount of contaminated solid materials. However, many hazardous compounds are resistant to microbial degradation due to complex chemical structure, toxicity and compound concentration that hardly support growth. Microbial growth is also affected by moisture, pH, inorganic nutrients and particle size. Because composting of hazardous wastes typically involves the bioremediation of contaminated substrate-sparse soils, support of microbial self-heating needs incorporation of proper amount of supplements. The hazardous compounds reported to disappear through composting includes aliphatic and aromatic hydrocarbons and certain halogenated compounds. The possible routes leading to disappearance of hazardous compounds include volatilization, assimilation, adsorption, polymerization and leaching (Hogan, 1998).

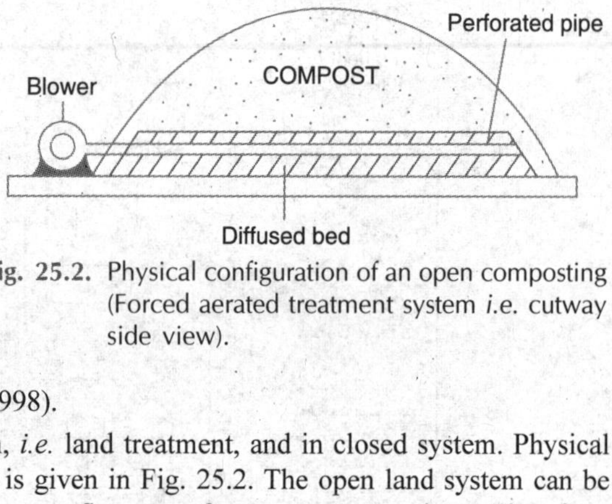


Fig. 25.2. Physical configuration of an open composting (Forced aerated treatment system *i.e.* cutway side view).

Composting can be done in open system, *i.e.* land treatment, and in closed system. Physical configuration of an open composting system is given in Fig. 25.2. The open land system can be inexpensive treatment method, but the temperature fluctuates from summer to winter. Therefore, rate of biodegradation of waste materials declines. Secondly, land treatment system may become

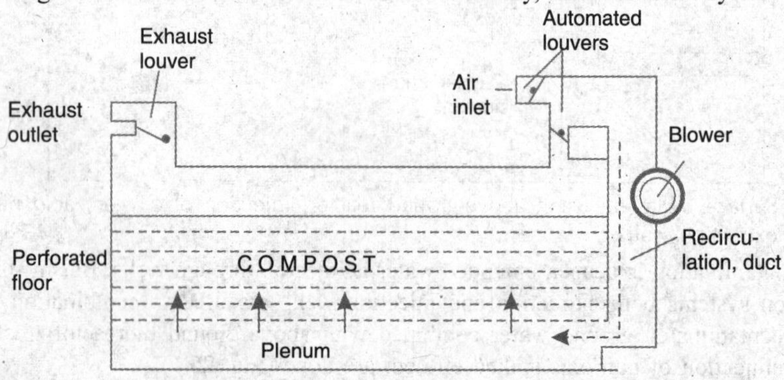


Fig. 25.3. Physical configuration of a closed composting system with forced ventilation (cutway side view).

oxygen limited, depending on amount of substrate, depth of waste, application, etc. However, efficiency of open treatment system can be increased by passing air (Fig. 25.2). This approach is referred to as *engineered soil piles* and *forced aeration treatment*. The *closed treatment system* is preferred over the open land treatment system because controlled air is supplied to maintain the microbial activity. As a result of microbial growth and volatilization of hazardous compounds, internal temperature gradually rises. Therefore, use of blowers for air circulation and exhaust for removal of toxic volatiles are set up in closed treatment system (Fig. 25.3). Ventilators supply oxygen and remove heat through evaporation of water.

***(ii)* Composting Process.** As composting is a solid-phase biological treatment, target compounds must be either solid or a liquid associated with a solid matrix. The hazardous compounds should be biologically transformed. To achieve this goal, the waste material should be suitably prepared so that biological treatment potential should maximise. This is done by adjustment of several physical, chemical and biological factors (Fig. 25.4). The hazardous wastes must be well solubilized so that they may be bioavailable. The hazardous compounds and soil organic matters serve the source of carbon and energy for microorganisms. Microbial enzymes secreted during growth phase degrade toxic compounds. However, proper maintenance of water, $O_2$, inorganic nutrients and pH increase the rate of decomposition.

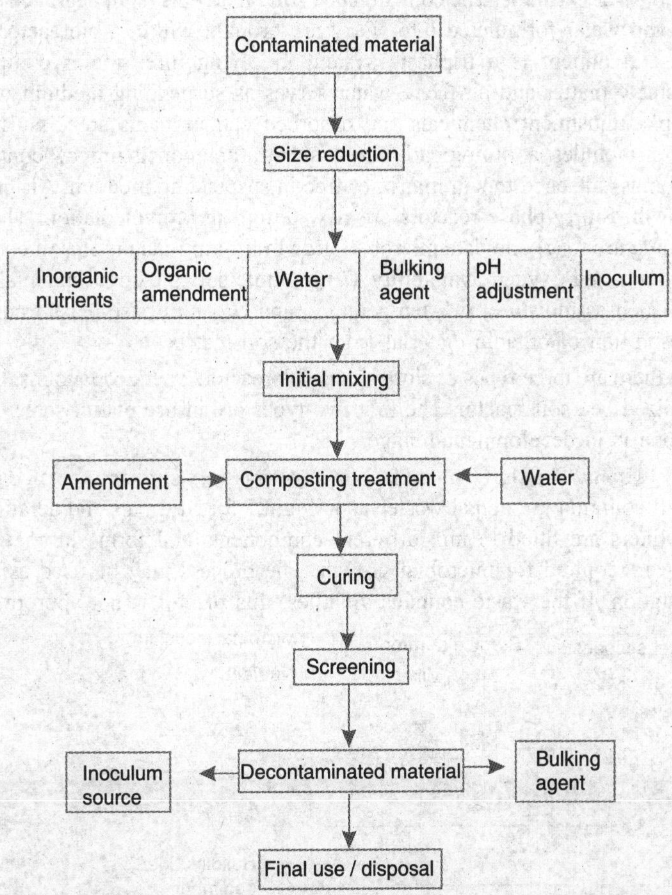


**Fig. 25.4.** Outline of composting treatment sequence (based on Hogan, 1998).

If there is low substrate-density or site-specific conditions, analogue or non-analogue, non-hazardous carbon sources that can stimulate microbial growth and enzyme production can be added to compost. Organic amendment also stabilize microbial population in inhibitory environment. Secondly, presence of sufficient amount of water enhances microbial growth. Addition of inorganic nutrients influences microbial growth and rate of decomposition of hazardous wastes. Under nitrogen limiting conditions *Phanerochaete chrysosporium* produces extracellular lignin peroxidase that degrades benzopyrene and 2, 4, 6-trinitrotoluene. It has also been noted that a pH range of 5.0–7.8 promoted the highest rates of degradation of hazardous wastes. But lignin degradation has been found the most rapid at pH of 3.0–6.5. This shows that optimal pH levels can be species, site and waste specific.

Gradual colonization of organic materials is done by indigenous microflora, but hazardous chemicals may inhibit microbial growth. Therefore, *bioaugmentation* (*i.e.* use of commercial or GEMs) of wastes is also recommended.

To provide experimental proof of biodegradation during composting, a common hazardous contaminant pesticide, $^{14}C$-labelled Carbaryl was added in sewage sludge-wood chip mixture at 1.3–2.2 ppm concentration. After 18–20 days in laboratory composting apparatus, 1.6–4.9 per cent of Carbaryl was recovered as $^{14}CO_2$ and remaining bound to soil organic matter.

**(*b*) Slurry-phase Treatment:** The contaminated solid materials (soil, degraded sediments, etc), microorganisms and water formulated into slurry are brought within a bioreactor *i.e.* fermenter. Thus slurry-phase treatment is a triphasic system involving three major components: water, suspended particulate matter and air. Here water serves as suspending medium where nutrients, trace elements, pH adjustment chemicals and desorbed contaminants are dissolved. Suspended particulate matter includes a biologically inert substratum consisting of contaminants (soil particles) and biomass attached to soil matrix or free in suspending medium. Air provides oxygen for bacterial growth. Slurry-phase reactors are new design in bioremediation. The objectives of bioreactor designing are to (*i*) alleviate microbial growth limiting factors in soil environment such as substrate, nutrients and oxygen availability, (*ii*) promote suitable environmental conditions for bacterial growth such as moisture, pH, temperature, and (*iii*) minimise mass transfer limitations and facilitate desorption of organic material from the soil matrix.

Biologically there are three types of slurry-phase bioreactors : aerated lagoons, low-shear airlift reactor, and fluidized-bed soil reactor. The first two types are in use of full scale bioremediation, while the third one is in developmental stage.

**(*i*) Aerated Lagoons.** The slurry-phase lagoon system is very similar to aerated lagoon used for treatment of small common municipal wastewater (Fig. 25.5). Nutrients and aeration are supplied to the reactor. Mixers are fitted to mix different components and form slurry, whereas surface aerators provide air required for microbial growth. The process may be used as single-stage or multistage operatation. If the waste contains volatiles, this reactor is not appropriate.

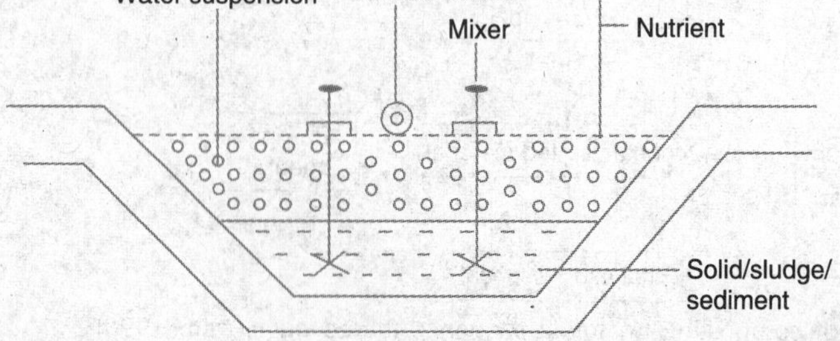


Fig. 25.5. Slurry-phase lagoon

**(*ii*) Low-shear Airlift Reactors (LSARs).** The LSARs are useful when waste contains volatile components; tight process control and increased efficiency of bioreactors are required. Fig. 25.6 shows a low-shear airlift slurry-phase bioreactor. LSARs are cylindrical tanks which is made up of stainless steel. In this bioreactor, pH, temperature, nutrient addition, mixing and oxygen can be controlled as desired. Shaft is equipped with impellers. It is driven by motor set up at the top.The rake arms are connected with blades which is used for resuspension of coarse materials that tend to settle on the bottom of the bioreactor. Air diffusers are placed radially along the rake arm. Airlift provides to bottom circulation of contents in reactor. Baffles make the hydrodynamic behaviour of slurry-phase bioreactors. Pre-treatment process includes size fractionation of solids, soil washing, milling to reduce particle size and slurry preparation. Certain surfactants such as anthracene, pyrene, perylene, etc. are added to enhance the rate of biodegradation. These act as co-substrate and utilize as carbon source. Co-substrates also induce the production of beneficial enzymes.

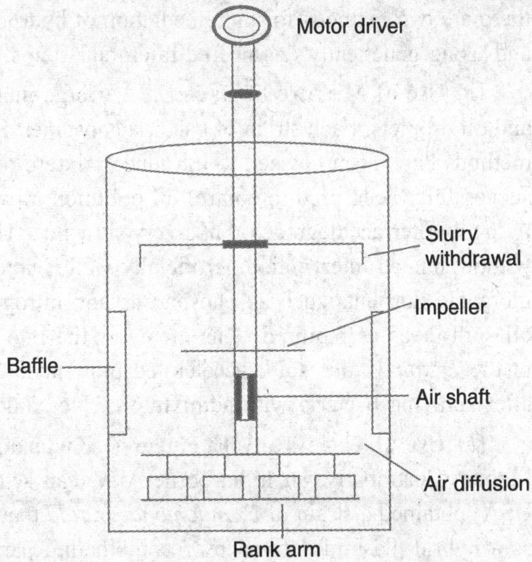


**Fig. 25.6.** A low shear airlift slurry phase bioreactor.

***(c)* Factors Affecting Slurry-phase Biodegradation:** Factors that affect slurry-phase biodegradation are *(i)* pH (optimum 5.5–8.5), *(ii)* moisture content, *(iii)* temperature (20–30°C), oxygen (aerobic metabolism preferred), *(iv)* aging, *(v)* mixing (mechanical and air mixing), *(vi)* nutrients (N, P and micronutrients), *(vii)* microbial population (naturally occurring microorganisms are satisfactory, genetically engineered microorganisms for layer compound may be added), and *(viii)* reactor operation (batch and continuous cultures).

## 3. BIOREMEDIATION OF HYDROCARBONS

Petroleum and its products are hydrocarbons. These two have much economic importance. Oils constitute a variety of hydrocarbons viz., xylanes, naphthalenes, octanes, camphor, etc. If present in the environment these cause pollution. For example, during cold war between Iraq and America, millions of gallons of petroleum was leaked into sea which resulted in fish mortality. In addition, oil and petrol leakage in marine environment is of usual phenomenon. For example, 11 million gallon oil spill from the supertanker *Exxon Valdez* ran around near Prince William Sound, Alaska in March 1989. It is interesting to note that in Pennasylvania about 27,000 litres (6,000 gallon) of petrol leakage occurred. It contaminated the underground water supplies. The input of oil to the environment can be ecologically devastating and cost of cleaning can go to several million dollars.

In toxic environment, microorganisms act only if the conditions *e.g.* temperature, pH and inorganic nutrients are adequate. Oil is insoluble in water and is less dense. It floats on water surface and forms slicks. It should be noted that in bulk storage tank microbial growth is not possible provided water and air are supplied. The microorganisms which are capable of degrading petroleum include pseudomonads, various corynebacteria, mycobacteria and some yeasts. However, there are two methods for bioremediation of hydrocarbons/oil spills, by using mixture of bacteria, and using genetically engineered microbial stains.

***(a)* Use of Mixture of Bacteria** : A large number of bacteria resides in interfaces of water and oil droplets. Each strain of bacteria consumes a very limited range of hydrocarbons, therefore, methods have been devised to introduce mixture of bacteria. Moreover, mixture of bacteria have successfully been used to control oil pollution in water supplies or oil spills from ships. Bacteria living in interface degrade oil at a very slow rate. The rate of degradation could not be accelerated without human intervention. Artificially well characterised mixture of bacterial strains along with inorganic nutrients such as phosphorus and nitrogen are pumped into the ground or applied to oil spill areas as required. This increases the rate of bioremediation significantly. For example, in the *Exxon Valdez* spill, accelerated bioremediation of oil washed upon beaches was noticed after spraying bacteria with admixture of inorganic nutrients.

***(b)* Use of Genetically Engineered Bacterial Strains** : In 1979, for the first time Anand Mohan Chakrabarty, an India borne American scientist (at the time with General Electric Co., USA) obtained a strain of *Pseudomonas putida* that contained the XYL and NAH plasmid as well as a hybrid plasmid derived by recombinating parts of CAM and OCT (these are incompatible and cannot co-exist as separate plasmids in the same bacterium). This strain could grew rapidly on crude oil because it was capable of metabolising hydrocarbons more efficiently than any other single plasmid. For more information *see* Chapter 7.

In 1990, the USA Government allowed him to use this superbug for cleaning up of an oil spill in water of State of Texas. Superbug was produced on a large scale in laboratory, mixed with straw and dried. The bacteria laden straw can be stored until required. When the straw was spread over oil slicks, the straw soaked up the oil and bacteria broke up the oil into non-polluting and harmless products.

## 4. BIOREMEDIATION OF INDUSTRIAL WASTES

A variety of pollutants are discharged in the environment from a large number of industries/mills. For example, textile industry alone contributes a significant amount of pollutants to water bodies such as enzymes, acids, alkali, alcohols, phenols, dyes, heavy metals, radionucliods, etc. Traces of zinc, cadmium, mercury, copper, chromium, lead are found in dyes.

Actinomycetes show a higher capacity to bind metal ions as compared to fungi and bacteria. In addition, uptake mechanism of living and dead cells differ. Due to these differences they have potential application in industries. The living microbial cells accumulate metals intracellularly at a higher concentration, whereas dead cells precipitate metals in and around cell walls by several metabolic processes. Dead biomass immobilised on polymeric membrane absorbs uranium well, and immobilised *Aspergillus oryzae* cells on reticulated foam particles have been used for Cd removal. *Aspergillus niger* biomass contains upto 30% of chitin and glucan. Chitin phosphate and chitosan phosphate of fungi absorb greater amount of U than Cu, Cd, Mn, Co, Mg and Ca.

***(a)* Bioremediation of Dyes** : There is limited study on microbial degradation of azo and reactive dyes. Maximum number of dyes undergo degradation through reduction. In 1981, M.G. Kulla discussed azo-reductase of *Pseudomonas* strains in the chemostat culture. This enzyme catalyses azo-linkage of the dye. During degradation process of azo, NAD(P) acts as electron donor. S.K. Srivastava and co-workers in 1995 observed degradation of black liquor pulp mill effluents by the strains of *Pseudomonas putida*. Some anaerobic bacteria, *Streptomyces* and fungi (*e.g. Phaenerochaete chrysosporium*) have been characterised for decolouration of chromogenic dyes. The enzymes involved in dye degradation are lignases (lignin peroxidase), Mn (II) dependent peroxidase and glyoxal oxidase. These enzymes are well associated with lignin degrading system.

***(b)* Bioremediation in the Paper and Pulp Industry:** Effluents from the pulp and paper industry contain chromophoric compounds and can be partly mutagenic and inhibitory to aquatic system. The presence of various pollutants produced during pulp and paper manufacturing necessitates the need for wastewater treatment before discharge.Bioremediation hosts promising in solving environmental problems in cost-effective ways. White-rot fungi have ability to process a variety of pollutants efficiently.

In 2002, L. Christove and B. Drissel have reviewed wastewater bioremediation in the pulp and paper industry. The paper and pulp industries discharge wastewater which contains halogenated (chlorogenated) organic materials (e.g. chlorolignins, chlorosyringols, chloroaliphatics, catechols, etc.) in high amount that can pose environmental problems.

***(i)* Role of Microorganisms:** Certain soil-inhabiting fungi, streptomycetes, bacteria and white-rot fungi (*Ganoderma lacidum, Coriolus (Trametes) versicolor, P. chrysosporium, Coprinus macrorhizus, Hericium erinaceus,* etc.) have been studied for their ability to decolourise chromophoric substracts. White-rot fungi produce a variety of lignin-degrading enzymes (e.g. proxidases and laccases) that degrade phenolic substances. These enzymes convert and biotransform phenolic compounds used to treat wastewater. Use of immobilised biomass or enzymes increases its stability.

***(ii)* Cultivation Strategies:** Several strategies for cultivation of white-rot fungi have been developed as briefly described below:

- **Mycor process :** In this process, *P. chrysosporium* is immobilised on the surface of rotating disks that can be enclosed for additional $O_2$ supply. A growth stage is necessary before the start of decolourisation so that nutrient nitrogen is depleted and the fungus becomes lignolytic. The Mycor process reduced colour (in an alkaline stage spent liquor) by 80% and further converted 70% of chlorophenols. Limitation with Mycor process is in constant with substrate to about 40% only.
- **Continuous flow system:** Three continuous flow laboratory decolourisation strategies were examined. *Method I* consisted of a surface area of 510 $cm^2$ for fungal cultivation. Mats of *Trichoderma* sp. were transferred to the reactor and decolourisation was carried out without reaction. *Method II* utilized in vessel in which height was increased by decreasing the surface area and four baffles were intorduced to divide the vessel into compartments. *Ttrichoderma* packed in synthetic wire bags were suspended in the middle of each zone and continuous $O_2$ was supplied. *Method III* was similar to the Mycor process where fungal mats were clasped between circular wires supported by outer control rings ans securely fixed in a middle flame. The disks were rotated continuously. It was the best method resulting in a total decolourisation of extraction stage effluent of more than 78% and COD reduction by 25%.
- **Fungal pellets :** Mycelial pellets of *C. versicolor* can be used to decolourise lignin-containing waste in an air-tight fermentor in a minimum cost. Addition of $Mg^{++}$ increases the rate of decolourisation because it acts as activator of many enzymes. *A colour* reduction by 92% with 67% reduction in COD may be obtained 7 days of using *C. versicolor.*
- **Immobilised culture :** The fungus immobilised in beads of calcium alginate gel can give the best result treated with water. In a case study, immobilised *C. versicolor* decolourised 80% after 3 days in the presence of sucrose. A fluidised bioreactor containing polyurethrane-immobilised *C. versicolor* reduced colour of a bleach.

*(c)* **Bioremediation of Heavy Metals :** Due to dicharge of industrial waste, metal pollution is generally increasing in the environment. Metals absorbed by organisms at the lower torphic level of food chain becomes gradually concentrated in those organisms which are at the higher levels where man is the ultimate to be victimized. Hence, the concentration of heavy metals gradually increases at each trophic level and enter in human body in higher amount (Fig. 25.7). Therefore, maximum damage is caused in the consumers of the last trophic level such as human. Several microorganisms have shown detoxification potential to deal with such environmental problem.

Various approaches have been made to detoxify and clean up these meals such as : use of certain chemical which in turn cause secondary pollution, and physical methods that requires large inputs of energy/expansion materials. Another option is the use of different type of microorganisms such as algae fungi and bacteria that romove metals from the solution.

**Metal-microbe Interactions and Mechanism of Metal Removal:** On the basis of localization site of the metal, microbial interactions with heavy metals can be classified into different categories such as extracellular, exocellular and intracellular where microorganisms mobilize, immobilize, transform, precipitate, accumulate, coordinate, exchange, adsorb the metals and can form complexes.

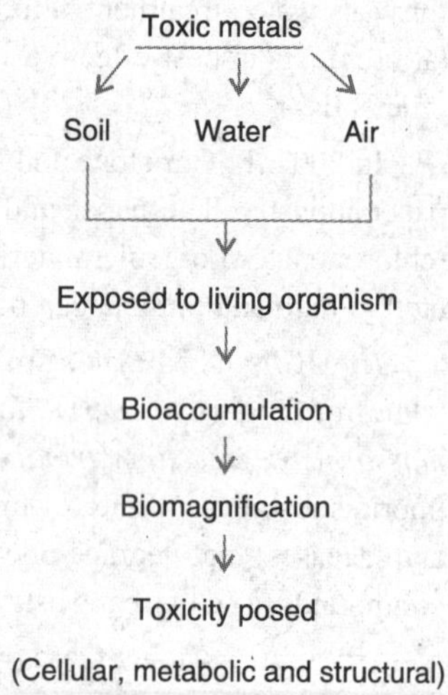


**Fig. 25.7.** Fate of toxic metals in biosystems.

Microorganisms remove metals by the following mechanisms, **adsorption** (negatively charged cell surfaces of microorganisms bind to the metal ions), **complexation** (microorganisms produce organic acids (e.g. citric acid, oxalic acid, gluconic acid, formic acid, lactic acid, malic acid) which chelate metal ions. Biosorption of metals also takes place due to carboxylic groups found in microbial polysaccharides and other polymers, **precipitation** (some bacteria produce ammonia, organic bases or $H_2S$ which precepitate metals as hydroxides or sulfates. For example, *Desulfovibrio* and *Desulfotomaculum* transform $SO_4$ to $H_2S$ which promotes extracellular precipitation of insoluble metal sulfieds. *Klebsiella aerogenes* detoxifies cadmium sulphate which precipitates on cell surface, **Volatilization** (some bacteria causes methylation of $Hg^{2+}$ and converts to dimethyl mercury which is a volatile compound), etc.

The species of *Chlorella, Anabaena inaequalis, Westiellopsis prolifica, Stigeoclonium tenue, Synechococcus* sp. tolerate heavy metals. However, several species of *Chlorella, Anabaena*, marine algae have been used for the removal of heavy metals. But the operational conditions limit the practical application of these organisms. Rai *et al*. (1998) studied biosorption of $Cd^{++}$ by a capsulated nuisance cyanobacterium, *Microcystis* both from field and laboratory. The naturally occurring cells showed higher efficiency for $Cd^{++}$ and $Ni^{++}$ as compared to laboratory cells.

Fungi also are capable of accumulating heavy metals in their cells. However, several mechanisms operate in them for removal of heavy metals from the solution, a few of these have been discussed below :

- **Metabolism-independent accumulation.** The positively charged ions in the solution are attracted to negatively charged ligands in cell materials. Biosorption of metal ion occurs on microbial cell surface. But composition of biomass and other factors affect biosorption. For example, in *Rhizopus arrhizus* adsorption depends on ionic radius of $Li^{3+}$, $Mn^{2+}$, $Cu^{2+}$, $Zn^{2+}$, $Cd^{2+}$, $Ba^{2+}$, $Hg^{2+}$ and $Pb^{2+}$. However, binding of $Hg^{2+}$, $Ag^{2+}$, $Cd^{2+}$, $Al^{3+}$, $Ni^{2+}$, $Cu^{2+}$ and $Pb^{2+}$ strongly depends on concentration of yeast cells.

- **Metabolism-dependent accumulation.** In fungi and yeast, heavy metal ions are transported into the cells through cell membrane. However, as a result of metabolic processes ions are precipitated around the cells, and synthesised intracellularly as metal-binding proteins. Energy-dependent uptake of $Cu^{2+}$, $Cd^{2+}$, $Co^{2+}$, $Ni^{2+}$, $Zn^{2+}$ by fungi has been demonstrated. Moreover, intracellular uptake is influenced by certain external factors such as pH, anions, cations and organic materials, growth phase, etc. Metal uptake by growing batch culture was found maximum during lag phase and early log phase in *Aspergillus niger, Penicillium spinulosum* and *Trichoderma viride.*
- **Extracellular precipitation and complexation.** Fungi produce several extracellular products which can complex or precipitate heavy metals. For example, many fungi and yeast release high affinity Fe-binding compounds that chelate iron. It is called siderophores. The $Fe^{3+}$ chelates which are formed outside the cell wall are taken up into the cell. In *Saccharomyces cerevisiae* removal of metals is done by their precipitation as sulphides *e.g.* $Cu^{2+}$ is precipitated as CuS.

Bacteria secrete extracellular polymers like capsule that have anionic character. External biofilm associated with cell, complexes with metals and accumulate substantially. Whole cell of *Bacillus subtilis* have shown to reduce gold from $Au^{3+}$ to $Au^0$ through extracellular enzymatic biotransformation. Under anoxic environmentals, sulphate-reducing bacteria (*Desulfovibrio* and *Desulfotomaculum*) oxidize organic matter using sulphate as an electron aceptor. *Citrobacter* sp. generates $HPO^{2-}_4$ after an en enzymatic cleavage of glycerol-1 -phosphate. Metal ions readily form complexes with $HPO_4^{2+}$ to form insoluble precipitates which remain bound to the cells.

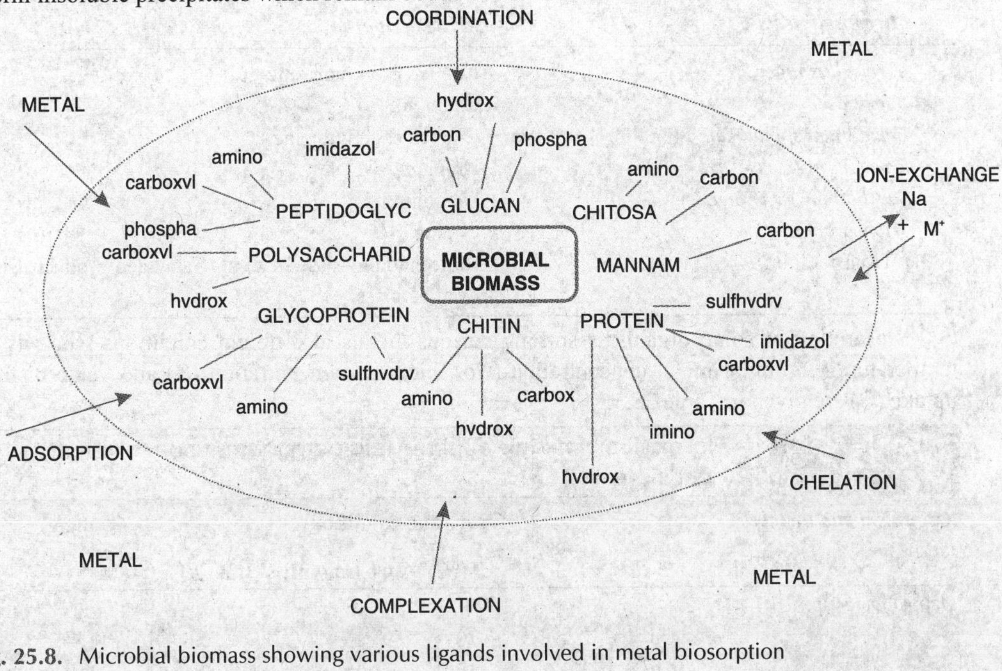


Fig. 25.8. Microbial biomass showing various ligands involved in metal biosorption (K.M. Panikar *et al.*, 2003).

(*i*) **Biosorption:** It is a passive metabolism-independent physico-chemical interactions between heavy metal ions and microbial surface. It is defined as '*a non-directed physico-chemical interaction that may occur between metal/radionuclide species and microbial cell*'. Most biosorption phenomena are combination of processes such as electrostatic interactions, ion exchange complexation, ionic band formation, precipitation, nucleation, etc. (Fig. 25.8).

The complexity of microbial surfaces and chemical/physical properties of metal ions help the interactions to occurs. For biosorption, active state of cells is not prerequisite. The process can occur even with inactivated/dead cells. Biosorption offers several advantages, for examples,

- Physiological constraints of microbial cells do not giveen the process. There is no need of costly nutrients for growth and aseptic operation of cells. It operates at a wider range of conditions such as pH , temperature and concentration of metal ions.
- Inactivated biomass works as an ion exchanger. Therefore, the prcess is rapid, for example *Streptomyces nouresei* carries out biosorption at a high speed (256 μ mol/g). Metals can be desorbed readily from the biosorbent and recovered thereafter.

(*ii*) **Organisms Involved in Biosorption:** There are several microorganisms such as bacteria (*Arthrobacter viscosus, Pseudomonas syryngae, P. aeruginosa, P. putida, Bacillus subtilis, E. coli, Streptomyces nouresei, S. pimprina,* etc.), fungi and yeast (*Saccharomyces* spp., *Aureobasidium pullulans, R. nigricans,* etc.) and algae (*Chlorella valgaris, Ascophyllum nodosum, Cladophora crispata,* etc.). Fungal mycelia (e.g. *Aspergillus* and *Penicillium*) also remove metals from wastewater and offer a good alternative for detoxification of effluents. Biosorption have shown that *Aspergillus oryzae* can remove cadmium efficiently from solution (Table 25.2).

**> Table 25.2. Fungi involved in metal removal from industrial wastewater (based on Bitton, 1999).**

| *Microorganisms* | *Metals removed* |
|---|---|
| *Aspergillus niger* | Copper, cadmium, zinc |
| *A. oryzae* | Cadmium |
| *Penicillium spinulosum* | Copper, cadmium zinc |
| *Rhizopus arrhizus* | Uranium |
| *Saccharomyces cerevisiae* | Uranium |
| *Trichoderma viride* | Copper |
| ATM-Bioclaim™ | Biotechnology-based use of granulated product derived form biomass |

Biosorption of some organisms sorbing various metals at different conditions (such as pH, temperature (T), initial metal concentration (Co), biomass concentration (B) and value of metal uptake (Q) are given in Table 25.3.

**> Table 25.3. Biosorption of metals by some microorganisms under different conditions.**

| *Organism and metals* | | Operating conditions | | | | Q (mg/g) |
|---|---|---|---|---|---|---|
| | | pH | T(°C) | Co (mg/ml) | B (g/1) | |
| *Arthrobacter* sp. | | | | | | |
| | Cd | 7 | 20 | 1 | 0.6 | 1.4 |
| | Cu | 3.5-6 | 30 | 180 | 0.4 | 1.48 |
| | Pb | 5-5.5 | 30 | 250 | 1.4 | 130 |

| | | | | | | |
|---|---|---|---|---|---|---|
| *Streptomyces nouresei* | Cd | 6 | 30 | 1-110 | 3.5 | 3.4 |
| | Cu | 5.5 | 30 | 0.6-65 | 3.5 | 3.4 |
| | Pb | 6.1 | 30 | 2-207 | 3.5 | 36.5 |
| | Ag | 6 | 30 | 1-100 | 3.5 | 38.6 |
| | Zn | 5.8 | 30 | 0.6-65 | 3.5 | 1.6 |
| *Rhizopus arrhizus* | Cd | 3.5 | 26 | 10-400 | n.a. | 25 |
| | Cr | 1-2 | 25 | 25-400 | n.a. | 4.5 |
| | Cu | 5.5 | 25 | 0.6-25 | n.a. | 9.5 |
| | Pb | 5 | 25 | 150 | 1 | 6.5 |
| | Zn | 6-7 | n.a. | 10-600 | 3 | 13.5 |
| *Penicillium digitatum* | Cd | 5.5 | 25 | 10-50 | 6.5 | 3.5 |
| | Cu | 5.5 | 25 | 10-50 | 6.5 | 3 |
| | Pb | 5.5 | 25 | 10-50 | 6.5 | 3 |

Source: K.M. Paknikar *et al.* (2003); n.a., not available.

***(iii)* Factor affecting biosorption of metals:** Biosorption depends on various factors like type of biomass, metal chemistry, temperature, pH, biomass concentration, initial metal concentration, composition of biomass, etc.

***(iv)* Metal bisorption technologies:** Several technoligies for metal removal have been commercialized and employed as briefly given below:

- **ATMBIOCLAIM™ process:** The Advance Mineral Technology (ATM) Inc. (U.S.A.) developed a wastewater treatment process with *Bacillus* sp. (immoblised in beads) pretreated with caustic solution. It is specific for metal cations in the order: $Cu^{2+} > Zn^{2+} > Cd^{2+} = Ni^{2+} > Pb^{2+}$.
- **AlgaSORB™ process:** Biorecovery systems, Inc. (U.S.A.) developed this proprietary based material which consists of several types of living and nonliving algae. Algal cultures are immobilised in silica gel in the form of beads. The desorption of metal is carried out.
- **BIO-FIX™ process:** The Bureau of Mines (U.S.A.) developed this process that consists of biomass immobilised in polysulfone. It consists of thermally killed biomass of *Sphagnum* peat moss, algae, yeast, bacteria and /or aquatic flora. The beads are suitable for practical application in stirred tank reactor, fixed and fluidized-bed columns.

***(d)* Bioremediation of Coal Wastes through VAM Fungi:** Bioremediation of coal waste land through VAM fungi is gaining importance in recent years. Selected VAM fungi are introduced through plants in coal mine areas. Extensive infection of most plant species colonizing coal waste has been observed in India and other countries. It has been found that VAM fungi improved the growth and survival of desirable revegetation species. Increased growth of red maple, maize, alfalfa and several other plants inoculated with VAM fungi growing in coal mine soil has been recorded.

## 5. BIOREMEDIATION OF XENOBIOTICS

Use of pesticides has benefited the modern society by improving the quantity and quality of the worlds' food production. Gradually, pesticide usage has become an integral part of modern agriculture system. Many of the artificially made complex compounds *i.e.* xenobiotics persist in environment and do not undergo biological transformation. Microorganisms play an important role in degradation of xenobiotics, and maintaining of steady state concentrations of chemicals in the environment. The complete degradation of a pesticide molecule to its inorganic components that can be eventually used in an oxidative cycle removes its potential toxicity from the environment. However, there are two objectives in relation to biodegradation of xenobiotics: (*i*) how biodegradation activity arises, evolves and transferred among the members of soil microflora, and (*ii*) to device bioremediation methods for removing or detoxifying high concentration of dangerous pesticide residues.

The characters of pesticide degradation of microorganisms are located on plasmids and transposons, and are grouped in clusters on chromosome. Understanding of the characters provides clues to the evolution of degradative pathways and makes the task of gene manipulation easier to construct the genetically engineered microbes capable of degrading the pollutants.

**(*a*) Microbial Degradation of Xenobiotics:** Biodegradation of pesticides occurs by aerobic soil microbes. Pesticides are of wide varieties of chemicals *e.g.* chlorophenoxyalkyl caboxylic acid, substituted ureas, nitrophenols, triazines, phenyl carbamates, organochlorines, organophosphates, etc. Duration of persistence of herbicides and insecticides in soil is given in Table 25.4. Organophosphates (*e.g.* diazion, methyl parathion and parathion) are perhaps the most extensively used insecticides under many agricultural systems. Biodegradation through hydrolysis of *p*-*o*-aryl bonds by *Pseudomonas diminuta* and *Flavobacterium* are considered as the most significant steps in the detoxification of organophosphorus compounds. Organomercurials (*e.g.* Semesan, Panodrench, Panogen) have been practised in agriculture since the birth of fungicides. Several species of *Aspergillus, Penicillium* and *Trichoderma* have been isolated from Semesan-treated soil. Moreover, they have shown ability to grow over 100 ppm of fungicide *in vitro*. The major fungicides used in agriculture are water soluble derivatives such as Ziram, Ferbam, Thiram, etc. All these are degraded by microorganisms.

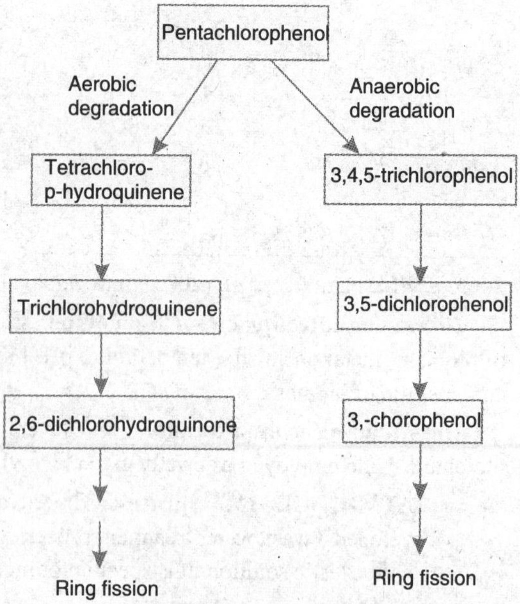


**Fig. 25.9.** Outline of aerobic and anaerobic degradation of pentachlorophenol.

Pentachlorophenol (PCP) is a broad-spectrum biocide which has been used as fungicide, insecticide, herbicide, algicide, disinfectant and antifouling agent. Bioreactors containing alginate immobilised + Polyurethane foam immobilized PCP degrading *Flavobacterium* (ATCC39723) cells have been used to remove PCP from contaminated water. Absorption of PCP by Polyurethane immobilized matrix plays a role in reducing the toxicity of PCP. Flavobacterium removed and detoxified PCP. In other experiment *P. chrysosporium* enzyme (ligninase) has been found to dehalogenate PCP. Steps of PCP degradation has been shown in Fig. 25.9.

**> Table 25.4.** **Duration of persistance of insecticides and herbicides in soil.** (*Source* : M.T. Madigan *et al.*, 1997).

| *Biocides 75-100% disappearance* | *Time taken for* |
|---|---|
| **Chlorinated insecticides** | |
| DDT (1,1,1-trichloro-2,2-bis-(p-chlorophenyl) ethane) | 4 years |
| Aldrin | 3 years |
| Chlordane | 5 years |
| Heptachlor | 2 years |
| Lindane (hexachloro-cyclohexane) | 3 years |
| **Organophosphate insecticides** | |
| Diazinon | 12 years |
| Malathion | 1 week |
| Parathion | 1 week |
| **Herbicides** | |
| 2,4-D (2,4-dichlorophenoxyacetic acid) | 4 weeks |
| 2,4,5-T | 30 weeks |
| Atrazine | 40 weeks |
| Simazine | 48 weeks |
| Propazine | 1.5 years |

Source : M.T. Madigan *et al.* (1197)

DDT (1,1,1-trichloro-2,2-bis (p-chlorophenyl) ethane) is an insecticide that persists in soil for four years. Degradation pathway of DDT involves an initial dechlorination of the trichloromethyl group to form 1,1-dichloro 2-ethane which then undergoes further dechlorination, oxidation and decarboxylation to form bis methane. Subsequent cleavage of one of the normal aromatic rings yields p-chlorophenyl acetic acid, which may also undergo ring cleavage. Microorganisms associated with DDT degradation are *Aspergillus flavus, Fusarium oxysporum, Mucor aternans, P. chrysosporium, Trichoderma viride,* etc. Environmental factors including pH, temperature, bioavailability, nutrient supply and oxygen availability affect biodegradation of pesticides.

(*b*) **Gene Manipulation of Pesticide-degrading Microorganisms:** Day-by-day the number of xenobiotic-degrading miroorganisms is increasing. However, pesticide-degrading genes of only a few microorganisms have been characterized. Most of genes responsible for catabolic degradation are located on the chromosomes, but in a few cases these genes are found on plasmids or transposons. The camphor degrading genes of *Pseudomonas putida* are located on plasmid. Soon after, naphthalene (NAH) degrading plasmid was isolated. Discovery of these genes made it possible to construct a new genetically engineered strain of *P. putida* that alone was potent to degrade camphor, naphthalene, xylene, toluene, octanes and hexanes. For detailed discission, *see* Chaper 17.

For the first time pesticide degradation through plasmid mediated genetically engineered microorganism was reported by A.M. Chakrabarty and coworkers in 1981. In 1993, S. Nagata and co-workers have also cloned and sequenced two genes involved in early steps of Y-HCH degradation in UT26. The *lin*A gene encodes Y-HCH dehydrochlorinase which converts Y-HCH to 1,2,4TCB via Y-PCCH and 1,4-TCDN. The *lin*B gene encodes 1,4-TCDN chlorohydrolase which converts 1,4-TCDN to 2,4-DDOL via 2,4,5-DNOL. This gene is a member of the haloalkanedehalogenase family with a broad range specificity for substrate. The genetically engineered *P. putida* comprises of both *lin*A and *lin*B genes.

The *opd* gene, initially isolated from *Flavobacterium* sp. ATCC27551 and *Pseudomonas diminuta*, has been well characterized. It is associated for degradation of pesticides such as parathion, methylparathion, etc. *Gliocladium virens* is a useful soil saprophyte which has shown a strong mycoparasitic activity against many fungal pathogens. Strains of *G. virens* are potential for use in the bioremediation of contaminated soil. Optimization of *opd* expression in bioremedially useful organism such as *P. chrysosporium*, *G. virens*, etc. holds a great promise in lessening the pesticide pollution.

## B. BIOAUGMENTATION (USE OF BLENDS OF MICROORGANISM)

Acceleration of biodegradation of specific compounds by inoculating bacterial cells is called **bioaugmentation.** Bacterial cells contian specific plasmid which encodes enzymes for degradation of those compounds. A variety of plasmids have been reported from *Alcaligenes. Acinetobacter, Arthrobacter, Beijeirinkia, Klebsiella, Flavobacterium* and *Pseudomonas.* Several genetically engineered strains have been developed exploiting *Pseudomonas.*

Production of microbial products for bioaugmentation.

### 1. Prinicples of Bioaugmentation

Microorganisms capable of degrading herbicides/other chemicals in industrial water are isolated from wastewater, compost, sludege, etc. Some of the strains. may be irradiated to enhance their ability and mutants are selected (Fig. 25.10). Before their use in the environment they are tested in laboratory for their biodegradation ability. Bioassays are also used to assess the toxicity of the wastewater for commercial preparation of microbial seeds. Selected strains are used in large fermentor to get mass culture. Then they are preserved through lyophilization, drying and freezing.

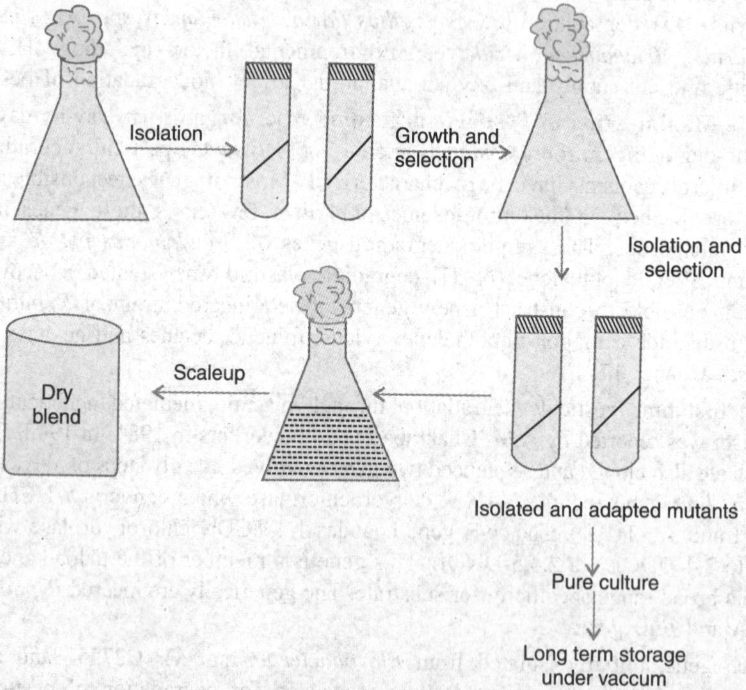


Fig. 25.10. Isolation and purification of microbial blends used for pollution control.

Commercial bioaugmentation products are single cultures of consortia of microorganisms with certain degradative properties or their desirable characters. At present most important users are the industrial wastewater treatment plants. The selected microorganism is added to a bioreactor so that potential for biodegradation of wastes must be maintained or enhanced. Due to trade secrets information on bioformulation of mixture of microbial cultures are not scanty.

Application of biaugmentation includes: (*i*) the increased BOD removal in wastewater treatment plants, (*ii*) reduction of sludge volume by about 30% after addition of selected miroorganisms, (*iii*) use of mixed cultures in sludge digestion, (*iv*) biotreatment of hydrocarbon wastes, and (*v*) biotreatment of hazardous wastes.

The use of added microorganisms for treating hazardous wastes usch as phenol, ethylene glycol, formaldehyde has been attempted. Bioaugmentation with parachlorophenol-degrading bacteria decomposed 96% parachlorophenol in 9 hours. Cells of *Candida tropicalis* have been used for removal of high concentration of phenol present in freshwater. Ability of a bioreactor to dechlorinate 3-chlorobenzoate was increased after addition of *Desulfomonile tiedjei* to a methanogenic upflow anaerobic granular sludge banket. Anoxygenic phototrophic bacteria have also been considered for the degradation of toxic compounds in wastes.

Some demerits of bioaugmentation are : (*i*) need of an acclimation period prior to onset of biodegradation, (*ii*) a short survival or lack of growth of microbial inocula in the seeded bioreactors, and (*iii*) some times negative or non-conclusion of some of commercial products.

### 2. Use of Enzymes

Many enzymes have been detected in wastewaters such as catalase, phosphate esterases and amino-peptidases. These enzymes can be added to fresh waters to improve biodegradation of xenobiotic compounds. For example, parathion hydrolases (isolated from *Pseudomonas* and *Flavobacterium*) have been used to cleanup the containers of parathion and detoxification of wastes containing high concentration of organophosphates. A little information is available on use of enzymes in wastewater treatment plants. This technology is applied to reduce the production of excessive amount of extracellular polysaccharides during wastewater treatment because overproduction of polysaccharides results in increased water retention with reduced rate of dewatering. Addition of enzyme can degrade these expolymers.

Some specific enzymes (e.g. horseradish peroxidase) can catalyse the polymerisation and precipitation of aromatic compounds (e.g. substituted aniline and phenols). Horseradish peroxidase catalyses the oxidation of phenol and chlorophenols by hydrogen peroxide. The extracellular fungal laccases (obtained from *Trametes versicolor* or *Botrytis cinerea*) can be used for the treatment of effluents generated by the pulp and paper industry because this enzyme can be useful for dechlorination of chlorinated phenolic compounds or oxidation or aromatic compounds even at adverse environmental conditions such as low pH, high temperature, presence of organic solvents, etc. Therefore, attention has now been paid to use **extremozyme** of microorganisms that can work at extreme environments also.

## C. BIOFILTRATION

Biofiltration is a new technology used to purify contaminated air evolved from volatile organic compounds by involving microorganisms. It is a low cost technology gradually becoming popular due to simple operational and waste-removal efficiencies. Biofiltration is the oldest biotechnological method for removal of undesired foul gas components from air. Since 1920, biofilters were used to remove odorous compounds from wastewater treatment plants or animal farming. It could be achived by digging trenches, laying an air distribution system and refilling the trenches with permeable soil, wood chips and compost.

In the 1960s, the first biofilters were built in the U.S.A.Between 1980s and 1990s, about 30 large and full-scale systems of about 1,000 $m^3$ capacity have been constructed. Biofiltration has more industrial success in Europe and Japan where over 500 biofilters are in operation. Moreover, biofiltration is not suitable for highly haloginated compounds (e.g. trichloroethylene, trichloroethane and carbon tetrachloride) due to its low aerobic degradation. Also the size of a biofiltration in inversely proprotional to the degradation rate.

***(i)* Biofilters:** Biofiltration is done by using biofilters. Biofilters are the packed-bed units in which gas is blown through bed of compost or soil covered by an active biofilm made by the natural microorganisms. Diagram of a biofilter is given in Fig. 25.11.

The microorganisms consume the gaseous organic pollutants and use as source of carbon and energy. Instead, it may contain an inner support where a special or pool of microorganisms are cultivated. The harmful compounds are degraded by an active biofilm covering the bed. The unwanted odorous organic compounds from gaseous phase are removed. They are absorbed or adsorbed on porous solid base of the biofilter, or dissolved into liquid phase and then oxidised by the microorganisms. Biofiltration is beneficial because it does not require large amount of energy during operation.

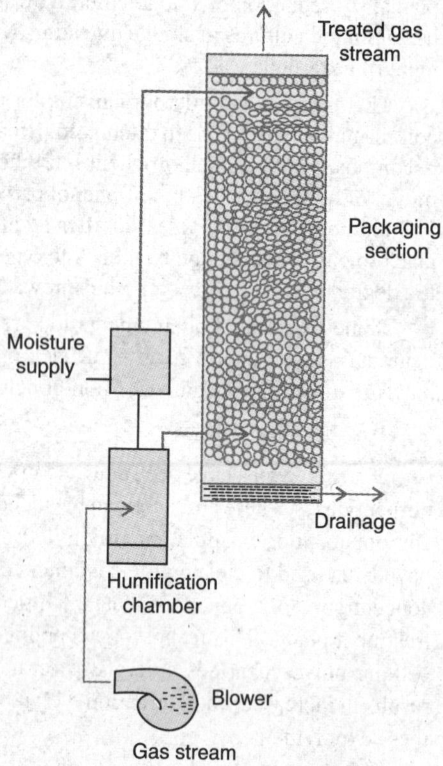


**Fig. 25.11.** Diagram of a biofilter.

***(ii)* Microroganisms used in biofilters:** Different aspects have been studied regarding the microbial potential of biofilters: (*i*) isolation and characterisation, (*ii*) use of pure cultures of bacteria or fungi, (*iii*) mixed microbial population, (*iv*) effect of enrichment culture including application of special strains, types of microorganisms and their metabolic activities, (*v*) effect of external conditions on microbial activity, and (*vi*) release of microorganisms from biofilters.

Biofilter for absorbption of contaminants from air.

Microorganisms present in biofilter are aerobic ones. Most of them are bacteria (mostly coryneforms and endospore formers), occasionally pseudomonads, protozoa, invertebrates and a few actinomycetes (mainly *Streptomyces* spp.) and some fungi (mostly *Alternaria, Aspergillus, Botrytis, Cladosporium, Fusarium, Mortierells, Rhizopus, Penicillium, Trichoderma*). Fungi form a large specific area which remain in direct contact of air flowing through filter.

Microorganisms are the most critical components of biofilters. Because they transform or degrade the contaminants. Naturally occurring microorganisms are available for the process because they get adapted to the contaminants. In some cases, a specific microbe or genetically engineered microorganisms may be used.

**(*iii*) Biofilter media:** The filter media must have some characteristics for performance of the biofilter. Because all the filter media allow the polluted air to interact closely with the degradative microorganisms, oxygen and water. Constitution of the physical media provides fine porous, large surface area and distribution of uniform pore size. Uniform pore size strognly defines the efficiency of biofilm. Inorganic bed material has a good flow properties and consists of a variety of metal oxides, glass or ceramics beads. PVC is commonly used as packing material.

The active microbial biofilm will adher onto the biofilter media. The amount of microorganisms present depends on the availability of surface which in turn increases efficiency of biofilters. Thus the suitable biofilter media have large surface area for both adsorption of contaminants and support for microbial growth. Syntheitc or inert media must be inoculated with soil, compost or sewage sludge. Because these materials have a big and complex population of microorganisms available to develop the proper microbial culture for the process. Pure culture can also be tested as inoculum. Suitable biofilter media have the ability to retain moisture to sustain biofilm layer and retain capacity of nutrient supply to microbes that form active biofilm. Some of the materials used as biofilter media are compost, peat, soil, activated carbon, wood chips or bark, perlite, vermiculite, lava rock and inert plastic material.

**(*iv*) Mechanism of biofiltration:** Removal of contaminated material is multistep process. The contaminants are converted into liquid phase and transported to bacterial cell in the biofilm and transferred across the cell membrane, where the compound is degraded and used in cell metabolism. The treatment process depends on two mechanisms: (*i*) direct adsorption in biofilm and degradation, (*ii*) adsorption on organic media and biodegradation, and (*iii*) dissolution in aqueous phase and degradation. After biodegradation, the contaminants are exhausted from the biofilter. The process can be expressed as below:

$$\text{Pollutants} + O_2 + \text{Microorganisms} \rightarrow \text{Microbial cells} + CO_2 + H_2O$$

**Immobilisation of microorganisms to the bedding materials in biofilters :** Immobilisation of microorganisms consists of two processes: (*i*) the self-attachment of cells to the filter bedding material, and (*ii*) the artificial immobilisation of microorganisms to the bedding material. Self-attachment of microorganisms to a surface depends on the microbial culture, *i.e. secretion of glycocalyx* (extracellular polysaccharide) and several forces such as electrostatic interaction, covalent bond formation, hydrophobic interaction, and partial covalent bond between imcroorganisms and hydroxyl groups on surfaces. Immobilisation of microorganisms at filter bedding is done by five methods such as: carrier bonding, cross-linking, entrapment, microcapsulation and membrane methods (for detail see Chapter 22)

# D. UTILIZATION OF SEWAGE, AND AGRO-WASTES

## 1. Production of Single Cell Protein (SCP) on Sewage Oxidation Ponds

*See* Chapter 18

## 2. Biogas from Sewage

*See* Chapter 24

## 3. Mushroom Production on Agro-wastes

*See* Chapter 18

## 4. Vermicomposting

Vermicomposting is the phenomenon of compost formation by earthworms. Obviously, earthworms play an important role in the cycling of plant nutrients, turnover of organic matter and maintenance of soil structure. They can consume 10–20% of their own biomass per day. The most important effect of earthworms in agro-ecosystems is the increase in nutrient cycling, particularly nitrogen. They ingest organic matter with a relatively wide C:N ratio and convert it to earthworm tissue with a lower C:N ratio. Thus, they affect the physico-chemical properties of soil.

It is a sustainable biofertilizer generated from organic wastes. Vermicompost is an excellent source of nutrients for vegetables ornamentals, fruit and plantation crops. Using vermicompost, one can get 10-15% more crop yield, besides improvements in quality of the product. For the first time, in 1970, vermicomposting was started in Ontario (Canada). In recent years, the U.S.A. , Japan and Philipines are the leaders of vermicompost producers. So far least attention has been paid in India. But is recent years, Government and non-government organisations (NGOs) are trying to popularise the vermicomposting process.

Through an NGO, Pithoragarh Municipality (Uttranchal) has started vermicomposting by using a thermotolerant earthworm *Eisenia foetida.* Vermicomposting is also being done at Shanti Kunj (Haridwar). One kg earthworm can consume 1 kg organic materials in a day. They secrete as casting which are rich in Ca, Mg, K, N and available P. Depending on substrate quality, vermicompost cosists of 2.5–3.0% N, 1–1.5% P and 1.5–2% K, useful microorganisms (bacteria, fungi actinomycetes, protozoa), hormones, enzymes and vitamins.

Earthworm makes soil rich in nutrients.

Earthworms make tunnels and mix soil. Thus they aerate the soil which promote the growth of bacteria and actinomycetes. Consequently, microbial activity of soil is increased due to increase in enzymatic and biological activity of earthworms. About 500 species of earthworms are known in India and over 3,000 in the world. The most common members of the earthworm to be used in vermicomposting include: *Eisenia andrie, E. coetida, Dravida willsii, Endrilus euginee, Lamito mauritii, Lubrieus rubellus* and *Perionyx excavatus.*

Vermicomposted soil ready to be used.

In several countries including India, significant work has been done. Scientists at Indian Institue of Sciences (Bangalore) have developed methods for frequent decomposition of coconut coir by using earthworms. Prof. B.R. Kaushal and coworkers at Kumaun University, Nainital have done significant work on earthworms, their food materials, food habit, organic matter turnover and established relationships between food consumption, changes in worm biomass, and casting activity of earthworms. They have also monitored the feeding and casting activity of *Amynthas alexandri* on corn, wheat leaves and mixed grasses in laboratory cultures. Casts were produced on surface and sides of the containers. Food consumption varied from 36 to 69 mg/g live worm/day. Cast production ranged from 4 to 6 mg/g live worm/day. Some of the known and potential waste decomposer (such as *Drawida nepalensis*, etc.) earthworms may be introduced in such places where they are absent.

(*a*) **Process of Vermicomposting:** Process of vermicomposting can be done in pits or concrete tanks, wells or wooden crates. A pit of $2 \times 1 \times 1$ m$^3$ dimension (1 m maximum depth) is dug under a shade to prevent the entry of water during rain (Fig. 25.12). Broken bricks or pebbles are spread on the bottom of pit followed by coarse sand to facilitate the drainage. It is covered by layer loamy soil which is moistened and inoculated by earthworms. It is covered by small lumps of fresh or dry cattle dung followed by a layer of hay or dry leaves or agro-wastes. Every day for about 20–25 days water is sprinkled over it to keep the entire set up moist. Until the pit is full dry and green leaves are put into the pit in each week. Vermicompost is ready after 40–45 days. Vermicompost appears soft, spongy, dark brown with sweet smelling. Then it is harvested and kept in dark. It is sieved and packed in polythene to retain 20% moisture content.

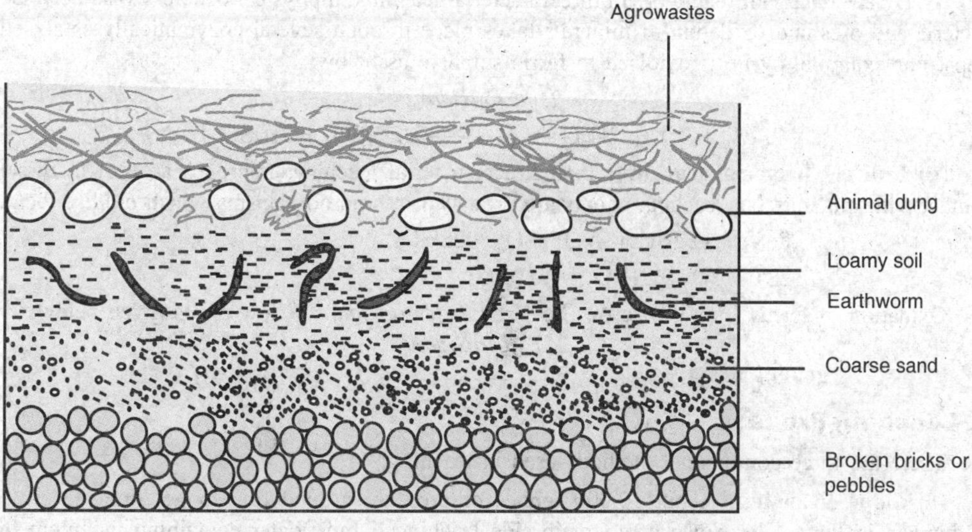


Fig. 25.12. Diagram of pit for operation of vermicomposting process.

## E. MICROBIAL LEACHING (BIOLEACHING, BIOMINING)

Microbial leaching is the process by which metals are dissolved from ore bearing rocks using microorganisms. For the last 10 centuries, microorganisms have assisted in the recovery of copper dissolved in drainage from water. Thus biomining has emerge as an important branch of biotechnology in recent years. Microbial technology renders helps in case of recovery of ores which cannot be economically processed with chemical methods, because they contain low grade metals. Therefore, large quantity of low grade ores are produced during the separation of high grade ores. The low grade ores are discarded in waste heaps which enter in the environment. The low grade ores contain significant amount of nickel, lead, and zinc ores which could be processed by microbial leaching. Bioleaching of uranium and copper has been widely commercialized. But large scale leaching process may cause environmental problems when dump is not managed properly. This results in seepage of leach fluids containing large quantity of metals and low pH into nearby natural water supplies and ground water.

Thus, biomining is economically sound hydrometallurgical process with lesser environmental problem than conventional commercial application. However, it is an inter-disciplinary field involving metallurgy, chemical engineering, microbiology and molecular biology. It has tremendous practical application. In a country like India, biomining has great national singnificance where there is vast unexploited mineral potential.

## 1. Miroorganisms used for Leaching

The most commonly used microorganisms for bioleaching are *Thiobacillus thiooxidans* and *T. ferrooxidans*. The other microorganisms may also be used in bioleaching viz., *Bacillus licheniformis, B. luteus, B. megaterium, B. polymyxa, Leptospirillum ferrooxidans, Pseudomonas fluorescens, Sulfolobus acidocaldarius, Thermothrix thioparus, Thiobacillus thermophilica*, etc.

**(*a*) Chemistry of Microbial Leaching:** *T. thiooxidans* and *T. ferrooxidans* have always been found to be present in mixture on leaching dumps. *Thiobacillus* is the most extensively studied Gram-negative bacillus bacterium which derives energy from oxidation of $Fe^{2+}$ or insoluble sulphur. In bioleaching there are two following reaction mechanisms:

**(*i*) Direct bacterial leaching:** In direct bacterial leaching, a physical contact exists between bacteria and ores and oxidation of minerals takes place through several enzymatically catalysed steps. For example, pyrite is oxidised to ferric sulphate as below:

$$2FeS_2 + 7O_2 + 2H_2O \xrightarrow{T.\ ferrooxidans} 2FeSO_4 + 2H_2SO_4$$

**(*ii*) Indirect bacterial leaching:** In indirect bacterial leaching, microbes are not in direct contact with minerals but leaching agents are produced by microorganisms which oxidise them.

$$FeS_2 + Fe_2(SO_4) \rightarrow 3FeSO_4 + 2S^o$$

$$2S^o + 3O_2 + 2H_2O \rightarrow 2H_2SO_4$$

Oxidation of ferrus ($Fe^{2+}$) to ferric ($Fe^{3+}$) by *T. ferrooxidans* at low pH is given below:

$$4FeSO_4 + 2H_2SO_4 + O_2 \xrightarrow{T.\ ferroxidans} 2Fe_2(SO_4)_3 + 2H_2O$$

## 2. Leaching Process

There are three commercial methods used in leaching:

**(*a*) Slope Leaching:** About 10,000 tonnes of ores are ground first to get fine pieces. It is dumped in large piles down a mountain side leaching dump. Water containing inoculum of *Thiobacillus* is continuously sprinkled over the pile. Water is collected at bottom. It is used to extract metals and generate bacteria in an oxidation pond.

**(*b*) Heap Leaching:** The ore is dumped in a large heaps called leach dump. Further steps of treatment is as described for slope leaching.

**(*c*) *In situ* Leaching:** In this process, ores remain in its original position in earth. Surface blasting of rock is done just to increase permeability of water. Thereafter, water containing *Thiobacillus* is pumped through drilled passage to the ores. Acidic water seeps through the rock and collects at bottom. Again from bottom water is pumped, mineral is extracted and water is reused after generation of bacteria.

## 3. Examples of Bioleaching

Bioleaching has been discussed with copper, uranium, gold, silver and silica.

**(*a*) Copper Leaching:** Throughout the world copper leaching plants have been widely used for many years. It is operated as simple heap leaching process or combination of both heap leaching and *in situ* leaching process. Dilute sulphuric acid (pH 2) is percolated down through the pile. The liquid coming out of the bottom of pile reach in mineral. It is collected and transported to precipitation plant, metal is reprecipitated and purified. Liquid is pumped back to top of pile and cycle is repeated. For removal of copper the ores commonly used are chalcocite ($Cu_2S$), chalcopyrite ($CuFeS_2$) or covellite (CuS). Several other metals are also associated with these ores. Chalcocite is oxidised to soluble form of copper ($Cu^{2+}$) and covellite by *T. ferrooxidans*.

$$Cu_2S + O_2 \rightarrow CuS + Cu^{2+} + H_2O$$

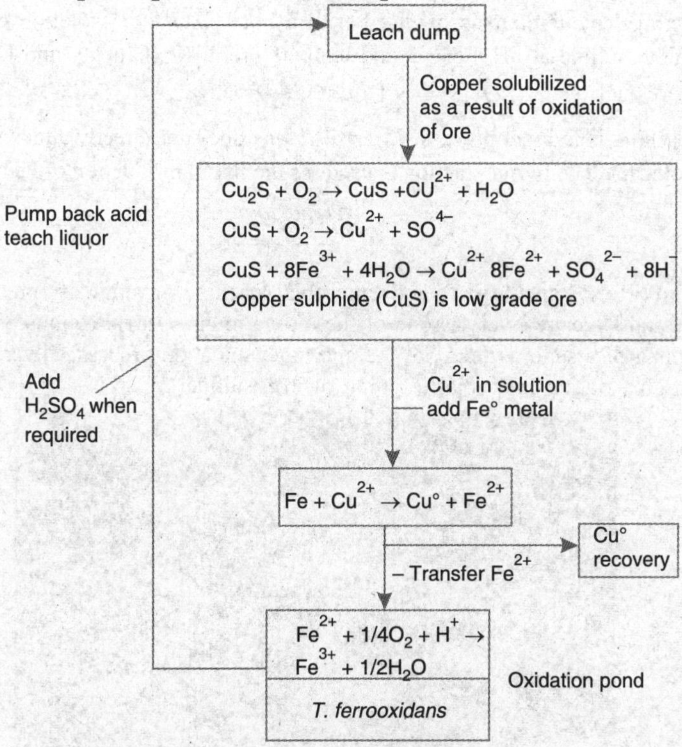


**Fig. 25.13.** Microbial leaching of copper.

Covellite is oxidised to copper sulphate chemically or by bacteria.

$$2CuFeS_2 + 8\frac{1}{2}O_2 + H_2SO_4 \rightarrow 2CuSO_4 + Fe_2(SO_4)_3 + H_2O$$

Thereafter, strictly chemical reaction occurs which is the most important reaction in copper leaching.

$$CuS + 8Fe^{3+} + 4H_2O \rightarrow Cu^{2+} + 8Fe^{2+} + SO_4^{2-} + 8H^+$$

Copper is removed as below:

$$Fe^o + Cu^{2+} \rightarrow Cu^o + Fe^{2+}$$

$Fe^{2+}$ is transferred in oxidation pond

$$Fe^{2+} + \frac{1}{4}O_2 + H^+ \xrightarrow{T.\ ferrooxidans} Fe^{3+} + \frac{1}{2}H_2O$$

The $Fe^{3+}$ ions produced is an oxidation of ores; therefore, it is pumped back to pile. Sulphuric acid is added to maintain pH. An outline of microbial leaching of copper is shown in the Fig. 25.13. Microbial leaching of copper has been widely used in the USA. Australia, Canada, Mexico, South Africa and Japan. In the USA 200 tonnes of copper is recovered per day.

(*b*) **Uranium Leaching:** Uranium leaching is more important than copper, although less amount of uranium is obtained than copper. For getting one tonne of uranium, a thousand tonne of uranium ore must be handled. *In situ* uranium leaching is gaining vast acceptance. However, uranium leaching from ore on a large scale is widely practised in the USA, South Africa, Canada and India.

Insoluble tetravalent uranium is oxidised with a hot $H_2SO_4/Fe^{3+}$ solution to make soluble hexavalent uranium sulfate at pH 1.5–3.5 and temperature 35°C (Crueger and Crueger, 1984).

$$UO_2 + Fe_2(SO_4)_3 \longrightarrow UO_2SO_4 + 2FeSO_4$$

Uranium leaching is indirect process. *T. ferrooxidans* does not directly attack on uranium ore, but on the iron oxidant. The pyrite reaction is used for the initial production of $Fe^{3+}$ leach solution.

$$2FeS + H_2O + 7\tfrac{1}{2}\,O_2 \xrightarrow{\textit{T. ferrooxidans}} Fe_2(SO_4)_3 + H_2SO_4$$

(*c*) **Gold and Silver Leaching:** Today's microbial leaching of refractory precious metal ores to enhance gold and silver recovery is one of the most promising applications. Gold is obtained through bioleaching of arsenopyrite/pyrite ore and its cyanidation process. Silver is more readily solubilized than gold during microbial leaching of iron sulfide.

Microbial leaching of copper ore by *T. ferrooxidans.*

(*d*) **Silica Leaching:** Magnesite, bauxite, dolomite and basalt are the ores of silica. Mohanty and coworkers in 1990 isolated *Bacillus licheniformis* from magnesite ore deposits. Later it was shown to be associated with bioleaching, concomitant mineralysis and silican uptake by the bacterium. It was concluded that silican uptake was restricted adsorption of bacterial cell surface rather than internal uptake through the membrane. The bioleaching technology of silica magnesite by using *B. licheniformis* developed at Bose Institute, Calcutta is being used for the first time for commissioning a 5 billion tonnes capacity of pilot plant at Salem Works of Burn, Standard Co. Ltd, Tamil Nadu, in collaboration with the Department of Biotechnology, Govt of India.

## F. HAZARDS OF ENVIRONMENTAL ENGINEERING

The genetically modified microorganisms which are purposefully released into the environment for bioremediation, plant growth stimulation or biocontrol of pathogens and pests of plants and animals either do well as hoped, influence the other harmful microorganisms or themselves changed are still in debate. However, some of performance of GMMs have been discussed in this context. There are several types of risks associated with release of GMMs into the environment (Table 25.5).

**> Table 25.5. Possible types of risks associated with the release of GMMs.**

| *Type* | *Mechanism* | *Results* |
|---|---|---|
| Risk for humans, animals, plants | - Transfer of drug resistance gene into clinical and/or symbiotic strains, | New forms of known diseases |
| | -GMMs useful for food production physiologically active | Toxins and/or unwanted |
| substances in food | | |
| Risk for environment | - GMMs will overgrew the indigenous strain | Disease of biodiversity |
| | - Disturbance of ecological balance pathogen | Activation of earlier unknown |
| | - Massive transfer of foreign genes into indigenous strains | Formation of new pathogen |
| Social risk | - Beneficial usage of the GMMs in industrial countries | Increasing of economical and social differences between industrial and developing countries |
| - territo-ries of developing countries for field tests of GMMs release | GMMs release for military and/ | Industrial countries will use or improper purpose |
| Ethical risk | Commercial secrecy concerning information about release of GMMs | Violation of consumers right |

*Source*: V. Velkov (1996).

## 1. Survival of Released Genetically Modified Microorganisms (GMMs) in the Environment

The GMMs after release into the environment are much influenced by multiple factors. Before release GMMs live in a very ideal conditions, but after release their survival is affected by both abiotic and biotic factors. Due to presence of poor quality of available nutrients their growth and multiplication declined. Substrates accessible to microorganisms in open nature are so limiting that within one year a typical soil microbe could go through 1–36 generations of growth. Therefore, under nutrient starvation conditions, non-differentiating microorganisms enter in a dormant viable state. During this course of time certain changes in redox potential, energy status and composition of proteins occur. It has been found that starvation induced proteins (unfoldases and chaperones), that protect non-growing cells of *E. coli* are produced.

After some days of starvation, microorganisms do not require energy to remain viable but enter into a viable but not culturable (VNC) state. During this period, they can remain viable for years, and retain their plasmid. According to physiological survival strategy of non-differentiating bacteria, there is a very high probability of entering of released GMMs into VNC state. In addition to physiological strategy of survival, the GMMs have genetical strategy of survival as discussed below:

***(a)* Adaptive Mutagenesis in GMMs:** The GMMs released into the environment do not grow even in the presence of potential substrates. It seems that the non-growing cells 'adapt' themselves to the potential substrates present in the environment. It may be assumed that in populations of microorganisms growing and persisting in open environment the random mutagenesis is increasing their biodiversity, and adaptive mutagenesis 'adapt' the non-growing cells to the substrates potentially present in the environment. However, if foreign genes from GMMs spread into the indigenous microbes it may result hazardous effects.

***(b)* The GMMs participate Gene Transfer:** There are four ways of gene transfer in microbial ecosystem *viz*., transformation by free DNAs, transduction by bacteriophage DNAs, conjugation by plasmids, and conjugal transfer by conjugative transposons (for detailed description *see A Text Book of Mirobiology* by Dubey and Maheshwari, 2012). Conjugal gene transfer is of the most considerable ecological importance. In an experiment, the GMMs and natural soil strains were released in sterile soil. Conjugal gene transfer from GMMs to soil strains occurred with a frequency of about $3 \times 10^{-6} - 3 \times 10^{-5}$ of transconjugants per donor. The addition of glucose and/or tryptone enhanced upto $2 \times 10^{-4}$ (Fray and Day, 1990). However, in non-sterile soil natural microorganisms inhibit conjugative gene transfer. On the other hand in the rhizosphere rich in nutrients, effective gene transfer among *Agrobacterium, Pseudomonas* and *Rhizobium* has been observed.

There are abundance of plasmids in aquatic microbial systems. In some lake waters upto 46 per cent of heterotrophs contain plasmids, but in river water 10–15 per cent. Majority of these plasmids are conjugative. The frequency of plasmid transfer in aquatic systems varies from $5 \times 10^{8}$ to $2.5 \times 10^{-3}$ conjugants per donor. Moreover, very effective plasmid transfer was obtained between GMMs and strains of epilithon (aquatic microbial community associated with stones), $10^{-2}$ transconjugants per recipient.

***(c)* Gene Transfer via Conjugative Transposons:** The conjugative transposons are self-transmissible elements that normally are integrated into a chromosome or plasmid but excise themselves and transfer by conjugation to recipients. The conjugative transposons are capable of transferring themselves as well as driving the tramsmission of other elements. For example, conjugative transposons of *Bacteroides* can mobilize co-resident plasmids either in *cis* or *trans* position. The indigenous strains have an efficient capability to transfer their plasmids into GMMs released in soil or aquatic ecosystems.

***(d)* Effect of Environmental Factors on Gene Transfer:** Soil and water polluted with heavy metals and organic xenobiotics are characterized by the presence of many strains with plasmids carrying genes encoding resistance to heavy metals and biodegradation. In such polluted ecosystem, selective pressure could be stimulative for conjugative transfer between GMMs and natural strains. However, the risk of spread of foreign genes by conjugation could be reduced if GMMs do not contain any conjugative plasmids.

## 2. Ecological Impact of GMMs Released into the Environment

The GMMs released into the environment disturb the ecological balance by inhibiting/ promoting growth of indigenous microflora or replacing the natural strains as diacussed below:

***(a)* Growth Inhibition of Natural Strains:** A genetically modified strain of *Klebsiella planticola* constructed to produce ethanol from organic wastes was found (*i*) to destroy mycorrhizal fungi, (*ii*) to reduce plant growth, and (*iii*) to increase the population of parasitic nematodes of plants. A genetically modified strain of *P. putida* pR103 degraded in soil microcosm the herbicide 2,4-D (2,4-dichlorophenoxyacetic acid) and resulted in accumulation of 2,4-DCP (2,4-dichlorophenol) which in turn is toxic metabolite. The 2,4-DCP was more toxic than 2,4-D as the former reduced the rate of $CO_2$ evolution of soil microflora as compared to 2,4-D.

***(b)* Growth Stimulation of Indigenous Strains:** The GMMs strain of *Erwinia carotovora* affected indigenous bacterial community. The total bacterial density significantly increased. This increase is attributed to an inoculum nutrient effect. The inoculated cells of *E. carotovora* died and became a source of nutrients for indigenous soil microorganisms. More interesting finding is that the genetically modified *P. cepacia* released in soil microcosm containing the pollutant increased the taxonomic diversity of soil microrganisms. This is because the intermediate metabolites of genetically modified *P. cepacia* produced after biodegradation of pollutant which stimulated the growth of static cells. This has resulted in increased taxonomic diversity. Stimulation of plasmid transfer and recombination might cause an increase in genetic diversity.

A strain of *Bacillus subtilis* used as biocontrol agent, inspite of decreasing the pathogens' inoculum, induced sexual reproduction in *Sclerotium rolfsii* which might help in survival of the pathogen. Similarly, *Trichoderma viride* has been found to induce oospore formation in *Phytophthora infestans*, the causal agent of late blight of potato.

**(*c*) Replacement of Natural Strains:** In 1992, H. Bolton and coworkers reported that a strain of *Pseudomonas* PC1 released in a wheat root system replaced efficiently the indigenous pseudomonads in their ecological niche. A genetically modified strain of *Agraobacterium tumifaciens*, with the deleted genes of tumour formation replaced the initial wild strain and protected plants from tumour formation. This ability of GMMs is useful in agricultural biotechnology for protection of plants against pathogens.

## 3. Monitoring of GEMs in the Environment

In a microbial community, diverse group of microorganisms are found. It is now possible to monitor individual microbe in a community. Sensitive and selectable trackling tools *e.g. ice*, *lac*ZY, *xyl*E, *lux* systems are used in monitoring studies. Nucleic acid hybridization is also useful method where rRNA specific probes can identify the specific microbial strains of interest. Bioluminiscent reporter gene has potential to identify the specific microbe, for example, enzyme luciferase acts on substrate luciferin. In addition, genes capable of giving resistance against certain antibiotics have also been used in environmental monitoring as these detoxify antibiotics such as gentamycin, ampicillin, tetracyclin, kanamycin, steptomycin, etc.

**(*a*) Risk Assessment of the GMMs Released into the Environment:** When GMMs are released in the environment, they get dispersed and it becomes difficult to eliminate them. Relatively few small scale field tests have been conducted with genetically modified bacteria (10 strains) and genetically modified viruses (5 strains). Most of the European countries, USA and Japan have Environmental Regulations that ban the deliberate release of GMMs. But the USA has permitted deliberate release of some of the GMMs. No country has defined what steps will be required to obtain permission for commercial use of GMMs for *in situ* bioremediation.

In July 1995, European parliament passed a resolution saying that "a legally binding international biosafety protocol is necessary". It must be negotiated by States partly on the United Nations' convention on biological diversity. Therefore, resolution of deliberate release of GMMs are being carried out in many developing countries for safe use of GMMs. Otherwise the whole biosphere will be at risk. India has formulated Biosafety Rules and Regulations for GMMs as discussed in Chapter 1. For stable development of Indian biotechnology, safety of environmental engineering is very actual.

There are two principal uncertainties of the assessment of risks associated with possible ecological consequences of release of GMMs: (*i*) total number of hazardous strains present in the environment as only 5% of total species are known and rest of 95% unknown, and (*ii*) absence of definite criteria for evaluation of similarity between microorganisms. However, estimates of similarity made by different experts may vary drastically. Moreover, there are certain questions unasked which may possibly help in assessment of risk arising from release of GMMs in the environment.

## PROBLEMS

1. What is bioremediation ? In what ways it is good tool for environmental clean-up ?
2. Write an extended note on *in situ* bioremediation with examples.
3. Write an essay on *ex situ* bioremediation with special reference to composting.
4. Discuss in detail slurry-phase treatments by using different approaches.
5. What are the industrial wastes ? Write in detail their bioremediation methods.
6. Write an essay on bioremediation of xenobiotics.
7. What do you know about bioleaching ? Write in detail bioleaching of copper by giving methods and mechanisms of bioleaching.
8. What is biogas ? In what ways biogas can be produced on sewage ?
10. Write an essay on production of single—cell protein on sewage water.
11. How will you utilize agro-wastes for production of edible mushrooms ?
12. What are the hazards associated with release of GMMs in the environment ? How will you monitor the GMMs in the environment ?
13. Write short notes on the following:

    (*i*) *In situ* bioremediation, (*ii*) *Ex situ* bioremediation, (*iii*) Intrinsic bioremediation, (*iv*) Engineered bioremediation, (*v*) Bioremediation of hydrocarbons, (*vi*) Bioremediation of xenobiotics, (*vii*) Composting, (*viii*) Vermicomposting, (*ix*) Microbial leaching (bioleaching), (*x*) Mushroom production on agro-wastes, (*xi*) Production of single-cell protein on sewage water, (*xii*) Hazards of GMMs released in environment, (*xiv*) GMMs monitoring in the environment, (*xv*) Bioremediation in paper and pulp industry, (*xvi*) Biosorption, (*xvii*) Bioaugmentation, (*xviii*) Biofiltration, (*xix*) Vermicomposting
14. What is biosorption ? Wirte organisms used in biosorption technologies.
15. Discuss in detail the bioaugmentation of industrial wastes.
16. Give a detailed account of biofiltration biofilter media and mechanism of biofiltration.

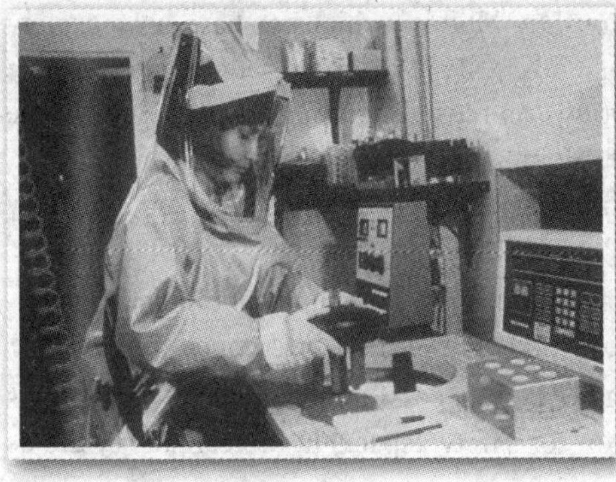

CHAPTER

# 26

# Biotechnology and Biosafety, Intellectual Property Right (IPR) and Protection (IPP)

There is a growing concern about the GMOs. The divers questions on safety, ethics and welfare are associated with genetic manipulation. Release of GMOs in the environment and their long-term effects have been well realised. Many countries and NGOs have opposed the release of GMOs.

The most important example of A.M. Chakrabarty's '*superbug*' may be quoted in this regard. He prepared genetically modified *Pseudomonas putida* and called superbug which had potential for degrading oil spills. Then a case was filed against him in the Court. After winning from Supreme Cout in 1992, the U.S. Government allowed him to treat oil spills by using superbug.

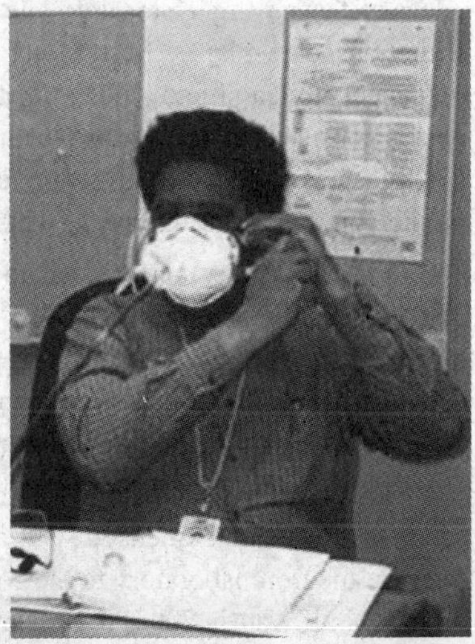

A device for respiratory protection.

## A. BIOSAFETY

Due to growing concerns arising from GMOs throughout the world the UNIDO/WHO/FAO/UNEP has built up an Informal Working Group on Biosafety. In 1991, this group prepared the *" Voluntary Code of Conduct for the Release of Organisms into the Environment"*. The ICGEB has also played an important role in issue related to biosafety and the environmentally sustainable use of biotechnology. The ICGEB organises annual workshops on biosafety and on risk assessment for the release of GMOs. It collaborates with the management of UNIDO's BINAS (Biosafety Information Network and Advisory Service), aimed at monitoring the global development in regulatory issues in biotechnology.

Since September 1998, the ICGEB has provided an on-line bibliographic data-base on biosafety and risk assessment for the environmental release of GMOs. This database which is accessible through the website of ICGEB also provides informations on biosafety to its Member States. The ICGEB is also assisting to its Member States in developing the national biosafety framework. Since February 1999, it has also adopted a legally binding biosafety protocols by the signatory countries (ICGEB, Activity Report, 1998).

Biotechnology is safe when practiced properly. Benefits deriving from biotechnological innovations will lead to major improvements in the health and well being of the world population.

The potential risks of biotechnology are manageable. For successful management, regulations have been constructed. The main areas of consideration for safety aspects in biotechnology are as below:

(*i*) Pathogenicity of living organisms and viruses (natural and genetically engineered) to infects humans, animals and plants to cause disease

(*ii*) Toxicity of allergy-associated with microbial production

(*iii*) Increasing number of antibiotic resistant pathogenic microorganisms

(*iv*) Problems associated with the disposal of spent microbial biomass and purification of effluents from biotechnological processes

(*v*) Safety aspects associated with contamination, infection or mutation of process strains

(*vi*) Safety aspects associated with the industrial use of microorganisms containing *in vitro* ricombinants.

## 1. Biosafety Guidelines and Regulations

Many countries have formulated the Biosafety Guidelines for rDNA manipulation with the aims: (*i*) to minimise the probability of occasional release of GMMs, and (*ii*) to ban the deliberate release of such organisms into the environment. There are high risks and benefits associated with transgenic plants. In 1999, a British medical journal published about the adverse effect of genetically modified (GM) potato (produced by Rowett Research Institute, Scotland). This potato consisted of snow drop lectin which posed adverse affect on small intestine of rats. The GM-potatoes stunted the growth and damaged immune system of the rat. This led to world-wide public concern about this issue. This sparked much controversy on the safety of GM-foodstuffs.

In India, there was much debate on cultivation of *Bt* transgenic cotton in Gujarat in 2001. The *Bt*-cotton was prepared for its cultivation mainly in Andhra Pradesh. But the crop of *Bt*- cotton grown in Gujarat was infected by bollworm which resulted in severe loss. The *Bt* cotton is a transgenic (*i.e.* genetically modified) plant which possesses a toxin producing gene (*Bt*) of *Bacillus thuringiensis.* Upon ingestion of this toxin, bollworms die.

Similarly, the attenuated virus (making virus less virulent by chemical treatment) are used as vaccines. Risks are associated with possible conversion of attenuated virus into virulent form and their escape from laboratory to the environment. Growing tension between America and Iraq ended with $2^{nd}$ Gulf War (in 2003) in spite of global protest. America and Britain emphasised for storage of biological weapon by Iraq in the form of millions of litres of botulism toxin, bacteria, etc. to which Iraq denied. However, if the microbial products are with Iraq or any other countries possibly prepared for 'warfare', it will certainly have a threat to the society for spread of *ha-voc.* Over the issue of GMOs and biotech a hot debate is taking place throughout the world.

In India, DBT has evolved "the recombinant DNA safety guidelines" to excercise powers conferred through the Environmental Protection Act 1986 (modified in December 1989) for the manufacture, use, import, export and storage of hazardous microorganisms/genetically engineered organisms, cell, etc.These guidelines are being implemented through the following three mechanisms: (*i*) the institutional biosafety committees (IBSCs) monitors the research activity at institutional level, (*ii*) the review committee on genetic manipulation (RCGM) functioning in the DBT which allows the risky research activities in the laboratories, and (*iii*) the genetic engineering approval committee (GEAC) of the Ministry of Environment and Forest has the power to permit large scale use of GMOs at commercial level, and open field trials of transgenic materials including agricultural crops, industrial products, health care products, etc. (DBT, Annual Report 1995-96).

Bt. Cotton.

Under the EPA 1986, the GEAC examines from the viewpoint of environmental safety and issues, clearance or the release of GMO/transgenic crops and products into the environment. The GRAS (generally regarded as safe) also grants permits to conduct experimental and large-scale field trials which are beyond the limit of 20 acres. In case of transgenic crops, applicants are required to seek clearance from the Ministry of Agriculture which will accord the final clearance of transgenic crop varieties for agricultural activites (Fig. 26.1).

In March 2002, *Bt* cotton was release for commercial cultivation following biosafety procedures laid down by the Govt. of India. Three cotton hybrids have been granted permission for field sowing and are currently in fields in six states: Maharastra, Gujarat, Madhya Pradesh, Andhra Pradesh, Karnataka and Tamil Nadu. Transgenic mustard hybrids are also in advanced stage of field trials.

## 2. Operation of Biosafety Guidelines and Regulations

All the institutions/industries working on genetic engineering activities have IBSCs. Moreover, with the permission of RCGM research activities are being carried out in the country using transgenic materials. A mechanism based on interaction between committees and different departments of the Govt. of India has been set up. (Fig. 26.1). Such materials will have to meet with the approval of the IBSC, RCGM and GEAC.

Some of the examples are given below:

(*i*) M/S Proagro PGS India Ltd had imported the transgenic mustard seeds expressing Barstar and Barnase genes from Belgium to evaluate the performance of seeds in Indian soils. This company has also imported transgenic seeds of tomato containing *B. thuringiensis* (Bt) CRYIA$^{(b)}$ gene to assess the resistance of cultivars containing transformed potato plants with the above gene to specific tomato fruitworms of India.

(*ii*) M/S MAHYCO, Mumbai imported seeds of transgenic cotton containing *Bt* gene to conduct trials in glasshouse by back crossing with Indian cotton lines and to evaluate resistance of the transgenic plants to bollworms in India.

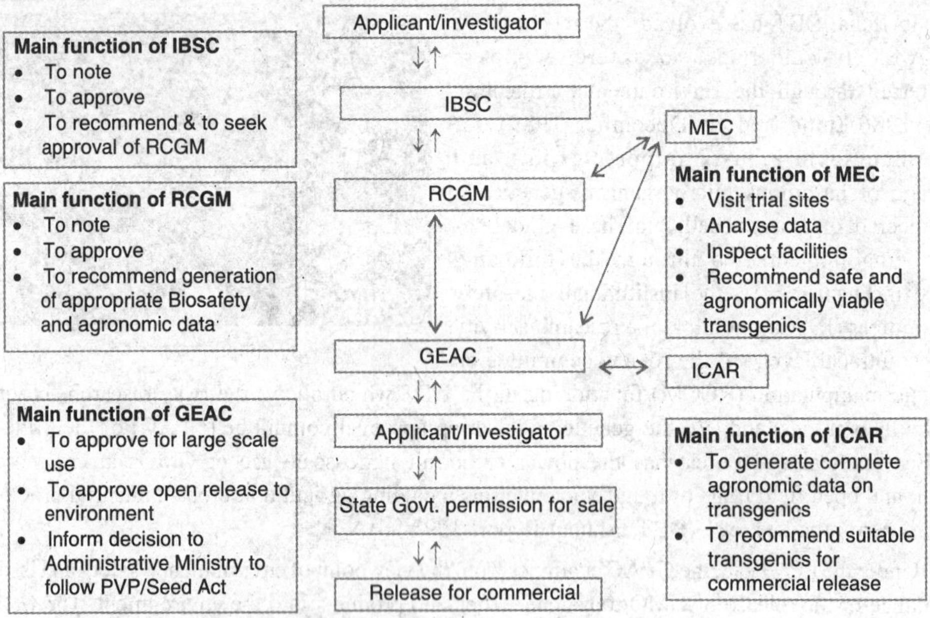


**Fig. 26.1.** Procedures involved in commercialization of indigenously developed transgenic crops (Source: Manju Sharma *et al.*, 2003).

(*iii*) M/S Transgene Vaccines Ltd, Hyderabad imported the recombinant strains of yeast expressing Hepatitis B surface antigen protein. The company was granted permission to carry out experiments for small scale production of vaccines at IMTECH, Chandigarh.

(*iv*) M/S Santha Biotechnics Ltd, Hyderabad conducted experiments for the production of recombinant Hepatitis B vaccines and human $\alpha$-interferon by using yeast and *E. coli*, respectively.

In Asia, India and China are the two major producers of genetically modified products. Currently India grows only GM cotton, while China produces GM varieties of cotton, poplar, petunia, tomato, papaya and sweet pepper. In China, GM crops go through three phases of field trials (pilot field testing, environmental release testing, and preproduction testing) before being submitted for assessment to the Office of Agricultural Genetic Engineering Biosafety Administration (OAGEBA).

The release of transgenic crops in India is governed by the Indian Environment Protection Act, which was enacted in 1986. The Institutional Biosafety Committee (IBSC), Review Committee on Genetic Manipulation (RCGM) and Genetic Engineering Approval Committee (GEAC) all review any genetically modified organism to be released, with transgenic crops that also need permission from the Ministry of Agriculture.

In October 2009, Indian regulators cleared the Bt brinjal (a genetically modified eggplant) for commercialisation. In February 2010, a moratorium was imposed on its release due to opposition from some scientists, farmers and environmental groups.

In 2011, some of the other Asian countries that grew GM crops were Pakistan, the Philippines and Myanmar. Japan requires labelling to help the consumers who may have choice between foods produced through genetically modified, conventional and organic farming methods.

## B. INTELLECTUAL PROPERTY RIGHT (IPR) AND PROTECTION (IPP)

Generally, the physical objects such as household goods or land are the properties of a person. Similarly, a country has its own property. The ownership and rights on the property of a person is protected by certain laws operating in the country. This type of physical property is tangible. On the other hand, the transformed microorganisms, plants and animals and technologies for the production of commercial products are exclusively the property of the intellectuals. The discoverer has the full rights on his property. It should not be neglected by the others without legal permision. The right of intellectuals must be protected and it does by certain laws framed by a country. However, it is important to distinguish between the physical property and intellectual property. For example, seed of a plant is tangible asset; it can be sold in market and money can be made from it. But the intellectual property is intangible asset. Legal rights or patents provide an inventor only a temporary monopoly on the use of an invention , in return for disclosing the knowledge to the others in a specification that is intended to be both comprehensive to, and experimentally reproducible by a person skilled in the art. Others in society may use the knowledge to develop further inventions and innovations.

The laws are formulated time to time at national and international levels. The USA has declared for adopting a strong and uniform IPR laws throughout the world. Development of crop varieties is the others intellectual property right. It is protected by 'plant breeders rights' (PBRs). The PBRs are available in developed countries but not in India. The principle of PBRs recognizes the pact that farmers and rural communities have contributed a lot to the creation, conservation, exchange and knowledge of genetic/species utilization of genetic diversity. The IPR and IPP granted by the government to plant breeders are to exclude others for about 15-20 years from producing or commercializing materials of a specific plant variety. But this variety should be new and never existing before.

Intellectual Property Rights.

Biotechnology has played a significant role in providing processing, designing and production of valuable commercial products utilizable in many area of the society as well as the country (*e.g.* agriculture, medical, health care, industry, environment, etc.). Technology transfer in biotechnology requires a minimum amount of technical and legal capability which the developing countries lack at present. Therefore, manpower in such country must be trained. The gene-rich/technology-poor developing countries must come out together and reach to an understanding to help in various mutual programmes.

### 1. Forms of Protection

The IPR is protected by different ways: patents, copyrights, trade secrets and trademarks, designs, geographical indications.

**(*a*) Patents:** Patent is a special right to the inventor that has been granted by the Government through legislation for trading new articles. A patent is a personal property which can be lisenced or sold by the person/organisation just like any other property. For example, Alexander Grahm

Bell obtained patent for his telephone. This gave him the power to prevent engine from making or using or selling a telephone elsewhere. In some European countries the monopoly rights were granted only to the inventors so that they may develop new articles beneficial for the society. In the USA the maximum limit of this monopoly is for 17 years. In India, the Indian Patent Act (1970) allows the 'process patents' but not the 'product patent', and the maximum duration of patent is for 5 years from the date of grant, and 7 years from the date of filing the patent application. The least duration between five and seven years is applicable for patents. The conditions of patents are given in the preceding section.

The patents in terms give the inventor the rights to exclude the others from making, using or selling his invention as disclosed in 'claims' of the patent. Obviously, it is difficult to keep secret the certain inventions such as the fermentation process.. Therefore, guidance should be obtained from a qualified patent attorney.

A patent consists of three parts; the grant, specifications and claims. The **grant** is filled at the patent office which is not published. It is a signed document which is actually the agreement that grants patent right to the inventor. However, the **specification** and claims are published as a single document which is made public at a minimum charge from the patent office. The specification part is narrative in which the subject matter of invention is described how the invention was carried out. The **claim** section specifically defines the scope of the invention to be protected by the patent to which the others may not practice.

The most important issue of discussion is the operation of State or Federal patent laws. For example, Food and Drug Administration of the US has regulatory purview on patented pharmaceuticals before permitting for clinical use. Similarly, the Environment Protection Agency of the USA working under the Federal Insecticide, Fungicide and Rodenticide Act permits the release of genetically engineered microbial pesticides. The local 'nuisance ordinances' can minimize the excessive use of genetically engineered inventions. In India, DBT has formulated 'the recombinant DNA safety guidelines' to excercise powers conferred through Environmental Protection Act (1986). The 'genetic engineering approval committee (GEAC) of the Ministry of Environment and Forest has the powers to allow large scale use of GEMs at commercial level, and open field trials of transgenic materials (*see* preceding section).

(*i*) **Reading a Patent:** In order to file a patent, the documents required should have a specialised structure. The inventors (s) must decide first that the patent is to be filed at National or International office. But normally, an applicant must file a patent first in his/her own country where he/she resides. Then it may be filed abroad on a later date in the international office. Thereafter, the specifications needs to be prepared according to the set guidelines. The structure of patents has two parts:

- **Description:** Basically it is the technical component which describes the: (*i*) field of invention, (*ii*) objectives, (*iii*) previous attempts made so far, (*iv*) solution solved through this possible variations and specific description. It is published as a single document and made public from the patent office with minimum charges.
- **Claims:** Claims are the legal documents. It defines the scope of invention to be protected through this patent so that the others may not use it. The scope of protection may be in terms of apparatus or instruments, prdoducts or process and their uses.

(*ii*) **Patenting strategies:** The application of a patent is prepared with a specific, clear and concise title. Novelty of product or process designed is mentioned. A patent attorney is appointed for the legal aspects of the patent. Then the patent is filed in the office of the Controller of Patents. After getting the grant, the patent comes into enforcement (Fig. 26.2). The patent is in the form of letter which contains the name of inventor, the name of patentee, a description of patent and the relevant claims.

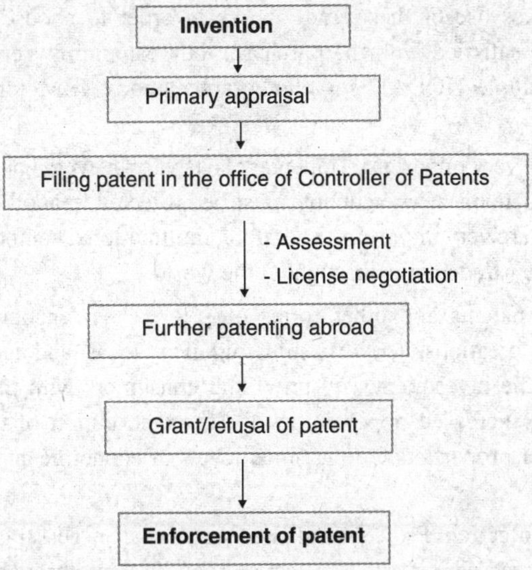


**Fig. 26.2.** Steps of patenting.

**(*b*) Copyrights:** The copyright protection is only a form of expression of ideas. One of the best example of copyrights is the books. The authors, editors, publishers or both publisher and author/editor have copyrights. The materials of the book cannot be reprinted or reproduced without written permission from copyright holders. However, it should be clear that patents and trade secrets provide protection of only basic knowhow, whereas copyrights protect the expressed materials viz., materials in printed, video-recorded or taped forms. Biotechnological materials subject to copyright include database of DNA sequence or any published forms, photomicrographs, etc.

Copyright for publication.

In India, Copyright Act 1957 was amended in 1994 and brought enforce in 1999. It includes computer programme, tables and databases. The computer programme is defined as ' set of works expressed in words, codes, schemes, or any other form including a machine readable medium capable of causing computer to perform a particular task or achieve a particular result'. In India, the Ministry of Human Resource and Development look after the Copyright Act.

**(*c*) Trade Secrets (Knowhow):** The private proprietary information that benefits the owners is called trade secret. It may be of any type, from process to product yield. The most popular example is Coca Cola that has covered its best kept secrets of its formula under this law. It is surprising to know that India does not have trade secrets.Therefore, it allows any company to register and protect the details of formulae. Usually, a patent runs out for 10-20 years, but under the law of trade secrets a company will have no obligation to reveal the trade secrets. In India, the limit of trade secrets is at least five years and two years in the USA. If the trade secrets become public before the granted period, the intellectual is paid compensation and unauthorised users are punished by the Court. The trade secrets in the area of biotechnology may comprise of hybridization conditions, cell lines, processing, designing, consumer's list, etc.

**(*d*) Trademarks :** A trade mark is an identification symbol which is used in the course of trade to enable the public to distinguish on trader's goods from the similar goods of the other traders. The public makes use of these trade works in order to choose whose goods they will have to buy. If they are satisfied with the purchase, they can simply repeat their order by using the trade mark, for example KODAK for photography goods, IBM for computers, Zodiac for readymade cloths, etc.

Trademark rights are very important. Trademark laws vary in every country. Through agreement it is ensured that the trademark of one country must be protected in another country. India's Trade and Merchandise Rights are very important. Therefore, multinational companies spend large amount of money to maintain their trademarks throughout the world.

The advantages of patents and other forms of IPR are: (*i*) encouraging and safeguarding intellectual and artistic creations, (*ii*) disseminating new ideas and technologies quickly and widely, (*iii*) promoting the investment, (*iv*) providing consumers with the result of creation and invention, (*v*) providing increased opportunities for the distribution of the above effects across the countries in a manner proportionate to national levels of economic and industrial development (OECD, Paris, 1989).

**(*e*) Plant Variety Protection:** Protection of plant varieties are necessary in the area of agricultural biotechnology. As discussed earlier, such protection may be done through 'Plant Breeders Right' (also called plant variety rights). The PBRs are available in developed countries but not in India. During 2001, the farmer's rights in relation to breeding of new varieties was realized. Therefore, the following two major steps were taken for consideration under PBRs:

(*i*) The Food and Agriculture Organisation (FAO) has an International treaty on plant genetic resources for food and agriculture. This treaty consists of a particular classes which refers to operation of farmer's rights.

(*ii*) The 'Plant Varietal Protection and Farmer's Rights' Act 2001 agrees for the right of farmers, breeders and researchers. It also protects the interest of public. The protection is provided by making compulsory licensing of rights, and inhibiting the import of plant varieties consisting of 'genetic use restriction technology' (GURT) e.g. terminator technology of Monsanto.

The principles of PBRs recognize the pact that the farmers and rural community have contirbuted a lot to the creation, conservation, exchange and knowledege of genetic/species utilization of genetic diversity. The IPR and IPP granted by the Government to plant breeders are for about 10–20 years so that the others may be excluded. The others could not produce and commericalize materials or a specific plant variety. But this variery should be new and never existing before. Through the national and international agreements the plant breeders are given limited monopoly rights over the varieties which they have created by the way of a registration system for plant varieties. Those who use the seeds or plant will pay the royalty to the breeders.

However, for granting protection the new varieties should fulfill some of the conditions: (*i*) the new variety must always be new *i.e.,* it should not have ever been exploited commercially, (*ii*) it must always be biologically *distinct* and possess different characters, (*iii*) the characters of the plants of new variety must be *uniform,* (*iv*) the distinguishing character of new variety must be *stable* from generations. If must be changed in any of generation, and (*v*) The new variety must have denomination. It must have taxonomic validity, *i.e.,* systematic position, generic and species names, etc.

In 1991, an international debate on 'Biotechnology and Farmer's Rights was organized in Armsterdam (the Netherlands). The recommendations made on the concept of farmers rights explains that: (*i*) PBRs must be protected, (*ii*) Patent legislation should not be extended to generate materials, (*iii*) There should be legal sanction, (*iv*) An 'International Gene Fund' must be established and the member country should pay to this fund, and (*v*) The new variety must have denomination. It must have taxonomic validity i.e. systematic position, generic and species names, etc.

## 2. World Intellectual Property Organisation (WIPO)

The WIPO is one of the specialized agencies of the United Nations. It has provided that the intellectual property shall include rights relating to the following:

(*i*) Literary, artistic and scientific works, performance of artists, phonograms, broadcast; innovation in all fields of human endeavor; scientific discoveries; trade marks; service marks and commercial names; industrial designs; protection against unfair competition and all other rights resulting from intellectual activity in the area of industrial, scientific, literary or artistic fields.

(*ii*) The intellectual property is protected by and governed by appropriate national legislation. The national legislation specifically describes the inventions which are the subject matter of protection and those which are excluded from a protection, for example, methods of treatment of humans or therapy and invention whose use would be contrary to law or invention which are injurious to public health are excluded from patentability in the Indian legislation.

## 3. General Agreement of Tariffs and Trade (GATT) and Trade Related IPRs (TRIPs)

The GATT was framed in 1948 by developed countries to settle the disputes among the countries regarding share of world trade. It is decided by tariffs rates and quantitative restrictions on imports and exports. For a long time benefits from GATT was achieved only by developed countries. In 1988, the US Congress enacted a law ' the Omnibus Trade and Competitiveness Act (OTCA). As a result of which the USA became powerful to investigate the laws related to trade and check them if not beneficial to its interest. After warning, if the investigated country does not change its law within the desired period, the US takes action against that country. In 1992, the US gave warning to India to change some of its laws of IPR, patents and copyrights. India had certain inhibitions to sign on GATT draft. Therefore, there was much debate throughout the country on this issue and bad intension of the US. Professionals, politicians and scientists argued that the total package of TRIPs must guarantee for economic and technological subjugation of the country.

The then Director General of GATT, A. Dunkel came to India for discussion on this issue. Certain provisions were suggested to include in GATT draft that India will not give any kind of subsidies for the production of oilseeds and pulses as the international price is more than that in India. India assured to change its patent laws by 2003. In the changed patent, it will introduce the 'product patent' and enhance patent duration. It will open the market to foreign patent holders also, and open the agriculture to patent technology. Even there are several groups and organisations that have rejected this draft of suggestion and opposed the decision taken by the government.

## 4. Patent Status - International Scenario

Patenting process varies form country to country. Therefore, it is effective only within the jurisdiction of that country which grants patent. For example, a biotechnology company Genetech applied for patent in the U.K. in 1989, for prodcution of human tissue plasminogen activator (tPA) by recombinant DNA process. It was rejected by the U.K. Patent Office. In contrast, Genetech was readily awarded a patent for human tPA in the U.S.A. and also in Japan. Benefits and disadvantages of patent systems are given in Table. 26.1.

> **Table 26.1.** Benefits and disadvantage of patent system.

| *Benefits* | *Disadvantage* |
|---|---|
| 1. The patent holder retains an absolute monopoly on product or process for the period of patent (up to 20 years in some cases) | 1. Knowledge is an public domain following expiry and could be valuable to competitors |
| 2. Administration of patent maintenance once has been obtained, it becomes relatively easy | 2. Litigation can be expensive |
| 3. There is least chance of misuse of the things which have been covered under the patent | 3. Things which have not been covered by patent could be misused |

In spite of such limitation, there has been an international cooperation over centuries. The cooperation is extended by means of international conventions. Conventions between the member countries regulate the formalities regarding patent operation. Some of the important conventions are discussed below:

(*a*) **Paris Convention:** Paris convention is one of the oldest convention which was signed in 1833. It is also called the 'International Convention for the Protection of Industrial Property'. The Majority of industrialized countries are the members of the Paris Union and the member accounts 151 Member States. As far as protection of the industrial property is concerned the member states equally treat the nationals of the other member states of the union and their own. The Paris Convention includes one of the important articles to which the member states agree. This actical recognizes the first filing of a patent application in any member state and permits the priority rights fromt the date of filing.

Similar rights will also be accorded to corresponding patent application in any other member state provided the patent application has been filed within 12 months of the first filing date. In Paris Convention 1988, the basic principles of equal treatment for domestic and foreign investors were established. Several modifications have been done in the text of Paris Convention. In 1979, the last convention was done at Stockholm.

(*b*) **Union International Pour Ia Protection des Obtentions Vegetales (UPOV):** The UPOV Convention was signed in 1961. International hromonization of Plant Breeders Rights laws have been achieved through the UPOV. At present there are 24 Member States of which India is not a member of UPOV.

(*c*) **Strasbourg Convention:** In November 1963, a convention was held at Strasbourg on the unification of certain points of substantive law on patents for invention. In this convention, plant and animal varieties have been excluded from patent protection. The patent laws formulated during their convention need the basic requirement before an invention is to be patented. For example, the invention must be: (*i*) *new,* (*ii*) *inventive,* and (*iii*) *industrially applicable.*

Moreover, some of the important features of this convention have been given in the 'European Patent Convention'.

(*d*) **Patent Co-operation Treaty (PCT):** The PCT was signed by the member states in 1970, but it came into effect from June 1978. At the same time, the European Patent Convention was implemented. Up to January 1999, it had 100 member states. Therefore, it has a scope internationally. However, the main instrument of international collaboration for intellectual property is the 'World Intellectual Property Oragnisation' (WIPO) based in Germany. The WIPO administers the PCT as well as Paris Covention. The patents granted by the PCT are treated as International patent. Because the patent applications are processed by the WIPO which is an intellectual body and one of the specialised agencies of the United Nations. The WIPO processes the patent application in an International phase'. The international phase has concerns with: (*i*) formal preliminaries, (*ii*) art search, and (*iii*) publication of patent application.

(*e*) **European Patent Convention (EPC):** The EPC was signed in October 1973 which started operation from June, 1978. The legal intity of EPC is called the European Patent Organisation (EPO) which is situated in Munich. The EPO has its two bodies: (*i*) the European Patent Office and an Administrative Council. The patent applications are submitted in any of the National Patent Offices of the member states. Finally, the EPO processes the patent applications filed by the member states.

After grant, the patent becomes the European Patent, and called as a 'bundle of national patents' in each member state as such. Then it is designated as the European Patent (U.K.), Euripean Patent (France), and so on. The EPC is credited to introduce to its patent statute the provisions for biotechnology inventions. The two important provisions are given as below:

(*i*) the need of 'Culture Collections' as patent depositories for the placement of microorganisms referred to in patent application, and

(*ii*) exclusion of certain inventions of plants and animal research through classical methods, from the list of patents.

(*f*) **Budapest Treaty:** The Budapest treaty was singned in 1977 and started operation form 1980. At present it has 45 member states. The purpose of convention of Budapest Treaty was to provide 'International Recognition of Microorganisms' for the purpose of patent procedure. If a new genus/species/strain of a microorganism is discovered or created, its culture is deposited at 'International Depository Authorities' in any member state. The strains which can serve the purpose of scientific community or society are accepted. These can be procured if required in future to work in industry or research. The strains which are deposited are alloted an accession number, accession date, source of isolation and specific function.

(*g*) **Organisation for Economic Cooperation and Development (OECD):** In December 1960, the OECD was set up under the convention in Paris to promote the policies designed:

(*i*) to achieve the highest sustainable economic growth and development and a rising standard of living in Member states (now 24 member), while maintaining financial stability and thus to contribute to the development of the world economy,

(*ii*) to conribute the sound economic expansion in Member as well as non-member countries in the process of economic development,

(*iii*) to contribute the expansion of world trade on a multilateral, non-discriminatory basis in accordance with international obligations.

## 5. Patenting of Biological Materials

As discussed earlier that different countries have different patent laws which are changed with advent of time. For example, Dr. Anand Mohan Chakrabarty (an India born American scientist) created a superbug by using a bacterium, *Pseudomonas* which eats upon oils. His superbug could not be patented because the existing US laws before 1980 did not permit to patent the liveforms. Lateron the patent laws were amended. In 1988 in the US, a patent was issued to genetically engineered mouse 'oncomouse' (containing human cancer gene) which is again a liveform. Dr. Chakrabarty filed a case in the US Supreme Court and won it. In 1990, the US government allowed him to treat oil spills by using *Pseudomonas*-based superbug.

The biological inventions are patented similar to the chemical ones. Following are the examples of biological inventions which can be patented:

(*i*) **Biological Processes:** Such processes producing useful products like biochemicals, secondary metabolites, proteins, enzymes, etc.

(*ii*) **The Products:** It includes chemical or new organisms isolated or genetically modified.

(*iii*) **Composition:** For specific uses products are formulated.

(*iv*) **New Applications:** It includes exploitation of microorganisms to produce antibiotics, etc.

(*v*) **Treatment Methods:** The methods for industrial products, plants, animals, etc.

Similarly, the examples of chemical inventions which can be patented are: (*i*) composition of new compounds developed for use in special cases (*e.g.* pesticides, pharmaceuticals, flavour chemicals, etc.), (*ii*) new compounds of practical application (*e.g.* dyes, additives, polymers, etc.), (*iii*) new uses of chemicals (*e.g.* herbicides, insecticides, fungicides, process claims, etc.), (*iv*) new process (to produce novel products through using reagents, catalysts, reaction conditions or integrating multiple steps), (*v*) new methods of treatment (of industrial products, non-therapeutics, etc.).

(*a*) **Products Patents- Its Importance to Inventors:** The products has much importance in the industrial sector. It may be exemplified by penicillin which was first discovered in Britain by Alexander Fleming in 1929. During World War II significant amount of penicillin was required. England had no money for penicillin production. Florey and Heatley, the co-workers with Fleming, shifted to the U.S.A. and granted aid from American Government for penicillin production. Thus the penicillin available during World War II for saving lives of soldiers was produced by Anglo-American collaborative research. The U.S.A. dominated the proprietary fermentation technology. Britain did not get any benefit form this collaborative effort.

Thereafter, 6-aminopenicillinic acid (6-APA) was discovered and patented by Beecham (U.K.). The 6-APA could be isolated by using the enzyme, penecillin G acylase. The 6-APA is used to prepare semi-synthetic penicillin, for example ampicillin. The product patented for 6-APA earned royalty for further development in industry.

(*b*) **Conditions for Patenting:** There has been a debate on the patentable articles and conditions related with them. It is not like that every discovery can be granted patents. Discovery cannot be patented because the discovered article is the product of nature. Yes, the process or techniques used to discover the nature's product may be granted patents. Therefore, patent laws differentiate between discovery and invention, and allow patenting of inventions but not discoveries. The European Patent Office (EPO) has given suggestions that the process developed to isolate the products from nature is patentable. If the product is new and does not have previously existing recognition (*e.g.* microbial metabolites, antibiotics, alcohols, organic acids, vitamins, enzymes, etc), it is patantable. Therefore, the specific conditions for patent application should be such that qualify for patent *i.e.* (*i*) the invention must have novelty and utility for the society, (*ii*) the product must be inventive *i.e.* skill has been applied to it, (*iii*) it must be reproducible (will give similar result after repetition) and disclosed, (*iv*) scope of protection should be in proportion to the invention, and (*v*) it must be patentable.

Before filing the patent application, the inventor must deposit a sample of officially approved material declaring that it is free from dispute of novelty and can be used by others when becomes legally free. Moreover, the application may be withdrawn before the grant of patent.

(*c*) **Patenting of Liveforms:** As discussed earlier, EPO has suggested to patent the genetically engineered liveforms. Also oncomouse is one of the examples of which initially the patent claims was rejected but on appeal the previous decision was overruled. Similarly, genetically engineered *E. coli* in which human genes for insulin, growth hormone, tPA, etc. have been introduced, have been patented in the USA. Likewise, transgenic herbicide- and bollworm-resistant cotton, and insect-resistant tobacco have been granted patents. Several countries (such as Japan, USA, Europe, etc.) have modified the patent laws stating that the transgenic plants and animals can be protected through patent claims.

## 6. Significance of Patents in India

The Indian Patents Act (1970) emphasises that any patentable commodity must possess novelty.

Government of India introduced the Patent Amendment Ordinance in January 1999 to meet the part of obligations under 'Trade Related aspects of Intellectual Property Rights' (TRIPS). The main features of the amendment were to provide for:

(*i*) a mail box facility to eanble filing of product patents in the areas of drugs, pharmaceuticals on or after 1st January 1995, and

(*ii*) an opinion for obtaining 'Exclusive Market Rights' (EMR) in drugs, pharmaceuticals and agro-chemicals.

Indian scientists now may first file their patent application anywhere in the world without first filing in India. After taking this priority further application may be filed in other countries including India.

Regarding patent system, the Government of India took significant decisions. On 7th December 1998, the Paris Convention and Patent Cooperation Treaty (PCT) was accepted by India. For the purpose of claiming priority of patents, on 28th December 1998 the Government notified 130 countries as '*Convention Countries*' and 20 '*non-Paris convention countries*'. For the purpose of claiming priority in patents and trademarks, the Government of India also notified 150 Paris Conventions Countries as '*Convention Countries*' on 30th November and 3rd December, 1998.

**(*a*) Benefits of joining Paris Convention and PCT:** There are many benefits to the Government and also to the inventors and industries for joining the Paris Convention and PCT as given below:

(*i*) India can reciprocate with 150 countries for patents and trademarks,

(*ii*) for joining other conventions and treaties it is a condition to be a member state of Paris Convention first,

(*iii*) collaborative research can be conducted between India and the other member states of Paris Convention,

(*iv*) technology transfer between the member states becomes easier,

(*v*) in research sector more foreign investment can be made,

(*vi*) a single international application for protection of patent can be filed by the member states,

(*vii*) at the time of filing 'national phase' application expenditure can be saved by droppintg the designated countries,

(*viii*) for individual inventors of the developing countries there is 75% reduction in fees,

(*ix*) it takes minimum time or the members countries to grant protection rights,

(*x*) in spite of being a member of Paris Convention and PCT, the Indian Patent Holders have to apply for granting IPP in Paris convention and PCT,

(*xi*) since India does not grant patent in the field of software and living organisms, the applicants should file PCT application for biotechnology invention.

**(*b*) Some Cases of Patenting:** The Chennai based patent office believes that South Indian delicacies like 'medu vadai', 'rava uppama', 'badam halwa', 'rice idli', 'rice pongal' and even green pea's 'masala' are the novel process. In 1973, patents were granted for these popular preparations to the Dasaprakas Hotel Chain. The Mumbai patent office has granted a process patent to Dilip Shantaram Dahanunkar for the preparation of 'tomato rasam' and custard chili jam spread used as pizza topping. The same person has been given a patent for an improved process for preparation of vitaminised sweet and sour lemon pickle rice and a process for manufacturing banana sauce.

In 1995, the USA had granted a patent to the Medical Centre, University of Missisippi (USA) for use of turmeric (*haldi*) powder as a wound healing agent. Council of Scientific and Industrial Research, New Delhi (India) objected this patent. Consequently, the patent grant was revoked following serious objection in an order passed on April 21, 1998. In 1997, the EPO has given a favourable interim judgment on the challenge of a European Patent on the fungicidal effect of neem oil owned by W. Grace & Co.The challenge to neem patent was done by Dr. Vandana Shiva Ms Magda Alvoet (M.P. of the European Parliament) and the other NGOs of neem campaign. This patent was also revoked in 2004 on the basis of evidence that in such aided property of neem has already been reported in Indian literature. Recenty, the US government has patented the Indian 'Basmati' rice as 'Ricetech'. The Government of India challenged on April 28, 2001. Consequently US PTO served a notice to M/S Rice Tech, Inc. for amendments in claims restricting to only three strains developed by it.

## PROBLEMS

1. What do you know about biosafety ? Write in detail the biosafety guidelines and regulations for release of genetically engineered microorganisms.
2. Write an essay on operation of biosafety guidelines in India.
3. What is intellectual property right ? Discuss in detail the different forms of its protection.
4. Write an essay on patenting the biological materials.
5. Write short notes on the following:

   (*i*) Patents, (*ii*) Copyrights, (*iii*) Trade secrets and trade marks, (*iv*) GATT, (*v*) TRIPs, (*vi*) WIPO, (*vii*) Patenting of liveforms, (*viii*) Biosafety, (*ix*) Operation of biosafety guidelines in India, (*x*) GMMs, (*xi*) Plant variety protection, (*xii*) International status of Patents.

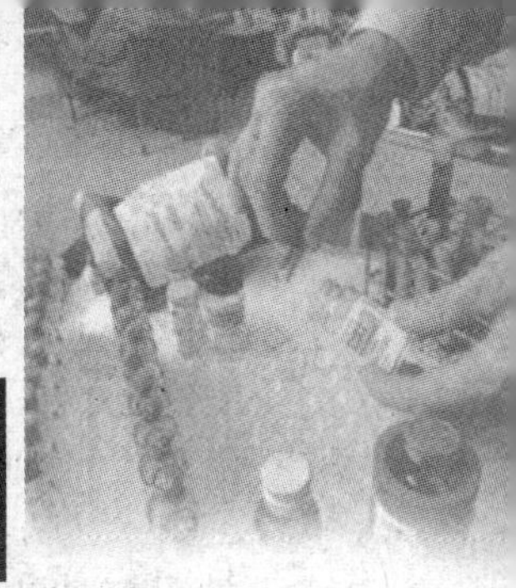

# Glossary

**3'-hydroxyl end:** The hydroxyl group attached to 3′ carbon atom of sugar of the terminal nucleotide of a nucleic acid.

**Acclimatization:** Adaptation of a living organism to a changed environment that subject it to environmental stress.

**Adaptor:** A synthetic single stranded oligonucleotide that produces a molecule containing cohesive ends and a restriction site. It provides a new restriction site to cloning vector.

**Agarose gel electrophoresis:** It is a process in which a matrix composed of a highly purified form of agar is used to separate large DNA and RNA molecules. *See* electrophoresis.

**Algalization:** The process of application of blue-green algae (cyanobacteria) in fields as biofertilizer.

**Allele:** One of pair of variant forms of a gene occurring at a given locus on a gene.

**Amplification:** Formation of many copies of a DNA segment of PCR.

**Amplified fragment length polymorphism (AFLP):** A combination of RFLP and RAPD very sensitive in detecting polymorphism throughout the genome.

**Analogous:** Feature of organisms which are functionally similar but have evolved in different ways.

**Androgenesis:** Development of haploid plants from the male gametophyte following a developmental pattern resembling embryogenesis, resulting from the culture of anthers or microspores.

**Annealing:** The process of healing (denaturing step) and slowly cooling (renaturing step) of double stranded DNA to allow the formation of hybrid DNA or DNA-RNA molecules.

**Antagonism:** Inhibitory relationships between microorganisms by involving amensalism, competition, and predation and parasitism.

**Antibiosis:** Growth inhibition of an organism by a substance of another organism.

**Antibiotic:** A biological substance produced by one organism inhibiting the growth (or killing) of another organism.

**Antibody:** A protein (immunoglobulin) synthesized by B lymphocytes that recognizes a specific site on an antigen.

**Antigen:** A compound that induces the production of antibodies.

**Antigenic determinants:** *See* epitope.

**Anti-parallel orientation:** Arrangement of two strands of DNA molecule oriented in opposite directions so that the 5′-$PO_4$ end of one strand is aligned with the 3′-OH end of the complementary strand.

**Antisense DNA:** The sequence of chromosomal DNA that is transcribed.

**Antisense RNA:** An RNA sequence complementary to all or part of a functional RNA.

**Antisense therapy:** *In vivo* treatment of a genetic disease by blocking translation of a protein with a DNA or RNA sequence that is complementary to a specific mRNA.

**Artificial seeds:** A synthetic seed consisting of a somatic embryo surrounded by nutrient medium which is protected by a thin membrane of chemical.

**Aseptic:** Free from all microorganisms.

**Assay:** A test to detect the presence of some substances in small amount in solution.

**Autoclave:** An instrument used for sterilization of glassware and culture media.

**Autonomously replicating sequence (ARS):** Any cloned DNA sequence that initiates and supports extrachromosomal replication of a DNA molecule in a host cell. It is often used in yeast cells.

**Autoradiography:** A technique that captures the image formed in a photographic emulsion due to emission of light or radioactivity from a labelled component placed together with unexposed film.

**Auxins:** A class of plant growth regulators that stimulate cell division, cell elongation, apical dominance and root initiation. They are used in plant cell and tissue culture e.g. 2, 4-D, IAA, NAA, etc.

**Axenic culture:** A pure culture of single propagule of an organism not contaminated by any other microorganism.

**Axillary bud:** A bud produced in the axis of leaves.

**B cells:** Lymphocytes derived from bone marrow cells that produce antibodies.

**BAC (Bacterial artificial chromosomes):** A cloning vector for isolation of genomic DNA constructed on the basis of F-factor.

**Bacteriophage:** A virus that infects bacteria.

**Bacterization:** A technique of seed-dressing with bacteria as water soluble suspension used for introduction of bacteria in soil.

**Bacteroids:** The pleomorphic forms of *Rhizobium* present inside root nodules that fix atmospheric nitrogen gas.

**Baculovirus:** A virus that infects arthropods including mostly insects.

**Base pairs (bp):** Pairs of nucleic acids in a double stranded helix e.g. pairing of A with T (or A with U in RNA) and G with C.

**Batch culture:** A suspension culture in which cells grow in a finite volume of nutrient medium following a sigmoidal pattern of growth.

**Batch fermentation:** Fermentation carried out by growing microbes in batch culture.

**Binary vector system:** A two-plasmid system in *Agrobacterium* for transferring a T-DNA region that carries cloned genes into plant cells. One plasmid contains the virulence genes and the other contains engineered T-DNA region.

**Bioaccumulation:** Concentration of a chemical agent (e.g. DDT) in the increasing amount in the organisms of a food chain.

**Biocontrol:** Any process using the living organism to inhibit the growth and development of the pathogenic organisms.

**Biodegradation:** The gradual breakdown of a compound to its constituents by a living organism.

**Biodiversity:** The variability among the living organism from all sources, soil, water, air, extreme habitat or associated with organisms.

**Biofertilizers (microbial inoculants):** Commercial preparation of microorganisms by using which nitrogen and phosphorus level and growth of plants increased.

**Bioinformatics:** The application of information sciences (mathematics, statistics and computer science) to increase our understanding of biology.

**Bioleaching:** Solubilization of metals in the ores by using a solution containing specific microorganisms.

**Biomass:** The cell mass produced by a population of living organisms.

**Biomining:** Solubilisation and recovery of metals from ores.

**Biopesticides:** Commercial preparation of microorganisms by using which insects, termite or nematode pests are killed or their level is minimized and crop yield increased.

**Biopolymer:** Any polymer such as protein, nucleic acids, lipids, polysaccharide, produced by a living organism.

**Bioreactor:** A vessel i.e. fermentor used for fermentation or other processes.

**Bioremediation:** The process of using living organisms to remove contaminants, pollutants or unwanted substances from soil or water.

**Biosensor:** An electronic device that uses biological molecules or cells to detect specific compounds.

**Biotic stress:** Stress resulting from living organisms which can harm the other plants such as virus, bacteria, fungi, weeds, harmful insects.

**Biotransformation or bioconversion:** The use of cultured cells to convert substrates into desired organic compounds by the virtue of an endogenous enzyme which catalyses the reaction.

**Blot transfer:** Transfer of blot from a gel to a membrane filter.

**Blot:** A spot on membrane filter gel made by adopting a solution.

**Blunt end:** The ends of a double stranded DNA that do not possess sticky end (single stranded bases).

**Blunt-end ligation:** Joining of nucleotides present at the ends of two DNA molecules.

**Brewer's yeast:** Strains of *Saccharomyces cerevisiae* used in beer production.

**Broad- host-range plasmid:** A plasmid that can replicate in a number of different bacterial species.

**Callus:** An unorganized mass of plant cells capable of cell division and growth *in vitro.*

**Cancatamer:** DNA segment made up of repeated sequences linked end to end.

**Capsid:** External protein coat of a virus particle.

**Carrier:** An inert material which is mixed with microbial inoculants for the production of biofertilizers and biopesticides.

**Caulogenesis:** Shoot induction from callus.

**cDNA clone:** A double stranded DNA complement of mRNA synthesised *in vitro* by using reverse transcriptase and DNA polymerase.

**cDNA library:** A collection of cDNA clones generated *in vitro* from the mRNA sequence of a single tissue or cell population.

**Cell culture:** Cell culture of growing cells *in vitro* in liquid medium.

**Cell line:** (1) Cell line derived from a primary culture and selected at the time of the first subculture, or derived from a cell population with particular attributes, (2) cells that acquire the ability to multiply indefinitely *in vitro*.

**Cell suspension:** Cells in liquid medium often used to describe suspension culture of a single cell and cell aggregates.

**Chelate:** Complex organic molecule that can combine with cation and does not ionise. They supply micro-nutrients to plants at slow and steady rates, casually used to supply iron to plant cells.

**Chemostat:** A open contiguous culture in which cell growth rate and cell density are held constant by a fixed rate of input of a growth limiting nutrients.

**Chimera:** (1) Recombinant DNA molecules containing sequences from different organisms; (2) a plant or animal that has population of cells with different genotypes.

**Chimeric DNA:** A recombinant DNA molecule containing unrelated genes.

**Chimeric protein:** Fusion protein.

**Chromosome walking:** A technique that identifies overlapping cloned sequences of about 40 kb long that form one continuous segment of a chromosome.

**Chromosome:** A physically distinct unit of the genome.

**Circularization:** A DNA fragment generated by digestion with a single restriction enzyme will have complementary $5' \rightarrow 3'$ extension (sticky ends). If these ends are annealed and ligated, the DNA fragment will have converted to a covalently closed circle or circularized.

**Cleave:** To break phosphodiester bonds of double stranded DNA, usually with a Type II restriction enzyme.

**Clonal propagation:** Vegetative propagation of plants considered to be physiologically and/ or genetically uniform which originated from a single individual or explant is called clonal propagation.

**Clone bank:** A population of organisms each containing a DNA molecule inserted through cloning vector.

**Clone:** A collection of genetically identical cells or organisms derived from a common ancestor where all members have similar genetic composition.

**Cloning site:** A location on a cloning vector into which DNA can be inserted.

**Cloning vector:** A small, self-replicating DNA molecule (plasmid or phage $\lambda$) into which foreign DNA is inserted in the process of cloning genes or other DNA sequence. It can carry inserted DNA and transferred into host cell.

**Cloning:** Incorporation of a DNA molecule into a chromosomal site or a cloning vector.

**Co-cultivation:** Incubation of host cells or tissues with *Agrobacterium* or any other organism acting as vehicle for a vector.

**Codon:** A set of three nucleotides in mRNA that specify a tRNA carrying specific amino acid which is incorporated into a polypeptide chain during protein synthesis.

**Cohesive ends:** Single stranded complementary nucleotides on the ends of double stranded DNA molecule.

**Colony hybridization:** A technique where a nucleic acid probe is used to identify a bacterial colony with a vector carrying a specific cloned gene(s).

**Complementary DNA (cDNA):** A DNA strand formed from mRNA by using the enzyme reverse transcriptase.

**Compost:** Fully decomposed organic material.

**Concatamer:** DNA segment made up of repeated sequences linked end to end.

**Conjugation:** The unidirectional transfer of DNA from donor bacterium to recipient bacterium through cell-to-cell contact.

**Consensus sequence:** The nucleotide sequence that is present in majority of genetic signals that perform a specific function.

**Contamination:** Presence of unwanted microorganisms in microbial, plant or animal cultures, food or any products in laboratory or industry.

**Contig:** A set of overlapping clones that provides a physical map of a portion of chromosome. It refers to contiguous map.

**Continuous culture:** A suspension culture held at constant volume and continuously supplied with nutrients by the inflow of fresh medium.

**Copy number:** The average number of molecules of a plasmid of a gene per genome present in a cell.

**COS sites (ends):** The 12-base, single strand, complementary extension of phage lambda (λ) DNA.

**Cosmid:** A plasmid vector which consists of the *COS* site of phage lambda (λ) and one or more selectable markers such as an antibiotic resistance gene.

**Crown gall:** A tumour occurring at the base of certain plants due to infection of the plant by *Agrobacterium tumifaciens*.

**Cryopreservation:** Culture preservation at ultra-low temperature of –196ºC.

**Cryoprotectant:** A chemical agent that inhibits freezing and thawing damage to cells.

**Cybrid:** A cytoplasmically hybrid cell with organelles from both parental sources (obtained through fusion of cytoplast with a whole cell) and a nucleus of only one cell. Nucleus of the other cell denatured.

**Cytokinins:** A class of plant growth regulators which cause cell division, cell differentiation, shoot differentiation and breaking of apical dominance e.g. BAP, kinetin, zeatin.

**Database:** A repository of sequences (DNA or amino acids) which provide a centralized and homogeneous view of its contents.

**de novo:** It means anew, afresh.

**Denaturation:** Separation of double stranded DNA molecule into single strands.

**Dextran:** A polysaccharide produced by certain bacteria.

**Diazotrophs:** The microorganisms that fix atmospheric nitrogen ($N_2$).

**Differentiation:** The process of biochemical and structural changes by which cells become specialized in form and function.

**Disarmed plasmid:** A plasmid from which some portion has been deleted.

**DNA amplification:** Multiplication of a piece of a DNA in a test tube into many thousand or million of copies by using PCR.

**DNA chip:** The solid supports of glass or silicon about the size of a microscope slide consisting of DNA attached in highly organized arrays. It is available commercially also e.g. GeneChip® Probe Array.

**DNA construct:** A suitable DNA which has been prepared for cloning purpose.

**DNA microarray:** *See* DNA chip.

**DNA polymerase:** An enzyme that catalyses the phosphodiester bond in the formation of DNA.

**DNA probe:** A radiolabelled ($^{32}P$) DNA segment used to hybridize the base pairing between the probe and a complementary base sequence in a DNA sample.

**DNA probe:** Isolated single radiolabelled $^{32}P$-DNA strands (oligonucleotides) used to detect the presence of the complementary (opposite) strands, and as a very sensitive biological detectors.

**DNA-delivery system:** Any procedure that introduce DNA into a recipient cell.

**Doubling time:** The time required to double the number of cells.

**Downstream processing:** Separation and purification of products(s) from a fermentation process.

**Downstream:** The stretch of nucleotides of DNA that lies in the 3′ direction of the site of initiation of transcription.

**Electrophoresis:** The method of separation of molecules such as DNA, RNA or protein based on their relative migration applying a strong electric field.

**Electroporation:** A technique where electric field is applied to facilitate protoplast fusion derived from two different plants.

**Embryo transfer:** Implantation of embryos from donor animals or generated by *in vitro* fertilization into the uterus of the recipient animals.

**Embryogenesis:** Process of development of embryo from a fertilized egg cell.

**Embryoids:** Embryo-like structure produced as a result of differentiation process such as embryogenesis and androgenesis.

**Embryonic stem (ES) cell:** Cells of early embryo that give rise to all differentiated cells including germ line cells.

**Encode:** To specify the sequence of amino acids in a protein.

**Endonucleases:** An enzyme that catalyses the cleavage of DNA at internal position, cutting DNA at specific sites.

**Endotoxin:** Cell wall portion (lypopolysaccharide) of Gram-negative bacteria that shows antigenic properties and trigger immune system.

**ENTREZ:** It is the integrated information database similarity search.

**Episome:** It is a plasmid which replicates within a cell independently of the chromosome and can integrate into the host bacterial DNA (e.g. F factor in *E. coli*).

**Epitope:** A specific chemical domain present on an antigen recognized by an antibody. Each epitope elicits synthesis of different antibodies.

**EST (expression tag sequence):** A partial gene sequence unique to the gene in question that can be used to identify and position the gene during genomic analysis. It is derived from cDNA molecules.

**Excision:** (1) Enzymatic removal of a DNA segment from a chromosome; (2) cutting (with scalpel, etc.) of tissue for culture.

**Exons:** The coding regions of a gene that are present in processed mRNA.

**Exonuclease III (ExoIII):** An enzyme from *E. coli* that removes nucleotides from 3'OH ends of double stranded DNA.

**Exonuclease:** An enzyme that digest DNA or RNA from the ends of strand; 5′ exonuclease requires a free 5′ end and degrades the molecule in 5′→3′ direction; 3′ exonuclease requires a free 3′ end and degrades the DNA in 3′→5′ direction.

**Explant:** The tissue taken from a plant or seed and transferred to a culture medium to establish a tissue culture system or regenerate a plant.

**Exponential phase:** A phase of cell growth where they undergo maximum rate of cell division.

**Expression vector:** A constructed cloning vector of which coding sequence is properly transcribed and the RNA is efficiently translated.

**F factor:** *See* episome.

**Fed-batch fermentation:** Microbial growth where nutrients are added into the fermentor without removing the product.

**Fermentation:** The process by which microorganisms turn raw materials such as glucose into products such as alcohol.

**Fermentor:** A large sized bioreactor/culture vessel used to produce products under controlled conditions.

**Fibroblast:** Flattened cells of connective tissue.

**Foreign DNA:** A DNA molecule which is incorporated into a cloning vector or chromosome.

**Friable:** Crubling or fragmenting callus. A friable callus easily dissected and readily digested into single cell or clump of cells in solution.

**Functional genomics:** The study of the function of all specific gene sequences and their expression in time and space in an organism.

**Fusion protein (hybrid protein):** Two or more coding sequences from different genes that have been cloned together act a single polypeptide chain after translation.

**Fusion protein:** A single protein molecule encoded by parts of two or more genes combined together as one unit.

**Fusogen:** A fusion inducing agent used for agglutination of protoplast in somatic hybridization.

**Gametoclonal variation:** Phenotypic variation apart from normal segregation occurring during gamete formation arising from the culture of gametophytic cells or tissues such as in anther culture or unfertilized ovule culture.

**Gene bank (gene library):** *See* clone bank.

**Gene cloning (molecular cloning):** Insertion of a gene into a vector to form a new DNA molecule that can be perpetuated into host cell.

**Gene gun (biolistic):** A device to accelerate particles for physically delivering the recombinant DNA, typically precipitated onto microprojectile, into a cell using an explosive propellant.

**Gene transfer:** The use of genetic or physical manipulations to introduce foreign genes into host cells to achieve desired characteristics in progeny.

**Gene:** (1) A hereditary unit; (2) a segment of DNA coding a specific protein.

**Genetic engineering:** *In vitro* DNA technologies used to isolate genes from an organisms, manipulate them in laboratory as per desire and insert them into another cell system for specific genetic trait. It is also called gene cloning.

**Genetic fingerprinting:** A technique of analysing DNA of an individual to reveal the pattern of repetition of particular nucleotide sequences through the genome. The unique pattern of DNA is identified by a PCR or Southern hybridisation technique.

**Genetic map (linkage map):** The linear array of gene on a chromosome based on recombination frequencies.

**Genetic marker:** A DNA sequence used to mark a specific location on a particular chromosome.

**Genetic polymorphism:** *See* polymorphisms.

**Genetically modified organism (GMO):** An organism that has been modified by the application of recombinant DNA technology.

**Genome:** (1) The entire complement of genetic material of an organism, virus or organelle; (2) the haploid set of chromosome of a eukaryotic organism.

**Genomic library:** A collection of clones containing sequences of genomic DNA of an organism. These molecules are propagated in bacteria or viruses.

**Genomics:** The study of molecular organization of genome, their information contents and the gene products they encode.

**Germ line cells:** The cells that produce gamete.

**Green manuring:** A farming practice where a leguminous plant which has derived benefits from its association with appropriate species of *Rhizobium* is ploughed in the soil and then non-legume is grown and allowed to take the benefits of the already fixed nitrogen.

**Hairpin loop:** A region of double helix found by base pairing within a single strand of DNA or RNA which has folded itself.

**Hardening:** Gradual acclimatization of *in vitro* grown plants to *in vivo* conditions just to adapt the outdoor conditions.

**Heterocysts:** The modified vegetative cells in cyanobacteria associated with $N_2$ fixation.

**Heterogeneous:** Large RNA molecules which are unedited mRNA transcripts found in nucleus of an eukaryotic cell nuclear RNA (hnRNA).

**Heterokaryon:** A cell containing two or more nuclei unlike genetic make up.

**Heterologous probe:** A DNA probe desired for one species and used to screen for a similar DNA sequence from another species.

**Homokaryon:** A cell with two or more identical nuclei as a result of fusion.

**Homologous:** Having the same evolutionary function or structure.

**Homopolymer:** A nucleic acid strand composed of one kind of nucleotide.

**Hybridoma:** A unique fused cell that produces quantities of a specific antibodies (immunoglobulins) and divide continuously.

**Hydrogen uptake (Hup) gene:** A gene found in $N_2$ fixing microbes that converts $H_2$ to $2H^+$ and harvests energy of the organisms.

**Immobilization:** Attachment of cells or protoplasts into a matrix.

**Immobilized enzymes:** An enzyme physically localized in a defined region enabling it to be reused in a continuous process.

**Immunoglobulin:** *See* antibody.

***In situ* hybridization:** A technique which detects and locates a given sequence of DNA directly with the cells or tissue by using labelled probes i.e. fluorescent antibodies.

***In situ*:** It means 'in original or natural place'.

***In vitro* translation:** Protein synthesis directed by purified DNA with bacterial extracts or mRNA that provide ribosomes, tRNAs and protein synthesis factors. The reaction mixture is supplemented with ATP, GTP and amino acids.

***In vitro*:** Any process carried out in sterile cultures or measurement of biological processes outside the intact organism such as enzyme reaction.

***In vivo*:** The natural conditions in which organisms live.

**Inoculum:** A culture containing viable cells of microorganism.

**Insert DNA:** A DNA molecule incorporated into a cloning vector.

**Interferon:** A glycoprotein produced by a virus-infected cells which protect another cell from attack of the virus.

**Intervening sequences:** *See* intron.

**Intron:** A gene segment that is transcribed from DNA which later on is excised from the primary transcript during processing to a functional RNA molecule.

**Isozyme:** Multiple molecular forms of an enzyme that exhibit similar catalytic properties.

**kb:** Abbreviation of kilobase pairs.

**Klenow fragment:** A product of proteolytic digestion of DNA polymerase I from *E. coli*. DNA Pol I has both polymerase and 3′-exonuclease activity (but not 5′ exonuclease activity).

**Label:** A radioactive compound attached with DNA/RNA to indicate the presence of a complementary DNA strand in a sample.

**Lag phase:** Time required by inoculated cells in fresh medium to adapt the new environment before the start of cell division.

**Leader sequence:** It is the non-translated sequence at the 5′ end of mRNA that precedes the initiation codon.

**Leghaemoglobin:** A globular protein containing haeme expressed in legume root nodules.

**Ligase:** An enzyme used by a genetic engineer to join the cut ends of the double stranded DNA.

**Ligation:** Joinig of two DNA molecules by formig phosphodiester bonds catalysed by T4 DNA ligase of *E. coli* DNA ligase.

**Lignocellulose:** The composition of woody biomass consisting of lignin and cellulose.

**Linkers:** A synthetic double stranded oligonucleotide carrying the sequence for one or more restriction sites.

**Liposomes:** Artificial vesicles of phospholipids.

**Log phase:** *See* exponential growth phase.

**Long terminal repeats (LTR):** Similar blocks of genetic information found at the end of the genomes of retroviruses.

**Lysogeny:** A phenomenon in which prophage survives within a host bacterium as a part of host genome or as extrachromosomal element and does not cause lysis.

**M13:** A single stranded DNA bacteriophage used as vector for DNA sequencing.

**Megabase (Mb):** A length of DNA consisting of $10^6$ bp. 1 Mb = $10^3$ kb = $10^6$ bp.

**Meristem:** Undifferentiated but determined tissue, the cells of which are capable of active cell division and differentiation into specialized and permanent tissue such as shoot and root.

**Messenger RNA (mRNA):** An RNA molecule transcribed from DNA that specifies during translation the amino acid sequences of a protein molecule.

**Metabolites:** The products produced through biochemical activities.

**Microarray:** *See* DNA chip.

**Microinjection:** A technique of injecting DNA or RNA into the nucleus of protoplast of cell with the help of a micropipette.

**Micropropagation:** Use of small piece of tissue such as meristem grown in culture to produce large number of plants.

**Modification:** Methylation of DNA to protect it from cellular restriction enzymes.

**Molecular markers:** The markers (DNA sequences) revealing variation at DNA level which can be detected and monitored in the subsequent generations e.g. RFLP, RAPD, SSR, etc.

**Monoclonal antibodies:** Anitibodies derived from a single clone of cells which recognize only one kind of antigen.

**Morphogenesis:** The development of a structure from an unorganized state to a differentiated and organized state.

**Mushrooms:** Edible fruiting bodies of ascomycetous and basidiomycetous fungi.

**Mutant:** A cell which has undergone a heritable change that has resulted in due to a change in its genes or chromosome.

**Mycoherbicide:** Fungal preparations that kill the abnoxious weeds/herbs.

**MycoRhiz®:** A commercial preparation of *Pisolithus tinctorius* mycelia produced by Abbott Lab (U.S.A.) used for raising tree seedling through tailoring of roots.

**NCBI (National Centre of Biology Information):** It was established in 1988 for development of information systems in molecular biology. It is the foremost repository of publically available genomic and proteomic data.

**Nick translation:** A procedure for labelling DNA *in vitro* using DNA Pol I.

***Nif* (nitrogen fixing) genes:** A 24 kb long nucleotide sequence organised in 7 oberons (transcriptional units) which take part in nitrogen fixation.

**Nitrogenase:** An enzyme expressed by *nif* genes that converts $N_2$ to $NH_4^+$ under $O_2$-tense conditions in prokaryotes.

**Nod genes:** Genes that encodes nodulins.

**Nodulation:** Formation of root nodules in legumes by *Rhizobium*.

**Northern blotting:** Hybridization of a labelled DNA probe to RNA fragments that have been transferred from an agarose gel to a nitrocellulose filter. It is also called Northern hybridization.

**Oligomer:** A molecule formed from a small number of monomers.

**Oligonucleotides:** A short synthetic single stranded DNA.

**Oncogene:** A gene that causes cancer.

**Open reading frame (ORF):** A nucleotide sequence of DNA molecule that encodes a polypeptide. This term is applied when the function of the encoded protein is unknown.

**Operator:** A site on DNA at which repressor or activator protein binds to control the transcription from initiating at an adjacent promoter.

**Operon:** A cluster of genes that are coordinately regulated.

**Opine:** The condensation product of an amino acid with *keto* acid or a sugar.

**Organ culture:** Culture of an organ *in vitro*.

**Organogenesis:** The process of initiation and development of a structure which show natural organ form and/or function, the initiation of which is temporally separated from the initiation of other organs.

**Origin for replication (ori):** A site of nucleotide sequence from which replication begins.

**Osmaticum:** Agents such as polyethylene glycol, mannitol, glucose, sucrose which maintains the osmotic potential of a nutrient medium equivalent to that of cultured cells.

**P1 clones:** A cloning system for isolating genomic DNA that uses elements from phage P1 i.e. *loxP* and *pax* sites.

**Palindromic sequence:** Complementary DNA sequences that are the same when each strand is read in the same direction (e.g. 5′→3′). These sequences act as recognition sites for Type II restriction endonuclease.

**Patent:** A legal document issued by a Government that permits the holder the exclusive right to manufacture, use of sell an invention for a defined period, which varies country to country.

**pH:** The negative logarithm of hydrogen ion concentration of any solution for measuring acidity or alkalinity.

**Phagemid:** A cloning vector that contains components derived from both phage DNA and plasmid.

**Plant growth promoting rhizobacteria (PGPR):** The bacteria that live in rhizosphere and promote plant growth through complex mechanisms.

**Plaque:** A clear area that is visible in a bacterial lawn formed on an agar plate due to lysis of bacterial cells by bacteriophage.

**Plasmid:** Extrachromosomal, self-replicating, circular double stranded DNA molecules containing some non-essential genes.

**Plating efficiency:** The percentage of cells plated on medium which develop colonies.

**Pluripotent:** *See* totipotent.

**Polyacrylamide gel electrophoresis (PAGE):** A method for separation of nucleic acids or protein molecules according to their size. The molecules migrate through the inert gel matrix under the influence of an electric field. *See* electrophoresis.

**Polyadenylation [poly(A) tailing]:** The addition of A residues to 3′ end of eukaryotic mRNAs.

**Polycistronic mRNA:** An mRNA that includes coding region of more than one gene.

**Polyclonal antibodies:** A serum containing antibodies that binds to different antigenic determinants of an antigen.

**Polylinker:** A segment of DNA that contains several restriction sites. It is also called multiple cloning sites.

**Polymerase chain reaction (PCR):** The action of the enzyme polymerase to produce many copies of a polynucleotide sequence of DNA at high temperature.

**Polymorphisms:** Occurrence of one or two alleles in a population.

**Polysomes:** Complexes of ribosome bound together by a single stranded mRNA molecule. It is also called polyribosome.

**Prebiotics:** They are non-digestible food ingredients that stimulate the growth and/or activity of bacteria in the digestive system and benefit the health.

**Primary culture:** A culture started from cell, tissue, organ taken directly from an organism. It is regarded as such until sub-cultured for the first time. Then it becomes cell line.

**Primer:** A short oligonucleotide that hybridizes the template strand and gives a 3′ –OH end for the initiation of nucleic acid synthesis.

**Primer:** It is a short sequence often RNA that is paired with one strand of DNA and provides a free –OH end at which a DNA polymerase starts synthesis of a deoxyribonucleotide chain.

**Probiotics:** The live microorganisms that may confer a health benefit on the host when administered in adquate amounts.

**Pro-embryo:** Early embryonic stage.

**Promoter –10 sequence:** It is the consensus sequence TATAATG centered about 10 bp before the start point of bacterial gene. It is involved in initial melting by DNA polymerase.

**Promoter –35 sequence:** It is consensus sequence centered about 35 bp before the start point of a bacterial gene. It is recognized by RNA polymerase.

**Promotor sequence:** A regulatory sequence of DNA that initiates the expression of a gene.

**Promotor:** A DNA segment lying upstream (5′ to) a gene onto which RNA polymerase gets attached.

**Propagule:** A part of plant that serve for propagation.

**Prophage:** An inactive state of a bacteriophage genome maintained in a bacterial cell as a part of host chromosome.

**Protein engineering (enzyme engineering):** Generation of protein with subtly modified structure that confer special properties (as present before) such as high *p*H stability, thermal stability, high catalytic specificity, etc.

**Proteome:** The entire collection of proteins that an organism can produce. The proteome is analyzed by two-dimensional electrophoresis followed by mass spectrometry in some case.

**Proteomics:** The study of proteome or the array of proteins that an organism produce.

**Protoplast:** Microbial or plant cells whose cell wall has been removed so that the cells assume a spherical shape.

**Pulse field gel electrophoresis (PFGE):** A procedure used to separate large DNA molecules by alternating the direction of electric current in a pulsed manner across a gel.

**Quantitative trait locus:** A locus that affects a quantitative trait.

**Random amplified polymorphic DNA (RAPD):** A PCR-based technique for detecting polymorphism at DNA level. A single short (usually 10-mer) synthetic nucleotide primers are used in PCR.

**Recognition site:** *See* restriction site.

**Recombinant DNA technology:** *See* gene cloning.

**Recombinant DNA:** The hybrid DNA produced by joining pieces of DNA at specific points.

**Recombinant:** An individual whose genes on a chromosome results from one or more cross-over events.

**Regeneration:** Development of new organs or plantlets from a tissue, callus culture or from a bud.

**Renaturation:** The re-union of two nucleic acid strands after denaturation.

**Repeated DNA sequence:** A sequence of nucleotides which occurs more than once in a genome. It may be present in a few to many millions of copies. The length of individual repeated sequence may be a few nucleotides to several kb.

**Repetitive DNA:** DNA sequences that are present in genome in multiple copies, sometimes million times or more.

**Replacement therapy:** The administration of metabolites, co-factors or hormones that are deficient due to genetic disease.

**Replica plating:** Transfer of cells from bacterial colonies growing on one Petri plate to the other plate corresponding to the location of the first plate (master plate).

**Replicon:** Synthesis of the genome in which DNA is replicated. It contains an origin of replication.

**Reporter gene:** A gene encoding a product that can readily be assayed during genetic transformation e.g. GUS, chloramphenicol transacetylase, etc. It may be connected to any promoter of interest so that it may express.

**Repression:** Inhibition of transcription by preventing RNA polymerase from binding to the initiation site for transcription.

**Restriction endonuclease (type II):** An enzyme that recognizes a specific duplex DNA and cleaves phosphodiester bonds on both the strands between specific nucleotides.

**Restriction fragment length polymorphism (RFLP):** DNA fragment of different length produced by cutting with restriction enzymes.

**Restriction map:** A linear array of sites on DNA cleaved by various restriction enzymes.

**Restriction site:** The sequence of nucleotide pairs of double stranded DNA that is recognized by a type II restriction endonuclease.

**Retroviruses:** The RNA viruses that contain double stranded genome by using reverse transcription. The double stranded DNA integrates to chromosomes of infected cells.

**Reverse transcriptase:** An RNA-dependent DNA polymerase that uses an RNA molecule as a template for the synthesis of complementary DNA strand.

**Rhizosphere:** The zone of immediate vicinity of growing plant roots that harbours increased microbial population.

**Ribonuclease:** An enzyme that hydrolyses RNA.

**Ribosomal binding site:** A nucleotide sequence near 5′ end of an mRNA that facilitates the binding of the mRNA to the small ribosomal sub-unit. It is also called Shine-Dalgarno sequence.

**Saccharification:** Enzymatic hydrolysis (by glucoamylase) of polysaccharides resulting in maltose and glucose.

**Scale-up:** (1) Expansion of laboratory experiments to full size industrial processes, (2) transition from small scale to large scale (industrial) production.

**Secondary metabolite:** The metabolite that is not required by a cell for its growth or metabolism but produced as a by-product for protection against microorganisms during stationary phase of growth.

**Secondary metabolites:** Metabolite produced as end products of primary metabolism and not involved directly in metabolic activity.

**Selectable:** Having a gene product that enables to identify and propagate a particular cell type. *See* also reporter gene.

**Selection:** A method to isolate or identify a specific organisms in a mixed culture.

**Sequence characterized amplified regions (SCARs):** A fragment of genomic DNA present at a single locus identified by PCR amplification using a pair of specific oligonucleotide primers.

**Shine-Dalgarno sequence:** *See* ribosomal binding site.

**Shuttle vector:** A plasmid cloning vector that can replicate in two different organisms due to presence of two different origin of replication i.e. $ori^{Euk}$ and $ori^{E.coli}$.

**SI nuclease:** An enzyme that degrades single stranded DNA.

**Siderophore:** A low molecular weight Fe-chelating protein synthesized by several soil microorganisms.

**Signal sequence:** A segment of about 15 to 30 amino acids at the N′ terminus of a protein that enables the protein to be secreted. The signal sequence is removed as the protein is secreted through the membrane. It is also called signal peptide and leader peptide.

**Single cell protein (SCP):** Cells or protein extracts of microorganisms produced in large quantities for use as human or animal protein supplements.

**Site directed mutagenesis:** The process of nucleotide changes in cloned genes by site specific mutagenesis.

**Somaclonal variation:** Variation that occurs in cell cultures and tissues which may be genetic or epigenetic.

**Somatic cell gene therapy:** Gene delivery to a tissue other than reproductive cells of an individual to correct gene defect.

**Sonication:** *See* ultrasonication.

**Southern blotting:** A method for transferring denatured DNA molecules separated by gel electrophoresis to a matrix (e.g. nitrocellulose membrane) on which hybridization can be done.

**Splicing:** Gene manipulation where one DNA molecule is attached to another.

**Split genes:** In eukaryotes structural, genes are split up by a number of non-coding regions called introns.

**Staggered cuts:** Symmetrically cleaved phosphodiester bonds that lie on both strands on DNA duplex not opposite to one another.

**Stem cell:** A precursor cell that undergoes division and gives rise to different lineages of differentiated cells.

**Sterilization:** The process for elimination of microorganisms.

**Sticky ends:** *See* cohesive ends.

**Stirred tank fermentor:** A bioreactor in which cells or microorganisms are mixed by mechanically driven impellers.

**Strain:** A genetic variant (to its parents) of multicellular organism or microorganism.

**Structural gene:** A DNA sequence that encodes a polypeptide molecule.

**Structural genomics:** The study of structure of all gene sequences in a fully sequenced genome.

**Sub-culture:** Sub-division of a culture and its transfer to a fresh medium.

**Substrate:** Food source for growing cells of an organism.

**Superbug:** A recombinant *Pseudomonas putida* prepared by Dr. A.M. Chakrabarty. *See* text.

**Suspension culture:** A culture consisting of cell aggregates initiated by placing callus tissue or some times seedling in an agitated liquid medium.

**Symbiosis:** A symbiotic relationship between two organisms for mutual benefits.

**Synbiotics:** They are the nutritional supplements combining probiotics and prebiotics in a form of synergism.

**Synchronous culture:** A culture in which a specified proportion of the cells are at the same indicated phase of the cell cycle, enter the same indicated phase of the cell cycle simultaneously, and/or exhibit the same approximate duration of cell cycle.

**T cells:** Lymphocytes that pass through thymus gland during maturation. They play a key role in immune response.

**T4 DNA ligase:** An enzyme produced by *E. coli* after infection by bacteriophage T4. It catalyses the joining of DNA duplex and repairs nick in DNA molecules. It uses ATP as source of energy.

**Tailing:** Addition of some nucleotides *in vitro* by the enzyme terminal transferase to 3′-OH ends of a duplex DNA molecule. It is also called homopolymer tailing.

**Tandem array:** A DNA molecule containing two or more identical sequences in series.

**Taq polymerase:** A heat stable DNA polymerase isolated from a thermophilic bacterium *Thermus aquaticus* and used in PCR; see PCR.

**T-DNA:** A fragment of Ti-plasmid of *Agrobacterium tumifaciens* that is transferred and integrated into chromosome of plant cells.

**Template strand:** The polynucleotide strand being used by DNA polymerase for determining nucleotide sequence during synthesis of a new strand.

**Template:** An RNA or single stranded DNA molecule on which a complementary strand is synthesised. It is also called antisense strand.

**Terminal transferase:** An enzyme that adds nucleotides to the 43′ terminus of DNA molecules.

**Thawing:** A slight melting of pre-cooled specimen before long-term storage at −196°C.

**Therapeutic agent:** A chemical compound used for treatment of a disease. It is also called as drug.

**Ti plasmid:** An extrachromosomal, double stranded, and self-replicating DNA molecules found in *Agrobacterium tumifaciens* that causes crown gall disease in plants.

**Tissue culture:** The rearing of cells or tissue in an artificial medium under controlled condition.

**Totipotency:** A property of normal cells that they have the genetic potential to give rise to a complete individual, except terminally specialized cells.

**Totipotent:** The ability of a cell to respond according to environmental stimuli and divide to form differentiated cell types.

**Transcription:** Synthesis of RNA molecule catalysed by RNA polymerase using a DNA strand as a template.

**Transfection:** DNA transfer into a eukaryotic cell.

**Transformation:** (1) Transfer of a DNA fragment into a cell and acquisition of new genetic marker by the recipient cell; (2) conversion of animal cells by various means in tissue culture from controlled to uncontrolled cell growth.

**Transgene:** A gene from one source incorporated into the genome of another organism.

**Transgenesis:** Introduction of a gene into animal or plant cells that is successively transmitted to onward generations.

**Transgenic animal:** A fertile animal consisting of a foreign gene in its germ line.

**Transgenic organisms:** Plants, microorganisms or animals in which novel DNA has been incorporated into their germ line cells.

**Transgenic plant:** A fertile plant consisting of a foreign gene in its germ line.

**Transient:** Of short duration.

**Translation:** Copying of mRNA into polypeptide.

**Transposition:** The process of insertion of a transposon into a new site on the same or another DNA molecule.

**Transposon tagging:** Insertion of transposable element into or nearby a gene, thereby making that gene with a known DNA sequence.

**Transposon:** A DNA segment that can jump from a locus and join a new locus on the DNA molecule by involving the enzyme transposase.

**Ultrasonication:** Disruption of cells or DNA molecules by high frequency sound waves.

**Upstream processing:** *See* upstream.

**Upstream:** (1) The stretch of DNA base pairs present in 5′ direction from the site of transcription. The first transcribed base is designated as +1 and the upstream nucleotide as –1, –10; (2) the stages of manufacturing process preceding the biotransformation step. It includes preparation of raw materials required for fermentation.

**Vectors:** (1) Vehicles for transferring DNA from one cell to another; (2) a DNA molecule that can carry inserted DNA and be perpetuated in a host cell.

**Vegetative propagation:** Reproduction of plants using a non-sexual process involving the culture of plant parts such as stem and leaf cuttings.

***vir* genes:** A set of genes present on Ti-plasmid which facilitate the transfer of T-DNA segment into a plant cell.

**VNTRs (minisatellites):** Polygenic markers.

**Western blot:** A technique in which protein is transferred from an electrophoretic gel to a cellulose or nylon support membrane following electrophoresis. A particular protein is then identified by probing the blot with a radiolabelled antibody which binds on the specific protein to which the antibody was prepared.

**Xenobiotic:** A chemically synthesized compound not produced by living organisms.

**YAC (A yeast artificial chromosome):** A vector of hundreds to kilobases long used for cloning of DNA fragment.

**YEP (Yeast episomal plasmid):** A cloning vector for yeast that uses 2 μm plasmid origin of replication (*ori*) and maintained as an extrachromosomal nuclear DNA molecule.

**Zooblot:** Hybridization of a cloned DNA from one species to DNA from the other organism to determine the extent to which the cloned DNA is evolutionary conserved.

# Bibliography

Alexander, M. 1977. *Introduction to Soil Microbiology,* John Wiley & Sons, New York.

Baxeevanis, A.D. and Quellette, B.F.F. 2001 *Bioinformatics : A Practical Guide to the Analysis of Genes and Proteins.* John Wiley & Sons, Inc.

Bernard, H.U. and Helinski, D.R. 1980. In *Genetic Engineering: Principles and Methods,* Vol. II, (eds. Setlow, J. and Hollaender, A.), pp. 133-168, Plenum Press : New York.

Bhojwani, S.S. and Bhatnagar, S.P. 1974. *The Embryology of Angiosperms,* Vikas, Delhi.

Bull, A.T. Holt, G. and Lilly, M.D. 1983. *Biotechnology : International Trends and Perspectives,* p. 84, Oxford & IBH Publ. Co. PVT. Ltd., New Delhi.

Butenko, R.G. 1985. In *Plant Cell Culture* (R.G. Butenko, translated from Russian by P.E. Chernilovskaya), pp. 11-34, MIR; Moscow.

Carlson, P.S. 1983. *Cell and Tissue Culture Techniques for Cereal Crop Improvement,* pp. 407-418, New York.

Chahal, D.S. 1982. In *Advances in Agricultural Microbiology,* (ed. Subba Rao, N.S.) pp. 551-584, Oxford & IBH Publ. Co., New Delhi.

Chang, S.T. and Hayes, W.A. 1978. *The Biology and Cultivation of Edible Mushrooms,* p. 819, Academic Press; New York and London.

Christodoulatos, C. and Koutsospyros, A. 1998. In *Biological Treatment of Hazardous Wastes* (eds. Lawandowski, G.A. and DeFlippi, L.J.) pp. 69-101, John Wiley & Sons.

Clement, G. 1968. In *Single Cell Protein* (eds. Mateles, R.I. and Tannebaum, S.A.) pp. 306-308, MIT Press, Cambridge, Massachusetts.

Collins, H.A. and Watts, M. 1983. In *Handbook of Plant Cell Culture: Techniques for Propagation and Breeding,* Vol. I (eds. Evans, D.A.; Sharp, W.R.; Ammirato, P.V. and Yamada, Y.), pp. 729-747, Macmillan, New York.

Constabel, F. 1978. In *Frontiers of Plant Tissue Culture,* (Ed. Thorpe, T.A.), pp. 141-149, Intern. Ass. Plant Tissue Cult. Calgary.

Crueger, W. and Crueger, A. 1984. *Biotechnology—A TextBook of Industrial Microbiology.* Sinaver Associates Inc., Sunderland, USA, pp. 284-287.

Dahl, H.H.; Flavell, R.A. and Grosveld, F.G. 1981. In *Genetic Engineering,* Vol. 2 (ed. Williamson, R.), pp. 50-127, Academic Press: London.

Dodds, J.H. and Roberts, L.W. 1985. *Experiments in Plant Tissue Culture,* 2nd Ed. Cambridge University Press, Cambridge.

Dubey, R.C. and Maheshwari, D.K. 2012. *A Textbook of Microbiology.* S. Chand & Co. Ram Nagar, New Delhi, India.

Elmerich, C.; Fogher, C.; Bozouklian, H.; Perroud, B.; Bandhari, S.K. and Nair, S.K. 1987. In *Biotechnology in Agriculture,* (Eds. Natesh, S.; Chopra, V.L. and Ramachandran, S.), pp. 31-37, Oxford & IBH Co., New Delhi.

Fokkema, N.J. 1981. In *Microbial Ecology of The Phylloplane,* (ed. Blackman, J.P.), pp. 433-454, Academic Press : London.

Freshney, R. I. 2002. *Culture of Animal cells*. John Wiley & Sons, Inc.

Glover, D.M. 1984. *Gene Cloning — The Mechanism of DNA Manipulation,* Chapman and Hall : London, p. 222.

Gronow, M.; Mullen, W.H.; Russel, I.J. and Anderson, D.J. 1988. In *Molecular Biology and Biotechnology* (eds. J.M. Walker and E.B. Gingold), pp. 323-247, Royal Soc. Chem. Burlington Houses, London.

Hogam, J.A. 1998. In *Biological Treatment of Hazardous Wastes.* (eds. Lewandowaski, G.A. and DeFlippi, L.J.) pp. 357-383, John Wiley & Sons.

Hu, C.Y. and Wang, P.J. 1983. In *Hand Book of Plant Cell Culture - Techniques for Propagation and Breeding,* Vol. I (Eds. Evans, D.A.; Sharp, W.R.; Ammirato, P.P. and Yamada, Y.), pp. 177-227, Macmillian, Publ. Co., New York.

Jahn, E.C. 1982. In *Progress in Biomass Conservation,* Vol. 3 (Eds. Sarkanen, K.V.; Tillman, D.A. and Jahn, E.C.), pp. 1-50, Academic Press, Inc., New York.

Johansson, B.C.; Nordlund, S. and Baltscheffsky, H. 1983. In *The Phototrophic Bacteria: Anaerobic life in the light* (Ed. Ormerod, J.G.), pp. 120-145, Balckwell Sci. Publications, Oxford.

Kado, C.I. and Tait, R.C. 1983. In *Genetic Engineering in Eukaryotes* (eds. Lurquin, P.F. and Kleinhofs, A.), pp. 103-110, NATO ASI Series A, Life Sciences Vol. 61, Plenum Press: New York, London.

Kornberg, A. 1974. *DNA Synthesis*, W.H. Freeman & Co., San Francisco.

Kull, H.G. 1981. In *Microbial Degradation of Xenobiotics and Recalcitrant Compounds.* (eds. Leisinger, T. Cook, DAM, Nuesch, J. and Hutter, T), pp. 387-399, Academic Press London.

Kwan S.; Yelton, D.E. and Scharff, M.D. 1980. In *Genetic Engineering —Principles and Methods,* Vol. 2, (eds. Setlow, J.K. and Hollaender, A.), pp. 31-46, Plenum Press: New York, London.

Lockwood, L.B. 1979. In *Microbial Technology - Microbial Process,* Vol. 1, 2nd Ed. (eds. Peepler, H.J. and Perlman, D.), pp. 355-387, Academic Press London.

Madigan, M.T. *et al.* 1997. *Brock Biology of Microorganisms.* 8th Ed. Prentice Hall Inc.

Maliga, P. 1980. In *Perspectives in Plant Cell Tissue Culture* (Ed. Vasil, I.K.), *Int. Rev. Cytol. Suppl.* 11A, pp. 225-250, Academic Press, New York.

Moo-Young, M. 1985. *Comprehensive Biotechnology.* Pergman Press, Oxford.

Nandi, S.K. and Palni, L.M.S. 1992. In *Microbial Activity in the Himalaya* (ed. R.D. Khulbe), pp. 419-428, Shree Almora Book Depot., Almora.

Narang, S.A.; Hsuing, H.M. and Brousseaun, R. 1979. In *Methods in Enzymology* (ed. Wu, R.), Vol. 68, pp. 90-98, Academic Press : New York, London.

Perlman, D. 1979. In *Microbial Technology*, 2nd Ed. Vol. 1, (eds. Peppler, H.J. and Perlman, D.), pp. 241-280, Academic Press, New York.

Peters, P. 1993. *Biotechnology— A Guide to Genetic Engineering*, WMC. Brown Publishers, Dubuwue, p. 253.

Prescott, L.M.; Harley, J.P. and Klein, D.A. 2003. *Microbiology.* McGraw- Hill High Education, New York.

Rao, D.V. 1989. *Ph.D. Thesis.* Univ. of Rajasthan, Jaipur, India.

Riviere, J. 1977. *Industrial Application of Microbiology,* (Translated and edited by M.O. Moss and J.E. Smith), Surrey University, Press, London.

Sasson, A. 1984. *Biotechnology : Challenges and Promises*, UNESCO, Paris.

Schieder, O. 1982. In *Plant Improvement and Somatic Cell Genetics* (eds. Vasil, I.K. Scowcroft, W.R. and Frey, K.J.), pp. 239-253, Academic Press, London.

Schlegel, H.S. 1986. *General Microbiology*, 6th Ed. (Translated, by M. Kogut), Cambridge University Press, London.

Scrimshaw, N.S. 1968. In *Single Cell Proteins* (eds. Mateles, R.I. and Tannenbaum, S.R.), pp. 3-7, MIT Press, Cambridge, Mass.

Senez, J.C. 1986. In *Perspectives in Biotechnology and Applied Microbiology,* (eds. Alam, D.I. and Mi Moo Young), pp. 33-48, Elsevier Applied Sci. Publ. in cooperation with Arab Bureau of Education for the Gulf States.

Steinkraus, K.H. 1983. In *The Filamantous Fungi*, Vol. 4, *Fungal technology* (eds. J.E. Smith and D.R. Berry), B.Kristiansen), pp. 171-189, Oxford & IBH Publ. Co. Calcutta.

Stewart-Tull, D.E.S. and Sussman, M. ed. 1992. *The Release of Genetically Modified Microorganisms,* REGEM2, Plenum Press, New York.

Subba Rao, N.S. 1982. Biofertilizers, In *Advances in Agricultrural Microbiology* (ed., Subba Rao. N.S.), pp. 219-242, Oxford & IBH Pub. Co., New Delhi.

Tilak, K.V.B.R. 1991. *Bacterial Biofertilizers*, ICAR, New Delhi, p. 66.

Trevan, M.D. 1980. *Immobilized Enzymes: An Introduction and Application in Biotechnology,* Chichester, John Wiley.

Trevan, M.D. 1987. In *Biotechnology : The Biological Principles* (eds. Trevan, M.D.; Boffey, S.; Goulding, K.H. and Stanbury, P.), pp. 155-228, (Indian Ed.), Tata McGraw Hill Publ. Co., New Delhi.

Vasil, I.K. 1982. In *Plant Improvement and Somatic Cell Genetics* (eds. Vasil, I.K.; Scowcroft, W.R. and Frey, K.J.), pp. 179-203, Academic Press: London.

Venkataraman, G.S. 1972. *Algal Biofertilizers and Rice Cultivation.* Today & Tomorrow Printers and Publ., New Delhi.

Yokotsuka, T. 1981. *The Quality of Food and Beverages*, 2: 171-196, Academic Press, New York.

# Index


Artificial insemination, 225

## C

## F



## G

## H

## N



## O



## P

## Q



## R



## S

## T